Machinery Malfunction Diagnosis and Correction

ISBN 0-13-240946-1

Hewlett-Packard Professional Books

Machinery Malfunction Diagnosis and Correction

Vibration Analysis and Troubleshooting for the Process Industries

Robert C. Eisenmann, Sr., P.E.

President — MACHINERY DIAGNOSTICS, Inc. — Minden, Nevada

and

Robert C. Eisenmann, Jr.

Manager of Rotating Equipment — HAHN & CLAY — Houston, Texas

http://www.hp.com/go/retailbooks

To join a Prentice Hall PTR internet mailing list, point to
http://www.prenhall.com/mail_lists/

Prentice Hall PTR
Upper Saddle River
New Jersey 07458

Library of Congress Cataloging in Publication Data

Eisenmann, Robert C. 1943–
    Machinery malfunction diagnosis and correction : vibration analysis and troubleshooting for
the process industries / Robert C. Eisenmann, Sr., and Robert C. Eisenmann, Jr.
        p.   cm -- (Hewlett-Packard professional books)
    Includes bibliographical references and index.
    ISBN 0-13-240946-1
    1. Machinery--Monitoring. 2. Machinery--Vibration. I. Eisenmann, Robert C., 1970– .
II. Title. III. Series.
TJ153.E355   1997
621.8'16—dc21   97-31974
                                                            CIP

Editorial/Production Supervision: *Joanne Anzalone*
Acquistions Editor: *John Anderson*
Cover Design Director: *Jerry Votta*
Cover Design: *WEE Design Group*
Manufacturing Manager: *Julia Meehan*
Marketing Manager: *Miles Williams*
Manager, Hewlett-Packard Press: *Patricia Pekary*

Published by Prentice Hall PTR
Prentice-Hall, Inc.
A Simon & Schuster Company
Upper Saddle River, New Jersey 07458

Prentice Hall books are widely used by corporations and government agencies
for training, marketing, and resale.
The publisher offers discounts on this book when ordered in bulk quantities.
For more information, contact: Corporate Sales Department, Phone: 800-382-3419;
FAX: 201-236-7141; E-mail: corpsales@prenhall.com
Or write: Corp. Sales Dept., Prentice Hall PTR,
1 Lake Street, Upper Saddle River, NJ 07458

Printed in the United States of America
10 9 8 7 6 5 4 3 2 1

ISBN 0-13-240946-1

Prentice-Hall International (UK) Limited, *London*
Prentice-Hall of Australia Pty. Limited, *Sydney*
Prentice-Hall Canada Inc., *Toronto*
Prentice-Hall Hispanoamericana, S.A., *Mexico*
Prentice-Hall of India Private Limited, *New Delhi*
Prentice-Hall of Japan, Inc., *Tokyo*
Simon & Schuster Asia Pte. Ltd., *Singapore*
Editora Prentice-Hall do Brasil, Ltda., *Rio de Janeiro*

To Mary Rawson Eisenmann,
Wife and Mother
Who Always Kept The Home Fires Burning
While The Boys Went Off To Play With Their Machines

Table of Contents

Preface

When my son graduated from Texas A&M University, he was understandably eager to start working, and begin earning a livable salary. He accepted a maintenance engineering position at a large chemical complex, and embarked upon learning about process machinery. In the months and years that followed, he and his colleagues had many questions concerning a variety of machinery problems. From my perspective, most of these problems had been solved twenty or thirty years ago. However, it was clear that the new engineering graduates were devoting considerable effort attempting to unravel mysteries that had already been solved.

The obvious question that arises might be stated as: *How come the new engineers cannot refer to the history files instead of reworking these issues?* A partial answer to this question is that the equipment files often do not provide any meaningful historical technical data. Major corporations are reluctant to spend money for documentation of engineering events and achievements. Unless the young engineers can find someone with previous experience with a specific malfunction, they are often destined to rework the entire scenario.

Although numerous volumes have been published on machinery malfunctions, there are very few technical references that address the reality of solving field machinery problems. This general lack of usable and easily accessible information was a primary force in the development of this text. The other significant driving force behind this book was the desire to coalesce over thirty-three years of experience and numerous technical notes into some type of structured order that my son, and others could use for solving machinery problems.

This is a book about the application of engineering principles towards the diagnosis and correction of machinery malfunctions. The machinery under discussion operates within the heavy process industries such as oil refineries, chemical plants, power plants, and paper mills. This machinery consists of steam, gas and hydro turbines, motors, expanders, pumps, compressors, and generators, plus various gear box configurations. This mechanical equipment covers a wide variety of physical characteristics. The transmitted power varies from 50 horsepower, to units in excess of 150,000 horsepower. Rotational speeds range from 128 to more than 60,000 revolutions per minute. There is a corresponding wide range of operating conditions. Fluid temperatures vary from cryogenic levels of minus 150°F, to values in excess of plus 1,200°F. The operating pressures range from nearly perfect vacuums to levels greater than 40,000 pounds per square inch. Physically, the moving elements may be only a few feet long, and weigh less than 100 pounds — or they may exceed 200,000 pounds, and cover the length of a football field. In virtually all cases, these process machines are assembled with precision fits and tolerances. It is meaningful to note that the vibration severity criteria for many of these machines are less than the thickness of a human hair.

In some respects, it is amazing that this equipment can operate at all. When the number of individual mechanical components are considered, and the potential failure mechanisms are listed, the probabilities for failures are staggering. Considerable credit must be given to the designers, builders, and innovators of this equipment. They have consistently produced machines that are constantly evolving towards units of improved efficiency, and extended reliability.

The majority of machinery problems that do occur fall into what I call the **ABC** category. These common problems are generally related to **A**lignment, **B**alance, and incorrect **C**learances (typically on bearings). Due to the continual appearance of these malfunctions, an entire chapter within this text has been devoted to each of these subjects. Machines also exhibit other types of failures, and a sampling of common plus unique problems are described within this book.

Some people might view this document as a textbook. Others might consider this to be a reference manual, and still other individuals might use this book for troubleshooting. It has also been suggested that this book be categorized as a *how to do it manual*. Since 52 detailed case histories are combined with numerous sample calculations and examples, each of these descriptions are accurate and applicable. In the overview, the contents of this book cover a variety of machinery malfunctions, and it engages the multiple engineering disciplines that are required to solve real world problems. Regardless of the perception, or the final application, this is a book about the mechanics, measurements, calculations, and diagnosis of machinery malfunctions. I sincerely hope that this text will provide some meaningful help for students, for new graduates entering this field, as well as provide a usable reference for seasoned professionals.

Finally, I would like to extend my deepest personal thanks to John Jensen of Hewlett Packard for the inspiration, encouragement, and opportunity to write this book. I am further indebted to John for his detailed and thorough review of much of the enclosed material. I would also like to thank Ron Bosmans, Dana Salamone, and Pamela Puckett for their constructive comments and corrections.

Robert C. Eisenmann, Sr., P.E.
October 1997

Introduction

*M*achinery development has been synonymous with technological progress. This growth has resulted in an evolutionary trend in industrial equipment that moves towards increased complexity, higher speeds, and greater sophistication. The water wheel has evolved into the hydroelectric plant, the rudimentary steam engine has grown into the gas turbine, and coarse mechanical devices have been replaced by elegant electronic circuits.

Throughout this evolution in technology, new industries and vocations have developed. In recent decades, the Machinery Diagnostician has appeared within most maintenance engineering organizations. These individuals generally possess an extensive knowledge of the machinery construction. They understand repair procedures, and they have a working knowledge of the peripheral equipment. This includes familiarity with the lube and seal oil system, the processing scheme, and the machine controls. Diagnosticians are generally knowledgeable of the machinery monitoring or surveillance instrumentation that covers everything from transducers to the data logging computers. Furthermore, when a problem does appear on a piece of equipment, it generally falls under the jurisdiction of the machinery diagnostician to resolve the difficulty, and recommend an appropriate course of corrective action. This requirement imposes another set of demands. That is, these individuals must be familiar with problem solving techniques and proven methodology for correcting the machinery malfunction.

Clearly, the diagnostician must be qualified in many technical disciplines. As depicted in the adjacent diagram, the basic areas of expertise include knowledge of *machinery*, knowledge of *physical behavior*, plus knowledge of *instrumentation*. The machinery background must be thorough, and it must allow the diagnostician to focus upon realistic failure mechanisms rather than esoteric theories. The category of physical behavior embraces technical fields such as: statics, dynamics, kinematics, mechanics of materials, fluid dynamics, heat transfer, mathematics, and rotordynamics. Knowledge in these areas must be fully integrated with the instrumentation aspects of the electronic measurements required to document and understand the machinery motion.

Competence in these three areas is only achieved by a combination of *knowledge* and field *experience*. Acquiring knowledge often begins with specific technical training. For instance, all academic institutions provide the mathematics and physics necessary to grasp many physical principles. A few universities provide an introduction to the world of analytical rotordynamics. Unfortunately, academia is often burdened by the necessity to obtain research grants, and generate complex general solutions for publication. Certainly the college level contributions to this field are significant, and the global solutions are impressive. However, the working machinery diagnostician often cannot use generalized concepts for solving everyday problems. To state it another way, integral calculus is absolutely necessary for success in the classroom, but it is reasonably useless for most activities performed on the compressor deck.

Within the industrial community, a variety of training programs are available. Instrumentation vendors provide courses on the application and operation of their particular devices. Similarly, machinery vendors and component suppliers have various courses for their clientele. Although these training courses are oriented towards solutions of field problems, they typically display shortcomings in three areas. First, the industrial courses are limited in scope to three or four days of training. This time frame is acceptable for simple topics, but it is inadequate for addressing complex material. Second, industrial training courses are restricted to the instruments or devices sold by the vendor providing the training. Although this approach is expected by the attendees, it does limit the depth and effectiveness of the training. The third problem with vendor training resides in the backgrounds of the training specialists. Although these people are usually well qualified to represent the products of the vendor, they often lack an understanding of the realities within an operating plant. Clearly, the smooth presentation of fifty computer generated slides has no relationship to the crucial decisions that have to be made at 2:00 AM regarding a shaking machine.

Another disturbing trend seems to permeate the specialized field of vibration analysis. Within this technical area, there have been long-term efforts by some vendors to train people to solve problems based entirely on simplistic vibratory symptoms. This is extraordinarily dangerous, and the senior author has encountered numerous instances of people reaching the wrong conclusions based upon this approach. Many problems display similar vibratory symptoms, and additional information is usually required to sort out the differences. In all cases, the measurements must be supplemented with the logical application of physical laws. In addition, the machinery construction and operation must be examined and understood in order to develop an accurate assessment of the malfunction.

Very few professional organizations provide a comprehensive and integrated approach targeted to the topic of machinery diagnosis. The text contained herein attempts to provide a pragmatic and objective overview of machinery malfunction analysis. The three fundamental areas of physical behavior, machinery, and instrumentation knowledge are integrated throughout this book. The structure of this text is directed towards developing a basic understanding of fundamental principles. This includes the applicability of those principles towards machinery, plus the necessary instrumentation and computational systems to describe and understand the actual behavior of the mechanical equipment.

It should be recognized that acquiring basic knowledge does not guarantee that the diagnostician will be qualified to engage and solve machinery problems. As previously stated, experience is mandatory to become proficient in this field. Although the preliminary knowledge may be difficult to obtain, the experience portion may be even harder to acquire. This is particularly true for the individual that works in an operating complex that contains a limited assortment of mechanical equipment. For this diagnostician, the ability to develop a well-rounded background may be hampered due to an absence of mixed machinery types, and associated problems. References such as the excellent series of books by Heinz Bloch[1] provide detailed machinery descriptions, procedures, and guidelines. If the diagnostician is not familiar with a particular machine, this is the one available source that will probably answer most mechanical questions.

In a further attempt to address the experience issue, this text was prepared with 52 field case histories interspersed throughout the chapters. These case studies are presented with substantial details and explanations. The logical steps of working through each particular problem are reviewed, and the encountered errors as well as the final solutions are presented. It is the author's hope that these field examples on major process machinery will provide additional insight, and enhance the experience level of the machinery diagnostician.

The equipment discussed in this text resides within process industries such as oil refining, pipeline, chemical processing, power generation, plus pulp and paper. The specific machines discussed include pumps, blowers, compressors, and generators that vary from slow reciprocating units to high speed centrifugal machines. The prime movers appear in various configurations from induction motors, to cryogenic and hot gas expanders, hydro-turbines, multistage steam turbines, and large industrial gas turbines. In some cases the driver is directly coupled to the driven equipment, and in other trains an intermediate gear box is included. Some of the discussed machinery was installed decades ago, and other mechanical equipment was examined during initial field commissioning.

It is an objective of this text to assist in understanding, and to demonstrate practical solutions to real world machinery problems. This book is not designed to be mathematically rigorous, but the presented mathematics is considered to be accurate. In all cases, the original sources of the mathematical derivations are identified. This will allow the reader to reference back to the original technical work for additional information. Significant equations in this text are numerically identified, and highlighted with an outline box such as equation (2-1). Developmental and supportive equations are sequentially numbered in each chapter. In addition, intermediate results plus numeric sample calculations are also presented. These examples are not assigned equation numbers. In essence, this book is structured to supplement a formal training presentation, and to provide an ongoing reference.

---

[1] Heinz P. Bloch, *Practical Machinery Management for Process Plants, Vol. 1 to 4* (Houston, TX: Gulf Publishing Company, 1982-1989).

MACHINERY CATEGORIES

It is organizationally advantageous to divide process machinery into three categories. Typically, these individual machinery categories are administered under a singular condition monitoring program since they share a common technology. However, the allocation of resources among the three segments varies in direct proportion to the process criticality of the mechanical equipment.

The first segment covers the large machinery within an operating plant. These main equipment trains are generally **critical** to the process. In most instances the plant cannot function without these machines. For example, the charge gas compressor in an ethylene plant, or a syngas compressor in an ammonia plant fall into this category. This equipment typically ranges between 5,000 and 50,000 horsepower. Operating speeds vary from 200 to 60,000 RPM, and fluid film bearings are normally employed. Most of the machinery problems presented within this text reside within this critical category.

Machines of this class are typically equipped with permanently installed proximity probe transducer systems for vibration and position measurements, plus bearing temperature pickups, and specialized transducers such as torque sensors. Historically, the field transducers are hard wired to continuous monitoring systems that incorporate automated trip features for machinery protection. These monitoring systems are also connected to process and/or dedicated computer systems for acquisition of static and dynamic data at predetermined sample rates. These data acquisition computer systems provide detailed information concerning the mechanical condition of the machinery.

The second major group of machines are categorized as **essential** units. They are physically smaller than the critical units, they normally have lower horsepower ratings, and they are usually installed with full backup or spare units. Machines within this category include trains such as product pumps, boiler feed water pumps, cooling water pumps, etc. Individual units in this category may not be critical to the process — but it is often necessary to keep one out of two, or perhaps two out of three units running at all times. It should be recognized that a particular service may be considered as essential equipment when a fully functional main and spare unit are in place. However, if one unit fails, plant operation then depends upon the reliability of the remaining train. In this manner, an essential train may be rapidly upgraded to the status of a critical unit.

These essential machinery trains are usually instrumented in a manner similar to the critical units previously discussed. Shaft sensing proximity probe systems, and thermocouples are hard wired to monitoring systems. These monitoring systems may be integrated with computerized trending systems. Due to the similarity of construction and installation of the critical and the essential machines, the text contained herein is directly applicable to essential units.

The third group of machines are referred to as **general purpose** equipment. These units are physically smaller, and they generally contain rolling element bearings. These machines are often installed with full backups, or they are single units that are non-critical to the process. Machines within this category have minimal vibration or temperature measuring instrumentation permanently installed. This equipment is often monitored with portable data loggers, and the information tracked with dedicated personal computer systems. In many instances, small machines are not subjected to detailed analytical or diagnostic procedures. An in-depth analysis might cost more than the original purchase price of the equipment. Although there are not many direct references to small machinery within this book, the techniques and physical principles discussed for large machines are fully appropriate for these smaller units.

The technology necessary to understand the behavior of process machinery has been evolving for many years. For example, dedicated machinery monitoring systems are being replaced by direct interfaces into Distributed Control Systems (DCS) for trending of general information. Detailed dynamic data is simultaneously acquired in a separate diagnostic computer system. This improvement in data trending and resolution allows a better assessment of machinery malfunctions. In addition, numerous developments in the areas of rotor dynamics, aerodynamics, blade design, cascade mechanics, metallurgy, fabrication, testing, plus optimizing bearing and support designs have all combined to provide a wealth of knowledge. Understanding these individual topics and the interrelationship between design parameters, mechanical construction, vibratory behavior, position between elements, and the array of electronic measurements and data processing can be an intimidating endeavor.

In support of this complex requirement for knowledge plus experience, this book has been prepared. To provide continuity through the chapters, various facets of several basic types of industrial machines are examined. It is understood that one text cannot fully cover all of the material requested by all of the readers. However, it is anticipated that the information presented within this text will provide a strong foundation of technical information, plus a source for future reference. The specific topics covered in this book are summarized as follows.

CHAPTER DESCRIPTIONS

The following chapter 2 on **dynamic motion** begins with a general classification of machinery vibration problems. A review of the fundamental concepts provides a foundation that extends into a description of a simple undamped mechanical system. The addition of damping, plus the influence of forced vibration are discussed. Although the majority of the emphasis is placed upon lateral motion, the parallel environment of torsional vibration is introduced. Finally, the theoretical concepts are correlated with actual measured machinery vibratory characteristics for lateral and torsional behavior.

Rotor mode shapes are discussed in chapter 3. This topic begins with a review of static deflection, followed by the influence of rotor mass, and the distribution of mass and supports. Various aspects of inertia of mechanical systems are discussed, and critical distinctions are identified. Next, system damping, and effective support stiffness are discussed, and their influence upon the deflected mode shapes are demonstrated. The physical transition of a rotor across a critical speed, or balance resonance region is thoroughly explained. These basic concepts are then extended into measured and calculated rotor mode shapes. In addition, the construction of interference maps are introduced, and a variety of illustrations are used to assist in a visualization of these important concepts.

Chapter 4 addresses **machinery bearings** and **supports** in rotating systems. This includes an introduction to oil film bearing characteristics, and some computational techniques. This is followed by proven techniques for determination of radial fluid film bearing clearances, plus the measurement of bearing housing coefficients. Fluid film thrust bearings are also discussed, and the characteristics of rolling element bearings are reviewed. Appropriate case histories are included within this chapter to assist in explanation of the main concepts.

Analytical rotor modeling is introduced in chapter 5. This is a continuation of the machinery behavior concepts initiated in the previous chapters. These concepts are applied to the development of an undamped critical speed analysis for lateral and torsional behavior. This is followed by the inclusion of damping to yield the damped response, plus a stability analysis of the rotating system. Further refinement of the machinery model allows the addition of dimensional forcing functions to yield a synchronous response analysis. This step provides quantification and evaluation of the transient and steady state vibration response characteristics of the machinery. Finally, the validity and applicability of these analytical techniques are demonstrated by six detailed case histories distributed throughout the chapter.

Chapter 6 provides a discussion of **transducer characteristics** for the common measurement probes. A traditional industrial suite of displacement, velocity, acceleration, and pressure pulsation probes are reviewed. The construction, calibration, and operating characteristics of each transducer type are subjected to a comprehensive discussion. In addition, the specific advantages and disadvantages of each standard transducer are summarized. Specialized transducers are also identified, and their general applications are briefly discussed. Finally, the topic of vibration severity and the establishment of realistic vibration limits is discussed.

Dynamic signal characteristics are presented in chapter 7. This section addresses the manipulation and examination of dynamic vibration signals with a full range of electronic filters. In addition, an explanation of combining time domain signals into orbits, and the interrelationship between the time and frequency domain characteristics are examined. Finally, common signal combinations such as signal summation, amplitude modulation, and frequency modulation are discussed. In all cases, appropriate examples are presented.

Chapter 8 covers **data acquisition and processing** in terms of the instrumentation systems required for accurate field data acquisition, plus the processing of the data into useful hard copy formats. Sample forms are included to facilitate documentation of field measurements. In addition, the functions and necessary compatibility issues between instruments and transducers are discussed, and operational guidelines are offered. This chapter concludes with an overview of the most useful machinery data presentation formats.

Based upon the concepts discussed in the previous sections, chapter 9 discusses the origin of many of the **common malfunctions** experienced by process machinery. The topics include synchronous (rotational speed) excitations such as unbalance, bowed shafts, eccentricity, and resonant responses. The influence of preloads, machinery stability, mechanical looseness, rubs, and cracked shafts are discussed. In addition, foundation considerations are reviewed from several perspectives. These general problems are applicable to all rotating machines, and several case histories are included to illustrate these fundamental mechanisms.

Chapter 10 addresses the **unique behavior** of different types of machinery. Excitations associated with gear boxes, electrical frequencies, and fluid excitations are included. In addition, the behavioral characteristics of traditional reciprocating machines, plus hyper compressors are reviewed. Although this group does not cover all of the potential sources of excitation, it does provide a useful summary of problems that occur with regularity on many types of machines. Again, a series of fully descriptive field case histories are distributed throughout the chapter.

Rotational speed vibration is the dominant motion on most industrial machines. Chapter 11 is devoted to an in-depth discussion of this synchronous behavior, and the direct application of these concepts towards **rotor balancing**. This chapter begins with the initial thought process prior to balancing, and the standardized measurements and conventions. The concept of combined balancing techniques are presented, and the machinery linearity requirements are identified. The development of balancing solutions are thoroughly discussed for single plane, two plane, and three plane solutions. In addition, static-couple solutions using two plane calculations are presented, and multiple speed calculations are discussed. The use of response prediction, and trim balance calculations are reviewed, and several types of supportive calculations are included. Again, field case histories are provided to demonstrate the applicability of the rotational speed analysis, and rotor balancing techniques on process machines.

The last portion of chapter 11 deals with shop balancing machines, techniques, and procedures. Although the fundamental concepts are often similar to field balancing, the shop balancing work is generally performed at low rotative speeds. This shop balancing discussion includes additional considerations for the various types of machinery rotors, and common balance specifications.

Machinery alignment persists as one of the leading problems on process machinery, and this topic is covered in chapter 12. Alignment is discussed in terms of the fundamental principles for casing position, casing bore, and shaft alignment. Each type of machinery alignment is discussed, and combined with explanations of several common types of measurements and calculations. This includes dial indicator readings, optical alignment, wire alignment, plus laser alignment, proximity probes, and tooling balls. The applicability of each technique is addressed, and suitable case histories are provided to demonstrate the field use of various alignment techniques.

The concepts of **applied condition monitoring** within an operating plant are discussed in chapter 13 of this text. This chapter was based upon a tutorial by the senior author to the Texas A&M Turbomachinery Symposium in Dallas, Texas. The first portion of this chapter describes the logic and evolution of condition monitoring, and the typical parameters involved. These concepts are illustrated with machinery problems detected during normal operation. The second part of this chapter reviews the turnaround checks and calibrations that should be performed on the machinery control and protection systems. The third portion of this chapter covers the application of condition monitoring during a post-overhaul startup of a machinery train. Again, case studies are used to illustrate the main points of the transient vibratory characteristics.

Chapter 14 address a **machinery diagnostic methodology** that may be used for diagnosis of complex mechanical problems. This chapter was based upon a paper prepared by the senior author for an annual meeting of the Vibration Institute in New Orleans, Louisiana. This topic discusses the fundamental tools, successful techniques, and the seven-step process used for evaluation of machinery problems. Again, specific field case histories are included to illustrate some of the germane points of this topic.

The final chapter 15 is entitled **closing thoughts and comments**, and it addresses some of the other obstacles encountered when attempting to solve machinery problems. This includes candid observations concerning the problems of dealing with multiple corporate entities, plus the politics encountered within most operating plants. In many instances, an acceptable solution is fully dependent upon a proper presentation of results that combine economic feasibility with engineering credibility.

The appendix begins with a **machinery diagnostic glossary** for the specialized language and terminology associated with this business. For reference purposes, a list of the **physical properties** of common metals and fluids, plus a table of **conversion factors** are included. The technical papers and books cited within this text are identified with footnotes, and summarized in a bibliography at the end of each chapter. In addition, a detailed **index** is provided in the last appendix section that includes technical topics, corporate references, and specific authors referenced throughout this book.

It is the authors' hope that the material included within this book will be beneficial to the machinery diagnostician, and that this text will serve as an ongoing technical reference. To paraphrase the words of Donald E. Bently (circa 1968), founder and owner of Bently Nevada Corporation ...*we just want to make the machinery run better...* To this objective, we have dedicated our professional careers and this manuscript.

BIBLIOGRAPHY

1. Bloch, Heinz P., *Practical Machinery Management for Process Plants, Vol. 1 to 4*, Houston, TX: Gulf Publishing Company, 1982-1989.

Dynamic Motion

*M*any mechanical problems are initially recognized by a change in machinery vibration amplitudes. In order to understand, and correctly diagnose the vibratory characteristics of rotating machinery, it is essential for the machinery diagnostician to understand the physics of dynamic motion. This includes the influence of stiffness and damping on the frequency of an oscillating mass — as well as the interrelationship between frequency, displacement, velocity, and acceleration of a body in motion.

MALFUNCTION CONSIDERATIONS AND CLASSIFICATIONS

Before examining the intricacies of dynamic motion, it must be recognized that many facets of a mechanical problem must be considered to achieve a successful and acceptable diagnosis in a timely manner. For instance, the following list identifies some of the related considerations for addressing and realistically solving a machinery vibration problem:

○ Economic Impact
○ Machinery Type and Construction
○ Machinery History — Trends — Failures
○ Frequency Distribution
○ Vibratory Motion Distribution and Direction
○ Forced or Free Vibration

The economic impact is directly associated with the criticality of the machinery. A problem on a main process compressor would receive immediate attention, whereas a seal problem on a fully spared reflux pump would receive a lower priority. Clearly, the types of machinery, the historical trends, and failure histories are all important pieces of information. In addition, the frequency of the vibration, plus the location and direction of the motion are indicators of the problem type and severity. Traditionally, classifications of forced and free vibration are used to identify the origin of the excitation. This provides considerable insight into potential corrective actions. For purposes of explanation, the following lists identify some common forced and free vibration mechanisms.

Forced Vibration Mechanisms   **Free Vibration Mechanisms**

○ Mass Unbalance   ❑ Oil Whirl
○ Misalignment   ❑ Oil or Steam Whip
○ Shaft Bow   ❑ Internal Friction
○ Gyroscopic   ❑ Rotor Resonance
○ Gear Contact   ❑ Structural Resonances
○ Rotor Rubs   ❑ Acoustic Resonances
○ Electrical Excitations   ❑ Aerodynamic Excitations
○ External Excitations   ❑ Hydrodynamic Excitations

Forced vibration problems are generally solved by removing or reducing the exciting or driving force. These problems are typically easier to identify and solve than free vibration problems. Free vibration mechanisms are self-excited phenomena that are dependent upon the geometry, mass, stiffness, and damping of the mechanical system. Corrections to free vibration problems may require physical modification of the machinery. As such, these types of problems are often difficult to correct. Success in treating self-excited problems are directly related to the diagnostician's ability to understand, and apply the appropriate physical principles. To address these fundamental concepts of dynamic motion, including free and forced vibration, the following chapter is presented for consideration.

It should be mentioned that much of the equation structure in this chapter was summarized from the classical textbook by William T. Thomson[1], entitled *Mechanical Vibrations*. For more information, and detailed equation derivation, the reader is encouraged to reference this source directly. The same basic equation structure is also described in his newer text entitled *Theory of Vibration with Applications*[2]. Regardless of the vintage, at least one copy of Thomson should be part of the reference library for every diagnostician.

FUNDAMENTAL CONCEPTS

Initially, consider a simple system consisting of a one mass pendulum as shown in Fig. 2-1. Assume that the pendulum mass M is a concrete block suspended by a weightless and rigid cable of length L. Further assume that the system operates without frictional forces to dissipate system energy. Intuitively, if the pendulum is displaced from the vertical equilibrium position, it will oscillate back and forth under the influence of gravity. The mass will move in the same path, and will require the same amount of time to return to any specified reference point. Due to the frictionless environment, the amplitude of the motion will remain constant. The time required for one complete oscillation, or cycle, is called the *Period* of the motion. The total number of cycles completed per unit of

---

[1] William Tyrell Thomson, *Mechanical Vibrations*, 2nd Edition, 9th Printing, (Englewood Cliffs, New Jersey: Prentice Hall, Inc., 1962), pp.1-75

[2] William T. Thomson, *Theory of Vibration with Applications*, 4th Edition, (Englewood Cliffs, New Jersey: Prentice Hall, 1993), pp. 1-91.

time is the *Frequency* of the oscillation. Hence, frequency is simply the reciprocal of the period as shown in the following expression:

$$Frequency = \frac{1}{Period}$$ (2-1)

The box around this equation identifies this expression as a significant or important concept. This same identification scheme will be used throughout this text. Within equation (2-1), period is a time measurement with units of hours, minutes or seconds. Frequency carries corresponding units such as Cycles per Hour, Cycles per Minute (CPM), or Cycles per Second (CPS or Hz). Understandably, the oscillatory motion of the pendulum is repetitive, and periodic. As shown in *Marks' Handbook*[3], Fourier proved that periodic functions can be expressed with circular functions (i.e., a series of sines and cosines) — where the frequency for each term in the equation is a multiple of the fundamental. It is common to refer to periodic motion as harmonic motion. Although many types of vibratory motions are harmonic, it should be recognized that harmonic motion must be periodic, but periodic motion does not necessarily have to be harmonic.

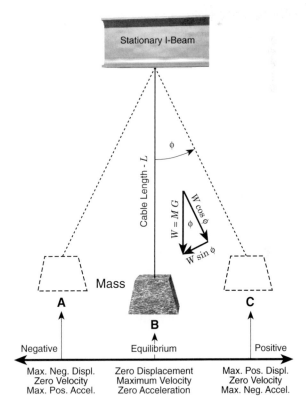

Fig. 2–1 Oscillating Pendulum Displaying Simple Harmonic Motion

[3] Eugene A. Avallone and Theodore Baumeister III, *Marks' Standard Handbook for Mechanical Engineers*, Tenth Edition, (New York: McGraw-Hill, 1996), pp. 2-36.

In a rotating system, such as a centrifugal machine, frequency is normally expressed as a circular rotational frequency ω. Since one complete cycle consists of one revolution, and one revolution is equal to 2π radians, the following conversion applies:

$$\omega = 2\pi \times Frequency = 2\pi \times F$$ (2-2)

Combining (2-1) and (2-2), the rotational frequency ω may be expressed in terms of the *Period* as follows:

$$\omega = \frac{2\pi}{Period}$$ (2-3)

The frequency units for ω in equation (2-3) are Radians per Second, or Radians per Minute. Again, this is dependent upon the time units selected for the period. Although these are simple concepts, they are continually used throughout this text. Hence, a clear and definitive understanding of period and frequency are mandatory for addressing virtually any vibration problem.

Returning to the pendulum of Fig. 2-1, a gravitational force is constantly acting on the mass. This vertical force is the weight of the block. From physics it is known that weight W is equal to the product of mass M, and the acceleration of gravity G. As the pendulum oscillates through an angular displacement ϕ, this force is resolved into two perpendicular components. The cosine term is equal and opposite to the tension in the string, and the sine component is the *Restoring Force* acting to bring the mass back to the vertical equilibrium position. For small values of angular displacement, $\sin\phi$ is closely approximated by the angle ϕ expressed in radians. Hence, this restoring force may be represented as:

$$Restoring \ Force \approx W \times \phi$$ (2-4)

Similarly, the maximum distance traveled by the mass may also be determined from plane geometry. As shown in Fig. 2-1, the cable length is known, and the angular displacement is specified by ϕ. The actual change in lateral position for the mass is the distance from A to B, or from B to C. In either case, this distance is equal to $L \sin\phi$. Once more, for small angles, $\sin\phi \approx \phi$ in radians, and the total deflection from the equilibrium position may be stated as:

$$Deflection \approx L \times \phi$$ (2-5)

This repetitive restoring force acting over the same distance has a spring like quality. In actuality, this characteristic may be defined as the horizontal stiffness K of this simple mechanical system as follows:

$$Stiffness = K = \frac{Force}{Deflection}$$ (2-6)

If equations (2-4) and (2-5) are substituted into (2-6), and if the weight W is replaced by the equivalent mass M times the acceleration of gravity G, the following expression is produced:

$$K = \frac{Force}{Deflection} \approx \frac{W \times \phi}{L \times \phi} = \frac{W}{L} = \frac{M \times G}{L} \qquad \textbf{(2-7)}$$

Later in this chapter it will be shown that the natural frequency of oscillation for an undamped single degree of freedom system is determined by equation (2-44) as a function of mass M and stiffness K. If equation (2-7) is used for the stiffness term within equation (2-44), the following relationship results:

$$\omega = \sqrt{\frac{K}{M}} = \sqrt{\frac{M \times G}{L} \times \frac{1}{M}} = \sqrt{\frac{G}{L}} \qquad \textbf{(2-8)}$$

Equation (2-8) is often presented within the literature for describing the natural frequency of a simple pendulum. A direct example of this concept may be illustrated by considering the motion of the pendulum in a grandfather's clock. Typically, the pendulum requires 1.0 second to travel one half of a stroke, or 2.0 seconds to transverse a complete stroke (i.e., one complete cycle). The length L of the pendulum may be determined by combining equations (2-3) and (2-8):

$$\omega = \frac{2\pi}{Period} = \sqrt{\frac{G}{L}}$$

If the period is represented in terms of the pendulum length L, the above expression may be stated as:

$$Period = 2\pi \times \sqrt{\frac{L}{G}} \qquad \textbf{(2-9)}$$

Equation (2-9) is a common expression for characterizing a simple pendulum. The validity of this equation may be verified in technical references such as *Marks' Handbook*[4]. For the specific problem at hand, equation (2-9) may be solved for the pendulum length. Performing this manipulation, and inserting the gravitational constant G, plus the period of 2.0 seconds, the following is obtained:

$$L = \frac{G \times Period^2}{4\pi^2} = \frac{(386.1 \, Inches/Second^2) \times (2.0 \, Seconds)^2}{4\pi^2} = 39.12 \, Inches$$

Thus, the pendulum length in a grandfather's clock should be 39.12 inches. This value is accurate for a concentrated mass, and a weightless support arm. In an actual clock, the pendulum is often ornate, and weight is distributed along the length of the support arm. This makes it difficult to accurately determine the location of the center of gravity of the pendulum mass. Nevertheless, even rough measurements reveal that the pendulum length is in the vicinity of 40 inches. In addition, clock makers normally provide a calibration screw at the bottom of the pendulum to allow the owner to adjust the clock accuracy. By turning this adjustment screw, the effective length of the pendulum may be altered. From the previ-

---

[4] Eugene A. Avallone and Theodore Baumeister III, *Marks' Standard Handbook for Mechanical Engineers*, Tenth Edition, (New York: McGraw-Hill, 1996), p. 3-15.

ous equations, it is clear that changing the pendulum length will alter the period of the pendulum. By moving the weight upward, and decreasing the arm length, the clock will run faster (i.e., higher frequency with a shorter period). Conversely, by lowering the main pendulum mass, the length of the arm will be increased, and the clock will run slower (i.e., a lower frequency with a longer period).

Although the grandfather clock is a simple application of periodic motion, it does provide a realistic example of the fundamental concepts. Additional complexity will be incorporated later in this text when the behavior of a compound pendulum is discussed. It should be noted that a compound pendulum is a mechanical system that normally contains two degrees of freedom. This additional flexibility might be obtained by adding flexible members such as springs, or additional masses to a simple system. In a two mass system, each mass might be capable of moving independently of the other mass. For this type of arrangement, each mass must be tracked with an independent coordinate system, and this would be considered as a two degree of freedom system.

The number of independent coordinates required to accurately define the motion of a system is termed the *Degree of Freedom* of that system. Process machinery displays many degrees of freedom, and accurate mathematical description of these systems increases proportionally to the number of required coordinates. However, in the case of the simple pendulum, only one coordinate is required to describe the motion — and the pendulum is a single degree of freedom system exhibiting harmonic motion. More specifically, this is an example of basic dynamic motion where the restoring force is proportional to the displacement. This is commonly referred to as *Simple Harmonic Motion* (SHM). Other devices such as the undamped spring mass (Fig. 2-7), the torsional pendulum (Fig. 2-25), the particle rotating in a circular path, and a floating cork bobbing up and down in the water at a constant rate are all examples of SHM.

Before expanding the discussion to more complex systems, it is desirable to conclude the discussion of the simple pendulum. Once again, the reader is referred back to the example of the oscillating pendulum depicted in Fig. 2-1. On this diagram, it is meaningful to mentally trace the position of the mass during one complete cycle. Starting at the vertical equilibrium position *B*, the displacement is zero at time equal to zero. One quarter of a cycle later, the mass has moved to the maximum positive position *C*. This is followed by a zero crossing at point *B* as the mass approaches the maximum negative value at position *A*. The last quarter cycle is completed as the mass returns from the *A* location back to the original equilibrium, or center rest point *B*.

Intuitively, the mass achieves zero velocity as it swings back and forth to the maximum displacement points *A* and *C* (i.e., the mass comes to a complete stop). In addition, the maximum positive velocity occurs as the mass moves through point *B* from left to right, combined with a maximum negative velocity as the mass moves through *B* going from right to left. Finally, the mass must de-accelerate going from *B* to *C*, and accelerate from *C* back to point *A*. Then the mass will de-accelerate as it moves from *A* back to the original equilibrium point *B* that displays zero lateral acceleration.

Another way to compare and correlate the displacement, velocity, and acceleration characteristics of this pendulum would be a time domain examination. Although a meaningful visualization of the changes in displacement, velocity, and acceleration with respect to time may be difficult — a mathematical description simplifies this task. For instance, assume that the periodic displacement of the mass may be described by the following fundamental equation relating displacement and time:

$$Displacement = D \times \sin(2\pi \times F \times t)$$ **(2-10)**

where: *Displacement* = Instantaneous Displacement
             D = Maximum Displacement (equal to pendulum position A or C)
             F = Frequency of Oscillation
             t = Time

In a rotating system, such as a centrifugal machine, this expression can be simplified somewhat by substituting the rotational frequency ω that was previously defined in equation (2-2) to yield:

$$Displacement = D \times \sin(\omega t)$$ **(2-11)**

The instantaneous velocity of this periodic motion is the time derivative of displacement. Velocity may now be determined as follows:

$$Velocity = \frac{d}{dt}Displacement = D \times \omega \times \cos(\omega t)$$

By converting the cosine to a sine function, expression (2-12) is derived:

$$Velocity = D \times \omega \times \sin(\omega t + \pi/2)$$ **(2-12)**

Note that velocity leads displacement by $\pi/2$ or 90°. Another way to state the same concept is that displacement lags behind velocity by 90° in the time domain. The same procedure can now be repeated to examine the relationship between velocity and acceleration. Since acceleration is the time rate of change of velocity, the first time derivative of velocity will yield acceleration. The same result may be obtained by taking the second derivative of displacement with respect to time to obtain acceleration:

$$Acceleration = \frac{d^2}{dt^2}Displacement = -D \times \omega^2 \times \sin(\omega t)$$

By adding π to the sine term, the negative sign is removed, and the following expression is obtained:

$$Acceleration = D \times \omega^2 \times \sin(\omega t + \pi)$$ **(2-13)**

Acceleration leads displacement by π or 180°, and it leads velocity by 90°. It may also be stated that displacement lags acceleration by 180° in time. The relationship between displacement, velocity, and acceleration may be viewed graphically in the polar coordinate format of Fig. 2-2. This diagram reveals that

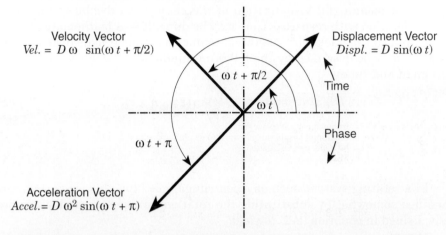

Fig. 2–2 Timing Relationship Between Displacement Velocity, and Acceleration

mechanical systems in motion do display a consistent and definable relationship between frequency, and the respective displacement, velocity, and acceleration of the body in motion.

Understanding the timing between vectors is mandatory for diagnosing machinery behavior. It is very easy to become confused between terms such as *leading* and *lagging*, and the diagnostician might inadvertently make a 90° or a 180° mistake. In some instances, this type of error might go unnoticed. However, during rotor balancing, a 180° error in weight placement might result in excessive vibration or even physical damage to the machine. This type of error is totally unnecessary, and it may be prevented by establishing and maintaining a consistent timing or phase convention.

From Fig. 2-2, it is noted that time is shown to increase in a counterclockwise direction. If this diagram represented a rotating shaft, time and rotation would move together in a counterclockwise direction. As discussed in succeeding chapters, phase is measured from the peak of a vibration signal backwards in time to the reference trigger point. This concept is illustrated in Fig. 2-3 that depicts a rotating disk with a series of angles marked off at 45° increments. Assume that the disk is turning counterclockwise on the axial centerline. If this rotating disk is observed from a stationary viewing position, the angles will move past the viewing point in consecutive order.

That is, as the disk turns, the angles progress in a 0-45-90-135-180-225-270-315° consecutive numeric order past the fixed viewing position. However, if the angles increased with rotation, the observed viewing order would be backwards. Since this does not make good physical sense, the direction of numerically increasing angles are *always* set against shaft rotation as in Fig. 2-3. This angular convention will be used throughout this text, and vector angles will always be considered as degrees of phase lag. This convention applies to shaft and casing vibration vectors, balance weight vectors, balance sensitivity vectors, plus all

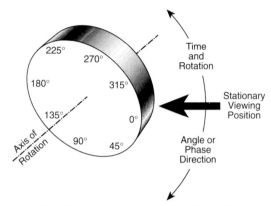

Fig. 2–3 Traditional Angle Designation On A Rotating Disk

analytically calculated vectors. In short, angles and the associated phase are measured against rotation based upon this physical relationship.

For proper identification, phase angles should be specified as a phase lag, or provided with a negative sign. In most cases, it is convenient to ignore the negative sign, and recognize that these angles are phase lag values. Using this convention, phase between the 3 vibration vectors in Fig. 2-2 may be converted by:

$$Phase_{displacement} = Phase_{velocity} + 90° \qquad (2\text{-}14)$$

$$Phase_{displacement} = Phase_{acceleration} + 180° \qquad (2\text{-}15)$$

$$Phase_{velocity} = Phase_{acceleration} + 90° \qquad (2\text{-}16)$$

If a velocity phase angle occurs at 225°, it is determined from (2-14) that the displacement phase angle is computed by: 225°+90°=315°. Similarly, the velocity phase may be converted to an acceleration phase from equation (2-16) as: 225°-90°=135°. If the phase lag negative sign is used, the angle conversions in equations (2-14) to (2-16) must also be negative (i.e., -90° and -180°). In either case, consistency is necessary for accurate and repeatable results.

In addition to phase, the vibration magnitude of an object may be converted from displacement to velocity or acceleration at a constant frequency. This requires a conversion of units within the motion equations (2-12) and (2-13). For example, consider the following definition of English units for these parameters:

D = Displacement — Mils,$_{peak\ to\ peak}$ = Mils,$_{p\text{-}p}$

V = Velocity — Inches/Second,$_{zero\ to\ peak}$ = IPS,$_{o\text{-}p}$

A = Acceleration — G's,$_{zero\ to\ peak}$ = G's,$_{o\text{-}p}$

F = Frequency — Cycles/Second (Hz)

Reinstalling $2\pi F$ for the frequency ω, and considering the peak values of the terms (i.e., sin=1), equation (2-12) may be restated as follows:

$$V = D \times \omega = D \times 2\pi \times F$$

Since velocity is generally defined as zero to peak $(_{o-p})$, and displacement is typically considered as peak to peak $(_{p-p})$, the displacement value must be halved to be consistent with the velocity wave. Applying the appropriate physical unit conversions, the following expression evolves:

$$V = \left(\frac{D}{2}\text{Mils} \times \frac{1\,\text{Inch}}{1,000\,\text{Mils}}\right) \times \left(2\pi\frac{\text{Radians}}{\text{Cycle}} \times F\frac{\text{Cycles}}{\text{Second}}\right)$$

Which simplifies to the following common equation:

$$\boxed{V = \frac{D \times F}{318.31}}$$ (2-17)

Next, consider the relationship between acceleration and displacement as described by equation (2-13), and expanded with proper engineering units to the following expression:

$$A = D \times \omega^2 = \left(\frac{D}{2}\text{Mils} \times \frac{1\,\text{Inch}}{1,000\,\text{Mils}}\right) \times \left(2\pi\frac{\text{Radians}}{\text{Cycle}} \times F\frac{\text{Cycles}}{\text{Second}}\right)^2$$

$$A = \left(\frac{D \times F^2}{50.661}\right)\frac{\text{Inches}}{\text{Second}^2}$$

Acceleration units for the above conversion are Inches/Second^2. Measurement units of G's can be obtained by dividing this last expression by the acceleration of gravity as follows:

$$A = \left(\frac{D \times F^2}{50.661}\right)\frac{\text{Inches}}{\text{Second}^2} \times \frac{1\,G}{386.1\,\text{Inches/Second}^2}$$

This conversion expression may be simplified to the following format:

$$\boxed{A = \left(\frac{D \times F^2}{19,560}\right) = D \times \left(\frac{F}{139.9}\right)^2}$$ (2-18)

The relationship between acceleration and velocity may be stated as:

$$A = V \times \omega = V \times 2\pi \times F$$

Expanding this expression, and including dimensional units, the following equation for converting velocity at a specific frequency to acceleration evolves:

$$A = \left(V\frac{\text{Inches}}{\text{Second}}\right) \times \left(2\pi\frac{\text{Radians}}{\text{Cycle}} \times F\frac{\text{Cycles}}{\text{Second}}\right) \times \left(\frac{1\,G}{386.1\,\text{Inches/Second}^2}\right)$$

Simplifying this expression, the following common equation is derived:

$$A = \frac{V \times F}{61.45} \qquad \textbf{(2-19)}$$

The last three equations allow conversion between displacement, velocity, and acceleration at a fixed frequency measured in Cycles per Second (Hz). A set of expressions for frequency measured in Cycles per Minute (CPM) may also be developed. Since machine speeds are measured in Revolutions per Minute (RPM), this additional conversion is quite useful in many instances. Performing this frequency conversion on equations (2-17), (2-18), and (2-19) produces the next three common conversion equations:

$$V = \frac{D \times RPM}{19,099} \qquad \textbf{(2-20)}$$

$$A = D \times \left(\frac{RPM}{8,391}\right)^2 \qquad \textbf{(2-21)}$$

$$A = \frac{V \times RPM}{3,687} \qquad \textbf{(2-22)}$$

The vibration units for equations (2-20), (2-21), and (2-22) are identical to the English engineering units previously defined. However, the frequency for these last three equations carry the units of Revolutions per Minute (i.e., RPM or Cycles per Minute).

The simultaneous existence of three parameters (i.e., displacement, velocity, and acceleration) to describe vibratory motion can be confusing. This is further complicated by the fact that instrumentation vendors are often specialized in the manufacture of a single type of transducer. Hence, one company may promote the use of displacement probes, whereas another vendor may strongly endorse velocity coils, and a third supplier may cultivate the application of accelerometers. The specific virtues and limitations of each of these types of transducer systems are discussed in greater detail in chapter 6 of this text. However, for the purposes of this current discussion, it is necessary to recognize that displacement, velocity, and acceleration of a moving body are always related by the frequency of the motion.

This relationship between variables may be expressed in various ways. For example, consider an element vibrating at a frequency of 100 Hz (6,000 CPM) and a velocity of 0.3 IPS$_{,o-p}$. From equation (2-17) the relationship between velocity and displacement may be used to solve for the displacement as follows:

$$D = \frac{318.31 \times V}{F} = \frac{318.31 \times 0.3\,\text{IPS}_{o-p}}{100\,\text{Hz}} = 0.955\,\text{Mils}_{p-p}$$

Similarly, the equivalent acceleration of this mechanical element may be determined from equation (2-19) in the following manner:

$$A = \frac{V \times F}{61.45} = \frac{0.3\text{IPS}_{\text{o-p}} \times 100\text{Hz}}{61.45} = 0.488 \ \text{G's}_{\text{o-p}}$$

Thus, the displacement and acceleration amplitudes for this velocity may be computed for any given frequency. Another way to view this interrelationship between parameters is to extend this calculation procedure to a large range of frequencies, and plot the results as shown in Fig. 2-4. Within this diagram, the velocity is maintained at a constant magnitude of 0.3 IPS,$_{\text{o-p}}$ and the displacement and acceleration amplitudes calculated and plotted for several frequencies between 1 and 20,000 Hz (60 and 1,200,000 CPM).

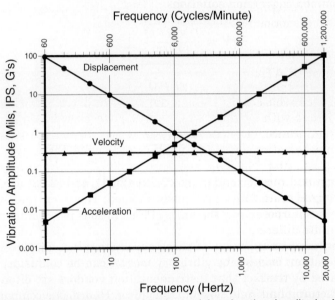

Fig. 2–4 Equivalent Displacement, Velocity, and Acceleration Amplitudes V. Frequency

Fig. 2-4 shows that displacement is large at low frequencies, and acceleration is larger at high frequencies. From a measurement standpoint, displacement would be used for lower frequencies, and acceleration would be desirable for high frequency data. Again, specific transducer characteristics must also be considered, and the reader is referred to chapter 6 for additional details on the actual operating ranges of transducers.

For purposes of completeness, it should be recognized that the circular functions previously discussed can be replaced by an exponential form. For instance, equation (2-23) is a normal format for these expressions:

$$Displacement = D \times e^{i\omega t} \tag{2-23}$$

In this equation, "i" is equal to the square root of minus 1 and "e" is the natural log base that has a value of 2.71828. This expression will satisfy the same equations, and produce identical results to the circular formats. However, it is sometimes easier to manipulate equations using an exponential form rather than a circular function. For reference, the relationship between the exponential and the circular function is shown as follows:

$$e^{i\omega t} = \cos(\omega t) + i \times \sin(\omega t) \qquad \text{(2-24)}$$

The "$\cos(\omega t)$" term is often referred to as the *Real*, or the *In-Phase* component. The "$i\ \sin(\omega t)$" term is the projection of the vector on the imaginary axis. This is normally called the *Imaginary*, or the *Quadrature* component. These terms are used interchangeably. It should be understood that the form, and not the intent of the equations has been altered. It should also be mentioned that both the *Real* and *Imaginary* (*In-Phase* and *Quadrature*) components must satisfy the equation of motion for the mechanical system.

VECTOR MANIPULATION

Many physical characteristics of machines are described with vectors. A magnitude is joined with a directional component to provide a parameter with real physical significance. These vector quantities are routinely subjected to various types of mathematical operations. More specifically, the addition, subtraction, multiplication, and division of vectors must be performed as an integral part of vibration and modal analysis, rotor balancing, analytical modeling, plus instrumentation calibration.

For reference purposes, it is necessary to define the methods used for vector manipulation. The different vector operations may be performed with a hand held calculator, they may be executed with the math tools incorporated in spreadsheets, or they may be included as subroutines into computer programs. In addition, some Dynamic Signal Analyzers (DSA) use vector math as part of the signal processing and computational capabilities. In all cases, these fundamental math operations must be performed in a consistent manner.

From an explanatory standpoint, the specific vector equations will be shown, and a numeric example will be presented for each type of operation. The examples will be performed with circular coordinates, however an exponential form will provide an identical solution. For consistency, the following pair of polar coordinate vectors will be used throughout this series of explanations:

$$\overrightarrow{V_a} = A\angle\alpha \qquad \text{(2-25)}$$

$$\overrightarrow{V_b} = B\angle\beta \qquad \text{(2-26)}$$

The first vector (2-25) has of a magnitude A, occurring at an angle α. Similarly, the second vector (2-26) has an amplitude of B, and an angle β. As previously discussed, these vectors may be represented in a Cartesian coordinate (X-Y) system by the following pair of equations:

$$\vec{V_a} = V_{a_x} + V_{a_y} = A \times \cos\alpha + A \times \sin\alpha$$

$$\vec{V_b} = V_{b_x} + V_{b_y} = B \times \cos\beta + B \times \sin\beta$$

Multiplying the amplitude by the cosine and sine of the associated angle will allow conversion from polar to rectangular coordinates. The cosine term represents the magnitude on the X-Axis, and the sine term identifies the amplitude on the Y-Axis. From the last pair of equations, the individual Cartesian amplitudes for each vector component may be summarized as:

$$V_{a_x} = A \times \cos\alpha \qquad\qquad (2\text{-}27)$$

$$V_{a_y} = A \times \sin\alpha \qquad\qquad (2\text{-}28)$$

$$V_{b_x} = B \times \cos\beta \qquad\qquad (2\text{-}29)$$

$$V_{b_y} = B \times \sin\beta \qquad\qquad (2\text{-}30)$$

This conversion of the initial vectors now provides the format to allow the addition and subtraction of two vector quantities. **Vector addition** is performed by summing the individual X and Y components, and converting from Cartesian back to polar coordinates. The summation of X-Axis components is achieved by adding equations (2-27) and (2-29) in the following manner:

$$\boxed{V_{add_x} = V_{a_x} + V_{b_x} = A \times \cos\alpha + B \times \cos\beta} \qquad (2\text{-}31)$$

Similarly, the Y-Axis summation component is obtained by addition of the previously described equations (2-28) and (2-30) as follows:

$$\boxed{V_{add_y} = V_{a_y} + V_{b_y} = A \times \sin\alpha + B \times \sin\beta} \qquad (2\text{-}32)$$

The X and Y summation components are now be converted back into polar coordinates of amplitude and angle as shown in equations (2-33) and (2-34):

$$\boxed{V_{add} = \sqrt{(V_{add_x})^2 + (V_{add_y})^2}} \qquad (2\text{-}33)$$

$$\boxed{\phi_{add} = \operatorname{atan}\left(\frac{V_{add_y}}{V_{add_x}}\right)} \qquad (2\text{-}34)$$

Vector addition is used with many different types of calculations. For instance, consider the installation of two weights into the balance ring on a turbine rotor. If the weights are both installed in the same hole, the effective weight correction would be the simple sum of the two weights. However, if the weights are screwed into two different holes, the effective balance correction must be

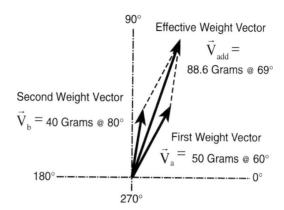

Fig. 2–5 Vector Addition
Of Two Balance Weights

determined by vector addition of the two individual weight vectors. For demonstration purposes, assume that 50 Grams was inserted into a hole at 60°, and 40 Grams was installed at the 80° hole as described in Fig. 2-5. The initial weight vectors are represented with equations (2-25) and (2-26) as:

$$\overrightarrow{V_a} = A\angle\alpha = 50 \text{ Grams } \angle 60°$$

$$\overrightarrow{V_b} = B\angle\beta = 40 \text{ Grams } \angle 80°$$

The summation of horizontal vector components in the X-Axis is determined with equation (2-31):

$$V_{add_x} = A \times \cos\alpha + B \times \cos\beta$$

$$V_{add_x} = 50 \times \cos 60° + 40 \times \cos 80°$$

$$V_{add_x} = 50 \times 0.500 + 40 \times 0.174 = 25.00 + 6.95 = 31.95 \text{ Grams}$$

Similarly, the summation of vertical vector components in the Y-Axis may be computed with equation (2-32) as follows:

$$V_{add_y} = A \times \sin\alpha + B \times \sin\beta$$

$$V_{add_x} = 50 \times \sin 60° + 40 \times \sin 80°$$

$$V_{add_x} = 50 \times 0.866 + 40 \times 0.985 = 43.30 + 39.39 = 82.69 \text{ Grams}$$

The calculated X and Y balance weights identify the combined effect of both weights in the horizontal and vertical directions. These weights are actually X and Y coordinates that may be converted to a polar coordinate magnitude using equation (2-33) in the following manner:

$$V_{add} = \sqrt{(V_{add_x})^2 + (V_{add_y})^2}$$

$$V_{add} = \sqrt{(31.95)^2 + (82.69)^2} = \sqrt{7,858} = 88.64 \text{ Grams}$$

Finally, the angle of the resultant vector may now be determined from equation (2-34) as shown:

$$\phi_{add} = \text{atan}\left(\frac{V_{add_y}}{V_{add_x}}\right)$$

$$\phi_{add} = \text{atan}\left(\frac{82.69}{31.95}\right) = \text{atan}(2.588) = 68.9°$$

Thus, the 50 Gram weight installed at 60° plus the 40 Gram weight at 80° are vectorially equivalent to 88.6 Grams at 69° (as shown in Fig. 2-5). The magnitude of this vector sum is the net effective weight that should be used for additional balancing calculations such as centrifugal force. The effective angle of this weight pair is necessary information for intermediate balancing response calculations, as well as the documentation of final results. For more information on this type of calculation, please refer to chapter 11 of this text.

The same basic approach is used for **vector subtraction**, with one significant difference. Instead of adding Cartesian coordinates, the X and Y components are subtracted. That is, by subtracting the B vector from the A vector, the X-Axis change is obtained by subtracting equation (2-29) from (2-27):

$$\boxed{V_{sub_x} = V_{a_x} - V_{b_x} = A \times \cos\alpha - B \times \cos\beta} \qquad \textbf{(2-35)}$$

In a similar manner, the Y-Axis component is obtained by subtraction of the previously identified equation (2-30) from (2-28) as follows:

$$\boxed{V_{sub_y} = V_{a_y} - V_{b_y} = A \times \sin\alpha - B \times \sin\beta} \qquad \textbf{(2-36)}$$

Calculation of the differential vector is achieved with equations (2-37) and (2-38) that are identical in form to the vector addition conversions:

$$\boxed{V_{sub} = \sqrt{(V_{sub_x})^2 + (V_{sub_y})^2}} \qquad \textbf{(2-37)}$$

$$\boxed{\phi_{sub} = \text{atan}\left(\frac{V_{sub_y}}{V_{sub_x}}\right)} \qquad \textbf{(2-38)}$$

This type of vector computation is extremely useful for performing routine tasks such as runout subtraction on proximity probe displacement signals. For instance, Fig. 2-6 displays a synchronous vibration vector at full operating speed of 2.38 Mils,$_{p-p}$ at an angle of 134°. Assume that the slow speed 1X runout was

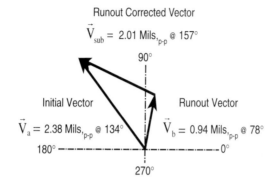

Runout Corrected Vector

$\vec{V}_{sub}$ = 2.01 Mils,$_{p-p}$ @ 157°

Initial Vector

$\vec{V}_a$ = 2.38 Mils,$_{p-p}$ @ 134°

Runout Vector

$\vec{V}_b$ = 0.94 Mils,$_{p-p}$ @ 78°

Fig. 2–6 Vector Subtraction Of Shaft Runout From Running Speed Vector

measured to be 0.94 Mils,$_{p-p}$, at 78°. Subtraction of the slow roll from the full speed vector yields a compensated, or a runout corrected vector.

Mathematically, the initial vibration at running speed may be identified as the A vector, and the slow roll runout may be represented by the B vector. Substitution of the defined vibration vectors into equations (2-25) and (2-26) provides the following vectors for subtraction:

$$\vec{V}_a = A\angle\alpha = 2.38 \text{ Mils}_{p-p}\angle 134°$$

$$\vec{V}_b = B\angle\beta = 0.94 \text{ Mils}_{p-p}\angle 78°$$

The difference between horizontal X-Axis vector components is determined with equation (2-35) in the following manner:

$$V_{sub_x} = A \times \cos\alpha - B \times \cos\beta$$

$$V_{sub_x} = 2.38 \times \cos 134° - 0.94 \times \cos 78°$$

$$V_{sub_x} = 2.38 \times (-0.695) - 0.94 \times 0.208 = -1.654 - 0.196 = -1.850 \text{ Mils}_{p-p}$$

Similarly, the difference of vector components in the vertical Y-Axis may be computed with equation (2-36):

$$V_{sub_y} = A \times \sin\alpha - B \times \sin\beta$$

$$V_{sub_y} = 2.38 \times \sin 134° - 0.94 \times \sin 78°$$

$$V_{sub_y} = 2.38 \times 0.719 - 0.94 \times 0.978 = 1.711 - 0.919 = 0.792 \text{ Mils}_{p-p}$$

The negative value for the horizontal component is perfectly normal, and acceptable. This negative sign, combined with the positive sign on the vertical component, identifies that the final vector will reside in the upper left polar quadrant (i.e., angle between 90° and 180°). The computed X and Y coordinates may now be converted to polar coordinates using equation (2-37) to determine the magnitude of the runout corrected vector:

$$V_{sub} = \sqrt{(V_{sub_x})^2 + (V_{sub_y})^2}$$

$$V_{sub} = \sqrt{(-1.850)^2 + (0.792)^2} = \sqrt{4.050} = 2.01 \text{ Mils}_{p\text{-}p}$$

The angle of the runout compensated vector may now be calculated from equation (2-38) as follows:

$$\phi_{sub} = \text{atan}\left(\frac{V_{sub_y}}{V_{sub_x}}\right)$$

$$\phi_{sub} = \text{atan}\left(\frac{0.792}{-1.850}\right) = \text{atan}(-0.428) = -23.2°$$

$$\phi_{sub} = -23.2 + 180° = 156.7°$$

The 180° addition to the angle is a quadrant correction. Thus, subtracting a runout of 0.94 Mils,$_{p\text{-}p}$ at 78° from the full speed vector of 2.38 Mils,$_{p\text{-}p}$ at 134° yields a runout compensated vector quantity of 2.01 Mils,$_{p\text{-}p}$ at 157°. This calculated result is in full agreement with the vector diagram shown in Fig. 2-6. This compensated vector represents the actual dynamic motion (i.e., vibration) of the shaft. For more information on runout compensation, please refer to chapters 6, 7, 8, and 11.

The major complexity associated with vector addition and subtraction is due to the necessity for converting from polar to Cartesian coordinates, performing a simple operation, and then converting from Cartesian back to polar coordinates. Fortunately, this multiple conversion is not required for vector multiplication and division.

Vector multiplication of two vector quantities may be executed by simply multiplying amplitudes, and adding the respective phase angles as follows:

$$\boxed{\overrightarrow{V_{mul}} = \overrightarrow{V_a} \times \overrightarrow{V_b} = (A \times B)\angle(\alpha + \beta)} \qquad \textbf{(2-39)}$$

This manipulation is easy to perform, and the only cautionary note resides with the value of the angle. In many cases, this may exceed 360°, due to the size of angles α and β. When a full circle has been exceeded (i.e., final angle greater than 360°), the size of the angle may be reduced by 360° to yield a physically meaningful angle between 0° and 360°.

Vector multiplication is necessary in the machinery diagnosis business. For example, consider the situation of determining the required balance weight to correct the 1X vibration response of a machine. Presuming that the unit has a properly defined balance sensitivity vector, the required balance weight and angle can be determined from equation (2-39). This requires a vector multiplication between the measured vibration, and the sensitivity vector. For demonstration purposes, assume that the measured vibration vector is 2.0 Mils,$_{p\text{-}p}$ at an angle of 40°. Further assume that the rotor balance sensitivity vector is equal to

150.0 Grams/Mil$_{,p-p}$ at an angle of 190°. Based on this data, the operable vectors for this vector manipulation are identified as:

$$\overrightarrow{V_a} = A\angle\alpha = 2.0 \text{ Mils}_{p-p}\angle 40°$$

$$\overrightarrow{V_b} = B\angle\beta = 150 \text{ Grams/Mil}_{p-p}\angle 190°$$

Multiplication of these two vectors is performed with equation (2-39) as:

$$\overrightarrow{V_{mul}} = (A \times B)\angle(\alpha + \beta)$$

$$\overrightarrow{V_{mul}} = (2.0 \text{ Mils}_{p-p} \times 150 \text{ Grams/Mil}_{p-p})\angle(40° + 190°) = 300 \text{ Grams}\angle 230°$$

This vector product indicates that the installation of a 300 Gram weight at an angle of 230° will balance the measured synchronous response of 2.0 Mils$_{,p-p}$ at 40°. Naturally, the accuracy of this value is dependent upon the correctness of the balance sensitivity vector.

As described in further detail in chapter 11, a vector summation between the calculated vibration from the weight, plus the current vibration vector will result in a predicted vibration vector with the weight attached. An additional vector summation with the shaft runout will produce an uncompensated 1X vector. For a perfectly linear mechanical system, this would be the vibration amplitude and phase displayed by a synchronous tracking filter. Although this discussion is somewhat premature within the sequence of this text, the main point is that vector calculations may involve a string of manipulations to achieve the necessary result.

Vector division represents the final category of vector math. Referring back to the initial vectors, equations (2-25) and (2-26), vector division is performed by dividing the amplitudes, and subtracting the angles as follows:

$$\overrightarrow{V_{div}} = \frac{\overrightarrow{V_a}}{\overrightarrow{V_b}} = \left(\frac{A}{B}\right)\angle(\alpha - \beta) \qquad\qquad \textbf{(2-40)}$$

This kind of manipulation is also easy to perform, and again a cautionary note resides with the final value of the angle. In many cases, this angle may drop below 0°, due to the relative size of angles α and β. When the zero point is crossed (i.e., negative angle), the size of the angle may be increased by 360° to yield a physically meaningful angle between 0° and 360°.

Vector division is widely used for various types of machinery calculations. For instance, the computation of a balance sensitivity vector requires the division of a calibration weight vector by a differential vibration response vector. The technical details associated with this calculation are in chapter 11. However, from a pure computational standpoint, consider the following initial pair of vectors for division.

$$\overrightarrow{V_a} = A\angle\alpha = 400 \text{ Grams}\angle 230°$$

$$\overrightarrow{V_b} = B\angle\beta = 5.00 \text{ Mil}_{p\text{-}p}\angle 60°$$

Division of these two vectors is performed with equation (2-40) as follows:

$$\overrightarrow{V_{div}} = \frac{\overrightarrow{V_a}}{\overrightarrow{V_b}} = \left(\frac{A}{B}\right)\angle(\alpha - \beta)$$

$$\overrightarrow{V_{div}} = \left(\frac{400 \text{ Grams}}{5.00 \text{ Mil}_{p\text{-}p}}\right)\angle(230° - 60°) = 80.0 \text{ Grams/Mil}_{p\text{-}p}\angle 170°$$

This calculation identifies a single balance sensitivity vector based upon a measured differential response vector of 5.00 Mils,$_{p\text{-}p}$ at an angle of 60°. This vector change in shaft vibration response was due to the installation of a 400 Gram weight at an angle of 230°. Vector division of the weight by the differential vibration vector yields the balance sensitivity vector of 80.0 Grams/Mil,$_{p\text{-}p}$ at an angle of 170°. This unbalance sensitivity vector may now be used to compute balance corrections in a manner similar to the earlier example of vector multiplication.

These simplified rules for vector multiplication and division may be verified by performing the same operations using exponential functions instead of the presented polar coordinates. The results will be identical, and this will reinforce the concept that the vector math may be successfully executed using either exponential or circular functions. In all cases, these vector manipulations are continually used throughout the field of machinery analysis, and these procedures must be mastered to allow progression to the real machinery topics.

UNDAMPED FREE VIBRATION

Expanding upon the concepts of the previous section, again consider the single mass pendulum of Fig. 2-1. Within this earlier mechanical system, the mass of the concrete block was identified as the only significant element in the system. If this concrete block remains constant, and if the weightless cable is replaced by a coil spring, the simple spring mass system of Fig. 2-7 is produced. Assume that the spring is suspended from a totally rigid I-Beam, and consider the mass to be confined to movement only in the vertical direction. Since damping is not involved, this is considered as an undamped mechanical system. In addition, there are no external forces applied to this system, so it must be classified as a system that exhibits free vibration when it is displaced, and allowed to oscillate in the vertical plane. The resultant motion is defined as undamped free vibration of this one degree of freedom mechanical system.

If this physical example is converted into a traditional physics diagram, the sketches shown in Fig. 2-8 evolve. The left diagram shows the main mechanical elements, and the right sketch displays the Free Body Diagram. Normally, this mechanical system would remain at rest (i.e., no motion). For this system to

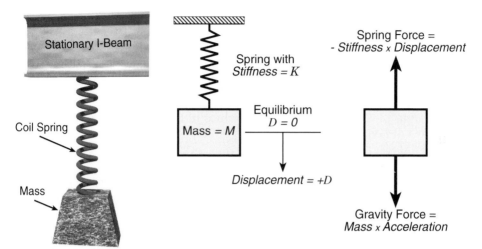

Fig. 2–7 Spring Mass
Mechanical System

Fig. 2–8 Equivalent Spring Mass Mechanical System And
Associated Free Body Diagram

move, some type of initial disturbance is required. Furthermore, when this
mechanical system is in motion, the free body diagram (Fig. 2-8) reveals two
active forces; a spring force, and the gravitational term. The general equation of
motion for this body is simply the equality of active forces as follows:

$$(Mass \times Acceleration) = (-Stiffness \times Displacement)$$

By rearranging terms, the following summation of forces is obtained:

$$(Mass \times Acceleration) + (Stiffness \times Displacement) = 0$$

Substituting a simpler alpha identification for each of the four variables,
the equation of motion for this simple spring mass system may be stated in the
manner that W. T. Thomson used:

$$(M \times A) + (K \times D) = 0 \qquad\qquad \textbf{(2-41)}$$

If equation (2-41) is divided by the mass, the resultant expression contains
a system mechanical constant (i.e., K/M), plus the interrelated acceleration and
displacement of the body:

$$A + \left(\frac{K}{M}\right) \times D = 0 \qquad\qquad \textbf{(2-42)}$$

Equation (2-42) can be satisfied by either of the previously discussed circular or exponential functions. For simplicity, assume that an exponential function
as defined in equation (2-23) is substituted into (2-42) to yield the following version of the equation of motion:

$$-D \times \omega^2 \times e^{i\omega t} + \left(\frac{K}{M}\right) \times D \times e^{i\omega t} = 0$$

Extracting the common terms from this equation, the following is obtained:

$$D \times e^{i\omega t} \times \left(-\omega^2 + \frac{K}{M}\right) = 0 \qquad (2\text{-}43)$$

Equation (2-43) is satisfied for all values of time t when the terms within the brackets are equated to zero:

$$\left(-\omega^2 + \frac{K}{M}\right) = 0$$

This may now be solved for the natural or critical frequency ω_c as follows:

$$\boxed{\omega_c = \sqrt{\frac{K}{M}}} \qquad (2\text{-}44)$$

Another common form of this expression is obtained by converting the rotational frequency ω_c units of Radians per Second to Cycles per Second in accordance with equation (2-2) to yield the following:

$$\boxed{F_c = \frac{1}{2\pi} \times \sqrt{\frac{K}{M}}} \qquad (2\text{-}45)$$

Clearly, the frequency of oscillation is a function of the spring constant, and the mass. This is the undamped natural frequency of the mechanical system. It is also commonly called the undamped critical frequency, and the subscript "c" has been added to identify frequencies ω_c and F_c. In all cases, following an initial disturbance, the mass will oscillate (or vibrate) at this natural frequency, and the amplitude of the motion will gradually decay as a function of time. This reduction in amplitude is due to energy dissipation within a real mechanical system.

Although this result is simple in format, it does represent an extraordinarily important concept in the field of vibration analysis. That is, the natural frequency of a mechanical resonance will respond to an alteration of the stiffness and the mass. Often, the diagnostician has limited information on the effective stiffness, or equivalent mass of the mechanical system. However, changes in stiffness or mass will behave in the manner described by equation (2-44). In many instances, this knowledge of the proper relationship between parameters will allow a respectable solution to a mechanical problem.

Initially, the existence of a unique natural frequency that is a function of the mechanical system mass and stiffness may appear to be only of academic interest. In reality, there are field applications of this physical relationship that may be used to provide solutions for mechanical problems. For instance, if a mechanical system is excited by a periodic force at a frequency that approaches a natural resonant frequency of the mechanical system — the resultant vibratory motion may be excessive, or even destructive. Three potential solutions to this type of problem were identified by J. P. Den Hartog[5], in his text *Mechanical Vibrations*. Quoting from page 87 of this book:

"...In order to improve such a situation, we might first attempt to eliminate the force. Quite often this is not practical or even possible. Then we may change the mass or the spring constant of the system in an attempt to get away from the resonance condition, but in some cases this is also impractical. A third possibility lies in the application of the dynamic vibration absorber, invented by Frahm in 1909...The vibration absorber consists of a comparatively small vibratory system k, m attached to the main mass M. The natural frequency $\sqrt{k/m}$ of the attached absorber is chosen to be equal to the frequency ω of the disturbing force. It will be shown that then the main mass M does not vibrate at all, and that the small system k, m vibrates in such a way that its spring force is at all instances equal and opposite to P_o sin ω t. Thus there is no net force acting on M and therefore the mass does not vibrate..."

In his text book, Den Hartog proceeds to derive a detailed equation set that supports the above statements. He also examines torsional systems, and damped vibration absorbers. Thomson[6] also discussed the utilization of both lateral and torsional vibration absorbers. However, for this discussion, the application of a simple lateral undamped spring mass vibration absorber will be reviewed. The fundamental engineering principles behind an absorber installation are illustrated with the following case history.

Case History 1: Piping System Dynamic Absorber

The mechanical system under consideration consists of a pair of product transfer pumps that were subjected to a modification of the discharge piping to span across a new roadway. These essential pumps were motor driven at a constant speed of 1,780 RPM. The pumps had a successful eight year operating history, with only minor seal problems, and one coupling failure. During a plant revision, the pump discharge piping was rerouted to a new pipe rack. Due to the design of the new rack, the discharge line was poorly supported, and problems began to appear on both pumps shortly after the piping modification.

Multiple seal failures were combined with repetitive bearing, and coupling failures. These two pumps that previously received maintenance attention only once or twice a year were now subjected to overhauls on a monthly basis. This increased maintenance passed unnoticed for a long time. Unfortunately, one night the main pump failed when the spare pump was out for repairs. This coincidence of mechanical failures forced a plant outage, and this event focused management attention upon the reduced reliability of these pumps.

Vibration analysis of the pumps and the associated piping revealed a domi-

---

[5] J.P. Den Hartog, *Mechanical Vibrations*, 4th edition, (New York: McGraw-Hill Book Company, 1956), p. 87.

[6] William T. Thomson, *Theory of Vibration with Applications*, 4th Edition, (Englewood Cliffs, New Jersey: Prentice Hall, 1993), pp. 150-159.

nant motion at the pump running speed of 1,780 RPM. Comparison with histori-cal data revealed 1X vibration amplitudes on the pump and motor were ten to twenty times higher than previously measured. This machinery abnormality was coincident with vertical vibration levels in excess of 25 Mils,$_{p-p}$ at the middle of the unsupported discharge line (i.e., midspan of the road crossing).

A temporary brace was fabricated, and placed below the discharge line. This support reduced the piping vibration, and also resulted in a drop in the pump synchronous motion. Considering the positive results of this test, and some preliminary calculations on the natural frequency of the piping span, it was concluded that the pump running speed was very close to a lateral natural frequency of the new discharge pipe.

Since a brace in the middle of the road was unacceptable as a long-term solution, other possibilities were examined and discarded. Finally, the applica-tion of a tuned spring mass vibration absorber was considered as a potential and practical solution. For this problem a simple horizontal cantilevered vibration absorber was designed to resemble the diagram in Fig. 2-9.

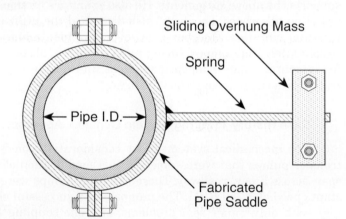

Sliding Overhung Mass

Spring

← Pipe I.D. →

Fig. 2–9 Typical Tuned
Spring Mass Vibration
Absorber Assembly
For Piping System

Fabricated
Pipe Saddle

This device consists of a fabricated pipe saddle that is securely bolted to the outer diameter of the discharge pipe. It is physically located at the point of high-est vibration (i.e., center of the piping span). Since the pipe vibrates vertically, the absorber is positioned horizontally so that the cantilevered weight may also vibrate vertically. In this case, the spring consists of flat bar stock that has the most flexible axis placed in the direction of the desired motion. The overhung mass is bolted to the flat bar stock spring, and it may be moved back and forth to allow adjustment of the natural frequency.

By inspection of this damper assembly, it is apparent that the stiffness and mass of the spring, plus the overhung mass are equivalent to a simple spring mass system. The problem in designing an appropriate vibration absorber is now reduced to a reasonable selection of physical dimensions to obtain a natural fre-quency of 1,780 CPM for this installed assembly.

Several approaches may be used to determine an acceptable set of absorber dimensions. For example, a Finite Element Analysis (FEA) could be performed. However, an FEA approach may become unnecessarily complicated and time consuming. Use of published beam natural frequency equations may also be considered. However, one must be careful of published canned equations where the assumptions and boundary conditions may not be clearly explained or understood. Fortunately, a practical approach for performing these calculations was presented by John D. Raynesford[7], in his *Hydrocarbon Processing* article on this subject. In this article, he considered the system as a simple spring mass assembly. The dimensions of the spring were combined with the overhung mass to provide the basic elements for the absorber design. Specifically, Raynesford considered the total static deflection Y_{total} of the vibration absorber to be associated with the weight W, mass M, gravitational constant G, and the spring constant K of the assembly in the following manner:

$$K = \frac{W}{Y_{total}} = \frac{M \times G}{Y_{total}} \qquad \textbf{(2-46)}$$

This is the same general stiffness relationship that was previously applied to the simple pendulum in equation (2-7). If equation (2-46) for stiffness is placed into the previously developed natural frequency equation (2-44), the following substitution and changes may be performed:

$$\omega_c = \sqrt{\frac{K}{M}} = \sqrt{\frac{M \times G}{Y_{total}} \times \frac{1}{M}} = \sqrt{\frac{G}{Y_{total}}}$$

Solving for the total deflection Y_{total}, the following equation is obtained:

$$Y_{total} = \frac{G}{\omega_c^2} \qquad \textbf{(2-47)}$$

The total end point deflection of the vibration absorber was presumed to be due to a combination of the uniformly distributed weight of the spring, plus the cantilevered mass on a weightless beam. Traditional deflection equations for these two elements may be extracted from various references. For example, deflection of a beam with a uniformly distributed load may be obtained from references such as Shigley[8], or Roark[9] as follows:

$$Y_{spring} = \frac{W_{spring} \times L^3}{8 \times E \times I} \qquad \textbf{(2-48)}$$

[7] John D. Raynesford, "Use Dynamic Absorbers to Reduce Vibration," *Hydrocarbon Processing*, Vol. 54, No. 4, (April 1975), pp. 167-171.

[8] Joseph E. Shigley and Charles R. Mischke, *Standard Handbook of Machine Design*, (New York: McGraw-Hill Book Company, 1986), pp 11.5-11.6.

[9] Warren C. Young, *Roark's Formulas for Stress & Strain*, 6th edition, (New York: McGraw-Hill Book Co., 1989), pp. 100-102.

where: Y_{spring} = End Deflection of Spring (Inches)
W_{spring} = Weight of Spring (Pounds)
L = Length of Spring (Inches)
E = Modulus of Elasticity (= 30 x 10^6 Pounds / $Inch^2$ for steel)
I = Spring Area Moment of Inertia ($Inches^4$)

Similarly, the deflection of a cantilevered mass on a weightless beam may be extracted from either Shigley, or Roark, as follows:

$$Y_{mass} = \frac{W_{mass} \times L^3}{3 \times E \times I}$$ (2-49)

where: Y_{mass} = End Deflection at Mass (Inches)
W_{mass} = Weight of Mass (Pounds)

The total deflection due to the weight of the spring plus the cantilevered mass is obtained by superposition (addition) of these well proven beam deflection equations as follows:

$$Y_{total} = Y_{spring} + Y_{mass}$$ (2-50)

Substituting equations (2-47), (2-48), and (2-49) into the total deflection equation (2-50) yields the following combined result:

$$\frac{G}{\omega_c^2} = \frac{W_{spring} \times L^3}{8 \times E \times I} + \frac{W_{mass} \times L^3}{3 \times E \times I}$$ (2-51)

At this point, the Raynesford article begins a trial and error solution to arrive at the vibration absorber dimensions. Another way to obtain a set of realistic dimensions is to pursue a further simplification of the equation. For instance, equation (2-51) may be solved for the weight of the overhung mass as:

$$W_{mass} = \left(\frac{3 \times G \times E \times I}{L^3 \times \omega_c^2} \right) - \left(\frac{3}{8} \times W_{spring} \right)$$ (2-52)

In equation (2-52), the area moment of inertia I for the flat bar stock used for the spring is determined by the next equation for a rectangular cross section:

$$I = \frac{b \times h^3}{12}$$ (2-53)

where: b = Width of Rectangular Spring (Inches)
h = Height of Rectangular Spring (Inches)

As always, the spring weight is calculated simply by multiplying volume by the material density as follows:

$$W_{spring} = b \times h \times L \times \rho \qquad (2\text{-}54)$$

where: ρ = Material Density (= 0.283 Pounds/Inches^3 for steel)

Equations (2-53) and (2-54) will now be substituted back into (2-52), and simplified to yield the following expression for the overhung mass:

$$W_{mass} = \left(\frac{G \times E \times b \times h^3}{4 \times L^3 \times \omega_c^2} \right) - \left(\frac{3 \times b \times h \times L \times \rho}{8} \right)$$

This expression contains the known quantities of the acceleration of gravity G, the modulus of elasticity E, the density of the spring material ρ. If a spring is constructed from flat stock that is 1 inch wide by 1/2 inch thick, then dimensions b and h are also defined. The undamped natural frequency of the system ω_c should be equal to the measured excitation frequency of 1,780 CPM. Performing these numerical substitutions into the last expression yields:

$$W_{mass} = \left(\frac{386.1 \frac{In}{Sec^2} \times 30 \times 10^6 \frac{Lb}{In^2} \times 1 In \times (0.5 In)^3}{4 \times L^3 \times \left(1,780 \frac{Cycle}{Min} \times 2\pi \frac{Rad}{Cycle} \times \frac{Min}{60 Sec} \right)^2} \right) - \left(\frac{3 \times 1 In \times 0.5 In \times 0.283 \frac{Lb}{In^3} \times L}{8} \right)$$

Performing these calculations, the following simplified result is obtained:

$$W_{mass} = \left(\frac{21.84}{L} \right)^3 Pound\text{-}Inch^3 - (0.053 \times L) \frac{Pounds}{Inch} \qquad (2\text{-}55)$$

Equation (2-55) correlates the weight of the overhung mass to the overhung length for the defined conditions. The graph shown in Fig. 2-10 is a plot of equation (2-55). It describes this specific relationship between the length of the spring and the magnitude of the overhung mass. From this plot it is obvious that the longer the spring, the less mass required. Conversely, as the spring is shortened, the overhung weight must be increased. For this particular piping problem, a spring length of 12 inches was selected with an overhung weight calculated from equation (2-55) of 5.4 pounds. This same weight value could also be extracted from the curve plotted in Fig. 2-10 for a spring length of 12 inches.

To allow fine tuning of the absorber resonant frequency, the spring was fabricated to be 15 inches long. This additional length does slightly violate the developed equation array, but the error is small. In addition, it must be recognized that the developed equations do not constitute a rigorous solution, but they do provide an acceptable solution. Thus, the extra spring length allows the ability to perform a final adjustment of the natural frequency to correct for variations in the calculations, the fabrication process, or the field attachment.

Normally, it is desirable to bench test the vibration absorber in the shop before installation, and perform most adjustments before installing the device in the field. In most instances, a simple hammer test with an accelerometer and

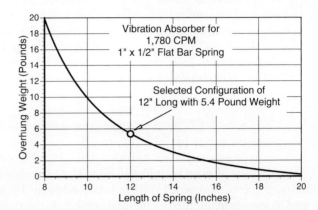

Fig. 2–10 Spring Length Versus Overhung Weight For Piping System Lateral Vibration Absorber

spectrum analyzer will identify the natural frequency of the absorber. If the natural frequency is low, then the overhung weight should be moved in towards the support. The opposite is also true. That is, if the measured natural frequency of the absorber in the shop test is on the high side, then the overhung weight should be moved away from the support.

In this case, the 5.4 pound weight was finally positioned at 12.5 inches from the base during the shop frequency response test. Another minor adjustment was made after the absorber was bolted into place on the discharge line. This device proved to be successful, and piping vibration was reduced from levels in excess of 25 Mils,$_{p-p}$, to a final condition at the pipe midspan of 1.5 to 2.0 Mils,$_{p-p}$. More significantly, the vibration amplitudes on the two transfer pumps returned to previous historical levels, and the failures ceased.

The article by Raynesford also offers the following two important rules regarding the attachment and fabrication of absorbers:

"1. Try to attach the absorber at the point of maximum vibration and in such a way as to vibrate in the same plane. That is, if the bearing housing vibrates in the horizontal plane, mount the absorber vertically so it can also vibrate in the horizontal plane. Adjust the weight in and out until minimum vibration on the unit (maximum on the absorber) is achieved.

2. A rigid attachment is essential-the wand must flex, not the attachment. Be careful when using welds. They are prone to failure in the heat effected zone. Make generous use of large radii at the juncture of the wand and the base or attachment..."

To this pair of recommendations, it would also be advisable to suggest that the absorber be shop tuned to the desired natural frequency. This is always easier to perform in the machine shop versus the field. In addition, the vibration absorber should be installed with a permanent safety chain loosely connecting the assembly with some adjacent rigid structure. If the support saddle or the attachment welds fail, this safety chain would restrain the spring mass assembly, and significantly minimize the potential for any personnel injury.

Overall, it must be recognized that a vibration absorber provides a cost-effective solution to some difficult problems, and it demonstrates a practical application of an undamped mechanical system. On the other hand, a vibration absorber is certainly not a universal solution for all machinery vibration problems. It is often ineffective when used to correct a rotor resonance, or an acoustic resonance problem. In most instances, this type of vibration absorber is useful for addressing certain types of structural resonance problems — and it should always be applied with good engineering judgment and common sense.

FREE VIBRATION WITH DAMPING

Now consider an expansion upon the concepts of the undamped system by including another type of vibration absorber. At this point, consider a single concrete block suspended from a rigid I-Beam by a spring, plus an automotive type shock absorber as shown in Fig. 2-11. Again, the weight is allowed to move in a vertical direction, and the equivalent spring mass damper system is depicted in Fig. 2-12. Since damping is now involved, this is a damped mechanical system. As before, there is no external force applied, and the behavior of this system must be classified as free vibration when it is allowed to oscillate.

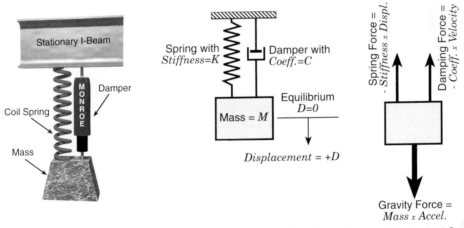

Fig. 2–11 Spring Mass Damper System

Fig. 2–12 Equivalent Spring Mass Damper Mechanical System And Associated Free Body Diagram

The shock absorber is a viscous damper that displays the fundamental property of a damping force that is proportional to velocity times the damping coefficient. This relationship is quite clear when the analogy of an old versus new shock absorber is considered. Specifically, an old worn out shock absorber will be quite loose, and the inner rod will move easily in and out of the main body. However, a new shock absorber will be tight, and extension or compression of the inner rod requires the application of a slow steady force. If an individual attempts to rapidly move the inner rod, they will find that this higher speed motion is resisted by a significantly larger force.

Functionally, the shock absorber or damper removes energy from the system. To state it another way, the damper provides the fundamental means of energy dissipation for the mechanical system. If this physical representation is converted into a traditional physics free body diagram, Fig. 2-12 evolves. Once more, the vertical motion must be initiated by an initial disturbance, and the system now reveals a spring force, a damping force, plus the necessary gravitational term. From this free body diagram, the force balance yields the following equation of motion for this damped mechanical system:

$$(Mass \times Accel) = (-Stiffness \times Displ) + (-Damping\ Coeff \times Velocity)$$

Moving all terms to the left side of the equation, the expression becomes:

$$(Mass \times Accel) + (Damping\ Coeff \times Velocity) + (Stiffness \times Displ) = 0$$

Substituting a simpler alpha identification for the six physical variables, the equation of motion may be stated as:

$$(M \times A) + (C \times V) + (K \times D) = 0 \qquad \textbf{(2-56)}$$

Again, in the manner used by W. T. Thomson, the periodic displacement of this damped spring mass system may be defined with an exponential function similar to equation (2-23). If the displacement is identified by D, and the time is specified by t, and S is a constant that has to be determined, an appropriate exponential equation would have the following form:

$$D = e^{St} \qquad \textbf{(2-57)}$$

As demonstrated earlier in this chapter, acceleration, velocity, and displacement are integrally related, and equation (2-56) may be rewritten in terms of displacement by the substitution of equation (2-57). Certainly the displacement term may be inserted directly. The velocity and acceleration terms are obtained by taking the first and second time derivatives of equation (2-57) to yield the following equation of motion for this damped single degree of freedom system:

$$(M \times S^2 \times e^{St}) + (C \times S \times e^{St}) + (K \times e^{St}) = 0$$

This expression may be simplified by factoring out the common exponential term, and dividing by the mass M to yield the next form of the motion equation:

$$e^{St} \times \left\{ S^2 + \left(\frac{C}{M}\right) \times S + \left(\frac{K}{M}\right) \right\} = 0 \qquad \textbf{(2-58)}$$

As discovered in the undamped case previously discussed, equation (2-58) may be satisfied for all values of time t, when the following occurs:

$$\left\{ S^2 + \left(\frac{C}{M}\right) \times S + \left(\frac{K}{M}\right) \right\} = 0$$

This last expression takes the distinctive form of a quadratic equation. From basic algebra, it is known that this expression may be solved for the constant S in the following traditional manner:

$$S_{1,2} = -\left(\frac{C}{2M}\right) \pm \sqrt{\left(\frac{C}{2M}\right)^2 - \left(\frac{K}{M}\right)} \qquad \text{(2-59)}$$

Two solutions are produced (± radical), and the general equation must be expanded to correspond with this dual root. Hence, the periodic displacement described by equation (2-57) is redefined in the following manner:

$$D = A \times e^{S_1 t} + B \times e^{S_2 t} \qquad \text{(2-60)}$$

Constants A and B depend on how the oscillation was started. The behavior of a damped system is dependent on whether the radical from equation (2-59) is real (+), imaginary (-), or zero (0). The simplest case is the zero value for the radical, and this term is defined as critical damping $\mathbf{C_c}$, as follows:

$$\left(\frac{C_c}{2M}\right)^2 - \left(\frac{K}{M}\right) = 0 \qquad \text{(2-61)}$$

By rearranging terms, the following intermediate result is obtained:

$$\left(\frac{C_c}{2M}\right)^2 = \left(\frac{K}{M}\right)$$

By taking the square root of both sides of the equation, and substituting equation (2-44), the following is obtained:

$$\frac{C_c}{2M} = \sqrt{\frac{K}{M}} = \omega_c \qquad \text{(2-62)}$$

For convenience, a damping ratio of ξ will be defined as the actual damping C divided by the critical damping C_c as follows:

$$\xi = \frac{C}{C_c} \qquad \text{(2-63)}$$

Combining the damping ratio from equation (2-63), and equation (2-62), the term $C/2M$ may be reconfigured as:

$$\frac{C}{2M} = \left(\frac{C}{2M}\right) \times \left(\frac{C_c}{C_c}\right) = \left(\frac{C_c}{2M}\right) \times \left(\frac{C}{C_c}\right) = \omega_c \times \xi \qquad \text{(2-64)}$$

Based on these derived expressions, the solution to the quadratic equation (2-59) may now be rewritten as:

$$S_{1,2} = -\omega_c \times \xi \pm \sqrt{(\omega_c \times \xi)^2 - (\omega_c)^2}$$

If the undamped natural frequency ω_c is factored out of this expression, the following result is obtained:

$$S_{1,2} = \omega_c \times \left\{ -\xi \pm \sqrt{\xi^2 - 1} \right\} \qquad \textbf{(2-65)}$$

Interestingly enough, the solution for constants S_1 and S_2 reveals a relationship between the undamped natural frequency ω_c, and the damping ratio ξ. The transition between oscillatory and non-oscillatory motion is referred to as critical damping. For this case, $C=C_c$, $\xi=1$, and equation (2-65) simplifies to:

$$S_{1,2} = -\omega_c \qquad \textbf{(2-66)}$$

Substituting this critical damping solution for S_1 and S_2 back into the general equation of motion, equation (2-60) produces the following result:

$$D = (A + B) \times e^{-\omega_c t} \qquad \textbf{(2-67)}$$

This function contains only one constant $(A+B)$, and the solution lacks the required number of independent constants to properly represent the general solution. In this case, an expression in the form of $t \times e^{-\omega_c t}$ will satisfy the equation. Upon substitution of this new form, the general solution of equation (2-67) can be correctly written in the following manner:

$$D = (A + B \times t) \times e^{-\omega_c t} \qquad \textbf{(2-68)}$$

The significance of a critically damped system is depicted in Fig. 2-13. In this diagram, the displacement of the mass is plotted against time. For numerical simplicity, the constants A and B were assigned values of +10 and -5 respec-

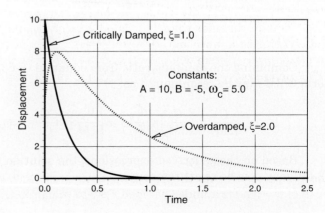

Fig. 2–13 Time Domain Amplitude Response Of Critically Damped, And Overdamped Mechanical Systems

tively, and the natural frequency ω_c was set equal to 5.0. The solid line represents a critically damped system. The resultant motion is aperiodic, and this critically damped system returns to rest in the shortest time without oscillation (vibration) of the mass. Stated in another way, a critically damped system contains the minimum amount of damping necessary for aperiodic motion.

If the system damping is greater than critical damping, the system is considered to be overdamped. Conversely, if the mechanical system has less than critical damping, the system is underdamped, and it will oscillate or vibrate with time. It should be noted that most process machines are underdamped, and sustained motion of the rotating or reciprocating elements is normal behavior.

For a better understanding of damping, consider an overdamped system (equivalent to a new shock absorber). In this case, the damping ratio would be greater than one ($\xi > 1$), and the S terms in the quadratic solution equation (2-65) may be specified as:

$$S_1 = \omega_c \times \left\{ -\xi + \sqrt{\xi^2 - 1} \right\}$$

$$S_2 = \omega_c \times \left\{ -\xi - \sqrt{\xi^2 - 1} \right\}$$

Combining these expressions with the general equation produces the following equation for the motion of an overdamped mechanical system:

$$D = A \times e^{\omega_c t \times \left\{ -\xi + \sqrt{\xi^2 - 1} \right\}} + B \times e^{\omega_c t \times \left\{ -\xi - \sqrt{\xi^2 - 1} \right\}} \qquad \textbf{(2-69)}$$

This function describing an overdamped mechanical system is plotted as the dotted line in Fig. 2-13. The displacement change with time is the sum of two decaying exponential functions, and system vibration is not maintained. Motion is aperiodic, and the body returns to rest without oscillation. It is also clear from this composite diagram that the overdamped system of equation (2-69) does not return to rest as rapidly as the previously discussed critically damped case.

Finally, consider the situation of a mechanical system with small damping (equivalent to a worn out shock absorber). This is generally referred to as an underdamped system where $\xi < 1$, and the radical of equation. (2-65) is imaginary. The constants S_1 and S_2 for this condition can be written as follows:

$$S_{1,2} = \omega_c \times \left\{ -\xi \pm \sqrt{1 - \xi^2} \right\} \qquad \textbf{(2-70)}$$

Using equation (2-70) to recompute the S terms, and then including these new expressions into the general equation, the following solution for the equation of motion for a under damped system was presented by W.T. Thomson as:

$$D = e^{-\xi\omega_c t} \times \left[A \times e^{\omega_c t \times \left\{ -i\sqrt{1-\xi^2} \right\}} + B \times e^{\omega_c t \times \left\{ -i\sqrt{1-\xi^2} \right\}} \right]$$

This expression may be simplified to a more understandable format as:

$$D = Y \times e^{-\xi\omega_c t} \times \left\{ \sin\left(\omega_c t \times \sqrt{1-\xi^2} + \varphi\right) \right\} \qquad \textbf{(2-71)}$$

By inspection, equation (2-71) consists of the superposition of an oscillating sine wave plus an exponential term. In most cases, the amplitude of the sine wave is decreased by the exponential function with increasing time. The variable Y in this equation represents the peak intersection between the exponential function and zero time. The φ term is the timing lag between the oscillatory curve and a zero time starting point. For demonstration purposes, a response curve for an under damped system is plotted in Fig. 2-14. A displacement value of 10.0 was assigned to Y, the timing offset φ was set equal to zero, and a constant value of 5.0 was used for the undamped natural frequency ω_c. This system exhibits an oscillatory motion with respect to time, and this is referred to as free vibration of the under damped mechanical system.

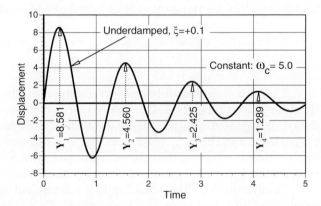

Fig. 2–14 Time Domain Amplitude Response Of An Under Damped Mechanical System

Another interesting point from examining (2-71) is that the term $\omega_c\sqrt{1-\xi^2}$ is multiplied by the time t to determine the number of radians. This suggests that the undamped natural frequency ω_c is altered by the damping ratio ξ to produce a new frequency. In fact, this is commonly identified as the damped natural frequency or $\omega_{damped\ critical}$ for the mechanical system, and it is defined in the following manner:

$$\boxed{\omega_{damped\ critical} = \omega_c \times \sqrt{1-\zeta^2}} \qquad \textbf{(2-72)}$$

This is a very important equation because it directly influences the impact of damping upon a resonance. As shown in the forthcoming Fig. 2-18 for forced vibration — variations in damping produce major changes in amplitude and phase through a resonance. However, there is also a subtle shift in the resonant frequency as the damping is varied. Thus, the machinery diagnostician must be fully aware of the fact that changes in system damping will alter the behavior through a resonance in the following three ways:

○ Significant changes in the peak amplitude at the resonance.
○ Significant variations in the phase angle change across the resonance.
○ Subtle change in the damped natural frequency (i.e., damped critical speed).

Fig. 2-14 for an under damped system shows that the oscillatory motion decays with time. Examination of a longer time record would reveal that the amplitude decrease is actually an exponential decay. The rate of this exponential decay may be quantified by the log decrement which is defined as follows:

$$Log\ Decrement\ =\ \delta\ =\ \ln\!\left(\frac{Y_1}{Y_2}\right) = \frac{1}{N} \times \ln\!\left(\frac{Y_1}{Y_{N+1}}\right) \qquad \textbf{(2-73)}$$

In equation (2-73), Y_1 and Y_2 represent any two successive amplitudes in the decaying dynamic signal. The natural logarithm of this ratio defines the damping as the log decrement δ. In some cases, particularly with lightly damped or short duration experimental data, it is necessary to examine multiple cycles of the decaying signal to determine the log decrement. For these situations, the right hand side of equation (2-73) may be used. Within this part of the expression, the initial peak amplitude is still specified by Y_1, and the amplitude following N number of cycles is identified as Y_{N+1}. The validity of this relationship may be checked by calculating the log decrement for different combinations. For instance, in Fig. 2-14, the first amplitude peak Y_1 has a magnitude of 8.581, and the second peak Y_2 is equal to 4.560. Using the first part of equation (2-73), the log decrement δ may be computed from these values in the following manner:

$$Log\ Decrement\ =\ \delta\ =\ \ln\!\left(\frac{Y_1}{Y_2}\right) = \ln\!\left(\frac{8.581}{4.560}\right) = \ln(1.882) = 0.632$$

The same calculation may be performed using the first three cycles in Fig. 2-14. For this case the third peak Y_4 has an amplitude of 1.289, and the log decrement δ may be computed with the right side of equation (2-73) as follows:

$$Log\ Decrement\ =\ \delta\ =\ \frac{1}{N} \times \ln\!\left(\frac{Y_1}{Y_{N+1}}\right) = \frac{1}{3} \times \ln\!\left(\frac{Y_1}{Y_{3+1}}\right)$$

$$\delta = \frac{1}{3} \times \ln\!\left(\frac{Y_1}{Y_4}\right) = \frac{1}{3} \times \ln\!\left(\frac{8.581}{1.289}\right) = \frac{\ln(6.657)}{3} = \frac{1.896}{3} = 0.632$$

The same result of $\delta=0.632$ has been reached using a single cycle and multiple cycles. Certainly this concept may be extended to the examination of various decaying dynamic data sets. It should also be mentioned that the log decrement δ may also be expressed in terms of the critical damping ratio ξ. It can be shown that the log decrement δ is accurately expressed as:

$$Log \ Decrement \ = \ \delta \ = \ \frac{2\pi \times \xi}{\sqrt{1 - \xi^2}} \qquad\qquad (2\text{-}74)$$

The decaying signal plotted in Fig. 2-14 was produced with a damping ratio of ξ=0.1. To check the validity of equation (2-74), this damping ratio may be used to calculate the log decrement as follows:

$$Log \ Decrement \ = \ \delta \ = \ \frac{2\pi \times \xi}{\sqrt{1 - \xi^2}} \ = \ \frac{2\pi \times 0.1}{\sqrt{1 - 0.1^2}} \ = \ 0.632$$

Once more the same value for the log decrement has been obtained. This provides confidence that equations (2-73) and (2-74) are compatible, and consistent. Depending on the available data, one expression may be easier to apply versus the other. Another usable format for these expressions is obtained by solving equation (2-74) for the damping ratio ξ to produce the following:

$$Damping \ Ratio \ = \ \xi \ = \ \frac{\delta}{\sqrt{4\pi^2 + \delta^2}} \qquad\qquad (2\text{-}75)$$

Equation (2-75) is useful for determining the damping ratio based upon experimental or analytical values of the log decrement. Further examination of equations (2-74) and (2-75) reveals that the damping ratio ξ and the log decrement δ are closely related. For instance, the polarity of the log decrement and the damping ratio must be the same. If the damping ratio is positive, the log decrement must also be positive. Similarly, if the log decrement is negative, then the damping ratio must be negative. The physical significance of negative damping is depicted in Fig. 2-15. This diagram is based upon equation (2-71) where Y was assigned a value of 1.0, the timing offset φ was equated to zero, a value of 5.0 was

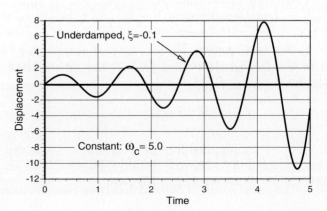

Fig. 2–15 Time Domain Amplitude Response Of An Unstable Mechanical System

used for ω_c, and the damping ratio ξ was minus 0.1. It is clear that the oscillatory motion of this system increases with time, and this is representative of an unstable mechanical system.

In essence, a large positive value for the log decrement is synonymous with a well damped system. Conversely, a small positive log decrement is indicative of a stable system with lower damping. The inclusion of the plus sign signifies a stable mechanical system that will exhibit decreasing amplitudes as a function of time. The appearance of a negative value for the log decrement defines a system that displays increasing amplitudes with time. This is descriptive of an inherently unstable system. In this type of mechanical system, the motion will continue to increase until physical damage, or eventual destruction occurs.

FORCED VIBRATION

The previous sections have addressed free vibration where the motion originates with an initial disturbance, and the energy of the resultant motion is dissipated by damping or friction. When the available energy is completely removed, the system returns to a rest condition of zero motion.

However, real mechanical systems are influenced by both external and internal forcing functions. These forces are often periodic, and they provide a continuous energy input into the system. In this situation, the mechanical system continues to vibrate, and does not return to a state of zero motion. A simple example of a periodic forcing function can be defined with an expression such as:

$$Force = F \times \sin(\omega t) \qquad \textbf{(2-76)}$$

Equation (2-76) defines a maximum force F that varies periodically (in this case sinusoidally) as a function of frequency ω, and time t. This force could be equated to the summation of damped spring mass forces described by W.T. Thomson in equation (2-56) to provide the following force balance:

$$(M \times A) + (C \times V) + (K \times D) = F \times \sin(\omega t) \qquad \textbf{(2-77)}$$

Equation (2-77) states that the weight, damping force, and spring force are equal to the applied forcing function. It is presumed that the body oscillates at the frequency of the forcing function, and that the resultant motion is identical to the forcing function. For this case of forced vibration, assume that the displacement is represented by the following circular function that is similar to the previous equation (2-11):

$$Displacement = D = Y \times \sin(\omega t - \varnothing) \qquad \textbf{(2-78)}$$

In this equation, Y represents the peak displacement, and $\varnothing$ is the phase lag between the applied force and the motion. The first and second time derivatives of equation (2-78) will yield the following two expressions for velocity and acceleration of the mechanical system:

$$Velocity = V = Y \times \omega \times \sin(\omega t - \varnothing + \pi/2) \tag{2-79}$$

$$Acceleration = A = Y \times \omega^2 \times \sin(\omega t - \varnothing + \pi) \tag{2-80}$$

Inserting equations (2-78), (2-79), and (2-80) back into the equation (2-77), yields the next expression for forced vibration of a spring mass damper system:

$$\{M \times Y \times \omega^2 \times \sin(\omega t - \varnothing + \pi)\} + \{C \times Y \times \omega \times \sin(\omega t - \varnothing + \pi/2)\} + \tag{2-81}$$
$$\{K \times Y \times \sin(\omega t - \varnothing)\} = F \times \sin(\omega t)$$

This expression defines four vector quantities that remain in a fixed relationship with respect to each other as depicted in Fig. 2-16. These four vectors rotate together at a constant frequency ω. The graphical representation of these four vectors reveals that the stiffness term includes displacement, the velocity term considers damping, and the acceleration vector incorporates the inertia force. As shown, the three vibration vectors are mutually perpendicular.

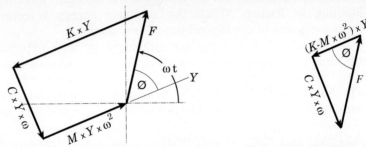

Fig. 2–16 Forced Vibration - Primary Vectors   **Fig. 2–17** Simplified Vector Triangle

This is consistent with the vector diagram previously shown in Fig. 2-2. In both cases, the 90° shifts between displacement, velocity, and acceleration are clearly evident. Since the displacement and acceleration terms are opposite in direction, the vector diagram of Fig. 2-16 may be redrawn into the simplified sketch shown in Fig. 2-17.

The simplified vector diagram contains the same elements as Fig. 2-16. However, the differential vector between displacement and acceleration terms is used instead of the separate and opposite vectors. This simplification provides a right triangle with the applied forcing function as the hypotenuse, the damping term as one leg of the triangle, and the differential stiffness-inertia vector as the third leg of the triangle. The maximum displacement Y may be determined from trigonometry by setting the square of the hypotenuse equal to the sum of the squares of the other two sides of this right triangle:

$$F^2 = \{(K - M \times \omega^2) \times Y\}^2 + \{C \times Y \times \omega\}^2$$
$$F = Y \times \sqrt{(K - M \times \omega^2)^2 + (C \times \omega)^2}$$

Solving for the displacement Y produces the following equation:

$$Y = \frac{F}{\sqrt{(K - M \times \omega^2)^2 + (C \times \omega)^2}} \tag{2-82}$$

The maximum angle between the forcing function and the displacement Ø may be determined from a trigonometric relationship in the following manner:

$$\tan\varnothing = \frac{C \times \omega}{K - M \times \omega^2} \tag{2-83}$$

If the numerator and denominator of equations (2-82) and (2-83) are divided by the stiffness K, the following intermediate results are obtained. Once again, these equations are consistent with the approach used by W.T. Thomson.

$$Y = \frac{\left(\dfrac{F}{K}\right)}{\sqrt{\left(1 - \dfrac{M \times \omega^2}{K}\right)^2 + \left(\dfrac{C \times \omega}{K}\right)^2}} \tag{2-84}$$

$$\tan\varnothing = \frac{\left(\dfrac{C \times \omega}{K}\right)}{\left(1 - \dfrac{M \times \omega^2}{K}\right)} \tag{2-85}$$

Equations (2-84) and (2-85) may be further simplified by converting these expressions into a non-dimensional format. This can be accomplished by incorporating the following previously defined terms:

From equation (2-44) the system natural frequency = $\omega_c = \sqrt{K/M}$

From equation (2-63) the damping ratio = $\xi = C/C_c$

From equation (2-62) the critical damping coefficient = $C_c = 2\,M\omega_c$

In addition, the following two new non-dimensional terms will be defined, and included in this conversion:

$$Critical\ Speed\ Frequency\ Ratio = \Omega = \frac{\omega}{\omega_c} \tag{2-86}$$

$$Amplitude\ Ratio = \frac{Y}{Y_c} \tag{2-87}$$

Where Y_0 is the zero frequency deflection of the spring mass damper system due to the application of the periodic forcing function F. In this context, Y_0 is equal to F/K in equation (2-84). By inserting the previously defined non-dimensional expressions into equations (2-84) and (2-85) the following are obtained:

$$Y = \frac{Y_o}{\sqrt{\left(1 - \dfrac{\omega^2}{\omega_c^2}\right)^2 + \left(2 \times \xi \times \dfrac{\omega}{\omega_c}\right)^2}}$$

$$\tan\varnothing = \frac{\left(2 \times \xi \times \dfrac{\omega}{\omega_c}\right)}{\left(1 - \dfrac{\omega^2}{\omega_c^2}\right)}$$

Finally, by incorporating equations (2-86), and (2-87) into the above expressions, the non-dimensional *Amplitude Ratio*, and *Phase Angle* $\varnothing$ are described by the resulting equations (2-88) and (2-89).

$$Amplitude\ Ratio = \frac{1}{\sqrt{(1 - \Omega^2)^2 + (2 \times \xi \times \Omega)^2}} \qquad\qquad (2\text{-}88)$$

$$\tan\varnothing = \frac{(2 \times \xi \times \Omega)}{(1 - \Omega^2)} \qquad\qquad (2\text{-}89)$$

From these two expressions, the amplitude ratio and phase angle are reduced to functions of the damping ratio ξ, and the critical speed ratio Ω This is an expected result since it is common knowledge that vibration amplitudes in the vicinity of a resonance are directly related to the frequency offset from the center frequency of the resonance. Furthermore, the damping ratio should also participate in determining the response characteristics since it is the fundamental indicator of energy dissipation.

This relationship is easier to understand in the graphic format of Fig. 2-18. Within this diagram, a family of five curves are plotted over a critical speed frequency ratio Ω between 0 and 3.0, and a damping ratio ranging from an underdamped system at $\xi=0.1$ to an overdamped condition of $\xi=2.0$. This general type of data display is referred to as a response plot or a Bode plot. In most cases, the synchronous 1X vibration amplitude and phase are plotted against rotative speed instead of the non-dimensional values used for Fig. 2-18. However, the concepts of tracking synchronous amplitude and phase as a function of speed are the same for a dimensional or a non-dimensional system.

In either case, an amplitude increase occurs at the natural or critical frequency ($\Omega=1$) of the system. This amplitude response is coincident with a substantial phase shift at the critical speed. It is also apparent that the magnitude of the amplitude response, and the amount of the phase shift are both directly related to the system damping.

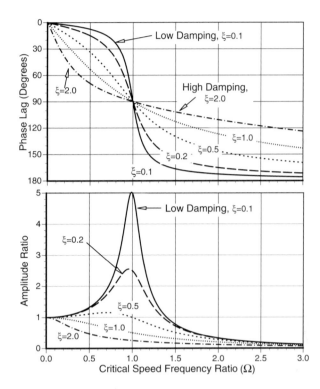

Fig. 2–18 Calculated
Bode Plot Of Forced
Response For A Simple
Mechanical System

Specifically, an under damped mechanical system (e.g., ξ=0.1) exhibits a large amplitude change through the resonant frequency, combined with a phase shift that approaches 180°. Conversely, a heavily damped system (e.g., ξ=2.0) displays no amplitude change through the critical speed region, and only a minor variation in phase angle to reveal the presence of the natural frequency.

It is meaningful to recall that the forcing function for equation (2-88) was based upon a constant driving force. That is, the same force is applied to the system at every frequency. In actuality, forced vibration is often excited by a force that varies with machine speed. For instance, centrifugal force due to unbalance varies as the square of the speed. Hence, it is desirable to consider revising the amplitude ratio of equation (2-88) to include a forcing function that incorporates an Ω^2 term. This would simulate the relationship between the amplitude response, and the mass unbalance forcing function. In the simplest format, it can be shown that the amplitude ratio for an unbalance response is described by the next equation that includes the speed squared Ω^2 term in the numerator:

$$Amplitude\ Ratio = \frac{\Omega^2}{\sqrt{\left(1 - \Omega^2\right)^2 + \left(2 \times \xi \times \Omega\right)^2}} \qquad \textbf{(2-90)}$$

For this forced unbalance response, the phase relationship remains identical to the previously presented equation (2-89). If the amplitude ratio and phase for a forced unbalance condition are plotted as a function of the critical speed ratio, Fig. 2-19 emerges. Note that the damping relationship remains consistent with the previous discussion, and the amplitude exhibits the most significant change. In this forced unbalance case, the amplitude ratio at low speeds approaches zero due to a small driving force (i.e., low unbalance force). At the critical speed of $\Omega=1$, the magnitude of the peak is governed by the damping. At frequencies above the critical speed, the amplitude and phase remain fairly constant for each value of damping ratio. Hence, above a resonance, it is normal to encounter a plateau region where synchronous amplitude and phase remain reasonably constant with increasing machine speed. This behavior will be demonstrated with actual examples of machinery vibration data throughout this text.

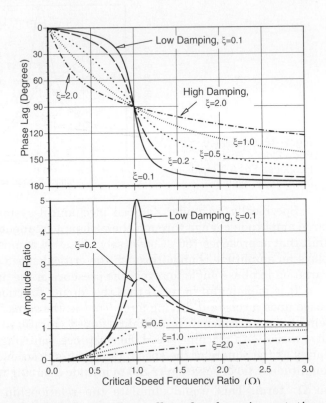

Fig. 2–19 Calculated Bode Plot Of Unbalance Response For A Simple Mechanical System

A calculated or a measured Bode plot is excellent for observing rotational speed vibration amplitude and phase as a function of speed (frequency ratio). However, there are conditions when a detailed examination of phase changes is required. Fortunately, the same vector information that is used to construct a Bode plot may also be viewed in terms of polar coordinates. In this type of data presentation, the vectors (amplitudes and angles) at each speed (or speed ratio) are plotted in a polar format. The heads of the vectors are then connected to form a continuous line. This type of diagram is generally referred to as a polar plot, and it has many applications within the domain of machinery malfunction analysis. A typical example of this type of data is shown in Fig. 2-20. In this diagram, the conditions of low damping (i.e., ξ=0.1 and 0.2) were extracted from Fig. 2-19, and these curves were replotted in a polar coordinate format. The peak of the resonance occurs at Ω=1, which is coincident with a 90° phase angle shift. As in the Bode plot, the high speed condition is identified as Ω=3.0.

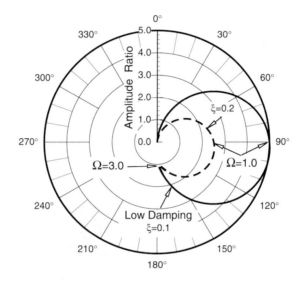

Fig. 2–20 Calculated Polar Plot Of Unbalance Response For A Simple Mechanical System

Vector angles on the polar plot are always plotted against rotation. This phase lag logic is directly associated with the vibration measurement systems used to analyze the machinery behavior. In virtually all cases, it is highly desirable (if not mandatory) to generate data plots that are physically representative of the machinery geometry. This topic will be discussed in much greater detail in the subsequent chapters 3, 5, 6, 7 and 11.

Before leaving the calculated Bode and polar plots for this forced response of a spring mass damper system, the magnitude of the vibration at the natural frequency should be examined in greater detail. From the previous definitions, it is clear that the critical speed frequency ratio Ω is equal to unity ($\omega/\omega_c=1$) at the natural frequency. Substituting this value of Ω=1 into either equations (2-88) or (2-90) yields identical results. Specifically, when a value of Ω=1 is placed into equation (2-88), the following result is obtained:

$$Amplitude\ Ratio\ =\ \frac{1}{\sqrt{(1-\Omega^2)^2+(2\times\xi\times\Omega)^2}}$$

$$Amplitude\ Ratio\ =\ \frac{1}{\sqrt{(1-1^2)^2+(2\times\xi\times1)^2}}\ =\ \frac{1}{\sqrt{0+4\times\xi^2}}$$

Thus, at the natural resonance of $\Omega=1$, the Amplitude Ratio is reduced to:

$$Amplitude\ Ratio_{at\ Resonance} = \frac{1}{2 \times \xi} \qquad \textbf{(2-91)}$$

The amplitude ratio for this simple system is a function of the critical damping ratio ξ. This relationship is easily tested by running some trial values. For instance, if the damping ratio $\xi=0.1$, then the amplitude ratio is equal to 5.0. Similarly, if the damping ratio $\xi=0.5$, then the amplitude ratio is equal to 1.0. These values are consistent with the plots presented on both Figs. 2-18 and 2-19.

The amplitude ratio computed with equation (2-91) is a useful quantity that is often referred to as the *Amplification Factor* for the mechanical system. This amplitude ratio, or amplification factor is also called the Q for the resonance. This dimensionless quantity provides a way to describe the severity of a particular resonance, or the magnitude of the damping ratio at a resonance. In all cases, a high Q is indicative of a system with minimal damping that exhibits large amplitudes at the peak of the resonance. Systems with low available damping may be easily excited, and may be susceptible to stability problems due to a lack of available system damping.

Conversely, systems with a small value for Q must be well damped by definition, and this type of system will exhibit low vibration amplitudes at the resonant frequency. Systems with higher damping will be more difficult to excite, and will be less susceptible to a variety of stability problems.

The amplification factor for a rotating machine passing through a specific resonance (critical speed) may be evaluated analytically from the damped critical speed calculations discussed in chapter 5. Based upon the real and imaginary portions of the complex Eigenvalue, the log decrement may be computed. The amplification factor for the resonance is determined by dividing π by the calculated log decrement.

From a measurement standpoint, various methods are used to determine the amplification factor based upon the vibration response data of a machine passing through the rotor resonance. A comparison of the three traditional methods are reviewed in chapter 3.

The theoretical model of mechanical system behavior is closely matched by the motion of actual rotating machines. For example, Fig. 2-21 depicts a Bode plot of a high speed compressor rotor mounted between bearings. In this diagram, the rotational speed in RPM is plotted on the horizontal axis, with synchronous 1X amplitude and phase lag presented on the dual vertical axes.

This field vibration data was measured with a shaft sensing proximity probe mounted close to the coupling end journal bearing. From this plot it is apparent that a resonance occurs at a speed of 6,100 RPM. This response is the first critical speed of the rotor. Amplitude response through this resonance is moderate, and the overall phase roll through the critical is approximately 110°. This data indicates that the mechanical system is underdamped, with a damping ratio in the vicinity of $\xi=0.2$. This type of transient speed behavior is normal and customary for many types of machines within the process industries.

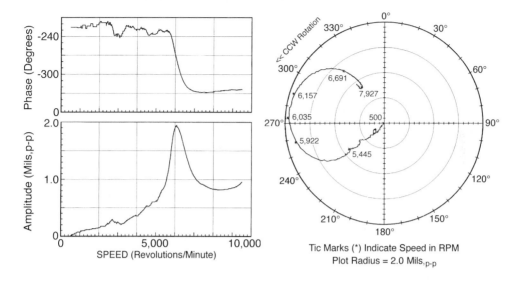

Fig. 2–21 Measured Bode Plot Of Actual Industrial Centrifugal Compressor

Fig. 2–22 Measured Polar Plot Of Actual Industrial Centrifugal Compressor

The same synchronous 1X vector vibration data may be plotted in the polar coordinate format of Fig. 2-22. This data is identical to the Bode plot, but the polar presentation provides improved resolution of phase changes. As viewed from the driver end of the train, this machine rotates in a counterclockwise direction. The angular reference system begins with 0° at the probe, and the phase angles increase in a direction that is counter to shaft rotation. This is the correct angular convention, and the specific logic for this phase convention will be discussed in succeeding chapters. Note that the origin of the polar plot is coincident with low speed, and the plot evolves in a clockwise direction as speed increases. Again, this is normal behavior for a rotor resonance where the phase continues to increasingly lag as the unit passes through the critical speed (balance resonance) region.

Generally, both Bode and the polar plots are required to accurately define resonances. This is applicable to rotor, structural, and secondary resonances. Although a Bode plot will allow accurate frequency identification of the critical, the type of resonance is often identified by the polar plots along the length of the rotor. A proper understanding of these transient speed plots is vital to a full comprehension of the transient vibration behavior of the machinery.

It should also be mentioned that this data is sensitive to any type of vector offset. This is particularly true for shaft measurements made with proximity probes. These displacement transducers are susceptible to shaft surface conditions such as scratches, surface imperfections, metallurgical variations, magnetized segments, and eccentricity of the observed shaft surface. These types of conditions produce erroneous signals that often appear as a substantial 1X vector at low speeds. For demonstration purposes, the transient data previously shown in Fig. 2-21 is replotted in Fig. 2-23 with the inclusion of a constant 1X runout vector of 0.83 Mils,$_{p-p}$ at 168°.

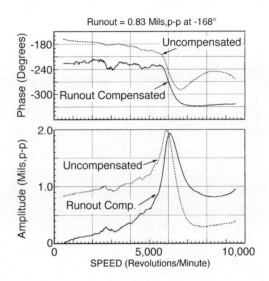

Fig. 2–23 Measured Bode Plot Of Actual Industrial Centrifugal Compressor With Shaft Runout

The solid lines on this Bode plot are identical to the data in Fig. 2-21, and the dotted lines for amplitude and phase show the influence of the 0.83 Mil,$_{p-p}$ runout vector. Clearly this slow roll vector influences the vibration signal throughout the entire speed domain. The uncompensated critical speed peak appears at 5,900 RPM instead of the actual resonance speed of 6,100 RPM. Vibration amplitudes at the operating speed of 9,500 RPM appear as 0.4 Mils,$_{p-p}$ with runout, versus the true magnitude of 0.95 Mils,$_{p-p}$. Finally, the phase shift through the resonance with runout included is about 50°, whereas the properly compensated vibration signal displays a more realistic 110° phase change through the resonance region.

It is apparent that the inclusion of a slow speed runout vector can result in serious data interpretation problems. Due to the potential implications of shaft runout, the origin and various corrective measures for shaft runout will be discussed in greater detail in subsequent chapters.

Overall, the relationship between the physical parameters of mass, stiffness, damping and the motion of a body including the displacement, velocity, acceleration, and frequency have been established in this chapter. When these fundamentals are clearly understood, complex mechanical vibration problems may be addressed, and successfully solved.

The previously discussed resonant response is quite typical for a piece of rotating machinery. In this common behavior, a rotor resonance is excited by a synchronous unbalance force during transient startup, and coastdown conditions. The coincidence of the excitation frequency (rotor speed) with the natural frequency (critical speed) results in a blossoming or amplification of the vibration amplitudes. It is also apparent that during normal operating conditions, the rotor resonance is normally dormant (inactive) due to the lack of any appreciable exciting force within the bandwidth of the resonant frequency.

In actuality, a variety of rotor and structural resonances exist on every machinery train. In most instances, the various major and minor resonances are inactive due to the lack of an appropriate stimulation. However, when a machine emits a discrete excitation that is coincident with a natural frequency, that particular natural frequency will become active. Similarly, when a process machine produces, or is subjected to a wide-band excitation, the resonant frequencies within the excitation bandwidth may likewise become active.

Case History 2: Steam Turbine End Cover Resonance

As an example of an excitation of a normally inactive resonance, consider a high pressure steam turbine driving a series of tandem centrifugal compressors. The operating speed for this train normally varied between 3,500 and 3,750 RPM. The eleven stage turbine is rated at 49,000 HP, and it had a history of low vibration amplitudes with minimal evidence of any abnormalities.

Prior to a maintenance turnaround, plant rates were increased to maximize production going to storage. Under this operating condition, the machine speed was temporarily increased to 3,800 RPM. Since the rated speed for this train was 3,930 RPM, the moderate speed increase was considered to be well within the performance envelope for this unit. Unfortunately, continuous operation at this higher speed resulted in an objectionable governor housing vibration. This axial casing vibration occurred at a frequency of 15,200 CPM.

With field measurements, it was readily determined that the high vibration amplitudes were confined to the bolted end cover of the governor housing. The maximum response was measured at the middle of the cover, and minimal amplitudes were evident within the bolt circle of the cover. It was clear that the measured frequency occurred at four times machine speed (4X), and it was discovered that a slight speed reduction resulted in the virtual elimination of the end cover vibratory motion. Naturally, the operations personnel were against any corrections that might reduce the plant production rates.

Hence, a solution other than slowing down the turbine was required. In an effort to determine the mode shape of the cover plate, a series of additional measurements were obtained. A grid pattern was established on the cover plate face with approximately 5 inches between grid lines. Vibration readings filtered at four times running speed were obtained at each grid intersection. In addition, a reference accelerometer was located at the middle of the cover plate, and the timing relationship to each intersection measurement point was visually determined on a digital oscilloscope.

In all cases, the readings were directly in phase, and the cover was moving back and forth in unison. The accelerometer field measurements were manually converted to casing displacement at the average frequency of 15,200 CPM. Plotting the amplitude at each grid intersection allows the construction of the three-

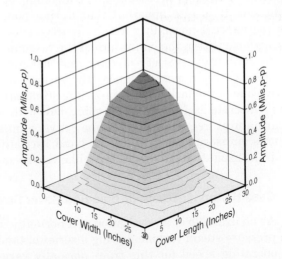

Fig. 2–24 Three-Dimensional Mode Shape Of Vertical Steam Turbine End Cover at 15,200 CPM

dimensional plate mode shape shown in Fig. 2-24.

This diagram suggests a simple drum mode, with maximum displacement at the center of the plate, and minimum motion at the plate perimeter. Since this appears to be a resonant plate, and since it behaves like a structural resonance; it would be desirable to confirm this hypothesis based upon the physical characteristics of the plate. An affirmation of this resonance may be obtained by performing a Finite Element Analysis (FEA). Alternatively, since this cover plate is a simple geometric structure, a set of manual calculations may be performed.

For instance, *Roark's Formulas for Stress & Strain*[10], includes Table 36 of Natural Frequencies of Vibration for Continuous Members. Case number 15, on page 717, within this table describes the equation for a rectangular flat plate with uniform thickness, and all four edges fixed. The natural frequency for this geometric configuration is presented as:

$$F = \frac{K}{2\pi} \times \sqrt{\frac{D \times G}{W \times a^4}} \tag{2-92}$$

where:
F = Plate Natural Frequency (Cycles / Second)
a = Plate Width and Length (Inches)
K = Constant based upon ratio of Length over Width. For a square plate, K = 36.
G = Acceleration of Gravity (= 386.1 Inches / Second^2)
W = Plate Unit Weight (Pounds / Inch^2)
D = Plate Flexural Rigidity (Pounds-Inch)

The turbine end cover was physically measured to be 30 inches square, with a thickness of 0.625 inches. The cover was attached to the turbine casing by a series of bolts that were located on a centerline of approximately 28.5 inches

---

[10] Warren C. Young, *Roark's Formulas for Stress & Strain,* 6th edition, (New York: McGraw-Hill Book Co., 1989), pp. 714-717.

square. Since the bolt pattern represented the point of attachment, and the zero motion perimeter, the cover dimension of 28.5 inches was used for the plate calculations. The plate unit weight W is computed by multiplying the plate density ρ times the plate thickness t as follows:

$$W = \rho \times t \tag{2-93}$$

where: ρ = Plate Density = 0.283 Pounds / Inch^3
         t = Plate thickness = 0.625 Inches

The plate unit weight may now be computed from equation (2-93) as:

$$W = \rho \times t = 0.283 \text{ Pounds/Inch}^3 \times 0.625 \text{ Inches} = 0.177 \text{ Pounds/Inch}^2$$

From Roark, page 714, the plate flexural rigidity D is determined from the modulus of elasticity E, Poisson's ratio v, and the plate thickness t as presented in equation (2-94):

$$D = \frac{E \times t^3}{12 \times (1 - v^2)} \tag{2-94}$$

where: E = Modulus of Elasticity (= 29.5 x 10^6 Pounds / Inch^2)
         v = Poisson's Ratio = 0.28 (Dimensionless)

Based on the plate thickness t, and the material constants (E and v), the Flexural Rigidity D may now be determined with equation (2-94) as:

$$D = \frac{E \times t^3}{12 \times (1 - v^2)} = \frac{(29.5 \times 10^6 \text{ Pounds/Inch}^2) \times (0.625 \text{ Inches})^3}{12 \times (1 - 0.28^2)}$$

$$D = \frac{7.202 \times 10^6}{11.059} = 651,200 \text{ Pounds-Inch}$$

From these computations of the unit weight W, and the flexural rigidity D, it is now possible to compute the natural frequency of the turbine cover plate. As previously noted, the constant K in the natural frequency equation is equal to 36 due to the equal length of the sides. Combining the various values, the natural frequency is computed from equation (2-92) in the following manner:

$$F = \frac{K}{2\pi} \times \sqrt{\frac{D \times G}{W \times a^4}} = \frac{36}{2\pi} \times \sqrt{\frac{(651,200 \text{ Pounds-Inch}) \times (386.1 \text{ Inches/Second}^2)}{(0.177 \text{ Pounds/Inch}^2) \times (28.5 \text{ Inches})^4}}$$

$$F = \frac{18}{\pi} \times \sqrt{\frac{251,428,000}{116,776 \text{ Second}^2}} = 265.8 \frac{\text{Cycles}}{\text{Second}} \times 60 \frac{\text{Seconds}}{\text{Minute}} = 15,950 \frac{\text{Cycles}}{\text{Minute}}$$

This calculated value of 15,950 CPM is higher than the measured frequency of 15,200. However, the 750 CPM difference represents only a 5% deviation. Considering that the material physical properties were estimated, and the plate dimensions were not precise, the 5% variation in frequencies is understandable.

Thus, the measured and calculated data point towards a plate resonance as the culprit. This was verified during operation by bolting a beam with a jack-screw across the plate. Application of the screw to the center of the cover plate immediately suppressed the plate resonant response. During the next scheduled turnaround, the temporary beam and jackscrew were removed, and two angle beams were welded across the inside of the cover plate. Subsequent operation at speeds up to 3,950 RPM revealed no reoccurrence of this problem.

In retrospect, this was not a serious mechanical problem, but it was an irri-tant to the Operations personnel since the vibratory response was audible on the compressor deck. Hence, sometimes problems must be corrected to satisfy the mechanics as well as the politics of the situation.

TORSIONAL VIBRATION

The previous sections of this chapter have considered the vibration of a mechanical system in general, and the lateral vibration of a rotating machine in particular. These same concepts are also applicable to the torsional behavior of a mechanical system. In fact, an analytical simulation of a new machine usually considers both the lateral and the torsional response of the proposed rotor sys-tem. Fortunately, the same concepts and equations developed for longitudinal vibration may be directly converted to torsional motion. For example, the four basic torsional vibration parameters are defined as follows:

D_{tor} = Torsional Displacement — Degrees,$_{\text{peak to peak}}$ = Deg,$_{\text{p-p}}$

V_{tor} = Torsional Velocity — Degrees/Second,$_{\text{zero to peak}}$ = Deg/Sec,$_{\text{o-p}}$

A_{tor} = Torsional Acceleration — Degrees/Second^2,$_{\text{zero to peak}}$ = Deg/Sec^2,$_{\text{o-p}}$

F = Frequency — Cycles/Second (Hz)

The torsional displacement D_{tor} is the angular twist ϕ due to a natural oscillation, or an applied torque. The first time derivative of torsional displace-ment yields the torsional velocity V_{tor}. Similarly, the second time derivative of torsional displacement provides torsional acceleration A_{tor}. These quantities are identical to their lateral counterparts, and a 90° timing shift exists between tor-sional vectors. As before, measurement of one torsional vibration amplitude and frequency will allow computation of the other two torsional vibration parame-ters. For instance, replacing $2\pi F$ for the frequency ω, and considering the peak values of the terms, equation (2-12) may be restated for a torsional system as:

$$V_{tor} = \frac{D_{tor}}{2} \times \omega = \left(\frac{D_{tor}}{2}\text{Degrees}\right) \times \left(2\pi\frac{\text{Radians}}{\text{Cycle}} \times F\frac{\text{Cycles}}{\text{Second}}\right)$$

Simplification produces the following equation:

$$V_{tor} = \pi \times D_{tor} \times F$$ (2-95)

Next, consider the relationship between torsional acceleration and displacement as originally described by equation (2-13), and modified as follows:

$$A_{tor} = \frac{D_{tor}}{2} \times \omega^2 = \left(\frac{D_{tor}}{2}\text{Degrees}\right) \times \left(2\pi\frac{\text{Radians}}{\text{Cycle}} \times F\frac{\text{Cycles}}{\text{Second}}\right)^2$$

Which simplifies to the following equation:

$$A_{tor} = 2\pi^2 \times D_{tor} \times F^2$$ (2-96)

A similar relationship exists between torsional acceleration and velocity as:

$$A_{tor} = V_{tor} \times \omega = \left(V_{tor}\frac{\text{Degrees}}{\text{Second}}\right) \times \left(2\pi\frac{\text{Radians}}{\text{Cycle}} \times F\frac{\text{Cycles}}{\text{Second}}\right)$$

Which is equal to the following expression:

$$A_{tor} = 2\pi \times V_{tor} \times F$$ (2-97)

These last three equations allow conversion between torsional displacement, velocity, and acceleration at a fixed frequency in Cycles per Second. For those unaccustomed to torsional vibration units, these expressions, and the resultant amplitudes may appear to be unusual. It should be noted that torsional displacement amplitudes are typically much smaller than one degree.

For demonstration purposes, consider the conversion of torsional vibration units on vibration data acquired on a 28,000 horsepower speed increasing gear box. In this case, a proximity probe was used to measure torsional displacement from the bull gear teeth. At the same time, a direct coupled torsiograph (velocity based torsional pickup) was attached to the outboard blind end of the gear. Operating at 100% load with a speed of 5,100 RPM (85 Hz), the proximity probe system revealed a torsional displacement of 0.0021 Degrees,$_{\text{p-p}}$. Simultaneously, the torsiograph system indicated a torsional velocity of 0.567 Degrees/Second,$_{\text{o-p}}$.

The installation of both types of torsional sensors was directed at providing a verification of the data by using two different and independent measurement systems. Hence, this steady state data can be directly compared if the units of one measurement are converted to the units of the other measurement. Using the previous equation (2-95), the measured torsional displacement may be converted to torsional velocity as follows:

$$V_{tor} = \pi \times D_{tor} \times F = \pi \times 0.0021\,\text{Deg}_{\text{o-p}} \times 85\frac{\text{Cycles}}{\text{Second}} = 0.561\ \text{Deg/Sec}_{\text{o-p}}$$

The measured torsional displacement of 0.0021 Degrees,$_{\text{p-p}}$ at 5,100 RPM (85 Hz) is thereby equal to 0.561 Degrees/Second,$_{\text{o-p}}$. This converted torsional displacement amplitude is in excellent agreement with the directly measured torsional velocity of 0.567 Degrees/Second,$_{o-p}$.

Torsional behavior of a mechanical system parallels the equations developed for a lateral system. Torsional vibration measurements are complex signals that usually require some type of post processing. The synchronous 1X rotational speed data is generally viewed as a vector quantity, and the concepts of springs, dampers, masses, and forcing functions are fully applicable. The names and units have changed somewhat, but the structure of the equations of motion remain the same.

The lateral stiffness K is equivalent to the torsional stiffness K_{tor}. The traditional units used for torsional stiffness are Inch-Pounds/Radian. Multiplication of the torsional displacement D_{tor} in Degrees, times the torsional stiffness K_{tor}, and a conversion of units yields the torsional stiffness term for the general equation of motion. This term will carry the final torque units of Inch-Pounds.

Lateral damping C is analogous to torsional damping C_{tor}. The customary units used for torsional damping are Pound-Inch-Second/Radian. Multiplication of the torsional velocity V_{tor} in Degrees/Second, times the torsional damping C_{tor}, and a conversion of units yields the torsional damping term for the general equation of motion. This term will carry the final torque units of Inch-Pounds.

Mass M in a lateral system is the comparable of the mass polar moment of inertia J_{mass} in a torsional system. The correct engineering units for polar inertia are Pound-Inch-Second^2/Radian. Multiplication of the torsional acceleration A_{tor} in Degrees/Second^2, times the inertia J_{mass}, and a conversion of units produces the torsional inertial term for the general equation of motion. This inertial term will also carry the torque units of Inch-Pounds.

A periodic forcing function $F\sin(\omega t)$ in a lateral system is the equivalent of an oscillating torque $T\sin(\omega t)$ in a torsional system. The traditional units used for torque are either Inch-Pounds, or Foot-Pounds. For the sake of completeness, it should be recognized that shaft torque in a process machine is directly associated with the shaft speed, and the transmitted power. The classical definition of power is the time rate of doing work or delivering energy. The inclusion of time into this definition is very important. For instance, a small motor can provide any amount of energy given a sufficient amount of time. However, producing a large amount of energy within a short period of time requires a large motor.

In a linear system, power is the product of force and velocity. Within a rotating system, power is determined by multiplying torque times speed. In either case, the expected engineering units for power would be in the format of Foot-Pounds/Minute or Inch-Pounds/Minute. Although these units are consistent with the previous discussion, these units are generally not used. In most instances, power is identified in units of horsepower for nonelectrical equipment, and units of kilowatts or megawatts for electric machinery. The traditional conversion expressions for computing steady state torque from both types of power units are presented in the following equations (2-98) and (2-99):

$$Torque(\text{Foot-Pounds}) = \frac{5,252 \times Horsepower}{RPM} \qquad \textbf{(2-98)}$$

$$Torque(\text{Foot-Pounds}) = \frac{7,043 \times Kilowatts}{RPM} \qquad \textbf{(2-99)}$$

The conversion factor between kilowatts and horsepower is 1.341, and that value provides the difference between the constants shown in equations (2-98) and (2-99). These expressions also define the relationship between speed, power, and torque. Thus, a machine at constant speed will exhibit periodic torque variations if the load or power oscillates. Conversely, if the power level remains constant, variations in operating speed (e.g., due to governor malfunction) will manifest as variations in torque.

The generalized equation of motion for a lateral single degree of freedom system subjected to a periodic forcing function was presented in equation (2-77). This expression equated the periodic forcing function to the summation of the spring, damping, and mass forces. This previous equation is also appropriate for a torsional system, and the following modification of equation (2-77) is presented to describe a forced torsional system:

$$\{J_{mass} \times A_{tor}\} + \{C_{tor} \times V_{tor}\} + \{K_{tor} \times D_{tor}\} = T \times \sin(\omega t) \qquad \textbf{(2-100)}$$

Equation (2-100) will respond in exactly the same way in a lateral or a torsional system. The same characteristic family of curves may be developed for amplitude ratio as a function of frequency ratio for various damping ratios. Although the physical system description has changed, the mathematical structure remains constant. Another proof of the mathematical similarity between lateral and torsional mechanical systems resides in a comparison of the engineering units applied on both types of systems. For example, Table 2-1 summarizes several mechanical parameters, and typical engineering units that may be used for a lateral, and a torsional system. From Table 2-1, it is noted that the lateral vibratory measurements are based on distance, versus an equivalent angular twist in the torsional analysis. The stiffness and damping coefficients are formulated in a similar manner with lateral displacement replaced by angular twist — and the applied lateral force is replaced by a torsional torque. Momentum or Impulse in a torsional system includes the moment arm, and work or energy carries the same units in either system.

Although the lateral and torsional systems display identical equation structures, the diagnostician must be aware of the physical differences between the two types of motion. For instance, torsional vibration problems are often transient excitations that drive one or more torsional resonances (e.g., during synchronous motor startup). It is unusual to encounter self-excited steady state torsional behavior. Although lateral vibration produces shaft deflection, torsional vibration results in shaft stress reversals. Hence, when a torsional problem is encountered, it is often characterized by mechanical failures (e.g. broken shaft).

To gain a better understanding of this type of twisting behavior, consider a simplification of the complex torsional system into an undamped system sub-

Table 2–1 Comparison Of Units For Lateral Versus Torsional Mechanical Systems

Parameter	Lateral	Torsional
Displacement	$Mils_{p-p}$	$Degrees_{p-p}$
Velocity	$\left(\dfrac{Inches}{Second}\right)_{o-p}$	$\left(\dfrac{Degrees}{Second}\right)_{o-p}$
Acceleration	$\left(\dfrac{Inches}{Second^2}\right)_{o-p}$	$\left(\dfrac{Degrees}{Second^2}\right)_{o-p}$
Stiffness	$\dfrac{Pounds}{Inch}$	$\dfrac{Inch-Pounds}{Radian}$
Damping	$\dfrac{Pound-Seconds}{Inch}$	$\dfrac{Pound-Inch-Seconds}{Radian}$
Mass — Inertia	$\dfrac{Pound-Second^2}{Inch}$	$\dfrac{Pound-Inch-Second^2}{Radian}$
Force — Torque	Pounds	Inch – Pounds
Momentum or Impulse	Pound – Seconds	Pound – Inch – Seconds
Work or Energy	Pound – Inches	Pound – Inches

jected to free vibration. The diagram in Fig. 2-25, describes such a mechanical system consisting of a disk supported by a circular rod. The top of the rod is rigidly attached to the stationary I-Beam, and the rod has a torsional stiffness of K_{tor}. The disk connected to the bottom of the rod has a mass polar moment of inertia identified as J_{mass}. Clearly, the stiffness and mass properties exhibited by the lateral spring mass system described in the earlier Fig. 2-7 are analogous to this simple undamped torsional system. It is intuitive that the torsional system shown in Fig. 2-25 will remain at rest until an initial disturbance is applied. In this case, if a torque, or twist is applied to the disk, the system will oscillate in accordance with the following simplification of equation. (2-100):

$$\{J_{mass} \times A_{tor}\} + \{K_{tor} \times D_{tor}\} = 0$$

Since the system has no damping, the torsional velocity term must be equal to zero. Also, the system is excited with an initial angular displacement, and a continuous forcing function does not exist. Thus, the forced $T\sin(\omega t)$ term is also equated to zero. The resultant expression may be rearranged as follows:

$$A_{tor} + \left\{\frac{K_{tor}}{J_{mass}}\right\} \times D_{tor} = 0 \qquad \textbf{(2-101)}$$

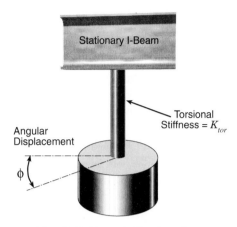

Fig. 2–25 Undamped
Simple Torsional System

Mass Polar Moment of Inertia = J_{mass}

This torsional expression is identical to the lateral equivalent presented in equation (2-42). Performing the same manipulation on equation (2-101) for the torsional system, the following expected solution evolves:

$$\omega_{c_{tor}} = \sqrt{\frac{K_{tor}}{J_{mass}}} \qquad (2\text{-}102)$$

Another common form of this expression is obtained by converting the rotational frequency ω_c units of radians per second to cycles per second in accordance with equation (2-2) to yield the following:

$$F_{c_{tor}} = \frac{1}{2\pi} \times \sqrt{\frac{K_{tor}}{J_{mass}}} \qquad (2\text{-}103)$$

This solution is the torsional natural frequency for the undamped disk hanging by a torsional spring (i.e., rod). As discussed in the lateral system, this natural frequency is a function of the torsional stiffness, and the mass polar moment of inertia. It is clear that following an initial angular disturbance, the disk will oscillate at this natural frequency, and the angular motion will gradually decay as a function of time.

From the previous discussion, it is apparent that the equation structure for lateral and torsional systems are virtually identical. However, the issue of measured torsional vibratory behavior should also be addressed. For machines such as reciprocating engines, pumps, or compressors, a synchronous torsional component will generally be present under all operating conditions. These types of machines often exhibit strong torsional resonances, and the clear identification of resonant frequencies is generally easily achieved.

However, on centrifugal machines, the torsional characteristics are generally quite small, and many of the torsional resonances are well damped by the process fluid. In addition, the torsional signals are often cluttered with extraneous excitations. Due to the sensitivity of the measurement, the influence from adjacent machines, electrical interference at 60, 120 and 180 Hz, transducer support resonances, and machining imperfections on the observed elements all contribute to noise in the torsional signal. Hence, the machinery diagnostician must be very careful during data acquisition and reduction. Special attention should be paid to the signal to noise ratio of the entire measurement system — and the use of low-pass and/or high-pass filters, plus a synchronous tracking filter may be necessary. These signal processing considerations will be discussed in much greater detail in the following chapters 6, 7, and 8 of this text.

Torsional vibration data may be steady state information acquired at a constant speed, and load. Data may also be obtained during transient conditions of machinery train startups or coastdowns. In most cases, the data processing techniques used for lateral vibration measurements may be successfully used for torsional data. However, there are significant differences in the vibration transducers used for lateral and torsional measurement. In most instances, the torsional measurements are delicate, and they must be handled carefully. For additional information on the characteristics of common transducers, and their signal conditioning requirements, chapters 6 and 7 should be reviewed.

An example of transient torsional vibration data is shown in the Bode plot of Fig. 2-26. This data depicts the transient response of a large, single helical, bull gear mounted in a speed increasing gear box. Note the clearly defined resonant response at 2,050 RPM, and the associated 160° phase shift. This data is analogous to the lateral response Bode plot previously presented in Fig. 2-21.

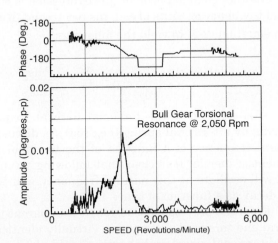

Fig. 2–26 Measured Torsional Bode Plot Of A Bull Gear In A Speed Increasing Gear Box

In the overview, lateral and torsional equations of simple systems are essentially identical. It is also clear that strong similarities exist between lateral and torsional vibration response characteristics. However, it must be understood that lateral and torsional measurements represent two entirely different modes of machinery behavior. The mechanical implications of torsional versus lateral vibration are quite different. For instance, torsional criticals are usually governed by coupling stiffness, and the major mass polar moment of inertia as described by equation (2-102). Lateral criticals on the other hand are generally controlled by the support and/or shaft stiffness combined with the rotor mass as described by equation (2-44). In most cases, the stiffness of the coupling torque tube (spacer) is significant to torsional behavior, whereas it is often inconsequential to the lateral response.

Furthermore, torsional vibration may not be detected by lateral vibration sensors, but lateral vibration may adversely influence the accuracy of torsional vibration measurements. It has been documented that a torsional failure may be in progress, and the machine might continue to exhibit low radial vibration amplitudes until a major component failure (e.g., broken shaft). For this type of machinery situation, the cross-coupling between torsional and lateral vibration may be virtually nonexistent. However, in other machinery trains, lateral and torsional modes are closely coupled so that vibration of a lateral resonance may excite a coupled torsional resonance, or vice versa. When in doubt, both the lateral and torsional characteristics of the machinery should be carefully examined as a routine part of the problem solving investigation.

BIBLIOGRAPHY

1. Avallone, Eugene A. and Theodore Baumeister III, *Marks' Standard Handbook for Mechanical Engineers*, Tenth Edition, pp. 2-36, 3-15, New York: McGraw-Hill, 1996.

2. Den Hartog, J.P., *Mechanical Vibrations*, 4th edition, p. 87, New York: McGraw-Hill Book Company, 1956.

3. Raynesford, John D., "Use Dynamic Absorbers to Reduce Vibration," *Hydrocarbon Processing*, Vol. 54, No. 4 (April 1975), pp. 167-171.

4. Shigley, Joseph E. and Charles R. Mischke, *Standard Handbook of Machine Design*, pp 11.5-11.6, New York: McGraw-Hill Book Company, 1986.

5. Thomson, William T., *Theory of Vibration with Applications*, 4th Edition, pp. 1-91, and 150-159, Englewood Cliffs, New Jersey: Prentice Hall, 1993.

6. Thomson, William Tyrell, *Mechanical Vibrations*, 2nd Edition, 9th Printing, pp.1-75, Englewood Cliffs, New Jersey: Prentice Hall, Inc., 1962.

7. Young, Warren C., *Roark's Formulas for Stress & Strain*, 6th edition, pp. 100-102, and 714-717, New York: McGraw-Hill Book Co., 1989.

Rotor Mode Shapes

The analysis of machinery vibration characteristics must be based upon a solid working knowledge of the dynamics associated with a rotating mechanical system. Items that influence rotor motion include the shaft construction, support locations, and the distribution of masses such as impellers, spacers, and couplings. Certainly the stiffness and damping characteristics of the bearings and machine support structure will play an important role in influencing rotor behavior. In addition, the relationship between the operating speed range, and the system criticals are very important.

These factors are parameters for establishing the response characteristics of the rotor system, and the associated dynamic rotor mode shapes. It is meaningful to recognize that many machine characteristics can be adequately described by applying the basic concepts of mass and support distribution, system viscous damping, plus stiffness of the major elements and support structure. It must also be recognized that mode shapes may be forced deflections based upon the active system forces. They may also be natural modes of vibration (resonances), plus a combination of the two to yield a forced resonant response.

MASS AND SUPPORT DISTRIBUTION

The distribution of weight and supports along the length of a rotor establishes the static deflections, plus the static bearing loads. For example, consider a constant diameter shaft that is simply supported between two points as shown in Fig. 3-1. This shaft will have a maximum deflection at the midspan, and each support location will carry one half of the shaft weight.

If one support is moved to the rotor midspan, the condition described in Fig. 3-2 will occur. In this case, the maximum deflection occurs at the unsupported end of the shaft. The load applied to each support will now be dependent upon the support characteristics. Specifically, if the left support is a free support, it will not have a static load. Under this condition, the center support will carry the entire weight, and the shaft will be balanced on this center pivot. On the other hand, if the left support is connected to the rotor (e.g., a bearing), it will produce a vertical reaction if the middle support is not perfectly centered. For this reason,

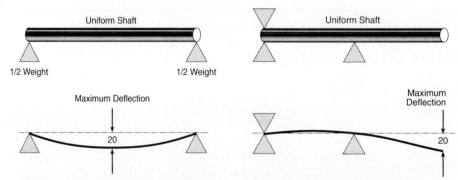

Fig. 3–1 Simply Supported Shaft With Static Deflection Due To Beam Weight

Fig. 3–2 Overhung Shaft With Static Deflection Due To Beam Weight

overhung machines such as power turbine rotors for dual shaft gas turbines, or overhung blowers must be carefully examined to determine the static and dynamic bearing loads, and directions.

Next consider the addition of a concentrated load (e.g., an impeller) at the middle of the rotor. The diagram presented in Fig. 3-3 represents the deflection associated with the additional force applied at the midspan of the simply supported shaft. Clearly the center deflection must increase when the additional load is applied. In addition, with the supports located at the shaft ends, it is reasonable to conclude that the total weight (shaft plus midspan load) will be equally shared between the two supports.

Fig. 3-4 illustrates the condition of an overhung rotor with the addition of a center weight. In this configuration, one support is located directly below the concentrated midspan load. In this case, the mode shape, and the maximum deflection are identical to Fig. 3-2 with zero external load. However, the force balance has been altered, and the center support must now carry the shaft weight plus the center load.

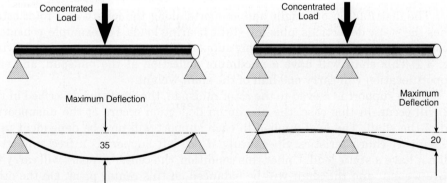

Fig. 3–3 Simply Supported Shaft Deflected By Beam Weight And Center Load

Fig. 3–4 Overhung Shaft Deflected By Beam Weight And Center Load

Next, consider the static mode shapes displayed in Figs. 3-5 and 3-6 that describe the influence of moving the concentrated load from the midspan to the end of the rotor. For the shaft simply supported between bearings (Fig. 3-5), the mode shape returns back to the initial condition (Fig. 3-1). The additional load is directly transmitted to the right hand support. Under this configuration, the deflected mode shape is dependent only on the shaft weight, but the support loads are clearly different.

Finally, for the overhung case of Fig. 3-6, the cantilevered load at the end of the shaft results in an increase in the maximum deflection. This type of behavior certainly makes intuitive sense, and it is representative of real overhung machines. It should also be recognized that the application of this load to the free end of the rotor will result in a downward vertical restraining force at the left end support. Again, this is consistent with the forces and moments encountered in machines such as power turbines and overhung blowers.

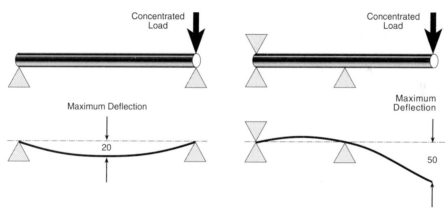

Fig. 3–5 Simply Supported Shaft Deflected By Beam Weight And End Load

Fig. 3–6 Overhung Shaft Deflected By Beam Weight And End Load

Overall, it is recognized that the static support forces (i.e., at the bearings), plus the location and magnitude of the maximum deflection are dependent upon the support characteristics. It is apparent that the addition of elements such as impellers, couplings, balance pistons, and spacers will directly influence the resultant support forces, and the associated maximum static deflection.

Simple mechanical systems can often be modeled as a uniform weight distribution for the shaft, combined with concentrated loads for the impellers. The static deflections can be calculated with beam theory, and the static bearing loads determined by summation of moments. Reference books such as *Roark's Formulas for Stress and Strain*[1] provide characteristic equations for many typical mechanical systems without resorting to detailed beam calculations.

For more complex rotors, it is necessary to divide the rotor into discrete and

---

[1] Warren C. Young, *Roark's Formulas for Stress & Strain*, Sixth Edition, (New York: McGraw-Hill Book Company, 1989).

definable segments. Based upon the dimensions and material density, it is possible to calculate the weight for each station. From this weight distribution, the static loads at both bearings can be computed. This approach identifies the mass distribution along the rotor, plus the resultant bearing forces in a static position.

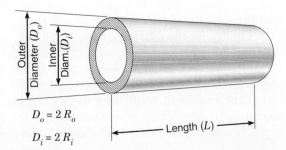

$D_o = 2\,R_o$

$D_i = 2\,R_i$

Fig. 3–7 Hollow Circular Cylinder With Physical Dimensions

The weight of each section or portion of a rotor is dependent upon the physical dimensions, plus the density of the shaft material. For example, the hollow circular cylinder depicted in Fig. 3-7 is dimensionally specified by an outer diameter D_o, an inner diameter D_i, and an overall length L. The shaft radius to the outer diameter is R_o, and the internal bore radius is identified as R_i. Based on these dimensions, a variety of necessary calculations may be performed. For instance, from plane geometry, the cross sectional area of this annulus is computed by subtracting circular areas in the following manner:

$$A_{annulus} = \pi \times R_o^2 - \pi \times R_i^2 = \frac{\pi}{4} \times (D_o^2 - D_i^2) \qquad \text{(3-1)}$$

The volume of the annulus may be determined by multiplying the cylinder length L times the cross sectional area $A_{annulus}$ as stated in the next expression:

$$V_{annulus} = L \times A_{annulus} = \frac{\pi \times L}{4} \times (D_o^2 - D_i^2) \qquad \text{(3-2)}$$

The weight of the annulus may be calculated by multiplying the cylinder material density ρ times the total volume $V_{annulus}$ as shown in equation (3-3):

$$W_{annulus} = \rho \times V_{annulus} = \frac{\pi \times L \times \rho}{4} \times (D_o^2 - D_i^2) \qquad \text{(3-3)}$$

For reference purposes, the densities of many common metals have been summarized in Appendix B of this text. This tabulation also includes the modulus of elasticity, the shear modulus, and the coefficient of thermal expansion for each metal. These fundamental properties are referred to throughout this book, and it is convenient to have typical values readily available. It should be mentioned that some materials do not have totally unique properties. For critical calculations the precise physical properties should be obtained from a metallurgical reference source, or specific tests of the metal.

For some calculations, it is necessary to know the mass of the mechanical part. For the annulus under discussion, the mass is easily determined by dividing the weight from equation (3-4) by the acceleration of gravity G as follows:

$$M_{annulus} = \frac{W_{annulus}}{G} = \frac{\pi \times L \times \rho}{4 \times G} \times (D_o^2 - D_i^2) \qquad \textbf{(3-4)}$$

In many instances, machinery shafts are not hollow, and they are fabricated of solid metal. If the inner diameter D_i is set equal to zero in equations (3-1) through (3-4), the following expressions for solid machine shafts with an outer diameter of D are easily developed:

$$A_{solid} = \frac{\pi \times D^2}{4} \qquad \textbf{(3-5)}$$

$$V_{solid} = \frac{\pi \times L \times D^2}{4} \qquad \textbf{(3-6)}$$

$$W_{solid} = \frac{\pi \times L \times \rho \times D^2}{4} \qquad \textbf{(3-7)}$$

$$M_{solid} = \frac{\pi \times L \times \rho \times D^2}{4 \times G} \qquad \textbf{(3-8)}$$

In order to be perfectly clear on the dimensional aspects of equations (3-1) to (3-8), the defined variables and their respective English engineering units are summarized as follows:

L = Cylinder Length (Inches)
D = Solid Cylinder Outer Diameter (Inches)
D_o = Hollow Cylinder Outer Diameter (Inches)
D_i = Hollow Cylinder Inner Diameter (Inches)
R_o = Hollow Cylinder Outer Radius (Inches)
R_i = Hollow Cylinder Inner Radius (Inches)
ρ = Cylinder Material Density (Pounds / Inch^3)
A = Cylinder Cross Sectional Area (Inches^2)
V = Cylinder Volume (Inches^3)
W = Cylinder Weight (Pounds)
M = Cylinder Mass (Pounds-Second^2 / Inch)
G = Acceleration of Gravity (= 386.1 Inches / Second^2)

These variables and associated units will be used in the next sections on inertia, plus throughout the remainder of this text.

Case History 3: Two Stage Compressor Rotor Weight Distribution

Consider the compressor rotor depicted in Fig. 3-8. This high speed rotating assembly consists of a shaft with various diameters, plus two midspan impellers, and a thrust collar. Since the impellers are mounted back to back, this particular design does not include a balance piston. As indicated on the diagram, the rotor weights 282 pounds, and there is a moderate overhang on the thrust end of the rotor. Normal operating speed varies between 15,000 and 17,500 RPM. This compressor is a drive through unit with a stream turbine coupled to one end, and another compressor coupled to the opposite (thrust) end. Historically, the bearing on the drive end of the rotor seldom exhibited any damage, but the journal bearing located at the thrust end was often found to be in a distressed or damaged condition. On other occasions, the bearings would fail during operation, and the rotor would be severely damaged. Depending on the severity of the failure, the shaft was often chrome plated. After severe failures, the entire rotating assembly was totally replaced.

This machine operates in a very dirty and corrosive environment. The rotor was constantly subjected to large unbalance forces due to the accumulation of

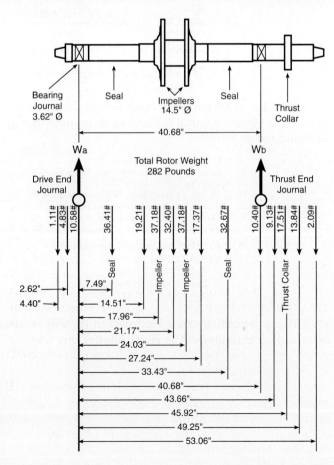

Fig. 3–8 Two Stage Centrifugal Compressor Rotor With Weight And Moment Diagram

material in the impellers, plus random corrosion of the impeller vanes. During a typical run, the compressor would startup smoothly after an overhaul. However, synchronous running speed vibration response would always deteriorate with the passage of time.

An initial step in the analysis of this problem required the determination of static bearing loads. To achieve this goal, the rotor was divided into fifteen sections, and the weight of each section was calculated. Since this was a solid shaft, equation (3-7) was used to compute the weight of each shaft section. The weight of the thrust collar was determined with (3-3), and the two impellers were weighed separately. Next, the thrust collar and wheel weights were combined with their respective shaft section weights, and the weight distribution summarized in Fig. 3-8.

This information was then combined with the distance from the center of the drive end bearing to the centroid of each rotor segment. The complete array of weights and distances are shown in Fig. 3-8. From this sketch, it is possible to perform a summation of moments around the center or transverse axis of the drive end journal in the following manner:

$$\boxed{\sum Moments_{ccw} = \sum Moments_{cw}}$$ (3-9)

$$17.54 \text{ Inch-Pounds} + W_b \times 40.68 \text{ Inches} = 6,768.74 \text{ Inch-Pounds}$$

$$W_b \times 40.68 \text{ Inches} = 6,751.20 \text{ Inch-Pounds}$$

$$W_b = 166 \text{ Pounds}$$

Since the total rotor weight is 282 pounds, the next force balance applies:

$$W_a + W_b = 282 \text{ Pounds}$$

$$W_a + 166 \text{ Pounds} = 282 \text{ Pounds}$$

$$W_a = 116 \text{ Pounds}$$

Thus, the drive end bearing has a 116 pound static load, and the thrust end journal carries 166 pounds. Although the differential force is only 50 pounds, it is an appreciable percentage difference. Ultimately, it was determined that the bearings were only marginally sized to accommodate the rotor weight. However, they were considerably undersized when the additional unbalance forces due to foreign objects were included. It was also determined that the available load capacity for the drive end was barely acceptable, whereas the load carrying capability for the thrust end journal bearing was unacceptable. Based upon these conclusions, both bearings were increased in size and load capacity.

The larger bearings reduced the number of machine failures per year, and overall reliability was substantially improved. Further improvements in machine longevity would require changes in the chemical plant process. Unfortunately, the required alterations to the processing scheme could not be economically justified. Hence, the compressor rotor was occasionally sacrificed to meet production quotas.

From this example, it is clear that even a simple analysis of rotor weight distribution and bearing static loads may be beneficial. In some cases this may solve a problem, or it may provide insight into prospective solutions. It must also be recognized that the dynamics of the rotating system must be considered. This includes the effects of mass unbalance that serves to deform the mode shape, plus the effects of inertia, stiffness, and damping of machine elements.

INERTIA CONSIDERATIONS AND CALCULATIONS

As shown in the previous example, the calculation of rotor weights is easily achieved for rotors that can be segmented into a series of cylinders. For a homogeneous material, the volume and weight of each section may be computed. The summation of individual cylindrical section weights yields the total rotor weight. In most instances, this is an acceptable and achievable computational procedure. However, problems occur with machines that have non-symmetric shafts, or are composed of multiple materials. For instance, the rotor weight of electric machines, such as motors and generators, may be indeterminate due to an unknown combination of iron, insulation, and copper within the rotor windings. In these situations, the rotor assembly can be weighed on a scale, and the distribution of weight between bearing journals estimated. Any additional elements such as a cooling fan, flywheel, or overhung exciter may then be added to determine the total rotating system weight. Hence, when accurate calculations cannot be performed, the machinery diagnostician can revert to the traditional experimental technique of weighing the rotor.

In addition to weight or mass, the inertia properties of a rotating system must also be considered during any dynamic analysis. Due to the complexity of inertia calculations, and the strong potential for confusion between area and mass moment of inertias, significant errors are possible. The physics definition of inertia is the property of a body to resist acceleration. This includes the tendency of a body at rest to remain at rest, or the tendency of a body in motion to stay in motion. This reference to a physical body implies mass, which suggests the presence of a three-dimensional object. Hence, an area moment of inertia is somewhat of a misnomer since a plane area has no depth, and therefore no mass, nor the potential for inertia. Nevertheless, area moments of inertia abound in the literature, and for the sake of completeness, they will be discussed.

It should also be mentioned that confusion regarding inertia is further compounded by some of the technical literature that transposes polar with the transverse inertias. In other cases, the authors' fail to identify the specific type of inertia calculations, much less the reference inertia axis. In an effort to circumvent these potential sources of confusion, the following discussion of inertia is offered for consideration.

It is desirable to begin with a review of the area moment of inertia. In Fig. 3-9, *Sketch A* shows an irregular plane surface with axes *x-x* and *y-y* passing through the area designated as *A*. These rectangular axes may be in any location

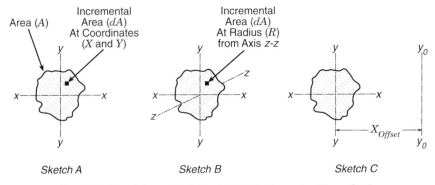

Fig. 3–9 Area Moment Of Inertia For An Irregular Plane Surface

with respect to the area. If the area was symmetric (e.g., circle or rectangle), and if the axes were axes of symmetry, they would be called the principal inertia axes. However, the general solution for inertia about each axis is determined from strength of materials integrals as:

$$I_{xx} = \int Y^2 \times dA \qquad\qquad (3\text{-}10)$$

$$I_{yy} = \int X^2 \times dA \qquad\qquad (3\text{-}11)$$

The X and Y terms in (3-10) and (3-11) are distances from each respective axis to the incremental area identified as dA. This definition of area moment of inertia is consistent with references such as *Roark*[2], or Shigley[3]. In both equations, inertia is calculated by multiplying each area by the square of the distance to the respective reference axis. The area is measured in inches squared, and this is multiplied by the square of the distance to the axis in inches. This product yields an area moment of inertia with English engineering units of Inches[4].

In *Sketch B* of Fig. 3-9, a third axis z-z has been added. This new axis is perpendicular to the plane of the area A, and it passes through the intersection of axes x-x and y-y. The distance from the intersection of axis z-z with the area A, to any incremental area dA is identified by the radius R. The inertia about this axis is termed the area polar moment of inertia J that is given by:

$$J_{zz} = \int R^2 \times dA \qquad\qquad (3\text{-}12)$$

From plane geometry it is known that the radius R, may be expressed in terms of X and Y coordinates as:

[2] Warren C. Young, *Roark's Formulas for Stress & Strain*, Sixth Edition, (New York: McGraw-Hill Book Company, 1989), p. 59.

[3] Joseph E. Shigley and Charles R. Mischke, *Standard Handbook of Machine Design*, (New York: McGraw-Hill Book Company, 1986), p. 9.13.

$$R^2 = X^2 + Y^2 \tag{3-13}$$

Equation (3-13) may be inserted into (3-12), and expanded as follows:

$$J_{zz} = \int (X^2 + Y^2) \times dA$$

$$J_{zz} = \int X^2 \times dA + \int Y^2 \times dA$$

Substituting (3-10) and (3-11) into the above expression, the polar inertia J_{zz} is equated to the summation of the transverse area inertias I_{xx} and I_{yy} as:

$$J_{zz} = I_{xx} + I_{yy} \tag{3-14}$$

For a symmetrical area such as a circle, the transverse area inertias I_{xx} and I_{yy} are equal. Thus, setting inertias $I_{xx}=I_{yy}=I$, equation (3-14) may be restated in the following simplified format:

$$\boxed{J_{zz} = I + I = 2 \times I} \tag{3-15}$$

Since most rotating shafts are circular, equation (3-15) is common within the machinery business. The applicability of this geometric simplification will be better appreciated during the forthcoming discussion of mass moment of inertia. However, before addressing that topic, the translation of the inertia axis should be reviewed. Specifically, *Sketch C* in Fig. 3-9 identifies a new axis y_o-y_o that is parallel to the previously defined vertical axis y-y. The constant distance between the two axes is identified as X_{offset}. It can be shown that the area moment of inertia about this new axis is given by the expression:

$$\boxed{I_{y_o y_o} = I_{yy} + A \times (X_{offset})^2} \tag{3-16}$$

The same argument may be made in the perpendicular plane for the area polar moment of inertia. Consider a new polar axis z_o-z_o that is parallel to the previously defined z-z axis. If the distance between the two axes is identified as R_{offset}, the area moment of inertia about this new polar axis is given by:

$$\boxed{J_{z_o z_o} = J_{zz} + A \times (R_{offset})^2} \tag{3-17}$$

The moment of inertia of an area about a particular axis may be converted to a mass moment of inertia by including the thickness, and the density of the body. This converts a two-dimensional into a three-dimensional problem, and the complexity of the associated equation structure increases proportionally. For explanation purposes, assume that the plane area shown in Fig. 3-9 is expanded by adding a finite depth. If a material density is applied to the resulting volume, the mass shown in Fig. 3-10 evolves. *Sketch A* in this diagram reveals that a two-dimensional *x-y* coordinate system is inadequate for locating a point within the

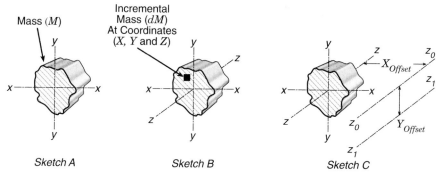

Sketch A   Sketch B   Sketch C

Fig. 3–10 Mass Moment Of Inertia For An Irregular Solid Object

mass M. Hence, the three-dimensional x-y-z coordinate system in *Sketch B* of Fig. 3-10 is necessary to locate an incremental mass dM within the boundaries of the object. The equations for mass moment of inertia around the three axes are defined by sources such as *Marks' Handbook*[4] as follows:

$$I_{xx} = \int (Y^2 + Z^2) \times dM \qquad (3\text{-}18)$$

$$I_{yy} = \int (X^2 + Z^2) \times dM \qquad (3\text{-}19)$$

$$I_{zz} = \int (X^2 + Y^2) \times dM \qquad (3\text{-}20)$$

The translation of one axis to another parallel axis also applies to mass inertia as well as to the previously discussed area moment of inertia. For example, the polar mass moment of inertia around axis z-z may be translated to the parallel axis z_o-z_o with the following common expression:

$$\boxed{J_{z_o z_o} = J_{zz} + M \times (X_{offset})^2} \qquad (3\text{-}21)$$

Within this equation, the distance X_{offset} represents the parallel offset along the x-x axis as shown in *Sketch C* of Fig. 3-10. If the offset of the new axis was further displaced from both the x-x and y-y axes, the polar moment of inertia along the new axis identified as z_1-z_1 would be given by the following:

$$\boxed{J_{z_1 z_1} = J_{zz} + M \times (R_{offset})^2} \qquad (3\text{-}22)$$

The distance between the z-z axis and the z_1-z_1 axis is defined by the radial

[4] Eugene A. Avallone and Theodore Baumeister III, *Marks' Standard Handbook for Mechanical Engineers*, Tenth Edition, (New York: McGraw-Hill, 1996), p. 3-9.

offset R_{offset}. From (3-13) it is known that $R_{offset}^2 = X_{offset}^2 + Y_{offset}^2$. The validity of this is demonstrated by allowing Y_{offset} to equal zero, and noting that (3-22) then reverts back to equation (3-21). In all cases, (3-21) and (3-22) are analogous to the parallel axis equation (3-17) for a plane area. In fact, this similarity in mathematical formats is the source for some of the confusion associated with the topic of inertia. In order to maintain a proper distinction between area and mass moment of inertia, the diagnostician should always pay close attention to the engineering units. Recall that the area moment of inertia carries English engineering units of Inches^4, and the mass inertia has units of Pound-Inch-Second^2.

The integrals shown on the last few pages are interesting calculus topics, but they do not help the field diagnostician until they are solved for specific geometric shapes. For instance, it is meaningful to develop the equations for a circular shaft cross section. In order to maintain continuity, the hollow circular

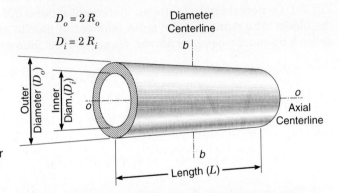

Fig. 3–11 Hollow Circular Cylinder With Principal Axes Identified

cylinder from Fig. 3-7 will be reused. For the calculation of inertia, the principal axes passing through the axial centerline *o-o*, and the midspan diametral centerline *b-b* have been shown in Fig. 3-11. For this geometric figure, the cross sectional area moment of inertia I_{area} along a diameter of the hollow circle is given by Spotts[5], or Harris[6] in the following common format:

$$I_{area_{annulus}} = \frac{\pi \times (D_o^4 - D_i^4)}{64} \qquad (3\text{-}23)$$

The area polar moment of inertia J_{area} is computed about an axis that is perpendicular to the circular cross section. On Fig. 3-11 this would be the centerline axis *o-o* of the cylinder. The polar inertia of this circular area is equal to twice the inertia along a diameter (i.e., $J=2xI$) as evident from (3-15). Multiply-

[5] M.F. Spotts, *Design of Machine Elements*, 6th Edition, (Englewood Cliffs, New Jersey: Prentice-Hall, Inc., 1985), p. 18.
[6] Cyril M. Harris, *Shock and Vibration Handbook*, Fourth edition, (New York: McGraw-Hill, 1996), p. 1.12.

ing equation (2-23) by two, the next expression for the area polar inertia evolves:

$$J_{area_{annulus}} = \frac{\pi \times (D_o^4 - D_i^4)}{32} \tag{3-24}$$

The validity of equation (3-24) is supported by Spotts, page 150, and others. Note that the last two expressions are plane area moments of inertia (i.e., no depth, no density, and no weight). However, in many machinery calculations it is mandatory to consider the mass of the element. When mass is included, the calculations become more complex. Specifically, if the hollow cylinder in Fig. 3-11 is considered to be a rotating machinery shaft, the mass polar moment of inertia along the axis of rotation (axis o-o) is defined by Gieck[7] and others. If these equations are placed into the nomenclature used in Fig. 3-11, the general expression for the mass polar moment of inertia of the cylinder may be stated as:

$$J_{mass_{annulus}} = \frac{M_{annulus}}{2} \times (R_o^2 + R_i^2) \tag{3-25}$$

Equation (3-25) is the mass inertia term of a rotating shaft that is often described as the WR^2 of the rotor. In actuality, this quantity should be identified as the mass polar moment of inertia, but common nomenclature sometimes supersedes technical accuracy. Nevertheless, if the diameters are used instead of the radii, and if (3-4) for the annulus mass is substituted into (3-25); the following manipulation may be performed to reach a common equation used for calculation of the mass polar moment of inertia for a hollow cylinder.

$$J_{mass_{annulus}} = \frac{M_{annulus}}{2} \times (R_o^2 + R_i^2) = \frac{M_{annulus}}{2} \times \left(\frac{D_o^2}{4} + \frac{D_i^2}{4}\right)$$

$$J_{mass_{annulus}} = \frac{M_{annulus}}{8} \times (D_o^2 + D_i^2) = \frac{\pi \times L \times \rho}{4 \times G \times 8} \times (D_o^2 - D_i^2) \times (D_o^2 + D_i^2)$$

Simplification of the last expression yields the following equation:

$$J_{mass_{annulus}} = \frac{\pi \times L \times \rho}{32 \times G} \times (D_o^4 - D_i^4) \tag{3-26}$$

This mass polar inertia is as important to a torsional analysis as the mass is necessary for a lateral analysis. Specifically, equation (2-102) identifies the utilization of the polar inertia to compute a torsional natural frequency in the same way that the mass is used in (2-44) to compute a lateral natural frequency. As shown in Table 2-1, the inertial term in the general equation of motion is governed by mass in a lateral system, and polar inertia in a torsional system. During the analysis or modeling of real machinery both the polar inertia and the

---

[7] Kurt Gieck and Reiner Gieck, *Engineering Formulas*, 6th edition, (New York: McGraw-Hill Inc., 1990), p. M3.

transverse inertia are utilized. Since (3-26) defines the mass polar moment of inertia J_{mass} through the axial centerline o-o, it is now reasonable to define the transverse mass Inertia I_{mass} of the hollow cylinder. Although any defined axis may be used for the calculations, the customary midspan diameter axis b-b depicted in Fig. 3-11 will be used for the following exercise. Again, extracting a standard mass inertia equation from Gieck, and modifying the terms to be consistent with Fig. 3-11, the general expression for the mass moment of inertia on the cylinder diameter axis b-b passing through the center of gravity is as follows:

$$I_{mass_{annulus}} = \frac{M_{annulus}}{12} \times (3R_o^2 + 3R_i^2 + L^2) \qquad (3\text{-}27)$$

If diameters are used instead of radii, and if the mass equation (3-4) is included, the transverse inertia equation (3-27), may be modified as shown:

$$I_{mass_{annulus}} = \frac{M_{annulus}}{12} \times (3R_o^2 + 3R_i^2 + L^2) = \frac{M_{annulus}}{12} \times \left(\frac{3D_o^2}{4} + \frac{3D_i^2}{4} + L^2 \right)$$

$$I_{mass_{annulus}} = \frac{M_{annulus}}{16} \times \left(D_o^2 + D_i^2 + \frac{4L^2}{3} \right)$$

$$I_{mass_{annulus}} = \frac{M_{annulus}}{16} \times (D_o^2 + D_i^2) + \frac{M_{annulus}}{12} \times L^2$$

$$I_{mass_{annulus}} = \frac{\pi \times L \times \rho}{16 \times 4 \times G} \times (D_o^2 - D_i^2) \times (D_o^2 + D_i^2) + \frac{\pi \times L \times \rho}{12 \times 4 \times G} \times (D_o^2 - D_i^2) \times L^2$$

$$I_{mass_{annulus}} = \left\{ \frac{\pi \times L \times \rho}{64 \times G} \times (D_o^4 - D_i^4) \right\} + \left\{ \frac{\pi \times L^3 \times \rho}{48 \times G} \times (D_o^2 - D_i^2) \right\}$$

If the mass polar moment of inertia from (3-26) is substituted into the last expression, the transverse inertia equation may be simplified as follows:

$$I_{mass_{annulus}} = \frac{J_{mass_{annulus}}}{2} + \left\{ \frac{\pi \times L^3 \times \rho}{48 \times G} \times (D_o^2 - D_i^2) \right\} \qquad (3\text{-}28)$$

If the length L of the annulus is small compared to the outer diameter, the influence of the far right hand term in equation (3-28) is significantly diminished, and the following approximation is often used:

$$I_{mass_{annulus}} \approx \frac{J_{mass_{annulus}}}{2} \qquad (3\text{-}29)$$

This result is consistent with the previous relationship described by equation (3-15) between the ratio of inertias for a plane circular cross section. In actual practice during the analytical modeling of a rotor system, the shaft station lengths are normally kept fairly short, and equation (3-29) may be a good approximation. For longer shaft sections, the more complex equation (3-28) must be used. Equation (3-29) is particularly useful for estimating the transverse inertia of wheels that have a large diameter, and a comparatively short length. For example, a centrifugal compressor impeller may be 24 inches in diameter, with a disk and a cover thickness of only 0.25 inches each. In this situation, one half of the mass polar moment of inertia J_{mass} will be very close to the detailed transverse mass moment of inertia I_{mass}.

As mentioned earlier in this chapter, the diagnostician must always be aware of the potential dilemma in the application of inertia within technical documents, and computer programs. For instance, one set of software uses weight inertia instead of mass inertia. The output inertia values from these programs carry inertia units of Pounds-Inches^2 instead of Pound-Inch-Second^2. Although the difference between the two inertia values is only the acceleration of gravity, the results can be confusing to the unprepared. In all cases, it is mandatory to be completely knowledgeable of all aspects of any inertia calculations. Within this text, inertias and their respective English engineering units are as follows:

I_{area} = Area Moment of Inertia on Diameter (Inches^4)
J_{area} = Area Polar Moment of Inertia (Inches^4)
I_{mass} = Mass Transverse Moment of Inertia (Pound-Inch-Second^2)
J_{mass} = Mass Polar Moment of Inertia (Pound-Inch-Second^2)

In many instances, machinery shafts are not hollow, and they are fabricated of solid metal. If the inner diameter D_i is set equal to zero in equations (3-23), (3-24), (3-26), and (3-28), the following inertia expressions for solid shafts with a diameter of D are easily developed:

$$I_{area_{solid}} = \frac{\pi \times D^4}{64} \qquad (3\text{-}30)$$

$$J_{area_{solid}} = \frac{\pi \times D^4}{32} \qquad (3\text{-}31)$$

$$J_{mass_{solid}} = \frac{\pi \times L \times \rho \times D^4}{32 \times G} \qquad (3\text{-}32)$$

$$I_{mass_{solid}} = \frac{J_{mass_{solid}}}{2} + \left\{ \frac{\pi \times L^3 \times \rho \times D^2}{48 \times G} \right\} \qquad (3\text{-}33)$$

Once again these expressions may be verified from various sources such as Spotts, *Marks'*, or Shigley. Please recall that the developed equations are based upon a circular cross section. If the cross sectional area is not circular, then the equations must be modified based upon the original integrals used to define inertia. Often this type of calculation is not practical due to the complexity of the mechanical part. In these cases, the diagnostician must resort to other techniques to determine the inertia properties of the machine element.

In the same way that a scale may be used to determine the weight of a rotor, there are experimental techniques that may be applied to determine inertia properties. For example, consider the machine part shown in Fig. 3-12. This could be a blank for a bull gear or a flywheel, or any other machine element for which a mass polar moment of inertia may be required. Due to the complexity of the part it may not be feasible to calculate the inertia directly, but it is possible to experimentally determine the inertia.

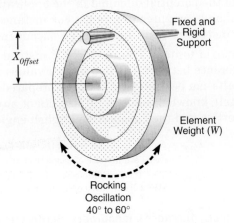

Fig. 3–12 Mechanical Arrangement For Rocking Test To Determine Mass Polar Moment of Inertia

One technique consists of suspending the part from a horizontal support as shown in Fig. 3-12. Ideally, this support member should be a knife edge, but more realistically, it will probably be a solid circular rod as depicted. If the mass polar moment of inertia through the axial centerline of the element is desired, then the distance between that centerline and the supporting pivot point X_{offset} must be accurately measured. In addition, the part should be weighed to determine the total weight W in pounds. The machine element may now be rocked back and forth as a compound pendulum. By measuring the period of the motion, the inertia about the pivot point may be determined. This offset inertia may now be translated back to the centerline of the machine part by applying the parallel axis equations previously developed.

In actual practice, the Inertia of this compound pendulum may be determined from *Marks' Handbook*[8]. Extracting the appropriate equation, and placing it in terms used within this text, the following equation for overall inertia about the pivot point may be stated:

$$J_{z_o z_o} = \frac{W \times X_{offset} \times Period^2}{4 \times \pi^2}$$ (3-34)

[8] Eugene A. Avallone and Theodore Baumeister III, *Marks' Standard Handbook for Mechanical Engineers*, Tenth Edition, (New York: McGraw-Hill, 1996), p. 3-16.

If equation (3-34) is inserted into the axis translation equation (3-21), the following is obtained:

$$J_{z_o z_o} = J_{zz} + M \times (X_{offset})^2$$

$$J_{zz} = J_{z_o z_o} - M \times (X_{offset})^2$$

$$J_{zz} = \frac{W \times X_{offset} \times Period^2}{4 \times \pi^2} - M \times (X_{offset})^2$$

$$J_{zz} = \frac{W \times X_{offset} \times Period^2}{4 \times \pi^2} - \frac{W}{G} \times (X_{offset})^2$$

Factoring out the common terms, the above may be simplified to:

$$J_{zz} = W \times X_{offset} \times \left\{ \left(\frac{Period}{2 \times \pi} \right)^2 - \left(\frac{X_{offset}}{G} \right) \right\} \qquad \textbf{(3-35)}$$

This is a useful expression since all of the variables are easily determined. Specifically, the weight W of the machine part can be measured on a scale, and the distance between the pivoting point and the geometric center of the element X_{offset} is easily measured. The acceleration of gravity G is constant, and the *Period* of the swinging motion is determined with a stopwatch in seconds. Normally, a series of runs are made to determine the average period of oscillation.

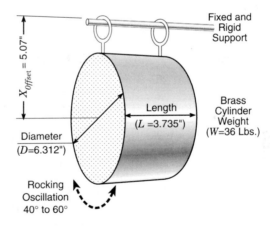

Fig. 3–13 Rocking Test To Determine Mass Polar Moment Of Inertia Of A Solid Brass Cylinder

Experimental techniques are always more credible if the validity of the equations, and the experimental procedures can be verified with a controlled test. With an inertia experiment, the object to be tested should be of simple geometry such as the solid cylinder shown in Fig. 3-13. As indicated on the diagram, the average cylinder diameter D is equal to 6.312 Inches, and the average length L is 3.735 Inches. A shop scale indicated that the weight W of the specimen was 36.0 ±0.1 Pounds. Since the cylinder was solid brass, the density was found in Appendix B of this text to be 0.308 Pounds/Inch^3. Two holes were drilled and tapped into the side of the cylinder, and two screw eyes were inserted to provide the pivot arm. Since the mass polar moment of inertia along the axial centerline is desired, the distance between the pivot point and the cylinder center line X_{offset} was measured to be 5.07 Inches.

Before starting the experiment, it is desirable to check the physical properties of the brass cylinder. For example, equation (3-7) may be used to compute the weight of the brass cylinder based upon the average dimensions, and the density of the material as follows:

$$W_{solid} = \frac{\pi \times L \times \rho \times D^2}{4}$$

$$W_{solid} = \frac{\pi \times 3.735 \text{ Inches} \times 0.308 \text{ Pounds/Inch}^3 \times (6.312 \text{ Inches})^2}{4} = 35.997 \text{ Pounds}$$

The calculated value shows excellent agreement with the measured weight of the cylinder, and 36.0 pounds may be confidently used for the ensuing inertia measurements. Some individuals might argue that this type of weight check is unnecessary. However, if the material contained inclusions, or if the density was wrong, or if one or more of the dimensions were incorrect — the weights would not match, and the experimental accuracy would be in jeopardy. So a simple calculation such as this weight check is desirable to insure that the physical parameters are in unison. The other calculation that should be made at this stage is the mass polar moment of inertia of the brass test specimen. From the previously developed equation (3-32) the polar moment of inertia of this brass cylinder is computed in the following manner:

$$J_{mass_{solid}} = \frac{\pi \times L \times \rho \times D^4}{32 \times G}$$

$$J_{mass_{solid}} = \frac{\pi \times 3.735 \text{ Inches} \times 0.308 \text{ Pounds/Inch}^3 \times (6.312 \text{ Inches})^4}{32 \times 386.1 \text{ Inches/Second}^2}$$

$$J_{mass_{solid}} = \frac{\pi \times 3.735 \times 0.308 \times 1,573.33}{32 \times 386.1} = 0.464 \text{ Pound-Inch-Second}^2$$

At this point the test calibration setup is established, the final answer is known, and the only remaining variable to be defined for equation (3-35) is the period of the oscillatory motion. Several preliminary swings of the brass cylinder revealed that the time for one complete cycle was less than a second. In addition, the friction between the two eye bolts and the support rod caused the oscillating mass to grind to a stop after only a few cycles. This problem was partially remedied by putting a tight plastic sleeve on the support rod, and then covering this surface with lithium grease. This friction reduction effort was rewarded by a substantial increase in the number of possible oscillatory cycles. However, it was then discovered that the horizontal support rod was not quite level, and the brass cylinder had a tendency to walk down the rod. This problem was corrected by re-leveling the support rod and the cylinder.

Following these test setup modifications, the actual test was conducted. The brass cylinder was displaced about 30° from the vertical centerline and released. The peak of the motion at one extremity was visually sighted, and a stopwatch was used to measure the time required for multiple back and forth cycles. The final measured test data is summarized as follows:

Rocking Run #1	7.83 Seconds for 10 cycles
Rocking Run #2	7.95 Seconds for 10 cycles
Rocking Run #3	7.98 Seconds for 10 cycles
Rocking Run #4	8.09 Seconds for 10 cycles
Rocking Run #5	7.79 Seconds for 10 cycles
Rocking Run #6	7.90 Seconds for 10 cycles
Rocking Run #7	7.89 Seconds for 10 cycles
Rocking Run #8	7.94 Seconds for 10 cycles
Rocking Run #9	7.93 Seconds for 10 cycles
Rocking Run #10	7.89 Seconds for 10 cycles
Total Time =	79.19 Seconds for 100 cycles

Average Time for 1 Cycle = 0.7919 Seconds = Period

It is easy to lose track of the cycle count, or miss a timing point, and negate the accuracy of a data set. These types of errors are evident during the data collection work, and erroneous times are identified and discarded. For instance, approximately twenty runs were made to collect the data in the above tabular summary. Ten of the timing runs were not used due to obvious errors in the data accumulation. The ten acceptable test runs reveal an average period of 0.7919 Seconds. This is considered to be a consistent value, and the experimental mass polar moment of inertia may now be computed from equation (3-35).

$$J_{zz} = W \times X_{offset} \times \left\{ \left(\frac{Period}{2 \times \pi} \right)^2 - \left(\frac{X_{offset}}{G} \right) \right\}$$

$$J_{zz} = 36.0 \text{ Pounds} \times 5.07 \text{ Inches} \times \left\{ \left(\frac{0.7919 \text{ Seconds}}{2 \times \pi} \right)^2 - \left(\frac{5.07 \text{ Inches}}{386.1 \text{ Inches/Second}^2} \right) \right\}$$

$$J_{zz} = 182.52 \text{ Pound-Inches} \times (0.01588 - 0.01313) \text{Second}^2 = 0.502 \text{ Pound-Inch-Second}^2$$

The experimental value for the polar inertia is 0.502 versus the calculated value for this brass cylinder of 0.464 Pound-Inch-Seconds^2. The difference of 0.038 represents an 8% error of the experimental versus the computed value. In some instances this level of deviation is perfectly acceptable. For example, if the part under test was a coupling hub that will be mounted on a power turbine with an inertia of 40.0 Pound-Inch-Seconds^2, the small differential of 0.038 Pound-Inch-Seconds^2 would be insignificant. However, if the part under test was one of eight impellers to be mounted on a slender shaft, the cumulative error may be unacceptable. In order to explain this error, it is necessary to re-examine the component equations used for the inertia test calculation. Specifically, (3-34) for the total inertia about the pivot point may be solved as follows:

$$J_{z_o z_o} = \frac{W \times X_{offset} \times Period^2}{4 \times \pi^2}$$

$$J_{z_o z_o} = \frac{36.0 \text{ Pounds} \times 5.07 \text{ Inches} \times (0.7919 \text{ Seconds})^2}{4 \times \pi^2} = 2.899 \text{ Pound-Inch-Second}^2$$

Substituting the test inertia from the above calculation back into equation (3-21), and performing the axis translation, the following result is obtained:

$$J_{z_o z_o} = J_{zz} + M \times (X_{offset})^2$$

$$2.899 \text{ Pound-Inch-Second}^2 = J_{zz} + \frac{W}{G} \times (X_{offset})^2$$

$$J_{zz} = 2.899 \text{ Pound-Inch-Second}^2 - \frac{36.0 \text{ Pounds}}{386.1 \text{ Inches/Second}^2} \times (5.07 \text{ Inches})^2$$

$$J_{zz} = (2.899 - 2.397) \text{ Pound-Inch-Second}^2 = 0.504 \text{ Pound-Inch-Second}^2$$

As expected, this result is identical to the previous answer obtained by using the composite equation (3-35). The interesting point of the above calculations is that inertia due to the axis translation is equal to 2.397, versus the overall test inertia of 2.899 Pound-Inch-Seconds^2. Hence, the cylinder inertia is only about 20% of the inertia due to the axis translation. This is not a desirable condition since the axis translation is the dominant term. The geometrical configuration displayed in Fig. 3-12 would not be as error prone since the X_{Offset} distance resides within the body of the element, and the axis translation term would not dominate the test. Hence, the diagnostician should always be concerned about trusting this type of experimental inertia test with a long X_{Offset} distance between the pivot axis and the desired principal polar moment of inertia axis. The other lesson to be learned is that simplified expressions such as (3-35) may not provide full visibility concerning the potential accuracy of the final results. In some cases it is necessary to revert back to the basic equations, and reexamine the entire calculation and/or experimental test procedure.

With respect to the brass cylinder, it is concluded that improvement of the test accuracy will require a reduction or elimination of the axis translation term. This could be accomplished with a test that consisted of suspending the mass from cables, and then measuring the period as the mass oscillated in a twisting manner (Fig. 3-14). Since the axis of oscillation is the axial geometric centerline of the element, there is no axis translation involved, and test accuracy should be improved. This type of inertia test is ideal for machine parts such as compressor impellers or turbine disks that contain complex geometrical cross sections. For

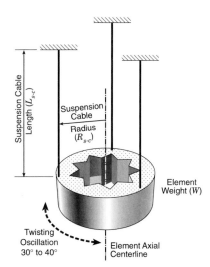

Fig. 3–14 Mechanical Arrangement For Twisting Test To Determine Mass Polar Moment of Inertia

instance, the internal star pattern shown in Fig. 3-14 might be difficult to model with an equivalent inner diameter for the machine part.

For this procedure, the machine element is suspended from three thin cables (3 points determine a plane), spaced at 120° apart. The test piece must be leveled as precisely as possible. If it is not level, then any induced twisting oscillations will cause the machine element to wobble during the test. This wobble not only negates the test accuracy, it can prove to be dangerous for parts with any appreciable physical size and weight. After leveling, the average suspension cable length L_{s-c} and the cable radius R_{s-c} are accurately measured and recorded. For best results, each of the suspension cable lengths should be equal, and the radius for all three cables should be identical. As before, the machine element to be tested is weighed on an accurate scale in English units of Pounds.

During execution of this test, the machine element is manually displaced in a twisting manner, and released. The machine part will torsionally twist back and forth, and the period of the twisting oscillations will be measured with a stopwatch. Since friction should not be major problem, the part will oscillate back and forth for many cycles. It is not unusual to observe thirty or more cycles resulting from one initial displacement. Based upon these measured parameters the mass polar moment of inertia may be computed as follows:

$$J_{zz} = \frac{W \times R_{s-c}^{2} \times Period^{2}}{4\pi^{2} \times L_{s-c}} \qquad \textbf{(3-36)}$$

The general form of (3-36) was extracted from the *Shock & Vibration Handbook*[9], and it was converted to the nomenclature used in this text. As previously

[9] Cyril M. Harris, *Shock and Vibration Handbook*, Fourth edition, (New York: McGraw-Hill, 1996), p. 38.5.

discussed on the rocking inertia test, it is mandatory to validate the test procedure with an actual test on a known geometric shape. For comparative purposes, the solid brass cylinder used for the rocking test will be used for this twisting inertia test as shown in Fig. 3-15. From this diagram it is noted that the average suspension cable radius $R_{s\text{-}c}$ was 3.00 Inches, and the average cable length $L_{s\text{-}c}$ was 33.73 Inches. As before, the total cylinder weight was 36.0 Pounds.

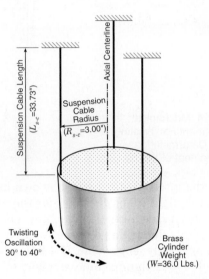

Fig. 3–15 Twisting Test To Determine Mass Polar Moment Of Inertia Of A Solid Brass Cylinder

The first test configuration used 48 Inch long suspension cables in an effort to increase the period of the oscillation, and improve the time measurement accuracy. Conceptually this was a good idea, but it turned out to be impractical since the long cables had a tendency to wrap around each other. This proved to be an unmanageable situation, and the support cable lengths were reduced to 33.73 Inches. During the acquisition of test data, the cylinder was twisted about 20° circumstantially from rest, and released. The peak of the motion at one extremity was visually sighted, and a stopwatch was used to measure the time required for complete back and forth cycles. The test data is summarized as follows:

Twisting Run #1	13.83 Seconds for 10 cycles
Twisting Run #2	14.23 Seconds for 10 cycles
Twisting Run #3	14.15 Seconds for 10 cycles
Twisting Run #4	13.94 Seconds for 10 cycles
Twisting Run #5	14.02 Seconds for 10 cycles
Twisting Run #6	13.89 Seconds for 10 cycles
Twisting Run #7	14.01 Seconds for 10 cycles
Twisting Run #8	14.11 Seconds for 10 cycles
Twisting Run #9	14.05 Seconds for 10 cycles
Twisting Run #10	13.95 Seconds for 10 cycles
Total Time =	140.18 Seconds for 100 cycles

Average Time for 1 Cycle = 1.4018 Seconds = Period

This is a much smoother test than the rocking inertia previously discussed. The number of miscounts and aborted runs were substantially reduced, and approximately fifteen runs were made to collect the data shown in the above tabular summary. Five of the timing runs were not used due to obvious errors in data accumulation. The ten consistent test runs reveal an average period of 1.4018 Seconds. This was considered to be a consistent value for this experimental procedure, and the mass polar moment of inertia may be computed from equation (3-36) as follows:

$$J_{zz} = \frac{W \times R_{s-c}^2 \times Period^2}{4\pi^2 \times L_{s-c}}$$

$$J_{zz} = \frac{36.0 \text{ Pounds} \times (3.00 \text{ Inches})^2 \times (1.4018 \text{ Seconds})^2}{4\pi^2 \times 33.73 \text{ Inches}} = 0.478 \text{ Pound-Inch-Second}^2$$

The experimental polar inertia from this twisting procedure of 0.478 is quite close to the previously calculated value of 0.464 Pound-Inch-Seconds^2. The 3% deviation is quite acceptable for most rotor dynamics calculations. This is particularity true for smaller components that are stacked on a shaft to achieve a final rotor assembly. It should be recognized that both the rocking and the twisting inertia tests have their own domain of application that is dependent on the size and geometry of the machine element.

Just as the weights of individual components are summed up to determine a total rotor weight, the inertia of the component pieces may be added to determine the overall rotor polar inertia. The origin of the inertia values may be from calculations of defined geometries, or from experientially determined inertia values. In any case, as long as the engineering units and the inertia axis are common, the numeric inertia values may be summed up to determine the mass polar moment of inertia for the entire rotating assembly.

In some instances, there is minimal opportunity to determine the inertia of rotor components since the unit cannot be disassembled or unstacked. In these situations, the general inertia characteristics may be estimated based upon available dimensions and probable materials of construction. In other cases, the complexity of the rotor may not allow a reasonable segmentation and estimation of inertia properties. This is particularly true for rotors that are constructed of multiple materials, plus rotors that contain complicated geometric configurations. In these instances, another experimental technique may be employed to determine the overall mass polar moment of inertia of the rotor.

This technique is based upon the familiar college physics experiment depicted in Fig. 3-16. In this diagram, a cylinder or drum is mounted in rigid bearings that allow rotation of the cylinder, but restrict any lateral or translation movement of the cylinder. A cord is wrapped around the cylinder at a shaft radius of R_s. It is assumed that this cord is of insignificant weight and diameter, and that it will not stretch with the application of axial tension. Next, a known weight (mass M) is attached to the end of the cord, and allowed to free fall. The time T required to fall a distance D is measured with a stop watch. The experiment normally consists of determining the cylinder mass polar moment of inertia J_{mass} based on the four known quantities of radius R_s, mass M, fall distance D, and the average fall time T.

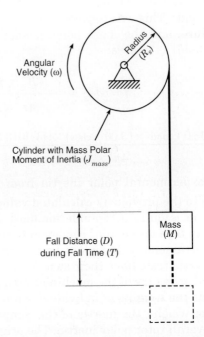

Fig. 3–16 Traditional Mechanical Arrangement For Polar Moment Of Inertia Experiment Based On Falling Mass Attached To A Rotating Cylinder

This basic physics problem may be solved by constructing free body diagrams of the cylinder and the falling mass, developing equations of motion, and then solving for the polar inertia term. Another way to achieve the same result is by performing an energy balance on the system shown in Fig. 3-16. Conservation of energy requires that the change in potential energy is equal to the change in kinetic energy. In this case, the change in potential energy is simply the elevation change in the mass ($M \times G \times D$). The overall change in kinetic energy is composed of a change in translational energy of the falling mass ($M \times V^2/2$), plus the change in the rotational kinetic energy of the rotor ($J_{mass} \, \omega^2/2$). These traditional physical concepts may be represented mathematically in the following manner:

$$M \times G \times D = \frac{M \times V^2}{2} + \frac{J_{mass} \times \omega^2}{2} \qquad \textbf{(3-37)}$$

The velocity V in equation (3-37) represents the average velocity of the falling mass M. In all cases, the falling weight W is equal to the mass M times the acceleration of gravity G. Substituting the quantity W/G for the mass M produces the following:

$$W \times D = \frac{W \times V^2}{2 \times G} + \frac{J_{mass} \times \omega^2}{2}$$

$$2W \times D = \frac{W \times V^2}{G} + J_{mass} \times \omega^2$$

$$J_{mass} \times \omega^2 = 2W \times D - \frac{W \times V^2}{G}$$

This may now be solved for the mass polar moment of inertia J_{mass} as:

$$J_{mass} = \frac{W}{\omega^2} \times \left\{ 2D - \frac{V^2}{G} \right\} \tag{3-38}$$

The average velocity V of the mass M during free fall may be determined from the fundamental equations of motion for rectilinear motion with constant acceleration. More specifically, all physics books agree that the average velocity multiplied times the drop time T will determine the fall distance D as follows:

$$D = \left(\frac{V_o + V}{2} \right) \times T \tag{3-39}$$

If the initial starting velocity V_o is equal to zero, then equation (3-39) may be simplified to represent the velocity V in terms of the drop distance D, and the total elapsed drop time T as shown in the next equation:

$$V = \frac{2D}{T} \tag{3-40}$$

The last conversion to be performed consists of an expression for the angular velocity ω of the cylinder in terms of the known experimental parameters. Since tangential velocity V divided by the radius R_s is equal to the angular velocity ω, equation (3-40) may be used to determine ω in the following manner:

$$\omega = \frac{V}{R_s} = \frac{2D}{R_s \times T} \tag{3-41}$$

Equation (3-38) for the polar inertia may now be clarified into the known experimental data by substituting equations (3-40) and (3-41) back into equation (3-38), and performing the following simplification of terms:

$$J_{mass} = \frac{W}{\omega^2} \times \left\{ 2D - \frac{V^2}{G} \right\}$$

$$J_{mass} = \frac{W}{(2 \times D/R_s \times T)^2} \times \left\{ 2D - \frac{(2 \times D/T)^2}{G} \right\}$$

$$J_{mass} = \frac{W \times R_s^2 \times T^2}{4 \times D^2} \times \left\{ 2D - \frac{4 \times D^2}{G \times T^2} \right\}$$

$$J_{mass} = W \times R_s^2 \times \left\{ 2D \times \frac{T^2}{4 \times D^2} - \frac{4 \times D^2}{G \times T^2} \times \frac{T^2}{4 \times D^2} \right\}$$

The cancellation of common terms results in equation (3-42).

$$J_{mass} = W \times R_s^2 \times \left\{ \frac{T^2}{2D} - \frac{1}{G} \right\}$$

(3-42)

where: J_{mass} = Mass Polar Moment of Inertia (Pounds-Inches-Seconds^2)
   W = Falling Weight (Pounds)
   R_s = Shaft Radius (Inches)
   D = Drop Distance (Inches)
   T = Drop Time (Seconds)
   G = Acceleration of Gravity (=386.1 Inches / Second^2)

With a few minor modifications, this traditional physics experiment may be used to determine the polar inertia of a complete rotor assembly. One implementation of this concept is the technique by Michael Calistrat[10] that is illustrated in Fig. 3-17. This diagram depicts a rotor resting in the rollers of a two bearing shop balancing machine. To minimize errors, the rotor assembly should be properly balanced, and the rollers in the balancing machine should be in good condition. As shown in Fig. 3-17, an overhead pulley is supported by an external structure, or the overhead crane that is used to move rotors in and out of the balancing machine. The diameter of the overhead pulley does not matter, but this pulley should be in good condition, and it should turn easily. The pulley redirects the gravitational force to pull upward on the rotor, but it does not diminish the effect of the falling mass. A plastic covered braided steel cable is attached to the shaft with Duct tape at a radius of R_{shaft}, and wrapped around the shaft for ten or fifteen turns. It's probably a good idea to keep the cable away from any of the rotor bearing journals, or the rotor sections observed by proximity probes. The thickness of this cable is generally small compared to the diameter of the shaft. Again, as shown in Fig. 3-17, the steel cable passes overhead, through the pulley, and then is secured with a pair of vertical weights. The first weight is a tare

---

[10] Michael M. Calistrat, *Flexible Couplings, their design selection and use*, (Houston: Caroline Publishing, 1994), p 464.

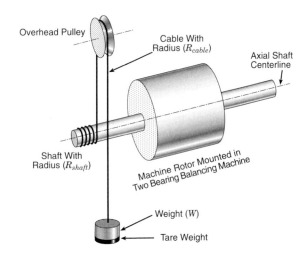

Overhead Pulley

Cable With
Radius (R_{cable})

Axial Shaft
Centerline

Shaft With
Radius (R_{shaft})

Machine Rotor Mounted in
Two Bearing Balancing Machine

Weight (W)

Tare Weight

Fig. 3–17 Mechanical Test Arrangement For Polar Moment Of Inertia Measurement Of Full Rotor Assembly Mounted In A Shop Balancing Machine, Technique by M. M. Calistrat

weight that is used to counterbalance the rotor, and keep the cable taunt. This tare weight should just barely allow the rotor to rotate, and overcome system friction.

The tare weight is determined by trial and error, and small changes in this weight will have a big influence on the rotor. Since this work is performed in a balancing machine, it makes sense to use balancing clay to get a good tare weight established. The second weight W is the experimental weight that actually turns the rotor. It may be compromised of balancing clay, a stack of washers, or any reasonable combination thereof. In all cases, the diagnostician must have the ability to accurately weigh this mass W at the end of the test.

As with the two previous inertia measurements, the majority of the time is spent in setting up the test. The test execution requires minimal time, and this rotor inertia procedure is no exception. Following the setup per Fig. 3-17, the cable is wound up on the shaft, and the weight is naturally raised in elevation. At this time the shaft is released, and the time T for the weight to fall through a predetermined distance D is measured with a stopwatch. The inertia calculation described by equation (3-42) may be used directly. However, improved accuracy will result by including the cable radius R_{cable} with the shaft radius R_{shaft} as recommended by Calistrat. This addition is not required for very thin cables, but it does improve accuracy for thicker cables, since the moment arm is really the sum of the shaft plus the cable radius. Hence, (3-42) may be rewritten as follows:

$$ J_{mass} = Weight \times (R_{shaft} + R_{cable})^2 \times \left\{ \frac{Time^2}{2 \times Distance} - \frac{1}{G} \right\} \qquad \textbf{(3-43)} $$

The nomenclature in equation (3-43) has been expanded to be more understandable, and yet maintain consistency with the general terms used within this text. Again, the experimental weight W is in pounds, the shaft radius R_{shaft}, the cable radius R_{cable}, and the fall *Distance* are all in inches. The measured *Time* for the free fall through the predetermined distance is defined as the time in seconds, and G is the gravitational constant.

As with any experiment, the validity should be checked with a test mandrel of known characteristics. For this test, a section of 4140 steel was selected with a density of 0.283 Pound/Inch^3. The length L of the specimen was 59.81 inches, and the ends were squared up in a lathe. In addition, the outer diameter D of the test piece was cleaned up to nominally 5.980 inches across the length of the element. Once more equation (3-7) is used to compute the mandrel weight:

$$W_{solid} = \frac{\pi \times L \times \rho \times D^2}{4}$$

$$W_{solid} = \frac{\pi \times 59.81 \text{ Inches} \times 0.283 \text{ Pounds/Inch}^3 \times (5.98 \text{ Inches})^2}{4} = 475 \text{ Pounds}$$

This calculated weight of 475 pounds was lower than the uncalibrated shop scale reading of 490 pounds. Although a better agreement in weights would have been comforting, the 15 pound deviation was considered to be within the measurement accuracy of the shop scale. Next, the mass polar moment of inertia of this steel shaft may be computed with equation (3-32) in the following manner:

$$J_{mass_{solid}} = \frac{\pi \times L \times \rho \times D^4}{32 \times G}$$

$$J_{mass_{solid}} = \frac{\pi \times 59.81 \text{ In.} \times 0.283 \text{ Pounds/In.}^3 \times (5.98 \text{ In.})^4}{32 \times 386.1 \text{ Inches/Second}^2} = 5.504 \text{ Pound-Inch-Second}^2$$

Prior to testing, the runout (eccentricity) along the length of the mandrel was confirmed to be less than 1.0 Mil. In addition, a check balance was performed, and the residual unbalance was minor. For the actual test runs, a fall distance of 60 Inches was established on a vertical reference stand. The shaft radius R_{shaft} was 2.99 Inches, and the cable diameter was 1/8", for a R_{cable} of 0.06 inches. The tare weight was found to be 0.696 Pounds (315.7 grams). The experimental weight W was adjusted until the free fall time was about 5 seconds. This required a total of 3.43 Pounds (1,556 grams). The timed test data was obtained by releasing the shaft and measuring the time required for the weight to fall 60 inches. A stopwatch was used to measure the time, and the test data is follows:

Falling Weight Run #1	4.52 Seconds
Falling Weight Run #2	4.54 Seconds
Falling Weight Run #3	4.65 Seconds
Falling Weight Run #4	4.54 Seconds
Falling Weight Run #5	4.56 Seconds
Total Time =	22.81 Seconds for 5 Drops

Average Time for 1 Drop of 60 Inches = 4.562 Seconds

Eight different test runs were made, and the obviously incorrect times were discarded. The above array was quite consistent, and there was good confidence in both the validity and the accuracy of this experimental data. Hence, equation (3-43) was now applied to compute the measured mass polar moment of inertia:

$$J_{mass} = Weight \times (R_{shaft} + R_{cable})^2 \times \left\{ \frac{Time^2}{2 \times Distance} - \frac{1}{G} \right\}$$

$$J_{mass} = 3.43 \text{ Pounds} \times (2.99 + 0.06 \text{ Inches})^2 \times \left\{ \frac{(4.562 \text{ Seconds})^2}{2 \times 60 \text{ Inches}} - \frac{1}{386.1 \text{ Inches/Sec}^2} \right\}$$

$$J_{mass} = 3.43 \times 9.302 \times (0.1734 - 0.0026) = 5.450 \text{ Pound-Inch-Second}^2$$

Note that there is excellent agreement between the experimental inertia of 5.450 and the calculated polar inertia of this steel shaft of 5.504 Pound-Inch-Second^2. The actual deviation of 1% is considered to be quite acceptable accuracy for this type of measurement. It is often desirable to perform this test with different weights, and different fall distances to verify the consistency of the procedure. In all cases, the test weight should be substantially larger than the tare weight, and please recall that the initial starting velocity is zero at the start time. For bigger and heavier rotors that have a tendency to sag within the balancing machine, consideration should be given to rolling the rotors for several hours before attempting this type of inertia test. In addition, dial indicator measurements should be made at the rotor midspan before beginning the inertia tests, and after the conclusion of the last run. Any appreciable sag to the shaft of the rotor under test might negate the test results, and require a repeat of the slow roll and the test procedure.

Finally, it should be recognized that other experimental procedures exist for determination of mass moment of inertia. In addition, many computation programs provide the capability for three-dimensional calculations of the inertia properties of complex bodies. As always, the machinery diagnostician should be aware of the engineering units, and test cases of known geometries should always be run to verify the technique.

DAMPING INFLUENCE

Three basic types of damping occur in a machinery system. These damping types are commonly referred to as *viscous* damping, *coulomb* damping, and *solid* or *structural* damping.

Viscous damping is encountered by solid bodies moving through a viscous fluid. In this type of damping, the resistance force is proportional to the velocity of the moving object. As an example of viscous damping, consider the situation of a cook stirring a pot of soup versus a pot of molasses. It is self-evident that stirring the molasses is considerably more difficult due to the thickness and higher viscosity of the molasses as compared to the thin soup. The required force is directly proportional to the velocity of the stirring spoon. In most cases the cook would stir the molasses at a much slower rate than he would the soup, simply because it would take too much strength or energy to stir it rapidly.

The same type of physical property, i.e., viscous damping, is encountered in the bearings and oil seals of large rotating machines. In this case, the damping is provided by the lubricating oil, and the rotating shaft is the rigid body moving through the viscous fluid. The process fluids handled by the machine also provide damping to the rotor system. For liquid handling machines such as pumps and hydraulic turbines, this is significant. However, for gas handling machines such as turbines or centrifugal compressors this is a minor consideration.

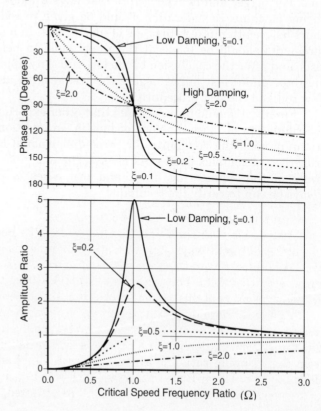

Fig. 3–18 Calculated Bode Plot Of Unbalance Response For A Simple Mechanical System With Variations In Damping

The next type of damping is **coulomb damping**, which arises from the sliding of one dry surface upon another (rub condition). The coulomb friction force is nearly constant, and it depends on the nature of the sliding surfaces, and the perpendicular pressure between the surfaces. This type of force is generally dominant in damped systems during the final stages of motion when other types of damping become negligible.

The third category of damping is often referred to as solid or **structural damping**. This is due to internal friction within the material, and it differs from viscous damping in that it is independent of frequency, and proportional to the maximum stress of the vibratory cycle. Since stress and strain are proportional in the elastic range, it can be stated that solid damping force is proportional to deflection. Structural damping in rotating machinery is small when compared with viscous damping, but it does exist.

The major contribution that positive damping makes to a rotating machinery system is the dissipation of energy. This influence is most dramatically illustrated when a mechanical system passes through a resonance as in Fig. 3-18. This calculated Bode diagram was duplicated from chapter 2. The unbalance response plot of frequency ratio versus amplitude ratio and phase of a damped system provides a good perspective of the actual influence of damping. The family of curves in this diagram are plotted with a damping ratio (or damping factor) extending from $\xi=0.1$ to 2.0. Recall that this ratio is the actual damping divided by the critical damping. Note, that with a lightly damped system of $\xi=0.1$, the response at the resonance is quite high. This translates to the fact that there is little energy dissipation under this condition. The system is under damped, and it is susceptible to instability due to a lack of an energy dissipation. Conversely, when the damping factor is large, $\xi=2.0$, the system is over damped, response through the resonance is restrained, and overall stability of the system is high.

In many instances of free vibration, rotor instability can be related to a lack of damping. The system damping may be assessed by examining the critical speed response on Bode plots (synchronous 1X vectors versus speed). Typically, the amplification factor Q through the resonance is used to quantify the severity of the resonance, plus the damping ratio. A large amplification factor is associated with a poorly damped, high amplitude resonance. Conversely, a low amplification factor is generally associated with a well damped resonance, that displays small amplitudes at the peak of the resonance.

Extraction of the amplification factor from the Bode plots may be performed in several different ways. Unfortunately, there is disagreement within the technical community as to the best manner to obtain this information from the measured vibration data. For the sake of completeness, three separate approaches for determination of this dimensionless amplification factor will be presented.

The first technique consists of visually comparing the measured vibration response data with a set of calculated curves, and estimating a damping ratio ξ. The rotor amplification factor for the specific resonance is then computed directly from the damping ratio. As a practical example of an industrial machine, consider the compressor Bode plot shown in Fig. 3-19 (same as Fig. 2-21). This variable speed vibration data is runout compensated, and it exhibits a clean transition through a single critical speed. If this data is viewed in conjunction with the computed family of curves shown in Fig. 3-18, a suitable damping ratio may be selected. Specifically, the shape of the amplitude versus speed, and the

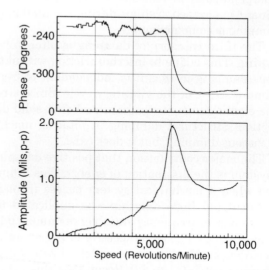

Fig. 3–19 Measured Bode Plot Of A Centrifugal Compressor Startup

phase versus speed are compared between the measured (Fig. 3-19) and the calculated plots (Fig. 3-18). It is reasonable to conclude that the calculated plot with a damping ratio ξ of 0.2 is the closest match to the measured machine response data. From the previously developed equation (2-91), the amplification factor (i.e., amplitude ratio) Q may be determined from the damping ratio as follows:

$$Q_{Curve\ Fit} = Amplification\ Factor = Amplitude\ Ratio = \frac{1}{2 \times \xi} \qquad (3\text{-}44)$$

Substituting the previously identified value for the damping ratio of $\xi=0.2$, the following result is obtained:

$$Q_{Curve\ Fit} = \frac{1}{2 \times \xi} = \frac{1}{2 \times 0.2} = \frac{1}{0.4} = 2.5$$

This visual comparison between curve shapes provides any easy way to estimate Q for a cleanly defined resonance. However, for more complex response characteristics, other techniques are available. For instance, the second approach for computation of the amplification factor is derived from electrical engineering terminology. This technique has also been adopted by various mechanical standards organizations such as the American Petroleum Institute. In this procedure, the center frequency of the resonance is divided by the resonance bandwidth at the *Half Power Point* in accordance with the following:

$$Q_{Half\ Power} = \frac{Center\ Frequency}{Frequency\ Bandwidth\ @\text{-}3dB} \qquad (3\text{-}45)$$

The half power point is equivalent to the amplitude that is -3 dB down from the peak of the resonance. In terms of linear scales, an amplitude change of -3 dB is equal to 0.707 times the peak vibration amplitude. This level defines the specific point on the resonance response curve where the frequency bandwidth is measured. Although this calculation procedure may sound complicated, the actual execution is fairly straightforward.

If the Bode plot from the last example is reconsidered, the amplification factor using this half power approach may be computed as shown in Fig. 3-20. Within this rendition of the Bode, various features have been identified to

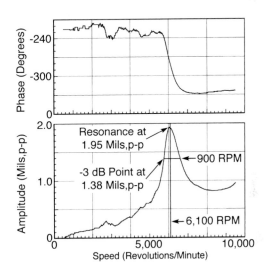

Fig. 3–20 Amplification Factor From A Bode Plot Using the Half Power Point Calculation Technique

enhance this discussion. Specifically, the center of the resonant peak (critical speed) has been identified at 6,100 RPM. The vibration amplitude at the peak of this resonance is 1.95 Mils,$_{p-p}$. The half power point is determined by multiplying 0.707 times the maximum of 1.95 Mils,$_{p-p}$ to obtain a -3 dB point of 1.38 Mils,$_{p-p}$. The response curve width at this amplitude is 900 RPM. Stated in another way, the frequency equal to an up slope amplitude of 1.38 Mils,$_{p-p}$ is 5,800 RPM. The frequency equal to a down slope amplitude of 1.38 Mils,$_{p-p}$ is 6,700 RPM. The bandwidth at this -3 dB amplitude is 6,700 minus 5,800, which is equal to 900 RPM. Substituting these values for center frequency and bandwidth into the previous equation (3-45) the next result is obtained:

$$Q_{Half\ Power} = \frac{Center\ Frequency}{Frequency\ Bandwidth\ @\text{-3dB}} = \frac{6,\,100\ Rpm}{900\ Rpm} = 6.8$$

The calculated amplification factor using this second approach (6.8) is considerably higher than the first method that produced a Q of 2.5. However, this second method may be used for complex resonant response conditions where multiple critical speeds appear close together. For instance, a machine that has a split critical (e.g., horizontal followed by a vertical mode) would be difficult to handle with the first approach. Whereas, the second scheme would allow a better quantification of the Q for each response peak.

One objection to the second approach is that Q varies with changes to the center frequency of the resonance. For instance, if the shape of the response curve is maintained, and the center frequency is reduced to 3,000 RPM, the Q drops to 3.3 (=3,000/900). By the same token, if the critical speed occurs at 12,000 RPM, the Q now increases to 13.3 (=12,000/900). This change in amplification factors with a constant shape to the response peak may be quite confusing, as well as contradictory to the desired definition.

The third approach for calculation of the amplification factor consists of dividing the amplitude at the resonance by the amplitude at a speed far above the resonance. Again referring to the same Bode plot example, this approach is illustrated in Fig. 3-21. As before, the magnitude of 1.95 Mils,$_{p-p}$ at the transla-

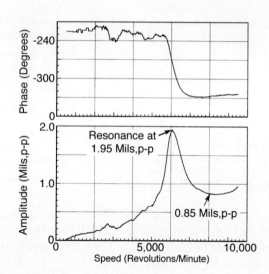

Fig. 3–21 Amplification Factor from a Bode plot Using A Simple Amplitude Ratio Between the Resonance and a Speed Far Above the Resonance

tional critical speed of 6,100 RPM is identified. Above this resonance, the phase and amplitude enter a plateau region where the 1X vector remains reasonably constant as speed increases. This is normal behavior, and the synchronous response above a critical is typically flat until some external force, or a higher order resonance influences the motion. Within the context of this example, the region at 8,000 RPM is sufficiently removed from the critical speed, and the rotational speed vibration amplitude in this region is 0.85 Mils,$_{p-p}$. The amplification factor determined with this third technique is computed by simply dividing the maximum vibration amplitude at the critical speed by the amplitude measured well above the resonance in accordance with the following expression:

$$Q_{Resonance/Above} = \frac{Amplitude_{At\ Resonance}}{Amplitude_{Above\ Resonance}} \qquad (3\text{-}46)$$

Using the vibration amplitudes identified on Fig. 3-21, the amplification factor determined by this method is easily computed as:

$$Q_{Resonance/Above} = \frac{Amplitude_{At\ Resonance}}{Amplitude_{Above\ Resonance}} = \frac{1.95\ \text{Mils}_{p-p}}{0.85\ \text{Mils}_{p-p}} = 2.3$$

This value of 2.3 is indicative of an adequately damped machine, and a stable mechanical system. Due to the differences between machines, it is difficult to precisely categorize the variations between good and bad Q factors. However, it is reasonable to identify several categories of amplification factors versus damping and stability characteristics. Summary Table 3-1 attempts to provide some realistic guidelines for these interrelated parameters:

Table 3–1 Shaft Amplification Factor Versus Damping And Anticipated Rotor Stability

Amplification Factor	Damping	Stability
Q = < 2	Well Damped	Extremely Stable
Q = 2 to 8	Adequately Damped	Normal Stability
Q = 8 to 15	Poorly Damped	Marginal Stability
Q = > 15	Insufficient Damping	Inherently Unstable

Machines that fall into the last category may be potentially dangerous, and may not survive a maiden startup. Machines within this group often require extensive modifications to bearings and/or seals to increase system damping. In some instances, modifications such as the installation of squeeze film damper bearings may be required to provide adequate damping for the system.

It should also be noted that the amplification factors referred to herein are associated with shaft vibration measurements where the oil film viscous damping is dominant (i.e., between the journal and bearing). Casing vibration measurements would typically be more receptive to structural damping, and not so sensitive to viscous damping in the bearings. Thus, casing measurements generally display Q's that are much higher than shaft measurements, simply due to the lack of damping within the casing and the support structure.

Another consideration that must be applied to any evaluation of variable speed data is the acceleration rate, or more specifically the rate of speed change of the rotating system. During machinery train coastdowns, there is usually minimal, if any, control of the deceleration. However, during startup, the rate of rotor acceleration is often controllable on variable speed drivers such as steam turbines. On machines with older control systems, the startup rate is often dependent upon the skill and knowledge of the operator handling the trip & throttle valve. On newer speed control systems, the startup rate is usually controlled by an electronic governor with predetermined startup speed ramps. Unfortunately, some electronic governors suppliers are not well versed in the acceptable startup rates for various types of machinery trains. These vendors often set abnormally fast startup rates in an effort to *snap* the rotor(s) through a critical speed range. This tendency has resulted in machinery damage, and more than one rotor failure. Hence, the end user should always verify that the proposed startup acceleration rates are reasonable for the machinery in question.

Intuitively, the passage of a rotor system through a critical speed region should be performed in a direct and knowledgeable manner. If the startup rate is inordinately slow, the machine may *hang up* in a resonance, and cause mechanical damage due to the high vibration levels. An example of this type of occurrence is briefly discussed in the turbine generator case history 39 in chapter 11. On the other hand, if the speed acceleration rate through a resonance is excessive, the machine may self-destruct after it reaches operating speed and attempts to rebalance itself about the mass center. Although this type of occurrence is rare, it is certainly avoidable, and totally unnecessary.

The speed transition rate through a resonance will alter the characteristics of the vibration response data. A slow startup will show a higher peak at the resonance, combined with a broader bandwidth. Conversely, a rapid startup will produce a lower peak amplitude at the critical speed, plus a smaller resonant bandwidth. This attenuated response characteristic has erroneously led many individuals into a false sense of security by ramping through critical speeds at a high rate. Hence, the diagnostician must be aware of the transition rate through the system critical speed(s), and any evaluation of the resonant characteristics (e.g., Q) should be weighed by this speed change rate.

It is impossible to fully quantify proper startup acceleration rates for all classes of machinery operating with various types of drives, and control systems. However, to provide some guidance in this area, Table 3-2 of acceptable startup rates is offered for consideration. This table summarizes field measurements on a variety of variable speed trains. In all cases, the general machine type is indicated, the rotor weight is shown, and the maximum machine speed is listed. The last column provides typical peak startup acceleration rates (generally through

Table 3–2 Various Machine Types Versus Acceptable Startup Acceleration Rates

Machine Type	Rotor Weight (Pounds)	Maximum Speed (RPM)	Acceleration Rate (RPM/Second)
Expander/Pinion	60	34,000	550 to 750
Pinion/Compressor - Small	300	14,000	350 to 450
Pinion/Compressor - Medium	1,500	11,000	80 to 100
Steam Turbine/Compressor	7,500	8,000	70 to 80
Gas Turbine/Compressor	25,000	5,400	20 to 30
Steam Turbine/Compressor	29,000	3,800	40 to 50
Steam Turbine/Compressor	54,000	4,000	35 to 40
Hydro Turbine/Generator	180,000	420	5 to 10

resonances). It should be recognized that other portions of the startup sequence often occur at much lower rates. Thus, the compressor that exhibits a 400 RPM per Second ramp through a critical may be preceded by a region with a rate of only 20 to 30 RPM per Second.

Table 3-2 represents conservative speed acceleration rates. The observed trend of light weight, high speed rotors that accelerate rapidly is considered to be reasonable. In addition, large rotors that run at slow speeds exhibit understandably lower rates. This table is only provided for reference purposes based upon a finite set of measurements. For specific recommendations of acceleration rates for defined machinery trains, the Original Equipment Manufacturer (OEM) must always be contacted.

Throughout this section on damping influence, the emphasis has been placed on the effects of damping upon a mechanical system. The changes of rotor amplification factor, and the changes in stability due to variations in damping are generally understood. For simple cases, it has been demonstrated that the theory and the actual machinery behavior are quite compatible. In all cases, the machinery diagnostician is encouraged to simplify problems to the greatest extent possible, and to strive towards straightforward explanations and solutions. Unfortunately, some mechanical systems may not cooperate with this approach, and the complexity of the problem plus the potential solution may very well exceed the explanations offered by simple systems.

When faced with difficult technical problems, the traditional engineering mentality often attempts to fully quantify and define all variables associated with the problem. With respect to damping characteristics, this is difficult to accomplish. For example, John Vance[11] states that *"Accurate values for the damping coefficients are usually difficult to obtain. The author has found from experiments that an all-steel rotor/shaft assembly with tight fits and no gear backlash will have about 1.5–2.0 percent of critical damping, not including friction between the rotor and ground."* This statement reinforces the difficulty in determination of actual damping coefficients. In many instances, the damping characteristics are either ignored or determined experimentally.

There is experimental information published on specific damping materials used for vibration and shock isolation. Technical references such as the *Shock and Vibration Handbook*[12] devote several chapters to isolators, and the characteristics and utilization of damping materials. This information is applicable to machines such as air blowers and refrigeration compressors that are installed on large buildings. For the comfort of the inhabitants of these buildings, the machines are normally isolated from their respective structures. However, on large industrial machines, the foundations and supporting structures are normally rigid, and external damping materials are not used. On large gas handling process machines, most of the damping is provided by the bearing lubricating oil,

---

[11] John M. Vance, *Rotordynamics of Turbomachinery,* (New York: Wiley-Interscience Publication, 1988), p 61.

[12] Cyril M. Harris, *Shock and Vibration Handbook*, Fourth edition, (New York: McGraw-Hill, 1996), chapters 32, 33, 36 and 37.

with only a minor contribution from the shaft oil seals. The viscous damping associated with the bearings normally dominates any coulomb and structural damping in the system. The engineering values for bearing damping are usually calculated with analytical programs, as discussed in chapter 5. Structural or support damping is often determined experimentally, and this topic is addressed in the following chapter 4.

Although quantification of discrete damping values is necessary from an initial design standpoint, it may not be particularly useful to the diagnostician attempting to solve a field problem. In most cases, a working knowledge of the influence of damping upon the machinery, plus an understanding of the contributing parameters that are involved in determination of the damping are the most important items. To this extent, it is mandatory for the diagnostician to understand that bearing damping varies with speed, applied load, bearing geometry, plus the lubricant characteristics. This knowledge may allow the development of a logical thought process during the analysis of a machinery problem. For instance, if a compressor exhibits symptoms of decreased damping, such as higher vibration amplitudes passing through a critical speed, one mechanism that might considered would be the influence of expanded bearing clearances upon the damping. Of course, increased bearing clearances might also produce significant changes in the bearing stiffness, which would probably change the center frequency of the measured critical speed. Hence, the influence of stiffness upon the machinery behavior must also be addressed, and the following section addresses the fundamentals of the stiffness aspects of machinery supports.

STIFFNESS INFLUENCE

The stiffness of a mechanical system is the spring like quality of mechanical elements to elastically deform under load. In essence, the application of a force measured in pounds will produce a deflection measured in inches. Thus, the stiffness of a mechanical element carries the traditional engineering units of Pounds per Inch. Mechanical properties such as the dimensions and weight of an object have a physical meaning for most individuals. However, a characteristic such as stiffness is commonly used, but it is seldom associated with anything of real physical significance. To provide a sense of magnitude to stiffness values, consider the following derivation of axial, radial, and torsional stiffness for a series of identical cylinders.

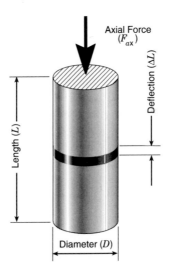

Fig. 3–22 Solid Cylinder Mounted On Infinitely Rigid Base And Subjected To Axial Compression

The diagram presented in Fig. 3-22 describes a solid metal cylinder with a diameter D, and a length L. Assume that the bottom of this cylinder is rigidly supported, and consider the application of a pre axial compressive force F_{ax}. Further assume that the applied force is sufficient to cause an elastic deformation of the cylinder, and that the amount of this deformation is described by the deflection ΔL. From fundamental strength of materials it is known that:

$$Strain = \frac{Deflection}{Length} = \frac{\Delta L}{L} \qquad (3\text{-}47)$$

$$Stress_{solid} = \frac{Force}{Area_{solid}} = \frac{F_{ax}}{\pi \times D^2/4} = \frac{4 \times F_{ax}}{\pi \times D^2} \qquad (3\text{-}48)$$

It is also known that within the elastic region for the material, stress and strain are related by the modulus of elasticity E as follows:

$$Modulus\ of\ Elasticity\ =\ E\ =\ \frac{Stress_{solid}}{Strain} \tag{3-49}$$

Substituting equations (3-47) and (3-48) into (3-49) produces the following:

$$E\ =\ \frac{\left\{\dfrac{4 \times F_{ax}}{\pi \times D^2}\right\}}{\left\{\dfrac{\Delta L}{L}\right\}}\ =\ \frac{4 \times F_{ax} \times L}{\pi \times D^2 \times \Delta L} \tag{3-50}$$

In accordance with the initial definition for stiffness, the axial stiffness of this metal cylinder may be determined by dividing the applied force F_{ax} by the deflection ΔL. Hence, it may be properly stated that:

$$K_{ax_{solid}}\ =\ \frac{F_{ax}}{\Delta L} \tag{3-51}$$

By combining (3-50) and (3-51), the axial stiffness may be included as:

$$E\ =\ \left\{\frac{4 \times L}{\pi \times D^2}\right\} \times \left\{\frac{F_{ax}}{\Delta L}\right\}\ =\ \left\{\frac{4 \times L}{\pi \times D^2}\right\} \times K_{ax_{solid}}$$

By rearranging terms, (3-52) for axial stiffness of a solid cylinder evolves:

$$K_{ax_{solid}}\ =\ \frac{\pi \times E \times D^2}{4 \times L} \tag{3-52}$$

Equation (3-52) makes sense from the standpoint of the contribution of the terms. Specifically, a large diameter cylinder will be stiffer than a smaller diameter cylinder, and a short cylinder should be stiffer than a tall one. To provide a physical representation of the meaning of various stiffness values, consider solving the last expression for the diameter as follows:

$$D\ =\ 2 \times \sqrt{\frac{L \times K_{ax_{solid}}}{\pi \times E}} \tag{3-53}$$

If the cylinder material is steel, the modulus of elasticity E would be in the vicinity of 30,000,000 Pounds/Inch^2. If the total length of the cylinder L is assumed to be 30 inches, then (3-53) may be simplified to the following format:

$$D\ =\ 2 \times \sqrt{\frac{30\ \text{Inches} \times K_{ax_{solid}}}{\pi \times 30 \times 10^6\ \text{Pounds/Inches}^2}}\ =\ \frac{\sqrt{K_{ax_{solid}}}}{886.2}$$

With this expression, the equivalent cylinder diameter may be computed for various stiffness values. For instance, if the axial stiffness of a solid cylinder is 1,000,000 Pounds/Inch, the required diameter is calculated as follows:

$$D = \frac{\sqrt{K_{ax_{solid}}}}{886.2} = \frac{\sqrt{1 \times 10^6}}{886.2} = 1.1284 \text{ Inches}$$

If this calculation is repeated for a series of stiffness values, the resultant diameters may be plotted against stiffness. Another way to view this information is shown in Fig. 3-23. Within this diagram, a series of consecutive cylinders are drawn that represent stiffness values ranging from 1,000,000,000 Pounds/Inch (i.e., 10^9) with an associated diameter of 35.682 inches — to a minimum stiffness of 100,000 Pounds/Inch (i.e., 10^5) with a computed diameter of 0.3568 inches.

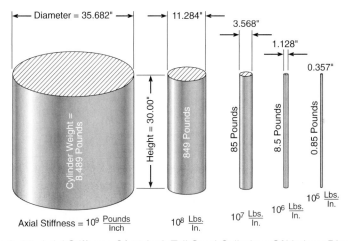

Fig. 3–23 Axial Stiffness Of 30 Inch Tall Steel Cylinders Of Various Diameters

Fig. 3-23 is a rendition of a similar explanation of stiffness presented by F.L. Weaver[13], in his paper entitled *Rotor Design and Vibration Response*. Although Weaver did not present the method for computing the stiffness of the cylinders — his diagram of spring gradients of 30 inch tall solid steel cylinders on an infinitely stiff support are equivalent to the drawing in Fig. 3-23.

If the cylinders subjected to axial compression are hollow instead of solid, the previously developed equation structure may be easily modified by incorporating an inner diameter into equation (3-48) as follows:

$$Stress_{annulus} = \frac{Force}{Area_{annulus}} = \frac{F_{ax}}{\pi \times \left(\dfrac{D_o^2}{4} - \dfrac{D_i^2}{4} \right)} = \frac{4 \times F_{ax}}{\pi \times (D_o^2 - D_i^2)} \qquad \textbf{(3-54)}$$

[13] F.L. Weaver, "Rotor Design and Vibration Response," *Proceedings of the First Turbomachinery Symposium*, Gas Turbine Laboratories, Texas A&M University, College Station, Texas, (1972), pp. 142-147.

In equation (3-54), the cylinder outer diameter is designated by D_o, and the hollow inner diameter is once more identified as D_i. The strain equation (3-47) may now be combined with equation (3-54) in the traditional stress—strain relationship described by equation (3-49) to produce the following result:

$$E = \frac{Sress_{annulus}}{Strain} = \frac{\left\{ \dfrac{4 \times F_{ax}}{\pi \times (D_o^2 - D_i^2)} \right\}}{\left\{ \dfrac{\Delta L}{L} \right\}} = \frac{4 \times F_{ax} \times L}{\pi \times (D_o^2 - D_i^2) \times \Delta L} \tag{3-55}$$

As before, the general expression for the spring constant ($F_{ax}/\Delta L$) may be inserted into (3-55) to establish an expression that includes the physical dimensions, the modulus of elasticity, plus the axial stiffness of a hollow cylinder:

$$E = \left\{ \frac{4 \times L}{\pi \times (D_o^2 - D_i^2)} \right\} \times \left\{ \frac{F_{ax}}{\Delta L} \right\} = \left\{ \frac{4 \times L}{\pi \times (D_o^2 - D_i^2)} \right\} \times K_{ax_{annulus}}$$

This equation may now be solved for the axial stiffness of a hollow cylinder:

$$\boxed{K_{ax_{annulus}} = \frac{\pi \times E \times (D_o^2 - D_i^2)}{4 \times L}} \tag{3-56}$$

Equation (3-56) may be used to compute the axial stiffness of a hollow cylinder. It is clear that if the internal diameter is set equal to zero, then (3-56) reverts back to (3-52) for a solid cylinder. From this pair of stiffness equations, it is noted that a hollow cylinder will always have an axial stiffness that is less than an equivalent solid cylinder of equal outer diameter.

This array of cylinders shown in Fig. 3-23 helps to visualize stiffness values for various physical dimensions. However, it should be recalled that these values are predicated upon a cylinder placed in axial compression. It is also meaningful to examine the radial or lateral stiffness for this same group of steel cylinders. For instance, assume that a cylinder is positioned horizontally as shown in Fig. 3-24. Further assume that the cylinder is simply supported at each end, and that a radial midspan force F_{rad} is applied. As before, the dimensions of the cylinder are the diameter D, and the length L. Furthermore, the maximum midspan radial deflection ΔR of the cylinder is identified on Fig. 3-24.

The radial cylinder stiffness is the applied force F_{rad} divided by the deflection ΔR. Even though Fig. 3-24 depicts a radial instead of an axial deflection, the same physical laws apply. Hence, it may be properly restated that the radial stiffness is equal to the applied force F_{rad} divided by the deflection ΔR as shown:

$$K_{rad} = \frac{F_{rad}}{\Delta R} \tag{3-57}$$

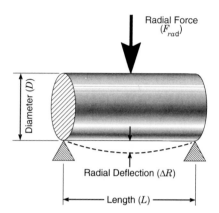

Fig. 3–24 Simply Supported
Solid Cylinder Subjected To
A Midspan Radial Force

The lateral deflection of this cylinder may be determined from simply supported beam formulas. From *Roark*[14], or Spotts[15], the maximum midspan deflection may be computed from the following common expression:

$$\Delta R = \frac{F_{rad} \times L^3}{48 \times E \times I_{area}} \tag{3-58}$$

Combining the previous equations (3-57) and (3-58), and solving for the lateral (radial) cylinder stiffness K_{rad}, the next expression is formed:

$$K_{rad} = \frac{F_{rad}}{\Delta R} = \frac{F_{rad}}{\left(\dfrac{F_{rad} \times L^3}{48 \times E \times I_{area}} \right)}$$

This is simplified to the following common format:

$$K_{rad} = \frac{48 \times E \times I_{area}}{L^3} \tag{3-59}$$

Equation (3-59) is often used for the calculation of the shaft stiffness. Since the element under load has a circular cross section, the area moment of inertia I_{area} for a solid may be determined from (3-30), and incorporated into the stiffness equation (3-59) to obtain the following result:

$$K_{rad_{solid}} = \frac{48 \times E \times I_{area_{solid}}}{L^3} = \left(\frac{48 \times E}{L^3} \right) \times \left(\frac{\pi \times D^4}{64} \right)$$

[14] Warren C. Young, *Roark's Formulas for Stress & Strain*, Sixth Edition, (New York: McGraw-Hill Book Company, 1989), p. 101.

[15] M.F. Spotts, *Design of Machine Elements*, 6th Edition, (Englewood Cliffs, New Jersey: Prentice-Hall, Inc., 1985), p. 27.

This simplifies to the next common equation for lateral shaft stiffness:

$$K_{rad_{solid}} = \frac{3 \times \pi \times E \times D^4}{4 \times L^3} \qquad (3\text{-}60)$$

If the length L remains at 30 inches, and the modulus of elasticity for steel is 30,000,000 Pounds/Inch^2, equation (3-60) may be solved for the stiffness of various cylinder diameters. For consistency, the previously calculated cylinder that had an axial stiffness of 10^6, was computed to have a diameter of 1.1284 inches. The lateral stiffness for this same cylindrical element is as follows:

$$K_{rad_{solid}} = \frac{3 \times \pi \times 30 \times 10^6 \text{ Pounds/Inch}^2 \times (1.1284 \text{ Inches})^4}{4 \times (30 \text{ Inches})^3} = 4,240 \text{ Pounds/Inch}$$

Repeating this calculation for cylinder diameters of 35.682, 11.284, 3.5682, and 0.3568 inches respectively, the summary diagram of Fig. 3-25 may be constructed. In this drawing, the cylinder dimensions are identical to the values previously determined from the axial load case. However, it is apparent that the radial stiffness magnitudes have changed significantly from the axial stiffness values shown in Fig. 3-23.

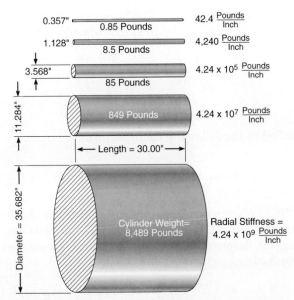

Fig. 3–25 Radial Stiffness Of 30" Long Steel Cylinders Of Various Diameters

For the case of a hollow cylinder, the radial stiffness is determined by installing the area moment of inertia for an annulus, equation (3-23), into the general radial stiffness equation (3-59) in the following manner:

$$K_{rad_{annulus}} = \frac{48 \times E \times I_{area_{annulus}}}{L^3} = \left(\frac{48 \times E}{L^3}\right) \times \left(\frac{\pi \times (D_o^4 - D_i^4)}{64}\right)$$

Simplifying and combining terms, the expression (3-61) for radial stiffness of a hollow cylinder may be easily produced:

$$K_{rad_{annulus}} = \frac{3 \times \pi \times E \times (D_o^4 - D_i^4)}{4 \times L^3} \qquad \text{(3-61)}$$

This is similar to equation (3-60) for a solid shaft. If the inner diameter of equation (3-61) is set equal to zero, the above expression will revert back to equation (3-60). Again, it is apparent that a solid shaft is always laterally stiffer than a comparable hollow shaft of equal outer diameter.

In addition to the radial and axial directions, machinery shafts are also subjected to twisting forces. For example, the normal movement of reciprocating machines produces synchronous torsional excitations in rotating elements such as crankshafts. Although centrifugal machines are less susceptible to torsional motion — failures do occur, and the torsional aspects of centrifugal machinery must be considered. For either class of machine, the torque on one end of a shaft

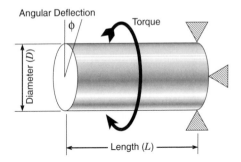

Fig. 3–26 Solid Cylinder, Fixed At One End, And Subjected To Torque

may result in an angular deflection along the shaft length. The amount of twist is directly related to the applied torque, and the torsional stiffness.

In general, torsional stiffness K_{tor} is analogous to the axial or lateral stiffness. However, the force is replaced by torque, and the linear displacement is replaced by an angular deflection ϕ. Performing this parameter substitution in (3-51), the following general definition for torsional stiffness is obtained:

$$K_{tor} = \frac{Torque}{\phi} \qquad \text{(3-62)}$$

A typical condition for a cylinder subjected to twisting deformation is depicted in Fig. 3-26. From strength of materials (*Roark*[16], or Spotts[17]), it is

---

[16] Warren C. Young, *Roark's Formulas for Stress & Strain*, Sixth Edition, (New York: McGraw-Hill Book Company, 1989), p. 346.

known that the angle of twist ϕ is related to the applied *Torque*, the member length L, the area polar moment of inertia J_{area}, and the material shear modulus G_{shear} in the following manner:

$$\phi = \frac{Torque \times L}{J_{area} \times G_{shear}} \tag{3-63}$$

Note that the shear modulus and the acceleration of gravity are both traditionally described by the letter G. To distinguish between these two common terms, the shear modulus will be identified by G_{shear} throughout this text. It should also be mentioned that the shear modulus is also known as the modulus of rigidity, the modulus of elasticity in shear, and the torsion modulus. In any case, combining equations (3-62) and (3-64), the following general expression for torsional stiffness is obtained:

$$K_{tor} = \frac{Torque}{\phi} = \frac{Torque}{\dfrac{Torque \times L}{J_{area} \times G_{shear}}} = \frac{J_{area} \times G_{shear}}{L} \tag{3-64}$$

The resultant JG/L expression for torsional stiffness is common within the literature. For instance, this same equation structure is used in the *ASM Handbook of Engineering Mathematics*[18]. Within (3-64), the area polar moment of inertia J_{area} is the value calculated through the center of the member (i.e., along the rotational axis). If the cylinder cross section is an annulus (hollow shaft) with an outer diameter of D_o, and an inner diameter of D_i, the area polar moment of inertia J_{area} was previously defined by equation (3-24). Substituting (3-24) into equation (3-64), the following torsional stiffness expression is formed:

$$K_{tor_{annulus}} = \left\{ \frac{G_{shear}}{L} \right\} \times \left\{ \frac{\pi \times (D_o^4 - D_i^4)}{32} \right\}$$

This is combined into the following common equation for the torsional stiffness of a hollow shaft (cylinder):

$$K_{tor_{annulus}} = \frac{\pi \times G_{shear} \times (D_o^4 - D_i^4)}{32 \times L} \tag{3-65}$$

The validity of equation (3-65) is again supported by identical expressions for torsional stiffness of hollow circular cylinders as shown in the *Shock and Vibration Handbook*[19]. If the shaft cross section is solid, then the inner diameter

---

[17] M.F. Spotts, *Design of Machine Elements*, 6th Edition, (Englewood Cliffs, New Jersey: Prentice-Hall, Inc., 1985), p. 150.

[18] William G. Belding and others, *ASM Handbook of Engineering Mathematics*, 4th printing, (Metals Park, Ohio: American Society for Metals, 1989), p. 319.

[19] Cyril M. Harris, *Shock and Vibration Handbook*, Fourth edition, (New York: McGraw-Hill, 1996), p. 1.11.

D_i is equal to zero, and equation (3-65) may be simplified to the following expression describing the torsional stiffness of a solid circular shaft:

$$K_{tor_{solid}} = \frac{\pi \times G_{shear} \times D^4}{32 \times L} \qquad \text{(3-66)}$$

Once more, this expression for torsional stiffness is common within the technical literature previously cited. From (3-66), it is evident that the torsional stiffness is a function of the cylinder diameter D, the cylinder length L, and the shear modulus G_{shear}. If the physical dimensions, and G_{shear} are known, then the torsional stiffness may be computed. However, in cases when the shear modulus is unknown, it may be computed from the modulus of elasticity E, and Poisson's ratio v. This relationship is defined by Shigley[20] as:

$$E = 2 \times G_{shear} \times (1 + v) \qquad \text{(3-67)}$$

If the steel cylinder dimensions previously used for computation of axial and radial stiffness are used for the calculation of torsional stiffness — the first step is to determine the shear modulus. As previously noted, the modulus of elasticity E for these steel cylinders was assumed to be 30,000,000 Pounds/Inch². If Poisson's ratio v for this material is equal to 0.3, then G_{shear} may be determined from equation (3-67):

$$G_{shear} = \frac{E}{2 \times (1 + v)} = \frac{30 \times 10^6 \text{ Pounds/Inch}^2}{2 \times (1 + 0.3)} = 11.5 \times 10^6 \text{ Pounds/Inch}^2$$

For a solid steel cylinder length L of 30 inches, and a diameter D equal to 1.1284 inches, the computed shear modulus may now be used to calculate the torsional stiffness from equation (3-66) in the following manner:

$$K_{tor_{solid}} = \frac{\pi \times 11.5 \times 10^6 \text{ Pounds/Inch}^2 \times (1.1284 \text{ Inches})^4}{32 \times 30 \text{ Inches}}$$

$$K_{tor_{solid}} = \frac{58,573,344}{960} \text{ Inch-Pounds} = 61,010 \frac{\text{Inch-Pound}}{\text{Radian}}$$

Note that the direct engineering units for this calculation are Inch-Pounds but the non-dimensional angular units of Radians have been included. This is inferred by the previous equation (3-62) of Torque/Angle — and it is customary nomenclature for torsional stiffness. For numerical comparison purposes, the torsional stiffness calculation is repeated for cylinder diameters of 35.682, 11.284, 3.5682, and 0.3568 inches respectively. The selected cylinder diameters are identical to the values previously determined from the axial load case. The axial, radial, and torsional stiffness values for each of the five cylinder diameters are now summarized in Table 3-3.

---

[20] Joseph E. Shigley and Charles R. Mischke, *Standard Handbook of Machine Design*, (New York: McGraw-Hill Book Company, 1986), pp. 10.6.

Table 3–3 Comparison of Calculated Axial, Radial, and Torsional Stiffness of 30 Inch Long Solid Steel Cylinders of Various Diameters

Cylinder Diameter (Inches)	Axial Stiffness (Pounds/Inch)	Radial Stiffness (Pounds/Inch)	Torsional Stiffness (Inch-Pounds/Radian)
35.682	1.00×10^9	4.24×10^9	6.10×10^{10}
11.284	1.00×10^8	4.24×10^7	6.10×10^8
3.5682	1.00×10^7	4.24×10^5	6.10×10^6
1.1284	1.00×10^6	4.24×10^3	6.10×10^4
0.3568	1.00×10^5	4.24×10^1	6.10×10^2

Clearly, the steel cylinder stiffness varies considerably from the axial to the radial direction. These values are independent of load, and they are a function of cylinder dimensions, and the modulus of elasticity. The significant point of this exercise is that individual machine elements seldom have singular elements that exceed the range of stiffness values presented in these physical examples. That is, machine parts will probably not have a stiffness much greater than 10^8 or 10^9 Pounds per Inch. These are enormously rigid elements, and stiffness values of 10^8 are seldom encountered in rotating machinery. At the other end of the scale, the skinny cylinder with a diameter of 0.36 Inches, and a height of 30 inches, displays an axial stiffness of only 100,000 Pounds per Inch. Again, it is hard to imagine many load carrying members within a rotating machine with this type of aspect ratio. Thus, it is logical to conclude that this stiffness of 10^5 Pounds per Inch is a realistic minimum value. For reference purposes, Table 3-4 of typical stiffness ranges (in Pounds/Inch) for various machine elements is presented.

Table 3–4 Summary of Typical Machinery Element Stiffness Values

Mechanical Element	Typical Stiffness Range (Pounds/Inch)
Oil Film Bearings	300,000 to 2,000,000
Rolling Element Bearings	1,000,000 to 4,000,000
Bearing Housing Support - Horizontal	300,000 to 4,000,000
Bearing Housing Support - Vertical	400,000 to 6,000,000
Shaft 1" to 4" Diameter	100,000 to 4,000,000
Shaft 6" to 15" Diameter	400,000 to 20,000,000

The concept of stiffness is applicable to shafts, bearings, cases, support structures, and foundations. In a real machine all of these stiffness parameters play a role in determination of the final vibration response characteristics. The relationship between effective system stiffness, and each of the individual elements is that the reciprocal of the overall system stiffness is equal to the sum of the reciprocals of each individual element. Thus, the change in any element (e.g., soft bearing support) can influence the entire system. It must also be recognized that the weakest (softest) member in the rotor support system will be the dominant element in establishing overall or effective system stiffness.

The previous discussion has centered around the variations in stiffness due to changes in geometry and materials of construction. Although this is an academically interesting topic, it is not particularly useful until it is applied within the realm of process machinery. One of the obvious influences of stiffness is the effect upon the natural frequency of machine elements. For discussion purposes, consider the axial, torsional, and lateral natural frequency of an undamped hollow cylinder based upon the previously developed equations. Specifically, the axial natural frequency may be determined from equation (2-45). The required mass of the cylinder may be specified by equation (3-4), and the axial stiffness of this annulus may be determined from equation (3-56). Combining these three expressions, the following result is obtained:

$$F_{c_{axial}} = \frac{1}{2\pi}\sqrt{\frac{K_{ax_{annulus}}}{M_{annulus}}} = \frac{1}{2\pi}\sqrt{\frac{\left\{\dfrac{\pi \times E}{4 \times L} \times (D_o^2 - D_i^2)\right\}}{\left\{\dfrac{\pi \times L \times \rho}{4 \times G} \times (D_o^2 - D_i^2)\right\}}} = \frac{1}{2\pi}\sqrt{\frac{\pi \times E}{4 \times L} \times \frac{4 \times G}{\pi \times L \times \rho}}$$

This equation for the axial natural frequency may be further simplified to:

$$\boxed{F_{c_{axial}} = \frac{1}{2\pi}\sqrt{\frac{E \times G}{\rho \times L^2}}} \tag{3-68}$$

This is an interesting result. The diameters have canceled out, and the only remaining physical dimension is the length L of the cylinder. Hence, the hollow cylinder diameters do not influence the axial natural frequency. Based on the material properties used for the steel cylinder examples, the axial natural frequency for a 30.0 Inch long member may be computed from equation (3-68) as:

$$F_{c_{axial}} = \frac{1}{2\pi}\sqrt{\frac{30\times10^6 \text{ Pounds/Inch}^2 \times 386.1 \text{ Inches/Second}^2}{0.283 \text{ Pounds/Inches}^3 \times (30 \text{ Inches})^2}} = 1,073\frac{\text{Cycles}}{\text{Second}}$$

$$F_{c_{axial}} = 1,073\frac{\text{Cycles}}{\text{Second}} \times 60\frac{\text{Seconds}}{\text{Minute}} = 64,400\frac{\text{Cycles}}{\text{Minute}}$$

The computed value of 64,400 CPM is a relatively high frequency, and would probably fall outside of the operating speed range for most machines. Similarly, if the shaft length was extended to 60 Inches, the axial resonant frequency would be 32,200 CPM, which is still quite a high frequency for a 5 foot long shaft. Due to this general relationship, the axial natural frequency of a shaft is seldom within the operating speed range. Hence, there is minimal opportunity to excite this resonance. Although axial vibration does occur in most machines, there is little evidence to suggest that the shaft axial resonant frequencies are commonly excited.

As a cautionary note, the diagnostician should not ignore other longitudinal or axial resonances within the rotor system. For instance, flexible disc or metallic membrane couplings will often have a lower frequency axial resonance that may be in the vicinity of the operating speed range. This is generally referred to as the Natural Axial Resonant Frequency (NARF) of the coupling assembly. It is common to have NARF values between 5,000 and 15,000 CPM for many configurations of axially compliant (soft) couplings. The computation of these frequencies becomes quite difficult due to the complex diaphragm stiffness, and the determination of the effective mass. Hence, the simplified format represented by equation (3-70) will probably not be adequate for calculating a coupling NARF. Furthermore, other machine elements such as wheels or thrust collars also have axial resonant frequencies that may have to be considered during a machinery analysis.

Moving on to the torsional behavior, the fundamental undamped torsional resonance of a hollow cylinder may be determined based upon equation (2-103). The mass polar moment of inertia may be specified by equation (3-26), and the torsional stiffness of this annulus may be determined from equation (3-65). Combining these three expressions, the following result is obtained:

$$F_{c_{tor}} = \frac{1}{2\pi}\sqrt{\frac{K_{tor_{annulus}}}{J_{mass_{annulus}}}} = \frac{1}{2\pi}\sqrt{\frac{\left\{\dfrac{\pi \times G_{shear}}{32 \times L}(D_o^4 - D_i^4)\right\}}{\left\{\dfrac{\pi \times L \times \rho}{32 \times G}(D_o^4 - D_i^4)\right\}}} = \frac{1}{2\pi}\sqrt{\frac{\pi \times G_{shear}}{32 \times L} \times \frac{32 \times G}{\pi \times L \times \rho}}$$

The general expression for the cylinder torsional natural resonant frequency may be further simplified as follows:

$$F_{c_{tor}} = \frac{1}{2\pi}\sqrt{\frac{G_{shear} \times G}{\rho \times L^2}} \tag{3-69}$$

Interestingly enough, the torsional resonance equation (3-69) has the same format as the axial resonance equation (3-68). The only difference is that the torsional equation uses the shear modulus G_{shear}, and the axial equation uses the modulus of elasticity E. Once more the cylinder diameter is canceled out, and equation (3-69) reveals that the torsional natural resonant frequency of the hollow cylinder is a function of the length L and the material properties. For comparative purposes, the torsional resonant frequency for the 30.0 Inch long steel cylinders may be computed as follows:

$$F_{c_{tor}} = \frac{1}{2\pi} \sqrt{\frac{11.5 \times 10^6 \text{ Pounds/Inch}^2 \times 386.1 \text{ Inches/Second}^2}{0.283 \text{ Pounds/Inches}^3 \times (30 \text{ Inches})^2}} = 664.5 \frac{\text{Cycles}}{\text{Second}}$$

$$F_{c_{tor}} = 664.5 \frac{\text{Cycles}}{\text{Second}} \times 60 \frac{\text{Seconds}}{\text{Minute}} = 39,870 \frac{\text{Cycles}}{\text{Minute}}$$

Since the shear modulus is smaller than the elastic modulus, the torsional resonant frequency is lower than the previously computed axial resonant frequency for the 30.0 inch long cylinder. This torsional frequency is probably outside of the normal operating speed range, and the torsional natural resonance of the shaft by itself is not the primary area of concern. For most machinery, the torsional resonance(s) of the system are governed by the torsional stiffness of the couplings, and the overall mass polar moment of inertia of the entire rotor assembly (i.e., not just the shaft). When these items are combined, the actual torsional resonance(s) may fall within the operating speed range. An example of this type of torsional resonance situation is presented in case history 11 of the torsional analysis of a power turbine and pump.

The third type of undamped radial or lateral resonance of a hollow cylinder may be determined with equation (2-45). The mass may be specified by equation (3-4), and the stiffness of this annulus may be determined from equation (3-61). Combining these three expressions, the following result is obtained:

$$F_{c_{rad}} = \frac{1}{2\pi} \sqrt{\frac{K_{rad_{annulus}}}{M_{annulus}}} = \frac{1}{2\pi} \sqrt{\frac{\left\{ \frac{3 \times \pi \times E}{4 \times L^3} \times (D_o^4 - D_i^4) \right\}}{\left\{ \frac{\pi \times L \times \rho}{4 \times G} \times (D_o^2 - D_i^2) \right\}}}$$

$$F_{c_{rad}} = \frac{1}{2\pi} \sqrt{\frac{\left\{ \frac{3 \times \pi \times E}{4 \times L^3} \times (D_o^2 + D_i^2) \right\}}{\left\{ \frac{\pi \times L \times \rho}{4 \times G} \right\}}} = \frac{1}{2\pi} \sqrt{\frac{3 \times \pi \times E}{4 \times L^3} \times \frac{4 \times G}{\pi \times L \times \rho} \times (D_o^2 + D_i^2)}$$

The expression for the cylinder radial natural resonant frequency may be further simplified as follows:

$$F_{c_{rad}} = \frac{1}{2\pi} \sqrt{\frac{3 \times E \times G \times (D_o^2 + D_i^2)}{\rho \times L^4}} \qquad (3\text{-}70)$$

The lateral natural resonance equation (3-70) includes the complete array of cylinder dimensions, plus the material constants. It is clear that the lateral resonant frequency will change in accordance with diameter changes. Obviously, this evolves as a problem of greater mechanical complexity than either the axial or the torsional natural resonance of the hollow cylinder. For example, assume that the cylinder inner diameter D_i is zero, and the outer diameter is D_o is 3.5682 inches (compatible with previous calculations). The resultant radial resonant frequency for the 30.0 inch long steel cylinders may be computed as follows:

$$F_{c_{rad}} = \frac{1}{2\pi}\sqrt{\frac{3 \times 30 \times 10^6 \text{ Pounds/Inch}^2 \times 386.1 \text{ Inches/Second}^2 \times (3.5682 \text{ Inches})^2}{0.283 \text{ Pounds/Inches}^3 \times (30 \text{ Inches})^4}}$$

$$F_{c_{rad}} = 221.1 \frac{\text{Cycles}}{\text{Second}} \times 60 \frac{\text{Seconds}}{\text{Minute}} = 13,270 \frac{\text{Cycles}}{\text{Minute}}$$

This frequency of 13,270 CPM is low enough to be a potential threat to the operating speed range of the machine. In actuality, the lateral critical speed of an entire rotor with this type of shaft geometry (i.e., 30.0 inches long, and 3.5682 inches in diameter), would be even lower due to the influence of the attached wheels, thrust collars, and couplings. Hence, based upon this simple example, it is clear that a machine generally has a greater propensity towards lateral critical speed problems, rather than axial or torsional resonant problems. This conclusion is strongly supported by actual industrial malfunctions where resonant problems are usually observed in the lateral directions, and infrequently encountered in the axial or torsional directions.

From the previously discussed static deflection curves in Figs. 3-1 through 3-6, the effect of mass distribution and support location should be appreciated. Now when the magnitude of the restraint or stiffness is included into the discussion, the dynamic response of the machine can be subjected to further evaluation. One way to envision the influence of stiffness upon vibration amplitudes is to consider the relationship of parameters with respect to an applied force. Specifically, the applied *Force* is equal to the *Response* (vibration) times the *Restraint* (stiffness). This concept is often stated in the following manner:

$$\boxed{Response = \frac{Force}{Restraint}} \tag{3-71}$$

The larger the restraint, the smaller the response for a given unit force input. Clearly, vibratory motion can be suppressed by increasing the stiffness, but there are penalties to be paid. For example, a rolling element bearing is stiffer than a fluid film bearing. However, the rolling element bearing may adversely alter the shaft mode shape. In addition, high stiffness dictates low displacement, which generally means small viscous damping. Under this case, the rotor may not make it through the critical speed range, and the stability will certainly suffer. Therefore, the machine that performed well for many years with sleeve bearings might self destruct with the addition of stiff ball bearings. The same argument often applies to tilting pad bearings. Over the years, many End Users have suffered the consequences of an inappropriate change from sleeve bearings to a tilt pad assembly with higher stiffness, and lower damping.

Another view of the stiffness influence is presented in Fig. 3-27 illustrating rotor mode shapes at three different critical speeds. The mode shapes on the left describe the rotor motion with compliant (soft) bearings. For comparison, the mode shapes on the right side of Fig. 3-27 depict the anticipated mode shapes with rigid rotor supports. This rigid support condition implies that the bearing clamps down the shaft motion, and produces a nodal point at each bearing. It is important to recognize that a lateral nodal point is a location of negligible motion along the shaft axis. This condition is not totally definitive of a node, and it necessary for the shaft motion on each side of a node to be nominally 180° out of phase. This means that the shaft on one side of the node is moving in the opposite direction of the shaft on the other side of the node. In other words, a true zero axis crossing nodal point is defined by a rocking motion.

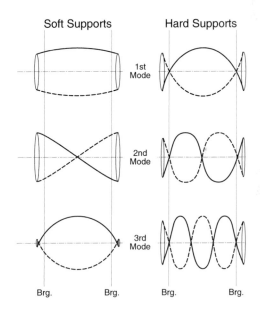

Fig. 3–27 Effect Of Radial Bearing Stiffness On Rotor Mode Shapes For The First Three Criticals

Note that the existence of a nodal point within an oil film bearing may be indicative of a serious situation. Specifically, a nodal point is a location of minimal motion. From previous discussions, low shaft vibration amplitudes within a bearing may manifest as low viscous damping, and this may result in low stability margins. Hence, a good rotor and bearing design may be rendered inoperative due to the coincidence of shaft nodal points and journal bearings.

Referring back to the rotor mode shapes shown in Fig. 3-27, it is noted that the machine describes a cylindrical mode at the first critical speed with soft supports. Relative motion at the bearings are significant, and phase measurements would indicate similar values between bearings. With a hard support, a nodal point occurs at each bearing, and shaft vibration measurements at the bearings would be nil. Furthermore, one can be fooled by taking simultaneous phase measurements inboard of one bearing, and outboard of the other bearing. This data would indicate a nominal 180° phase difference, and the diagnostician might believe that the unit was going through a second mode, rather than a first mode with hard supports.

The second critical describes the shaft mode shapes at the pivotal balance resonance. This is the critical speed during which shaft motion pivots through the centerline of the rotor producing a zero axis crossing nodal point within the rotor span. With a soft support stiffness, a pure conical mode is observed. It is evident that increasing stiffness will clamp down at the bearing locations, and cause the formation of two more nodal points. Again, the vibration measurement planes are extremely important, and a machine could be interpreted as passing through a first critical, when it is really transcending through a well restrained second mode. As with any type of modal analysis, the use of transducers at additional lateral locations along the shaft may be quite informative, and in some cases absolutely necessary.

The third critical speed with soft bearings is similar in shape to a translational first mode with hard supports. Again, the diagnostician is cautioned about premature mode shape speculations based upon partial data. The third mode with hard bearings exhibits a predictable shape with four nodal points. For many rotors this third critical speed is a rotor bending mode, and shaft alterations are usually required to appreciably change this mode shape or the associated natural frequency. Conversely, the first two modes are normally bearing dependent. The mode shapes and natural frequencies of the first and second criticals can often be altered by changing the stiffness characteristics of the radial journal bearings and/or the stiffness of the bearing housings.

CRITICAL SPEED TRANSITION

The nominal 180° phase change across a shaft nodal point is generally understandable. The concept of a shaft reversing direction (i.e., pivoting) across a nodal point makes physical sense, and extensive technical explanations are not required. However, the amplitude change, and the ideal 180° phase shift associated with a speed transition through a critical speed (balance resonance) do not lend themselves to direct intuitive logic. Certainly the analytical models exhibit these characteristics (chapters 2 and 5), and the field vibration measurements (chapters 7, 8, 11, etc.) also display vector changes across a rotor resonance. However, an instinctive, and universal understanding of this behavior by mechanically inclined individuals is not a common trait.

This is not a new topic, and references date back to the post civil war era. For instance, the 1869 technical paper by W.A. Rankin[21] initially addressed this subject. By 1882, Carl Gustaf de Laval of Sweden introduced a flexible shaft, single stage steam turbine. The German teacher and researcher August Föppl demonstrated in 1895 that a rotor could operate above a critical speed. In 1919, the English investigator H.H. Jeffcott[22] published his classic paper *The Lateral Vibration of Loaded Shafts in the Neighborhood of a Whirling Speed: The Effect of Want of Balance.* Fifteen years later, McGraw Hill Book Company began pub-

---

[21] W.A. Rankin, "On the Centrifugal Force of Rotating Shafts, *Engineer (London)*: 27, (1869).

[22] H.H. Jeffcott, "Lateral Vibration of Loaded Shafts in the Neighborhood of a Whirling Speed - The Effect of Want of Balance," *Philosophical Magazine*, Vol 37, (1919), pp. 304-314.

lishing the J.P. Den Hartog[23] textbook *Mechanical Vibrations*. In 1948, Prentice-Hall published the first edition of the W.T. Thomson[24] textbook also titled *Mechanical Vibrations*. The academic explanations continue with many papers and textbooks such as the 1988 *Rotordynamics of Turbomachinery* by John Vance[25], followed by the 1993 *Turbomachinery Rotordynamics, Phenomena, Modeling, and Analysis* by Dara Childs[26]. Obviously, these explanations of fundamental rotor behavior have gone on for well over a century.

Interestingly enough, Jeffcott corrected the earlier work of Rankin. Some historians believe that the Jeffcott work should be credited to de Laval or Föppl. It appears that Den Hartog, and Thomson both used the Jeffcott model, but they apparently neglected to reference the earlier work of Jeffcott. The books by Vance and Childs do recognize Jeffcott, and his contributions to the field of rotordynamics. However, many good machinery engineers have expressed the opinion to the senior author that ... *the physical understanding of the critical speed phenomena still seems to be clouded by the lack of a direct physical explanation of the amplitude change and the phase shift through the resonance.*

Since this is such an important concept to the field of rotating machinery, another explanation will be attempted on the following pages. This interpretation of the phenomena is based upon the Jeffcott model, and comments by Den Hartog and Thomson. This explanation will not be as mathematically rigorous or as extensive as the discussions by either Vance or Childs. However, the derivation presented herein is simply directed at a physical explanation of rotor behavior through a critical speed region. It is hoped that the following discussion does make intuitive sense to the machinery diagnostician.

To begin this discussion, it is appropriate to consider the traditional diagram of a Jeffcott rotor as depicted in Fig. 3-28. The bearings in this model are mounted at the ends of the shaft, and they are considered to be rigid, and frictionless. Initially, the damping will intentionally be set to zero, and influences from any fluids or other sources will be ignored. The shaft is uniform, massless, and flexible (i.e., elastic), and the shaft will have a uniform radial stiffness defined by K.

At the midspan of the shaft, a thin, flat disk is mounted. This disk will have a residual, or concentrated mass unbalance that is identified as M. This single mass unbalance is located on the mass centerline identified as G (not to be confused with the acceleration of gravity, or the modulus of rigidity). As shown in Fig. 3-28, the bearing centerline B, and the shaft and disk centerline S are identified. When the rotor is at rest, the bearing, shaft, and disk centerlines are coincident. The distance between the shaft and disk centerline S, and the mass centerline G, will be defined as the eccentricity e. This is not the journal offset

[23] J.P. Den Hartog, *Mechanical Vibrations*, (New York: McGraw-Hill Book Company, 1934).

[24] William Tyrrell Thomson, *Mechanical Vibrations*, (Englewood Cliffs, New Jersey: Prentice-Hall, Inc., 1948).

[25] John M. Vance, *Rotordynamics of Turbomachinery* (New York: Wiley-Interscience Publication, 1988).

[26] Dara Childs, *Turbomachinery Rotordynamics – Phenomena, Modeling, and Analysis*, (New York: Wiley-Interscience Publication, 1993).

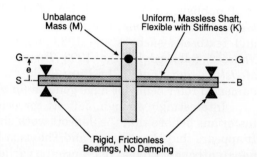

Fig. 3–28 Jeffcott Rotor Configuration

within a bearing, or an eccentric surface. It is merely the distance between the center of the shaft and disk **S**, and the unbalance mass **M** (and associated mass centerline **G**). Throughout this discussion, the geometry of the disk and the mass remains fixed, therefore the eccentricity distance **e** remains constant.

At very slow speeds, unbalance forces are negligible. The shaft turns around the bearing centerline **B**, and all rotating elements are concentric. This condition is depicted in the end view of Fig. 3-29. A dial indicator or a proximity probe mounted anywhere along the length of the shaft will indicate zero motion.

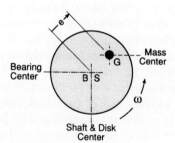

Fig. 3–29 Jeffcott Rotor At Very
Low Shaft Rotational Speed

As rotor speed increases, the straight shaft will deflect into the predictable mode shape shown in Fig. 3-30. The translational mode shapes within this discussion exhibit zero motion at the bearings, and a maximum deflection at the shaft midspan. The only driving force in the system is the centrifugal force due to the unbalance mass **M**. The following series of diagrams are consistently annotated with the previous centerline designations (**B**, **S**, and **G**), and the mass eccentricity **e** (offset) from the deflected shaft center. In time honored tradition, the maximum bending deflection of the shaft (distance between **B** and **S**) is identified by **r**. Furthermore, the rotational speed is indicated by ω in all of the diagrams and equations.

By inspection of these diagrams, it is clear that the shaft and disk are rotating at the operating speed ω Simultaneously, the deflected shaft is whirling in the bearings at this speed, and it carries the disk along with it as it moves. This motion is often referred to as synchronous whirl. The mechanism driving this whirl is the centrifugal force generated by the eccentric mass on the disk. As

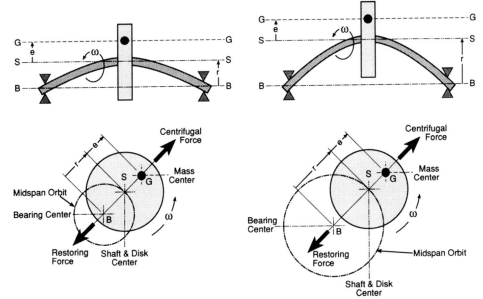

Fig. 3–30 Jeffcott Rotor At Moderate Speed Well Below the Balance Resonance

Fig. 3–31 Jeffcott Rotor Approaching The Critical Balance Resonance Speed

rotor speed increases, the outward force increases in accordance with the normal centrifugal force equation:

$$Centrifugal \ Force \ = \ M \times (r + e) \times \omega^2 \qquad \text{(3-72)}$$

In this expression, the total radius of the mass unbalance M is composed of the shaft bending r, plus the eccentricity of the mass with respect to the shaft centerline e. If the disk was perfectly balanced (i.e., M=0) there would be no deflection of the shaft, and no resultant whirl. With respect to real machines, there is always some amount of shaft bending, and some level of residual unbalance in the rotor that produces synchronous whirl. The physical influence of the centrifugal force defined in equation (3-72) may be graphically depicted in Figs. 3-30 and 3-31. The first drawing (Fig. 3-30) describes a moderate speed below the balance resonance (critical) speed. The condition described in Fig. 3-31 depicts a higher speed that is approaching the critical speed. Note that the two drawings are identical except for the increased midspan deflection r caused by the higher operating speed, and the associated larger centrifugal force.

The end views of the midspan disk in Figs. 3-30 and 3-31 show the bearing center B, shaft center S, and mass G. Note that the centrifugal force is in the same line as B, S, and G. These points are collinear. That is, they lie along a common straight line. The logic behind this statement is based on a force balance. Specifically, the centrifugal force acts outward from the bearing center B through the mass center G. The only opposition to this radial centrifugal force is the restoring force of the shaft. The restoring force is simply the shaft spring constant K multiplied by the shaft deflection r as described in the next expression:

$$\boxed{Restoring\ Force\ =\ K \times r} \tag{3-73}$$

The restoring force is the elastic pull of the shaft that attempts to straighten the shaft, and resist any deflection. It acts from the shaft center S, back to the bearing center B. In order for the centrifugal and the restoring forces to be in equilibrium, they must be equal in magnitude, and opposite in direction. Hence, for the simple case, points B, G, and S must be collinear. Based upon this logic, it is reasonable to equate the two opposing forces as follows:

$$Restoring\ Force\ =\ Centrifugal\ Force \tag{3-74}$$

Substituting (3-72) and (3-73) into equation (3-74) the following is obtained:

$$\boxed{K \times r\ =\ M \times (r + e) \times \omega^2} \tag{3-75}$$

From this expression, a mechanical dilemma is immediately apparent. The left side of the expression that represents the restoring force is linear with shaft deflection. However, the right side of this expression that defines centrifugal force, varies as the speed squared. Hence, at some rotational speed, the shaft displacement must become infinite to restrain the disk and attached unbalance mass. This point of infinite displacement is aptly termed the critical speed, or the balance resonance speed of the rotating assembly.

Another way to demonstrate this concept of infinite amplitude is based upon an expansion, and re-configuration of (3-75) in the following manner:

$$K \times r\ =\ M \times r \times \omega^2 + M \times e \times \omega^2$$

$$K \times r - M \times r \times \omega^2\ =\ M \times e \times \omega^2$$

$$r \times (K - M \times \omega^2)\ =\ M \times e \times \omega^2$$

This expression may now be solved for the bending deflection r as follows:

$$r\ =\ \frac{M \times e \times \omega^2}{K - M \times \omega^2}\ =\ \frac{e \times \omega^2}{(K/M) - \omega^2} \tag{3-76}$$

From equation (2-41), it was shown that (K/M) is equal to the undamped natural frequency ω_c (critical speed) squared. Substituting this expression into equation (3-76), the deflection r may now be restated as:

$$r\ =\ \frac{e \times \omega^2}{\omega_c^2 - \omega^2} \tag{3-77}$$

The eccentricity e between the shaft center S, and the mass centerline G remains constant. In addition, the natural frequency ω_c remains fixed due to an invariable mass M and spring constant K. Hence, the shaft deflection r is a function of the operating speed ω. Furthermore, if the numerator and denominator of (3-77) are divided by the critical speed squared ω_c^2, the following is obtained:

$$r = \frac{e \times (\omega/\omega_c)^2}{1 - (\omega/\omega_c)^2} \qquad (3\text{-}78)$$

In an effort to further simplify this expression, recall that the undamped critical speed frequency ratio was defined in equation (2-86) as:

$$Critical\ Speed\ Frequency\ Ratio = \Omega = \omega/\omega_c$$

Substituting this speed ratio into (3-78) yields the next relationship:

$$r = \frac{e \times \Omega^2}{1 - \Omega^2} \qquad (3\text{-}79)$$

As the shaft rotational speed increases from some slow roll condition, the value of the critical speed ratio Ω increases, and the mid span shaft deflection r increases. If the eccentricity e between the mass centerline G, and the shaft centerline S is assumed to be some realistic value such as +2, Fig. 3-32 may be plotted. As expected, the midspan deflection becomes quite large as the critical speed ($\Omega=1$) is approached.

Stated in another way, as the ratio of ω/ω_c approaches unity, the denominator of equation (3-79) becomes smaller, and the value of r becomes increasingly large. When the shaft speed ω is equal to the critical speed ω_c, the $\Omega = \omega/\omega_c=1$, and the denominator of equation (3-79) becomes zero. Obviously, division by zero will result in infinity. This is consistent with the previous logic, and the definition of an undamped system.

The response plot in Fig. 3-32 was constructed between Ω values of 0 and 0.94 to describe the amplitude characteristics below the resonance. If the plot

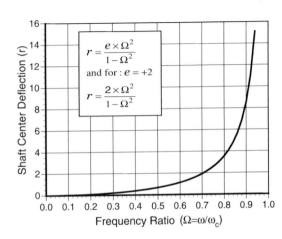

Fig. 3–32 Plot Of Jeffcott Rotor Approaching The Critical Balance Resonance Speed of $\Omega=1$

range is extended from 0 to 3.0, the diagram shown in Fig. 3-33 evolves. Again, a numeric value of +2 was selected for the eccentricity e, and it is noted that the midspan shaft deflection indeed moves off towards infinity. In a real machine, infinitely large shaft deflections are not possible. Only two possibilities for the Jeffcott (or any other) rotor are feasible. The first option is for the midspan amplitudes to increase to the point where the machine destroys itself (only choice for an undamped system). The other alternative for a real machine is to have the displacement amplitudes at the critical speed restrained by damping.

In accordance with the earlier discussion within this chapter, positive damping is an energy dissipater. It will limit the vibration amplitude through the resonance. For a system with low damping, such as a structural resonance, the amplitude at the resonance will be high, and the resonance bandwidth will be small (high Q). Conversely, for a system with large damping, such as a fluid film bearing with viscous damping, the resonance peak will be lower, and the bandwidth will be wider (small Q). Thus, without damping in the system, the machine could not survive a resonance. This discussion also identifies the logic associated with the amplitude increase at the critical speed. It is hopefully clear from the preceding explanation, and the general equations, that the displacement at the resonance must increase. Furthermore, the amount of the vibration increase at the critical is dependent upon the available system damping.

Unfortunately, the characteristics of the amplitude response through the resonance are not completely defined, because one other peculiarity must be reconciled. Specifically, Fig. 3-33 reveals a midspan amplitude r that migrates off towards positive infinity ($+\infty$) as the resonance is approached. In true mathematical fashion, the amplitude above the critical returns from negative infinity ($-\infty$). That is particularly disturbing when it is also recognized that the amplitudes above the resonance are all negative. For example, if $e=2$, and $\Omega=2$, then by equation (3-79), r is computed to be minus 2.67. Hence, the plotted curve is mathematically correct, but it does not describe a true physical situation. That is, a negative vibration amplitude is incomprehensible.

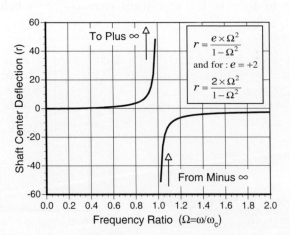

Fig. 3–33 Plot Of Jeffcott Rotor Passing Through The Critical Balance Resonance Region of $\Omega=1$

In retrospect, equation (3-79) used for computation of this deflection *r* only contains two terms. The eccentricity *e*, plus the frequency ratio between the running speed and the natural frequency $\Omega = \omega/\omega_c$. Above the resonant frequency ω_c it is obvious that this speed ratio must be a positive number that is greater than one. Hence, there is no variation of the frequency ratio Ω that would reverse the sign of the midspan deflection *r*.

The only other alternative resides with the eccentricity *e*. As previously stipulated, the magnitude of *e* is fixed by the geometry of the disk, and the location of the mass on the disk. However, the original definition of the Jeffcott rotor did not restrict the direction (i.e., the ± sign) of this eccentricity *e*. Hence, if the direction of *e* was reversed, the plotted response curve should flip over into the positive domain.

In fact, that is exactly what happens with a real machine, as well as the Jeffcott model. To prove this point, the response data above the resonance will be replotted in Fig. 3-34, and *e*=+2 will be replaced by *e*=-2. In order to maintain reasonable amplitudes, a frequency ratio range extending from 1.08 to 3.0 will be used. It is noted that the resultant plot illustrates the proper positive deflection *r*, and it also is indicative of normal behavior on the back slope of a resonance.

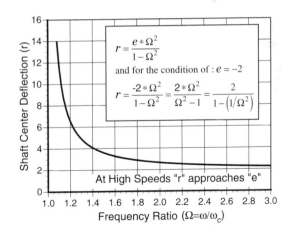

Fig. 3–34 Plot Of Jeffcott Rotor Leaving Critical Balance Resonance Speed Region of Ω=1 with e=-2

The sign reversal of the eccentricity *e* is physically equivalent to a reversal in the positions of the shaft *S*, and mass *G* centerlines. For instance, compare the Jeffcott rotor running below the critical in Figs. 3-30 and 3-31, with the rotor operating above the critical as shown in Figs. 3-35 and 3-36. Below the critical, the mass *M* was on the outside of the rotor. Whereas, above the critical speed, the mass *M* is tucked away underneath the curvature of the shaft — and the light side of the rotor is now on the outside. This reversal of *heavy side out* to *light side out* is the mechanism behind the nominal 180° phase shift across a resonance.

Another interesting point from the response plot in Fig. 3-34 is that the value of *r* steadily diminishes with an increasing frequency ratio Ω. At high shaft speeds, the magnitude of *r* approaches the magnitude of *e*. Physically, this characteristic is described in Fig. 3-36 of the Jeffcott rotor at a high speed condition.

In this state (i.e., well above the critical speed), the values of *e* and *r* are essentially equal. Since *e* carries a negative sign (above the critical), the centerline for the mass *G* is now coincident with the bearing centerline *B*. This direction reversal eliminates the centrifugal force, since the radius to the mass is now equal to, or very close to zero, and the ***rotor has self balanced itself***.

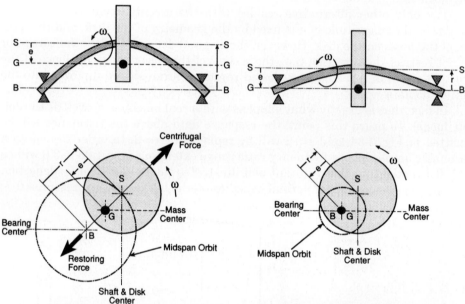

Fig. 3–35 Jeffcott Rotor Operating Slightly Above The Critical Balance Resonance Speed

Fig. 3–36 Jeffcott Rotor Operating In A Self Balanced Condition Well Above The Critical Balance Resonance Speed

This self-balancing characteristic is the mechanism that allows so many machines to operate successfully at speeds in excess of the rotor balance resonance, or critical speed. In this high speed condition, the system is in equilibrium, and the shaft actually rotates around the mass center *M*, which is equivalent to the centerline of gravity *G* for the rotor. For further discussions within this chapter, and especially the balancing chapter 11, reference will be made to rotation about the geometric axis below the critical speed, and rotation about the mass or inertial centerline above the critical. These continuing references are based upon the fundamental behavior and understanding of the traditional Jeffcott rotor.

At the beginning of this discussion, it was stated that the damping was intentionally set at zero. That was a convenient and necessary presumption to maintain simplicity of the analytical model. The inclusion of damping into the system does substantially complicate the response of the Jeffcott rotor. For instance, consider Fig. 3-37 of this rotor with damping. Note that the centerlines ***B***, ***S***, and ***G*** are no longer collinear. An angular phase change ϕ has been invoked upon the mass centerline ***G***. This implies that the centrifugal force, and the shaft

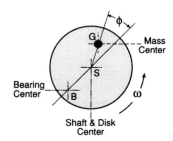

Fig. 3–37 Jeffcott Rotor With Damping

spring restoring force are no longer equal and opposite. Actually, a new force must now be included to achieve a force balance. This new force will be a damping term that may be tangential to the disk, and may vary in magnitude with the shaft surface velocity.

The addition of this damping does not invalidate the previous discussion, but it does complicate the scenario. For simplicity, the damping force might be assumed to be viscous, and the resultant force equal to the tangential velocity times the damping coefficient. This type of force could be included, and the previous analysis repeated. However, this inclusion would not necessarily improve the overall understanding of the critical speed phenomena. Furthermore, damping characteristics of real machines are not a simple linear function of rotor velocity, and a proper analysis would be substantially more intricate.

In essence, it must be recognized that all rotors have some amount of damping. The oil film in the bearings, oil in the shaft seals, or the process fluid itself may provide the damping. In all cases, the presence of damping will influence the shaft behavior. There will be positive contributions to the addition of damping, such as lower vibration amplitudes through the critical speed region. However, damping will also open the door to a variety of mechanisms such as nonsynchronous whirl. In all cases, the real world is always more complicated than the models that we can build to explain physical events. Fortunately, the undamped Jeffcott model may be used to explain the fundamental characteristics associated with the balance resonance or critical speed phenomena.

MODE SHAPE MEASUREMENT

Previous portions of this chapter have briefly covered the deflection of static beams with variations in weight distribution, and support location. In addition, the influence of damping, and the effect of rotor support stiffness have been addressed. Although these are important concepts that directly influence the behavior of rotating machinery, there is a strong argument to be made for direct observation of the deflected rotor mode shapes.

A visual observation of the rotor mode shapes would provide substantiation of the previously discussed theory, and allow further study into the complexities and subtleties of rotor dynamics. Fortunately, a method exists to provide this visualization by the application of proximity probe transducers, and electronic display instrumentation. For example, consider Fig. 3-38 of a simple rotor that consists of a shaft plus a center mounted disk. This is similar to the Jeffcott rotor. For this model, assume that the rotor and disk are initially well balanced, and that system damping is minimal. Further presume that a single concentrated mass unbalance is placed on the disk.

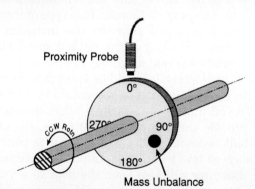

Fig. 3–38 Simple Rotor With Concentrated Mass Unbalance And Midspan Radial Proximity Probe

The outer diameter of the disk is observed by a vertical proximity displacement probe that measures the distance between the probe tip and the disk outer surface. In many ways, this type of vibration transducer may be considered as the electronic equivalent of a mechanical dial indicator. Detailed discussions of the characteristics of this type of probe are presented in chapters 6 and 7. However, for the current discussion, the reader should accept that this probe will accurately measure the distance between the probe tip and the observed surface.

The conditioned output from this proximity probe transducer system is a voltage sensitive signal that may be observed on an oscilloscope. The typical scale factor for this type of transducer is 200 millivolts per Mil. Thus, a 1 Mil (0.001 Inch) change in distance between the probe and the target surface will produce an electronic signal equal to 200 millivolts. If the distance between the rotor surface and the probe remains constant at all angular positions, the oscilloscope will display a straight line as the shaft is rotated. However, if the observed surface is eccentric (for whatever reason), then the probe will display a sine wave on the oscilloscope. The frequency of this sine wave will be equal to the speed of rotation, and the amplitude of the resultant sine wave will be dependent upon the magnitude of the eccentricity.

An angular coordinate system must be established and maintained. This angular coordinate system must allow a definitive and repeatable relationship between the rotating system, and a stationary reference point. Normally, this is achieved by another proximity probe that observes a notch (keyway) or projection (key) on the rotating shaft. During shaft rotation, this timing or Keyphasor® transducer produces a synchronous pulse as described in later chapters of this text. The angular location of this pulse is determined by stopping the machine, and physically lining up the trigger point on the notch or projection with the timing probe. When this physical alignment occurs at zero speed, it is equivalent to the trigger point of a pulse signal during rotation.

Again, specific details of this trigger arrangement are discussed in chapters 6, 7, 8, and 11. For the purposes of this current discussion, accept the fact that a trigger/reference arrangement does exist. Furthermore, the vibration sensing probe is always located at zero degrees for all phase measurements. In the example rotor shown in Fig. 3-38, the vibration sensing proximity probe is shown at the top vertical position above the disk. The rotation is specified as counterclockwise, the angular coordinate system on the disk begins at 0° at the probe, and the angles increase in a clockwise direction. This same logic will be used throughout this text.

Also, it must be understood that the proximity probe cannot directly identify the angular location of an effective mass unbalance. The proximity probe can only measure distances. That is, the probe can measure the change in distance between transducer and rotor around the entire circumference, but it cannot directly identify the location of a mass unbalance. This effective, or equivalent, or lumped mass unbalance location is normally identified as the *Heavy Spot* for the disk or rotor system. The circumferential point that the proximity probe does identify is the high point of the observed surface. This is the point of peak vibration that is identified by the synchronous vector phase angle. This physical location is generally referred to as the *High Spot*.

Consider a closer examination of the wheel on the simple rotor described in Fig. 3-38. The diagram presented in Fig. 3-39 represents the motion of this wheel at low rotative speeds. For purposes of discussion, assume that the rotor is operating at a speed that is well below the first critical speed (i.e., translational balance resonance) of the system. In this drawing, the center of rotation is shown to be coincident with the geometric center. A residual or effective mass unbalance is shown as the *Heavy Spot*.

If this rotor was perfectly balanced, then the *Mass Center* would be identical to the *Geometric Center*. However, if a lumped unbalance is placed on the disk, it is clear that a shift in the *Mass Center* must occur. It is also logical to recognize that the new *Mass Center* must reside on a line between the *Geometric Center* and the *Heavy Spot*. Furthermore, as the rotor turns, centrifugal force will cause the disk to deflect radially in the direction of the *Heavy Spot*. Under a low speed condition, the *High Spot* occurs at the same angular location as the *Heavy Spot*. Stating it another way, the minimum distance between the disk and the probe occurs at the *High Spot*. Clearly, this is coincident with the *Heavy Spot* at speeds well below the critical speed (this is exactly true only in the complete absence of damping).

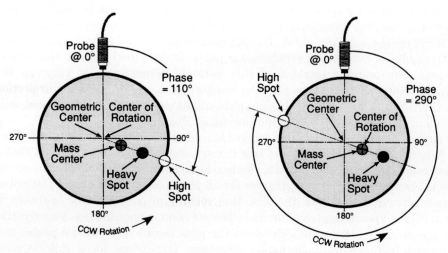

Fig. 3–39 Simple Rotor Operating Well Below The Shaft Critical Speed

Fig. 3–40 Simple Rotor Operating Well Above The Shaft Critical Speed

If rotation of this example system is counterclockwise, the angles would be measured against rotation, or clockwise from the probe. Thus, the *Heavy-High Spot* angular position is shown at a physical location of 110° in Fig. 3-39. If the shaft is turning, the phase angle obtained from a Vector Filter would be 110°. On the vibration waveform, this would be equivalent to the time lag between the peak of the vibration signal, and the trigger point of the Keyphasor® pulse. If the shaft is not rotating, and the timing notch is positioned under the Keyphasor® probe, then a counter rotation angle of 110° from the vertical vibration probe would locate the *High Spot*, and the associated *Heavy Spot*. Balancing of this rotor would require removing weight at 110° or adding weight at 290°. This is the logic behind the traditional proximity probe balancing rule of:

At speeds well below the critical,
remove weight at the phase angle,
or add weight at the phase angle plus 180°

Next, consider what happens at rotational speeds well above the first critical. This condition is described in Fig. 3-40. In this drawing, the center of rotation is now coincident with the actual mass center of the rotor. That is, rotation occurs around the mass center, or principal inertia axis instead of the geometric axis. This is identical to the behavior described by the Jeffcott rotor in the self balanced condition in Fig. 3-36.

The residual mass unbalance or *Heavy Spot* remains in the same angular location as the slow speed case. Intuitively, this must be true. That is, the relationship between the probe and the unbalance must remain constant, or the rotor could never be balanced. For instance, an unbalanced and loose impeller on a shaft would have a different phase angle during every runup. This wheel could not be balanced because the angle between the probe and the unbalance changes. A rotor can only be balanced when the angle between the stationary probe and the *Heavy Spot* on the rotor remains constant from run to run.

At high speed conditions, the new center of rotation manifests as a new *High Spot*. The rotor has self-balanced itself through the critical speed region, and the new center of rotation about the mass center produces an eccentric rotation axis. This eccentricity of rotation about the mass center results in a new *High Spot* that is 180° away from the *Heavy Spot*. Stating it another way, the minimum distance between the disk and the proximity probe occurs at the identified *High Spot*. This location is exactly opposite the *Heavy Spot* at rotational speeds well above the critical (completely true in the case of no damping).

As previously mentioned, the *Heavy Spot* remains at 110°, and the new location of the *High Spot* is shown at a physical location of 290°. If the shaft is turning, the phase angle obtained from a DVF would be 290°. On the vibration waveform, this would be equivalent to the lag between the peak of the vibration signal, and the Keyphasor® pulse. If the shaft is not turning, and the timing notch is positioned under the Keyphasor® probe, then a counter rotation angle of 290° would locate the *High Spot*. The *Heavy Spot* would remain at 290° minus 180° or at 110°. Balancing of this rotor would require adding weight at 290° or removing weight at 110°. This is the logic behind the proximity probe balancing rule of:

> *At speeds well above the critical,*
> *add weight at the phase angle,*
> *or remove weight at the phase angle minus 180°*

Obviously the weight corrections required below and above the critical speed would be performed in exactly the same angular locations. This is the property that allows slow speed balancing machines to correct the mass unbalance characteristics of high speed flexible rotors operating over one or more critical speeds. However, more important to the current topic of rotor mode shapes; it is clear that techniques exist to measure dynamic shaft motion, and to identify shaft critical speeds. It should also be mentioned that this example was presented as a rotor with minimal damping. As previously mentioned, damping will influence the response, and the identification of the true location of the residual rotor unbalance, i.e., the *Heavy Spot*. This determination is usually made by the computation of balance sensitivity vectors as presented in chapter 10. In a highly damped mechanical system, the true vertical and horizontal vibration response characteristics must be carefully examined, and the previously stated balancing *rules of thumb* judicially applied.

Up to this point, the discussion has considered vibratory motion only in the vertical plane. This is acceptable if the horizontal support characteristics are identical to the vertical. However, in many cases there is a definable difference between vertical and horizontal rotor support. This difference may be attributable to variations in oil film characteristics, variations in bearing housing stiffness from vertical to horizontal, or a combination of both. In order to accommodate this asymmetry, it is customary to install mutually perpendicular probes to observe both the vertical and horizontal vibration response characteristics. The installation of orthogonal proximity probe transducers provides the capability to measure total motion of the shaft within the bearing.

In addition, the measurement of shaft mode shapes should be considered as a three-dimensional proposition. This is due to the differences between horizontal and vertical restraints, damping, and excitations along the length of the rotor. An informative approach towards quantification of three-dimensional mode shapes consists of measuring shaft orbits at various points along the rotor. These orbits may be combined into an isometric view of the shaft with the respective orbits constructed at each measurement location.

The diagram in Fig. 3-41 illustrates such a presentation. In this example, a 3/8 inch diameter shaft Rotor Kit is configured with two sleeve bearings plus a midspan mass. Relative shaft vibration is observed with X-Y proximity probes mounted at six different lateral positions. The shaft orbits are plotted at each probe location, and the Keyphasor® dots are connected to describe the deformed rotor mode shape at a specific speed.

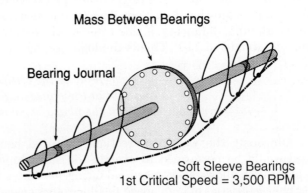

Mass Between Bearings

Bearing Journal

Fig. 3–41 Mode Shape Of Rotor Kit With Soft Sleeve Bearings And Center Mass

Soft Sleeve Bearings
1st Critical Speed = 3,500 RPM

This drawing depicts the rotor kit operating at the vertical translational critical speed of 3,500 RPM. These small rotor kits are quite sensitive to support characteristics. Hence, it is normal to observe a horizontal critical speed, followed by a vertical critical with a similar mode shape. At the measured critical speed of 3,500 RPM, the center of the shaft is considerably more deformed than at the bearings, and a typical first translational resonance for this type of machine is noted.

Using the same rotor kit, and changing the sleeve bearings to ball bearings (higher stiffness), the resultant orbital patterns and associated mode shapes are presented in Fig. 3-42. The system resonant frequency has increased to 6,300 RPM due to the higher support stiffness, and the midpoint shaft deflection has greatly increased. It is also noted that a zero axis crossing nodal point appears at both bearings. This is analogous to the hard support case presented in Fig. 3-27 where a definite phase change occurs, and the orbits are completely reversed passing through a nodal point.

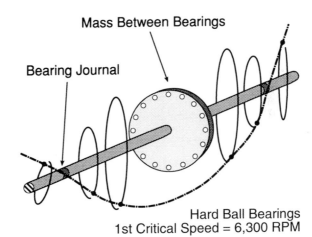

Mass Between Bearings

Bearing Journal

Fig. 3–42 Mode Shape Of Rotor Kit With Hard Ball Bearings And Center Mass

Hard Ball Bearings
1st Critical Speed = 6,300 RPM

Since the shaft motion at the bearings is quite small, it must be recognized that bearing damping is minimal with this configuration. The absence of damping increases the amplification factor through the critical, and increases the susceptibility of the machine to a variety of instability mechanisms. For stiff bearings, it should also be noted that the unbalance energy of the rotor is closely coupled to the supporting structure. In this configuration, casing or structural vibration is likely to be higher than the shaft relative motion due to the direct transmissibility of energy from the rotating shaft to the stationary casing.

On large turbines or compressors that contains very stiff bearings, the designers often include a circumferential squeeze film damper to provide additional damping. These squeeze film dampers consist of a non-rotating, loose fitting annulus, around the bearing outer diameter. The cavity between the bearing outer diameter and the inner diameter of the damper is filled with lube oil. The minor motion (velocity) of the bearing housing with respect to the stationary damper provides a means of energy dissipation through the viscous oil film. Hence, this type of device can be used to provide more damping for a poorly damped system.

Another popular machine configuration consists of an overhung wheel with a short inboard bearing span. This arrangement is used by machines such as overhung blowers, or power turbines on dual shaft gas turbines. The behavior of this type of mechanical system is dominated by the mass and the gyroscopics of the overhung wheel. To demonstrate the modal differences of a mass between bearings versus an overhung mass, the rotor kit was re-configured to the arrangement shown in Fig. 3-43. Hard ball bearings were again used, and orbital measurements were obtained at 5 locations. Note that stiff ball bearings produce nodal points with a 180° phase change across each node (bearing). Also observe the large excursions at the outboard end of the rotor, and the new critical speed of 7,000 RPM. Clearly, the dynamic motion outboard of the bearings describes the anticipated conical mode shape. This result is to be expected, and it is fully consistent with the static beam diagrams presented in Figs. 3-2, 3-4, and 3-6.

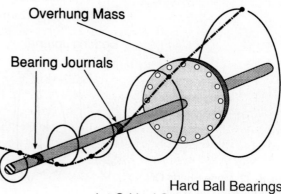

Overhung Mass

Bearing Journals

Fig. 3–43 Mode Shape Of Rotor Kit With Hard Ball Bearings And Overhung Mass

Hard Ball Bearings
1st Critical Speed = 7,000 RPM

Although the orbital technique is extremely informative, it is sometimes graphically difficult to draw a meaningful isometric diagram of the rotor system. This is often combined with the difficulty of installing mutually perpendicular probes at each desired location. An alternate approach consists of installing a single proximity probe at each location. This information provides a single vector quantity at each measurement point, which could be plotted in an isometric manner as previously achieved with the orbits. However, this again can be difficult from a graphics standpoint.

Another technique for handling the vector data is to resolve it into Cartesian coordinates and plot the sine or cosine component at each measurement position. This approach will provide a single quantity, either positive or negative, as a function of position along the rotor. This technique provides acceptable mode shapes, and the mechanics of plotting the two-dimensional data is usually quite straightforward. It should also be mentioned that this technique, as well as the other proximity probe measurements, must be corrected for slow roll runout at each probe location. Failure to perform this basic runout compensation may easily destroy the validity of the dynamic shaft mode shape measurements.

Case History 4: Vertical Generator Mode Shape

Application of any measurement technique is always predicated upon the availability of accessible measurement planes. Within the refining and petrochemical industries, rotating shafts are almost totally enclosed, and the installation of additional vibration transducers is difficult if not impossible. However, within the power generation industry, the mechanical equipment is significantly larger, and exposed shaft surfaces are often available.

This accessibility to the rotating elements allows the direct application of the previously discussed measurement technique. For example, consider the vertical generator rotor displayed in Fig. 3-44. This generator is driven by a hydro turbine at a normal operating speed of 277 RPM. Physically, the turbine extends downward from the generator, and this diagram primarily depicts the three guide bearings plus the generator shaft. The weight of the entire rotating system approaches 100 tons, and the vertical span between upper and lower bearings is approximately 27 feet. Although the majority of the hydro turbine rotor is under water — large portions of the generator shaft are externally visible, and all three radial guide bearings are accessible.

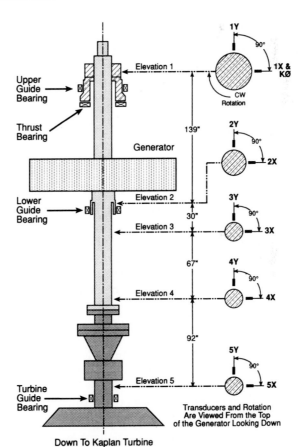

Fig. 3–44 Machinery And Vibration Transducer Arrangement For A Vertical Generator Mounted On Top Of A Hydro Turbine

The machine is normally monitored with a pair of proximity probes mounted below the lower generator guide bearing. For the purposes of this test, relative X-Y proximity probes were temporarily installed at five different elevations as shown in the machinery arrangement diagram, Fig. 3-44. The displacement probes mounted at elevations 1, 2, 3, and 5 were attached to each respective bearing housing with rigid brackets. The X-Y probes installed above the coupling assembly (Elevation 4) were quite difficult to mount since the machine had no stationary elements in the vicinity of the coupling. To address this situation, a uni-strut frame was constructed within the confines of the turbine pit. This aluminum structure was attached to the walls, and it was reinforced with many cross braces to provide a rigid support for the vibration transducers observing the generator shaft above the coupling assembly.

At all five elevations, the structural motion of the bearing housings and the uni-strut frame provided a potential error source for the shaft displacement measurements. In order to quantify the magnitude, frequency content, and timing relationships of the structural vibration, a series of ten low frequency accelerometers were installed on the machine. One accelerometer was mounted adjacent to each radial proximity probe, and the relative shaft motion from the displacement probe was acquired simultaneously with the casing absolute vibration sensed by the accelerometer. In most cases, the casing acceleration signals were insignificant, and the majority of the vibration was relative motion between the shaft and the casing. However, under some test conditions, the casing motion had to be included, and the techniques discussed in chapter 6 were applied to determine the overall, or absolute shaft vibration.

Under most test conditions, the largest error source was attributed to surface imperfections in the observed shaft surfaces. All of the shaft surfaces were pitted and/or rusted, and the radial vibration signals had to be runout compensated to minimize the shaft surface imperfections. The normal operating speed of 277 RPM required that slow roll data be obtained at a much lower speed, and a slow speed of 40 RPM was selected. Although this slow roll speed was a substantial percentage of the normal running speed, it was observed that minimal changes occurred between 60 and 20 RPM during coastdown. Hence, 40 RPM was selected as the nominal slow roll speed since the Digital Vector Filter (DVF) used for this data could correctly process this low speed value.

This particular machinery train always experienced a speed increase during load rejection. Depending on the actual generator load at the time of the trip, the rotor speed of the unit might accelerate from the normal value of 277 to as high as 450 RPM, and then coastdown to a stop. Dynamic vibration data could be acquired throughout the speed range, and slow roll runout vectors may be subtracted from the database to obtain runout compensated synchronous 1X shaft displacement. For example, consider the array at runout compensated 1X vectors at 410 RPM presented in Table 3-5.

These compensated vectors carry the units of Mils,$_{p-p}$, and the phase angles are referenced to each individual probe location. By inspection of this tabular summary two immediate conclusions may be drawn. First, the motion at each

Table 3–5 Vertical Generator - Summary Of Initial Runout Compensated X-Y Shaft Vibration Vectors At Maximum Load Rejection Speed Of 410 RPM

Location	Y-Probe	X-Probe
Elevation 1	10.25 Mils,$_{p-p}$ @ 138°	7.54 Mils,$_{p-p}$ @ 215°
Elevation 2	4.48 Mils,$_{p-p}$ @ 292°	5.55 Mils,$_{p-p}$ @ 337°
Elevation 3	10.81 Mils,$_{p-p}$ @ 275°	8.37 Mils,$_{p-p}$ @ 12°
Elevation 4	21.73 Mils,$_{p-p}$ @ 271°	27.41 Mils,$_{p-p}$ @ 357°
Elevation 5	10.10 Mils,$_{p-p}$ @ 255°	11.09 Mils,$_{p-p}$ @ 350°

elevation is forward, and essentially circular. Second, it is apparent that the bottom four elevations are generally in phase, and the top elevation is out of phase with respect to the other four measurement planes. Based on these observations, it is reasonable to compute an average radial motion at each elevation (basically circular response). Also recognize that a sign change (indicative of a phase change) must occur between the top two planes (Elevations 1 and 2).

The vector array from Table 3-5 may be manipulated in various ways to determine an average value for the generally circular orbits. One way would be to rotate the initial Y-Axis probe angles by 90° to be in the same angular reference position as the X-Axis probes. Performing this simple addition of 90° to each of the Y-Axis probe angles, Table 3-6 is easily generated. The In-Phase magni-

Table 3–6 Vertical Generator - Summary Of Initial Runout Compensated X-Y Shaft Vibration Vectors With Common Angular Reference At Load Rejection Speed Of 410 RPM

Location	Y-Probe	X-Probe
Elevation 1	10.25 Mils,$_{p-p}$ @ 228°	7.54 Mils,$_{p-p}$ @ 215°
Elevation 2	4.48 Mils,$_{p-p}$ @ 22°	5.55 Mils,$_{p-p}$ @ 337°
Elevation 3	10.81 Mils,$_{p-p}$ @ 5°	8.37 Mils,$_{p-p}$ @ 12°
Elevation 4	21.73 Mils,$_{p-p}$ @ 1°	27.41 Mils,$_{p-p}$ @ 357°
Elevation 5	10.10 Mils,$_{p-p}$ @ 345°	11.09 Mils,$_{p-p}$ @ 350°

tudes of each runout compensated shaft displacement vector may now be determined by multiplying each amplitude by the cosine of the associated angle (i.e., *In-Phase=A* cos Ø). Performing this manipulation on the vectors in Table 3-6, and calculating a simple arithmetic average, Table 3-7 may be produced.

The average In-Phase amplitudes from Table 3-7 may now be combined into a rotor mode shape. If these respective plus or minus magnitudes are plotted at each elevation, and if this is performed on an overlay of the rotor drawing, the mode shape shown in Fig. 3-45 evolves. Further embellishment to the graphics was performed by sweeping this shape over a complete.

Table 3–7 Vertical Generator - Summary Of In-Phase Components At 410 RPM

Location	Y-Probe In-Phase	X-Probe In-Phase	Average In-Phase
Elevation 1	-6.86 Mils,$_{p-p}$	-6.18 Mils,$_{p-p}$	-6.52 Mils,$_{p-p}$
Elevation 2	+4.15 Mils,$_{p-p}$	+5.11 Mils,$_{p-p}$	+4.63 Mils,$_{p-p}$
Elevation 3	+10.77 Mils,$_{p-p}$	+8.19 Mils,$_{p-p}$	+9.48 Mils,$_{p-p}$
Elevation 4	+21.73 Mils,$_{p-p}$	+27.37 Mils,$_{p-p}$	+24.55 Mils,$_{p-p}$
Elevation 5	+9.76 Mils,$_{p-p}$	+10.92 Mils,$_{p-p}$	+10.34 Mils,$_{p-p}$

From Table 3-7 and Fig. 3-45, it is noted that the bottom four In-Phase components are positive, and the top elevation 1 carries a negative sign. This change in sign signifies a zero axis crossing between the upper guide bearing, and the lower guide bearing. The resultant nodal point is visible directly above the middle of the generator in Fig. 3-45. This same general mode shape was originally evident during load rejection, and at the normal operating speed of 277 RPM.

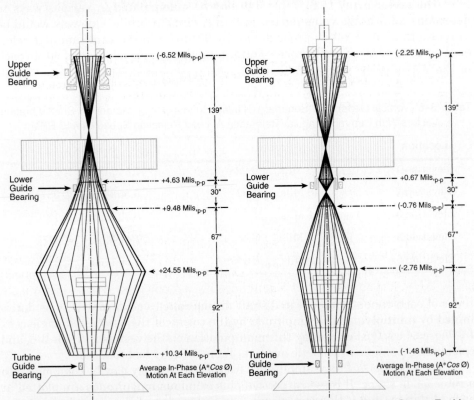

Fig. 3–45 Measured Mode Shape For Vertical Generator At The Maximum Load Rejection Speed Of 410 RPM

Fig. 3–46 Measured Mode Shape For Vertical Generator At The Normal Machine Operating Speed Of 277 RPM

After some deliberation, it was finally concluded that a constant mode shape under all operating conditions was unusual. This was combined with abnormal temperature characteristics of the bearing pads at each guide bearing. It was well documented that the upper generator guide bearing and the turbine guide bearing were functioning at temperatures that were nominally 20°F higher than the lower generator guide bearing. This inconsistency appeared for many years, and no explanation was available for this behavior.

Eventually, the machinery train was rebuilt by mechanics that were employees of the utility company instead of contractors hired by the OEM. During reassembly of this unit, the utility company mechanics discovered that the OEM procedure for setting the lower generator guide bearing clearances produced excessive radial clearances. This assembly procedure was modified by the utility company mechanics, and the final results were verified with swing checks of the vertical rotor. The ensuing test run was very successful, and bearing temperatures were now consistent at all three guide bearings. It was concluded that the lower generator guide bearing had been essentially ineffective in restraining the middle of the generator rotor for many years. This change in restraint at the generator lower guide bearing was clearly reflected in the shaft mode shapes. Specifically, the runout compensated In-Phase components at the normal operating speed of 277 RPM are summarized in Table 3-8.

Table 3–8 Vertical Generator - Summary Of In-Phase Components At 277 RPM

Location	Y-Probe In-Phase	X-Probe In-Phase	Average In-Phase
Elevation 1	-2.14 Mils,$_{p-p}$	-2.35 Mils,$_{p-p}$	-2.25 Mils,$_{p-p}$
Elevation 2	+0.62 Mils,$_{p-p}$	+0.72 Mils,$_{p-p}$	+0.67 Mils,$_{p-p}$
Elevation 3	-0.78 Mils,$_{p-p}$	-0.73 Mils,$_{p-p}$	-0.76 Mils,$_{p-p}$
Elevation 4	-2.68 Mils,$_{p-p}$	-2.85 Mils,$_{p-p}$	-2.76 Mils,$_{p-p}$
Elevation 5	-1.39 Mils,$_{p-p}$	-1.56 Mils,$_{p-p}$	-1.48 Mils,$_{p-p}$

The average X and Y-Axis In-Phase components are used to plot the shaft mode shape shown in Fig. 3-46. The influence of an active generator lower guide bearing is quite apparent, and a new shaft nodal point was clearly introduced below this midspan bearing. It should also be mentioned that the synchronous 1X vectors were reduced, and the vibration amplitude scaling on Fig. 3-46 at 277 RPM is five times larger than the adjacent Fig. 3-45 at 410 RPM. This change in scaling was considered to be appropriate to allow full visibility of the modified mode shape at 277 RPM.

It was concluded that the excessive bearing clearance associated with the incorrect OEM assembly procedure contributed to many of the mechanical problems on this machinery train. The validity of the end user modified bearing clearance adjustment was demonstrated by a new consistency in guide bearing temperatures, a reduction in synchronous vibration amplitudes, and logical changes in shaft mode shapes. In addition, the overall vibration characteristics were less susceptible to changes in electrical load.

Although the presented hydro turbine generator mode shape data was discussed at only two speeds, the entire 1X response during startup or coastdown should be checked. This data should be examined for the presence of rotor and structural resonances. The number of critical speeds, combined with the general rotor configuration, and the generic bearing type (i.e., soft versus hard) will go a long way towards definition of the expected mode shapes. In addition, the examination of runout compensated polar plots across a machine should confirm the inphase behavior of a translational resonance, and the out-of-phase motion of a pivotal mode. Analysis of this type of data will be discussed in greater detail in subsequent chapters of this text.

Overall, it is reasonable to conclude that there are techniques available for measurement and presentation of the dynamic shaft mode shapes. Some companies apply the techniques described in this section to produce on-line three-dimensional mode shapes that are continuously displayed and monitored by both operations and maintenance personnel. This is particularly true in the hydro power generation industry where multiple access points to the rotating shaft are readily available. On other machines, it is impossible to mount probes all along the length of the shaft. However, in some cases it is possible to attach proximity probes inboard and outboard of the bearings, and outboard of the shaft seals. Further information may be obtained from probes in the coupling area. This is particularly true for machines with solid (inflexible) couplings. Based on this type of additional shaft motion data, it is often possible to obtain a realistic estimate of the deformed shaft mode shape. Again variable speed data must be considered, and the diagnostician must have an awareness of the expected type of mode shape for the machine under examination.

ANALYTICAL RESULTS

There are situations when direct mode shape measurements are not practical due to a lack of probe locations, or during the initial design stages of a machine. In these cases, mathematical modeling of the rotor system is not only an extremely informative tool, it may be the only available option. These calculations provide the ability to preview machine response before construction of the machine, and it allows examination of many parameters associated with the system behavior. In addition, if problems do develop during operation, the computer models can provide considerable insight into the anticipated operating mode shape of the system. In addition, mode shape changes due to various excitations, or mechanical abnormalities can be modeled and examined in the computer.

There are three fundamental types of shaft mode shape calculations. The simplest form consists of undamped critical speed calculations, and the associated mode shapes. This type of analysis is based upon mass and stiffness properties of the system. The undamped computations yield calculated natural frequencies, and dimensionless mode shapes. The second type of mode shape calculation evolves from a damped stability analysis of the mechanical system. The inclusion of damping allows the computation of stability, as well as the entire array of forward and reverse modes. This damped stability analysis also produces dimensionless shaft mode shapes.

The third major type of computed mode shape evolves from a forced response analysis of the system. Whereas the previous two types of calculations do not include synchronous forcing functions (e.g., unbalance), the forced response does incorporate input forces, and it produces dimensional mode shapes. Hence, rotor displacement amplitudes at any location and any speed may be calculated. Chapter 5 on analytical modeling will address these three types of mode shape calculations in greater detail. However, for the purposes of this current chapter on rotor mode shapes, the discussion will remain at the fundamental level of undamped modes.

Case history 3 at the beginning of this chapter on the two impeller rotor forms the basis of performing undamped critical speed calculations. In all cases, the rotor model consists of a series of circular cylinders with particular inner and outer diameters, length, density, and modulus of elasticity. Lumped weights and inertias are included for elements such as impellers or couplings. The support stiffness characteristics are also included as part of the input. This type of analytical program for undamped critical speeds performs an iterative solution for each of the natural frequencies. It also provides a simple mode shape for each critical speed. The specific results that can be obtained with this type of analytical computation are best illustrated with the following example of a flexible compressor rotor.

Case History 5: Eight Stage Compressor Mode Shape Change

For this case study, consider the compressor rotor described in Fig. 3-47. This is an eight stage centrifugal compressor operating at 10,500 RPM, and consuming 7,000 horsepower. The outboard is at the thrust end of the compressor, and a large balance piston is mounted at the outboard discharge end of the machine. This particular machine was operated in surge for an extended period of time. When the unit was returned to normal operation, it was discovered that probe gap voltages had increased, and radial vibration had increased to levels between 3.5 and 4.0 Mils,$_{p-p}$. Attempts to slow the machine down were met with even higher vibration amplitudes, and operations personnel were understandably perplexed. It is common practice within many plants to slow down a

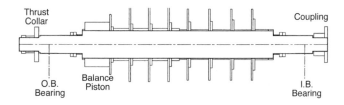

Fig. 3–47 Eight Stage Centrifugal Compressor Rotor Configuration

machine when it starts to vibrate excessively. In some companies, this is a normal instruction given to all compressor and turbine operators. Hence, when this time honored approach of slowing down the machine to get away from a point of high vibration results in even greater vibration amplitudes, the operations personnel find that their options are severely limited.

Examination of the full speed vibration data revealed that the machine was apparently operating in a pivotal mode, that became more severe as the speed was decreased. This was considered to be unusual since the compressor normally operated above the first, but below the second critical speed. In addition, the probe gap voltages indicated a measurable babbitt loss at both journal bearings (i.e., increased clearances). Fortunately, an undamped critical speed analysis has been performed, and this information was available for examination in conjunction with the vibration data. Fig. 3-48 displays the calculated mode shapes at the translational first critical with three different bearing stiffness values. Note that at a stiffness of 200,000 Pounds/Inch (top plot), a typical cylindrical mode is predicted with a critical speed of 3,350 RPM. On the middle plot, the stiffness was increased to 500,000 Pounds/Inch, and this change has little modal effect, but it

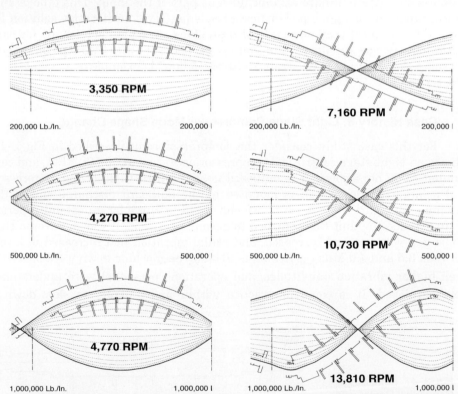

Fig. 3–48 Mode Shapes For Compressor Rotor Translational First Critical Speed

Fig. 3–49 Mode Shapes For Compressor Rotor Pivotal Second Critical Speed

does increase the first critical speed to 4,270 RPM. A further jump in bearing stiffness to 1,000,000 Pounds/Inch reduces motion (vibration) at the bearings, and raises the critical speed to 4,770 RPM. Also note the similarity between these plots and the previously discussed Fig. 3-27.

The next set of mode shape diagrams in Fig. 3-49 describes the pivotal balance resonance, or second critical speed. Again, the top plot shows the 200,000 Pounds/Inch stiffness case, and an expected conical mode at a calculated critical speed of 7,160 RPM. The middle plot reveals the change in mode shape as the stiffness is increased to 500,000 Pounds/Inch. At this stiffness the critical has been increased to 10,730 RPM, and the compressor could be operating on the front slope of this second mode. A final step of increasing stiffness to 1,000,000 Pounds/Inch has an additional clamping influence upon the mode shape, and the critical has been raised to 13,810 RPM.

Due to the abnormal operation of this compressor in surge, it is probable that some mechanical damage was inflicted on the bearings. It was postulated that the potential bearing damage resulted in a reduction in bearing stiffness, and this allowed the second critical speed to drop below operating speed. In essence, the unit was presumed to be running above the second critical speed.

For the sake of completeness, the third critical was also examined, and the resultant mode shapes are presented in Fig. 3-50 as bearing stiffness is increased through the three previous steps. Note that changes in bearing stiffness has minimal effect on the deflected shaft mode shapes. As expected, the fre-

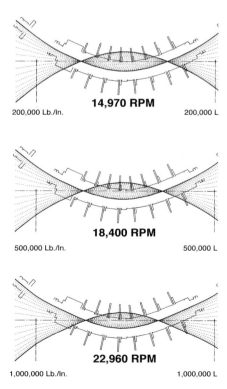

Fig. 3–50 Mode Shapes For Compressor Rotor Bending Third Critical Speed

quency continues to increase with a third critical speed at 14,970 RPM at 200,000 Pounds/Inch, and 22,960 RPM for the stiffest case of 1,000,000 Pounds/Inch. It is also worthwhile to note that the third critical is a sensitive area from a rotor fatigue standpoint. This is due to the reverse flexure acting within the rotor. This mode is sensitive to outboard masses. For instance, the installed coupling must be extremely well balanced to allow the machine any success of survival in this mode. In actuality, most compressors of this general configuration would be designed to avoid this mode. With respect to the current mechanical problem on this rotor, there is no evidence to suggest any influence from this third critical speed upon the measured high shaft vibration.

The information from a series of undamped critical speed calculations is often summarized into a family of curves that is typically called a critical speed map. More specifically, this type of data array is defined as an undamped critical speed map. In this type of presentation the combined effects of bearing stiffness, critical speed, and shaft mode are displayed on a log-log scale. The format of this diagram facilitates the examination of a large amount of information in a reasonably concise manner.

For instance, if the calculated critical speeds from the eight stage compressor rotor are plotted against bearing stiffness, the undamped critical speed map displayed in Fig. 3-51 may be constructed. In order to provide improved definition of the curve shapes, two additional data points at 100,000 and 2,000,000 Pounds/Inch have been added. Hence, each curve consists of five calculated critical speeds corresponding to five different stiffness values. It is significant to note that changes in stiffness alone can determine the mode shape of the system. The lower order modes are generally controlled by bearing stiffness, and the higher order modes are typically governed by shaft stiffness. The controlling parameter is normally determined by the distribution of strain energy within the undamped critical speed calculations.

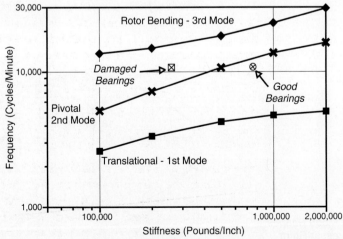

Fig. 3–51 Undamped Critical Speed Map For Eight Stage Compressor Rotor

However, the slope of the various critical speed curves may also provide an indication of the controlling physical parameters. For instance, if the critical speed changes significantly with stiffness variations, then the rotor response is probably dependent upon the bearing stiffness. Under this condition, changes in the journal bearings would change the resultant shaft critical. However, if the critical speed curve is fairly flat, then shaft stiffness is probably the controlling parameter. In that case, the rotor would have to be physically modified to alter the shaft critical speed and/or associated mode shape.

Returning to the case history under discussion, it was concluded from the critical speed map (Fig. 3-51), that operation at the normal speed of 10,500 RPM could result in the compressor experiencing a change in operating modes due to a reduction in bearing or support stiffness. Hence, the machine that ran above the first critical speed may be altered into operating above the second mode. Any speed reduction would then raise shaft vibration amplitudes due to the rotor entering the pivotal balance resonance. This predicted behavior was certainly consistent with the previous experiences of the operating personnel. This conclusion is also consistent with the strong potential of bearing damage manifesting as a reduction in support stiffness and effective bearing damping.

In the final scenario, the unit was tripped, and it did pass through both a pivotal and a translational resonance during coastdown. Both journal bearings were found to be damaged with expanded clearances, and the shaft seals reflected the bearing damage. Fortunately, the rotor journals were not scored, and all bearings and seals were replaced. The post overhaul compressor startup was normal, and the full speed vibratory characteristics returned to previously documented values. Hence, the damaged bearings (due to extended operation of compressor in surge) proved to be responsible for the increased vibration amplitudes, and the change in operating shaft mode shapes.

In summary, this mode shape analysis, and the associated critical speed map is informative during the design stages of a machine. This information is also quite useful during troubleshooting of a field problem. In most situations, the field problem solving exercise does not allow sufficient time to perform this type of analysis. Hence, it is always desirable to perform this type of detailed critical speed and mode shape analysis before the machine starts to shake. It is also desirable to apply some of the more advanced techniques discussed in the following chapters to examine the influence of bearing characteristics, torsional resonances, damped analysis, plus dimensional simulation of synchronous response characteristics.

BIBLIOGRAPHY

1. Avallone, Eugene A. and Theodore Baumeister III, *Marks' Standard Handbook for Mechanical Engineers*, Tenth Edition, pp. 3-9, 3-16, New York: McGraw-Hill, 1996.

2. Belding, William G. and others, *ASM Handbook of Engineering Mathematics*, 4th printing, p. 319, Metals Park, Ohio: American Society for Metals, 1989.

3. Calistrat, Michael M., *Flexible Couplings, their design selection and use*, p. 464, Houston: Caroline Publishing, 1994.

4. Childs, Dara, *Turbomachinery Rotordynamics – Phenomena, Modeling, and Analysis*, New York: Wiley-Interscience Publication, 1993.

5. Den Hartog, J.P., *Mechanical Vibrations*, New York: McGraw-Hill Book Company, 1934.

6. Gieck, Kurt and Reiner Gieck, *Engineering Formulas*, 6th edition, p. M3, New York: McGraw-Hill Inc., 1990.

7. Harris, Cyril M., *Shock and Vibration Handbook*, fourth edition, pp. 1-11, 1-12, 38-5, and chap. 32, 33, 36, and 37, New York: McGraw-Hill, 1996.

8. Jeffcott, H.H., "Lateral Vibration of Loaded Shafts in the Neighborhood of a Whirling Speed - The Effect of Want of Balance," *Philosophical Magazine*, Vol. 37 (1919), pp. 304-314.

9. Rankin, W.A., "On the Centrifugal Force of Rotating Shafts, *Engineer (London)*: 27, 1869.

10. Shigley, Joseph E. and Charles R. Mischke, *Standard Handbook of Machine Design*, pp. 9.13, and 10.6, New York: McGraw-Hill Book Company, 1986.

11. Spotts, M.F., *Design of Machine Elements*, 6th Edition, pp. 18, 27, and 150, Englewood Cliffs, New Jersey: Prentice-Hall, Inc., 1985.

12. Thomson, William Tyrell, *Mechanical Vibrations*, Englewood Cliffs, New Jersey: Prentice Hall, Inc., 1948.

13. Vance, John M., *Rotordynamics of Turbomachinery*, p 61, New York: Wiley-Interscience Publication, 1988.

14. Weaver, F.L., "Rotor Design and Vibration Response," *Proceedings of the First Turbomachinery Symposium*, Gas Turbine Laboratories, Texas A&M University, College Station, Texas (1972), pp. 142-147.

15. Young, Warren C., *Roark's Formulas for Stress & Strain*, Sixth Edition, pp. 59, 101, and 346, New York: McGraw-Hill Book Co., 1989.

Bearings and Supports

*B*earings provide the primary interface between the moving and the stationary parts of a machine. Although the seals and process fluids (or magnetic fields) coexist, the bearings provide the majority of the stiffness and damping for the moving assembly. It is understandable that dynamic forces developed on the moving parts are transmitted to the stationary parts across these main support bearings. The forces may be static radial loads due to rotor weight, or they may be dynamic forces due to mechanisms such as mass unbalance. In either case, the radial bearings must carry the applied loads, or the machine will fail.

Many machines are also equipped with thrust bearings to restrain the axial loads imposed by differential fluid pressures across stages of compression or expansion. Motors, generators, some gear boxes, and some double flow compressors do not contain thrust bearings. On other trains, a single thrust bearing may accommodate several machines connected via a series of hard couplings. In all cases, if the thrust bearing(s) fail, the machinery train will cease to function.

It is reasonable to conclude that bearings are one of the most vulnerable machinery elements. In many cases, bearings are the scapegoat for other malfunctions. It has been repeatedly demonstrated that bearings are often redesigned to handle higher and higher loads, when the problem actually originates within some other part of the machine. In these situations it is desirable to return to basic engineering principles, and examine the behavior of the entire mechanical system, including the interaction with the main bearings. The influence of loads, physical dimensions, bearing clearances, lubricant properties, and various geometric configurations will be reviewed in this chapter. The primary emphasis of this discussion will be on fluid film radial bearings. However, general comments for thrust bearings and roller element bearings will be included.

As noted in the preface to this text, excessive bearing clearances represent a major category of common machinery malfunctions. Hence, various methods for accurate measurement of bearing clearances will be presented. In addition, the measurement of bearing housing support characteristics will also be addressed within this chapter.

FLUID FILM RADIAL JOURNAL BEARINGS

In the development of analytical machinery models, there is always a temptation to begin the project with a detailed rotor analysis. However, the machinery diagnostician soon discovers that bearing characteristics must be defined and included within the rotor model. In some cases, such as the optimization of a mechanical system, many different bearing types may be considered with different rotor models. Hence, it is often an iterative procedure to determine the proper combination of bearings and rotor configuration.

One of the easiest, and informative starting points is the determination of the static shaft loading upon the planar area of the bearing. This calculation consists of dividing the static journal load by the plane area of the bearing as shown in the following equation (4-1):

$$Bearing\ Unit\ Load\ =\ BUL\ =\ \frac{Journal\ Weight}{Length\times Diameter} \qquad (4\text{-}1)$$

where: $\quad$ *Bearing Unit Load* $\ =\ $ Static Bearing Loading (Pounds/Inch^2)
$\qquad\qquad$ *Journal Weight* $\ =\ $ Static Load on Bearing (Pounds)
$\qquad\qquad\qquad$ *Length* $\ =\ $ Bearing Length (Inches)
$\qquad\qquad\quad$ *Diameter* $\ =\ $ Bearing Diameter (Inches)

If the weight of a rotor is evenly distributed between two bearings, then the Journal Weight is equal to 50% of the rotor weight. The product of bearing length times diameter yields the planar area of the bearing. That is, a top view of the bottom half of a bearing will have a projected area that is determined by the bearing dimensions. For example, consider a 14,400 pound rotor supported by two bearings that are each 6.0 inches long and 8.0 inches in diameter. The static bearing unit load may be computed from (4-1) as follows:

$$BUL\ =\ \frac{Journal\ Weight}{Length\times Diameter} = \left\{ \frac{(14,400/2)\ \text{Pounds}}{6.0\ \text{Inches}\times 8.0\ \text{Inches}} \right\}$$

$$BUL\ =\ \frac{7,200\ \text{Pounds}}{48.0\ \text{Inches}^2} = 150\ \text{Pounds/Inches}^2$$

A loading of 150 Psi is acceptable for most industrial machines. Generally, the range of allowable bearing loads for fluid film bearings varies between 100 and 300 Psi. For lightly loaded bearings (i.e., <100 Psi), the rotor bearing system may be susceptible to instability. In addition, the bearing size with respect to the shaft would violate common sense. If the above example had a loading of 50 Psi, and the rotor weight plus bearing diameter remained constant — then the bearing length would have to be 18.0 inches, which is ludicrous.

Conversely, for heavily loaded bearings (i.e., >300 Psi), the bearing may fail prematurely due to the excessive radial loads. Also, the bearing size with respect to the shaft would not make sense. If the original example had a loading of 600 Psi, and the rotor weight plus bearing diameter remained constant — then the bearing length would have to be 1.5 inches. This is likewise unreasonable for a seven ton rotor with an 8.0 inch diameter shaft.

In addition to these static loads, forces due to unbalance, misalignment, fluids, gears, etc., must be considered in the bearing design. These cyclic forces do not lend themselves to the simple analysis that was just performed. In order to evaluate these parameters, plus the oil film characteristics, the bearing analysis must be significantly expanded. Many excellent papers on dynamic bearing characteristics have been published by investigators such as Edgar Gunter[1], Jim McHugh[2], Paul Allaire and Ron Flack[3], and Dana Salamone[4].

In most cases, determination of machine support coefficients is generally a two step effort. The first part consists of the computation of oil film characteristics, and the second task requires the measurement of the bearing structural support. The overall or effective rotor support is based upon a combination of these individual, yet interrelated parameters.

Bearing oil film coefficients can be determined with two different types of computer programs. The first type is a bearing *look up* program that outputs principal stiffness K_{xx}, K_{yy} and damping C_{xx}, C_{xx} coefficients, plus the cross-coupling coefficients of K_{xy}, K_{yx}, C_{xy}, and C_{yx}. These programs also display Sommerfeld numbers with associated speeds, loads, and journal eccentricities. The bearing parameters can be calculated for specific conditions, or over a defined speed domain. Programs of this type run quite rapidly, and are useful for examining multiple cases prior to the final definition of parameters, and the detailed bearing calculations.

This second type of bearing coefficient program computes the equilibrium position, plus the stiffness and damping coefficients for a defined bearing geometry. Programs exist for cylindrical bearings, multi-lobe bearings, pressure dam bearings, and tilting pad bearings. These programs are often sophisticated finite element solutions that allow variable oil viscosity within the bearing, accept oil turbulence, plus the application of vertical and horizontal external forces, and variations in preload (where appropriate).

Typical program output data includes the principal stiffness K_{xx}, K_{yy} and damping C_{xx}, C_{xx} coefficients, plus the cross-coupling coefficients K_{xy}, K_{yx}, C_{xy}, and C_{yx} as required. These results are usually displayed graphically, as illustrated in Fig. 4-1. This plot describes dimensional stiffness and damping coefficients as a function of rotating speed. The presented data was calculated for a five (5) shoe tilting pad bearing. This bearing has a length/diameter (L/D) ratio of 0.4, a 60° arc length, load on pad (LOP), a 50% offset, and a 0.25 preload. The

---

[1] Edgar J. Gunter, "Dynamic Stability of Rotor Bearing Systems," NASA Report SP-113, 1966.

[2] James D. McHugh, "Principles of Turbomachinery Bearings," *Proceedings of the Eighth Turbomachinery Symposium*, Gas Turbine Laboratories, Texas A&M University, College Station, Texas (November 1979), pp. 135-145.

[3] Paul E. Allaire, and Ronald D. Flack, "Design of Journal Bearings for Rotating Machinery," *Proceedings of the Tenth Turbomachinery Symposium*, Turbomachinery Laboratories, Texas A&M University, College Station, Texas (December 1981), pp. 25-45.

[4] Dana J. Salamone, "Journal Bearing Design Types and Their Applications to Turbomachinery," *Proceedings of the Thirteenth Turbomachinery Symposium*, Turbomachinery Laboratories, Texas A&M University, College Station, Texas (November 1984), pp 179-188.

applied bearing static load was 1,750 pounds, and diametrical clearance was 9.0 Mils (0.009 inches). Since this is a tilting pad bearing, the cross-coupling coefficients are zero, and only the principal (xx and yy) coefficients are shown. It should also be mentioned that due to the extended amplitude range, this type of information is normally plotted with a log-log scale. Based on the relative magnitudes of the coefficients, the stiffness parameters appear on the top half, and the damping curves are towards the bottom portion of the plot.

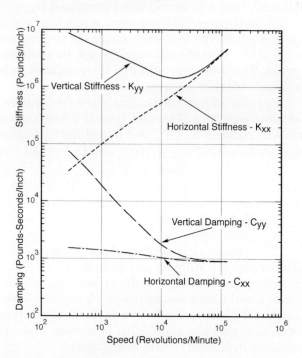

Fig. 4–1 Dimensional Oil Film Bearing Stiffness And Damping Coefficients Versus Rotor Speed

In many instances, the bearing coefficients are presented in a non-dimensional format. The customary form used for non-dimensional stiffness coefficients may be expressed by:

$$K_{NonDim} = K_{Dim} \times \left\{ \frac{C_b}{W} \right\}$$ (4-2)

where: K_{NonDim} = Non-dimensional Stiffness
          K_{Dim} = Stiffness (Pounds/Inch)
          C_b = Bearing Radial Clearance (Inches)
          W = Static Bearing Load (Pounds)

Similarly, the non-dimensional bearing damping coefficients are calculated by:

$$C_{NonDim} = C_{Dim} \times \left\{ \frac{\omega \times C_b}{W} \right\} \qquad (4\text{-}3)$$

where: C_{NonDim} = Non-dimensional Damping
         C_{Dim} = Damping (Pounds-Seconds / Inch)
         ω = Shaft Rotative Speed (Radians / Second)

Non-dimensional coefficients are often used to define a particular bearing design, and they permit direct comparison between bearings. Non-dimensional coefficients also allow reasonably easy conversions between test cases for variations in bearing clearances, loads, or speed. Hence, the designer may evaluate changes in parameters without repeating the full array of bearing calculations.

Another non-dimensional parameter that is computed for fluid film bearings is the Eccentricity ratio. In bearing terminology, eccentricity is defined as the distance between the bearing center, and the shaft centerline position. Dividing this distance by the bearing clearance yields the dimensionless quantity of eccentricity ratio. This physical location represents the calculated equilibrium position of the journal. It may be presented as a vector, or as a horizontal and a vertical eccentricity (or offset), from the bearing center.

These terms are easier to understand if an example of a fluid film bearing is examined. For instance, consider a shaft journal rotating within a cylindrical bearing as illustrated in Fig. 4-2. In this example, the bearing clearance circle is

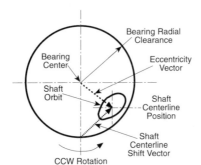

Fig. 4–2 Journal Eccentricity Position Within A Plain Circular Fluid Film Bearing

shown, and the bearing geometric center is identified. The shaft is defined as rotating in a counterclockwise direction, and the shaft orbit is indicated in the lower right quadrant of the bearing. The center of the shaft orbital motion is commonly referred to as the shaft centerline position. The physical distance between this shaft centerline position and the geometric center of the bearing is defined as the shaft eccentricity. As previously noted, eccentricity may be stated as a vector quantity, or as X-Y Cartesian coordinates.

The eccentricity ratio consists of the eccentricity magnitude divided by the bearing clearance. Most analysts' use radial bearing clearance to compute the eccentricity ratio. As a precautionary point, eccentricity should not be confused with the shift in shaft centerline position from some initial rest position (e.g., at the bottom of the bearing). Although these two vectors are directly related via the bearing clearance, the eccentricity vector is a calculated parameter based upon the bearing center. Whereas, the centerline position vector is measured with shaft sensing orthogonal proximity probes, and it is generally referenced to the bottom of the bearing (for a horizontal machine). Ideally, these two vectors should terminate at the same point within the clearance circle.

Since the eccentricity ratio is associated with the minimum oil film thickness, it is important information for the bearing designer, as well as the machinery diagnostician. It should be recognized that each particular bearing type or configuration displays a unique shaft centerline position of the journal within the bearing. This running position is a function of physical parameters such as bearing geometry, operating speed, shaft weight, and lubricant characteristics. The actual running position may be influenced by the application of shaft preloads originating from normal sources such as gear contact forces, or abnormal forces such as coupling misalignment.

In many machinery analysis problems, it is difficult to separate normal versus abnormal forces acting on the shaft. The dynamic motion of the shaft (vibration) is altered, and the running position of the journal within the bearing is influenced. Hence, one must evaluate the dynamic as well as the static information. This type of evaluation is often predicated upon a comparison between normal behavior and the current motion and/or position characteristics of the shaft position within the journal bearing. More specifically, the diagnostician must be aware of normal shaft position characteristics in order to identify an abnormal position. For instance, Fig. 4-3 describes the normal shaft centerline running position for three different types of common industrial journal bearings.

The plain journal bearing shown on the left side of Fig. 4-3 is a typical bearing installed in many types of horizontal machines. On smaller machines, this type of bearing may consist of upper and lower thin bearing liners restrained by a heavy bearing housing. On larger machines, the babbitt bearing surface may be integral with the bearing housing. In either case, this type of bearing generates an oil wedge in the lower right bearing quadrant (with CCW rotation). The shaft is supported at the minimum oil film, and journal weight is supported by the hydrodynamic forces within the bearing. In most cases, the shaft centerline vector pivots up from 20° to 40° above the bottom horizontal plane of the bearing.

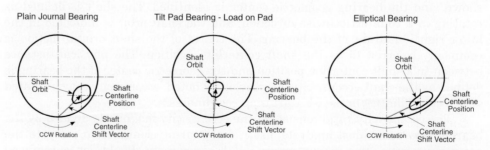

Fig. 4–3 Shaft Centerline Position With Three Different Types Of Fluid Film Bearings

The tilt pad bearing displayed in the middle of Fig. 4-3 consists of a series of floating pads that surround the journal. A common configuration for a horizontal machine consists of three pads in the bottom half of the bearing, and two pads located in the top bearing half. The three bottom pads are usually configured with one pad directly below the shaft (6 o'clock position). This physical pad location is commonly referred to as a *Load on Pad* (LOP) arrangement. If the bearing pads were repositioned or rotated on the shaft circumference to allow the two bottom pads to straddle the true vertical centerline, this would be considered as a *Load Between Pad* (LBP) configuration.

For a LOP arrangement, the shaft supporting oil wedge is established and maintained on the bottom pad. Due to the location of this oil wedge, the shaft rises essentially straight up into the normal running position. In most cases, the shaft centerline vector pivots up from 80° to 100° above the bottom horizontal plane of the bearing. The normal shaft running position is slightly offset from the true vertical centerline. Usually this offset is in the direction of rotation as noted in the center diagram of Fig. 4-3.

Shaft centerline position for a lemon bore or elliptical bearing are shown in the diagram located on the right side of Fig. 4-3. Within this type of fixed lobe bearing the horizontal clearances are much larger than the vertical clearances. Typically, a ratio of 1.5:1 or 2:1 is maintained between horizontal and vertical clearances. This physical configuration allows the rotating shaft to *slide over* into lower right bearing quadrant for a counterclockwise rotating shaft (as shown), or the lower left bearing quadrant for a clockwise shaft rotation.

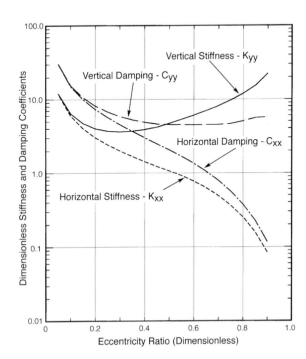

Fig. 4–4 Non-Dimensional Stiffness And Damping Coefficients Versus Journal Eccentricity Ratio

Although there are many types and configurations of journal bearings, the calculation of expected journal location within a bearing may be compared with the running position as determined by DC measurements with proximity probes. Again, it should be mentioned that the probes measure shaft position from an initial point such as the bottom of the bearing. Thus, the vector algebra for the probe calculation is based upon the rest point of the shaft in the bearing, whereas the analytical calculations are referenced to the bearing center. Significant deviations between computed and measured shaft centerline positions may be useful in the identification of a machinery problem. Conversely, if the radial position calculations and measurements agree, the validity of the computations are reinforced, and the diagnostician should consider looking into other aspects of abnormal behavior of the machinery.

Refer back to the five shoe tilt pad bearing data from Fig. 4-1, and the non-dimensional stiffness and damping parameters defined by equations (4-2) and (4-3). Clearly, the dimensional coefficients may be converted into non-dimensional values. It is common to plot these dimensionless parameters against the non-dimensional eccentricity ratio as shown Fig. 4-4. Since this is a tilt pad bearing, the shaft will rise vertically from the bottom pad, and the angle associated with the eccentricity vector will be in the vicinity of 90°. For this common bearing, it is noted that vertical stiffness increases as the minimum oil film decreases (i.e., larger eccentricity ratio). Hence, the computed results are consistent with intuitive logic and the expected behavior for this type of mechanical system.

Bearing analytical programs also compute the dimensionless Sommerfeld number based upon the inlet viscosity, speed, length, diameter, load, and clearance. This parameter is widely used as a characteristic number for journal bearing performance. Typical values for the Sommerfeld number vary from 0.01 to 10.0. The common format for the Sommerfeld number calculation is presented in the following expression:

$$N_{So} = \left\{ \frac{\mu \times \omega \times L \times D}{W} \right\} \times \left\{ \frac{R}{C_b} \right\}^2 \qquad \textbf{(4-4)}$$

where N_{So} = Sommerfeld Number (Dimensionless)
    μ = Absolute or Dynamic Oil Viscosity (Pounds-Seconds / Inch^2)
    ω = Shaft Rotational Speed (Radians / Second)
    L = Bearing Length (Inches)
    D = Shaft Diameter (Inches)
    R = Shaft Radius (Inches)
    W = Load on Bearing (Pounds)
    C_b = Bearing Radial Clearance (Inches)

It should be mentioned that other forms of the Sommerfeld number are used. Although the general intent of the expression remains intact, individual designers may use different values for this dimensionless number. For instance, the rotational speed may be stated Revolutions per Second instead of Radians per Second. In all cases, it is mandatory to identify the parameters and engineering units when attempting to compare the Sommerfeld numbers generated by

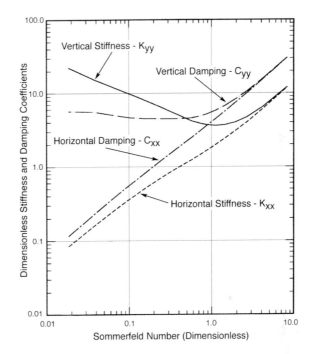

Fig. 4–5 Non-Dimensional Stiffness And Damping Coefficients Versus Sommerfeld Number

two or more bearing designers.

A typical plot of Sommerfeld number versus the non-dimensional stiffness and damping is presented In Fig. 4-5. This plot is based upon the same five shoe tilting pad bearing depicted in Figs. 4-1 and 4-4. Generally, the bearing designer will consider the characteristics displayed by the Sommerfeld plots, and the previously discussed eccentricity ratio plots, plus the dimensional plots of bearing coefficients versus speed. Bearing designers also examine plots of Sommerfeld numbers versus other non-dimensional quantities such as whirl or speed ratios. Clearly, individual design techniques and computer programs yield a large assortment of plot formats.

Another common calculation is the Reynolds number at the minimum oil film. This non-dimensional number is the ratio of inertia to viscous forces, and it may be computed by:

$$N_{Re} = \left\{ \frac{\rho \times H \times \omega \times R}{\mu \times G} \right\} \tag{4-5}$$

where: N_{Re} = Reynolds Number (Dimensionless)
   ρ = Oil Density (Pounds / Inch^3)
   H = Minimum Oil Film Height (Inches)
   ω = Shaft Rotational Speed (Radians / Second)
   R = Shaft Radius (Inches)
   G = Acceleration of Gravity (386.1 Inches / Second^2)
   μ = Absolute or Dynamic Oil Viscosity (Pounds-Seconds / Inch^2)

The Reynolds number allows characterization of the oil flow at the minimum oil film. This is useful in determination of the fluid flow regime. In most instances, laminar flow through the minimum oil film is encountered. Cavitation often occurs above the journal, but within the load carrying bottom half of the bearing, laminar flow is the normal and desired situation. In most cases, if the minimum oil film Reynolds number is less than 1,000, then laminar flow should be expected. Conversely, if this value exceeds 1,000, then turbulent flow would be a concern. The computational software should be able to handle both types of flow regimes, and provide meaningful bearing coefficients, plus proper equilibrium position and force balance.

As noted on the Sommerfeld number, various forms of these equations are in use, and dimensional analysis should always be performed to verify the consistency of units. In fact, some analytical computer programs do not yield true non-dimensional values for parameters such as the Reynolds or Sommerfeld numbers. Often a residual unit remains that alters the magnitude of the number. Again, to avoid confusion, the machinery diagnostician should make sure that dimensionless numbers are truly non-dimensional, or at least consistent between comparative cases.

A non-dimensional quantity that most bearing designers agree upon is the Preload factor for lobed or segmented bearings (e.g., tilt pad). These types of bearings display a pad curvature that is greater than the shaft curvature. This physical configuration forces the oil to converge close to the middle of each pad due to the reduced clearance. In essence, preload produces or forces an oil wedge in the bearing pad. For these types of bearings, the preload may be determined by the ratio bearing clearance C_b, and the pad clearance C_p as follows:

$$Preload = \frac{C_p - C_b}{C_p} = 1 - \left\{ \frac{C_b}{C_p} \right\} \qquad \textbf{(4-6)}$$

Either radial or diametrical clearances may be used for (4-6), but both variables must be the same. That is, C_b and C_p must be both radial, or both diametrical clearances. Another way to express the shaft preload is to convert equation (4-6) into equivalent diameters. By substitution in the previous expression, it can be easily shown that the Preload may also be calculated with the following:

$$Preload = \left\{ \frac{D_p - D_b}{D_p - D_s} \right\} \qquad \textbf{(4-7)}$$

where: D_p = Diameter of Pad Curvature (Inches)
D_b = Diameter of Bearing Clearance (Inches)
D_s = Diameter of Shaft (Inches)

For a positive preload, the shaft diameter D_s is the smallest number, and the pad diameter D_p is the largest number. If the pad and bearing clearances are equal, then the preload is zero. The bearing is circular, with the pad and bearing sharing the same center of curvature. At the other extreme, if preload is equal to 1.0, then the bearing clearance is zero, and the shaft is in direct contact with the bearing pad. In practice, bearing preloads are typically found to be in the range of 0.1 to 0.5. As preload increases, the bearing stiffness increases, and the damping often decreases. This relationship between the principal coefficients and the bearing pad preload may be used to optimize the stiffness and damping characteristics of the bearing. Conversely, if the bearing pad is damaged during installation, or the babbitt is scraped by a well intentioned millwright, the preload may be seriously altered, and the bearing characteristics totally corrupted.

The normal preload is a positive number, indicating that the bearing pad radius of curvature is greater than the bearing radius. If this situation was reversed, and the bearing pad displayed a radius that was smaller than the bearing, a negative preload would result. Physically, this means that the shaft is riding on the pad edges (i.e., bearing pad is edge loaded), and premature failure of the bearing is a certainty. On a questionable installation, bluing on the shaft may be used to determine the actual contact area between the journal and the bearing pads.

Additional visibility of bearing characteristics may be obtained from the calculated pressure and temperature profiles for each design. For example, consider Fig. 4-6 of radial pressure distribution within an elliptical bearing. This type of radial bearing is also referred to as a two lobe, or lemon bore bearing with 25 Psi oil supply pressure.

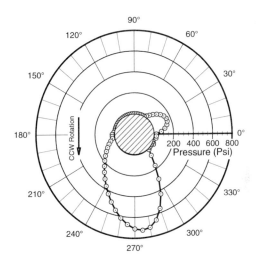

Fig. 4–6 Oil Pressure Distribution Around An Elliptical Journal Bearing

The polar coordinate plot displays the circumferential pressure distribution within the bearing at normal load and speed. Note that the maximum developed pressure occurs at the bottom of the bearing. This point is slightly upstream of the minimum oil film, and is consistent with expected bearing behavior and theory. Further examination of this diagram reveals another positive pressure buildup at an angle of 25°. This pressure buildup is due to the development of another converging oil wedge between the journal and the top half of the bearing. This secondary oil wedge is due to the fixed lobe bearing geometry. In some cases, the influence of forces originating in the top half of the bearing may be responsible for driving an instability, or providing a positive stabilizing force (e.g., pressure dam).

The difference between the physical location of the maximum oil film pressure, and the minimum oil film thickness points out an interesting fact in the world of rotors and bearings. Specifically, a non-rotating shaft with a vertical load will deflect downward in the direction of the applied load. Now assume that the shaft is turning at a constant speed, and that an oil wedge has developed. In this condition, the application of a downward force on the shaft will be greeted by a vertical shift in the direction of the applied vertical load, plus a horizontal shift of the shaft within the bearing. The horizontal shift will occur in the direction of rotation. That is, a shaft rotating in a counterclockwise direction will move to the right, and a shaft turning clockwise will slide to the left. This cross-coupling mechanism across the oil film is responsible for many bearing behavioral characteristics including the self-excited instabilities discussed in chapter 9 of this text.

It should also be mentioned that most fluid film bearings are constructed with a steel base or backing, and a babbitt coating that provides the actual bearing surface. The babbitt may be either tin or lead based, and various compositions are in common use. Since babbitt is softer than the steel journal, it is the first sacrificial element in a bearing assembly. Ideally, during a bearing failure, the babbitt will sustain the majority of the distress, and the steel journal will not be damaged. Thus, the bearings may be replaced, and the rotor may be reused without any repairs to the journals. Of course, during a major failure, the steel journals may contact the steel backing on the bearings, and substantial damage may be inflicted on both the journals as well as the bearings.

The babbitt thickness on journal bearings may range from micro-babbitt thickness of 0.005 to 0.015 Inches (5 to 15 Mils) as a minimum, to 0.050 or 0.060 Inches (50 or 60 Mils) as a maximum. The thick babbitt will be a better choice for conditions of dirty lube oil, or anticipated wear on the bearings. Unfortunately, thick babbitt layers are susceptible to damage from impact loads, various bearing instabilities, and shaft misalignment. In many instances, a malfunction can break off a chunk of thick babbitt, and carry it around the entire bearing, with disastrous consequences. Babbitt is also a poor conductor of heat, and a hot bearing will generally result in a premature fatigue failure. On the other hand, the thin micro-babbitt bearings will transmit heat to the backing material more readily, and this type of bearing is more resistant to impact loads, and other dynamic forces. However, the oil system must be maintained in a very clean condition. Any dirt or foreign objects in the oil may seriously damage a micro-babbitt coating. The oil film dynamics of thin versus thick babbitt bearings are essentially the same. The size and the geometry of the bearing is often more important than the babbitt thickness. However, the diagnostician should not ignore this important bearing parameter.

Case History 6: Shaft Position In Gas Turbine Elliptical Bearings

In most cases, it is technically difficult (if not impossible) to directly check the validity or accuracy of the computed bearing coefficients. However, each calculation must conclude with a force balance, plus a position balance of the journal within the bearing clearance. It is reasonable to believe that if the calculated eccentricity position is correct, then the other computed parameters are also representative of the bearing characteristics. Since journal position within an oil film bearing can be measured directly with proximity probes, it is logical to perform a check of the analytical predictions versus actual machine data.

For this case history, consider a group of four single shaft gas turbines that operate between 5,000 and 5,350 RPM. These units are rated at 40,000 HP, and they are used to drive high pressure centrifugal compressors through a single helical gear box. The shaft sensing proximity probes are mounted at ±45° from the true vertical centerline as shown in Fig. 4-7. At the turbine inlet end #1 bearing, the probes are mounted above the shaft. Conversely, at the exhaust end #2 bearing, the probes are located below the shaft.

Fig. 4–7 Angular Arrangement Of Radial Proximity Probes On A Single Shaft Gas Turbine

Inlet Bearing

Exhaust Bearing

The eight inch diameter journals are supported in elliptical bearings. These bearings have an average vertical diametrical clearance of 16 Mils (0.016 inches), and a normal horizontal diametrical clearance of 32 Mils (0.032 inches). These physical dimensions are consistent with a nominal 2:1 clearance ratio previously mentioned in this chapter.

The shaft centerline position for these machine journals was determined by measuring the proximity probe DC gap voltages at a stop condition, and at full speed. The difference between these DC voltages is divided by the transducer scale factor to determine the position change in the direction of each transducer. This X-Y change in radial position may be plotted on a graph that displays the bearing clearance, plus the calculated journal position in the X and Y directions.

Fig. 4-8 depicts the radial journal positions for the turbine inlet bearings. Shaft centerline locations for the A unit were obtained on different dates, and at slightly different speeds varying between 5,100 and 5,340 RPM. Three additional machines identified as the B, C, and D units are also included in this survey. Speeds for these last three units varied between 5,010 and 5,350 RPM. It is noted that excellent agreement has been achieved between the calculated position at 5,340 RPM, and the six sets of field data.

The same position information for the exhaust end #2 bearing is contained in Fig. 4-9. Notice that the scatter of data is much greater at this bearing, and

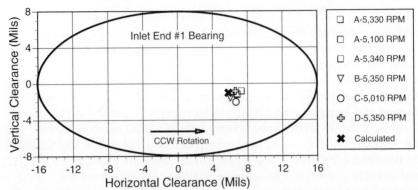

Fig. 4–8 Shaft Centerline Position On Four Gas Turbines At Inlet End #1 Bearing

the deviations from the calculated position are substantial. Initially, it might be concluded that the theory does not support the actual machinery behavior. However, a partial explanation for these aberrations resides within the characteristics of the proximity probe measurements. Specifically, the early vintage of both the proximity probes, and the companion drivers are sensitive to operating temperature. The temperature limit specification for this specific probe and cable was 350°F; and the oscillator-demodulator operating limit was specified as 150°F for a standard unit, or 212°F for an extended temperature range version.

As shown in Fig. 4-7, the exhaust end probes are mounted outboard of the #2 Bearing, and below the horizontal centerline. These probes are subjected to a high temperature environment that can easily heat the transducers to temperatures in excess of 200°F. The oscillator-demodulators are mounted in an explosion proof housing. Although a heat shield is installed between the turbine exhaust and this box, the electrical components often operate at temperatures above 130°F. Thus, the exhaust end probes, cables, and drivers are all exposed to elevated temperatures that affect the calibration curve slope.

For many years, the instrumentation vendors have recognized that operat-

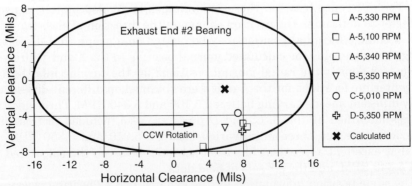

Fig. 4–9 Shaft Centerline Position On Four Gas Turbines At Exhaust End #2 Bearing Based On Direct Probe Gap Measurements Without Temperature Correction

ing temperature will influence probe calibration. For instance, Fig. 4-10 depicts the variation in calibration curves at temperatures of 75, 200, and 350°F. This

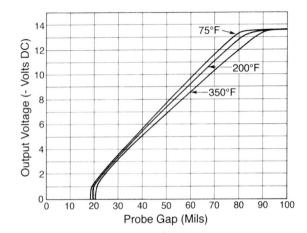

Fig. 4–10 Influence Of Operating Temperature On Proximity Probe System Output Voltage

data was published by the manufacturer of the proximity probes installed on these particular gas turbines. The plotted data is for a 0.300 inch diameter probe. Larger excursions are normally exhibited by smaller diameter probe tips. From this family of curves, it is clear that the calibration curve will bend downward as temperature increases. At 200°F, the calibration curve is nominally 0.5 volts below the normal curve for gaps in the vicinity of -9.0 to -10.0 volts DC. Hence, for a given distance between the probe and shaft, the output DC voltage from the Proximitor® is reduced by about 0.5 volts. Since the measurement system operates with a negative bias, the gap voltages are likewise negative.

The correction for this temperature behavior requires adding the incremental voltage to the Proximitor® output voltage. Thus, the measured output DC voltage should be corrected by -0.5 volts DC to yield a temperature compensated value. Specifically, Table 4-1 summarizes the cold (at stop) gap voltages, plus the hot (running) gap voltages for the B machine. The differential gap voltages are merely the cold minus the hot gap voltages at the turbine exhaust bearing. Dividing the Y-Axis (vertical) probe differential gap voltage by the normal transducer sensitivity of 0.2 Volts/Mil (200 mv/Mil) yields a displacement change of 2.15 Mils towards the probe. Similarly, the X-Axis (horizontal) transducer exhibits a -1.24 volt change, which is equivalent to a 6.20 Mil position shift away from the probe. This is equivalent to an overall shaft vector shift of 6.56 Mils at 26°

Table 4–1 Direct Proximity Probe Gap Voltages At Turbine Exhaust End #2 Bearing

Probe and Angular Location	Cold Gap Voltage	Hot Gap Voltage	Differential Gap Voltage	Differential Position
Y-Axis @ 315°	-9.66 volts DC	-9.23 volts DC	+0.43 volts DC	+2.15 Mils
X-Axis @ 225°	-9.23 volts DC	-10.47 volts DC	-1.24 volts DC	-6.20 Mils

from the cold to the hot position.

However, if the measured hot probe gap voltages are corrected by -0.5 volts DC to compensate for the transducer temperature sensitivity, the results are shown in Table 4-2. The initial cold gap voltages (zero speed) remain the same as

Table 4–2 Corrected Proximity Probe Gap Voltages At Turbine Exhaust End #2 Bearing

Probe and Angular Location	Cold Gap Voltage	Hot Gap Voltage	Differential Gap Voltage	Differential Position
Y-Axis @ 315°	-9.66 volts DC	-9.73 volts DC	-0.07 volts DC	-0.35 Mils
X-Axis @ 225°	-9.23 volts DC	-10.97 volts DC	-1.74 volts DC	-8.70 Mils

before. The displacement shift is again determined by dividing the differential gap voltages by 0.2 Volts/Mil to determine the distance shift. For the Y-Axis probe, this yields a displacement change of 0.35 Mils away from the probe. The X-Axis transducer now displays a -1.74 volt change, which is equal to an 8.70 Mil position shift away from the transducer. The total shift of the journal centerline position is therefore equivalent to a vector shift of 8.71 Mils at 47° (cold to hot position). Thus, the temperature correction reveals that the shaft is really riding higher in the bearing than the uncorrected data revealed.

Certainly the accuracy of this correction may be improved by detailed temperature sensitivity calibration of each transducer on each machinery train. However, that type of information is often not available, or the expense of producing and maintaining this database might be cost prohibitive. Hence, the use of a reasonable voltage correction is considered to be adequate and acceptable for this situation.

Correcting each of the hot gap voltages from the initial shaft centerline diagram produces the journal positions presented in Fig. 4-11. Again, the exhaust end probes are mounted on the bottom of the shaft, and the corrected DC voltages reveal a shaft rise. It is evident that agreement between the calculated and

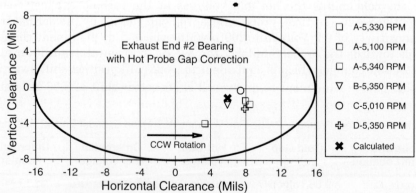

Fig. 4–11 Shaft Centerline Position On Four Gas Turbines At Exhaust End #2 Bearing Based On Temperature Corrections To The Proximity Probe Gap Voltages

measured journal position has been significantly improved by this simple probe gap temperature correction. The remaining deviations in measured radial position between both ends of the turbine may now be attributed to the presence of external loads, moments, or other influences acting upon the shaft.

Since the Inlet end #1 Bearing is adjacent to the accessory coupling, very little torque is transmitted during normal operation. Thus, the presence of external forces, and misalignment loads are minimal at the front end bearing. As previously observed, the measured positions agree very well with the theoretical calculations that consider only the load due to the applied journal weight.

However, at the gas turbine exhaust end bearing, the full power output from the turbine is transmitted across the load coupling. Dependent upon coupling type, alignment position and associated external forces, the actual journal location would probably deviate from the predicted eccentricity that was computed with only the journal weight. In fact, the reverse statement might also be appropriate. That is, since the exhaust end shaft centerline position agrees with the computed location, the influence of external forces may be considered to be minimal (i.e., indicative of a well-aligned Load coupling).

Overall, the eccentricity calculations at both ends of the turbine appear to be realistic and representative of average machine behavior. This correlation between the measured journal positions, and the computed equilibrium location is considered to be supportive of the accuracy of the analytical fluid film bearing calculations. Similar measurements and comparisons with calculated results may be performed at other speeds or different oil supply conditions. In most cases there should be a respectable correlation between the measured and the calculated shaft centerline position. This technique may also be used as a diagnostic tool. For example, if the measured shaft operating position is substantially different from the calculated position, the diagnostician should give strong consideration to the presence of internal or external shaft preloads.

FLUID FILM RADIAL BEARING CLEARANCE MEASUREMENTS

Assuming a proper bearing design, constant mechanical configuration, and the availability of a suitable lubricant at the required flow, temperature, and pressure — most of the variables shown in the bearing equations migrate towards constant values. The one parameter that generally does not remain constant is the bearing clearance. Although fluid film bearings are often touted as *lifetime* bearings, the reality is that these babbitt bearings are subject to physical damage whenever the oil film collapses. This could be caused by heavy shock loads on the bearings, loss of lubricant, or the detrimental long-term effects from excessive unbalance or misalignment. Many other physical mechanisms will also produce attrition in bearing babbitt thickness. In all cases, it is necessary to monitor journal position at each radial bearing with X-Y proximity probes, and to compare and trend this data within accurate bearing clearance diagrams.

The total diametrical clearance between the stationary bearing and the rotating shaft may appear to be an easily measurable value. Unfortunately, it is often quite difficult to accurately determine the true assembly clearance of a journal bearing. For a simple bearing configuration such as a plain circular bore, the clearance is the difference between the shaft diameter and the inner diameter bore of the bearing. Dependent on the length of the journal, the shaft diameter is normally measured at two or more axial locations (minimum fore and aft positions). Each axial location is typically measured at three to five different diameters. Besides providing the necessary average shaft diameter, this data checks for any gross diameter variations, or taper across the length of the journal. If the bearing is assembled (without the shaft), the inner diameter of the bearing should be measured in a manner similar to the shaft. The average diametrical clearance for a plain circular bore bearing is therefore:

$$C_{d_{plain}} = 1,000 \times (D_b - D_s) \qquad\qquad \textbf{(4-8)}$$

where: $C_{d\text{-}plain}$ = Average Diametrical Bearing Clearance (Mils)
D_b = Average Bearing Inner Diameter (Inches)
D_s = Average Shaft Outer Diameter (Inches)

If the shaft is resting solidly in the bottom half of the bearing, Plastigage may be placed on top of the shaft, and the upper bearing half installed, bolted down, unbolted, and then removed. Comparison of the deformed width of the Plastigage against the *Width Chart* supplied on each package of Plastigage will identify the diametrical clearance. Care should be taken to insure that the correct thickness of Plastigage be used for the bearing clearance measurement. The common colors and measurement ranges are as follows:

Green Plastigage ....1 to 3 Mils

Red Plastigage ....2 to 6 Mils

Blue Plastigage ....4 to 9 Mils

If these ranges are not appropriate, or if Plastigage is not available, then lead wire (or soft solder) may be used. For this measurement, the lead wire may be placed on top of the shaft, the top half of the bearing installed, bolted down, unbolted, and then removed. The thickness of the lead wire may then be measured with a 0 to 1 inch micrometer. The resultant thickness will correspond to the diametrical bearing clearance.

For long journal bearings, a strip of Plastigage or lead wire should be placed at either end of the bearing (i.e., fore and aft). Ideally, the clearances should be the same at both ends of the bearing. If variations do appear, the journal should be checked for a possible taper, and the bearing should be checked for wear or any evidence of a conical bore. In addressing this type of incongruity, it might be desirable to run a strip of Plastigage or lead wire axially along the top of the shaft. This would help to identify if the dissimilar bearing clearances vary uniformly along the length of the journal, or if some type of step change in clearance has occurred somewhere within the bearing.

For a fixed two lobe bearing such as an elliptical or lemon bore bearing, the previous techniques may be used to measure the vertical bearing clearance. Horizontal clearances are somewhat more difficult to determine. One approach is to measure across the assembled width of the bearing to determine the horizontal bearing diameter. Since these types of bearings are often provided with axial oil supply grooves at the splitline, it is important to measure the distance from *above* one groove to *below* the opposite groove as shown in Fig. 4-12. Two mea-

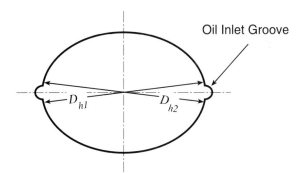

Fig. 4–12 Horizontal Diameter Of Elliptical Bearing With Axial Oil Grooves

surements are obtained at each end of the bearing, and they are averaged to determine the horizontal bearing inner diameter. Subtracting the shaft diameter, and multiplying by 1,000 will yield the average horizontal clearance in Mils. This should be compared with the vertical clearance to verify that a proper ratio exists. Typically, a large steam turbine will have a horizontal to vertical clearance ratio of 1.5:1, and an industrial gas turbine will generally be in the vicinity of 2:1. Another way to measure horizontal clearance on an elliptical bearing is to use feeler gages between the shaft and the bottom half of the bearing (top half removed). Measurements must be made on both sides of the journal, and their sum is a good approximation of the horizontal diametrical bearing clearance.

As bearing complexity increases, the techniques used to measure bearing clearances become more sophisticated. For stationary multi-lobe bearings, devices such as custom taper gauges, various multipoint measurement devices, or profile measurement machines may be used to determine bearing dimensions. Once again, when the minimum inner bearing dimensions are determined, subtraction of the shaft diameter yields the effective diametrical clearance. In essence, an expression similar to equation (4-8) may be used to determine the running clearance of fixed pad bearings. For pressure dam bearings, care should be taken to insure that the primary clearance measurement is based upon the dam lip, and not the pressure dam depth.

A further complication is introduced when tilt pad radial bearings are considered instead of fixed geometry bearings. In these assemblies, the clearance is influenced by pivoting of the bearing shoes. For instance, Fig. 4-13 depicts a five shoe bearing with an internal shaft journal. This is the same type of bearing that was used for computation of the Fig. 4-1 data. For illustration purposes, the journal in Fig. 4-13 is drastically undersized to allow an improved graphical visualization of the pad motion. If the five bearing pads are uniformly positioned

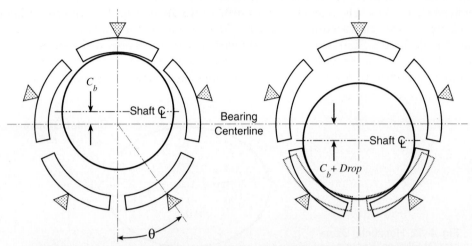

Fig. 4–13 Total Vertical Clearance In A Five Pad LBP Tilting Pad Journal Bearing

around the journal (i.e., no tilt), and the journal center is coincident with the bearing center — the distance from the journal surface to the center of each pad is equal to the radial bearing clearance C_b. If the journal is lifted vertically upward (left side of Fig. 4-13), the shaft will stop at the center of the upper pad. The total vertical travel from the bearing center will be equal to the radial bearing clearance C_b. If the shaft is now allowed to sink into the bottom half of the bearing, the condition shown on the right side of Fig. 4-13 will occur. In this diagram, the shaft will sink below the physical bearing clearance circle due to the rotation of the two bottom pads. The amount of the vertical shift will be equal to the radial bearing clearance C_b, plus an additional *Drop* due to the pad pivot.

It should be noted that this *Drop* only occurs in the static shaft condition depicted in Fig. 4-13. During machine operation, the diametrical bearing clearance is twice the radial bearing clearance C_b, and the static *Drop* does not occur. However, if bearing clearance is to be extracted from the static journal *Lift*, the pad *Drop* must be subtracted. Stated in another way, if a dial indicator is used to measure the total vertical *Lift* of the shaft within the bearing, the indicator reading will exceed the bearing clearance. Clearly, the *Lift* value must be reduced by the pad *Drop* in order to determine the actual diametrical bearing clearance.

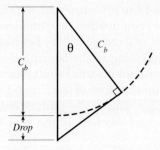

Fig. 4–14 Shaft Drop In A
Radial Tilting Pad Bearing
Due To Pad Pivot

A direct approach for determining bearing clearance from shaft *Lift* was presented in the 1994 papers by Nicholas[5], plus Zeidan and Paquette[6]. These authors used diagrams similar to Figs. 4-13 and 4-14 to explain this characteristic of radial tilting pad bearings. Specifically, within Fig. 4-14, the dotted line describes the radial bearing clearance C_b, and θ is the angle between the vertical centerline and the pad pivot point. The trigonometric relationship within this right triangle is as follows:

$$\cos\theta = \left(\frac{C_b}{C_b + Drop}\right)$$

or

$$C_b + Drop = \left(\frac{C_b}{\cos\theta}\right)$$

From Fig. 4-13, the total shaft *Lift* for this tilt pad bearing is the summation of the movement in the top half plus the bottom half of the assembly. This may be stated and combined with the last expression as:

$$Lift = (Top\ Half\ Clearance) + (Bottom\ Half\ Clearance)$$
$$Lift = (C_b) + (C_b + Drop)$$

$$Lift = (C_b) + \left(\frac{C_b}{\cos\theta}\right) = C_b \times \left(1 + \frac{1}{\cos\theta}\right)$$

The diametrical clearance C_d may be determined by solving the last equation for the radial bearing clearance C_b, and multiplying by 2 to yield:

$$\boxed{C_{d_{odd}} = 2 \times C_b = \frac{2 \times Lift}{\left(1 + \dfrac{1}{\cos\theta}\right)}} \qquad \textbf{(4-9)}$$

A sub-subscript $_{odd}$ has been added to the diametrical clearance solution of equation (4-9). This identifies the fact that this solution is for a tilt pad bearing with an odd number of pads. The assembly diagram shown in Fig. 4-13 represents a load between pad (LBP) configuration. However, if this bearing was turned upside down, a load on pad (LOP) arrangement would be depicted. Thus, equation (4-9) is correct for either a LOP or a LBP tilt pad bearing with an odd number of pads. For an even number of pads with LOP configuration, $\theta = 90°$,

[5] John C. Nicholas, "Tilting Pad Bearing Design," *Proceedings of the Twenty-Third Turbomachinery Symposium*, Turbomachinery Laboratory, Texas A&M University, College Station, Texas (September 1994), pp. 179-194.

[6] Fouad Y. Zeidan, and Donald J. Paquette, "Application of High Speed and High Performance Fluid Film Bearings In Rotating Machinery," *Proceedings of the Twenty-Third Turbomachinery Symposium*, Turbomachinery Laboratory, Texas A&M University, College Station, Texas (September 1994), pp. 209-233.

and the cosine of 90° is equal to 1. Thus, equation (4-9) may be simplified as:

$$C_{d_{LOP-even}} = \frac{2 \times Lift}{\left(1 + \dfrac{1}{\cos\theta}\right)} = \frac{2 \times Lift}{\left(1 + \dfrac{1}{1}\right)} = \frac{2 \times Lift}{2} = Lift \qquad (4\text{-}10)$$

The final common configuration for a radial tilt pad bearing would be a LBP with an even number of pads. This bearing type would display excessive clearance or *Drop* in both the upper, and the lower halves, and the total *Lift* would be:

$$Lift = (Top\ Half\ Clearance) + (Bottom\ Half\ Clearance)$$
$$Lift = (C_b + Drop) + (C_b + Drop)$$
$$Lift = \left(\frac{C_b}{\cos\theta}\right) + \left(\frac{C_b}{\cos\theta}\right) = \frac{2 \times C_b}{\cos\theta}$$

Once more, the diametrical clearance C_d may be determined by solving for the radial bearing clearance C_b, and multiplying by 2 to produce:

$$C_{d_{LBP-even}} = 2 \times C_b = \frac{2 \times Lift \times \cos\theta}{2} = Lift \times \cos\theta \qquad (4\text{-}11)$$

For symmetrical tilting pad bearings with an odd number of pads, the angle between the vertical centerline and the pad pivot point θ is fixed. The following values are typically used for this bearing angular dimension:

3 Pads ....θ = 60°
4 Pads ....θ = 45°
5 Pads ....θ = 36°
6 Pads ....θ = 30°
7 Pads ....θ = 25.7°
8 Pads ....θ = 22.5°

For purposes of simplification and easy reference, the developed equations (4-9) and (4-11) are combined with the standard bearing pad pivot angles, and a diametrical clearance to lift ratio calculated for each configuration. These results are summarized in Table 4-3, and the trivial case represented by equation (4-10)

Table 4–3 Tilt Pad Bearing Diametrical Clearance[a] Based On Shaft Lift Measurements

Load	3 Pads	4 Pads	5 Pads	6 Pads	7 Pads	8 Pads
LBP	0.667x*Lift*	0.707x*Lift*	0.894x*Lift*	0.866x*Lift*	0.948x*Lift*	0.924x*Lift*
LOP	0.667x*Lift*	*Lift*	0.894x*Lift*	*Lift*	0.948x*Lift*	*Lift*

[a]Diametrical Clearance = Numerical Factor x *Lift*

has also been included. If the shaft *Lift* in a radial tilt pad bearing is measured with a dial indicator or a proximity probe, the measured *Lift* may be converted to a diametrical clearance based upon the factors in Table 4-3 for the specific bearing configuration. For example, if a *Lift* of 14 Mils was measured, and the bearing was a five pad assembly, the diametrical bearing clearance is determined from Table 4-3 as follows:

$$C_{d_{odd}} = 0.894 \times Lift = 0.894 \times 16 \text{ Mils} = 14.3 \text{ Mils} \approx 14 \text{ Mils}$$

Table 4-3 may also be applied to the situation where a lift check of the journal within the bearing is not physically possible. In these cases, a separate mandrel may be used to measure the allowable motion within the bearing (i.e., the *Lift*). The shaft bearing journal diameter must be accurately measured as previously discussed, and a suitable mandrel cut on a precision lathe to exactly the same dimensions. Depending on the bearing configuration, the mandrel and assembled bearing may be mounted either vertically or horizontally. Typically, the mandrel is fixed, and the bearing housing is physically moved back and forth across the pads. For a fixed mandrel, a dial indicator is used to measure the overall motion of the bearing housing about the mandrel.

Conversely, if the bearing housing is mounted in some rigid fixture, the fabricated mandrel may be moved between pads, and the overall motion of the mandrel measured with the dial indicator. In either case, care must be exercised to insure that the stationary element remains fixed, and that the mandrel and bearing housing are collinear (i.e., the axial centerline of the mandrel is parallel to the bearing axial centerline). In addition, an accurate dial indicator reading to tenths of a Mil should be used for these measurements. The resultant *Shift* or *Lift* may then be multiplied by the appropriate geometric factor from Table 4-3 to determine the diametrical bearing clearance of the tilt pad bearing.

When checking tilt pad bearing clearances with a mandrel and an assembled bearing, it is desirable to check clearances in more than one direction. For instance, a four pad bearing should be checked at orthogonal diameters. That is, the up–down, and the left–right pads should be measured to verify that uniform clearances exist in both directions. For a five pad bearing, a total of three positions should be checked to insure that clearances are measured with respect to each pad. Obviously, the same concept may be extended to bearings with a larger number of pads.

This discussion of bearing *Lift* checks was predicated upon the assumption that a vertical *Lift* or *Shift* measurement could be made directly at the bearing. Obviously, this is an ideal condition. In many instances it is physically impossible to both mount a dial indicator next to the bearing housing, and position the indicator on top of the shaft. A more common condition is shown in Fig. 4-15. This diagram depicts a three stage rotor, horizontally supported between two radial journal bearings. In the sketch at the top of Fig. 4-15, a vertical dial indicator is located close to the coupling end of the rotor. The axial distance between the adjacent bearing and the indicator is identified as Z_{b-i}. The span or axial distance between bearing centerlines is specified as Z_b.

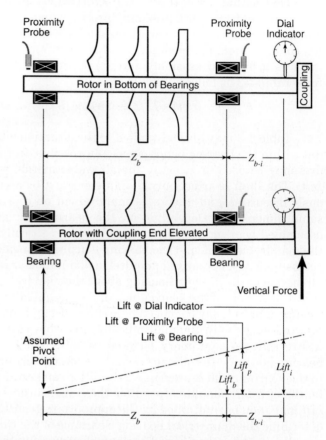

Fig. 4–15 Field Lift Check Of Coupling End Bearing On A Horizontal Rotor Mounted Between Two Journal Bearings

In addition, a vertical proximity probe is located on the outside of each bearing in Fig. 4-15. If an upward vertical force is applied at the coupling end of this rotor, the shaft will move towards the top of the coupling end bearing. It is assumed that the left end shaft remains reasonably stationary at the bottom of the outboard end bearing. This may be verified by another dial indicator at the outboard bearing, or the DC gap of the proximity probe at this bearing.

If no interference occurs, and if the rotor remains rigid, a straight lift should occur between the zero motion pivot point at the outboard end of the rotor, and the lifting point. This elevated rotor position is shown in the middle sketch of Fig. 4-15. Additionally, the vertical change in shaft centerline along the length of the rotor is presented at the bottom of Fig. 4-15. In this diagram, the vertical axis is expanded for clarity. It is noted that a series of similar right triangles are present in this ideal lift diagram. By simple proportion of these right triangles, the following expression evolves:

$$\frac{Lift_b}{Z_b} = \frac{Lift_i}{Z_b + Z_{b-i}}$$

This equation of proportions may be solved for the bearing $Lift_b$, as follows:

$$Lift_b = \frac{Z_b \times Lift_i}{Z_b + Z_{b-i}} = \frac{Lift_i}{\left(1 + \dfrac{Z_{b-i}}{Z_b}\right)} \tag{4-12}$$

Thus, the bearing shaft $Lift_b$ may be determined based upon a dial indicator $Lift_i$ obtained at a different axial location on the shaft. Next, the bearing diametrical clearance may be determined by applying the appropriate correction factor from Table 4-3 for a tilt pad bearing, or by equating the $Lift$ to the vertical clearance for a fixed pad bearing. If a vertical proximity probe is mounted adjacent to the bearing (e.g., Fig. 4-15), the change in DC gap voltages may also be used to determine the lift as shown in equation (4-13):

$$Lift_p = \left\{5.0\frac{\text{Mils}}{\text{Volt}}\right\} \times \{DC\,Gap_{rest} - DC\,Gap_{elevated}\}\ \text{Volts} \tag{4-13}$$

In many cases, the vertical shaft lift measured by a proximity probe $Lift_p$ may be very close to the shaft shift within the bearing $Lift_b$. This is due to the short axial distance between the bearing and the probe location (e.g., the configuration shown in Fig. 4-15). In fact, it is highly desirable to compare the corrected dial indicator readings from equation (4-12) with the differential probe gap readings computed with equation (4-13). This logic also applies to the opposite end of the machine. For instance, in Fig. 4-15, the outboard bearing is the assumed pivot point for lifting the rotor. At this location a vertical dial indicator should show zero motion as the shaft is lifted. In many cases, this *non-motion* is taken for granted, and an indicator is seldom positioned at the bearing opposite the unit subjected to lift check. However, proximity probes are often installed, and these probes should be monitored to verify that the shaft is not moving at the opposite end of the rotor. In practice, the DC gaps at this opposite bearing should not change as the shaft is raised.

On many installations, the machinery is equipped with the preferable combination of X-Y proximity probes. Often these transducers are mounted at ±45° from the vertical centerline, and a true vertical proximity probe does not exist. In this situation, the distance changes with respect to each probe should be vectorially summed to determine the overall shaft lift at that location. The specific steps are outlined in the following case history 7.

Case History 7: Expander Journal Bearing Clearance

A 5,000 HP hot gas expander operates at 8,016 RPM with 4.00 inch diameter journals mounted in tilting pad bearings. The bearings are four pad with a load between pad (LBP) configuration. This machine is equipped with X-Y proximity probes adjacent to each bearing at ±45° from vertical. A dial indicator was mounted 11 inches from the bearing, and the distance between bearing centerlines was measured to be 53 inches. The shaft was lifted with a pry bar, and the indicator showed a vertical lift of 10.5 Mils. The probe gap voltages measured during the lift are summarized in Table 4-4:

Table 4–4 Summary Of Probe Gap Voltages During Lift Check

Shaft Physical Position	Y-Axis	X-Axis
Probe Location	45° Left of Vertical	45° Right of Vertical
At Rest - Bottom of Bearing	-10.73 Volts DC	-9.69 Volts DC
Elevated - Top of Bearing	-9.96 Volts DC	-8.51 Volts DC

The lift at the bearing may be calculated based upon the external lift measurement, and the axial distances between bearings and indicator position. Using equation (4-12), it is easily determined that:

$$Lift_b = \frac{Lift_i}{\left(1 + \frac{Z_{b-i}}{Z_b}\right)} = \frac{10.5 \text{ Mils}}{\left(1 + \frac{11 \text{ Inches}}{53 \text{ Inches}}\right)} = \frac{10.5 \text{ Mils}}{(1 + 0.208)} = 8.69 \text{ Mils}$$

This mechanical result should now be compared with the lift measurements obtained with the shaft proximity probes. Applying equation (4-13) for each transducer, the shaft shift detected by each probe may be computed in the following manner:

$$Lift_{p_{Y-Axis}} = \left\{5.0\frac{\text{Mils}}{\text{Volt}}\right\} \times \{(-10.73) - (-9.69)\} \text{ Volts} = -5.20 \text{ Mils}$$

and

$$Lift_{p_{X-Axis}} = \left\{5.0\frac{\text{Mils}}{\text{Volt}}\right\} \times \{(-9.96) - (-8.51)\} \text{ Volts} = -7.25 \text{ Mils}$$

The negative signs indicate that the shaft movement was towards the probes. If a standard coordinate system is used, the true horizontal axis would be at 0°, and true vertical would be at 90°. Within this coordinate system the X-Axis probe would be located at 45°, and the Y-Axis transducer at 135°. If the measured shifts are considered as vectors towards each probe, the overall motion may be expressed as the following two vectors:

$$\overrightarrow{V}_y = A\angle\alpha = 5.20\ \text{Mils}\angle135°$$

$$\overrightarrow{V}_x = B\angle\beta = 7.25\ \text{Mils}\angle45°$$

The sum of horizontal vector components are determined with (2-31):

$$V_{add_{horiz}} = A\times\cos\alpha + B\times\cos\beta$$

$$V_{add_{horiz}} = 5.20\times\cos135° + 7.25\times\cos45° = -3.68 + 5.13 = 1.45\ \text{Mils}$$

Similarly, the sum of vertical vector components are computed with (2-32):

$$V_{add_{vert}} = A\times\sin\alpha + B\times\sin\beta$$

$$V_{add_{vert}} = 5.20\times\sin135° + 7.25\times\sin45° = 3.68 + 5.13 = 8.81\ \text{Mils}$$

From these shaft position changes it is noted that the shaft did not come straight up in the bearing. The horizontal shift of nominally 1.5 Mils indicates that the shaft moved sideways. This is not a surprising result since the pry bar used for the lift was not completely level, and some horizontal force was probably applied to the rotor. For bearing clearance purposes, the vertical lift of 8.8 Mils should be used for further calculations. However, before addressing the bearing clearances, it is desirable to conclude the vector addition computations of the shifts measured by the proximity probes. If equation (2-33) is used to determine the combined magnitude shift, the following result is obtained:

$$V_{add} = \sqrt{(V_{add_{horiz}})^2 + (V_{add_{vert}})^2} = \sqrt{(1.45)^2 + (8.81)^2} = \sqrt{79.72} = 8.93\ \text{Mils}$$

Note that the vector sum of 8.9 Mils is very close to the vertical shift of 8.8 Mils determined in the previous group of calculations. Finally, the angle of the shaft lift is determined from equation (2-34) as:

$$\phi_{add} = \text{atan}\left(\frac{V_{add_{vert}}}{V_{add_{horiz}}}\right) = \text{atan}\left(\frac{8.81}{1.45}\right) = \text{atan}(6.076) = 80.6° \approx 81°$$

Ideally, the lift angle should be 90°. Since some horizontal shift was imposed, a slight variation in angles does occur. If the lift angle is between 75° and 105° the total vertical lift error will be less than 4%. In many cases, it is more convenient to add the shift vectors on a handheld calculator rather than go through the detail required in the previously outlined steps. For this application, the diagnostician should make sure that the calculator is capable of easily performing vector addition (e.g., HP 48SX).

The vertical lift readings based upon the dial indicator should be close to the values measured by the proximity probes (assuming that the probes are mounted next to the bearing). If the deviation between the two values is greater than approximately 5 to 10% — then there is something wrong, and the entire measurement scenario should be re-examined. In this case, the calculated lift from the X-Y probes (8.8 Mils) should be compared with the mechanical lift as measured with the dial indicator (8.7 Mils). Since the probes are mounted outboard of the bearing, the indicated vertical lift from the probes is slightly greater than the mechanical lift corrected to the center of the bearing. It is reasonable to conclude that the vertical bearing lift is equal to 8.7 Mils. Based on this information, Table 4-3 or equation (4-11) may be used to determine the vertical diametrical bearing clearance as follows:

$$C_{d_{LBP-even}} = Lift \times \cos\theta = 8.7 \times \cos 45° = 8.7 \times 0.707 = 6.2 \text{ Mils}$$

The final step is to verify the general validity of this measurement. Typically, a bearing clearance ratio (*BCR*) is calculated as follows:

$$BCR = \frac{Bearing\ Diametrical\ Clearance \text{(Mils)}}{Journal\ Diameter \text{(Inches)}} \qquad \textbf{(4-14)}$$

Since this expander had 4.00 inch journals, the *BCR* is simply:

$$BCR = \frac{Bearing\ Diametrical\ Clearance \text{(Mils)}}{Journal\ Diameter \text{(Inches)}} = \frac{6.2 \text{Mils}}{4.00 \text{Inches}} = 1.6 \text{ Mils/Inch}$$

A clearance to diameter ratio of 1.6 makes good sense for this bearing configuration in a horizontal machine. Table 4-5 describes the general behavior of key parameters as the *BCR* is varied. Most bearing designers agree that a *BCR* of 1.0 is generally on the tight side. Small bearing clearances result in high oil film stiffness, and this is accompanied by low shaft vibration, and potentially high bearing temperature. If the *BCR* is increased to 2.0, the stiffness and damping will decrease, vibration will increase, and the bearing would probably run cooler. In addition, the machine with larger bearing clearances would be more susceptible to a variety of instability mechanisms. In most horizontal industrial machines, the *BCR* is seldom less than 1.0, and it generally does not exceed 2.0. In specialized applications, with exotic mechanical designs and metallurgy, these traditional limits may be extended. However, in most cases, the *BCR* runs between 1.0 and 2.0.

On large vertical machines, the radial bearing loads are low, and the weight of the rotating element is supported by a massive thrust bearing that is usually located at the top of the machine. On these units, the radial bearing clearances

Table 4-5 General Trends Of Key Bearing Parameters With Variations In Bearing Clearance Ratio (BCR) For Horizontal Machines Mounted In Fluid Film Bearing

Bearing Clearance Ratio (BCR)	Oil Film Stiffness	Oil Film Damping	Shaft Vibration	Bearing Temperature
1.0 Mil/Inch	Increases	Increases	Decreases	Increases
1.5 Mils/Inch	*Nominal*	*Nominal*	*Nominal*	*Nominal*
2.0 Mils/Inch	Decreases	Decreases	Increases	Decreases

are much tighter, and Table 4-5 is not applicable. For these vertical machines, the bearings are basically a flooded oil bath, with diametrical clearances that generally vary between 10 and 20 Mils (0.010 and 0.020 inches). These are typically referred to as *guide bearings*, and their fundamental function is to keep the shaft running in a vertical position. The clearance of these bearings are normally obtained by physically swinging the rotor back and fourth in orthogonal directions (e.g., North-South and East-West). In this case, the upper thrust bearing becomes the pivot point, and bearing clearance is measured with dial indicators at each bearing. For vertical machines equipped with tilt pad bearings, the individual pads are often radially adjustable in position to provide the capability to change the overall bearing clearance. On fixed geometry bearings, the proper clearance has to be built into the bearing based upon actual diameter.

In any lift measurement on assembled machines, consideration must be given to physical configurations or conditions that could cause measurement errors. For instance, close clearance seals, or a long balance piston might restrict the rotor lift, and appear as reduced bearing clearances. On gear boxes, if an element is partially supported by a mating gear, the lift check will be erroneous since the starting point will not be at the bottom of the bearing. Similarly, installed couplings, governor drive gears, and engaged turning gears will all inhibit the shaft lift, and may be incorrectly interpreted as reduced bearing clearances.

Conversely, excessive clearances in other machinery parts associated with the bearings may look like large clearances. Loose hold down bolts, or housing attachment bolts can produce inordinate shaft lift readings. On electric machines such as motors or generators, the bearings are normally insulated with some type of non-conducting material. This electrical insulation isolates the rotor voltage from passing to ground through the machine bearings. These insulating blocks are usually installed with zero clearance. However, clearances can expand with time and excessive vibration, with an overall reduction in support stiffness. The same argument applies to the fit between the bearing assembly and the housing. Although wide variations may be encountered for this dimension, most machines operate somewhere between an interference fit, or crush, of 1 or 2 Mils; and a clearance of 1 or 2 Mils. Clearly, excessive crush can distort the bearing assembly resulting in premature failure, whereas excessive clearance will reduce the support stiffness. This stiffness reduction may allow a rotor resonance that normally resides above operating speed to creep back into the operating speed domain. When this occurs, shaft vibration increases, and the propensity towards early failure of the bearing increases.

It is generally advisable to refer to the OEM specifications for guidance in establishing the proper clearances between the bearing assembly and the bearing cap. If this information is not available, then a zero to 1 Mil clearance should be used as a reasonable starting point. Determination of this clearance may be difficult due to the possibility of a zero clearance. If Plastigage or lead wire in installed between machine parts that have essentially no clearance, the measurement media becomes smeared, and essentially useless. The solution to this situation resides in providing an initial, or reference, clearance at the split line.

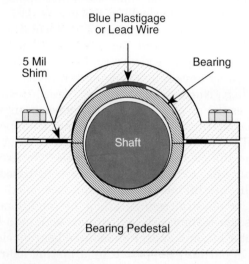

Fig. 4-16 Typical Clearance Measurement Between Bearing Liner And Bearing Housing

For example, a 5 Mil shim has been installed at the housing split line shown in Fig. 4-16. This shim elevates the entire upper half of the bearing cap by 5 Mils, and allows the use of Blue Plastigage (4 to 9 Mil range) to measure the remaining clearance. If the Plastigage shows a 4 Mil clearance, then subtraction of the 5 Mil shim reveals an interference fit of 1 Mil. Conversely, if the Plastigage indicates a 6 Mil clearance, then subtraction of the 5 Mil split line shim results in a bearing to cap clearance of 1 Mil. For clearances that exceed the measurement range available from Plastigage, lead wire may be used. In either case, when the measurement checks are completed, the Plastigage (or lead wire) remnants, plus the split line shims must be removed before final assembly of the housing.

If excessive cap to bearing clearances are encountered, the best permanent solution is to re-machine the offending stationary element(s) to restore proper clearances. In some cases, this is not a viable option due to production or maintenance demands. In this situation, a temporary stainless steel shim may be installed between the cap and the bearing to tighten up the assembly. If this correction technique is used, then the machine history records should clearly indicate the installation of this shim.

In all cases, the success of the lift check is highly dependent upon the method used to mechanically lift the shaft. For light rotors, a simple pry bar is quite adequate for this task. For heavier rotors, a screw jack, or an overhead chain hoist might be used. On very heavy rotors, a hydraulic jack may be necessary to lift the rotor. It must be recognized that this is a potentially dangerous practice. Rotors have been permanently bent, and bearing housings have been cracked or broken due to the aggressive use of a hydraulic jack. This type of lift should be performed carefully, and with full knowledge of the expected clearances. Multiple dial indicators might be installed axially on the shaft to verify that a linear (straight line) lift is occurring. This information would help to minimize potentially bending the shaft. It might also be desirable to mount a separate dial indicator on the outboard end of the bearing housing. This vertical indicator would be used to reveal any tendency towards a vertical lift of the housing. This information would help to minimize any damage to the bearing housing from the vertical hydraulic jack under the shaft.

BEARING SUPPORTS — MEASUREMENTS AND CALCULATIONS

The previously discussed bearing characteristics are associated with the properties of the oil film between the rotating shaft, and stationary bearings of different configurations. This is an acceptable description of the rotor support system if the bearings are rigidly supported. Industrial machines with heavy cases, and light rotating elements fall within this category. Barrel compressors with internal bearings, rigid gear boxes, high pressure pumps, and many older pieces of equipment operate with structural stiffness that are substantially greater than the oil film stiffness.

However, this is not the case for many other machines that have flexible supports and/or foundations. Units such as induced draft or forced draft fans, steam or gas turbines, horizontally split centrifugal compressors, and pumps with external bearings are just a few examples of machines that operate with flexible supports. For these types of machines the remainder of the mechanical system must be included. In a general case, the effective support stiffness for a typical rotor on a flexible support may be defined by equation (4-15) that describes the relationship as a group of springs in series:

$$\frac{1}{K_{eff}} = \frac{1}{K_{oil}} + \frac{1}{K_{hsg}} + \frac{1}{K_{base}} + \frac{1}{K_{fnd}} + \frac{1}{etc} \qquad \textbf{(4-15)}$$

where: K_{eff} = Effective Rotor Support Stiffness (Pounds/Inch)
K_{oil} = Oil Film (Bearing) Support Stiffness (Pounds/Inch)
K_{hsg} = Bearing Housing Support Stiffness (Pounds/Inch)
K_{base} = Baseplate Support Stiffness (Pounds/Inch)
K_{fnd} = Foundation Support Stiffness (Pounds/Inch)

This expression will be subjected to substantial modification if the support structure is in a resonant condition, or if the support is highly flexible. However, these are rare occurrences, and the above equation (4-15) is considered to be generally representative of the normal rotor support parameters.

Quantification of the structural support terms in equation (4-15) is a formidable technical feat. The calculation of these individual stiffness terms is difficult at best, and in some cases it is virtually impossible. The most reasonable approach for determination of the support coefficients is a direct measurement of the dynamic stiffness of the support structure. This measurement requires the application of a defined force to the structure, and the determination of the resultant movement. In the simplest case, a dial indicator is used to measure the displacement in thousands of an inch, and a calibrated hydraulic jack provides the force. Division of the applied force by the total movement yields a static stiffness in Pounds/Inch. This is a *zero frequency* technique that often fails to provide the correct structural stiffness since the characteristics vary with frequency.

It is possible to measure variable frequency structural stiffness by exciting the system with an appropriate device, and measuring the response with a vibration transducer. A frequency response function (FRF) measurement (a.k.a., transfer function) may be performed between the signal emitted by a force transducer, and the resultant displacement motion signal. This FRF measurement should include the amplitude relationship between force and motion at each frequency bin, plus phase and coherence information. The force applied to the structure would typically be measured in Pounds, and the structural response would be measured in Inches. The vibration or motion measurement could be made with a proximity probe mounted on an isolated stand, or with a seismic transducer that is integrated to displacement. In most cases, the field motion measurements are obtained with an accelerometer, and this signal is double integrated to obtain casing displacement. The engineering units for the frequency response function are Pounds/Inch, and this measurement is commonly referred to as Dynamic Stiffness.

The device used to excite the structure may vary from an electromechanical shaker to an impact hammer. The use of an electromechanical shaker provides a highly controllable excitation source, whereas an impact hammer is easily applied in a variety of situations. The physical installation of any shaker is often hampered by limited access to the assembled machinery bearing housings. In some cases, an electromechanical shaker with a stinger attached may be used to reach specific mechanical elements. In other situations, the selection of an impact hammer provides the necessary size and flexibility to excite a machine bearing housing with an acceptable and definable impact force. A typical arrangement for measuring the horizontal stiffness of a bearing housing with an

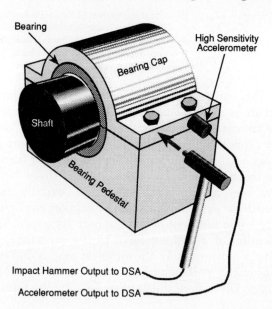

Fig. 4–17 Typical Test Arrangement For Bearing Housing Horizontal Impact Test

impact hammer is shown in Fig. 4-17.

Manual impact hammers come in various sizes for different testing applications. For instance, small hammers weigh between 1 and 2 Grams, and exhibit a frequency limit of nominally 900,000 CPM (15 KHz). These miniature impact hammers are used for static testing of items such as turbine blades. At the other end of the scale a 12 pound sledge hammer, or an instrumented battering ram may be used for low frequency tests on large structures such as foundations or buildings. For bearing housing measurements, a typical impact hammer weighs between 0.3 and 3 Pounds, and it is capable of producing a concentrated 5,000 pound force upon the test element. The dynamic force produced by the hammer is generally measured with an integral piezoelectric force transducer. Frequency response characteristics typically vary from 300 CPM (5 Hz), to a usable maximum frequency of about 60,000 CPM (1,000 Hz). Thus, the dynamic characteristics of a small to medium sized force hammer adequately cover the operating speed range of most machines.

During structural impact tests, the casing response is usually measured with an accelerometer attached to the bearing cap. The accelerometer should have a frequency range that is compatible with the force transducer. In addition, the accelerometer signal must be double integrated to convert acceleration to casing displacement. This double integration may be performed in an external analog device, or by application of wave form math in the DSA. In either case, the final FRF measurement between the applied force (Pounds) and the resultant displacement response (Inches) yields an equivalent support stiffness for the bearing housing (Pounds/Inch). In all cases, the validity of the frequency response data is checked with the coherence function, and the relative phase between signals should be examined.

For measurements of structural natural frequencies, the test setup is identical to the dynamic stiffness measurements, but the transducers are reversed. In this type of test, a FRF is performed between the measured acceleration divided by the input force. Double integration of the acceleration signal is neither required, nor desirable — and the FRF output units are typically G's/Pound. This type of data is commonly referred to as Inertance, and it should be performed whenever a structural resonance is suspected.

Case History 8: Measured Steam Turbine Bearing Housing Stiffness

For demonstration purposes, a typical data set is presented in Fig. 4-18. This information was obtained with a field instrumentation setup identical to Fig. 4-17. This test was performed on the exhaust bearing housing of an 8,000 HP steam turbine that normally operates at 8,520 RPM. The data was acquired with a 3 pound impact hammer, and a high sensitivity accelerometer mounted in a horizontal plane. Both the impact hammer and the accelerometer were directly connected to an HP-35665A Dynamic Signal Analyzer (DSA). The power source within the DSA was used to drive both piezoelectric transducers, and the resultant data was stored on a floppy disk. The data was later examined on an HP-35670A, and the data displayed in Fig. 4-18 committed to hard copy format with an HP-5L LaserJet printer. The FRF yields the dynamic stiffness plot at the bottom of Fig. 4-18. Since this data covers a wide amplitude range, a log scale was used for the stiffness. It is noted that a reasonably flat region exists between 4,800 and 13,000 CPM. At the normal operating frequency of 8,520 RPM the FRF reveals a dynamic stiffness value of 1,210,000 Pounds/Inch. This is judged to be a realistic value for the heavy cast steel bearing housing. The FRF also shows a substantial drop in stiffness at frequencies of 660 and 16,560 CPM.

If the center phase plot in Fig. 4-18 is examined, the large phase shift at 16,560 CPM might be interpreted as a structural resonance. However, when the coherence plot at the top of Fig. 4-18 is considered, it is evident that coherence between force and motion signals has dropped to below 0.2 at 16,560 CPM. This indicates that the FRF data is not valid, and the significance of the change at 16,560 CPM should be removed from further consideration.

At the turbine speed of 8,520 RPM the computed coherence was 0.97. Generally, coherence values greater than 0.9 are indicative of acceptable FRF data. Hence, the information in the vicinity of the turbine running speed is considered

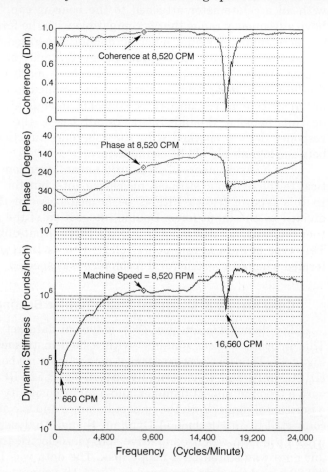

Fig. 4–18 Frequency Response Function (FRF) Of Steam Turbine Bearing Housing Horizontal Dynamic Stiffness

to be excellent data. When coherence drops to levels below 0.9, the FRF data should be cautiously applied. If coherence drops below 0.7, the FRF data should generally be ignored.

The data array shown in Fig. 4-18 is easily acquired, and rapidly processed. From the previous discussion it is summarized that the dynamic stiffness at turbine speed of 8,520 RPM was obtained directly from the FRF plot, and verified by the coherence. The validity of the amplitude and phase change at 16,560 CPM was found to be highly questionable due to the low coherence. However, the drop in FRF amplitudes at low frequencies was not fully explained. For an improved understanding of this behavior, and examination of the component force and displacement signals is required. This supplemental data is presented in Fig. 4-19 over the same frequency range used for the FRF data in Fig. 4-18.

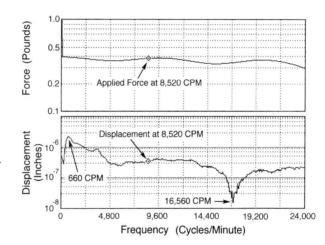

Fig. 4–19 Force And Displacement Data Used to Develop Steam Turbine Frequency Response Function (FRF) Dynamic Stiffness Plot

The upper diagram in Fig. 4-19 is the force (in Pounds) applied across the frequency domain of 0 to 24,000 CPM. If this same data was viewed in the time domain, a sharp initial pulse would be observed. Within the frequency domain, this pulse provides a reasonably uniform excitation across the selected analysis bandwidth. Hence, it may be properly concluded that the low frequency drop off of the FRF data is not due to any significant variations in the applied force.

However, the measured displacement presented in the bottom diagram of Fig. 4-19 reveals a large increase in the response at 660 CPM. There might be a tendency to consider the 660 peak as a resonance, but this conclusion is not supported by the differential phase data of Fig. 4-18. Furthermore, structural resonances have a narrow bandwidth, and the 660 CPM peak show in Fig. 4-18 does not display this characteristic. In all probability, the 660 CPM peak is due to a measurement anomaly. More specifically, an accelerometer was used to make the bearing housing response measurements. The acceleration signal was double integrated to obtain displacement. This conversion is accomplished within the DSA by dividing the acceleration signal by frequency squared.

At the boundary condition of zero frequency, the integrated displacement would have a value of infinity. This does not appear in the data because there is virtually no measurable acceleration output until the vibration transducer becomes active around 180 CPM (3 Hz). However, it is a fundamental fact that displacement amplitudes at low frequencies may be abnormally amplified due to the double integration process. This is further complicated by any noise in the acceleration signal that might also be erroneously amplified during double integration. Hence, the mechanical significance of the 660 CPM peak is eliminated. In addition, the displacement drop at 16,560 CPM is not meaningful information due to the previously mentioned low coherence at this frequency.

Since the dynamic stiffness FRF consists of force divided by displacement, the increased displacement at low frequencies produces a reduction in the dynamic stiffness. This is common behavior in all of these measurements, and low frequency data is generally ignored. This is acceptable since the low speed stiffness is considerably less important than the housing stiffness within the operating speed domain.

From a measurement standpoint, it should be mentioned that the casing displacement resulting from an impact hammer test is very small. For example, the peak displacement at 8,520 CPM on Fig. 4-19 is only 3.36×10^{-7} Inches,$_{o-p}$. This value is equivalent to 0.000672 Mils,$_{p-p}$. Using equation (2-21), this displacement converts to an accelerometer output of 0.000693 G's,$_{o-p}$ at the machine frequency. Fortunately, a high sensitivity accelerometer was used for this data, and the scale factor of 10,000 mv/G resulted in a signal strength of only 6.93 millivolts,$_{o-p}$. If the DSA is set for a full scale range of 1.0 volt,$_{o-p}$, the acceleration signal would appear at -43 dB. Although this is a small voltage, it is still within the range of most analyzers.

As an alternate scenario, if the response accelerometer had a scale factor of only 100 mv/G, the electrical signal would be proportionally reduced. In this situation, the analyzer would have to accommodate a low level signal of -83 dB. Unfortunately, many instruments do not have an adequate dynamic range to handle this variation in amplitudes. In all cases, it is recommended that a high sensitivity accelerometer (1,000 or 10,000 mv/G) be employed for this type of measurement. In addition, the data should be processed with a DSA that has a sufficiently large dynamic range (e.g., HP-35670A).

If an electromechanical shaker was substituted for the impact hammer, the applied force would be greater, and the resultant casing motion measured by the accelerometer would also increase. Thus, the measurement problems would diminish. However, the larger signal amplitudes must be compared with the potential difficulty and time required to properly mount an electromechanical shaker in the field. Regardless of the excitation source, the data processing and examination techniques are essentially the same. Overall, the user must be fully aware that this type of measurement is subject to a variety of errors, and all aspects of the FRF must be validated.

In retrospect, the primary objective of this exercise is directed at a measurement of bearing housing dynamic stiffness. As discussed, the bottom plot of Fig. 4-18 depicts the variation of this parameter with frequency. In many cases it is desirable to develop an equation that describes this behavior. By using appropriate curve fitting software within the DSA, a suitable polynomial equation may be defined that relates frequency to stiffness. This is an important consideration during the accurate modeling of rotating equipment as discussed in chapter 5. However, there are situations when this type software is not available, and another approach must be used to develop the characteristic equation.

Case History 9: Measured Gas Turbine Bearing Housing Stiffness

The data presented in Fig. 4-20 was acquired on the inlet bearing housing of a natural gas fired 40,000 HP gas turbine. This machine operates at 5,300 RPM, and casing stiffness information was required to enhance the accuracy of the analytical rotor model. Data was obtained in the vertical direction ($_{Y-Y}$) with a vertical accelerometer, and a vertical impact hammer excitation. Information was also acquired in the horizontal plane ($_{X-X}$) with a horizontal accelerometer, and a horizontal impact. For consistency with previous examples, the horizontal FRF data is shown in Fig. 4-20. This information was derived with a three pound

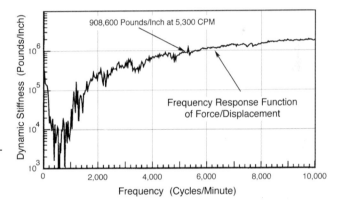

Fig. 4–20 Horizontal Stiffness Measurement Of Gas Turbine Inlet End #1 Bearing Housing

force hammer directly connected to an HP-3560A portable analyzer. Casing motion was measured with a high sensitivity accelerometer that was subjected to analog double integration prior to DSA processing. A comparison of this FRF with the previous example reveals a somewhat jagged curve in Fig. 4-20. After examining the various supplemental plots, it was concluded that the deviations are attributed to the analog double integration. Fortunately, coherence was above 0.9 at all frequencies above 1,000 CPM. Hence, the FRF data was considered to be acceptable, but a polynomial equation of this FRF was still required. Since the HP-3560A does not have curve fitting capabilities, the FRF data was exported to a Microsoft® Excel spreadsheet, and dynamic stiffness values were listed at 30 CPM intervals. This data was then subjected to a sixth degree polynomial curve between frequencies of 1,000 and 6,000 CPM, and the following characteristic equation was generated:

$$K_{S_{xx}} = 4,317 \times \left(\frac{Cpm}{1,000}\right)^6 - 86,566 \times \left(\frac{Cpm}{1,000}\right)^5 + 678,621 \times \left(\frac{Cpm}{1,000}\right)^4$$

$$-2.6269 \times 10^6 \times \left(\frac{Cpm}{1,000}\right)^3 + 5.2612 \times (Cpm)^2 - 5,165 \times (Cpm) + 2.448 \times 10^6$$

Within the specified frequency range, this polynomial expression may be used to calculate the horizontal structural stiffness of this inlet end #1 gas turbine bearing housing as a function of speed. As an example, the horizontal stiffness at the machine running speed of 5,300 RPM may be computed in the following manner.

$$K_{S_{xx}} = 4,317 \times \left(\frac{5,300}{1,000}\right)^6 - 86,566 \times \left(\frac{5,300}{1,000}\right)^5 + 678,621 \times \left(\frac{5,300}{1,000}\right)^4$$

$$-2.6269 \times 10^6 \times \left(\frac{5,300}{1,000}\right)^3 + 5.2612 \times (5,300)^2 - 5,165 \times (5,300) + 2.448 \times 10^6$$

$$K_{S_{xx}} = \{95.684 - 362.015 + 535.464 - 391.085 + 147.787 - 27.347 + 2.448\} \times 10^6$$

$$K_{S_{xx}} = 0.909 \times 10^6 = 909,000 \text{ Pounds/Inch}$$

This calculated value of 909,000 Pounds/Inch is consistent with the FRF measurement in Fig. 4-20. This type of curve fitting may also be used to determine the oil film stiffness as a function of rotating speed. The calculated stiffness versus speed curves (e.g., Fig. 4-1) may be converted to a polynomial equation, and the resultant expression used within the rotor response programs. In most cases, a third or fourth degree polynomial is sufficient to describe the stiffness curves, but the use of fifth or sixth degree equations are common.

The oil film stiffness, and various structural stiffness are combined in a reciprocal manner as described in equation (4-15). However, the field FRF tests essentially combine the stationary structural elements into a single housing stiffness. Thus, it is reasonable to simplify the overall or effective rotor support equation into the following common format:

$$\boxed{\frac{1}{K_{eff}} = \frac{1}{K_{oil}} + \frac{1}{K_{hsg}}}\qquad \text{(4-16)}$$

In this expression, the oil film stiffness is identical to the previous discussion, and the housing stiffness reflects the results of the field impact or shaker test. That is, the measured housing stiffness incorporates the flexibility of all of the bearing housing support elements. From a practical side, it is reasonable to assume that baseplate and foundation stiffness values are much greater than the bearing housing stiffness. Hence, the lower bearing housing stiffness is the dominant or controlling structural stiffness. This simplification provides the familiar format for the effective stiffness for two springs in series. It is understandable that the effective support stiffness upon the rotor is always lower than either the oil film stiffness, or the housing support stiffness.

The diagnostician should always be cautious in the application and interpretation of structural dynamic stiffness measurements. The excitation source, the elements influenced by the excitation source, the specific measurement points, method of accelerometer double integration, plus the final signal windowing and processing within the DSA can all contribute to substantial errors. These informative measurements should be carefully performed, and thoroughly understood. As previously mentioned, these FRF housing measurements are combined with the speed dependent oil film coefficients. Both of these coefficients are used within the stability and damped critical speed analysis, plus the forced response analysis. This combination of parameters provides a substantially improved model of the rotor support system across the operating range.

BEARING HOUSING DAMPING

The foregoing discussion has centered on the determination of bearing housing support stiffness. Although stiffness is a major consideration, it is not fully definitive of the complete mechanical system. In chapter 2 of this text, a minimum mechanical system description consisted of mass and damping in addition to the stiffness. In most cases, the mass may be considered as the weight of the bearing housing assembly. However, the determination of structural damping is considerably less defined.

Technical papers by Gunter and Kirk[7], Barrett and Nicholas[8], and Nicholas, Whalen and Franklin[9] generally include some type of support damping. The most common form recommended for structural damping consists of 10% of the critical damping for the housing or pedestal. By expanding equation (2-62) and dividing by 10, the structural damping for the bearing housing may be estimated with the following expression:

$$C_{hsg} \approx 0.2 \times \sqrt{\frac{K_{hsg} \times W_{hsg}}{G}} \qquad (4\text{-}17)$$

where: C_{hsg} = Housing Damping (Pounds-Seconds / Inch)
K_{hsg} = Housing Stiffness (Pounds / Inch)
W_{hsg} = Housing Weight (Pounds)
G = Acceleration of Gravity (386.1 Inches / Second^2)

---

[7] Edgar J. Gunter, and R.G. Kirk, "The Effect of Support Flexibility and Damping on the Dynamic Response of a Single Mass Flexible Rotor in Elastic Bearings." Report No. ME-4040-106-72U, University of Virginia, 1972.

[8] L.E Barrett, and John C. Nicholas, "The Effect of Bearing Support Flexibility on Critical Speed Prediction," *ASLE Transactions*, 1984 Joint Lubrication Conference, San Diego, California, February 1984 (revised June, 1984).

[9] John C. Nicholas, John K. Whalen, and Sean D. Franklin, "Improving Critical Speed Calculations Using Flexible Bearing Support FRF Compliance Data," *Proceedings of the Fifteenth Turbomachinery Symposium*, Turbomachinery Laboratory, Texas A&M University, College Station, Texas (November 1986), pp. 69-78.

The housing stiffness in (4-17) is the measured value from the FRF measurements at a particular speed. In practice, the support stiffness polynomial equations are modified in accordance with equation (4-17), and plots of housing damping versus speed are generated. Alternatively, structural damping plots may be subjected to a polynomial curve fit to develop the equations for support damping as a function of speed. It should be noted that the housing damping values are quite small when compared to the oil film characteristics. In one case, the structural damping was in the range of 200 to 300 Pounds-Seconds/Inch, whereas the oil film damping coefficients were calculated to vary between 6,000 to 20,000 Pounds-Seconds/Inch. Although the support damping is a minor correction, inclusion of this parameter should improve the analytical model accuracy.

FLUID FILM THRUST BEARINGS

In addition to the radial journal bearings, most industrial machines contain some type of thrust bearing to restrain axial motion of the shaft. As the journal bearings support the static rotor weight plus dynamic loads — the thrust bearings must contain constant thrust loads plus various axial dynamic forces. The constant thrust loads are primarily due to differential pressure across wheels in fluid handling machines (turbines and compressors), or axial components of gear contact forces. Additional dynamic axial forces are caused by items such as cocked wheels, bent rotors, or misaligned shafts. It should be noted that electric machines such as motors or generators do not contain thrust bearings. The magnetic forces across the air gap center the rotor within the stator.

In addition to restraining axial forces, the thrust bearing must also hold the rotor in a fixed axial position with respect to the stationary casing. This is necessary from a machine efficiency standpoint. For instance, in a steam turbine the critical clearance between the first stage wheel and the nozzle block is maintained by the thrust bearing. Similarly, the axial position between wheels and diaphragms in a centrifugal compressor is maintained by the thrust bearing assembly. It should be recognized that axial clearances in most machines are relatively small, and the thrust bearing is required to maintain the axial position, and prevent catastrophic contact between the rotating and the stationary parts.

In all cases, the thrust bearing assembly must fulfill this requirement for maintaining axial *position*, plus axial *float* within the thrust bearing. These two basic requirements demand two adjustments to the thrust bearing assembly. In many machines, solid shims are installed behind the thrust shoes to establish these critical axial dimensions. These shims are precision ground to tight tolerances, and they are normally segmented into two 180° sections. Typically, the axial position shim located behind the active thrust shoes is ground to establish the axial location of the rotor. Then the shims behind the inactive shoes are ground to establish the total thrust float within the bearing. Typical installations of thrust shims are shown in Figs. 4-21 and 4-22. It should also be mentioned that other mechanical schemes are used on process machines. However, in virtually all applications, one adjustment must be available for the rotor *position*, and a second adjustment must be available to set the thrust *float*.

On journal bearings, maximum shaft surface velocities are often limited to 200 or 250 feet per second. By comparison, the larger diameter thrust collars may exhibit peripheral speeds in the vicinity of 450 to 500 feet per second. These higher speeds demand proper lubrication to cool the higher temperatures associated with the increased surface velocities. In most machines, thrust bearings consist of segmented floating pads that are often used in conjunction with directed lubrication between the pads (e.g. Kingsbury type bearing). The stationary thrust bearing pads are assembled in a circular pattern, and they operate with an oil film between the stationary pads and a rotating thrust collar. As shown in Fig. 4-21, a machine normally contains a set of active thrust shoes, plus a set of inactive shoes. Under normal operating conditions, the machine runs on the active shoes. During process or mechanical upsets, the rotor may slam up against the inactive thrust shoes. To protect the machine during this type of transition, the inactive thrust bearing is often identical to the active thrust bearing in terms of mechanical construction, and load carrying capability.

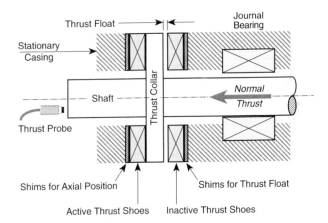

Fig. 4–21 Typical Single Thrust Collar Assembly On Horizontal Machinery

The diagram in Fig. 4-21 is representative of the thrust bearing arrangement in most centrifugal compressors, many large pumps, and small turbines. This configuration consists of a removable thrust collar that is mounted on the shaft with an interference (shrink) plus a keyed fit. If the thrust collar becomes damaged, it may be replaced with a spare collar. The active and inactive thrust shoe assemblies are rolled into the casing on either side of the single thrust collar. Precision ground shims are installed behind the shoes to establish the proper axial position of the rotor with respect to the stator, and to establish the thrust float indicated on Fig. 4-21. For fluid film thrust bearings, the thrust or axial float is normally in the vicinity of 15 Mils (0.015 Inches). Thrust floats of less than 10 Mils (0.010 Inches) or greater than 20 Mils (0.020 Inches) are seldom encountered on fluid film thrust bearings.

Due to the critical nature of the rotor thrust position, a proximity probe system is employed to monitor the relative rotor axial position via changes in the probe DC gap voltage. As shown in Fig. 4-21, an axial probe is used to observe the end of the shaft. During bearing assembly and setup, a dial indicator is also positioned axially, and changes in probe gap readings are compared with axial movements measured with the dial indicator. For system redundancy during operation, two thrust probes are normally installed, and the readings from both probes are compared with the dial indicator. In virtually all cases, all three readings (2 probes plus 1 indicator) must compare within ±1.0 Mil. Although this might seem like a tight tolerance, it is almost 7% of a normal float zone. Hence, this is a precision measurement that must be accurately calibrated and setup in a static condition. If this is not done properly, then the information obtained from the electronic probes when the machine is running will always be questionable.

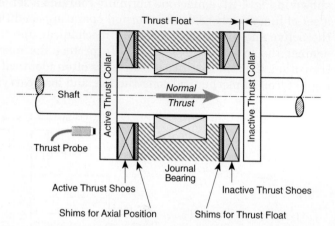

Fig. 4–22 Typical Dual Thrust Collar Assembly On Horizontal Machinery

A thrust bearing configuration commonly used on large steam and gas turbines is depicted in Fig. 4-22. In this arrangement, dual thrust collars are employed. These collars are generally integral with the shaft forging. Hence, if these thrust collars are damaged, the rotor or stub shaft might have to be scrapped and replaced. The journal bearing in Fig. 4-22 is mounted between the thrust shoes, and the stationary casing supports the bearings, as well as the shims mounted behind each set of thrust shoes. Once again, one or more axial proximity probes are used to measure rotor thrust position. Due to the differences in thrust bearing configurations, the axial probe shown in Fig. 4-22 will display an increase in gap voltage as the thickness of the active thrust pads diminish under normal thrust loads. However, the thrust bearing shown in Fig. 4-21 will display a decrease in axial probe gap voltage as the thickness of the active pads diminish under load. Depending on the thrust bearing configuration, the direction of the normal thrust loads, and the location of the axial proximity probes — the probe gap voltage may either increase or decrease as thrust bearing attrition occurs. Both situations can occur, and the machinery diagnostician must have full documentation of the mechanical setup in order to be confident in the on-line DC probe gap information.

Thrust bearings in large horizontal machines nominally contain 40 to 80 Mils (0.040 to 0.080 inches) of babbitt. In most cases, the babbitt is considered as the sacrificial element. During a thrust failure, the machine should coastdown on the available babbitt, and not allow contact between other parts of the rotor and the casing. The specific machine and probe setup for proper thrust monitoring is presented in chapter 6.

Another common application for a fluid film thrust bearing occurs on large vertical machines such as motors or generators. These vertical units are usually directly coupled to pumps or turbines, and the entire weight of the machinery train is supported by a single thrust assembly at the top of the upper machine. A typical installation is shown in Fig. 4-23. The radial guide bearings in this type of machinery often consist of a series of tilt pad bearings, and the thrust bearing usually consists of segmented pads. The thrust bearing may be a fixed geometry or a tilting pad configuration. In either case, an oil bath is provided for the thrust bearing, and high pressure oil lifting jets are often used to supplement the oil bath during startup and shutdown.

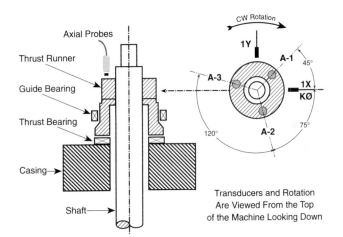

Fig. 4–23 Typical Fluid Film Thrust Bearing On Large Vertical Machinery

Vertical load support for this machine occurs at the thrust runner shown in Fig. 4-23. Shaft weight is essentially *hung* from the center of the thrust runner, and this mechanical element is positioned over the fluid film thrust bearing. Since smooth operation of this machine often depends upon proper behavior of the thrust runner, it is an ideal location to measure vibration and position. As noted in Fig. 4-23, X and Y probes are mounted in a true horizontal direction to observe radial motion of the thrust runner. A Keyphasor® probe is installed at the same angular location as the X axis probe for train timing measurements.

A group of three axial probes are mounted in the true vertical direction observing the top of the thrust runner. On the horizontal machines previously discussed, only one axial thrust position probe was required for thrust position measurement. Two thrust probes were normally installed on horizontal machines to provide full redundancy for the axial measurement. This is particularly necessary to support the voting logic used on monitoring systems setup for automatic trip. However, on a large vertical machine, there are very few installations that incorporate automatic trip based upon changes in axial position. For

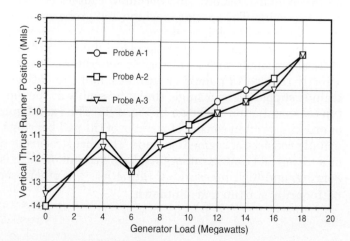

Fig. 4–24 Variation In Vertical Thrust Position Of Thrust Runner As A Function Of Generator Load

these vertical machines, three probes are installed above the thrust runner to determine the relative level of the thrust runner. Since three points determine a plane, any significant deviation in differential gap voltages from the three axial probes is easily interpreted as a *tilting* of the thrust runner.

For instance, Fig. 4-24 describes the thrust position changes measured by three axial probes mounted on a unit configured like Fig. 4-23. In this case, the three proximity probes provide consistent information. Clearly, the level of the thrust runner remains constant as load is varied on this turbine generator set. The same transducers may also be used to examine the axial vibration of the thrust runner. For example, running speed 1X vectors were obtained at slow roll speeds of 40 to 50 RPM, and at synchronous speed of 277 RPM at 14 megawatts. The full speed vectors were corrected for slow roll, and the runout compensated 1X vectors are presented in Table 4-6.

The probe angle corrections represent the angular spacing between the three probes. These static angles may be used to correct to dynamic vectors to a common location. In this case, the probe A-1 was selected as the common reference point, and the vectors from A-2 and A-3 were adjusted to the equivalent angular position of A-1. The resulting three corrected vectors are totally consistent in terms of both amplitude and angle. The final 1.5 Mil synchronous vector is considered to be quite acceptable for this class of machine.

Table 4–6 Runout Compensated Axial 1X Vectors With Probe Angle Correction At 14 MW

Axial Transducer Identification	Runout Compensated 1X Vector	Probe Angle Correction	1X Vector with Static Angle Correction
Probe A-1	1.49 Mils,$_{p-p}$ @ 307°	0°	1.49 Mils,$_{p-p}$ @ 307°
Probe A-2	1.43 Mils,$_{p-p}$ @ 65°	240°	1.43 Mils,$_{p-p}$ @ 305°
Probe A-3	1.49 Mils,$_{p-p}$ @ 188°	120°	1.49 Mils,$_{p-p}$ @ 308°

ROLLING ELEMENT BEARINGS

Most large process machines rely on fluid film bearings to support the rotating or reciprocating elements. When fluid film bearings are properly designed, installed, and operated, their life expectancy is quite long. However, another class of bearings are normally installed on smaller machines, or machines that require close tolerances, or minimal relative motion between parts (e.g., precision machine tools such as lathes or milling machines). Categorically, these bearings are identified as rolling element bearings, but they are commonly referred to as ball bearings, roller bearings, tapered roller bearings, etc. These bearings all contain a series of internal elements that roll between an inner and an outer race. They may depend on external lubrication, or they may be internally packed with grease. Bearings of this type run with very close clearances, and they provide high stiffness combined with low damping to the rotor system. The application of high loads, shock loads, or bearing attrition due to extended run times will expand the internal clearances, and result in bearing failure. Due to these fundamental characteristics, this class of bearings is considered to have a finite lifespan.

Since rolling element bearings have a limited life, it is desirable to monitor these bearings for early indication of impending failures. It has been historically demonstrated that vibration analysis is an effective tool for this type of monitoring. In addition, vibration analysis is useful for analysis of bearing faults. In many instances, the machinery diagnostician may be able to successfully predict the remaining life of a bearing. This allows maintenance to be performed in a scheduled and cost-effective manner rather than in reaction to a failure.

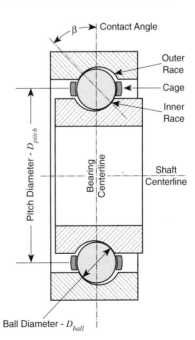

Fig. 4–25 Typical Rolling Element Bearing Configuration

Within any bearing, the primary dynamic excitation occurs at shaft rotational frequency. Analysis of the vibratory behavior of a shaft in a fluid film bearing often centers around the running speed vibration, and harmonics of this primary excitation. However, in a rolling element bearing, the fundamental shaft vibration is supplemented by the mechanics inherent with the additional moving bearing parts. For instance, consider Fig. 4-25 that depicts a typical ball bearing. In this example, assume that the outer race remains stationary, and that the inner race rotates at shaft frequency. It is clear that the supplemental frequencies emitted by this type of bearing must consider the geometry of the bearing in addition to the fundamental shaft rotational speed RPM. This would include the number of rotating balls N, the ball diameter D_{ball}, and the ball pitch diameter D_{pitch}. The contact angle between the balls and the races is also significant in the determination of the fundamental defect frequencies. This contact angle is identified as angle β in Fig. 4-25. Intuitively, the running position of a rolling element bearing is dependent upon the radial and axial forces applied to the bearing. If this ratio is changed due to variations in either the radial or the axial forces, the load contact angle across the bearing will be influenced.

In addition, the basic model for this bearing requires a tight clearance fit, with the internal balls rolling (and not sliding) in the raceways as the shaft turns. If the balls are sliding, this is indicative of excessive bearing clearance, and the general equations for specific defects are no longer applicable.

The traditional equations for the repetition rate of various defects were formulated in the 1960s. Based upon the bearing geometry, and the rotational speed, it can be shown that a defect in the bearing outer race will generate a frequency F_{ord} that may be computed with the following expression:

$$F_{ord} = \left\{\frac{N \times RPM}{2}\right\} \times \left\{1 - \left(\frac{D_{ball}}{D_{pitch}}\right) \times \cos\beta\right\} \qquad (4\text{-}18)$$

For bearing defects on the inner race, the emitted frequency F_{ird} may be determined with the next equation:

$$F_{ird} = \left\{\frac{N \times RPM}{2}\right\} \times \left\{1 + \left(\frac{D_{ball}}{D_{pitch}}\right) \times \cos\beta\right\} \qquad (4\text{-}19)$$

From a vibration measurement standpoint, the outer race defects normally appear at higher amplitudes than the inner race defects. These types of flaws are often combined with vibration at the ball spin frequency that is coincident with the frequency for a ball defect. This ball defect frequency F_{bd} may be calculated with equation (4-20):

$$F_{bd} = \left\{\frac{RPM}{2}\right\} \times \left\{\left(\frac{D_{pitch}}{D_{ball}}\right) - \left(\frac{D_{ball}}{D_{pitch}}\right) \times (\cos\beta)^2\right\} \qquad (4\text{-}20)$$

The fundamental train frequency, or more commonly the cage defect frequency F_{cd} may be determined by:

$$F_{cd} = \left\{\frac{RPM}{2}\right\} \times \left\{1 - \left(\frac{D_{ball}}{D_{pitch}}\right) \times \cos\beta\right\}$$ (4-21)

where: F_{ord} = Frequency of Outer Race Defect (Cycles / Minute)
F_{ird} = Frequency of Inner Race Defect (Cycles / Minute)
F_{bd} = Frequency of Ball Defect (Cycles / Minute)
F_{cd} = Frequency of Cage Defect (Cycles / Minute)
N = Number of Contained Balls or Rollers (dimensionless)
RPM = Rotational Speed of Bearing Inner Race (Revolutions / Minute)
D_{ball} = Diameter of Ball or Roller (Inches)
D_{pitch} = Pitch Diameter of Balls or Rollers (Inches)
β = Bearing Load Contact Angle (Degrees)

The defect frequencies computed with equations (4-18) through (4-21) carry the engineering units of Cycles per Minute (CPM). If frequencies in Cycles per Second (Hz) are desired, the results from the last four equations may be divided by 60. Once again, these equations are for a stationary outer race, and a rotating inner race, with rolling and *not* sliding balls or rollers. It must always be recognized that ball bearings in real machines with reasonable loads do actually slip, and the measured and calculated frequencies will probably not be identical.

Clearly, the frequencies computed with these equations are non synchronous. For a typical 100 HP motor with less than 20 balls, the *outer race defect* frequency computed by (4-18) will be a high frequency component in the vicinity of 40 to 45% of rotating speed times the number of balls ($\approx 0.45 \times RPM \times N$). The *inner race defect* frequency from equation (4-19) will also be a higher frequency component at about 55 to 60% of speed times the number of balls ($\approx 0.6 \times RPM \times N$). The ball spin or *ball defect* frequency will be in the order of 3 to 4 times running speed ($\approx 3.5 \times RPM$), and the *cage defect* frequency will be approximately 45% of running speed ($\approx 0.45 \times RPM$).

For a stationary inner race and a rotating outer race, equations (4-18) through (4-20) remain the same. The expression for a cage defect with a fixed inner race is given by equation (4-22). In this situation, the frequency of the cage defect will be greater that one half of running speed ($\approx 0.55 \times RPM_{out}$).

$$F_{cd\_fir} = \left\{\frac{RPM_{out}}{2}\right\} \times \left\{1 + \left(\frac{D_{ball}}{D_{pitch}}\right) \times \cos\beta\right\}$$ (4-22)

where: F_{cd_fir} = Frequency of Cage Defect with Fixed Inner Race (Cycles / Minute)
RPM_{out} = Rotational Speed of Bearing Outer Race (Revolutions / Minute)

Bearing dimensional data is often available directly from the bearing manufacturer. Some suppliers publish lists of these fault identification frequencies for their bearings. Whether tabular lists or discrete calculations are used, the machinery diagnostician should recognize that a new bearing will probably exhibit some, or all of these frequencies at very low amplitudes. As defects occur, the amplitudes at the associated defect frequencies will increase. As the defects continue to grow with time, the observed frequencies will often shift as the load distribution changes, and the balls (or rollers) begin sliding instead of rolling. In this condition, overall vibration levels are probably unacceptable, and the unit should be shutdown for bearing replacement.

The commonly observed types of damage on rolling element bearings result in craters or spalls on the raceways. As the rolling elements pass over these indentations, impact or shock pulses are generated. In some cases, the defect frequencies identified by the previous equations may not be visible in a frequency spectrum. However, as noted by John Mitchell[10], if this housing vibration data is observed in the time domain, the repetitive pulses are easily distinguished. In time domain analysis, the period of the four defect frequencies should be used to identify the origin of the pulse patterns.

For the sake of completeness, it should be mentioned that other transducers, measurements, and data processing techniques are used to examine the vibratory behavior of rolling element bearings. Methods such as spike energy, envelope detection, shock pulse, and various demodulation techniques are commonly employed. From a practicality standpoint, the diagnostician should evaluate any bearing analysis tool on the basis of performance. Specifically, the following question should always be asked: *can the instrumentation successfully and consistently identify mechanical failures on the rolling element bearings?*

BEFORE CONSIDERING BEARING REDESIGN

In the opinion of the senior author, the Original Equipment Manufacturers (OEMs) do a credible job of designing, building, and installing bearings. Within the industrial end user community there seems to be an overt tendency to redesign, and continually attempt to improve on the OEM bearings. In multiple cases, the bearing redesign has rendered the machinery inoperative. In other situations, the operating speed range of the machinery has been severely limited following the installation of presumably improved bearings. In all cases, the machinery diagnostician must approach bearing problems carefully, and conduct a methodical engineering analysis of the problem. Specifically, the following items should always be thoroughly examined and in some cases re-examined:

- ❍ Check that the oil console or reservoir contains the correct lubricant.
- ❍ Check the oil quality for proper density, viscosity, water content, etc.
- ❍ Check the oil for the presence of any foreign materials.
- ❍ Check for proper oil supply pressure, temperature, and system control.
- ❍ Check the oil flow rate to each bearing, and verify that orifices are properly installed, and that orifice diameters are both reasonable and correct.

---

[10] John S. Mitchell, *Introduction to Machinery Analysis and Monitoring*, second edition (Tulsa, OK: Pennwell Publishing Company, 1993), pp. 241-249.

○ Check the oil drain temperatures, and relative flow rates.

○ Check that the bearing is properly installed with respect to shaft rotation.

○ Check that anti-rotation pins are properly installed with respect to rotation.

○ Check that the shaft to bearing clearance is correct.

○ Check that the bearing to housing clearance is correct.

○ Check that the bearing liner is not distorted or warped.

○ Check that the bearing splitline is not sealed with RTV, silicone, or other incompressible sealants. Use a thin grade of Permatex® sealant for this job.

○ Check for other mechanical changes in the train that would influence bearing load (e.g., changing a gear coupling to a large diaphragm coupling).

○ Check rotor balance records, and the last set of transient startup data.

○ Check coupling alignment for proper cold offset and hot running position.

○ Check for proper temperatures from imbedded thermocouples or RTD's.

○ Check bearing temperature trends (day to night, week to week, etc.).

○ Check to be sure that shaft is level when hot and running.

○ Check bearings, seals, and couplings for evidence of electrical discharge.

○ Check pads and backing for evidence of wear, cracking, or fretting.

○ Check bearings for evidence of edge wear across bearings and machines.

○ Check for proper position of the journal within the bearing with prox probes.

○ Check shaft vibration for normal 1X running speed vibration vectors.

○ Check shaft vibration for any abnormal frequency components.

○ Check the attachment of the bearing housing to the casing and/or baseplate.

○ Check grout condition, and the attachment of baseplate to foundation.

If these checks are followed, and all identified problems corrected, the necessity to redesign or continually replace bearings will be greatly reduced. Bad habits seem to develop over time, and both operations and maintenance personnel have a tendency to get complacent. In many instances this will allow small oversights to turn into major problems. Hence, before jumping into a major redesign effort, the use and abuse of the current bearings should be examined.

There are situations when the bearings really do require an upgrade. If rotor or coupling changes are to be implemented, if the process loads or the lube and seal oil system are to be modified, or if greater reliability is required — then the existing bearings should be audited for potential areas of improvement. The addition of ball and socket bases for tilt pad bearings, the use of micro-babbitt, a change in bearing metallurgy for improved heat transfer, or providing directed lubrication are all common modifications that may benefit a particular bearing installation. In some applications, the installation of new bearing designs such as the *Flexure Pivot Bearings* described by Zeidan and Paquette[11] may be highly beneficial. In other cases, an additional five gallons per minute of oil flow may be all that is required. Once again, the machinery diagnostician is advised to proceed with logic, and proper engineering discipline.

---

[11] Fouad Y. Zeidan, and Donald J. Paquette, "Application of High Speed and High Performance Fluid Film Bearings In Rotating Machinery," *Proceedings of the Twenty-Third Turbomachinery Symposium*, Turbomachinery Laboratory, Texas A&M University, College Station, Texas (September 1994), pp. 209-233.

BIBLIOGRAPHY

1. Allaire, Paul E. and Ronald D. Flack, "Design of Journal Bearings for Rotating Machinery," *Proceedings of the Tenth Turbomachinery Symposium*, Turbomachinery Laboratories, Texas A&M University, College Station, Texas (December 1981), pp. 25-45.

2. Barrett, L.E and John C. Nicholas, "The Effect of Bearing Support Flexibility on Critical Speed Prediction," *ASLE Transactions*, 1984 Joint Lubrication Conference, San Diego, California, February 1984 (revised June 1984).

3. Gunter, Edgar J., "Dynamic Stability of Rotor Bearing Systems," *NASA Report SP-113*, 1966.

4. Gunter, Edgar J. and R.G. Kirk, "The Effect of Support Flexibility and Damping on the Dynamic Response of a Single Mass Flexible Rotor in Elastic Bearings." *Report No. ME-4040-106-72U*, University of Virginia, 1972.

5. McHugh, James D., "Principles of Turbomachinery Bearings," *Proceedings of the Eighth Turbomachinery Symposium*, Gas Turbine Laboratories, Texas A&M University, College Station, Texas (November 1979), pp. 135-145.

6. Mitchell, John S., *Introduction to Machinery Analysis and Monitoring*, second edition, pp. 241-249, Tulsa, OK: Pennwell Publishing Company, 1993.

7. Nicholas, John C., "Tilting Pad Bearing Design," *Proceedings of the Twenty-Third Turbomachinery Symposium*, Turbomachinery Laboratory, Texas A&M University, College Station, Texas (September 1994), pp. 179-194.

8. Nicholas, John C., John K. Whalen, and Sean D. Franklin, "Improving Critical Speed Calculations Using Flexible Bearing Support FRF Compliance Data," *Proceedings of the Fifteenth Turbomachinery Symposium*, Turbomachinery Laboratory, Texas A&M University, College Station, Texas (November 1986), pp. 69-78.

9. Salamone, Dana J., "Journal Bearing Design Types and Their Applications to Turbomachinery," *Proceedings of the Thirteenth Turbomachinery Symposium*, Turbomachinery Laboratories, Texas A&M University, College Station, Texas (November 1984), pp. 179-188.

10. Zeidan, Fouad Y. and Donald J. Paquette, "Application of High Speed and High Performance Fluid Film Bearings In Rotating Machinery," *Proceedings of the Twenty-Third Turbomachinery Symposium*, Turbomachinery Laboratory, Texas A&M University, College Station, Texas (September 1994), pp. 209-233.

Analytical Rotor Modeling

*M*achinery shaft vibration characteristics reflect the combined interaction of rotating assemblies with various fluids, and stationary machine elements. Often these characteristics can be segregated and quantified with appropriate measurement techniques. However, there are situations where the required data cannot be obtained due to other restraints. For instance, the critical speed of a rotor cannot be determined due to the economic impact of shutting down the unit. In another case, the machine running speed cannot be increased to investigate the effects of higher order modes. Similarly, changing mechanical parts, such as bearings or couplings, cannot be realistically evaluated based upon hardware trial and error substitutions. From a problem solving standpoint, the machinery diagnostician may begin an investigation by comparing the predicted machinery behavior with the measured vibratory characteristics. In these types of situations, the development and utilization of mathematical models to simulate the mechanical systems may be mandatory.

During the latter half of the twentieth century, various computational techniques have been developed and refined into working tools. Techniques such as Transfer Matrix, Computational Fluid Dynamics (CFD), and Finite Element Analysis (FEA) provide significant capability for modeling physical systems. In order to provide any overview of some analytical techniques and the generic types of available software for the evaluation of rotating machinery behavior, the following descriptions and examples of machinery calculations are presented.

MODELING OVERVIEW

The mathematics associated with simple mechanical systems were presented in chapter 2. For a single degree of freedom system consisting of a undamped mass hanging from a spring, it was concluded that the system natural frequency was a function of stiffness and effective mass. Paraphrasing equation (2-44), it can be stated these three variables are related in the following manner:

$$Natural\ Frequency \approx \sqrt{\frac{Stiffness}{Mass}} \tag{5-1}$$

From this expression it is clear that stiff elements have high natural frequencies, and flexible parts have lower resonances. Similarly, heavy elements will display low natural frequencies, and lighter components will exhibit higher values. For example, the natural frequency for a large steel girder may be 5 Hz, and a small tuning fork may emit a tone equal to a frequency of 500 Hz. In either case, the geometry and mechanical configuration define a combination of stiffness and mass that yield a discrete natural frequency.

As system complexity increases, the intricacies of the descriptive equations also expand. The simple expression of equation (5-1) is replaced by a matrix solution, and items such as inertia and rotational forces are included. It is apparent that mass, and the distribution of that mass is critical to the solution. Furthermore, shaft stiffness must be determined, and combined with the mass properties. When these elements are defined, the calculation of undamped critical speeds may be performed. These calculations do not include damping, they do not allow asymmetric stiffness, and they do not consider specific forcing functions. However, an undamped analysis provides an overview of the natural frequencies associated with a mass distribution at a selected support stiffness, plus the shaft mode shapes for each resonance and stiffness combination.

The next level of analytical modeling programs incorporates asymmetrical stiffness coefficients, plus damping from the bearings, foundation, or process fluid. The oil film coefficients are calculated for the bearing configuration, and support coefficients are normally measured. It is important to include damping into the calculations. This energy dissipater allows the examination of damped critical speeds, the computation of rotor stability, plus damped mode shapes. Although this calculation refinement does a credible job of finding the Eigenvalues (natural frequencies and damping), it does not accept actual forcing functions such as unbalance, skewed wheels, or bowed rotors.

Adding forced vibration mechanisms requires another evolution of the program structure. Within forced synchronous response programs, dimensional forces are used to compute rotor response in displacement units. Hence, the anticipated motion (vibration) at any speed, and at any position along the rotor may be computed. The accuracy of these calculations is often determined by a comparison with measured shaft vibration data at specific locations. This verification of calculations by measured vibration response characteristics is an often ignored step. In actuality, the verification of results is vital to the development of confidence in the calculations. It also helps to define areas where the custom analytical programs require improvement or modification.

Clearly, the construction of a successful analytical model requires the integration of numerous calculations into a cohesive set of results. A single computer program does not contain the entire model. In fact, many calculations are performed in separate environments from the rotor dynamics calculations. For instance, cross-sectional inertias may be computed in a mathematical program, and rotor dimensional configuration may be initially established in a spreadsheet program. In most cases, several different programs are required to perform the full array of calculations. In the remainder of this chapter, the primary rotor dynamics programs will be discussed, and illustrated with field examples.

UNDAMPED CRITICAL SPEED

In order to understand the fundamental behavior of a rotor system, it is mandatory to determine the frequency of the system critical speeds, and the associated mode shapes. One of the easiest tools to begin such an investigation is an analysis of the undamped critical speeds. To appreciate the current array of computations it is meaningful to briefly review the origin of these calculations.

Historically, one of the earliest procedures for critical speed calculations was developed by A. Stodola circa 1925. This graphical construction technique required the estimation of a rotor deflection curve, followed by the computation of relative kinetic and potential energy. By equating the kinetic to the potential energy, a first critical speed was approximated. This technique assumed rigid bearings, gyroscopic effects were ignored, affects of coupling weights could not be determined, and higher order modes could not be successfully addressed.

In 1944 Myklestad[1] published a paper on calculating natural modes of airplane wings and other beams. That was followed in 1945 by Prohl[2] and his paper on critical speeds of flexible rotors. By 1954, Prohl and Myklestad initiated development of calculation techniques for lateral critical speeds. This was followed by the combined work of Holzer, Prohl and Myklestad for torsional critical speeds. In both cases, the transfer matrix method was applied, and this provided a significant improvement in the ability to predict rotor resonant behavior. Although the required matrix calculations are quite complex, the final results reflect the sophistication of the model. Evolution of these techniques have progressed through various stages. During the 1980s, progress has been closely associated with the development of smaller and faster computers that can adequately process the matrix calculations.

The undamped critical speed program (CRITSPD) used in this text was developed by Edgar J. Gunter[3]. In the formulation of this program, the transfer matrix is divided into a point matrix containing the mass, inertia, and bearing properties, plus a massless field matrix containing the shaft properties. The use of a massless field transfer matrix for shaft properties has been used by Lund[4] in rotor stability and unbalance computer codes, and it is also described by Thomson[5]. Although the massless field matrix is a considerable improvement over the hyperbolic continuum formulation, it suffers from numerical difficulties when large numbers of stations are employed. Gunter incorporated a unique automatic scaling procedure to minimize the errors, and to allow the successful mod-

---

[1] N.O. Myklestad, "A New Method of Calculating Natural Modes of Uncoupled Bending Vibration of Airplane Wings and Other Types of Beams," *Journal of the Aeronautical Sciences,* Vol. 11, No. 2 (April 1944), pp. 153-162.

[2] M.A. Prohl, "A General Method for Calculating Critical Speeds of Flexible Rotors," *Journal of Applied Mechanics,* Vol. 12, Transactions of the ASME, Vol. 67 (September 1945), pp. A142-148.

[3] E. J. Gunter and C. Gareth Gaston, "CRITSPD-PC, Version 1.02," Computer program in MS-DOS® by *Rodyn Vibration, Inc.,* Charlottesville, Virginia, August, 1987.

[4] J.W. Lund, "Modal Response of a Flexible Rotor in Fluid Film Bearings," *Transactions American Society of Mechanical Engineers,* Paper No. 73-DET-98 (1973).

[5] William T. Thomson, *Theory of Vibration with Applications,* 4th Edition, Prentice-Hall, Englewood Cliffs, New Jersey, 1993.

eling of 100 station rotors on desktop computers.

The CRITSPD program includes the primary effects of rotor mass, shaft and bearing flexibility, plus transverse and polar inertia. In addition, the undamped analysis can incorporate synchronous or non-synchronous gyroscopic effects, shear deformation, plus variable bearing and seal stiffness. Couplings, impellers, thrust disks, and shaft spacers can be included by several different modeling schemes. Hollow shafts plus variable material densities, and various boundary conditions for each end of a rotor are accommodated. It is possible to compute synchronous critical speeds, planar modes, plus order tracking. The program calculates, and compares total kinetic and potential energy for each mode, plus the undamped critical speeds, and associated rotor mode shapes. In addition, the strain energy distribution, deflection, slope, moment, and shear at each station is computed and presented. Graphical outputs include the rotor cross section, a summary mode shape diagram for all criticals, and a mode shape for each individual critical speed, as illustrated in Fig. 5-1.

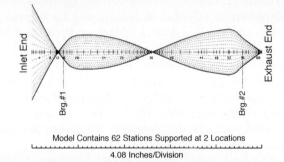

Fig. 5–1 Undamped Steam Turbine Mode Shape Output From CRITSPD Program

Model Contains 62 Stations Supported at 2 Locations
4.08 Inches/Division

Although undamped critical speed analysis can provide significant insight into the behavior of the machinery, it does have inherent limitations. For example, it does not include the effects of forces such as mass unbalance or internal synchronous mechanisms such as shaft bows. This analysis does not consider damping, or cross-coupled influences from bearings, seals, or aerodynamics. Undamped critical speed calculations are a simplification of the system mathematical model, which in turn is a further simplification of the real mechanical system. The parameters neglected in an undamped analysis can alter system resonances. These parameters can control the amplitudes at the resonant frequencies, and they may be responsible for instabilities. Hence, the results from an undamped analysis must be considered in the proper context.

One of the main utilizations for undamped critical speed calculations resides in the ability to quickly compare the change in natural frequencies as a function of support stiffness. As mentioned in chapter 3, the undamped critical speeds may be computed for a large range of stiffness values, and the results plotted as a family of curves. This type of summary plot is normally referred to as an undamped critical speed map. For instance, Fig. 5-2 depicts this type of plot for a 22,500 pound gas turbine rotor. Calculations were performed at 1, 2,

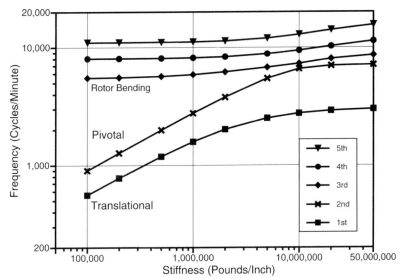

Fig. 5–2 Undamped Critical Speed Map For Single Shaft Gas Turbine

and 5 intervals to allow for even spacing of points on the logarithmic stiffness scale. The stiffness calculation range began at 100,000 Pounds/Inch that is certainly less than any machine stiffness, and it extends to 50,000,000 Pounds/Inch that is greater than any potential machine element stiffness.

The first five critical speeds for each stiffness were plotted, and the points connected for each natural frequency. In this manner, a log-log plot of stiffness versus each critical speed is produced. The first two modes reveal a variation of critical speeds with stiffness. It is reasonable to conclude that these are bearing dependent modes. This observation is confirmed by the detailed calculations that show the majority of the strain energy contained within the bearings. For the higher order criticals, and the stiffer portion of the 1st and 2nd modes, the natural frequencies display minimal variation with support stiffness. These conditions are indicative of resonant modes that are primarily controlled by shaft stiffness. This is important information, since in the first scenario, bearing changes could alter the rotor critical speed(s). This type of mechanical change is reasonably inexpensive to perform. However, in the second situation, shaft modifications would be required to change the critical speeds, and this type modification can be expensive as well as technically complicated.

The undamped critical speed map is also used to examine the relationship between the calculated resonant frequencies, and the operating speed range as shown in Fig. 5-3. In this diagram, the rotor support stiffness is shown for three different conditions. First, the minimum or soft condition of 500,000 Pounds per Inch is shown. Second, the horizontal bearing stiffness K_{xx} is plotted, followed by the third line of vertical bearing stiffness K_{yy}. The support stiffness values help to define the potential operating range of the machine. This information allows

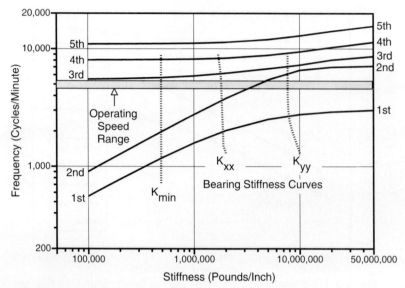

Fig. 5–3 Undamped Critical Speed Map For A Single Shaft Gas Turbine With Principal Bearing Stiffness Curves And Normal Operating Speed Range Identified

further quantification of the anticipated turbine characteristics.

The computation of natural resonant frequencies must also be compared with the potential excitation frequencies. It is perfectly understandable that natural frequencies will remain essentially dormant until some type of excitation coincides with these critical speed(s). The applied excitation may be a broad band forcing function, such as a steam turbine subjected to a slug of water. The excitation may also be a discrete frequency such as rotational speed mass unbalance (1X), or an excitation due to specific geometry within the machine (e.g., pump vane passing activity).

This type of evaluation is performed on an interference plot that is commonly referred to as a Campbell diagram. Fig. 5-4 depicts such a diagram based upon the previous critical speed map. In this diagram, the natural frequencies are plotted on the vertical axis, and the excitation frequencies are shown along the horizontal axis. For instance, the five critical speeds at normal operating conditions were extracted from the critical speed map, and they are shown as horizontal lines in the Campbell Diagram.

Excitations at running speed 1X unbalance, 2X misalignment, plus 5X, and 10X excitations are presented as the slanted lines. The intersection between the horizontal natural frequency lines, and the slanted excitation lines represents a potential case for the appearance of the specific resonance. Naturally, this type of diagram is often expanded to include higher frequency excitations, and other resonant frequencies such as structural or torsional critical speeds.

The number of resonant frequencies that exist on most large machinery trains can be staggering. This is particularly true for bladed machines such as

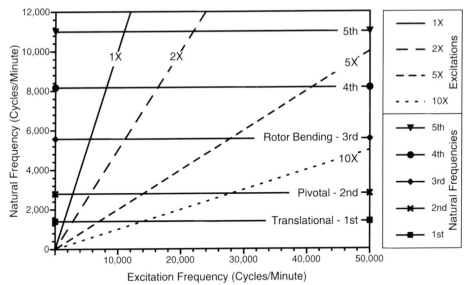

Fig. 5–4 Campbell Diagram For Single Shaft Gas Turbine At Design Stiffness Values

steam or gas turbines where the individual blades and the segmented groups of blades exhibit a variety of tangential and axial modes of vibration. Other complications, such as separate and distinct horizontal and vertical rotor balance resonances, disk or impeller resonances, plus external resonances can substantially increase the number of potential resonant frequencies.

Furthermore, when the sources of excitation are examined, the problem becomes even more complicated. For instance, harmonics of the fundamental excitations may be generated within the machine. In addition, specific frequencies may interact to form distinct beat or modulation frequencies. When all of the possible excitation frequencies are considered, the potential for exciting the expanded group of resonances increases dramatically.

In essence, an interference diagram such as the Campbell plot of Fig. 5-4 may become congested with all of the inherent excitations and resonances. It might even be concluded that the machinery cannot be operated at any reasonable speed due to the coincidence of excitations and natural frequencies. Obviously, this is not an acceptable conclusion, and it is not representative of the varied array of operating machinery trains.

In order to address this dilemma, it is suggested that machinery behavior be examined in two different categories. The first category would consist of the major rotor balance resonances (lateral critical speeds), and the potential low frequency excitations. This part of the analysis follows the scheme presented in the Campbell plot of Fig. 5-4, but additional detail is necessary to determine the severity of the interference points.

For example, the rotor model should be expanded to include a forced response analysis, as discussed later in this chapter. By varying the definable excitations (e.g., rotor bow, unbalance at various rotor locations, disk skew, etc.), the machinery diagnostician should be able to evaluate the vibration severity for anticipated forcing functions. This approach will identify the major or significant resonances, and allow other interference points to be discounted.

The second category of machinery behavior considers the higher frequency characteristics associated with turbine blades or compressor wheels. In this complex mechanical domain, the traditional two-dimensional Campbell diagram should be expanded into a three-dimensional SAFE[6] diagram (acronym for Singh's Advanced Frequency Evaluation). This analytical tool combines the two-dimensional Campbell plot with a third dimension of nodal diameters or mode shapes. The three-dimensional intersection of natural resonant frequencies, excitation frequencies, and nodal diameters are then used to identify potential resonant conditions. The inclusion of the blade mode shape allows the diagnostician to ignore the majority of the interfere points, and identify the frequencies and modes of greatest potential vibration.

Case History 10: Mode Shapes For Turbine Generator Set

Undamped critical speed calculations are relatively easy to setup and run. As noted, they do not include synchronous forcing functions, and the support stiffness characteristics represent a simple condition. However, these calculations can provide significant visibility into the behavior of rotating systems.

For rotors supported between bearings, and for overhung assemblies, the mode shapes discussed in chapter 3 make intuitive sense. Armed with the knowledge of the general rotor configuration, and the relative bearing stiffness, the anticipated mode shapes may be estimated. Even though the frequencies may not be calculated, the mode shapes for simple systems can be deduced. However, for more complicated systems, the shaft mode shapes may not be obvious.

For example, consider the turbine generator set depicted in Fig. 5-5. This is a three bearing machine that runs at 3,600 RPM. The combined weight for both rotors is 21,400 pounds, and a solid coupling is used between the two shafts. The turbine is an extraction unit, with a surface condenser at the exhaust. The synchronous generator has collector rings mounted at the outboard end, and a sepa-

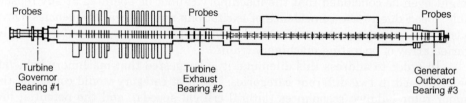

Fig. 5–5 Rotor Arrangement For Three Bearing Turbine Generator Set

[6] Murari P. Singh and others, "SAFE Diagram - A Design and Reliability Tool for Turbine Blading," *Proceedings of the Seventeenth Turbomachinery Symposium*, Turbomachinery Laboratory, Texas A&M University, College Station, Texas (November 1988), pp. 93-101.

rate exciter. The entire machinery train is mounted on a mezzanine deck, and the structure plus the bearing supports are compliant with stiffness values approaching a minimum value of 600,000 Pound/Inch.

Over the operating history of this machinery train various problems have occurred. The majority of the difficulties have been traced to generator unbalance problems, or high eccentricity at the solid coupling. Although successful field balance corrections have been performed on both the turbine and the generator, the logic behind some of the weight corrections was not fully understood.

In an effort to resolve some of the issues, the train was retrofitted with X-Y proximity probes as shown in Fig. 5-5. During startup, these transducers revealed critical speeds that were in direct contradiction with historical conclusions. For example, the local personnel believed that the T/G Set had a critical speed that began at 1,000 RPM, and lasted until well above 2,000 RPM. This behavior was considered to be inconsistent with any expected response through a single critical speed. In addition, the three planes of installed X-Y proximity probes provided additional contrasting information.

In order to address these anomalies, the system was eventually subjected to an undamped critical speed analysis. The computed mode shapes for the first three critical speeds are presented in Fig. 5-6. The first mode was calculated to

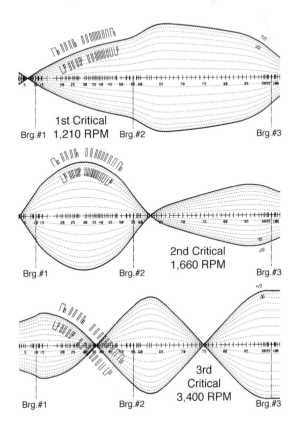

Fig. 5–6 Undamped Shaft Mode Shapes For The First Three Critical Speeds On A Three Bearing Turbine Generator Set

be 1,210 RPM, which agreed with the measured value of 1,250 RPM. This critical was always visible by the proximity probes installed at the #2 and #3 Bearings. The probes at the #1 bearing are close to a nodal point, and this translational resonance is not particularly visible at the front turbine bearing.

The measured second critical speed occurred at 1,650 RPM. This value is virtually identical with the calculated second critical of 1,660 RPM. In addition, the proximity probes mounted at bearings #2 and #3 always displayed an out of phase behavior. This was not fully understood until the analysis was performed, and the calculated pivotal mode shapes produced. From this second critical mode shape, it is clear that the probes at #2 and #3 bearings are on opposite sides of a shaft node, and a phase reversal must exist. Again, the analytical calculations are consistent with the field shaft vibration response measurements.

Finally, eccentricity problems at the coupling always caused high vibration amplitudes at slow speeds, and at speeds just below 3,600 RPM. The reason for this behavior is evident from the calculated mode shape plots where a large deflection in the coupling area is visible for the first critical at 1,210 RPM, plus the third critical speed at a computed value of 3,400 RPM. This resonance was fully corroborated by the proximity probe transient data that exhibited a resonance at 3,440 RPM.

Additional supporting evidence concerning the behavior of this turbine generator set was documented when balance weights were placed on each end of the generator. The anticipated modal response with the balance weights was consistent with the shaft mode shapes described in Fig. 5-6. There are other correlations that may be extracted from this data set. However, the main point is that undamped critical speed calculations provide an analytical tool that is directly applicable to existing machinery. In many situations, it can provide the diagnostician with significantly more insight into the dynamic behavior of the rotating machinery, and it also provides valuable modal information for field balancing.

Case History 11: Torsional Analysis of Power Turbine and Pump

The same general techniques used for undamped lateral calculations may also be applied towards the computation of undamped torsional frequencies and mode shapes. As stated in chapter 2, the basic equations for lateral and torsional characteristics are similar. However, the calculation scheme, and interpretations of results are somewhat different. In a lateral system, stiffness is expressed as Pounds per Inch, and mass carries the units of Pound-Seconds^2 per Inch. Within a torsional analysis, the torsional stiffness is a torque per unit angle, with common units of Inch-Pounds per Radian. Polar inertia carries the units of Pound-Inch-Seconds^2 per Radian, and this is analogous to mass in a lateral analysis.

Within a lateral analysis, the rotor stations with maximum motion are significant (as in the previous case history). During a torsional analysis, the nodal points are meaningful since stress reversals occur across each torsional node. Furthermore, in a lateral analysis, the mass and stiffness properties are typically confined to a single rotor, or rotor system with hard couplings. The influence of the flexible couplings between rotors are seldom included in a lateral analysis. During a torsional analysis the inertia and torsional stiffness properties of the entire train are considered. Since inertia elements are essentially fixed, the mechanical element used for alteration of torsional natural frequencies often reverts to the coupling spool piece between machines.

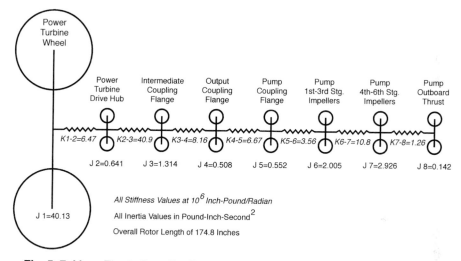

Fig. 5–7 Mass Elastic Data For Power Turbine, Drive Through Gear Box, And Pump

For example, consider the mass elastic data shown in Fig. 5-7. This diagram describes the lumped inertia at eight major machinery sections, plus the seven interconnecting torsional springs. This data was based upon OEM specified values, plus independent calculations of cylindrical sections with the equations presented in chapter 3 of this text. The actual machinery in Fig. 5-7 represents a processed water injection pump that is gas turbine driven via a straight through gear box. The gas turbine is a two shaft unit, with no mechanical connection between the gas generator and the power turbine. Hence, the drive end of this train begins with the power turbine wheel.

Considering the relative size and mass of the machine elements, it is understandable that a majority of the system inertia is contained within this turbine drive wheel. As noted, a gear box was attached to the turbine output shaft. Since turbine and pump speeds were compatible, the gear box consisted of a single drive through shaft element, with no speed change. The horizontally split pump was originally a six stage unit that was de-staged to five stages to meet process demands. The expected operating speed range for this pump varied from 5,180 to 6,800 RPM.

Various standards (e.g., API 617) recommend a 10% separation between any torsional resonance and the operating speed range. Application of this criteria expands the above speed range to include a minimum torsional frequency of 4,660 RPM, combined with a maximum of 7,480 RPM. Before performing any extensive calculations, it would be reasonable to estimate the first torsional frequency based upon the available mass elastic data. For instance, the major inertia occurs at the gas turbine wheel (40.13 Pound-Inch-Sec.^{2}/Radian), and the main coupling stiffness (6.67x10^6 Inch-Pound/Radian) would normally be varied to control the torsional resonance frequency. If these values are placed in equation (2-103), a first torsional critical speed may be estimated as follows:

$$F_{c_{tor}} \approx \frac{1}{2\pi} \times \sqrt{\frac{K_{tor}}{J_{mass}}} \approx \frac{1}{2\pi} \times \sqrt{\frac{6.67 \times 10^6 \text{ Inch-Pound/Radian}}{40.13 \text{ Pound-Inch-Second}^2/\text{Radian}}}$$

$$F_{c_{tor}} \approx \frac{1}{2\pi} \times \sqrt{166,210} = 64.89 \frac{\text{Cycles}}{\text{Second}} \times 60 \frac{\text{Seconds}}{\text{Minute}} = 3,890 \text{ CPM}$$

This estimated speed of 3,890 CPM does fall below the 4,660 to 7,480 RPM exclusion range, but the differential is uncomfortably small. Obviously, a full set of undamped torsional resonance calculations are required to obtain sufficient precision in the torsional natural frequency calculations. The most significant results of these computations are in Fig. 5-8. At the top of this diagram, the torsional mode shape at the calculated first critical speed of 5,030 RPM is shown. This frequency is much higher than the simple model estimate of 3,890 RPM. In addition, the computed first critical speed falls well within the exclusion speed range of 4,660 to 7,480 RPM. Clearly, this deviation demonstrates that a very simple model may not properly represent the actual mechanical system.

The second undamped torsional critical speed appears at 16,990 RPM, as indicated at the bottom of Fig. 5-8. This frequency is considerably higher than

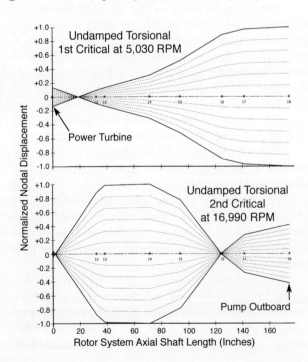

Fig. 5–8 Undamped First and Second Torsional Mode Shapes For Power Turbine, Drive Through Gear Element, and Coupled Five Stage Centrifugal Pump

the operating speed range, and it is beyond excitation by twice rotational speed oscillation (2 x 7,480=14,960 CPM). Hence, this torsional resonance should not cause any distress. Although higher order torsional frequencies are not shown, they should also be computed, and compared against potential excitations with a Campbell or a SAFE diagram.

From this data, it is evident that the major problem resides with the coincidence of the first torsional critical and the previously described exclusion range (operating speed range ±10%). This predicted relationship is unacceptable, and physical changes must be implemented to correct the deficiency. Unfortunately, the machinery under discussion included existing equipment that was in the process of re-configuration to accommodate new operating conditions. Although changes in these existing rotors were feasible, the economic considerations voted heavily against any significant changes to the turbine or pump rotors. The drive through gear box shaft was identical to other units at the same facility. Hence, there was reluctance to change this drive shaft to a one-of-a-kind assembly.

The last candidate for modification was the load coupling to the pump. The available couplings all contained hollow spool pieces with outer diameters that varied between 4.16, and 5.20 Inches. The torsional stiffness values for these couplings ranged from 6.67×10^6 to 7.00×10^6 Inch-Pound/Radian. This coupling stiffness range was previously judged unacceptable due to the frequency interference. Hence, a new coupling must be provided that is both torsionally softer, and still able to transmit the torque with acceptable stress values.

After the examination of potential stiffness changes, a nominal value of 2.00×10^6 Inch-Pound/Radian was selected as acceptable. The complete array of undamped torsional calculations were repeated, and the results summarized in Table 5-1. With this reduction in torsional stiffness, the calculated undamped first critical was reduced from 5,030 to 4,160 RPM. This is approximately 20% below the minimum operating speed, and outside of the exclusion speed range. The higher order torsional resonances were influenced by this reduction in load coupling stiffness, but the variations were insignificant. More importantly, there was minimal interference between higher order excitations (e.g., pump vane passing) and the undamped torsional criticals.

The final mechanical design by the coupling manufacturer included a solid coupling spool piece with an overall length of 16.80 Inches, and an outer diameter of 2.27 Inches. The flanges at each end of this spacer were 0.53 Inches wide, and 8.25 Inches in diameter. It is always a good idea to check the torsional stiffness provided by the vendor. In the vast majority of the cases, the stiffness pro-

Table 5–1 Comparison Of Undamped Torsional Critical Speeds For Two Different Couplings

Coupling Type	Stiffness (Inch-Lb/Rad)	1st Mode (RPM)	2nd Mode (RPM)	3rd Mode (RPM)	4th Mode (RPM)
Original Hollow Spool	6.67×10^6	5,030	16,990	27,220	30,680
New Solid Spool	2.00×10^6	4,160	16,450	25,530	29,630

vided by the coupling vendor is accurate. However, a quick check of this critical parameter is always warranted. The torsional stiffness of this solid spool piece can be closely approximated by computing the stiffness of the main torque tube, and the mounting flanges. These values are summed in a reciprocal manner to determine torsional stiffness of the spool piece. Since this is a solid assembly, equation (3-66) may be used to compute the spool stiffness as follows:

$$K_{tor_{spool}} = \frac{\pi \times G_{shear} \times D^4}{32 \times L} = \frac{\pi \times (11.9 \times 10^6 \text{ Pounds/Inch}^2) \times (2.27 \text{ Inches})^4}{32 \times (16.80 - 2 \times 0.53 \text{ Inches})}$$

$$K_{tor_{spool}} = \frac{992.66 \times 10^6}{503.68} = 1.97 \times 10^6 \text{ Inch-Pound/Radian}$$

Similarly, the torsional stiffness of each end flange is calculated as:

$$K_{tor_{flange}} = \frac{\pi \times G_{shear} \times D^4}{32 \times L} = \frac{\pi \times (11.9 \times 10^6 \text{ Pounds/Inch}^2) \times (8.25 \text{ Inches})^4}{32 \times (0.53 \text{ Inches})}$$

$$K_{tor_{flange}} = \frac{173.19 \times 10^6}{16.92} = 10.21 \times 10^9 \text{ Inch-Pound/Radian}$$

These individual torsional stiffness values may now be combined in a reciprocal manner to determine the overall or effective torsional stiffness of the entire solid spool piece in the following manner:

$$\frac{1}{K_{tor_{eff}}} = \frac{1}{K_{tor_{spool}}} + \frac{1}{K_{tor_{flange}}} + \frac{1}{K_{tor_{flange}}}$$

$$\frac{1}{K_{tor_{eff}}} = \frac{1}{1.97 \times 10^6} + \frac{1}{10.21 \times 10^9} + \frac{1}{10.21 \times 10^9}$$

or

$$K_{tor_{eff}} = 1.97 \times 10^6 \text{ Inch-Pound/Radian}$$

From these calculations, it is clear that the spool piece torsional stiffness is governed by the center torque tube. The end flanges are very stiff, and the most flexible member (center tube) controls the effective stiffness. This coupling torsional stiffness value is consistent with the required 2.00×10^6 Inch-Pound/Radian determined from the undamped analysis. Overall, this proved to be a mechanically acceptable field retrofit that performed with good reliability.

STABILITY AND DAMPED CRITICAL SPEED CALCULATIONS

The inclusion of damping into the analysis significantly expands the usefulness of the analytical calculations. This manifests as improved correlation between the rotor dynamics computations, and the real machinery behavior. In many ways, a damped analysis is similar to undamped calculations, with the important inclusion of damping and cross-coupled stiffness from bearings, seals, and aerodynamic and/or fluid interactions. Most programs allow the consideration of non symmetric bearing and support coefficients, plus an output scheme that displays representative vertical and horizontal motion. This type of analysis also determines stability, and it provides a comparison of stability between modes. Some programs, such as ROTSTB by Gunter[7] use a complex matrix transfer method, and other programs such as DYROBES by Chen, Gunter, and Gunter[8] are based upon finite element analysis (FEA) numerical methods. It should be noted that the matrix transfer method may skip roots that are closely spaced, but this problem does not occur with FEA.

Damped calculations may be considered as somewhat of a *pure* analysis due to the fact that these programs calculate all lateral critical speeds of the mechanical system, including all potential forward, reverse, and mixed modes. In addition, the damped analysis determines the stability characteristics of each mode. This is quite significant since reverse modes and stability are not computed by other methods. For a rotor mounted between bearings with small wheel diameters, this feature may not be particularly important. However, for a rotor with large overhung wheels, the potential of various reverse modes is significant, and this behavior must be visible to the diagnostician. On the limitation side, this type of program does not include external forces from mass unbalance, shaft bows, or other external excitations.

Damped critical speed programs include multiple gyroscopic effects, shear deformation, rotary inertia, plus a full set of eight bearing and support coefficients at defined speeds. Couplings, impellers, thrust disks, shaft spacers, and hollow shafts are accommodated. Hysteretic shaft damping may be included as well as aerodynamic cross-coupling effects. These programs allow variable density of rotor materials, and various boundary conditions for the rotor. Forward, backward, or mixed criticals are computed, plus the complex Eigenvalue for each resonance. The Eigenvalues consist of *real* and *imaginary* portions, and the normal output units are Radians per Second. The *real* portion of the Eigenvalue is the modal damping or growth factor. The *imaginary* portion of the Eigenvalue represents the damped natural frequency in Radians/Second. Multiplication by $30/\pi$ converts this value from Radians/Second to RPM as shown in equation (5-2):

---

[7] Edgar J. Gunter, "ROTSTB, Stability Program by Complex Matrix Transfer Method - HP Version 3.3," Computer Program in Hewlett Packard Basic by *Rodyn Vibration, Inc.*, Charlottesville, Virginia, March, 1989, modified by Robert C. Eisenmann, Machinery Diagnostics, Inc., Minden, Nevada, 1992.

[8] W.J. Chen, E. J. Gunter, and W. E. Gunter, "DYROBES, Dynamics of Rotor Bearing Systems, Version 4.21," Computer Program in MS-DOS® by *Rodyn Vibration, Inc.*, Charlottesville, Virginia, 1995.

$$Damped \; Critical \; Speed \; = \; \frac{30 \times Imag}{\pi} \qquad \textbf{(5-2)}$$

This class of analytical program also computes the *Log Decrement* for each resonance to allow an evaluation of rotor stability. The log decrement is determined by multiplying -2π times the ratio of real to imaginary portions of the Eigenvalue as follows:

$$Log \; Decrement \; = \; \delta \; = \; \frac{-2\pi \times Real}{Imag} \qquad \textbf{(5-3)}$$

A positive log decrement identifies a stable system, whereas a negative value signifies an unstable mode. Since the log decrement is a direct indication of the damping and the decay rate through a resonance, it may also be used to determine the amplification factor of the resonance. Dividing π by the log decrement will result in the amplification factor Q as shown in the next expression:

$$Amplification \; = \; Q \; = \; \frac{\pi}{Log \; Dec} \; = \; \frac{\pi}{\delta} \qquad \textbf{(5-4)}$$

On machines with split criticals, each individual resonance will be calculated, and the dominant direction will be identifiable from the mode shape plots. For instance, consider Fig. 5-9 that describes the non-dimensional damped vertical and horizontal mode shapes (Eigenvectors) of a gas turbine rotor. Based on the relative amplitudes of the vertical versus the horizontal mode shapes, it is self-evident that the described resonance is predominantly a vertical mode. On some programs, two levels of normalization are provided for each non-dimensional mode shape. Within these programs, the peak displacement for both orthogonal directions is always 1.0, and there is no visibility of any dominant motion in either the vertical or the horizontal directions. The data presented in Fig. 5-9 contains only one level of normalization, and the dominant direction of the computed motion is maintained. From this diagram, it is noted that the

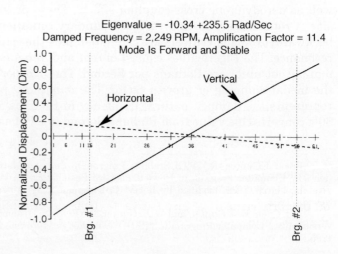

Eigenvalue = -10.34 +235.5 Rad/Sec
Damped Frequency = 2,249 RPM, Amplification Factor = 11.4
Mode Is Forward and Stable

Fig. 5-9 Damped Gas Turbine Mode Shape

Eigenvalue for mode was calculated to be:

$$Eigenvalue = -10.34 + 235.5 \text{ Radians/Second}$$

The real portion of this Eigenvalue is the first term of -10.34, and the imaginary part is the second term, or +235.5. Based on these values, the last three equations may be applied to determine the natural frequency in RPM, the log decrement for this mode, and the associated amplification factor. The damped critical speed is determined by converting units with (5-2) as follows:

$$Damped\ Critical\ Speed = \frac{30 \times Imag}{\pi} = \frac{30 \times 235.5}{\pi} = 2,249 \text{ RPM}$$

This value is identical to the damped frequency on the mode shape plot Fig. 5-9. Next, consider the calculation of the log decrement with equation (5-3):

$$Log\ Decrement = \frac{-2\pi \times Real}{Imag} = \frac{-2\pi \times (-10.34)}{235.5} = +0.276$$

This positive log decrement indicates a stable mode, and the value of the log decrement may now be used to determine the amplification factor of the resonance with equation (5-4) in the following manner:

$$Amplification = \frac{\pi}{Log\ Dec} = \frac{\pi}{0.276} = 11.4$$

The influence of bearing clearance upon the damped critical speeds is always a question to be addressed. By calculating bearing oil film coefficients under various clearance conditions, and combining this information with the support coefficients, the anticipated machinery response may be computed. For instance, consider the data presented in Table 5-2.

Table 5–2 Gas Turbine Damped Critical Speeds Versus Bearing Clearance

Journal Bearing Clearance	1st Mode Translational	2nd Mode Pivotal	3rd Mode Bending
Minimum	1,146 RPM	1,877 RPM	5,732 RPM
Average	1,104 RPM	1,826 RPM	5,727 RPM
Maximum	1,073 RPM	1,760 RPM	5,719 RPM

These damped critical speed calculations were performed at average, minimum, and maximum allowable bearing clearances. These computations were performed for a 40,000 HP gas turbine at 5,100 RPM. They reveal that the bearing dependent 1st and 2nd modes are moderately influenced, but the frequency of the 3rd critical is insensitive to bearing clearance variations. Generally, changes in journal bearing clearances are not a major factor in the resonant behavior of these machines. However, for long-term operation, it is always desirable to begin with the minimum bearing clearances to allow room for babbitt attrition during the run. Another perspective of the machine characteristics may be obtained by examining the variations in log decrement of each mode at each bearing clearance condition. For this gas turbine, Table 5-3 summarizes these stability parameters for the first three critical speeds.

Table 5–3 Gas Turbine Log Decrement Versus Bearing Clearance

Journal Bearing Clearance	1st Mode Translational	2nd Mode Pivotal	3rd Mode Bending
Minimum	0.418	0.497	0.041
Average	0.415	0.519	0.044
Maximum	0.338	0.496	0.046

If the log decrement is positive, vibration amplitudes will decay with time. Conversely, if the log decrement carries a negative sign, then the mode is unstable, and amplitudes will increase with time. Within Table 5-3, all values are positive. That indicates stable modes within the operating speed domain of the gas turbine. The magnitude of the log decrement describes the rate of oscillation decay. Specifically, a large positive log decrement delineates a well damped system that will rapidly attenuate vibratory motion. A well damped resonance will display a low amplification factor, and will persist over a broader frequency range. On the other hand, a small log decrement identifies a poorly damped resonance, with a higher amplification factor, and a smaller bandwidth.

With respect to the log decrements presented in Table 5-3, it is clear that the second pivotal critical speed exhibits the largest values. As such, this second mode would be the most difficult to excite, and the resultant motion would be quickly suppressed (i.e., damped out). The first translational mode has somewhat lower log decrement values. This critical would be slightly easier to excite, and the resultant motion would continue for a longer time. Finally, the rotor first bending mode (3rd critical) has the lowest log decrement. This resonance is the easiest to excite, and the motion would decay at a slower rate. A low log decrement is indicative of a high amplification factor at the resonance. This manifests as rapidly increasing vibration amplitudes with minimal phase change as the skirt of the resonance is approached.

Case History 12: Complex Rotor Damped Analysis

This particular rotor consists of an overhung hot gas expander wheel, a pair of midspan compressor wheels, and three stages of overhung steam turbine wheels[9] as described in Fig. 5-10. A series of axial through bolts are used to connect the expander stub shaft through the compressor wheels, and into the turbine stub shaft. This type of assembly is similar to many gas turbine rotors. However, in this machine, the rotor must be built concurrently with the inner casing. Specifically, the horizontally split internal bundle is assembled with the titanium-aluminum compressor wheels, stub shafts, plus bearings and seals. The end casings are attached, the expander wheel is bolted into position, and the turbine stages are attached with another set of through bolts.

The eight rotor segments are joined with Curvic® couplings, identified as #1 through #7 on Fig. 5-10. Even with properly ground and tight fitting Curvics®, there is potential for relative movement of rotor elements. Although each of the rotor segments are component balanced, any minor shift between elements will produce a synchronous force. Since this unit operates at 18,500 RPM, a few grams of unbalance, or a Mil or two of eccentricity will result in excessive shaft vibration, and strong potential for machine damage. Furthermore, the distribution of operating temperatures noted on Fig. 5-10 reveals the complexity of the thermal effects that must be tolerated by this unit. The 1,250°F expander inlet is followed by compressor discharge temperatures in excess of 430°F. The steam turbine operates with a 700°F inlet, and a 160°F exhaust.

By any definition, this must be considered as a complicated and difficult rotor. On the positive side, this machine is a compact design that yields a high thermal efficiency. Hence, when the unit is properly assembled, and balanced, it is very cost-effective to operate.

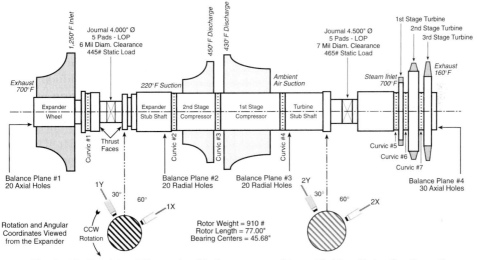

Fig. 5–10 Combined Expander-Air Compressor-Steam Turbine Rotor Configuration

[9] Robert C. Eisenmann, "Some realities of field balancing," *Orbit*, Vol.18, No. 2 (June 1997), pp 12-17.

A double overhung rotor with an appreciable midspan mass has the potential for multiple resonances with either forward or reverse modes. In order to better understand the behavior of this machine, various historical data sets were reviewed. It was noted that reverse orbits appeared around 7,000, and 17,000 RPM. Field balancing activities on this machine were generally successful when a two step correction was used. The first step consisted of an intermediate balance based on transient data as the machine passed through 14,000 RPM. This initial balance was accomplished using the outboard planes #1 and #4. This was followed by a trim at 18,500 RPM on the inboard planes #2 and #3 located next to the compressor wheels. It was evident that if the rotor was not adequately balanced at 14,000 RPM, it probably would not run at 18,500 RPM.

Further examination of historical data revealed that vibration severity changed in accordance with the machinery operational state. For instance, the peak vibration amplitudes occur at a rotor critical that appears between 7,600 and 8,100 RPM. This resonance displays the following variable characteristics:

❍ Cold Startup to 14,500 RPM — Peak Response of 2.0 to 5.0 Mils,$_\text{p-p}$
❍ Warm Coastdown from 14,500 RPM — Peak Response of 4.0 to 5.0 Mils,$_\text{p-p}$
❍ Hot Crashdown from 18,500 RPM — Peak Response of 6.0 to 8.0 Mils,$_\text{p-p}$

These amplitude variations are combined with changes in the amplification factor through the resonance (potential change in damping). Clearly, this information must be supplemented by an examination of the variable speed vibration data — plus an understanding of the rotor critical speeds, and mode shapes.

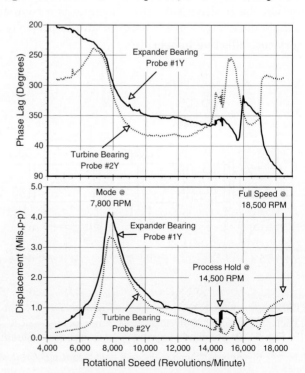

Fig. 5–11 Bode Plot Of Shaft Y-Axis Proximity Probes During A Typical Machine Startup

A typical startup Bode plot of the Y-Axis response from each measurement plane is shown in Fig. 5-11. Both plots are corrected for slow roll runout at 1,000 RPM, and the resultant data is representative of the true dynamic shaft motion at each lateral measurement plane. The major resonance appears at 7,800 RPM. A process hold point occurs at 14,500 RPM, and the unit displays various amplitude and phase excursions at this speed. Some of this behavior is logically due to the heating of the rotor and casing, plus variations in settle out of the operating system (i.e., pressures, temperatures, flow rates, and molecular weights).

The Bode also exhibits additional vector changes between 14,500 and 18,500 RPM. Some of these changes are due to the influence of a backward mode around 17,000 RPM. Other changes appear as the machine approaches the normal operating speed of 18,500 RPM. This higher speed data is difficult to fully comprehend in the Bode plot, but it becomes more definitive when replotted in the polar format of Fig. 5-12.

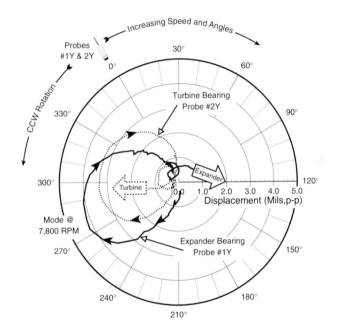

Fig. 5–12 Polar Plot Of Shaft Y-Axis Proximity Probes During A Typical Machine Startup

The point of major interest on Fig. 5-12 is that at full speed, the turbine end shaft is moving towards the 9 o'clock direction, and the shaft at the expander is heading towards 4 o'clock. This behavior indicates a couple, and the presence of some type of pivotal mode occurring at a frequency above the normal running speed of 18,500 RPM. In many cases, this type of response would not be unusual. However, for this unit, the machinery files had no indication of a resonance around the operating speed. Due to the measured response of midspan balance weights (planes #2 & #3) at 18,500 RPM, it was clear that the vibration data was correct. This also implies that the historical undamped mode shapes were not fully representative of actual machinery behavior.

As previously noted, there are only two lateral vibration measurement planes along the entire length of this rotor. Since there were no other feasible locations for shaft probes, additional measurement options were eliminated. Hence, the only viable approach resided with a proper analytical model of this rotor system. A 65 station damped model was constructed. This computer model included bearing stiffness and damping that varied with speed, plus flexible bearing supports. Damped natural frequencies, direction of each mode, and the log decrement for each mode are summarized in Table 5-4.

Table 5–4 Summary Of Calculated Damped Natural Frequencies

Mode Description	Damped Frequency	Mode Direction	Log Decrement
Stiff Shaft - Pivotal	5,040 RPM	Backward	1.61
Stiff Shaft - Pivotal	5,470 RPM	Forward	1.39
Stiff Shaft - Translational	5,910 RPM	Forward	1.92
Stiff Shaft - Translational	6,310 RPM	Forward	1.47
Shaft Bending - 2 Nodes	7,080 RPM	Backward	0.166
Shaft Bending - 2 Nodes	7,840 RPM	Forward	0.0753
Shaft Bending - 3 Nodes	17,710 RPM	Backward	0.263
Shaft Bending - 3 Nodes	21,930 RPM	Forward	0.263

The first four modes are stiff shaft pivotal and translation shapes with high log decrements. These modes did not appear in the vibration data due to the high damping for each mode. A backward mode was detected at 7,080 RPM. This mode was not visible in the startup plots, but it briefly appears in some of the hot coastdown data. The most active forward mode within the operating speed range occurs at a damped frequency of 7,840 RPM, and the calculated mode shapes for this resonance are presented in Fig. 5-13. From this diagram, it is noted that the

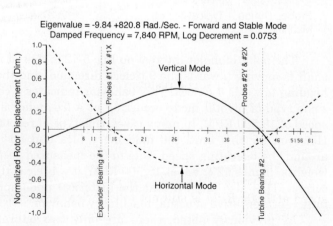

Eigenvalue = -9.84 +820.8 Rad./Sec. - Forward and Stable Mode
Damped Frequency = 7,840 RPM, Log Decrement = 0.0753

Fig. 5–13 Calculated Damped Mode Shapes Of Main Rotor Resonance At Nominally 7,800 RPM

normalized deflections at both bearings are quite small. This indicates minimal motion of the journals within their respective bearings. With small relative motion, the velocity is low, and bearing damping is minimal. This behavior is reflected in the low 0.0753 log decrement for this mode.

The validity of the analytical model is supported by correlation of the computed resonant frequency of 7,840 RPM (from Fig. 5-13), with the measured resonance of 7,800 RPM (Fig. 5-11). It is also clear from Fig. 5-13, that the rotor balance response at this resonance can be effectively controlled by corrections at the modally effective end planes #1 and #4.

With increasing speed, the damped analytical model reveals another backward mode at 17,710 RPM. The vertical and horizontal shapes for this reverse mode are presented in Fig. 5-14. This pivotal mode is often visible as reversed

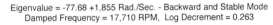
Eigenvalue = -77.68 +1,855 Rad./Sec. - Backward and Stable Mode
Damped Frequency = 17,710 RPM, Log Decrement = 0.263

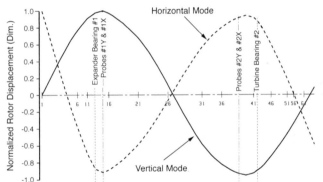

Fig. 5–14 Calculated Damped Mode Shapes Of Backward Resonance At Nominally 17,710 RPM

orbits on the transient vibration data. Immediately above the normal operating speed of 18,500 RPM, a damped mode was calculated at a frequency of 21,940 RPM as in Fig. 5-15. This forward mode has the same deflection characteristics as the previously backward mode at 17,710 RPM. In both cases, the inboard balance planes #2 and #3 are the most modally effective correction planes for this speed domain. It is concluded that weight corrections adjacent to these compressor wheels (planes #2 and #3) should be out of phase. This is due to the fact that a nodal point exists at the middle of the rotor. The validity of this conclusion was field tested on the machine. The installation of a pair of weights at the middle planes at the same angle resulted in excessive vibration. However, a couple shot proved to be smooth, and supportive of the analytical mode shape at high speed.

Finally, the existence of a pivotal resonance at slightly above running speed was previously noted on the polar plot, Fig. 5-12. The damped mode shape presented in Fig. 5-14 supports this observation. Once again, the vibration measurements, and the analytical tools are combined to explain the behavior of a complex machine. Although the variable behavior through the main critical speed at 7,800 RPM is still not totally clear, it is speculated that a loosening or relaxation of the segmented rotor occurs with elevated temperatures.

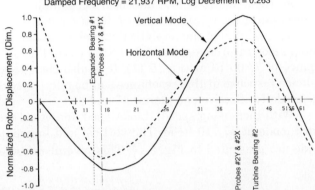

Fig. 5–15 Calculated
Damped Mode Shapes
Of Rotor Resonance
Occurring Above The
Normal Operating Speed

This work provided an improved understanding of the shaft response and damped mode shapes, plus a better appreciation of the process influence. Armed with this information, the rotor was balanced at the intermediate speed by using transient data acquired during cold startups at 14,000 RPM. This intermediate step was a two plane balance with weight corrections at the outboard planes #1 and #4. This allowed the rotor to run at full speed of 18,500 RPM, and a final two plane trim balance was performed on the interior planes #2 and #3 after a full heat soak. The synchronous shaft vibration amplitudes were significantly reduced. The magnitude of the final running speed vibration vectors ranged from 5 to 8% of the diametrical bearing clearance. The suitability of this balance state is demonstrated by an extended process run on this machine. Additional details on the unbalance response of this machine are presented in case history 36.

FORCED RESPONSE CALCULATIONS

The programs used to calculate undamped critical speeds, stability, plus damped critical speeds, all display the final results as non-dimensional amplitudes. This is adequate for determining mode shape geometry, and identifying the stations of maximum deflection. However, these basic concepts must be significantly extended to duplicate actual rotor behavior.

This desirable simulation of rotor motion is addressed by a forced response analysis of a damped rotor system. Various forcing functions such as mass unbalance, skewed disks, or shaft bows are allowed in this type of analysis. In addition to previous rotor modeling capabilities, the configuration for a forced response analysis typically includes rigid plus flexible disks. This type of program accepts constant coefficients for bearings and supports, or coefficients that vary as a function of speed. Usually, the most accurate results are obtained by employing fifth or sixth degree polynomial functions that describe the stiffness and damping coefficients for the bearing oil film, and the housing as a function of rotational speed. The program computes Bode and polar plots, elliptical orbits, two and three-dimensional mode shapes, plus bearing forces. Some programs, such as UNBAL by Gunter[10] use a Complex Matrix Transfer method, and other programs such as DYROBES by Chen, Gunter, and Gunter[11] are based upon Finite Element Analysis (FEA) numerical methods.

Since support coefficients may be calculated as a function of rotational speed, shaft displacement response vectors may be computed with minimal discontinuity. One of the obvious applications for this information would be the development of synchronous 1X vectors in Bode plots as shown in Fig. 5-16. The

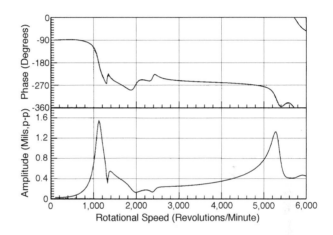

Fig. 5–16 Calculated
Gas Turbine Bode Plot

bottom half of the Bode plot displays 1X vibration amplitudes in Mils,$_{p-p}$ as a function of rotative speed in RPM. The data in the top half of Fig. 5-16 depicts the phase lag in Degrees. The same characteristics are shown in the polar plot of 1X vectors in Fig. 5-17. In both cases, the data is observed from a true vertical or horizontal perspective. The angular starting point for the phase angles begins at the probe location, and the convention follows standard phase lag logic with the angles progressing against rotation.

These analytical data presentation are designed to be analogous to the Bode and polar plots measured by proximity probe systems. The same 1X synchronous vectorial data is presented on both types of plots. The Bode plot displays excellent visibility of amplitude and phase changes with respect to speed, and the polar plot enhances the variations with respect to phase. In addition, the computer solution is not limited to the physical restrictions imposed on the physical installation of the proximity probes. In fact, the analytical model allows the generation of Bode and polar plots at any rotor station over any speed domain.

Since the entire rotor motion has been computed at numerous speeds, it is possible to construct both two and three-dimensional mode shapes of the scaled

---

[10] Edgar J. Gunter, "UNBAL, Unbalance Response of A Flexible Rotor - HP Version 4," Computer Program in Hewlett Packard Basic by *Rodyn Vibration, Inc.*, Charlottesville, Virginia, July, 1988, modified by Robert C. Eisenmann, Machinery Diagnostics, Inc., Minden, Nevada,1992.

[11] W.J. Chen, E. J. Gunter, and W. E. Gunter, "DYROBES, Dynamics of Rotor Bearing Systems, Version 4.21," Computer Program in MS-DOS® by *Rodyn Vibration, Inc.*, Charlottesville, VA, 1995.

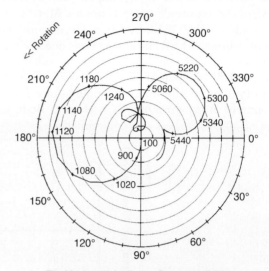

Fig. 5–17 Calculated
Gas Turbine Polar Plot

Tic Marks (*) Indicate Speed in RPM
Plot Radius = 1.8 Mils,p-p with 0.2 Mils,p-p/Div.

rotor behavior. For example, Fig. 5-18 depicts a two-dimensional rotor mode
shape superimposed upon an outline of the gas turbine rotor. The solid lines represent the predicted vertical vibration, and the dotted lines depict the horizontal
shaft motion. In all cases, scaling is provided via the left hand axis. This type of
scaled mode shape allows the comparison of anticipated displacement amplitudes with the actual machine clearances. On some machines, this type of information may not be particularly useful. However, on industrial turbines with
close tip clearances on the axial blades, this type of displacement data along the
rotor may be extraordinarily important.

The calculated shaft mode shape may also be viewed as a three-dimensional plot as shown in Fig. 5-19. In many cases this type of display is visually
more informative than the two-dimensional plot. This three-dimensional plot is

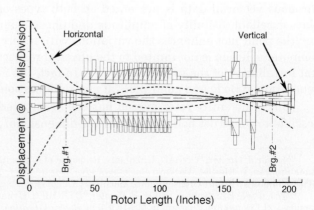

Fig. 5–18 Calculated Two-
Dimensional Plot for Horizontal And Vertical Gas Turbine Rotor Mode Shapes

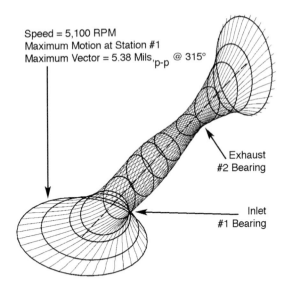

Speed = 5,100 RPM
Maximum Motion at Station #1
Maximum Vector = 5.38 Mils$_{\text{p-p}}$ @ 315°

Exhaust
#2 Bearing

Inlet
#1 Bearing

Fig. 5–19 Calculated
Three-Dimensional Gas
Turbine Mode Shape At
Normal Operating Speed

composed of shaft orbits at various locations, and it is scaled from a maximum vector that is listed on each diagram. The bearing locations are identified, and the speed is listed. This presentation provides an additional level of visibility to the calculated rotor mode shapes. The entire display may be rotated to observe or enhance the characteristics of a particular plot. The determination of calculated amplitude at specific rotor stations is easier with the two-dimensional plot of Fig. 5-18. However, the three-dimensional diagram of Fig. 5-19 does provide a better physical rendition of the shaft modal behavior.

The shaft orbits at any location and any rotative speed may also be extracted from the calculations and presented separately. Fig. 5-20 depicts a typical pair of calculated orbits from opposite ends of the turbine. The orbits are

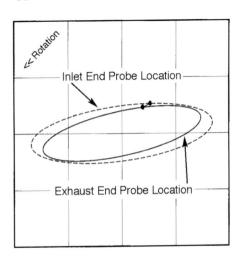

Rotation

Inlet End Probe Location

Exhaust End Probe Location

Fig. 5–20 Calculated
Gas Turbine Shaft Orbits
At Proximity Probe Mea-
surement Stations

viewed from the inlet end of the turbine, and the calculated orbits are oriented to be consistent with the measured shaft vibration data. The specific data used for Fig. 5-20 is representative of the anticipated shaft vibration at the actual probe locations. By changing the forcing function, such as various levels of unbalance at different locations within the turbine, the affect upon the overall mode shape, and resultant loads plus shaft vibration at the journal bearings may be calculated. Similarly, the impact of skewed wheels or bent rotors may be examined on paper before the machine is ever built.

Overall, it is evident that the computation of anticipated vibratory behavior along the length of a rotor provides useful information regarding the behavior of the machinery. If it can be demonstrated that this computational information is correct, and consistent with shaft vibration measurements, then a significant tool is available for the machinery designer as well as the diagnostician.

Case History 13: Gas Turbine Response Correlation

At the conclusion of a set of analytical calculations, the issue of verification of the results must be addressed. This is not an easy topic since comparison of analytical computations with field vibration measurements is seldom performed. As such, specific items of comparison are rarely defined, and tradition evaluations are often filled with generalities. Within the context of this chapter, it seems appropriate to perform a comparison on the basis of both qualitative and quantitative parameters. Specifically, synchronous characteristics of a single shaft gas turbine will be reviewed, and definable items during an unbalance response test will be correlated.

The 40,000 horsepower machine under discussion contains a 22,500 pound rotor that normally operates between 5,000 and 5,300 RPM. The unit is equipped with elliptical journal bearings, and a double acting Kingsbury type thrust bearing. This rotor contains seventeen stages in the axial flow air compressor, and two turbine stages as depicted in Fig. 5-21.

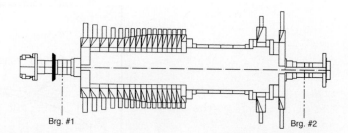

Fig. 5–21 Gas Turbine Rotor Configuration   Brg. #1   Brg. #2

In nearly all situations, the measured shaft vibration is elliptical, with the horizontal motion exceeding the vertical. In many cases, the orbit is tilted in the direction of rotation. At normal speeds, the phase angles between inboard and outboard orbits are almost identical. The same general behavior is noted in the analytical computations. For example, the orbit plots presented in Fig. 5-20 were extracted from a model of this machine. Hence, the general motion described by the analytical model reflects the measured field vibratory characteristics.

Transient speed characteristics are often difficult to understand due to the potential variation, and distribution of residual unbalance across the length of the rotor. For example, a mid span weight will drive the first critical, but will have minimal influence upon the pivotal mode. By the same token, a couple unbalance may be sufficient to excite the pivotal second critical, but the first translational mode may experience minimal excitation. Thus, the modal weight distribution will influence the critical speeds stimulated, and the amount of measured excitation. Although it is difficult to duplicate amplitude response through a series of resonant frequencies, it is reasonable to compare the measured critical speeds with the calculated natural frequencies. For instance, Table 5-5 compares the results for the first three critical speeds of this turbine rotor. The top row of calculated values displays the damped critical speeds with flexible supports. This analysis identifies a split horizontal (H) and vertical (V) critical for the first and second modes. The second row of calculated critical speeds were extracted from a forced response analysis that includes an improved definition of bearing and support characteristics. Finally, the bottom row summarizes the range of measured criticals for several different machines during multiple startup and coastdown field vibration data sets.

Table 5–5 Comparison Of Calculated Versus Measured Critical Speeds

Origin of Critical Speed	1st Mode Translational (RPM)	2nd Mode Pivotal (RPM)	3rd Mode Bending (RPM)
Calculated Damped	1,100 (H) 1,400 (V)	1,830 (H) 2,250 (V)	5,730
Calculated Forced	1,100 (H) 1,300 (V)	1,800 (H) 2,360 (V)	5,600
Field Measured	1,000 to 1,450	1,900 to 2,300	5,600+

Note that the first critical displays excellent agreement between the damped natural frequencies, the forced response criticals, and the measured critical speeds. Similarly, the pivotal second mode also shows excellent agreement between the calculated and measured resonant frequencies. The bending third critical is visible in the calculations, but is somewhat elusive in the field measurements. Since this mode is above the normal operating speed range, it can only be reached during over speed runs. These runs are usually of rather short duration, and the resonance generally has minimal time to respond. Overall, this agreement between the calculated and measured critical speeds provides increased confidence in the validity of the computations.

Another check on the accuracy of the model may be performed by installing an easily definable excitation on a real machine, and adding the same excitation to the model. A direct comparison of measured versus calculated vibration response characteristics should provide a suitable test of the model. For this test, consider the addition of unbalance calibration weights to each end of the turbine.

These weights would alter the 1X synchronous response, and the results should be visible in the vibration measurements, plus the analytical computations. For the purposes of this response test, an unbalance calibration weight of 77 Gram-Inches at 230° was added to the inlet coupling. The centrifugal force from this weight at 5,100 RPM was 125 Pounds (0.6% of the rotor weight). At the exhaust coupling, an unbalance of 234 Gram-Inches was attached at 275°. This weight produced a centrifugal force of 381 Pounds at 5,100 RPM (1.7% of rotor weight). Since the rotor residual unbalance was low, the vertical shaft vibration amplitudes were also small. Hence, the most meaningful data was extracted from the horizontal proximity probes.

Sequentially, an initial data set was obtained at 5,100 RPM without any extra unbalance. Next the machine was shutdown, the 77 Gram-Inch weight was installed at the inlet, and a second data set acquired. The turbine was again shutdown, and the inlet weight was removed. Next, the 234 Gram-Inch exhaust end weight was added, and a final data set was acquired at 5,100 RPM. The 1X vectors from the horizontal probes were runout compensated, and the results are summarized in Table 5-6.

Table 5–6 Measured X-Axis Vibration Response Vectors With Unbalance Weights

Weight Condition	Inlet Bearing #1	Exhaust Bearing #2
No Weight Installed	0.85 Mils,$_{p-p}$ @ 32°	1.01 Mils,$_{p-p}$ @ 346°
Weight at Inlet End	1.04 Mils,$_{p-p}$ @ 15°	1.38 Mils,$_{p-p}$ @ 330°
Weight at Exhaust End	1.06 Mils,$_{p-p}$ @ 350°	1.65 Mils,$_{p-p}$ @ 337°

It is obvious that the small weight installed at the inlet end of the turbine produced only minor changes, whereas the exhaust end weight resulted in a significantly larger change in shaft vibration. Since the initial synchronous 1X vectors are quite small, a comparable analytical case was developed with minimal shaft bow, and low residual unbalance. Specifically, a midspan shaft sag of 0.2 Mils (0.4 Mils TIR), was combined with a residual unbalance at the first compressor stage of 100 Gram-Inches. Another 100 Gram-Inch residual was located at the second stage turbine wheel. Calculations were performed at 5,100 RPM

Table 5–7 Calculated Horizontal Vibration Response Vectors With Unbalance Weights

Weight Condition	Inlet Bearing #1	Exhaust Bearing #2
No Weight Installed	0.93 Mils,$_{p-p}$ @ 281°	1.09 Mils,$_{p-p}$ @ 290°
Weight at Inlet End	1.10 Mils,$_{p-p}$ @ 260°	1.26 Mils,$_{p-p}$ @ 261°
Weight at Exhaust End	1.56 Mils,$_{p-p}$ @ 265°	1.67 Mils,$_{p-p}$ @ 274°

with the initial shaft bow, and the two residual unbalance locations. Two additional cases were run with the previously identified coupling unbalance weights. The computed 1X vibration vectors from these runs are presented in Table 5-7.

The initial rotor bow and residual unbalance vectors were selected to match the initial measured shaft vibration vector magnitudes. Since two sets of unbalance weights were used, both the direct and the cross-coupled balance response vectors may be calculated. The specific equations for these calculations are listed in chapter 11 of this text. For a two plane correction, equations (11-13), through (11-16) may be used. For example, the measured shaft vibration data at the turbine exhaust bearing may be used to calculate the balance sensitivity vectors from equation (11-16) as follows:

$$\overrightarrow{S_{22}} = \left\{ \frac{\overrightarrow{W_2}}{\overrightarrow{B_{22}} - \overrightarrow{A_2}} \right\} = \left\{ \frac{234 \text{ Gram Inches} \angle 275°}{1.65 \text{ Mils}_{p-p} \angle 337° - 1.01 \text{ Mils}_{p-p} \angle 346°} \right\}$$

$$\overrightarrow{S_{22}} = \left\{ \frac{234 \text{ Gram Inches} \angle 275°}{0.67 \text{ Mils}_{p-p} \angle 323°} \right\} = 349 \text{ Gram-Inches/Mil} \angle 312°$$

If the same calculations are performed for each set of primary and cross coefficients, the measured versus calculated balance sensitivity vectors may be generated as shown in Table 5-8. Note that the calculated sensitivity angles were adjusted by 45° to correct for the true horizontal orientation of the analytical calculations versus the +45° location of the proximity probe. Thus, the tabulated vectors in Table 5-8 are directly comparable in terms of angular position.

The similarity between measured and calculated sensitivity vectors in Table 5-8 lends further credibility to the validity of the analytical calculations. Certainly the analytically derived balance sensitivity vectors are not of sufficient accuracy to perform a refined field trim balance. However, they exhibit magnitudes that reflect the vibration response measurements, with reasonably consistent vector angles. Again, it is concluded that the analytical model does an excellent job of simulating the field dynamic behavior of the gas turbine rotor.

Table 5–8 Comparison Of Measured Versus Calculated Balance Sensitivity Vectors

Vector Identification	Measured Sensitivity	Calculated Sensitivity
Inlet Bearing - S_{11}	229 Gram-Inches/Mil,$_{p-p}$ @ 263°	190 Gram-Inches/Mil,$_{p-p}$ @ 340°
Inlet Bearing - S_{12}	329 Gram-Inches/Mil,$_{p-p}$ @ 338°	328 Gram-Inches/Mil,$_{p-p}$ @ 346°
Exhaust Bearing - S_{21}	156 Gram-Inches/Mil,$_{p-p}$ @ 294°	126 Gram-Inches/Mil,$_{p-p}$ @ 344°
Exhaust Bearing - S_{22}	349 Gram-Inches/Mil,$_{p-p}$ @ 312°	339 Gram-Inches/Mil,$_{p-p}$ @ 342°

Case History 14: Charge Gas Compressor with Internal Fouling

The centrifugal compressor depicted in Fig. 5-22 operates in cracked gas service. A low stage double flow compressor is coupled to the discharge end of this machine, and a high stage compressor is coupled at the thrust end of the subject compressor. This train is steam turbine driven at a maximum operating speed of 5,400 RPM. As noted, the rotor weighs 3,520 pounds, and it has a span of 107 inches between bearing centers. This compressor contains six impellers, and they are equally divided between the 2nd and 3rd process stages.

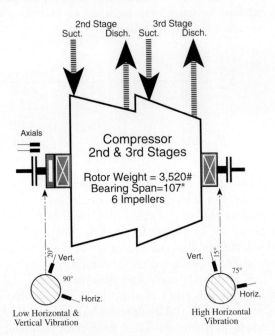

Fig. 5–22 Charge Gas Centrifugal Compressor Case Configuration

The compressor had been operating smoothly for an extended period of time when the horizontal vibration at the discharge end began to increase. The trend plot in Fig. 5-23 documents the vibration change over a four month period. At the beginning of this data, the machine displayed low and acceptable vibration amplitudes from all radial probes. A power outage in February resulted in an increase of vibration amplitudes at the discharge end. Approximately one week later, a problem with a seal pot float mechanism occurred, and vibration levels increased again. The amplitudes remained fairly constant throughout March, and then began a gradual downward trend towards the end of May.

It should be mentioned that these data points were acquired manually with a portable data collector on a weekly route. Changes or variations between these periodic samples are not visible. Hence, the transition between the low vibration condition on or about May 28, and the 6.0 Mil$_{p-p}$ value displayed on June 5 was unknown. Furthermore, the constituent parameters of rotational speed vectors, and radial position data was not available. The high vibration amplitude of 6.0

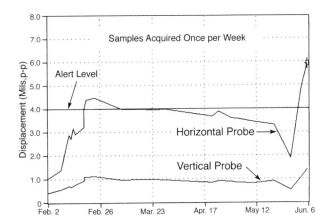

Fig. 5–23 Charge Gas Compressor Vibration Amplitude - Four Month Trend Plot

Mils,$_{p-p}$ on June 5th continued to increase until the horizontal probe exhibited an unfiltered amplitude of 6.9 Mils,$_{p-p}$. This behavior was documented in the orbit and time base plots in Fig. 5-24. Simultaneously, the suction end displayed low vibration amplitudes (1.3 Mils,$_{p-p}$), and this data is shown in Fig. 5-25. It was clear that the discharge journal was moving horizontally across the entire bearing clearance. That is, the 7 Mil vibration, plus 1 or 2 Mils for the oil film thickness is equivalent to the total diametrical bearing clearance of nominally 9 Mils.

The 1X vibration was reduced by unloading the compressor to allow operation at a lower speed. A further drop in vibration was achieved by reducing the oil supply temperature 7°F to increase the damping. This temperature reduction was accomplished by adding cold firewater to the water side of the oil cooler. Since this is a dirty gas service, the issue of coke buildup should always be considered. In this case, it was understood that a drop in efficiency had occurred during the past few months, but the specific decrease was not quantified. In retrospect, the plant personnel performed machinery efficiency calculations based on a heat and material balance. This was a poor method to determine compressor efficiency, and it turned out to be extraordinarily inaccurate. The only realistic approach to determine process machinery efficiency is to begin with an accurate measurement of the input shaft torque as discussed in chapter 6 of this text.

Wash oil rates were increased, with no measurable improvement. Based upon the available evidence, it was initially concluded that the discharge bearing was damaged. In addition, the suction end historical data was inconsistent. Specifically, the orbits in Figs. 5-24 and 5-25 describe a pivotal behavior across the compressor. The suction end phase had changed several times, and motion of this compressor was considered to be abnormal. Finally, it was agreed to shutdown the machinery, and prepare for a rotor, bearing, and seal change.

Following an orderly shutdown, the subsequent disassembly and inspection of the compressor resulted in several surprises. First, the discharge journal bearing was not damaged. In fact, the disassembly clearances were similar to the previous installation clearances. Second, the suction end bearing displayed babbitt

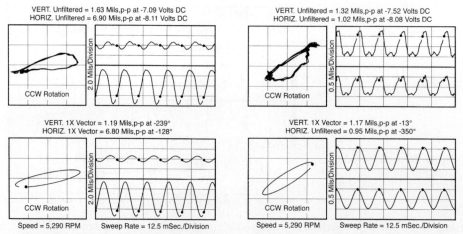

Fig. 5–24 Compressor Discharge End Bearing Shaft Radial Vibration

Fig. 5–25 Compressor Suction End Bearing Shaft Radial Vibration

damage on the bottom pads. Third, the compressor had a major accumulation of coke on the stationary, and the rotating elements. Inlet guide vanes, diaphragms, and return bends all exhibited various levels of coke deposits. In addition, the last stage wheel revealed major coke clusters at random locations within the impeller.

The compressor rotor contains three impellers for each process stage (2nd and 3rd). In both sections of the compressor, the inlet wheel for the respective stage was reasonably clean, and coke buildup increased progressively on the next two wheels. This is typical for a cracked gas machine to display increasing coke deposits as the heat of compression increases across the wheels that form the particular process stage. However, the amount of buildup on the last wheel in each process stage was substantial. Further examination of the casing revealed that most of the interstage labyrinths, and the balance piston labyrinths were completely filled with coke. At six locations, the mating surfaces on the rotor were highly polished, and the evidence of close contact between the rotor and the *filled-in* labyrinths was clear and unmistakable.

The condition of three of these surfaces is documented in Fig. 5-26. This photograph of the third process Stage shows the relatively clean inlet wheel on the left, and the heavily coked discharge end wheel on the right side. The polished shaft surfaces on the rotor are coincident with the coke filled interstage labyrinths. On the back side of the last impeller, the rotor balance piston resides. Although photographic evidence of this element is not as clearly defined, the balance piston displayed most of the same characteristics as the coke filled interstage labyrinths. The physical interpretation of this unique mechanical condition was hypothesized as a machine that was operating with a series of internal bearings. Specifically, the two external oil film tilt pad bearings were supplemented by six internal dry bearings. Five of these internal bearings were associated with

Fig. 5–26 Compressor Third Process Stage With Internal Coke Deposits Producing Midspan Pseudo Bearings

interstage labyrinths, and the sixth was at the discharge end balance piston.

The hypothesis of the development of six new internal bearings was examined in greater detail to determine if this could be responsible for the compressor high vibration problems. The only viable method to approach this problem would be with an analytical simulation of the machinery. The arrangement of shaft, impellers, spacers, and couplings for a normal rotor is depicted in Fig. 5-27. This machinery sketch identifies the proximity probe locations, and the radial journal bearings at each end of the rotor. Stiffness and damping coefficients for the oil film portion of these tilt pad bearings were computed. At a speed of 5,300 RPM, the calculated horizontal oil film stiffness K_{xx} was 350,000 Pounds/Inch. The vertical stiffness K_{yy} was computed to be 2,050,000 Pounds/Inch. The calculated horizontal oil film damping C_{xx} was 1,100 Pounds-Seconds/Inch. Finally, the vertical damping C_{yy} was 3,000 Pounds-Seconds/Inch. Since these are tilt pad bearings, cross-coupling coefficients do not exist. The journal and thrust bearing housing weight was approximately 200 Pounds, and the horizontal and vertical support stiffness (K_{sxx} & K_{syy}) for this housing were estimated at 2,000,000 Pounds/Inch.

Housing damping was calculated at 10% of the critical damping to be 200 Pounds-Seconds/Inch for the suction end bearing housing (C_{sxx} & C_{syy}) as per equation (4-17). The discharge end housing contains only a journal bearing, and the weight of this housing was estimated at 100 pounds. The vertical stiffness of this housing was set at 2,000,000 Pounds/Inch, and the horizontal stiffness was slightly reduced to 1,500,000 Pounds/Inch. The estimated vertical and horizontal damping values were proportionally reduced in accordance with the changes in stiffness and housing weight.

The normal model included a residual unbalance of 30 Gram-Inches at 175° on the suction end, and another 30 Gram-Inches at 145° on the discharge end. This total residual unbalance was set to be somewhat less than the normal bal-

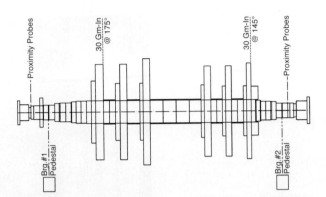

Fig. 5–27 Normal Compressor Rotor Configuration With Normal Bearings And Typical Residual Unbalance Levels

ance tolerance (113.4W/N) of 74 Gram-Inches for the entire rotor. This initial model allowed examination of the normal synchronous vibration response between 500 and 5,500 RPM. The computed response at operating speed provided an acceptable duplication of normal machine behavior. In addition, the transient calculations accurately predicted the first critical speed region centered at 2,500 RPM. Thus, the initial model (Fig. 5-27) successfully duplicated the historical machinery behavior. It was now reasonable to extend this model to the abnormal condition of a heavily coked compressor as shown in Fig. 5-26.

The rotor removed from the compressor was check balanced, and the residual unbalance determined at each end of the rotor. At the suction, the residual was 488 Gram-Inches at 218°. A much higher unbalance was discovered at the discharge end of the rotor with a measured 2,074 Gram-Inches at 201°. This synchronous excitation data was loaded into the model in conjunction with a 0.25 Mil midspan rotor sag. The support condition for the abnormal case required a minor modification of the existing bearings, plus the addition of the new internal bearings at the filled laby locations. The previous bearing housing characteristics were held constant. Similarly, the tilt pad bearing oil film coefficients at the discharge end were retained without modification. However, the suction end journal bearing coefficients were modified to reflect the demonstrated higher loads at this location. Horizontal stiffness K_{xx} at this location was increased to 1,500,000 Pounds/Inch, and the vertical stiffness was held at 2,050,000 Pounds/Inch. Finally, the oil damping at the suction bearing was held constant.

Internal compressor bearings were placed at each of the locations where the labyrinths were filled with coke, and there was obvious physical evidence of close clearance contact between the shaft and the laby areas. These internal bearings are identified as Brg. #2 through Brg. #7. The normal rotor journal bearings are shown as Brg. #1, and Brg. #8 on this model. The photograph in Fig. 5-26 shows the three internal bearings associated with the 3rd process stage as Brg. #5, Brg. #6, plus Brg. #7 at the balance piston. These internal bearing locations are identical to the locations on the model diagram presented in Fig. 5-28.

Computation of internal bearing coefficients was difficult due to the various unknowns associated with the internal behavior of this unit. Using short bearing

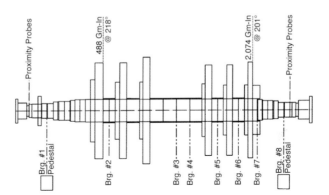

Fig. 5–28 Abnormal Compressor Rotor Configuration With Pseudo Internal Bearings And Measured Residual Unbalance

theory, calculations were performed at various clearances, and hydrocarbon gas viscosities. The resulting minimum stiffness values varied between 85,000 and 325,000 Pounds/Inch. On the high side of the potential stiffness envelope, values of 5,000,000 to 18,000,000 Pounds/Inch were computed. These maximum support stiffness would be reduced by the actual structural rigidity of the casing itself. Hence, the effective internal bearing stiffness would probably fall within the range of 400,000 to 1,500,000 Pounds/Inch.

In the final assessment, it was clear that a direct computation of support stiffness would not be attainable. A compromise value of 600,000 Pounds/Inch was selected for the vertical and horizontal support coefficients at internal bearings #2 through #6. The balance piston displayed less contact than the shaft labys, and a stiffness of 400,000 Pounds/Inch was used for this location. Cross-coupling coefficients, and all damping coefficients for these internal bearings were set to zero. Although this represents a simplistic model, the available mechanical data allows no other realistic alternative.

The forced synchronous response calculations were repeated for this abnormal case of multiple internal bearings, plus high unbalance. The results of these calculations at 5,300 RPM are presented in Fig. 5-29. The computed shaft vibration is compared with the measured 1X shaft orbits extracted from the previously discussed Figs. 5-24 and 5-25. In both sets of orbits, the same scaling of 2.0 Mils/Division was used, and both orbital sets display a true vertical/horizontal orientation.

Note that both discharge orbits in Fig. 5-29 are elliptical, and primarily horizontal. Also note that the horizontal magnitudes are similar, and the Keyphasor® dots for the discharge orbits reside in the same quadrant. Vertical magnitudes between the measured and computed discharge orbits show some variation; and the suction end orbits are rotated approximately 90° between the calculated and measured orbits. Nevertheless, the correlation between the measured and computed vibration response is considered to be acceptable.

Based on this ability to analytically duplicate the field machine behavior, it is reasonable to conclude that the compressor experienced the physical changes that were imposed upon the analytical model. Specifically, the high vibration lev-

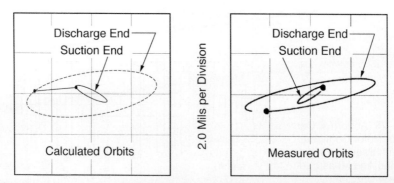

Fig. 5–29 Comparison Of Calculated Versus Measured Compressor Shaft Orbits

els at the discharge plus the low suction end vibration amplitudes were attributable to the combined effect of an internal coke buildup on the stationary internals (manifesting as internal bearings), plus large unbalance due to coke accumulation on the rotor. This combination of abnormalities resulted in a heavily loaded suction end bearing with low vibration (and pad damage), combined with a generally unloaded (and undamaged) discharge end journal that migrated across the available bearing clearance.

Once more, an analytical approach provides an acceptable simulation of a mechanical abnormality on a centrifugal machine. In this case, the physical evidence was used to develop a model that explained the abnormal behavior detected by the shaft sensing proximity probes. In all cases, it should be recognized that measurement and calculation technologies are coexistent resources that can provide significantly improved understanding of mechanical behavior.

Case History 15: Hybrid Approach To A Vertical Mixer

As demonstrated in the last two case histories, analytical solutions may be effectively combined with field vibration measurements to examine the machinery behavior from two different perspectives. This combination of techniques provides confidence in the individual technologies, plus the accuracy of the final results. It is clear that a comparison of calculated versus computed lateral vibration behavior makes good engineering sense. However, some physical situations cannot be properly examined by exclusively using only one technique. It these situations, it is necessary to combine the computational techniques with the physical measurements to arrive at a solution. This type of *hybrid approach* is not a common practice, but it does provide a way to get the job done with acceptable technical credibility.

As an example of this type of problem, consider the vertical mixer rotor displayed in Fig. 5-30. Charles Jackson would probably classify this assembly as the proverbial *mud ball on a willow stick*. The long and slender shaft is supported by two bearings at the top end, and two mixer wheels are located at the bottom of the rotor. A 30 inch elevation difference exists between the upper and lower mix-

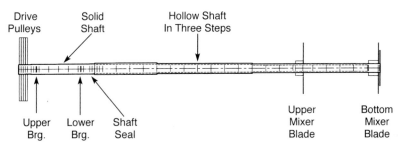

Fig. 5–30 Physical Configuration Of Vertical Mixer Rotor

ing blades. The distance between bearings is approximately 18 inches, and the vertical length of unsupported shaft approaches 116 inches. This rotor is driven by a variable speed motor via a belt and pulley configuration at the top of the assembly. In operation, the mixer is used in a batch process where the rotor is totally immersed in the process fluid, and pulley rotational speed is normally between 900 and 1,200 RPM. The radial bearings are rolling element units, and a mechanical seal is used to contain the process fluids.

The dual mixer blades have an outer diameter of 20 inches, and an average thickness of 0.188 inches. Various perforations and raised lips are fabricated into the blades to provide the necessary agitation action. This blade design was empirically based, and proven successful over many years of operation. However, due to process revisions, it would be necessary to install a thicker pair of mixer blades for future mixtures. The maximum anticipated thickness for the new blades was 0.488 inches. This blade thickness increase could add an additional 54 pounds to the rotor assembly. Since the initial rotor weight was 615 pounds, the additional blade weight represented a nominal 9% increase in the assembly weight. In addition, this extra blade weight represented an appreciable increase in the overhung mass.

During startup of this mixer with thin blades, it was observed that a critical speed existed between 250 and 300 RPM. Since this frequency was considerably below the operating speed range of 900 to 1,200 RPM, there was no interference between the resonance and normal running speed excitation. However, there was concern that the heavier mixer blades might have a detrimental influence upon the rotor critical speeds (especially the higher order modes). There was no information regarding rotor natural frequencies in the machinery files, and there was limited opportunity for traditional vibration response testing. As displayed in Fig. 5-30, the entire rotor is suspended from the two top bearings. During operation of the mixer, the only possible vibration measurements must be made from the exterior of the bearing housing. Obviously, this type of rotor will exhibit a variety of cantilevered modes, and vibratory motion at the bearings will be minimal under most conditions. Thus, direct casing vibration measurements will not be beneficial in solving this problem.

The undamped natural frequencies of the mixer rotor could be computed as discussed earlier in this chapter. Unfortunately, internal shaft diameters were not known, and the shaft material properties were reasonably undefined. Hence, a direct calculation of the critical speeds could not be attempted due to a lack of the fundamental mechanical information on the rotor.

The time honored *bump test* technique of *hit the stationary rotor with a 4x4 timber and measure the vibration response* could be used, but this approach leaves much to be desired. Although one or more natural frequencies would be excited, there is minimal ability to determine accurate mode shapes for each resonance, and virtually no way to separate out closely spaced or coupled modes.

From many aspects, a realistic engineering solution to this problem might seem to be unattainable. However, if the question is approached with multiple tools instead of a single technique, a logical *hybrid approach* may be developed. In this particular case, the initial step consisted of accurately measuring the static mode shape of the non-rotating shaft using an HP-35670A Dynamic Signal Analyzer plus an accelerometer, and a modally tuned impact hammer. The accelerometer was mounted close to the bottom mixer blade. The force hammer was used to impact the shaft at twelve different elevations at 10 inch increments up the length of the shaft. Frequency response functions (FRF) were then acquired between the accelerometer and each hammer location (acceleration / force). The data was checked for proper phase shifts, plus acceptable coherence as discussed in chapters 4 and 6. At this point, the FRF vectors at the various resonances could then be extracted and used to construct representative mode shapes.

Performing the above tasks manually can be a time consuming process. Handling a dozen FRFs is not impossible, but it is clear that a complex three-dimensional model may prove to be quite challenging. Hence, it is appropriate to consider methods of automating the field test, plus the associated calculations and animation of the resultant mode shapes. Historically, this type of work has been performed with large instrumentation systems operating under computer control. These types of measurement and data processing systems are complicated to set up and operate. In many cases, the field environment will not tolerate the time or expense associated with large scale modal tests.

A much more attractive approach resides in operating the DSA with software that is dedicated to modal analysis. In this specific case, the DSA was controlled with Hammer-3D[12] software that runs directly on the HP-35670A and eliminates the need for external devices. Within this software, the test element geometry and transducer array are physically defined. FRFs were acquired between the accelerometer and each hammer location as previously noted. Following a validity check of the averaged FRFs, curve fitting was applied to each of the first four resonant frequencies. The individual modes were then assembled, scaled, and presented as animated mode shapes on the DSA. Since this is a simple and symmetrical rotor, the Hammer-3D software was used in a single plane mode. The resultant mode shapes from these impact tests were committed to hard copy, and the first two modes are presented in Fig. 5-31.

As expected, the measured first mode was a pure overhung cantilever mode

[12] David Forrest, "Hammer-3D Version 2.01," Computer Program in Hewlett Packard Instrument Basic by Seattle Sound and Vibration, inc., Seattle, Washington, 1997.

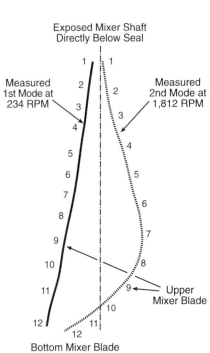

Fig. 5–31 FRF Measured Static Mode
Shapes Of Vertical Mixer Rotor

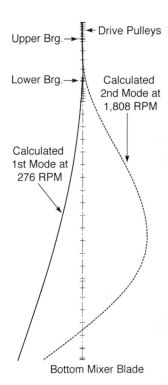

Fig. 5–32 Calculated Planar Mode
Shapes Of Vertical Mixer Rotor

that appeared at a frequency of 234 RPM. As shown in Fig. 5-31, the second mode displayed a zero axis crossing between the mixer blades, and it had a measured natural frequency of 1,812 RPM. This measured FRF data was obtained only on the exposed shaft sections below the shaft seal. There was no opportunity to acquire any meaningful FRF data in the vicinity of the bearings. Again, this is static mode shape data with a non-rotating shaft.

The next step consisted of generating an appropriate analytical model to simulate the measured behavior. This was a difficult task since the specific shaft material was unknown, and the internal hollow shaft diameters were likewise unknown. However, the total rotor weight was known to be 615 pounds, and the external shaft dimensions were easily measured. It was also noted that the top portion of the mixer shaft underneath the pulleys and bearings was solid. The hollow portion of the shaft was in three steps with decreasing diameters of 4.5, 4.0, and 3.5 inches. The weight of the pulleys and the mixer blades were measured on a shop scale, and the shaft material density was assumed to be 0.283 pounds per cubic inch. This density of steel was used since the shaft was magnetic, and therefore it was not any type of aluminum or stainless steel.

A simple model of the shaft was then constructed on a Microsoft® Excel spreadsheet. The external shaft dimensions were combined with the known component weights, plus the density of steel previously mentioned. It was assumed that the wall thickness for each of the three sections of hollow shaft were constant. This wall thickness for the hollow sections was then varied until the overall rotor weight matched the total physical weight of 615 pounds. This match occurred with a wall thickness of 0.5 inches, which seemed to be a reasonable value for this rotor assembly.

The dimensional rotor data from the spreadsheet was then loaded into the undamped critical speed program CRITSPD previously referenced in this chapter. In this software, a *planar* analysis was performed that consisted of setting the polar inertia terms to zero. Basically this is used to simulate a stationary non-rotating shaft. The bearing stiffness were then varied between 400,000 and 1,000,000 pounds per inch. As expected, this had little influence upon the calculated mode shapes or resonant frequencies. Certainly, this is a reasonable result since better than 95% of the strain energy was contained in the shaft, and less than 5% of the strain energy was in the bearings. Hence, the shaft properties controlled the natural resonant frequencies, plus the associated mode shapes.

The final piece of unknown data for performing the CRITSPD calculations was the modulus of elasticity E for the shaft material. Initially, the value for steel of 30,000,000 Psi was used. This produced planar modes that did not match the measured FRF results. A series of repetitive runs were made, and the value of E was incrementally reduced for each run. At a level of 21,000,000 Psi for E, the calculated planar results closely matched the measured FRF modes. Specifically, the first two computed modes are shown in Fig. 5-32.

The similarities between the measured FRF modes in Fig. 5-31 and the calculated CRITSPD modes in Fig. 5-32 are self-evident. The frequencies for both first and second modes are consistent, and the comparable mode shapes are virtually identical. The analytical model covers the entire rotor up through the drive pulleys, whereas the measured static model only addresses the exposed shaft. The largest deviation occurs in the frequency of the first critical. The measured FRF data provided a value of 234 RPM, and the calculated planar mode revealed a speed of 276 RPM for this first mode. Although the 42 RPM differential is an appreciable percentage of the resonant frequency, it is still well below the normal operating speed range.

Since the zero speed planar model matches the static FRF results, it is concluded that the analytical model is an acceptable representation of the mixer rotor. The next step requires activating the polar moment terms in the CRITSPD program, and performing a normal synchronous analysis. This run indicated that the first mode of the rotating shaft would occur at 277 RPM, and the second critical would increase to 1,841 RPM. The predicted first critical of 277 RPM was consistent with the plant observations of a resonance between 250 and 300 RPM. Furthermore, the calculated frequency of the second critical was considerably above the normal running speed range of 900 to 1,200 RPM.

At this point, the analytical model provided a good representation of the real machine. This similarity gave confidence to pursue the final step of increasing the thickness of the two mixing blades from 0.188 to 0.488 inches. This provided additional weight to the rotor, plus additional inertia due to the 20 inch

Table 5–9 Summary Of Measured And Calculated Natural Frequencies For Vertical Mixer

| Rotor Resonance | Original Thin Mixer Blades | | | Thick Blades |
	Static FRF Measurement	Planar Calculation	Synchronous Calculation	Synchronous Calculation
1st Mode	234 RPM	276 RPM	277 RPM	244 RPM
2nd Mode	1,812 RPM	1,808 RPM	1,841 RPM	1,705 RPM

diameter of these blades. This change dropped the first mode to 244 RPM, and it lowered the second critical to 1,705 RPM. Again, these frequencies are considerably removed from the operating speed range, and it is concluded that the additional blade thickness will not adversely influence the natural frequency characteristics of this vertical mixer.

For comparative purposes, the entire array of measured and calculated natural frequencies of this vertical rotor are summarized in Table 5-9. Additionally, the calculated mode shapes for the vertical mixer with the thicker mixer disks are presented in Figs. 5-33 and 5-34 for the first and second modes respectively.

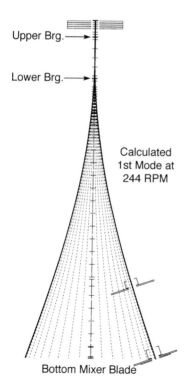

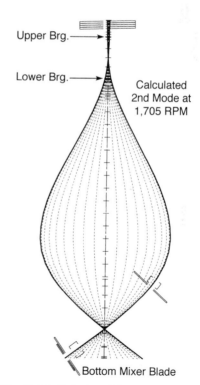

Fig. 5–33 Calculated Synchronous First Mode With Thicker Mixer Disks

Fig. 5–34 Calculated Synchronous Second Mode With Thicker Mixer Disks

In many ways, this vertical mixer case history is a simplistic example of interleaving measurements and calculations to achieve a realistic engineering solution. At all times, the machinery diagnostician must be cognizant of the fact that direct solutions are not always possible, and indirect or hybrid approaches are sometimes necessary to solve a problem. Furthermore, the accuracy of the final result does not have to extend to the third decimal point. In many cases, if you are within 5%, or perhaps 10% of the exact solution, that answer is fully acceptable within the field environment. You do not want to be inaccurate in your work, but then again, you do not want to try and attain some unrealistic measure of accuracy or precision.

BIBLIOGRAPHY

1. Chen, W.J., E. J. Gunter, and W. E. Gunter, "DYROBES, Dynamics of Rotor Bearing Systems, Version 4.21," Computer Program in MS-DOS® by *Rodyn Vibration, Inc.*, Charlottesville, Virginia, 1995.

2. Eisenmann, Robert C., "Some realities of field balancing," *Orbit*, Vol. 18, No. 2 (June 1997), pp. 12-17.

3. Forrest, David, "Hammer-3D Version 2.01," Computer Program in Hewlett Packard Instrument Basic by Seattle Sound and Vibration, inc., Seattle, Washington, 1997.

4. Gunter, Edgar J., "ROTSTB, Stability Program by Complex Matrix Transfer Method - HP Version 3.3," Computer Program in Hewlett Packard Basic by *Rodyn Vibration, Inc.*, Charlottesville, Virginia, March, 1989, modified by Robert C. Eisenmann, Machinery Diagnostics, Inc., Minden, Nevada, 1992.

5. Gunter, Edgar J., "UNBAL, Unbalance Response of A Flexible Rotor - HP Version 4," Computer Program in Hewlett Packard Basic by *Rodyn Vibration, Inc.*, Charlottesville, Virginia, July, 1988, modified by Robert C. Eisenmann, Machinery Diagnostics, Inc., Minden, Nevada, 1992.

6. Gunter, E. J. and C. Gareth Gaston, "CRITSPD-PC, Version 1.02," Computer program in MS-DOS® by *Rodyn Vibration, Inc.*, Charlottesville, Virginia, August, 1987.

7. Lund, J.W., "Modal Response of a Flexible Rotor in Fluid Film Bearings," *Transactions American Society of Mechanical Engineers*, Paper No. 73-DET-98 (1973).

8. Myklestad, N.O., "A New Method of Calculating Natural Modes of Uncoupled Bending Vibration of Airplane Wings and Other Types of Beams," *Journal of the Aeronautical Sciences*, Vol. 11, No. 2 (April 1944), pp. 153-162.

9. Prohl, M.A., "A General Method for Calculating Critical Speeds of Flexible Rotors," *Journal of Applied Mechanics,* Vol. 12, Transactions of the ASME, Vol. 67 (September 1945), pp. A142-148.

10. Singh, Murari P. and others, "SAFE Diagram - A Design and Reliability Tool for Turbine Blading," *Proceedings of the Seventeenth Turbomachinery Symposium*, Turbomachinery Laboratory, Texas A&M University, College Station, Texas (November 1988), pp. 93-101.

11. Thomson, William T., *Theory of Vibration with Applications*, 4th Edition, Englewood Cliffs, New Jersey: Prentice-Hall, 1993.

Transducer Characteristics

*P*revious chapters have discussed the fundamental motion characteristics of machinery, including the inter-relationship between displacement, velocity, acceleration, and frequency. Rotor mode shapes and the role of mass distribution, support characteristics, stiffness, and damping have also been reviewed. These physical characteristics have been expanded to examine some of the analytical computations available. It is understandable that mathematical modeling techniques provide some powerful tools for predicting the behavior of rotating equipment. However, it must be recognized that the real world is always more complicated than the models that are developed to explain physical phenomena. At best, the mathematical models are approximations of the actual physical system, and improvement or refinement of the model is often dependent upon correlations with field observations, and direct machinery measurements.

Over the years, many knowledgeable individuals have stated and restated that *one good measurement is worth a thousand expert opinions*. Hence, it is not only desirable, it is mandatory that machinery measurements be considered as a major problem solving tool. This topic will be explored from the standpoint of static measurements, such as position measurements versus machine clearances. In addition, the complex dynamic measurements associated with machinery vibratory behavior will be addressed throughout this text. Within this chapter, specific consideration will be given to the dynamic measurement transducers, and their associated signal characteristics.

In most cases, mechanical motion cannot be adequately quantified with human senses. Although the human eye can observe objects vibrating at amplitudes of 10 Mils,$_{\text{p-p}}$, and the ear can detect frequencies of 10,000 Hz, and differential acceleration levels of 0.1 G's,$_{\text{o-p}}$ are normally considered to be unpleasant, there are substantial differences in perception and threshold levels between individuals. In order to have some type of measurement quantification and repeatability, an interface device must be provided between the operating machinery and the diagnostician. The devices used for this task are electronic sensors, or transducers. These transducers convert numerous types of mechanical behavior into proportional electronic signals. The transducer outputs are usually converted into voltage sensitive signals that may be recorded and processed with various electronic instruments. Within this chapter, the operational characteristics of the major types of industrial transducers will be examined. In addition, the calibration methods, plus the fundamental advantages and disadvantages of each type of common industrial sensor will be reviewed.

BASIC SIGNAL ATTRIBUTES

Industrial transducers used for measurement of dynamic characteristics typically fall into three distinct categories. This includes the shaft sensing proximity probes, the mechanical motion velocity coils, and the solid state piezoelectric devices. Each of these three groups are generally used for measurement of displacement, velocity, and acceleration respectively. Each of the transducer types exhibits an array of strengths combined with a set of limitations. At this point in technology, there is no *universal sensor* that can be used for all measurements, on all machines, under all conditions. Thus, the machinery diagnostician must be intimately familiar with the characteristics of each type of transducer, and apply these devices to their best advantage.

Dynamic transducers measure events that occur in very small time increments. For instance, a thermocouple in process service may be used to measure temperatures that vary in minutes or even hours. Dynamic transducers used for vibration measurements must be able to detect phenomena that occur in fractions of a second. The resultant electronic signals are complex traces that contain significant information, and this data is generally quantified in terms of the following fundamental parameters:

- ❍ **Amplitude** (Magnitude or Severity)
- ❍ **Frequency** (Rate of Occurrence)
- ❍ **Timing** (Phase Relationship)
- ❍ **Shape** (Frequency Content)
- ❍ **Position** (from Proximity Probes Only)

The **amplitude** of the dynamic signal is generally proportional to the severity of the vibratory motion. That is, large amplitudes are directly related to high levels of vibration, and this is generally associated with the occurrence of a mechanical problem. Conversely, low vibration amplitudes are typically associated with proper behavior of the machinery, and the absence of mechanical difficulties. In most cases, the degradation of a piece of mechanical equipment is evident by increasing vibration amplitudes. There are exceptions to this general rule, and occasionally a machine will exhibit decreasing vibration amplitudes as mechanical condition degenerates (e.g., case history 50).

However, in most circumstances, amplitude is a direct indicator of vibration severity, and general machinery condition. The terminology used to define amplitude has originated from different sources, with several different meanings. For instance, consider Fig. 6-1 of a sine wave in the time domain. In this diagram three different types of amplitude measurements are identified. The total magni-

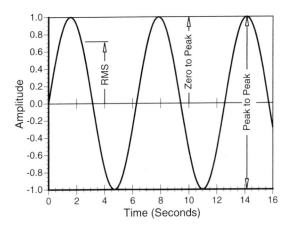

Fig. 6–1 Three Types Of
Common Sine Wave
Amplitude Designations

tude of the curve is represented by the *Peak to Peak* value. As its name implies, this measurement of amplitude extends from the lowest portion of the dynamic signal (i.e., bottom peak) to the highest portion of the signal (i.e., top peak). From Fig. 6-1, the minimum amplitude is -1.0, and the maximum or peak amplitude is equal to +1.0. Therefore, the total or peak to peak amplitude is equal to the total signal height of 2.0. Occasionally, historic references will be made to *double amplitude* measurements, which are synonymous with peak to peak values.

Shaft vibration measurements are generally expressed as peak to peak amplitudes. In the USA, units of Mils are used for displacement measurements (1 Mil = 0.001 Inches). This combination of engineering units and amplitude selection is abbreviated as Mils$_{p-p}$. It is convenient to use peak to peak displacement to relate the severity of shaft motion against the total bearing clearance. For instance, assume that a shaft radial vibration amplitude is 1.0 Mil$_{p-p}$, and the total bearing clearance is 10 Mils. The vibratory motion is easily determined to be 10% of the available diametrical clearance. Intuitively, this is an acceptable and comfortable level. However, if the shaft vibration was 9.0 Mils$_{p-p}$, the motion would be 90% of the available bearing clearance. This magnitude of vibration with respect to bearing clearance would be considered as unacceptable.

The second common type of amplitude measurement shown in Fig. 6-1 is the *Zero to Peak* value. This measurement extends from the middle of the dynamic signal to the highest portion of the signal (i.e., top peak). From the diagram, the maximum or peak is +1.0, and the zero to peak amplitude is 1.0. Casing vibration measurements are normally expressed as a zero to peak value. In the USA, the units of Inches per Second (IPS) are used for velocity measurements, and the acceleration of gravity in G's are used for acceleration. This combination of amplitude measurements and engineering units are generally abbreviated as IPS$_{o-p}$ and G's$_{o-p}$ respectively. Clearly, for a simple voltage signal, the relationship between these peak based amplitude measurements may be accurately expressed as:

$$Amplitude_{peak\ to\ peak} = 2 \times Amplitude_{zero\ to\ peak} \qquad \textbf{(6-1)}$$

The third common type of amplitude measurement is the *Root Mean Square* (RMS) value. As shown in Fig. 6-1, the RMS amplitude is lower than the zero to peak value. For a pure sine wave, the actual reduction is equal to $\sqrt{2}/2$, or a numerical value of 0.7071. The following conversions may be used to relate all three types of common amplitude measurements:

$$Amplitude_{rms} = 0.7071 \times Amplitude_{zero\ to\ peak} \qquad \textbf{(6-2)}$$

$$Amplitude_{rms} = 0.3536 \times Amplitude_{peak\ to\ peak} \qquad \textbf{(6-3)}$$

RMS values are used in Europe for vibration measurements. However, in the USA this type of magnitude measurement is seldom used for machinery vibration. The largest utilization of RMS occurs with electrical devices such as voltmeters. For instance, if a digital multimeter is used to measure AC voltage from a wall outlet, a reading of 115 volts may be observed. This voltage of 115 Volts,$_{rms}$, is equivalent to 325 Volts,$_{p-p}$ based upon equation 6-2. Hence, the common household power outlet really has 325 volts of electrical potential difference.

Other types of voltage measurements such as *Average* amplitudes are occasionally used. However, within this text, vibration amplitudes will be expressed in accordance with the previous discussion. Specifically, vibration amplitude measurements within this book will be stated as:

Displacement:	Mils,$_{peak\ to\ peak}$	=	Mils,$_{p-p}$
Velocity:	IPS,$_{zero\ to\ peak}$	=	IPS,$_{o-p}$
Acceleration:	G's,$_{zero\ to\ peak}$	=	G's,$_{o-p}$

Conversion of vibration units at a fixed frequency may be achieved with equations (2-17) through (2-22). Using these expressions, one vibration unit may be converted into another. For instance, a casing velocity measurement in IPS,$_{o-p}$ may be converted to a casing displacement measurement in Mils,$_{p-p}$. This type of conversion is useful when shaft versus casing measurements must be correlated. In all cases, the frequency of vibration must be included in the conversion calculation. Furthermore, the timing or phase relationship between vibration signals should also be considered.

The signal **frequency** is determined by the reciprocal of the period. To use a consistent example, the previously discussed sine wave is reproduced in Fig. 6-2. In this diagram, three different time ranges are identified on the curve. Basically, all three ranges are all identical, and they each define one cycle. In one case, a zero axis crossing is used to identify a complete cycle. In the other two ranges, the time between two consecutive peaks, or two consecutive valleys are identified. It does not matter what part of the cycle is used to determine the period, as long as the measurement point is identical from one cycle to another. For a repetitive signal, the time required to complete one cycle will remain con-

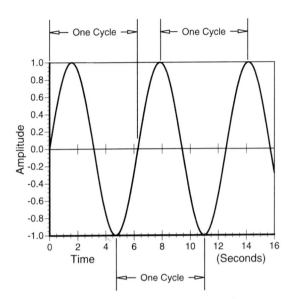

Fig. 6–2 Period Of A Sine Wave Measured At Three Different But Repetitive Locations

stant. The associated frequency may be determined from equation (2-1), and this expression is repeated as follows:

$$Frequency = \frac{1}{Period}$$

(6-4)

Period is measured in time units of Seconds or Minutes, and it identifies the length of time necessary to complete one cycle. If the *Period* is measured in Seconds, then the reciprocal frequency must carry the engineering units of Cycles per Second (CPS). This is also know as a *Frequency* in hertz (Hz). Similarly, if the *Period* is measured in Minutes, then the *Frequency* will carry units of Cycles per Minute (CPM), or Revolutions per Minute (RPM) if a machine speed is identified. The relationship between these frequency units are as follows:

$$Frequency \text{ (CPM)} = 60 \times Frequency \text{ (CPS or Hz)}$$

(6-5)

The example sine wave shown in Fig. 6-2 has a period of 6.28. If the time scale is in seconds as noted on this plot, then the fundamental period of the signal is 6.28 seconds. The frequency may be determined from the previous equations (6-4) as follows:

$$Frequency = \frac{1}{Period} = \frac{1 \text{ Cycle}}{6.28 \text{ Seconds}} \times 60 \frac{\text{Seconds}}{\text{Minute}} = 9.55 \frac{\text{Cycles}}{\text{Minute}}$$

For comparative purposes, now assume that the time scale on Fig. 6-2 is in milliseconds instead of seconds. One second contains 1,000 milliseconds, hence a conversion factor of 1,000 exists between the two scales. With this new scale, the fundamental Period of the sine wave would be 6.28 x 10^{-3} seconds. The frequency may again be determined from equation 6-4:

$$Frequency = \frac{1}{Period} = \frac{1 \text{ Cycle}}{0.00628 \text{ Seconds}} \times 60 \frac{\text{Seconds}}{\text{Minute}} = 9,550 \frac{\text{Cycles}}{\text{Minute}}$$

From this example, it is clear that low frequency motion has a long period (e.g., an earthquake). High frequency vibration is necessarily associated with a short period (e.g., a turbine rotational speed). This basic relationship is not only important during the analysis of mechanical behavior, it must also be addressed during the selection of an appropriate vibration transducer suite, and the associated diagnostic instrumentation.

Before considering the details of any vibration transducer, it is necessary to discuss the **timing** between events. This is a vitally important point, since very few mechanical problems are solved based upon one measurement with a single transducer. In most instances, multiple transducers are applied, and the signal outputs are examined for relative amplitudes, frequency content, and the timing or phase between signals.

This concept of timing may be expressed in various ways, and it is often confused by different triggering schemes. In the simplest format, timing is just the time delay between two signals as depicted in Fig. 6-3. The solid line represents the same sine wave that has been used for the two previous examples. The dotted line describes another sine wave that is offset in time by one quarter of a cycle from the solid line. On a rotating machine, one complete cycle is equal to one revolution, or 360 degrees. Hence, one cycle is 360°, and a quarter of a cycle is 90° as shown in Fig. 6-3.

The relative timing between the two sine waves may be described in two different ways. With time progressing from left to right, it may be properly stated that the solid curve leads the dotted curve by 90°. It may also be stated that the dotted curve lags the solid curve in time by 90°. Both statements are correct, and it can be very confusing if the same parameter is described in more than one way.

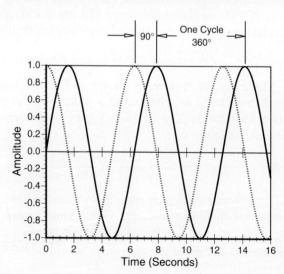

Fig. 6–3 Phase Or Timing Differential Between Two Sine Waves

To alleviate any potential of misinterpretation, the following six rules for phase angles shall be consistently applied throughout this text:

1. All phase measurements require a minimum of two signals. One signal will be the dynamic motion signal (e.g., vibration), and the other signal must be the trigger signal (e.g., Keyphasor®).
2. Phase measurements require that both the dynamic motion signal and the trigger occur at the same frequency.
3. Phase measurements must be made within the same time cycle.
4. The positive peak of the filtered motion signal (e.g., vibration) shall always be considered as the zero degree (0°) reference point for that transducer.
5. Looking backwards in time, the phase angle shall be the angular distance between the positive peak of the motion signal, and the first trigger point.
6. Phase angles shall be expressed as degrees of phase lag. The term *lag* implies a negative angle, i.e., against rotation, and backwards against time.

These rules apply for all measured vibration, pressure pulsation, force signals, etc. Vectors derived from the measured dynamic signals shall maintain the same rules, and all analytical calculations shall be configured to conform to similar criteria. Finally, the location of balance weights, and other references to specific angular locations on a rotor shall be consistent with the above rules.

To illustrate this phase convention, Fig. 6-4 was prepared. This sketch describes the reference sine wave, combined with two different timing marks (dots). The depicted events are filtered at the same frequency, and the peak of the motion signal is identified as the zero degree (0°) reference point. If the trigger point was coincident with this positive peak of the vibration signal, the signal phase angle would be 0°. If the trigger signal occurred one quarter of a cycle earlier, the resultant signal phase angle would be a 90° lag.

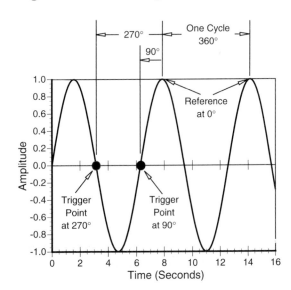

Fig. 6–4 Sine Wave Combined With Two Different Phase Reference Points

The second trigger point on Fig. 6-4 is located at three quarters of a cycle before the peak of the motion (or vibration) signal. The phase angle for three quarters of a cycle (3/4 times 360°) is equal to 270°, and this trigger point is also identified. In a similar manner, all phase angles will be determined, and reported in consistent units of degrees of *phase lag*.

One of the popular misconceptions on phase measurements is that the phase angle is referenced or determined from the angular location of the timing or Keyphasor® transducer. It must be recognized that the timing probe is used to provide an angular reference point between the rotating system and the stationary mechanical system. As discussed throughout this text, the timing probe identifies the angular or rotative position of the rotor at the exact instant in time when the trigger pulse is produced. The actual phase measurements are measured against rotation from the angular location of the vibration transducer. To state it another way, the angular position of the vibration probe always represents zero degrees (0°), irrespective of the timing probe location.

The trigger point in Fig. 6-4 is shown as a single dot. In actuality, the dot originates from analog oscilloscope utilization techniques where a pulse shaped timing signal is connected to the scope Z-Axis input. A negative going pulse input would produce a *blank* followed by a *bright* spot as shown in Fig. 6-5. The Z-Axis oscilloscope input of the timing signal is also referred to as *blanking*. For most oscilloscopes, this *blank-bright* sequence is superimposed upon the main dynamic motion signal. The portion of the trigger signal with a negative (downhill) slope produces a void or *blank* spot on the main motion signal. Conversely, the portion of the trigger signal with a positive (uphill) slope produces an intensified *bright* spot upon the motion signal. If a digital oscilloscope is used instead of an analog scope, the dot intensification does not occur, and only a *blank* spot

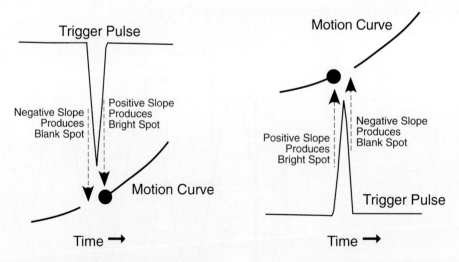

Fig. 6–5 Negative Trigger Pulse Signal And Associated *Blank-Bright* Sequence

Fig. 6–6 Positive Trigger Pulse Signal And Associated *Bright-Blank* Sequence

appears on the display. The width of the digital scope *blank* spot is equal to the width of the timing pulse.

A negative going trigger pulse is typical for a proximity probe observing a notch or a hole drilled into a shaft (e.g., Fig. 6-5). For a probe triggering off a projection such a shaft key, or for an optical transducer, a positive going trigger signal will be generated. This type of positive pulse is shown in Fig. 6-6. It is noted that the positive slope of the trigger still produces a *bright* spot, and the negative slope still shows up as a *blank* spot on the dynamic motion curve. The sequence of these events are now reversed to be *bright* then *blank* due to the directional characteristics of the timing pulse. Since all oscilloscopes have a sweep that displays signals from left to right on the screen, the actual sequence for any type of pulse on any oscilloscope can be easily determined by visual observation of the signals in the time domain. Again, this type of *blank-bright* display only applies to an analog oscilloscope such as a Tektronix 5110. A digital scope with blanking such as the HP-54600B will only display the *blank* spot.

The synchronizing trigger on an oscilloscope, plus the trigger on a tracking filter or DSA generally require the user to select a positive or a negative slope. Often these devices are set to trigger at about 50% of the slope (halfway between the upper and lower voltage). On many instruments, it is also possible to manually set the trigger at any other part of the pulse slope. In all cases, it is highly recommended that the input Keyphasor® signal be observed in the time domain, and the trigger point verified. On some signals it may be necessary to manually adjust the instrument trigger point to obtain a consistent trigger signal. From a mechanical interface standpoint, the physical trigger point is discussed in further detail in chapters 7, 8 and 11 of this text.

Many computerized data acquisition and processing systems have mimicked this analog oscilloscope logic into the computer output displays. Hence, the *blank-bright* or *bright-blank* analog timing convention remains as an integral part of the data presentation. This ability to precisely measure the timing between events proves to be very useful for incidents occurring at the same frequency. However, most industrial machines display more than a single discrete running speed frequency.

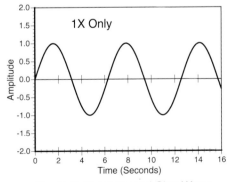

Fig. 6–7 Fundamental Sine Wave

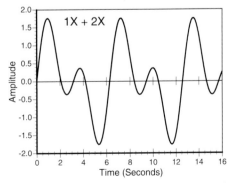

Fig. 6–8 Fundamental Plus 2nd Harmonic

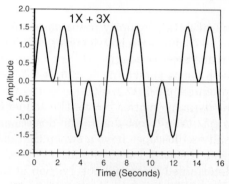

Fig. 6–9 Fundamental Plus 3rd Harmonic   **Fig. 6–10** Fundamental Plus 4th Harmonic

For instance, consider the time domain plots in Figs. 6-7 through 6-10. In each plot, the fundamental 1X amplitude, frequency, and timing are identical to the sine wave originally used in Fig. 6-1. However, it is apparent that the overall amplitudes, and the **shape** or general appearance of Figs. 6-8 through 6-10 have significantly changed from the initial sine wave. These changes are due to the addition of higher order harmonics to the fundamental. In all three cases, the additional harmonic components are set to be even multiples of the fundamental, and the amplitudes are equal in magnitude to the fundamental. The resultant time domain curves each have distinctive shapes or patterns that may be further altered by changing the respective amplitudes and/or the timing relationships of the higher order harmonics. The inclusion of non-synchronous frequencies will produce additional pattern variations. Specifically, case histories on signal summation, amplitude modulation, and frequency modulation are presented in chapter 7.

The complexity of time domain signals manifests as a wide variety of potential shapes and patterns. It is easy to visually recognize some of the simple signal combinations, but the more complex signals may not be comprehensible in the time domain. For these situations, additional signal processing techniques and instruments are required.

Various types of filters and frequency analyzers may be employed to dissect these time domain signals into quantifiable and manageable portions. These electronic devices and their specific attributes are reviewed in chapter 8. However, prior to any further discussion of signal manipulation, it would be desirable to examine the characteristics of the dynamic transducers that are used for the majority of the industrial measurements on process machinery.

PROXIMITY DISPLACEMENT PROBES

Shaft sensing proximity probes are used to obtain relative displacement measurements of rotating or reciprocating shaft surfaces. These are non-contacting transducers that are mounted on a reasonably stationary mechanical structure (e.g., bearing housing). From the mounting point they observe the static and dynamic displacement behavior of the moving machinery element. Proximity probes are supplied in a wide variety of sizes, shapes, and configurations. For example, the photograph in Fig. 6-11 depicts four different sizes and configurations of Bently Nevada Corporation proximity probes.

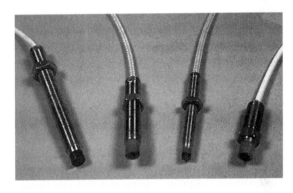

Fig. 6–11 Typical Configurations Of Proximity Probes Manufactured by Bently Nevada Corporation

The left hand probe in this photo consists of a fully threaded stainless steel body with an 8 mm Ryton® probe tip. The protective probe tip may also be constructed of durable high performance plastics such as polyphenylene sulfide (PPS) that is capable of withstanding harsh physical and/or chemical environments. For less severe applications, the two middle probes in Fig. 6-11 contain fiberglass tips of 8 mm and 5 mm diameters mounted in fully threaded bodies. Finally, a different body configuration is shown in the probe at the right side of the photograph. This is a reverse mount probe that screws into a mounting assembly (stinger) that can be cut to length to accommodate a variety of installation requirements.

A probe tip diameter of 5 mm (0.0197 inches) is typically mounted in a stainless steel body with an external thread of 1/4x28 UNF. Similarly, the standard 8 mm (0.315 inches) diameter probe tip is usually mounted in stainless steel body with external 3/8x24 UNF threads. Smaller diameter probe tips (e.g., 1/8") with lower dynamic ranges, and larger diameters (e.g., 2") for extended range measurements are also commercially available. However, the vast majority of the industrial applications use 5 mm and 8 mm probes.

Regardless of the physical configuration, all eddy current proximity probes consist of the same basic components. For example, consider the sketch of a typical probe assembly in Fig. 6-12. In this diagram, a flat wound coil located close to the probe tip is connected by two wires to a coaxial cable that runs between the probe and Proximitor®. This coaxial cable must be electrically tuned to a specific length in order to maintain the proper impedance between the probe and the

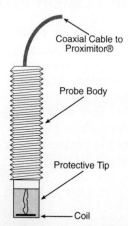

Coaxial Cable to
Proximitor®

Probe Body

Protective Tip

Fig. 6–12 Typical Components of A Proximity Probe

Coil

Proximitor®. If the interconnecting cable length is altered from the correct value, transducer calibration will be influenced.

All external wiring connections are terminated at the Proximitor®. Typically this is a fully shielded three conductor cable. The cable shield is normally grounded at the monitor rack, and the field end of the shield is allowed to float. This prevents the development of ground loops in the transducer wiring. Within the three wire cable, one wire is used for the output signal (white), the second wire provides a common ground (black), and the third wire is the power input (red). A power input of -24 volts DC is applied to the Proximitor® from a monitor, or a regulated DC power supply. The Prox contains an internal oscillator that converts some of the input energy into a radio frequency signal in the megahertz range. This high frequency signal is directed to the probe coil via the coaxial cable. The Proximitor® is also know generically as an Oscillator-Demodulator.

The flat pancake coil at the tip of the probe broadcasts this radio frequency signal into the surrounding area as a magnetic field. If a conductive material does not intercept the magnetic field, there is no power loss in the radio frequency signal. However, if a conductive substance intercepts the magnetic field, eddy currents are generated on the surface of the material, and power is drained from the radio frequency signal. As the conductive material approaches the probe tip, additional power is consumed by the eddy currents on the surface of the conductor. When the probe is in contact with the conductive material, the majority of the power radiated by the probe tip is absorbed by the material. As the power loss varies, the output signal from the Proximitor® also exhibits a change in voltage. In all cases, a small gap produces a small output voltage, and a large gap results in a large output voltage from the Proximitor®.

The relationship between distance and output voltage is achieved by a combination of electronic circuits within the Proximitor®. For instance, a demodulator removes the high frequency carrier signal, and a linearization circuitry provides a reasonably flat curve over a typical range of 80 to 100 Mils (0.080 to 0.100 Inches). These signal output characteristics may be easily checked by performing a simple distance versus voltage calibration. This type of calibration is generally performed with a spindle micrometer device as shown in Fig. 6-13. In this configuration, the proximity probe remains in a fixed position, and the spindle micrometer is used to accurately move a circular target back and fourth against the probe. Other fixture designs may be used that consist of a stationary target combined with a moveable probe mount. In either case, the probe to target distance must be both adjustable, and accurately measurable.

Fig. 6–13 Proximity Probe Static Calibration Fixture - Spindle Micrometer From Bently Nevada Corp. TK-3

As previously mentioned, when the probe tip is in physical contact with the observed conductive surface, the transducer output is at a minimum voltage level. As the distance between the probe and the target material increases, the output voltage increases in a proportional manner. A typical calibration procedure requires the tabulation of output voltage versus the physical distance between the probe and the target. For accuracy, these readings are generally performed at 5 Mil (0.005 Inch) increments, and the results are normally presented in a graphical format such as Fig. 6-14.

Within this calibration plot, the vertical axis displays the Proximitor® DC output voltage. Since a minus 24 volt DC power supply was supplied to the Proximitor®, the output voltages are also negative voltages. It is understandable that the final output voltage remains less than the power input level. The horizontal axis in this plot presents the physical distance between the probe tip and the target surface in Mils. For this particular calibration plot, a range of 10 to 110 Mils was spanned. The actual coordinates for each plotted point are listed next to the curve. This information not only provides good documentation of the calibration curve, it also allows the easy computation of transducer sensitivity. In all cases, proximity probe sensitivity is defined as the slope of the calibration curve as in the following calibration expression:

$$Proximity\ Probe\ Sensitivity\ = \frac{Differential\ Voltage}{Differential\ Gap} \qquad (6\text{-}6)$$

It is presumed that the linear portion of the calibration curve is a straight line. If deviations from a straight line are encountered, there is probably something wrong with the transducer system or the calibration setup. Once a proper straight line is established for the linear portion of the curve, the transducer sensitivity may be determined by computing the slope of the line. If voltage output

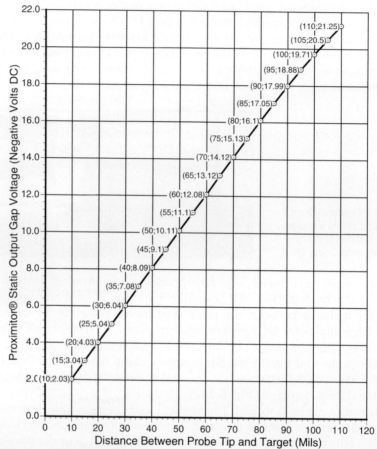

Fig. 6–14 Proximity Probe Static Calibration Curve On 4140 Target Material

values for 20 and 100 Mils are extracted from the calibration plot in Fig. 6-14, the probe sensitivity may be determined as follows:

$$Sensitivity = \frac{19.71 - 4.03 \text{ Volts}}{100 - 20 \text{ Mils}} = \frac{15.68 \text{ Volts}}{80 \text{ Mils}} \times 1,000 \frac{\text{milliVolts}}{\text{Volt}} = 196 \frac{\text{milliVolts}}{\text{Mil}}$$

The normal voltage sensitivity for this class of transducer is 200 millivolts/ Mil ±5%. Thus, values between 190 and 210 millivolts/Mil are acceptable, and the calculated slope of 196 millivolts/Mil falls within this acceptance range. In many cases, the proximity probe calibration is simply used to verify compliance with the standard sensitivity used for system calibration, trending and analysis. However, in situations requiring additional measurement accuracy, the actual curve slope will be used. For numerical computations it is common to perform a curve fit of the calibration curve. For instance, a linear curve fit on the probe calibration plot in Fig. 6-14 produces the following expression for determining output voltage based upon a gap for this particular proximity probe:

$$Voltage = 0.1957 \times Gap + 0.2395$$

This expression may be used for various calculations, or it may be incorporated into computerized monitoring or trending programs. Since the calibration curve is nominally a straight line, the 0.2395 value is the zero gap intercept point. Also, the 0.1957 multiplier is the slope of the curve that corresponds to the manually calculated value of 0.196 Volts per Mil (196 mv/Mil). It is useful to recognize that the reciprocal of the standard sensitivity is 5.0 Mils per Volt. This format is sometimes easier to remember, and easier to convert units. For example, if a transducer exhibits a 2.0 volt change in DC gap voltage, this is equivalent to a 10.0 Mil shift in gap between the stationary probe and the observed surface (i.e., 2.0 Volts x 5.0 Mils/Volt = 10.0 Mils).

Since proximity probe requires an electrical coupling to the observed surface, it is reasonable to conclude that variations in the target will influence the transduction. In actuality, anything that distorts the crystal lattice of the conductor will influence probe sensitivity. Changes in electrical resistivity, magnetic permeability, residual magnetism, or localized stress concentrations will appear as noise on the signal. Dependent on their magnitude, they may also change the scale factor. Material changes will have a significant effect upon the sensitivity. In the preceding discussion, sensitivity was assumed to be 200 mv/Mil. If this was calibrated on 4140 steel, the same transducer system would exhibit output sensitivities with other common metals as summarized in Table 6-1.

Table 6–1 Variation Of Proximity Probe Voltage Sensitivity With Different Metals

Material	Probe Voltage Sensitivity
Copper	380 millivolts/Mil
Aluminum	370 millivolts/Mil
Brass	330 millivolts/Mil
Tungsten Carbide	290 millivolts/Mil
Stainless Steel	250 millivolts/Mil
Steel — 4140 or 4340	200 millivolts/Mil

These variations in signal voltage sensitivity may be used directly for computation of displacement. In addition, any monitors, computers, or analytical instrumentation connected to the probe output may be programed with the precise sensitivity for the observed material. Alternatively, the Proximitor® may be electrically tuned to the new material, and the output scale factor adjusted to obtain a consistent output of 200 mv/Mil. Either approach may be successfully applied. However, it is highly recommended that a uniform and consistent approach within the entire operating complex be established and maintained.

Regardless of the specific calibration for a transducer, it is clear that the proximity probe provides a direct means for converting distances into electronic voltages. As such, this type of transducer has been referred to as an *electronic micrometer*. This characteristic is widely used in many industrial applications. For instance, measurement of average axial (thrust) position of rotating elements with respect to their stationary casings has evolved into a fundamental machinery protection measurement.

For thrust position measurements, the machine is assembled with knowledge of the forward and reverse rub points of the rotor against the stationary elements. Thrust bearing position is shimmed to obtain a specific axial clearance within the machine (e.g., 1st stage nozzle clearance in a steam turbine). Float shims are then ground to maintain the correct float of the thrust collar(s) within the bearing. As illustrated in Figs. 4-21 and 4-22, the physical configuration of the thrust bearing, the actual probe location, plus the direction of normal thrust loads must be considered during setup of a thrust probe installation.

When the mechanical clearances are properly established and verified, the thrust (axial) proximity probes are installed. Typically, two probes are mounted at each thrust bearing to provide full redundancy. These thrust probes are usually connected to a dual voting thrust position monitor. In order to have full confidence in the thrust monitor readings, it is necessary to have physical verification of the mechanical thrust setup, and full correlation throughout the system. This is achieved by acquiring and comparing several different measurements. Normally, a dial indicator is positioned at the end of the shaft, and zeroed with the rotor sitting hard against the active shoes. A digital multimeter (voltmeter) is connected to the Proximitor® output to allow measurement of the output voltage. Normally, a second digital multimeter is connected to the monitor input to verify that excessive line losses have not occurred between the Proximitor® output and the monitor input. In addition, the meter reading from the thrust position monitor must be obtained. If a computerized trending system is installed, the thrust position indication, plus the digitized gap voltage readings must be compared against the actual transducer output values.

With the various readings established, the rotor is physically bumped or moved back and forth between the active and the inactive thrust shoes. Depending on the philosophy of the particular operating company, the zero point on the thrust monitor may be set for the active thrust shoes, or the middle of the float zone. Initially, this is an arbitrary point, but once established, it must be maintained for all future measurements on the machine. Furthermore, all machines within a plant complex should be set up with the same reference scheme. It is unnecessarily confusing to all parties to have some machines with a zero thrust position at the center of the float zone, and other units with zero equal to the rotor positioned hard against the active thrust shoes. Pick one scheme, and stay with it for all of the machines within an operating complex.

During a routine setup, the probes are initially set to a gap voltage at the middle of the curve (e.g., approximately -12 volts DC). The rotor is then manually thrust back and forth between the active and inactive thrust shoes. Based upon the proximity probe calibration curve, the desired zero point, and the actual thrust bearing float zone — the probes are then reset to realistic gap voltages.

Once the probes are set and locked into position, the rotor is bumped back and forth at least three times to compare and verify all readings. Normally, it is good practice to obtain a set of *soft* readings when the rotor stops at the thrust shoes. Next, a heavy pull on the pry bar will usually result in additional movement to a *hard* thrust position. The difference between the *soft* and the *hard* thrust is due to compression of any springiness in the thrust assembly. With full thrust loads, the machine should typically run in the *hard* thrust positions.

When all of the readings are consistent, the rotor is rolled 180° (half a turn), and three more sets of readings are obtained. Each group of three readings must indicate consistent *hard* thrust points. In addition, the 180° shaft rotation should produce similar results. If the values are significantly different, the thrust bearing should be disassembled and carefully reexamined. Variations can be due to anything from a steel burr, to a cocked thrust collar, to a warped bearing housing, or a mismatch between upper and lower halves. In any case, the reasons for the thrust inconsistencies must be identified, and corrected before

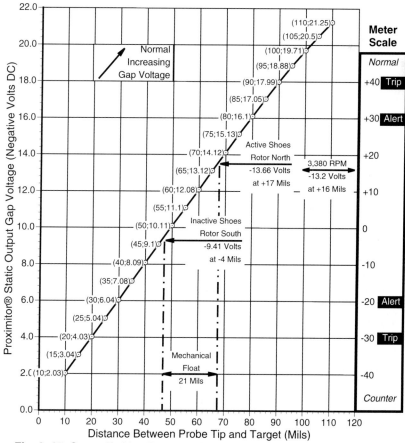

Fig. 6–15 Steam Turbine Proximity Probe Thrust Position Calibration Curve

final assembly, and startup of the machine.

At the conclusion of the field setup, the acquired information should be documented in a usable format. Generally, the probe calibration curve should be used to summarize the thrust setup. For example, consider the chart presented in Fig. 6-15 depicting the thrust calibration on a large steam turbine. This curve describes one of two thrust channels. The characteristics exhibited by the companion thrust channel display very similar behavior to this example curve. In this machine, the axial probe installation results in a gap voltage characteristic that increases with normal turbine thrust position. Hence, increasing gap voltages are identified as *Normal*, and reverse thrust positions are identified as *Counter* on the graph. The mechanical float zone measured with a dial indicator is shown as 21 Mils. With the rotor positioned *hard* on the active shoes, the output gap voltage is shown as -13.66 volts DC. As the rotor is bumped back *hard* against the inactive thrust shoes, the gap voltage decreases to -9.41 volts DC. Based upon a normal sensitivity of 200 millivolts per Mil (or 5.0 Mils per Volt), the thrust float measured by the probe voltage is determined as follows:

$$Thrust\ Float_{probe\ volts} = (13.66 - 9.41\ \text{Volts}) \times \left(5.0 \frac{\text{Mils}}{\text{Volt}}\right) = 4.25 \times 5.0 = 21.2\ \text{Mils}$$

This probe gap differential agrees with the mechanical float of 21 Mils. In addition, the thrust monitor meter scale shown to the right of the graph displays the following variation across the float zone:

$$Thrust\ Float_{monitor} = (+17\ \text{Mils}) - (-4\ \text{Mils}) = 21\ \text{Mils}$$

Once more, proper agreement is achieved between the mechanical float, probe voltage float, and the indicated monitor float zone. Hence, there is good confidence in the accuracy of this thrust position measurement, and the relationship between the electronics and the actual mechanical system. The calibration plot in Fig. 6-15 also indicates the *Normal* and *Counter* setpoints for the thrust *Alert* and *Trip*. In all cases, the *Alert* or *Alarm* point should set well within the available babbitt thickness of the thrust bearing. The second setpoint of *Trip* or *Danger* should be set with the idea of saving the rotor. This may allow damage or even destruction to the thrust bearing. Certainly the *Trip* point must be less than the forward and reverse axial rub points to allow the machine to coastdown without rotor contact to the stationary casing.

The running position of this turbine at 3,681 RPM is shown on the right hand side of Fig. 6-15. This steady state operating position at full process load is reasonably close to the *hard* position of the rotor on the active shoes. This is the normally trended position measurement, and it should be recognized that the actual reading will be dependent upon many other elements. Consideration should always be given to items such as the actual thrust load, compressibility of the thrust shoes, thrust balancing scheme for the machine, condition of interstage labyrinths, probe temperature sensitivity, ambient conditions, etc.

One final point on Fig. 6-15 should be mentioned. Note that the normal trip setpoint is located at +40 Mils, and the counter trip is at -30 Mils. Actually, the thrust float zone is set about 8 to 10 Mils high on the calibration curve. The available transducer operating range would be better utilized if the probe cold gap voltage had been set in the vicinity of -12.0 volts DC.

These position measurements with proximity probes are not limited to the axial direction. In fact, radial measurements of journal position are very important during the diagnosis of any piece of mechanical equipment. Machinery axial measurements are limited to one dimension (forward and reverse). However, radial or lateral position changes can occur in two dimensions (vertical and horizontal). To accommodate this additional degree of freedom, radial position measurements must be performed in two directions. This requirement drives the need for two orthogonal, or perpendicular proximity probes for each journal bearing. Orientation of these probes does not matter from a motion detection standpoint, but it is extraordinarily significant for proper determination of the actual lateral movement, and shaft position change.

Traditional transducer identification often attempts to relate the angular orientation to a true vertical, or a true horizontal direction. The vertical direction is often termed the Y-Axis, and the horizontal direction is called the X-Axis. As shown in the upper left hand shaft **A** of Fig. 6-16, this is an accurate description

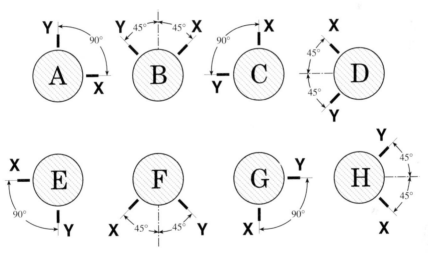

Fig. 6–16 Radial Proximity Probe Angular Identification

when the vertical probe is located directly above the shaft at 12 o'clock, and the horizontal pickup is on the right side of the shaft at the 3 o'clock position. If gap voltages from these probes are measured with a digital multimeter, the differential changes can be directly converted to plus and minus position changes in the true vertical and horizontal directions. Similarly, if the probe outputs are wired directly to a DC coupled oscilloscope, the X-Y movement of the dot on the CRT will accurately describe changes in radial shaft centerline position.

Unfortunately, proximity probes are seldom located at true vertical and horizontal locations. In many instances, the probes are offset at ±45° from the vertical centerline as shown in the **B** shaft diagram in Fig. 6-16. This configuration avoids the horizontal bearing splitline, and it provides top access for oil supply lines, thermocouples, seal oil piping, etc. This configuration has been adopted as a standard orientation by many organizations such as the American Petroleum Institute[1]. Fig. 6-16 also displays six additional potential configurations for radial proximity probes (diagrams **C** through **H**). These sketches were generated by indexing the original probe configuration by 45° increments around the shaft, and maintaining a fixed relationship between the X-Axis and Y-Axis probes. Obviously, the orthogonal probes may be installed at any angle, but the proper identification of X and Y is mandatory for correct physical interpretation of the resultant transducer data.

In many cases, the probes may be installed above and below the horizontal splitline. This is common on industrial gear boxes where the transducers are mounted towards the outside of the box. Thus, the bull gear probes may be mounted at ±45° from the horizontal splitline on one side of the box. The pinion probes are mounted at ±45° from the opposite horizontal splitline on the other side of the gear box. This type of variation in transducer mounting locations is often necessitated by the physical construction of the machine.

An example of this type of gear box proximity probe installation is shown in Fig. 6-17. This is a down mesh, speed reducing, double helical gear box. The input pinion shows the Y-Axis probes mounted above the horizontal splitline; whereas the output bull gear displays the X-Axis probes mounted above the horizontal splitline. It has been argued that this difference in probe orientation is due to the difference in rotation direction between the two shafts. In fact, this is

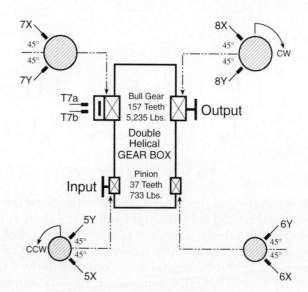

Fig. 6–17 Radial And Thrust Position Proximity Probes Mounted On A Speed Decreasing Double Helical Gear Box

[1] "Vibration, Axial Position, and Bearing Temperature Monitoring Systems — API Standard 670, Third Edition," *American Petroleum Institute*, (Washington, D.C.:American Petroleum Institute, November 1993).

not the case, and the direction of shaft rotation has nothing to do with correct identification of radial probe angular orientation.

In all cases, the diagnostician must maintain a consistent transducer orientation that will interface with traditional analog instruments, and provide proper signal polarity (i.e., direction). Since most computer-based instruments and analysis systems are patterned after their analog predecessors, the same transducer orientation rules normally apply. The diagram shown in Fig. 6-18 depicts the standard probe orientation scheme that will work under all condi-

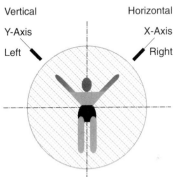

Fig. 6–18 Identification Procedure For All Radial Vibration Transducers

View from the Machinery Drive End

tions for clockwise and counterclockwise shaft rotation. First, the individual should assume a viewing position from the drive end of the machinery train. Next, the individual should view the probes from the center of the shaft. Then, when looking directly between the two radial probes (i.e., put your nose between the probes), the left hand probe will always be the *Vertical* or the *Y-Axis*. The right hand probe will always be the *Horizontal*, or the *X-Axis* transducer. Regardless of the angular position of the orthogonal probes this simple procedure provides the correct identification. This technique was used to identify transducers within the array of eight different probe orientations in Fig. 6-16. In addition, the gear box probes shown in Fig. 6-17 were correctly identified using this same method. For consistency, this technique should also be used for the identification of casing transducers such as velocity coils and accelerometers.

In practice, the Vertical, Y-Axis, Left Hand transducer is always connected to the Vertical or Y-Axis deflection of analytical instrumentation such as an oscilloscope. Similarly, the Horizontal, X-Axis, Right hand probe is connected to the Horizontal or X-Axis of the oscilloscope.

In order to observe the final data with respect to a true vertical and horizontal coordinate system, the data must be rotated by the angular offset of the probes from the desired coordinates. This can be accomplished by an angular rotation of the graphical display. The amount of rotation will be equal to the angle between the true vertical centerline, and the centerline for the Y-Axis probes. For instance, in Fig. 6-18, the Y-Axis probe is 45° counterclockwise from the true vertical centerline. Data from this probe arrangement would be rotated 45° counterclockwise to provide the correct physical viewing of the data.

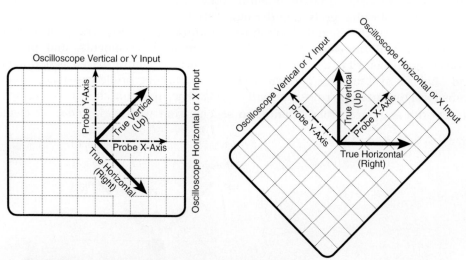

Fig. 6–19 Angular Vibration Transducer Position And Oscilloscope Signal Orientation

This concept may not be intuitively obvious, and an additional perspective is provided in Fig. 6-19. In this diagram, the left sketch depicts the normal orientation of an oscilloscope screen. As shown, the Y-Axis transducer is connected to the scope vertical voltage amplifier. Variations in voltage at this input will result in vertical deflections of the trace. If the scope is operated in an X-Y mode, and the X-Axis transducer is directed to the horizontal amplifier, any voltage change will produce horizontal deflections of the trace. Furthermore, assume that the probes are physically oriented at ±45° from true vertical as depicted in Fig. 6-18. In this orientation, it is clear that true vertical is located halfway between the probes. Obviously, in the left sketch of Fig. 6-19, the true vertical (up) direction must also be halfway between the oscilloscope vertical and the horizontal axis. By simple deduction, it is also reasonable to conclude that the true horizontal (right) direction is an axis that points down towards the lower right hand corner of the grid.

If the oscilloscope is physically rotated 45° counterclockwise, and the Vertical or Y-Axis of the scope is placed coincident with the Y-Axis probe, the sketch shown at the right side of Fig. 6-19 will evolve. In this arrangement, the principal oscilloscope axes line up with the respective transducers, and a proper Up–Down and Left–Right view is obtained. Although this might seem like a trivial exercise, it is vitally important to maintain proper direction and orientation of all of the machinery response measurements. This applies to dynamic motion (i.e., vibration) as well as static shaft centerline position data.

Proper manipulation of static data acquired with a Digital Multimeter (voltmeter) requires that the installed probe angle be included within the calculations. For instance, consider a bearing with proximity probes installed at ±45° from the true vertical centerline as shown in Fig. 6-20. Further assume that the probe gap voltages were obtained when the machine was stopped, and again at

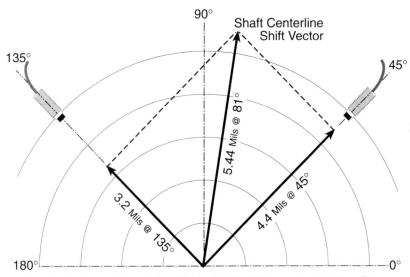

Fig. 6–20 Vector Calculation Of Radial Shaft Centerline Position Change

normal operating speed. If the left hand Y-Axis probe displayed a cold gap at stop of -9.58 volts DC, and -8.94 volts DC at full speed, the overall change was +0.64 volts toward the probe. Similarly, if the right hand, X-Axis probe had a cold gap of -9.44 volts DC, and a hot running gap of -8.56 volts DC, the change would be +0.88 volts toward the probe. The positive sign associated with both of the differential voltages indicates that shaft displacement was towards the probes. This is also evident by the fact that the full speed gap voltages decreased from the zero speed values — indicating that the shaft moved closer to both of the probes.

Based upon a sensitivity of 200 millivolts per Mil (or 5.0 Mils per Volt), the Y-Axis distance change is calculated by: 0.64 Volts x 5.0 Mils/Volt = 3.2 Mils. In the orthogonal X-Axis, the distance change with respect to the probe is determined in the same manner: 0.88 Volts x 5.0 Mils/Volt = 4.4 Mils. These calculated X-Y changes in shaft position are individual vector quantities where the angles are governed by the physical orientation of the probes. If the true right horizontal axis is designated as 0°, and the top vertical axis is identified as 90°; then the X probe would be located at 45°, and the Y transducer would be positioned at an angle of 135°. Combining these angles with the previously calculated magnitudes, the following shift vectors with respect to each probe may be defined:

$$\text{Y Probe Vector} = \vec{Y} = A \angle \alpha = 3.2 \, \text{Mils} \angle 135°$$

$$\text{X Probe Vector} = \vec{X} = B \angle \beta = 4.4 \, \text{Mils} \angle 45°$$

These two vector quantities may be summed to determine the centerline shift of the shaft. This vector summation may be performed on a pocket calculator suitable for vector math (e.g., HP-48SX), or the necessary result may be obtained by using the vector addition equation structure previously discussed in chapter 2 of this text. This vector summation may be performed in the true horizontal direction by using equation 2-31 as follows:

$$Horiz_{add} = A \times \cos\alpha + B \times \cos\beta$$

$$Horiz_{add} = 3.2 \times \cos 135° + 4.4 \times \cos 45°$$

$$Horiz_{add} = 3.2 \times (-0.707) + 4.4 \times 0.707 = (-2.263) + 3.111 = 0.848 \text{ Mils}$$

Similarly, a vector summation may be performed in the true vertical direction by applying equation 2-32 in the following manner:

$$Vert_{add} = A \times \sin\alpha + B \times \sin\beta$$

$$Vert_{add} = 3.2 \times \sin 135° + 4.4 \times \sin 45°$$

$$Vert_{add} = 3.2 \times 0.707 + 4.4 \times 0.707 = 2.263 + 3.111 = 5.374 \text{ Mils}$$

The true vertical and horizontal coordinates may now be plotted on graph paper, or they may be converted to polar coordinates. The magnitude of the centerline vector shift is calculated from equation 2-33 as:

$$Shift_{add} = \sqrt{(Horiz_{add})^2 + (Vert_{add})^2}$$

$$Shift_{add} = \sqrt{(0.848)^2 + (5.374)^2} = \sqrt{29.599} = 5.44 \text{ Mils}$$

The vector angle associated with this magnitude shift may now be computed with the arctangent equation 2-34 from chapter 2 as follows:

$$\phi_{add} = \text{atan}\left\{\frac{Vert_{add}}{Horiz_{add}}\right\}$$

$$\phi_{add} = \text{atan}\left\{\frac{5.374}{0.848}\right\} = \text{atan}\{6.337\} = 81°$$

Clearly, the changes in X-Y proximity probe gap voltages allows the determination of a shaft centerline shift. This information may be presented as a Cartesian coordinate position of 0.848 Mils horizontally to the right, and 5.374 Mils vertically upward. This same change in position may be expressed as a vector quantity of 5.44 Mils at an angle of 81°. The vector summation may also be performed graphically as shown in Fig. 6-20. This diagram also describes the physical representation of these shaft centerline shift calculations. It is apparent that the final results are identical no matter what calculation or plotting technique is applied. The diagnostician should apply the most appropriate method based upon the accuracy required, and the resources available for the calculation.

Knowledge of the radial shaft centerline position is an important diagnostic tool. Both the magnitude of the centerline shift and the associated angle are significant in evaluation of the machinery behavior. For instance, the previous

example revealed a shift vector of 5.44 Mils at 81°. If this occurred on a machine
with a load-on-pad tilt pad bearing, and a 15.0 Mil diametrical clearance, the
results would be indicative of normal operating position. This type of bearing
usually exhibits a steep attitude angle indicative of a vertical rise of the shaft
from the bottom pad. The magnitude of the shift is also reasonable with respect
to the diametrical clearance for this type of radial bearing.

However, the same shift vector of 5.44 Mils at 81° would be quite worrisome
for a machine equipped with plain sleeve bearings and a 10.0 Mil diametrical
clearance. With this mechanical configuration the journal would be positioned
close to the center of the bearing, and there would be a strong potential for instability of the machine. This centered rotation of a cylinder within a cylinder (i.e.,
shaft within a sleeve bearing) will be discussed in further detail in chapter 9.

Up until this point, the discussion has centered around the static measurements that may be obtained with shaft sensing displacement proximity probes.
These transducers also have the capability to detect vibratory motion of the
observed surface. In fact, the majority of the industrial applications for these
probes are based upon their ability to accurately measure relative shaft vibration. The easiest way to understand the transducer operation when observing a
vibrating surface is to consider the behavior around the calibration curve as
depicted in Fig. 6-21.

In this diagram, the distance between the probe tip and the target material
is oscillating in a uniform fashion. As stated in chapter 2, this type of repetitive
movement is referred to as Simple Harmonic Motion (SHM), and it may be conveniently described with a sine or cosine wave. This oscillating motion is trans-

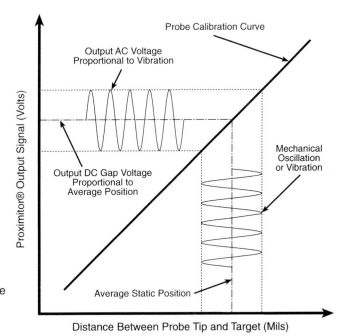

Fig. 6–21 Proximity Probe
Voltage Output Signal
With An Oscillating
Observed Surface

lated by the probe calibration curve into an oscillating voltage. More specifically, this is commonly referred to as an alternating, or an AC voltage.

The time domain characteristics of both curves are identical. The maximum voltage is coincident with the peak gap distance. Similarly, the minimum output voltage matches the point of the closest gap. The average distance between the probe and the target is referred to as the Average Shaft Position, and this is directly measured as the previously discussed DC Gap Voltage. Hence, the Proximitor® signal output consists of two interrelated parameters, the AC voltage proportional to vibration, and the DC voltage that indicates the average distance of the oscillating target with respect to the stationary probe.

Both the vibration and position characteristics use the probe calibration curve to translate the mechanical motion into a voltage output. For a standard 200 mv/Mil system, the gap voltage will change by 1 Volt for every 5 Mils of distance change within the linear portion of the curve. The oscillating or vibratory movement will be converted into an AC voltage signal at the transducer sensitivity of 200 mv/Mil (i.e., curve slope). Since the fundamental displacement measurement is expressed in terms of peak to peak motion, the vibration units will likewise be in Mils,$_{\text{p-p}}$.

In most instances, vibration measurements are based upon peak detection, and conversion via the appropriate scale factor. For instance, if the previous diagram displayed a top peak of the Sine wave at -10.0 volts DC, and a bottom peak of -8.0 volts DC, the vibration may be determined by:

$$Vibration_{peak\ to\ peak} = (10.0 - 8.0\ \text{Volts}) \times (5.0\ \text{Mils/Volt})$$

$$Vibration_{peak\ to\ peak} = 2.0\ \text{Volts,}_{\text{p-p}} \times 5.0\ \text{Mils/Volt} = 10.0\ \text{Mils,}_{\text{p-p}}$$

The same result can be achieved if the peak to peak voltage amplitude is extracted from an oscilloscope and multiplied by the scale factor in Mils/Volt, or divided by the scale factor in Volts/Mil. Another way to determine the vibration amplitude is to convert the scope voltage sensitivity to the transducer sensitivity. For instance, if the proximity probe calibration is 200 mv/Mil, or 0.2 Volts/Mil, the oscilloscope voltage amplifier may be set at 200 mv/Division or more commonly 0.2 Volts/Division. Dividing the scope amplifier setting by the probe scale factor yields a conversion factor of 1.0 Mil per Division. Thus, if the resultant vibration signal covers six vertical divisions on the oscilloscope, the vibration is determined to be 6.0 Mils,$_{\text{p-p}}$ by a direct visual observation.

For a pure sine wave, the conversion equation 6-3 may be applied to compute the peak to peak magnitude of a 0.4245 Volts,$_{\text{rms}}$ value obtained from a Digital Multimeter as follows:

$$Voltage_{peak\ to\ peak} = \frac{Voltage_{rms}}{0.3536} = \frac{0.4245}{0.3536} = 1.20\ \text{Volts}_{\text{p-p}}$$

$$Vibration_{peak\ to\ peak} = 1.20\ \text{Volts}_{\text{p-p}} \times 5.0\ \text{Mils/Volt} = 6.0\ \text{Mils,}_{\text{p-p}}$$

Again the same results are obtained, and it is clear that conversion between voltage and vibration amplitudes may be accomplished in different, but consistent ways. It is also self-evident that proximity probes may be used to measure average position, and vibration in both radial and the axial directions. This is necessary from a machinery analysis standpoint, since real process machinery does translate and vibrate in the lateral and axial directions, and the transducer suite must be able detect this overall movement.

Another application of proximity probes resides in the realm of providing timing signals. Typically, these are synchronous, once per revolution pulses that may be used for accurate speed measurements. They are also employed for phase measurements and determination of precession when combined with other probes on the machinery train. These 1X timing pulses are typically referred to as Keyphasor® signals, and the transducers are called Keyphasor® probes. These proximity probe timing sensors are often positioned over a shaft notch or drilled hole. With this arrangement, the probes produce a negative going pulse as the shaft indentation passes beneath the timing probe (e.g., Fig. 6-5).

In some instances, a raised surface such as the top of a shaft key is observed by the Keyphasor® probe. In these cases a positive going pulse is generated (e.g., Fig. 6-6). The shape of these timing pulses, and the instrumentation trigger points are discussed in greater detail earlier in this chapter, and also at the beginning of chapter 11 on balancing.

The radial, axial, and Keyphasor® probes are combined in many different combinations. In most instances, a pair of mutually perpendicular radial probes are installed at each journal bearing (X-Y probes). A pair of axial probes are typically mounted at each thrust bearing, and a radial Keyphasor® probe will usu-

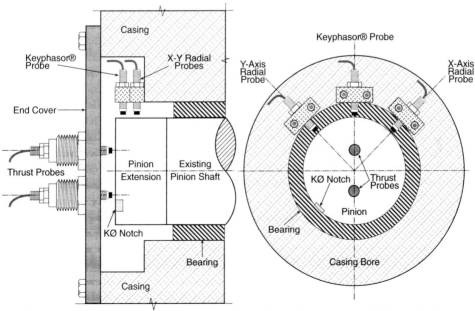

Fig. 6–22 Side View Of Proximity Probe Installation On A High Speed Pinion

Fig. 6–23 End View Of Proximity Probe Installation On A High Speed Pinion

ally be installed for each shaft speed. A typical installation that involves all three basic applications for proximity probes is presented in Figs. 6-22 and 6-23.

The machine under consideration is a speed increasing, double helical gear box. The unit is motor driven at the bull gear input, and the pinion output is coupled to a centrifugal compressor. Figs. 6-22 and 6-23 describe the probe installation at the outboard, or blind end of the pinion. Installation of probes on this unit was hampered by a short shaft that extended only 1/8" past the radial bearing. To provide measurable surfaces, shaft extensions for the bull gear and the pinion were fabricated from 4140. These extensions were threaded into the respective gear elements, and locked into position with a pair of countersunk cap screws.

The pinion drawings describe the location of the two thrust probes, and the surface observed by these axial transducers. Typically, the probes should be spaced a minimum of two probe diameters from any potential source of interference. In this case, the axial probes had to be located to avoid the Keyphasor® notch, the center punch on the shaft, as well as the axial cap screws. The radial vibration probes were positioned to have a clear view of the pinion extension, plus they must not be influenced by the timing notch milled into the outboard end of the extension. Thus, there is no chance for cross talk or interference with the thrust probes or the Keyphasor®. Although the side view places the X-Y probes fairly close to the timing probe, the end view in Fig. 6-23 reveals the ±45° angular offset between each of the radial probes and the Keyphasor®. Hence, the radial probes have minimal potential for signal interference.

In virtually all cases, the Keyphasor® probe should be oriented in a radial direction to maintain a uniform pulse shape and size during all operating conditions and speeds. Placement of the Keyphasor® transducer in an axial direction will often jeopardize the consistency of the timing pulse. This is due to variations in thrust position that manifest as substantial changes in the pulse signal. The observed notch (or projection) should normally be 40 to 60 Mils deep (40 to 60 Mils high for projection), and the width should be one and a half (1.5) times the probe diameter. For instance, a 1/4" diameter Keyphasor® probe would produce an acceptable pulse signal with a 3/8" wide slot that is 50 Mils deep (or 50 Mils high for a projection). In the final output, the Keyphasor® signal should have a pulse height between 5 and 15 volts.

The necessity for a strong and consistent Keyphasor® signal cannot be overstated. Many mechanical malfunctions and the entire array of normal balancing activities are totally dependent upon a good once-per-revolution trigger signal. Furthermore, most dynamic data acquisition systems and analytical instruments require a rotational speed trigger signal to allow full utilization of the capabilities of the instrumentation. Since the availability of this timing signal is so important to the business of machinery diagnostics, the manipulation and proper utilization of this signal will be discussed in further detail in subsequent chapters 7, 8, and 11.

In the overview, the proximity probe transducer suite is applicable to a wide range of process machinery. Due to the ability to measure relative position changes as well as relative vibration measurements, the potential industrial applications are substantial. As with any signal transducer, proximity probes exhibit a variety of advantages, but they also have a series of disadvantages. For purposes of comparison with other vibration transducers, the following two summaries of proximity probe features are presented for consideration:

Proximity Probe Advantages

○ Measures Shaft Dynamic Motion
○ Measures Shaft Static Position
○ Excellent Signal Response Between DC and 90,000 CPM (1.5 kHz)
○ Flat Phase Response Throughout Transducer Operating Range
○ Simple Calibration
○ Solid State Electronics
○ Rugged and Reliable Construction
○ Available in Many Physical Configurations
○ Suitable for Installation in Harsh Environments
○ Multiple Machinery Applications for the Same Transducers

Proximity Probe Disadvantages

❑ Sensitive to Surface Imperfections and Magnetism
❑ Sensitive to Material Properties
❑ Shaft Surface must be Conductive
❑ Low Dynamic Signal Response Above 90,000 CPM (1.5 kHz)
❑ External Power Source Required
❑ Correct Probe to Proximitor® Cable Impedance Must Be Maintained
❑ Minor Temperature Sensitivity in pre-1990 Probes
❑ Sensitive to Interference from Adjacent Proximity Probes
❑ Sensitive to Probe Mounting Bracket Resonance(s)
❑ Potentially Difficult to Install

In the majority of applications, proximity probes are used for permanent monitoring and machinery protection measurements on units with fluid film bearings. In many process plants this would be the critical and essential machinery trains. The probes would typically be connected to dedicated monitors for functions such as radial position and vibration, axial position and vibration, rotational speed, differential expansion, and eccentricity. The monitors provide power to the Proximitors®. These monitors also provide additional signal conditioning, various display functions, plus dedicated alarm capabilities. Furthermore, many monitoring systems are commonly interfaced with Distributed Control Systems (DCS), and dedicated dynamic data acquisition and processing systems such as a Transient Data Manager® system.

VELOCITY COILS

Velocity transducers represent one of the earliest forms of vibration probes. References from the 1950s often refer to these transducers as *vibrometers*. They are used to obtain absolute velocity measurements of stationary machinery elements. These are fully contacting probes that are mounted directly on a mechanical structure (e.g., bearing housing); and they measure the dynamic motion of that structure. Velocity coils are supplied in a several basic configurations. For example, the photograph in Fig. 6-24 depicts a typical industrial velocity pickup.

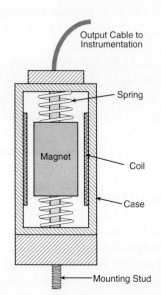

Fig. 6–24 Typical Industrial Velocity Coil   **Fig. 6–25** Velocity Coil Cross Section

This type of vibration transducer generally has an output sensitivity of 500 millivolts/IPS. The usable frequency range extends from a bottom end of 600 to 900 CPM (10 to 15 Hz), to an upper limit of approximately 90,000 CPM (1,500 Hz). Special velocity coils are built specifically for low frequency measurements; and although the lower end of the frequency response curve is significantly improved, the physical size and weight of the transducer grows substantially. For instance, some low frequency applications use velocity coils that weigh in excess of 200 pounds.

Regardless of the physical configuration, velocity coils consist of the same fundamental components. For example, consider the cross section of a typical velocity probe shown in Fig. 6-25. The transducer casing is directly attached to the vibrating surface. The method of attachment could be a screwed connection between the transducer mounting stud and a drilled and tapped hole at the measurement point. Another common approach would be to screw the velocity pickup into a high strength double bar magnet, and use the magnet for attachment of the pickup to the vibrating surface. This approach provides good mobility to the probe. However, the rigidity of the attachment should always be checked before believing the resultant vibration data. In all cases, rigid mounting of the probe to the vibrating surface insures that the transducer will move in unison with the measurement point. The transducer senses the vibrating surface via a circumferential electrical coil that is attached to the inside of the housing. This coil must move together with the casing due to the physical attachment between elements. Hence, the coil motion is presumed to be virtually identical to the vibratory motion of the attached surface.

Located within the center of the electrical coil is a permanent magnet mounted on very soft springs. This spring supported magnet is confined to oscillate in the principal axis of the transducer (e.g., up and down for a vertical probe). The combination of a heavy magnet (mass), plus a soft spring yields a low natural resonant frequency for the assembly. In actual operation, the transducer case and coil vibrate in sympathy with the attached surface, and the spring mounted magnet tends to remain stationary. This relative motion between the essentially stationary magnet, and the vibrating coil results in the generation of a coil voltage that is proportional to the velocity of the transducer outer casing.

Since this type of vibration probe produces a signal without the necessity of an external power source, the transducer is considered to be self-generating. This feature simplifies the field installation of a velocity transducer system. For instance, a velocity probe is commonly referred to as a *Geophone* within the field of seismic testing. This category of testing is applied to seismic measurements for earthquake detection, and petroleum exploration. In these applications a self-generating probe is highly desirable.

The natural frequency of the spring supported magnet typically falls in the range of 300 to 600 CPM (5 to 10 Hz). The severity of this fundamental system resonance might easily dominate the resultant output signal, and render the data unusable. In order to restrain this resonant spring mass response, the out-

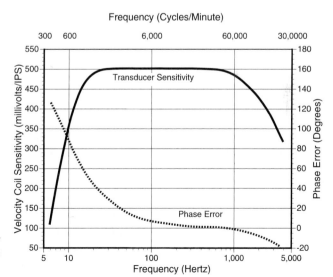

Fig. 6–26 Typical Velocity Coil Frequency Response Characteristics

put is either damped electronically, or the mechanical system is over damped with a viscous fluid surrounding the magnet. The damped response of the spring mass system is subject to the same physical laws that were previously discussed in chapter 2. Since the transducer system is over damped, the natural resonance is fully suppressed as shown in the calibration plot of Fig. 6-26.

The characteristics displayed in this velocity coil calibration plot are for a 600 CPM transducer. By definition, this transducer should display an attenuation in output sensitivity of -3 dB at 600 CPM. In fact, this probe does exhibit 354 mv/IPS (= 500 x 0.707) at the resonant frequency of 600 CPM (10 Hz). In addition, it should be noted that damping not only suppresses amplitude sensitivity, it also influences the output signal phase. This results in a 90° phase error at 600 CPM. Even at 6,000 CPM, the phase error is still 10°. Thus, the diagnostician should be cautious during any low frequency applications of velocity pickups. When accurate amplitude and phase characteristics are required for these transducers, each pickup must be individually calibrated to determine the specific characteristics for each velocity coil.

Full calibration of a velocity transducer requires the integration of several instruments as depicted in Fig. 6-27. The dual channel Dynamic Signal Analyzer (DSA) is used to generate a white noise output that attempts to simultaneously produce all frequencies within the analysis bandwidth. Another way to visualize a white noise source is to consider a signal that contains a uniform distribution of energy across the analysis bandwidth. In most instances, this type of excitation allows a faster calibration when compared against a single frequency oscillator that is used to sweep the transducer operating range.

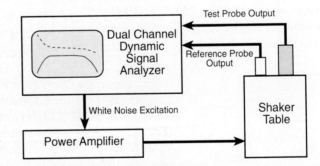

Fig. 6–27 Typical Instrumentation Required For Proper Calibration of Seismic Vibration Transducers

The white noise signal is used as an input to the Power Amplifier. This amplifier actually drives the vertically oscillating Shaker Table in accordance with the frequency content of the excitation signal. Since the shaker table is a mechanical device, the upper frequency will be governed by the design of the particular table. However, for the calibration of velocity coils, the testing devices will generally have the capability to exceed the frequency response characteristics of the test object (i.e., the velocity coil).

A calibrated reference transducer is mounted on the shaker table next to the test probe. This reference transducer is usually an accelerometer that has a flat amplitude and frequency response across the bandwidth of the velocity pickup. Output signals from the reference transducer, and the probe under test are directed back to the two channel DSA. At this time, a frequency response function (also known as a transfer function) is performed between the two signals (test probe to reference probe). The resultant amplitude and phase characteristics versus frequency are documented in a plot similar to Fig. 6-26.

General characteristics for any particular type of velocity transducers will govern the average results of this type of transducer testing. In most cases, deviations between transducers will be quite apparent. Hence, for any critical work with velocity coils (especially low frequency), this type of detailed response testing is considered to be mandatory.

Since velocity coils contain internal springs and moving parts, they are subject to fatigue failures as a function of time. Hot service in a high vibration environment will shorten the life span of these elements. Conversely, moderate operating temperatures combined with low vibratory surfaces will allow the longest transducer life. In many cases, the condition of the velocity coils and their suitability for service may be determined with a routine calibration check. Generally, significant deviations from the initial calibration curves would be grounds for refurbishing (i.e., rebuilding) or replacing the velocity pickup.

The velocity coil calibration curve and the discussed calibration check both presume that the transducer is securely bolted to the vibrating surface. In many instances, magnets or extension stingers are used between the velocity pickup and the measurement surface. These additional interface devices degrade the performance of the probe, and they may result in mounting resonances between 500 and 3,000 Hz (30,000 and 1800,000 CPM). Thus, the type of mounting fixture should always be considered. If possible, the velocity coil mounting device (e.g., magnet) should be incorporated into the calibration setup to determine the frequency response characteristics of the entire transducer system.

Based upon the calibration curve and the fundamental operation of the velocity pickup, it is clear that position measurements (i.e., DC or zero frequency) cannot be made with velocity coils. In all applications the orientation rules, and the interaction with the Keyphasor® timing probes are identical to the behavior previously discussed with the proximity probes.

It should also be restated that velocity amplitudes are zero to peak values, whereas displacement measurements are expressed as peak to peak amplitudes. Velocity signals may be integrated with respect to time to obtain displacement, and they may also be differentiated with time to obtain acceleration. These manipulations of the raw velocity signals may be performed with electronic circuits dedicated to these integration or differentiation functions; or they may be performed more effectively in a digital format within a DSA.

For velocity amplitudes at a given frequency, the equivalent displacement or acceleration amplitudes may be computed with equations 2-17 through 2-22 from chapter 2. In many cases, the velocity amplitudes at a single frequency are defined as vector quantities. The conversion equations allow the computation of proper amplitudes, but the diagnostician must convert the vector phase angles by 90°. The angle conversions are performed in accordance with equations 2-14 through 2-16. For velocity measurements, the angle conversions are as follows:

$$Phase_{displacement} = Phase_{velocity} + 90°$$

$$Phase_{acceleration} = Phase_{velocity} - 90°$$

Conversion of vibration units can be quite useful. For instance, if a shaft sensing proximity probe is attached to a bearing housing, and a velocity coil is mounted directly in line with the proximity probe, a variety of measurements are possible. The proximity probe can measure position and vibration of the shaft with respect to the housing. The casing mounted velocity coil can measure absolute vibration of the bearing housing. A summation of shaft plus casing vibration will result in shaft absolute motion. Fig. 6-28 depicts this type of dual transducer installation on a machine with a fluid film bearing.

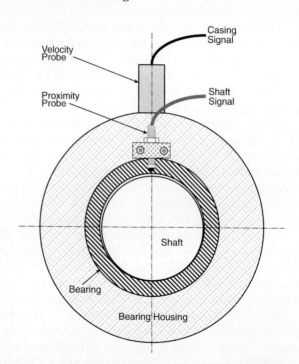

Fig. 6–28 Simultaneous Measurement Of Absolute Casing Velocity And Relative Shaft Displacement

In this type of arrangement, the casing velocity data may be integrated to casing displacement, and the resultant signal electronically summed with the proximity probe output. This type of signal manipulation is performed with a voltage summing amplifier. A differential voltage amplifier may also be used, but one of the signals must be inverted to obtain the proper result. The complex summed signal from the amplifier may then be processed in a variety of manners with additional diagnostic instrumentation.

Alternately, the velocity signal may be digitally integrated to displacement in one channel of a DSA. If the shaft displacement signal is connected to another channel of the DSA, the integrated velocity may be added to the shaft signal within the DSA. This summed signal may then be viewed as time or FFT data.

Another approach that is quite convenient for signals that are dominated by a single frequency component is a direct vector summation of the output from each transducer. For example, consider the following calculations for 1X vectors emitted by a 3,600 RPM synchronous motor:

$$\text{Shaft Relative Displacement} = 2.85 \text{ Mils,}_{\text{p-p}} \angle 165°$$

$$\text{Casing Absolute Velocity} = 0.24 \text{ IPS,}_{\text{o-p}} \angle 31°$$

Converting the casing velocity amplitude into the equivalent casing displacement amplitude may be accomplished with equation (2-20) as follows:

$$\text{Casing Absolute Displacement} = \frac{19,099 \times V}{Rpm} = \frac{19,099 \times 0.24}{3,600} = 1.27 \text{ Mils,}_{\text{p-p}}$$

The phase of the casing displacement vector is determined from the velocity phase angle using equation (2-14) as follows:

$$\text{Case Phase}_{displ} = \text{Case Phase}_{velocity} + 90° = 31° + 90° = 121°$$

These two conversions define the casing displacement vector, which may now be added to the relative shaft displacement vector to determine the shaft absolute vector. Using the equations (2-31) to (2-34) for vector addition, the summation of these two running speed vectors yields the following:

$$\text{Shaft Absolute Vector} = 2.85 \angle 165 + 1.27 \angle 121 = 3.86 \text{ Mils,}_{\text{p-p}} \angle 152°$$

This is an important solution for machines with flexible supports. In many cases, neither the shaft relative nor the casing absolute vibration vectors are fully descriptive of the overall shaft motion. For these types of machines, the shaft absolute measurement may be mandatory. Furthermore, during vibration analysis of machinery with flexible supports, the relationship of shaft to casing to absolute motion is often a key ingredient in the determination of the mechanical element(s) responsible for the vibration problem. It should also be mentioned that support flexibility might occur in the bearing housing assembly, or the soft members might be the machine foundation or even the subsurface soil.

As shown in the previous example, the absolute motion was computed from the transducer outputs for simple single frequency signals. Alternatively, the information may be obtained by electronic summation of the signals, followed by synchronous filtration. In either method, the shaft displacement signal should be corrected for shaft runout to obtain the best possible representation of the absolute shaft motion. For complex transducer signals containing multiple frequencies, the shaft plus casing data must be handled by electronic summation, plus further data processing such as FFT analysis.

Due to the self-generating signal characteristics, and the general ease of installation of these probes on different mechanical elements, the velocity coil has been a popular transducer. As with any signal transducer, velocity coils exhibit several advantages, but they also have a series of disadvantages. For purposes of comparison with other vibration transducers, the following two summaries of velocity coil features are presented for consideration:

Velocity Coil Advantages

○ Measures Casing Absolute Motion
○ Easily Attached to Machinery Externals, Piping, Baseplates or Structures
○ Good Signal Response Between 900 and 90,000 CPM (15 and 1,500 Hz)
○ Self-Generation Signal Electronics
○ No Special Wiring Required
○ Available in Several Configurations

Velocity Coil Disadvantages

❑ Sensitive to Mounting Fixture and Transducer Orientation
❑ Unable to Measure Shaft Vibration or Position
❑ Difficult Calibration Check
❑ Poor Signal Response Below 900 CPM (15 Hz) Above 90,000 CPM (1.5 KHz)
❑ Amplitude and Phase Errors at Frequencies below 1,800 CPM (30 Hz)
❑ Operates Above Transducer Natural Frequency of 600 CPM (10 Hz)
❑ Potential for Failure due to Fatigue of Moving Internal Parts
❑ Temperature Sensitive, Typical Upper Limit of 250°F, Lower Limit of 30°F
❑ Difficult to Install in Cramped Areas

During recent years the cost of velocity coils has increased, and the price of accelerometers has substantially decreased. This cost change, combined with the previously listed disadvantages, has eliminated the use of velocity coils in many locations. For applications where velocity remains the preferable measurement, the traditional mechanical velocity coil has been replaced by an accelerometer with integral electronics to provide a velocity output from the accelerometer.

PIEZOELECTRIC ACCELEROMETERS

Accelerometers are versatile vibration transducers for absolute measurement of stationary machinery elements or structures. These devices are fully contacting probes that are mounted directly on a mechanical element (e.g., bearing housing). They are available in a variety of configurations, and they may be designed to cover a wide range of operating and environmental conditions. For example, the photograph in Fig. 6-29 depicts four different sizes and configurations of standard PCB® Piezotronics high frequency ICP® accelerometers.

The ICP® designation is a registered trademark of PCB® Piezotronics, Inc., and it stands for Integrated Circuit Piezoelectric. This type of probe contains much of the necessary signal conditioning electronics within the body of the transducer. The fundamental flexibility of the basic transduction scheme allows

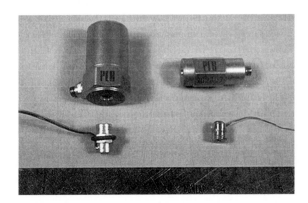

Fig. 6–29 Typical Configurations Of Accelerometers Manufactured By PCB® Piezotronics, Inc.

significant interchangeability of transducers with common external power supplies. In addition, many instruments (e.g., HP-35670A) contain a constant current power supply that allows the DSA to power the accelerometer directly.

Although the four accelerometers shown in the Fig. 6-29 may be driven by the same power supply, the characteristics of the accelerometers are significantly different. For comparative purposes, the accelerometer shown in the upper left hand portion of Fig. 6-29 weighs 60 grams. This pickup displays a ±5% operating range from 60 to 180,000 CPM (1 to 3,000 Hz). It has a mounted natural frequency of nominally 1,680,000 CPM (28.0 kHz), and an output sensitivity of 100 millivolts per G. Pickups of this type are generally used for measurement of low to medium frequency behavior.

In the upper right hand corner of Fig. 6-29, the displayed accelerometer is smaller with a total weight of 25 grams. It has an extended operating range (±5%) of 60 to 600,000 CPM (1 to 10,000 Hz). This transducer has a mounted resonant frequency in the vicinity of 2,100,000 CPM (35.0 kHz), and an output sensitivity of 10 millivolts per G. At the bottom of Fig. 6-29, miniature accelerometers are shown. These devices weight 2 grams and 1 gram respectively. Due to the small size and weight of these accelerometers the cables are integral with the transducers. The probe in the lower left hand corner has a ±5% operating range between 60 and 1,200,000 CPM (1 and 20,000 Hz). The mounted resonant frequency resides in the vicinity of 4,620,000 CPM (77.0 kHz), and the output scale factor is 10 millivolts per G. The smallest transducer shown in the lower right hand corner has a ±10% operating range of 180 to 1,800,000 CPM (3 to 30,000 Hz). The mounted resonant frequency for this accelerometer is 7,200,000 CPM (120.0 kHz), and the output scale factor is 5 millivolts per G.

Thus, it is apparent that as accelerometers decrease in size, the transducers exhibit lower scale factors combined with extended frequency response characteristics. These smaller accelerometers are used for measurements on small mechanical elements such as turbine blades and circuit boards. Conversely, larger accelerometers have higher output sensitivities plus a reduced frequency response range. For instance, one particular ICP® transducer that is suitable for low frequency seismic testing has a scale factor of 10,000 millivolts per G. The

±5% operating range on this pickup runs from 9 to 6,000 CPM (0.15 to 1,000 Hz), and the rated output is only ±0.5 G's. If the examination of lower frequencies are required, then piezoresistive accelerometers may be used, and they allow measurement of frequencies down to DC (zero frequency). These specialized piezoresistive transducers are more commonly used for static plus dynamic pressure measurements, and additional discussion of this sensing element will be provided in the following section of this chapter.

Although piezoelectric accelerometers are manufactured in a wide variety of physical configurations to address a large range of applications, the fundamental internal elements for this class of transducer remains fairly consistent. For instance, consider the cross section of a typical ICP® industrial accelerometer as presented in Fig. 6-30. This type of vibration transducer depends on the

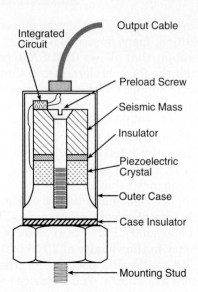

Fig. 6–30 Cross Section Of A Typical Industrial ICP® Accelerometer

electromechanical properties of the piezoelectric crystal. Specifically, the crystal will emit an electrical charge when a mechanical load or stress is applied. Conversely, when an electrical charge is applied to the crystal, it will physically deform in a direct relationship to the magnitude of charge. This concept is applied in many electronic devices varying from computers, to communication equipment, to accelerometers. Although various materials exhibit piezoelectric properties, the majority of the industrial accelerometers employ either natural quartz crystals, or man-made polycrystalline ceramics. Each type of material displays specific features, and the final selection of the piezoelectric crystal depends on the eventual application of the transducer.

The diagram in Fig. 6-30 depicts a compression accelerometer that includes a stainless steel base that supports the crystal, plus a seismic mass. Full contact between the mass and crystal is insured by the preload screw that joins the mass, crystal, and base into an essentially solid structure. When this assembly is bolted onto a vibrating surface, the seismic mass imposes a definable force upon the crystal. In accordance with the second law of motion by Sir Isaac Newton (1643 to 1727), force is equal to mass times acceleration. Within an accelerometer, the crystal is subjected to a force from the mass, and the output charge is proportional to the acceleration. Obviously, small forces will produce low acceleration levels, and large forces will manifest as high acceleration.

The charge output from the crystal in Fig. 6-30 is wired directly to an internal Integrated Circuit (IC). This electronics package provides the necessary signal conditioning to convert the crystal charge signal (picocoulomb/G) to a voltage sensitive signal (millivolts/G). The presence of this IC within the transducer normally limits the maximum operating temperature to approximately 250°F. For applications at higher temperatures, the signal conditioning electronics may be located in a separate Charge Converter that is placed in a cooler environment. Removing this IC from the accelerometer allows the transducer to be designed for effective operation at elevated temperatures. Standard transducers may be purchased for operation up to 600°F, and custom accelerometers have been built to withstand temperatures in excess of 1,200°F. This type of transducer is commonly referred to as a Charge Mode accelerometer. It is sensitive to cable whip, and electrical interference of the wiring between accelerometer and Charge Converter. Once the signal is converted to a voltage sensitive signal, the wiring downstream of the Charge Converter may be safely directed to the measurement or recording instrumentation with normal coaxial cable.

For most industrial applications, the internal IC plus a piezoelectric crystal are combined into an ICP® transducer that is suitable for use up to 250°F. The accelerometer in Fig. 6-30 represents an upright compression configuration. Accelerometers are also built in an inverted compression, an isolated compression, a shear mode, and a flexural mode configuration. Each particular model exhibits different technical characteristics for various measurement applications, combined with a range of transducer prices. As with any technical selection, the individual must arrive at an equitable balance between performance and cost. From an operational standpoint, the previous explanation remains generically applicable throughout this suite of transducers.

As previously noted during the accelerometer transducer review, the natural frequency of an accelerometer resides above the operating range. Hence, accelerometers function as a rigid system, and phase excursions do not appear within the transducer operating frequency range. In virtually all cases, acceleration phase angles may be converted to displacement angles by adding or subtracting 180°. The amplitude response of an accelerometer is somewhat more complicated due to the low frequency roll off of the transducer, plus the mounting resonance located well above operating frequency range.

Typical frequency response characteristics of an accelerometer are presented in the calibration plot of Fig. 6-31. In this diagram, the accelerometer output sensitivity in millivolts per G is plotted against frequency. This particular transducer exhibits the anticipated low frequency attenuation below 900 CPM (15 Hz), and the mounted resonance at a frequency of about 1,800,000 CPM (30,000 Hz). Clearly, measurements made in the vicinity of the resonant frequency would be influenced by the amplification associated with this resonance.

The output sensitivity between 1,200 and 600,000 CPM (20 and 10,000 Hz) remains reasonably flat, and an average scale factor of 100 millivolts per G is appropriate to apply within this region.

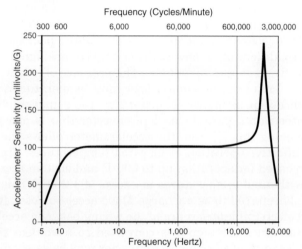

Fig. 6–31 Typical Frequency Response Characteristics For An Industrial Accelerometer

It must be mentioned that accelerometer calibration plots are generally performed with a stud mounted transducer. If the accelerometer is used in the field with an alternate mounting technique, the mounted resonant frequency may drift into the normal operating range due to a reduction in the attachment stiffness. This reduction in the mounted natural frequency serves to reduce the effective operating range of the accelerometer. For example, Table 6-2 summarizes the reduction in the maximum operating frequency with various types of accelerometer mounting techniques:

The *direct stud mount* presumes a smooth and flat mounting surface on the structure to be measured that will interface with the machined bottom of the accelerometer. The *direct adhesive mount* is typically a *super glue* type of material that is used to attach small one and two gram accelerometers. This type of

Table 6–2 Influence Of Mounting Technique On Frequency Response Characteristics

Type of Accelerometer Mount	Maximum Frequency Response
Direct Stud Mount	15,000 to 20,000 Hz
Direct Adhesive Mount	8,000 to 16,000 Hz
Epoxy Glue Base	8,000 to 12,000 Hz
Strong Double Bar Magnet	3,000 to 5,000 Hz
Mechanical Clamp	500 to 1,500 Hz
Hand Held	250 to 1,000 Hz

adhesive provides a good solid mount for small accelerometers. At the conclusion of the test, the accelerometer may be pried or twisted off of the mounting surface. The epoxy *glue bases* provide one of the most convenient methods for field mounting accelerometers. These *glue bases* are made of anodized aluminum, with a drilled and tapped center hole to accept the accelerometer mounting screw. These bases are attached to the measurement surface with a quick setting (5 to 10 minute cure time) two-part epoxy cement. One surface of the glue base is finished flat for contacting the accelerometer, and the other side is normally grooved to accept the epoxy.

Strong *double bar magnets* may be used to attach accelerometers, but the overall frequency response is significantly reduced. The use of weak magnets is discouraged since the frequency response characteristics of these mounts is generally unacceptable. A variety of *mechanical clamps* or attachments may be used to mount accelerometers. Again, a further reduction in overall frequency response should be anticipated. The final category in Table 6-2 of *hand held* accelerometers may be used if no other form of attachment is available.

From this summary, it is clear that an accelerometer may be rendered ineffective for high frequency measurements simply due to the method of transducer attachment. This characteristic may also be used to control the frequency response of the final measurement. For instance, assume that low frequency information is required on a machine that emits significant high frequency excitations. In this situation a mechanical isolation pad between the measurement surface and the accelerometer may be used to eliminate or suppress the high frequency components. Whenever possible, the calibration procedure should employ the same accelerometer mounting technique that will be used in the field. Within the low frequency domain, variations in accelerometer mounting techniques will not cause appreciable differences in the data. However, the diagnostician is always encouraged to utilize a substantial and rigid mount to eliminate any concerns or potential data corruption associated with a poor transducer attachment.

The ideal accelerometer mounting surface should be both smooth and flat. Historically, transducer manufacturers have specified surface finishes that are often unattainable on cast machinery structures. Spot facing of accelerometer mounting locations in the field, or even machine shop finishing generally results in surfaces that do not meet the ideal vendor requirements. In recent years, these stipulations have been somewhat relaxed, and an emphasis has been placed on using a thin coating of silicone grease or acoustic couplant between the accelerometer base and the mounting device. This approach allows an improved contact between the transducer and the mounting surface by allowing the grease to fill in the voids in both metallic surfaces. The appropriateness of this technique has been verified with shaker table tests, and extensions of the upper frequency limit of 5% to 20% have been documented on several occasions.

In addition to accelerometer sensitivity to the method of attachment, the machinery diagnostician must also secure the coaxial cables leaving the accelerometer. This is particularly important in high vibratory environments were the cable motion could provide addition strain to the accelerometer, and result in false and/or erratic electronic signals. To prevent the occurrence of this potential signal interference problem, it is generally desirable to tape, clamp, or otherwise secure the signal cable within 2 to 4 inches from the accelerometer. In addition, a slight bend or relief of the cable should be maintained.

The same type of shaker table calibration techniques used for velocity coils may generally be applied for accelerometers (e.g., Fig. 6-27). Since top end frequencies are higher with accelerometers, particular attention must be paid to the mounting techniques used for the test as well as for the reference accelerometer. Whenever possible, the accelerometer should be attached to the test surface (e.g., shaker table) in the same manner that it will be mounted in the field. Single point, swept frequency, as well as multi-frequency test systems are used for accelerometer calibration. Other common techniques such as gravitational *drop tests* may also be used for certain types of transducers. In all cases, the downstream electronics should be included in the calibration test to allow documentation of frequency response characteristics for the entire system.

Due to the solid state reliability, and the extended frequency response range of most accelerometers, the applications are substantial. As with any transducer, accelerometers exhibit a variety of advantages and disadvantages. For purposes of comparison with other vibration transducers, the following two summaries of accelerometer features are presented for consideration:

Accelerometer Advantages

- ○ Measures Casing or Structural Absolute Motion
- ○ Easily Attached to Machinery, Piping, Baseplates or Structures
- ○ Good Signal Response Between 900 and 600,000 CPM (15 >10,000 Hz)
- ○ Flat Phase Response Throughout Transducer Operating Range
- ○ Solid State Electronics with Rugged and Reliable Construction
- ○ Operates Below Mounted Natural Resonant Frequency
- ○ Same ICP® Signal Conditioning Usable with Various Transducers
- ○ Special Units Available for High Temperature Applications (>1,200°F)
- ○ Available in Many Configurations
- ○ Small Transducer Size, Easiest to Install in Cramped Areas

Accelerometer Disadvantages

- ❏ Sensitive to Mounting Technique and Surface Condition
- ❏ Unable to Measure Shaft Vibration or Position
- ❏ Difficult Calibration Check
- ❏ External Power Source Required
- ❏ Low Dynamic Signal Response Below 600 CPM (10 Hz)
- ❏ Transducer Cable Sensitive to Noise, Motion, and Electrical Interference
- ❏ Temperature Limitation of 250°F for ICP® Transducers
- ❏ Extended Frequency Range Often Requires Signal Filtration
- ❏ Double Integration Often Suffers from Low Frequency Noise

The characteristics of piezoelectric transducers revolve around the relationship between force, mass, and acceleration. It is reasonable to deduce that any other physical phenomena that can apply a force or load upon a piezoelectric crystal may also be measured with this technology. In fact, the following sections addresses some of these similar dynamic transducers.

PRESSURE PULSATION TRANSDUCERS

One of the most useful devices for analyzing fluid behavior is the dynamic pressure pulsation transducer, as depicted in the photograph of Fig. 6-32. The unit at the bottom of the photo is the actual pressure pulsation transducer, and the assembly shown at the top of the photograph portrays an identical pressure transducer mounted in a 1/2" NPT bushing.

Fig. 6–32 Typical Configuration of Pressure Pulsation Probes Manufactured By PCB® Piezotronics, Inc., Plus Installation Pipe Plug

These small transducers contain an Invar diaphragm that is exposed to the process fluid, and it is connected to the piezoelectric crystal. Application of pressure to the diaphragm causes the crystal to produce a charge that is converted to a voltage sensitive signal by the internal integrated circuit. For the pressure transducers depicted in Fig. 6-32, the output sensitivity is equal to 1.0 millivolt per Psi. These particular transducers are rated for a maximum pressure of 15,000 Psi, and the transducer operating temperature range generally extends between -100°F and +275°F. These pressure probes exhibit a large frequency range, with a high resonant frequency of 28,500,000 CPM (475,000 Hz).

Other configurations of static and dynamic pressure pulsation transducers are available. There are entire families of transducers dedicated specifically for high frequency measurements of shock waves, ballistic phenomena, engine combustion, acoustics, explosive blasts, and machinery related pressure pulsations. Although the physical transducer configurations and performance do vary, the fundamental transducer concepts remain constant.

Although the pressure pulsation transducer technology is fairly straightforward, the application and field utilization of these probes can become quite complicated. In most instances, the complexity is associated with the mounting location of the transducer versus the fluid stream to be investigated. Ideally, the pressure probe would be installed within the fluid to obtain the best possible indication of pressure fluctuations. However, this mounting location is often physically impossible to achieve. The next thought might be to mount the pressure probe flush with the pipe inner diameter. This sounds plausible, but in many instances the desired dynamic data is masked by the fluid boundary layer effects at the pipe wall.

In many applications, the pressure transducer is simply attached to the atmospheric side of an available vent or drain valve. This type of location is physically accessible, but the fluid pressure variation is now attenuated by the intermediate nipples, fittings, and valve(s). In addition, the dynamic signal may also be adversely influenced by a standing acoustic wave within the measurement piping (blowing across an empty bottle effect). In some cases, it may be desirable to *fool* the fluid mechanics of the system by externally increasing the length of the measurement cavity. Specifically, a fixture may be constructed with several feet of stainless steel tubing connected to the transducer measurement chamber. This tubing is generally rolled into a coil to be physically manageable, and the end of the tubing is plugged. Since the acoustic resonant frequency is inversely proportional to the passage length, an increased physical length will result in a significant reduction in the value of the acoustic frequency. A typical field application of this concept is shown in Fig. 6-33.

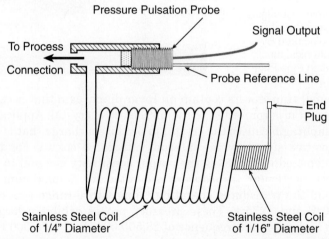

Fig. 6–33 Measurement Fixture For Extending Pressure Pulsation Probe Passage Length

This assembly is connected to the process stream through whatever nipples or valves are required. The 1/4" diameter coil is connected directly to the measurement chamber, and a smaller coil with a tube diameter of 1/16" is connected to the end of the large coil. This functions as a long tube that is open at one end (process end), and closed at the opposite end (plugged end). One quarter of a standing acoustic wave is contained within this device, and the fundamental wavelength is equal to four times the total passage length. The acoustic natural frequency is determined by dividing sonic velocity of the process fluid by the fundamental wavelength. A more detailed explanation of this phenomena is described in chapter 10. However, for this discussion it should be recognized that the acoustic resonant frequency can be adjusted, within physical limits, to a frequency that is outside of the frequency domain of interest.

If the pressure sensor installed in the measurement fixture consists of a sealed assembly containing a diaphragm plus attached piezoelectric crystal, the output will be dynamic pressure pulsation via a coaxial cable (signal plus shield). However, other types of pressure probes are in common use. Manufacturers such as Endevco Corporation and Kulite Semiconductor Products, Inc. offer miniaturized solid state devices that consist of a four arm Wheatstone bridge mounted on the surface of a silicon diaphragm. This type of pressure transducer is known as a piezoresistive transducer. In some probes, a pair of silicone strain gages are combined with a pair of fixed resistors in the transducer bridge. A higher sensitivity is obtained by using four active gages in the Wheatstone bridge circuit. In either case, the primary advantage of a piezoresistive transducer is the ability to measure down to zero frequency (DC). With respect to pressure measurements, this provides the unique and desirable capability of simultaneously measuring the static pressure as well as the dynamic pressure pulsation. The ability of piezoresistive pressure pickups to measure static pressures also allows an easy calibration with a simple hydraulic dead weight tester. These types of pressure transducers are easily recognizable by the four wire output cable configuration that is often combined within an external shield for noise suppression.

Another distinguishing characteristic of some piezoresistive pressure transducers is the presence of a reference line or vent tube protruding from the probe. An example of this thin wall tubing is shown in Fig. 6-33 as the Probe Reference Line. This tubing is connected to the back side of the diaphragm, and allows several different operating modes for the pressure pickup. If the tubing is open to the atmosphere as shown in Fig. 6-33, the transducer will be referenced to atmospheric pressure, and the resultant output will be a relative pressure (Pounds/Inch^2 Gage). If the reference tube is evacuated in a vacuum chamber, and the tube crimped shut, the transducer output will be an absolute pressure (Pounds/Inch^2 Absolute).

Alternatively, the probe reference line may be connected to other pressure sources to measure a differential pressure. This same capability may be employed to eliminate the static pressure from the final output signal. For instance, if the reference line shown in Fig. 6-33 is connected to the end of the small coil, the static pressure will be equalized across the diaphragm. This would effectively AC couple the output signal, and allow enhanced observation or amplification of the dynamic pressure pulsation portion of the electronic signal.

Clearly, there are many items to be considered in the proper selection and application of pressure pulsation transducers. A properly engineered installation produces meaningful information, and a poorly conceived installation generates more questions than answers. As with most dynamic transducers, the machinery diagnostician must carefully consider all aspects associated with the installation and utilization of pressure pulsation probes.

SPECIALIZED TRANSDUCERS

The piezoelectric transducer concept is not only applicable to accelerometers and pressure probes, it is also used for other types of dynamic transducers such as force and shock probes. For ICP® devices, the end user enjoys the convenience of interfacing with identical power supplies. Hence, a variety of ICP® transducers may be driven by the same power supplies, or a Dynamic Signal Analyzer (DSA) such as an HP-35670A.

Another common application of piezoelectric transducers resides in the domain of force measurement. Force sensors are used to measure compression, tension, and impact forces involved with a wide variety of manufacturing processes. There are also many applications for load or force transducers for measurement of radial or axial bearing loads on machines. Some of these installations are for temporary test measurements, and other installations are designed for long-term continuous monitoring of machinery forces.

In the realm of structural testing, piezoelectric force transducers are used in devices such as impact or impulse hammers to deliver a short duration force pulse to a structure. The force transducer measures the output characteristics of the force pulse, and a separate accelerometer is used to measure the resultant structural response. The clarity of these signals are verified in the time domain, and a frequency response function (FRF) is performed with a Dynamic Signal Analyzer between acceleration and force. The resultant frequency response curve is used to identify structural resonances plus damping characteristics.

In critical applications, it may be desirable to perform a system calibration that involves the accelerometer, the impact hammer, plus the data processing instrumentation. This apparently difficult task may be accomplished by the simple test fixture depicted in Fig. 6-34. In this diagram, a steel cylinder is suspended from a single stationary point using a two point hitch. This type of

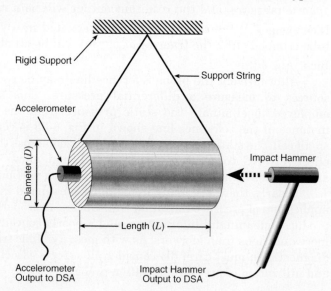

Fig. 6–34 Typical Test Setup Used For Mass Calibration of Piezoelectric Force Hammer And Accelerometer

suspension is critical for obtaining proper accuracy. This cylinder represents a mass at rest. If the mass is excited with an input force, the mass should accelerate in proportion to the applied force. More succinctly, this is known as Newton's *second law of motion*, and it is customarily stated as follows:

$$\boxed{Force = Mass \times Acceleration} \qquad\qquad \textbf{(6-7)}$$

In the test setup shown in Fig. 6-34, the mass is constant. Assume that the input force is measured with a force transducer that is integral with the impact hammer, and acceleration of the mass is measured with the axial accelerometer. Intuitively, mass should be determinable with a FRF of force over acceleration. As an example, consider a solid steel cylinder that is 1.50 inches in diameter, and 3.5 inches long. The weight is computed with equation (3-7) as follows:

$$W = \frac{\pi \times L \times \rho \times D^2}{4} = \frac{\pi \times 3.50 \text{ Inches} \times 0.283 \text{Pounds/Inch}^3 \times (1.50 \text{ Inch})^2}{4} = 1.75 \text{ Pounds}$$

To check these calculations, the steel cylinder was weighed on a digital laboratory scale at 790 grams. This metric mass unit may be converted to equivalent English units of pounds in the following manner:

$$W = 790 \text{ Grams} \times \frac{1 \text{ Pound}}{453.6 \text{ Grams}} = 1.74 \text{ Pounds}$$

Hence, the calculated and measured weights are equal. The next step consists of connecting the impact hammer force transducer to channel 2, and the accelerometer to channel 1 of an HP-35670A analyzer, and performing an FRF during several sample hits with the hammer. The results of this test are shown in Fig. 6-35 for a total of eight samples. At the midscale of 48,000 CPM (800 Hz), the amplitude is equal to 1.744 pounds. Obviously this is excellent agreement with the calculated and the measured weight of this steel cylinder. If significant differences appear between the actual weight and the FRF weight, the calibration of the force transducer or the accelerometer might be questionable. This could be resolved by individual calibration of both piezoelectric transducers. Alternatively, a percentage deviation could be established between the actual weight and the FRF weight. This constant offset could then be applied to all future FRFs with the DSA math functions.

For reference purposes, the test setup shown in Fig. 6-34 and the resultant FRF presented in Fig. 6-35 were performed prior to a turbine blade test. The overall test weights were fairly small, and a 1 gram accelerometer was combined with a miniature force hammer for this test. The same results may be achieved with physically larger systems, but the diagnostician should be very careful with the physical setup as well as the DSA operation. The transducers should be connected directly to the DSA, and powered by the internal ICP® power source. A suitable force window should be used for the hammer input on channel 2, and a proper exponential window selected for the accelerometer on channel 1. The hammer should be fixed to a constant pivot point so that it strikes the mass at essentially the same point for every sample. Also, the mass must be stationary

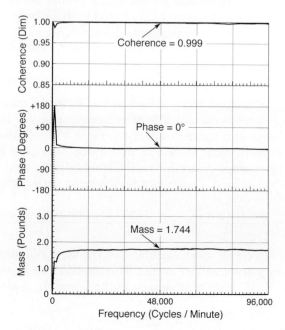

Fig. 6–35 Frequency
Response Function (FRF)
Used For Mass Calibration
Of ICP® Force Hammer
And Accelerometer

before every hit. This is the only way to guarantee comparable data for each of the FRF samples. Finally, if the mass line is not straight, or if the phase differential is higher than ±2°, or if the coherence falls below 0.99, then there is something wrong with the test setup or the DSA.

This same data processing scheme may be used for examining force divided by an acceleration signal that is double integrated to displacement. The resultant FRF of force over displacement provides a measurement of dynamic stiffness versus frequency. Certainly this is important parameter that may be necessary to properly describe or model the behavior of a particular machine and/or associated structure. This technique was discussed in chapter 4, and case histories 8 and 9 describe the support stiffness measurements on a steam turbine and a gas turbine bearing housing support.

Further explanations and discussions of these various piezoelectric transducers are beyond the current scope of this text. The machinery diagnostician should recognize that these devices do exist, and they may provide essential information on specific mechanical problems. In addition, there are other specialized transducers that are commonly available within the industrial community. For instance, lasers are used for static machinery alignment measurements as discussed in chapter 12. Lasers are also used for non-contacting lateral and torsional vibration measurements on various machines and structures. Although the laser based instruments are both expensive and complex, the capabilities are significant. In some cases, laser based transducers provide the only method for direct measurement of a vibrating surface.

Another type of optical transducer is the timing probe shown in the photo-

Fig. 6–36 Typical Optical Transducer Used For Machinery Timing Measurements

graph in Fig. 6-36. This device is used to observe a piece of reflective tape, or other variable contrast media attached to a rotating shaft or a reciprocating surface (e.g., compressor drive rod). The coincidence of the reflective tape with the optical probe produces a pulse signal that may be used as a Keyphasor® for field balancing or malfunction diagnosis. This type of transducer is ideally suited to machines that cannot tolerate drilled holes or milled slots in exposed shaft surfaces. High speed machines are often candidates for this type of timing pickup. Unfortunately, the use of reflective tape for the shaft timing mark is usually limited by the surface speed of the shaft. For cases of high surface velocities, the use of black spray paint or layout bluing may be used to uniformly darken the entire shaft. A strip of reflective paint may then be applied to the shaft surface to act as the trigger mark. The signal output from the optical driver may then be observed on an oscilloscope, and the suitability of the resultant pulse signal evaluated. Obviously, corrections may be performed to the reflective mark to improve the clarity and consistency of the final optical Keyphasor® signal.

Other dynamic transducers such as strain gages may be successfully used on mechanical structures as well as on rotating shafts. There are many available techniques for extracting this type of information, including direct wiring, telemetry, brush contacts, and some newer optical transmission techniques. It should be recognized that strain gage application, installation, and proper operation are almost an independent branch of measurement technology. Strain gages may be installed individually to measure strain in one direction, or three gages may be mounted in rectangular or various rosette configurations to determine the principal strains. From the strain data, the principal stresses may be computed by incorporating the modulus of elasticity E for the material. The resultant stress levels may then be compared with the elastic strength of the material under test. Typically, various safety factors are also included to form the final assessment of measured versus allowable stress limits.

The proper application of strain gage technology is a complex endeavor that should categorically be classified as an experimental stress analysis technique. Certainly there are occasions when it is necessary to examine the stress characteristics of a mechanical element. In this situation, the machinery diagnostician is encouraged to examine additional technical references, such as chapter 17 of the *Shock and Vibration Handbook*[2], or *Practical Strain Gage Measurements* by Hewlett Packard[3]. More detailed discussion of strain gage characteristics and specific application notes may be obtained from the vendors of these devices. Companies such as Measurements Group, Inc.[4] provide excellent technical literature as well as extensive training materials.

Strain gages are used in devices such as piezoresistive accelerometers and pressure transducers. These probes have the ability to measure constant acceleration or pressure levels (via DC output) combined with high frequency behavior or long duration shock pulses. Strain gages may also be applied for the measurement of tension and torsion in a machine shaft. This is a simple measurement on a non-rotating shaft. However, when the shaft is turning, the complexity of the strain measurement increases significantly. For temporary tests on rotating shafts, the radio telemetry system offered by Binsfeld Engineering (Maple City, Michigan) works quite well. In this system, the shaft mounted strain gages are powered by a nine volt battery and a transmitter that are attached to the shaft with strapping tape. A stationary circumferential antenna picks up the strain output from the rotating transmitter, and directs the signal to a Receiver Demodulator. After this device demodulates the signals, traditional instruments may be used to examine the static and dynamic strain signals.

For permanent or continuous monitoring of shaft torque, the non-contacting TorXimitor® by Bently Nevada Corporation provides an attractive offering. Within this system, a balanced strain gage bridge is mounted on a machine coupling or spool piece. The strain gages are combined with inductive power receivers, and FM signal transmission electronics. Adjacent to the coupling, a stationary module provides inductive energy to the rotating coupling electronics, and receives the strain gage signals broadcasted by the coupling transmitter. Finally, the FM signals are demodulated and conditioned in a separate module to provide an electrical signal that is proportional to torque. If a Keyphasor® signal is included, the transmitted horsepower is determined from equation (2-98).

In addition to the diversity of specialized dynamic transducers that are available, there are many variations of the standard displacement, velocity, and acceleration probes. For instance, extended range proximity probes exist that have linear operating ranges that easily exceed one inch. These types of probes generally have low sensitivity values in the range of 5 to 10 millivolts per mil. Other proximity probes are designed to be waterproof, resistive to a variety of industrial chemicals, and capable of withstanding significant operating pressures and temperatures.

Magnetic probes may be used to measure repetitive events when they are

---

[2] Cyril M. Harris, *Shock and Vibration Handbook*, Fourth edition, (New York: McGraw-Hill, 1996), chap. 17.
[3] *Practical Strain Gage Measurements, Application Note 290-1*, (Hewlett Packard, printed in USA, 1987).
[4] *Strain Gage Technology-Technical Reference Binder*, (Raleigh, North Carolina: Measurements Group, Inc., 1996 update).

positioned over conductive materials with discontinuities. These types of transducers are often used as primary sensors in turbine speed control systems. Typically, they are mounted in a radial direction over multi-tooth gears, and they produce an electrical pulse as each gear tooth passes underneath the probe. In control systems, three probes are used over the same gear to provide full redundancy and backup for the measurement. Dividing the pulse passing frequency by the number of gear teeth will yield the shaft rotative speed. It should be mentioned that the electrical signal output from these transducers is dependent upon the size and shape of the gear teeth, the peripheral speed, and the distance between the magnetic probe tip and the top of the gear teeth. In many applications these pickups are gapped at only 10 to 15 Mils (0.010 to 0.015 inches) from the gear teeth. This close distance makes the probes susceptible to damage.

Magnetic pickups or proximity probes are also used for the measurement of torsional vibration. For this application, the transducer is positioned to observe a set of gear teeth or a precision slotted wheel. The resultant tooth (or slot) passing frequency is radially influenced by the torsional motion. In this application, the transducer output is a frequency modulated (FM) signal. The high frequency carrier signal is the tooth (or slot) passing frequency, and the lower frequency modulating signal is proportional to the torsional vibration. Thus, demodulation of the FM signal produces the torsional vibration. Typically, two non-contacting pickups, spaced 180° apart, are used to observe the slotted wheel. Both of the signals are fed into a device such as the Bently Nevada TK17 Torsional Vibration Signal Conditioner. Within this instrument the two opposed signals are used to cancel out radial vibration, and thereby minimize measurement noise. The final output of the TK17 is the conditioned and demodulated signal proportional to torsional vibration. This signal carries engineering units of volts per degree, and it may be processed or recorded with any traditional array of diagnostic instrumentation.

Historically, torsional velocity coil transducers have been installed on the blind end of gear shafts for short-term unit testing. These devices operate similar to casing velocity pickups, with the significant difference that the output signals from the velocity pickup must be obtained through slip rings. This requirement limits the total useful operating life of this type of transducer to a few hours. Furthermore, the machinery train must be shutdown for removal of the torsional velocity coil at the end of the test. This may not be a particular problem for the small blower installation, but it can be a significant expense to shutdown and restart a large turbine generator set.

In the overview, a wide variety of static and dynamic measurement transducers are available in the industrial marketplace. The machinery diagnostician must define the specific test measurements required. Based on these requirements, the best possible transducer suite should be selected. There may be distinct differences in the transducers temporarily installed for a field test versus the pickups normally used for permanent monitoring and machinery protection. This applies to the type and quantity of the transducers as well as to the accuracy of the calibration of the various devices. In all cases, direct measurements are preferable to implied measurements, and correlation of multiple measurements remains as a highly desirable objective.

ASPECTS OF VIBRATION SEVERITY

Machinery vibration measurements are often viewed as quality control, or machinery condition measurements. Although this may be an oversimplification, most people consider low vibration levels to be indicative of proper machinery behavior. When vibration amplitudes are high, the general tendency is to believe that the machinery is experiencing some type of distress or mechanical malfunction. Ideally, this concept should be predicated upon the existence of a definitive method, engineering specification, or set of guidelines that allow a clear cut definition of acceptable versus unacceptable vibration amplitudes. In reality, the only statement that applies to all machines, under all conditions, is as follows:

...There Are No Universal Vibration Severity Limits...

Over the years many organizations and corporations have tried to establish vibration severity standards. Although partial success has been attained with some machines, and some specific applications, the establishment of universal vibration tolerances has not been achieved. When the measurement complexities are combined with the intricacies and variations between machines, it is self-evident that development of universal vibration severity limits may be beyond any reasonable expectation.

Even though it is doubtful that a universal criteria can be established, that still does not eliminate the need for a methodology to define workable vibration limits. Within the context of this book, it is reasonable to examine the various parameters that influence the measured vibration, and provide some guidance for addressing this complex question. The first issue that should always be considered is the fundamental accuracy of the measurement. The vibration transducer, cables, interface devices, signal converters, and final readout must be checked and calibrated as a system. If a problem exists in any part of the measurement chain, the validity of the entire measurement is compromised. Hence, routine calibration checks are mandatory.

The next consideration is the proper operation and calibration of any diagnostic instrumentation that is used to analyze the vibratory behavior. This equipment must also be rigorously and periodically checked for proper operation and calibration. Fortunately, many modern digital instruments are both self-checking and self-calibrating. This is a comforting feature that provides an improvement in data consistency and accuracy. However, with any instrumentation system, the diagnostician must continually compare the final hard copy output data with the transducer output signals to verify that some anomaly has not occurred within the data processing system.

For many measurements, a more fundamental question is the utilization of the correct vibration transducer for a specific measurement task. In chapter 2 of this text it was shown that displacement, velocity, and acceleration of an element are integrally related. If one of the three vibration parameters and the frequency was known, the other two vibration values could be easily calculated with equations (2-20) through (2-22). Phase difference between the three fundamental motion properties was also defined in equations (2-14) through (2-16). Hence, it might appear that any vibration value or vector could be easily converted into any other convenient set of engineering units. Indeed, this was the case presented in Fig. 2-4 where a constant velocity of 0.3 IPS,$_{o-p}$ was converted to equivalent displacement and acceleration amplitudes for a wide range of frequencies (1 to 20,000 Hz). This earlier plot considered nothing more than the pure transformation of vibration parameters.

However, as demonstrated within this chapter, the individual vibration transducers have specific areas of application, and definite limitations of frequency response characteristics. If the data from Fig. 2-4 is replotted with some realistic limitations on the actual frequency operating ranges of the vibration pickups, the following Fig. 6-37 may be drawn.

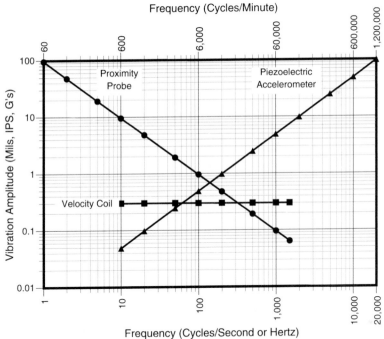

Fig. 6–37 Typical Frequency Operating Ranges For Traditional Vibration Transducers

This log-log plot reinforces the fact that proximity probes are more suitable for the lower frequency measurements around machine operating speed. Accelerometers are the correct transducer for high frequency measurements such as blade passing or gear meshing characteristics. Furthermore, the relative position and vibration capabilities of proximity probes makes them eminently appropriate for measurements on machines with fluid film bearings. On machines equipped with rolling element bearings, or units that normally emit high frequency excitations, the external mounting of casing accelerometers makes the most engineering sense.

Direct velocity measurements have often been touted as the most informative type of vibration data. This reputation is based upon the fact that frequency and displacement are combined into one value, for example, equation (2-20). Although a velocity coil has a limited frequency range (as shown in Fig. 6-37), the concept of using specific velocity amplitudes for severity evaluation has been popular in many applications, and for many years.

For bearing housing vibration measurements, it is common to assign specific velocity amplitudes to varying levels of mechanical condition. For instance, Fig. 6-38 identifies nine different levels of probable machinery condition versus the associated casing velocity amplitudes. This varies from a category of *excellent* condition at 0.002 IPS,$_{o-p}$ to *danger* at 1.0 IPS,$_{o-p}$. This plot converts the velocity amplitudes into bearing housing displacement amplitudes with associated engineering units of Mils,$_{p-p}$.

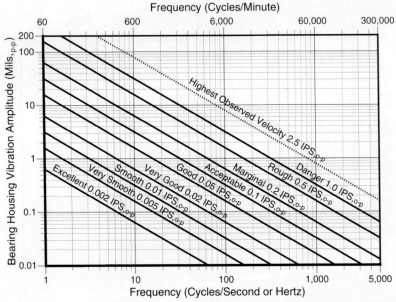

Fig. 6–38 Typical Vibration Severity Chart For Properly Supported Bearing Housings

Understandably, the highest amplitudes are tolerated at low frequencies, and the acceptable vibration level decreases with increasing frequency. This behavior of allowable vibration amplitude versus frequency is reasonable, and it is incorporated on virtually all severity charts. Many organizations use vibration severity charts that are similar to Fig. 6-38. Although the names of the categories will change, and the assigned velocity amplitudes will vary, the general chart configuration will resemble Fig. 6-38.

Although the concept of setting vibration limits based upon constant velocity sounds like a reasonable approach, it does not work in many cases. For instance, Fig. 6-38 shows a line of constant velocity at 2.5 IPS,$_{o-p}$. This line is labeled as *highest observed velocity*. Casing vibration amplitudes at this level have been observed by the senior author on several occasions. Although failures did eventually occur, the machinery operated for a considerable length of time. In two other cases, speed increasing gear boxes displayed bearing cap vibration levels in excess of 2.0 IPS_{o-p}. These gear boxes sustained large casing vibration amplitudes for many years with no adverse influence upon the gear teeth or the bearings. At the opposite end of the severity chart, mechanical failures have been documented with casing vibration levels of less than 0.05 IPS_{o-p}.

Clearly, the correlation between casing vibration and the excitation associated with various types of failure mechanisms is neither a definable, nor a constant quantity. For these types of measurements, the machinery diagnostician must recognize that different machines, operating in different services, with different mechanical components are always installed with a variety of foundation and piping systems. It is only logical to conclude that different machines must exhibit different casing vibration characteristics. In practice, casing vibration measurements are normally applied on small general purpose equipment trains equipped with rolling element bearings. Since universal casing vibration limits cannot be accurately quantified, the typical approach for this type of machinery is to perform trend analysis as a function of time. Although routine variations in casing amplitudes and frequencies will appear, the envelope of normal behavior should eventually become visible, and reasonable alert and danger levels may be established for each machine.

For larger machinery trains, the main bearings are normally fluid film bearings as discussed in chapter 4. For these critical units, the additional flexibility and damping at the journal bearings makes a severity chart such as Fig. 6-38 even less appropriate. On these types of machines, the primary vibration measurements are made with non-contacting proximity probes mounted at each bearing housing. As discussed earlier in this chapter, the proximity probes have the ability to measure journal position within the bearing, plus relative shaft vibration. If the peak to peak shaft vibration is compared against the available bearing diametrical clearance, a measure of severity may be deduced. For instance, the values listed in Table 6-3 are considered to be appropriate for a variety of machine trains equipped with fluid film bearings.

The proper utilization of Table 6-3 requires knowledge of the radial bearing clearance. If this diametrical clearance is not available, the machinery diagnosti-

Table 6–3 Machinery Condition Based Upon Runout Compensated Shaft Vibration As A Percentage Of The Total Available Diametrical Journal Bearing Clearance

Machine Condition	Peak to Peak Shaft Vibration As A Percentage of Diametrical Bearing Clearance	Appropriate Action
Normal	Less Than 20% of Available Clearance	Continue Monitoring
Alert	40% to 60% of Available Clearance	Initiate Corrective Action
Danger	More than 70% of Available Clearance	Shutdown Machine

cian might estimate the bearing clearance based upon the shaft diameter. As discussed in chapter 4, a bearing clearance ratio (BCR) of 1.5 Mils per inch of diameter is quite common on large machines. Thus, if a 6.0 inch diameter shaft journal is encountered, a normal diametrical clearance would be in the vicinity of 9.0 Mils. Applying the percentages shown in Table 6-3 would indicate that normal runout compensated shaft vibration should be less than 1.8 Mils,$_{\text{p-p}}$. By the same token, a danger or shutdown level would be at 70% of the available bearing clearance, or a trip setpoint of 6.3 Mils,$_{\text{p-p}}$.

These are reasonable amplitudes for shaft vibration of a 6 inch journal. However, they do not directly address the speed or load characteristics of the bearing. It is implied that large shafts run at slower speeds with larger absolute clearances, and smaller diameter shafts run at higher speeds with proportionally smaller bearing clearances. The work by Jim McHugh[5] incorporated the static rotor load on the bearings. For a simplified case of synchronous 1X shaft vibration in fluid film bearings, McHugh offered the following empirical equations:

$$Acceptable\ Level \le \left(\frac{25}{BUL}\right) \times D_c + 1 \qquad \text{(6-8)}$$

$$Alarm\ Level = \left(\frac{50}{BUL}\right) \times D_c \qquad \text{(6-9)}$$

$$Danger\ Level = \left(\frac{100}{BUL}\right) \times D_c \qquad \text{(6-10)}$$

where: *Acceptable Level* = Acceptable Shaft Vibration Amplitude (Mils,$_{\text{p-p}}$)
          Alarm Level = Setpoint for First Vibration Alarm (Mils,$_{\text{p-p}}$)
          Danger Level = Setpoint for Second Vibration Alarm (Mils,$_{\text{p-p}}$)
          D_c = Bearing Diametrical Clearance (Mils)
          BUL = Bearing Unit Load (Pounds / Inch^2)

The bearing unit load (BUL) in the above three expressions is simply the shaft weight upon the bearing divided by the plane area of the bearing. This concept was previously discussed in chapter 4, and the BUL is easily calculated with equation (4-1). McHugh includes a +1.0 Mil,$_{\text{p-p}}$ additional amplitude in equation (6-8) to allow for runout, or for measurements at other than the mid-plane of the bearing. Of course, a runout vector could be directly opposed to the synchronous vibration vector, and this would result in a decrease in uncompensated shaft vibration. It could also be argued that equations (6-9) and (6-10) should also include some type of runout correction. Nevertheless, these three equations do reinforce the concept that bearing load is an important consideration in establishing vibration severity limits. In essence, machines with high static loads are less tolerant to excessive vibration than machines with low bearing loads.

---

[5] James D. McHugh, "Setting Vibration Criteria for Turbomachinery," *Proceedings of the Eighteenth Turbomachinery Symposium*, Turbomachinery Laboratory, Texas A&M University, College Station, Texas (October 1989), pp. 127-135.

For a typical case of a bearing unit load of 150 Pounds/Inch^2 and a diametrical clearance of 9.0 Mils, the acceptable vibration amplitude from equation (6-8) would be 2.5 Mils,$_{p-p}$. Similarly, the danger or trip level may be determined from equation (6-10) as 6.0 Mils,$_{p-p}$. These values are in general agreement with the bearing clearance percentages presented in Table 6-3.

Another approach incorporates machine operating speed into the severity evaluation. For example, consider Fig. 6-39 that plots relative shaft vibration amplitude in Mils,$_{p-p}$ versus machine running speed. Clearly, this diagram is similar in construction to other vibration severity charts. In this case, the magnitudes and severity categories are based upon a myriad of measurements by the senior author on a wide variety of industrial machines. These vibration levels are from proximity probes mounted adjacent to the bearings, and the amplitude readings are corrected for shaft runout

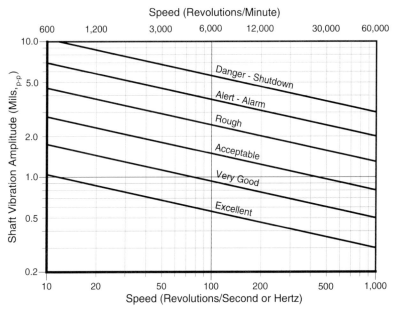

Fig. 6–39 Runout Compensated, Relative Shaft Vibration Severity Chart

Once again, it is noted that vibration tolerance decreases with operating speed. This is both reasonable and expected behavior. At a fixed speed, it is presumed that deterioration on a machine will propagate as increased vibration amplitudes, and a shutdown will eventually be required for problem correction. It can be argued that this graph is conservative in some areas, and it is liberal in other regions. Some individuals might disagree with the categories or the specific amplitudes displayed on Fig. 6-39, but the overall concept is generally sound from a machinery management standpoint. Please note that any of the vibration severity charts presented in this text (or from any other common source), should only be considered as general information. These vibration severity charts must not be construed as absolute or specific guidelines.

It is often difficult to draw a line of demarcation between acceptable and dangerous vibration amplitudes. The fundamental issue always comes down to the situation of shutting down the machine before a catastrophic failure occurs, versus the potential for unnecessary shutdowns due to normal changes or variations of the machinery. On large process machines many factors influence the shaft vibratory behavior, and the following list summarizes typical considerations for severity evaluation, and the establishment of suitable vibration limits.

○ Bearing Diametrical Clearance
○ Machinery Maximum Continuous Operating Speed
○ Bearing Static Load (that is, shaft weight upon journal)
○ Bearing Dynamic Load (loads due to gear contact forces, unbalance, etc.)
○ Actual Stress/Strain Levels Imposed by the Static plus Dynamic Loads
○ Shaft Centerline Position
○ Bearing Temperature at Minimum Oil Film
○ Modal Location of Measurement versus Physical Bearing Location
○ Shaft Electrical Runout
○ Shaft Mechanical Runout or Eccentricity
○ Frequency Distribution of Shaft and/or Casing Vibration
○ Journal Bearing Configuration
○ Bearing Housing Support Flexibility
○ Foundation Support Flexibility
○ Casing to Rotor Weight Ratio
○ Potential Influence from Attached Large Bore Piping
○ Potential Influence from Adjacent Machinery or Other External Excitations
○ Sensitivity to Process and/or Load Variations
○ Sensitivity to Variations between Day and Night
○ Sensitivity to Variations between Summer and Winter

New machines are often started with somewhat loose vibration limits due to the unknown behavior in many of the above categories. Although computer simulations can provide meaningful information of some mechanical conditions (e.g., unbalance), the final installed field vibratory behavior must always be measured and evaluated. Over a period of time, the vibration limits may be decreased as the operating experience increases. Again, the objective is always to protect the personnel, protect the machinery, and minimize false alarms.

With respect to new plants and machinery installations, one of the issues that continually reappears is the allowable vibration limits on piping. Acceptable piping vibration amplitudes are not well defined within the process industries. Technical organizations and specifications often bypass this topic with statements such as *...in cases of excessive piping vibration, the problem shall be corrected by adjusting supports, dampers, and snubbers accordingly...* Although that may be acceptable in a design specification, it may be inordinately difficult to evaluate and correct in the field.

Generally, there are three acceptable methods for evaluating piping vibration severity. The first approach is the direct technique of measuring strain, and multiplying by the modulus of elasticity to determine stress. Comparison of the stress against established endurance limits results in an evaluation of the potential for high cycle fatigue. In many cases of carbon steel piping, a measured level of 100 microstrain is considered to be acceptable. This value of 100 microstrain includes stress concentration plus safety factors. Hence, experimental stress analysis may be used to determine the suitability of vibrating pipe.

The second approach for evaluating the suitability of a piping system is to measure the peak velocity of the piping, and apply a constant velocity criteria to all frequencies. Historically, acceptable levels varying from 0.5 to 2.0 IPS,$_{o-p}$ have been suggested as maximum piping vibration limits. Again, this depends upon the specific technical application, and method of pipe support.

The third method for evaluating piping systems is also based upon piping vibration measurements. In this technique, vibration amplitudes are measured in displacement, and the amplitudes are compared against frequency for various severities. The empirical results by J.C. Wachel[6] have been widely published, and a rendition of this piping severity chart is shown in Fig. 6-40. In addition to the

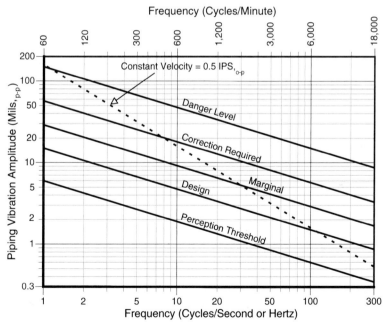

Fig. 6–40 Typical Piping Vibration Severity Chart By J. C. Wachel and J.D. Tison

---

[6] J.C. Wachel and J.D. Tison, "Vibrations In Reciprocating Machinery and Piping Systems," *Proceedings of the Twenty-Third Turbomachinery Symposium*, Turbomachinery Laboratory, Texas A&M University, College Station, Texas (September 1994), pp. 243-272.

five severity levels, a line of constant velocity corresponding to 0.5 IPS,$_{o-p}$ has been included. Again it must be recognized that this type of information may be directly applied on some installations, and in other cases it provides a reference point to begin a more detailed evaluation of the installed mechanical system.

In order to be perfectly clear, it should be restated that ...*There Are No Universal Vibration Severity Limits*...Tolerances will vary by machinery type, configuration, application, installation, vibration transducer, and industry. The only real answer to the establishment of vibration severity criteria lies in careful measurement and proper engineering evaluation of each specific machinery installation. Finally, the machinery diagnostician should not be surprised when some malfunctions result in a decrease in measured vibration. Although this only occurs under unique conditions, it is a true physical reality.

BIBLIOGRAPHY

1. Harris, Cyril M., *Shock and Vibration Handbook*, Fourth edition, chap. 17, New York: McGraw-Hill, 1996.
2. McHugh, James D., "Setting Vibration Criteria for Turbomachinery," *Proceedings of the Eighteenth Turbomachinery Symposium*, Turbomachinery Laboratory, Texas A&M University, College Station, Texas (October 1989), pp. 127-135.
3. *Practical Strain Gage Measurements*, *Application Note 290-1*, (Hewlett Packard, printed in USA, 1987).
4. *Strain Gage Technology - Technical Reference Binder*, (Raleigh, North Carolina: Measurements Group, Inc., 1996 update).
5. "Vibration, Axial Position, and Bearing Temperature Monitoring Systems - API Standard 670, Third Edition," *American Petroleum Institute*, (Washington, D.C.: American Petroleum Institute, November 1993).
6. Wachel, J.C. and J.D. Tison, "Vibrations In Reciprocating Machinery and Piping Systems," *Proceedings of the Twenty-Third Turbomachinery Symposium*, Turbomachinery Laboratory, Texas A&M University, College Station, Texas (September 1994), pp. 243-272.

Dynamic Signal Characteristics

Dynamic signals may contain a myriad of discrete frequencies, associated harmonics, interactions between components, plus various types of noise and signal interference. As the transducer frequency range is expanded, the signal complexity increases. For instance, a proximity probe observing a clean radial surface on bull gear shaft might exhibit the bull gear rotational frequency, plus an excitation at the pinion speed. By comparison, a casing mounted accelerometer might pick up the running speed components, the gear mesh frequency, the frequencies associated with box structural resonances, blade passing excitation from the lube oil pump, plus excitations from an adjacent machinery train. Obviously, the acceleration signal will be considerably more complex than the displacement signal.

In order to attempt any type of mechanical assessment of machinery behavior, it is necessary to dissect the dynamic signals into discrete and understandable portions. To accomplish this goal, a wide variety of electronic instruments are commercially available. In virtually all cases, these devices incorporate some fundamental types of electronic filters to assist in examination of the dynamic signals. These electronic filters may be analog, digital, or a combination of both.

ELECTRONIC FILTERS

The array of electronic filters may initially appear to be staggering, but in essence there are only two fundamental types of filters. These basic filters are the low-pass, and the high-pass filters. All other filter types and configurations are merely combinations of the two basic types. Hence, if the characteristics of the basic filters are understood, then the more sophisticated combinations of filters may be comprehended, and these electronic devices properly applied for dynamic signal enhancement and examination.

In order to explain filter characteristics in a consistent manner, a series of examples have been prepared. In the following cases, a Hewlett Packard 35665A Dynamic Signal Analyzer (DSA) was used to generate a broad band 1.0 volt signal that maintained the same amplitude over a wide frequency range. This white noise signal was used as the input into the various filters as shown in Fig. 7-1.

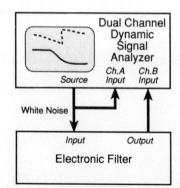

Fig. 7–1 Instrumentation
Arrangement For Frequency
Response Testing Of An
Electronic Filter

The broad band noise signal was also used as a direct input into channel *A* of the DSA. The noise signal was filtered by the external electronic filter, and the filtered output signal was used as an input into channel *B* of the DSA. The actual influence of the electronic filter was determined by performing a frequency response function (FRF) between the filter output and the noise source input.

In this type of testing, it is mandatory to consider the filter effect upon the signal amplitude as well as the influence upon the phase angle. Although this information could be generated with an oscilloscope and a function generator, the DSA performs this output/input comparison much faster, and with greater accuracy. As an example, consider the frequency response function (FRF) shown in Fig. 7-2 of a low-pass filter set at 50 Hz (3,000 CPM).

For this data, a Krohn-Hite, model 3323 dual channel analog filter was used (-24 dB/Octave per channel). This class of filter passes the low frequencies (i.e., low-pass), and it rejects the high frequencies. By definition, this filter has a single transmission band extending from the lower frequency limit of the device to some finite upper cutoff frequency. The filter frequency is coincident with an amplitude attenuation of -3 dB. In this case, the filter was manually set at 50 Hz (3,000 CPM). From the test data in Fig. 7-2 it is noted that the 1.0 volt input signal was reduced to 0.702 volts at 50 Hz (3,000 CPM). This is very close to the filter design value of 0.707 volts (i.e., -3 dB = 0.707).

Note that the signal amplitude varies with frequency. Even though the filter was set at 50 Hz (3,000 CPM), amplitude attenuation occurs down to a frequency of 30 Hz (1,800 CPM). At the upper end of the frequency scale, it is clear that 10% of the voltage signal is still passed through the filter at 90 Hz (5,400 CPM). Hence, it must be recognized that the crossover between acceptance and rejection of a signal is not an instantaneous event. It does occur over a finite frequency range. By increasing the sharpness to -48 dB/octave, the filter characteristics are imposed over a much narrower frequency range.

The presented phase data in Fig. 7-2 for the low-pass filter also reveals significant changes across the examined 100 Hz (6,000 CPM) frequency range. This is normal behavior for this type of device; and failure to consider this characteristic can easily corrupt the final data interpretation. In all cases, it must be acknowledged that filters may be successfully applied to reduce the complexity,

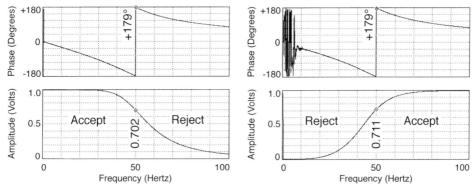

Fig. 7–2 Low-Pass Filter Set At 50 Hz   **Fig. 7–3** High-Pass Filter Set At 50 Hz

or *clean up* a dynamic signal. However, the resultant amplitude and phase data may be distorted. It is mandatory that the diagnostician be fully aware of the characteristics of any applied electronic filters.

A low-pass filter is quite useful for eliminating high frequency interference. For instance, a proximity probe signal may be passed through a low-pass filter to minimize the influence of shaft scratches or other surface imperfections. This is particularly significant when X-Y signals are displayed as a shaft orbit. A clean pair of vibration signals will result in a shaft orbit that is definitive of the journal centerline motion. Conversely, an orbit that includes multiple scratches might be totally illegible (e.g., ball of string display).

The low-pass represents one of the basic types of electronic filters. The companion filter that exhibits the opposite frequency characteristics is a high-pass filter. This type of filter only passes the higher frequencies (i.e., high-pass), and it rejects the lower frequencies. By definition, this filter has a single transmission band extending from a defined finite lower cutoff frequency (amplitude at -3 dB) to the upper frequency of the device. The diagram in Fig. 7-3 depicts the behavior of a 50 Hz (3,000 CPM) high-pass filter. Again, the filter was manually set at 50 Hz, and the 1.0 volt input signal was reduced to 0.711 volts at the set frequency. This is very close to the predicted value of 0.707 volts (i.e., -3 dB = 0.707).

Once more it is noted that the filtered amplitude varies with frequency. Even though the filter was set at 50 Hz (3,000 CPM), amplitude attenuation occurs up to a frequency of 80 Hz (4,800 CPM). At the bottom end of the frequency scale, it is noted that 10% of the signal is still passed through the filter at 28 Hz (1,680 CPM). The phase data for the high-pass filter also reveals significant variations across the 100 Hz (6,000 CPM) frequency range. This is normal behavior for this type of electronic filter. Once more, the machinery diagnostician must consider this characteristic during data analysis.

A high-pass filter is quite useful for eliminating low frequency interference. For example, double integration of an acceleration signal often produces substantial levels of low frequency noise. This condition may drive the significant vibration data down into the noise floor of the analytical instrumentation due to the overwhelming level of the low frequency components. One way to avoid this problem is to use a high-pass filter to eliminate the low frequencies from the acceleration signal before integration. In some instances, a double high-pass filter might be necessary. One stage of the high-pass filter would be used on the acceleration signal prior to integration, and the second stage would be used on the velocity signal before integration to displacement.

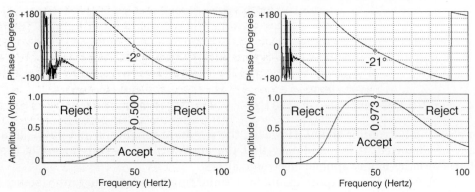

Fig. 7–4 High-Pass Filter Set At 50 Hz Plus Low-Pass Filter Also Set At 50 Hz

Fig. 7–5 High-Pass Filter Set At 30 Hz Plus Low-Pass Filter Set At 70 Hz

Low and high-pass filters may be considered as the building blocks for other filter types. For instance, the 50 Hz (3,000 CPM) low-pass and the 50 Hz (3,000 CPM) high-pass filter may be consecutively applied to the same signal as shown in Fig. 7-4. In this combination, the manually tuned frequency of 50 Hz will be subjected to a -3 dB attenuation from the low-pass, and another -3 dB reduction from the high-pass filter. This combined attenuation of -6 dB is equivalent to a 50% voltage ratio, and that is exactly equal to the measured value of 0.500 volts at 50 Hz (3,000 CPM). The phase change of the combined filter at 50 Hz is only -2°. However, increased or decreased frequency deviation will result is significantly greater phase errors. Obviously, this 50 Hz low-pass and high-pass combination is an unrealistic filter configuration due to the overall signal attenuation at all frequencies.

A separation of the tuned frequencies to a 30 Hz (1,800 CPM) high-pass plus a 70 Hz (4,200 CPM) low-pass yields a much more useful filter combination. This combination is displayed in the FRF in Fig. 7-5. This dual filter provides a comfortable range for passing a band of frequencies with minimal signal reduction. Of course phase errors are encountered, and the magnitude of the phase errors increase as the frequency varies from the center value of 50 Hz (3,000 CPM). This type of filter characteristic is commonly referred to as a band-pass filter. In essence, a band of frequencies are passed by the filter, and the remainder of the signal frequencies are rejected. A filter of this configuration is very useful for examination of a specific frequency component.

With respect to rotating machinery, the frequency of primary interest is usually the machine running speed. Although many frequency components can and do appear on process machines, the rotational speed motion should always be scrutinized. Due to the potential influence of other vibratory sources, the bandwidth of accepted frequencies should be reduced to allow examination of only the running speed vibration. For instance, Fig. 7-6 displays a 2 Hz (120 CPM) band-pass filter from a Bently Nevada, Digital Vector Filter (DVF). This data was produced by directing the DSA white noise signal into the DVF, and connecting the DVF filtered output back into the second channel of the DSA as shown in Fig. 7-1. The DVF was manually tuned to 3,000 RPM (50 Hz), and a frequency response function (FRF) performed between the filtered output signal and the broad band noise input.

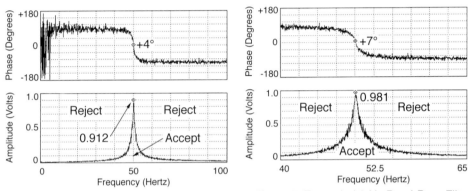

Fig. 7–6 Band-Pass Filter Set At 2 Hz   **Fig. 7–7** Expanded 2 Hz Band-Pass Filter

This filter provides a close fit around a specific frequency component, plus a minimal phase error at the center frequency. In addition, the DVF is designed to allow the center frequency of the filter to be automatically tuned to coincide with a Keyphasor® pulse. This *speed tracking* characteristic of the DVF band-pass filter is extraordinarily valuable for measuring synchronous behavior during transient speed conditions. Some of the fundamental applications for this type of filter include the generation of variable speed Bode and polar plots, plus the examination of synchronous 1X response at a constant speed.

Amplitude accuracy is important for this type class of electronic filter. However, the 2 Hz (120 CPM) DVF band-pass filter presented in Fig. 7-6 exhibits a maximum amplitude of 0.912 volts at the center frequency (i.e., 50 Hz). Since the input signal is 1.000 volt, the apparent error of this filter approaches 9%. This error would generally be considered as unacceptable, and the source of the deviation should be examined in greater detail. Fig. 7-7 represents another view of this same DVF band-pass filter. In this diagram, the 0 to 100 Hz (0 to 6,000 CPM) frequency scale has been narrowed to a 25 Hz (1,500 CPM) bandwidth extending from 40 to 65 Hz (2,400 to 3,900 CPM). This simple translation in frequency was performed within the DSA, and it has produced a fourfold improvement in frequency resolution (100 Hz versus 25 Hz bandwidth). Please note that the four times improvement in resolution is directly coupled with a fourfold increase in sample time.

Examination of the 2 Hz (120 CPM) band-pass filter characteristics by zooming in (Fig. 7-7) reveals an amplitude of 0.981 volts at the center of the filter. This value suggests an amplitude error of less than 2%. However, the filter was not adjusted, only the frequency analysis range was changed. This behavior suggests a difference in the DSA handling of the 100 Hz versus the 25 Hz analysis bandwidth. In actuality, the difference of a 2% versus a 9% amplitude error is directly attributable to the coincidence of the 50 Hz DVF filter frequency with a digital filter within the DSA. Also, a minimum of ten FFT frequency bins must be contained within the half power bandwidth of the filter under test. If this requirement is satisfied, then the amplitude accuracy will be 2% or better. If less filter bins are contained within the half power bandwidth, the amplitude accuracy will suffer. Additional digital filter characteristics will be discussed in further detail later in this section. Prior to that discussion, it is desirable to conclude the current review of band-pass filters.

As previously mentioned, the fundamental purpose of a band-pass filter is to provide a narrow bandwidth filter around a specific frequency such as machine running speed. The 2 Hz (120 CPM) bandwidth is quite adequate at higher rotational speeds, but at low speeds, the fixed bandwidth might be excessive. For instance, a machine operating at 12,000 RPM (200 Hz) would be properly observed with a 120 CPM (2 Hz) band-pass filter. However, if the machine was running at 200 RPM (3.33 Hz), the 120 CPM (2 Hz) filter width would not provide the desired clean signal. In fact, the filter would include a good portion of the frequency domain adjacent to the machine operating speed range.

To minimize this influence, the DVF and similar instruments are equipped with a 12 CPM (0.2 Hz) bandwidth filter. The frequency response function (FRF) shown in Fig. 7-8 depicts the characteristics of this filter between 0 and 100 Hz

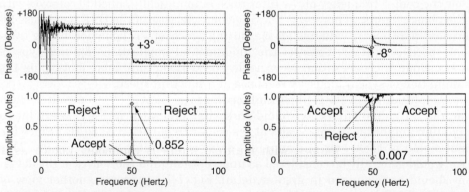

Fig. 7–8 Band-Pass Filter Set At 0.2 Hz   **Fig. 7–9** Band-Reject Filter Set at 2 Hz

(0 and 6,000 CPM). It is clear that the signal acceptance region is much tighter, and the phase transition is much steeper than the 2 Hz (120 CPM) band-pass filter. Although it is not obvious from this data, the filter settling time for these two filters varies by a factor of ten. That is, the 12 CPM (0.2 Hz) filter will require ten times longer to settle on to a definitive amplitude and phase angle as a 120 CPM

(2 Hz) filter. Thus, vectors that change rapidly with respect to time should be processed with a 120 CPM (2 Hz) bandwidth filter.

Generally, the narrow 12 CPM (0.2 Hz) filter should be employed at speeds below 1,000 RPM, and the wider 120 CPM (2 Hz) filter should be used for machine operating speeds above 1,000 RPM. Obviously, there are exceptions to this general rule, and the specific characteristics of the machinery should always be considered when selecting filter bandwidths.

One final variation to the standard suite of electronic filters is presented in Fig. 7-9. This 2 Hz (120 CPM) band-reject filter performs the opposite function of a band-pass filter. That is, instead of passing only a small frequency range, this type of filter rejects a small frequency range. For instance, the example reveals that the tuned frequency of 50 Hz (3,000 CPM) has reduced the 1.000 volt input level to 0.007 volts. In addition, the phase and amplitude effects upon the neighboring frequencies are minimal. This band-reject or notch filter is very useful for eliminating a particular frequency from a dynamic signal. For instance, if a machine is experiencing some minor subsynchronous motion, it may be advisable to filter out (i.e., band-reject) the running speed vibration component. This filtration would allow better visibility of the sub-harmonic activity for detailed determination of the orbital precession plus other signal characteristics.

A band-reject filter is also quite useful for reducing the level of interference on a dynamic signal. For example, if a measurement system displayed a 60 Hz line frequency interference, a band-reject or notch filter might be used to attenuate or effectively remove the 60 Hz component. As with all filters, this type of procedure does eliminate data. Thus, the application of a 60 Hz band-reject filter might be totally appropriate for examining steady state data on a 10,000 RPM turbine. However, a 60 Hz notch filter would be a poor selection for investigating the full load behavior of a 3,600 RPM synchronous motor.

Electronic filters are an integral part of the tools used by the machinery diagnostician. The previously discussed Krohn-Hite low and high-pass filters are typical of the tunable analog filters available within the marketplace. This same vendor also produces a line of digital filters that exhibit improved performance over the analog variety. The automatic tracking band-pass and band-reject filters available in the Bently Nevada DVF represent a digital filter configuration that is mandatory for virtually any type of machinery analysis. However, when the simultaneous examination of multiple frequencies is required, the diagnostician must also employ an instrument with multiple filters.

An instrument equipped with multiple filters is commonly known as a Dynamic Signal Analyzer (DSA), or a Fast Fourier Transform (FFT) analyzer. In years past, this type of instrument was also referred to as a spectrum analyzer. The device originally appeared in a single channel analog configuration that was physically large, and weighed in excess of 150 pounds. Since 1970, it has evolved into a smaller and lighter digital instrument. Currently, this device is available as a portable, battery powered, 2 channel field instrument that weighs 5 to 10 pounds (e.g., HP-3560A). It is also common to find stand-alone 2 to 4 channel instruments with substantial capabilities that weigh in the vicinity 20 to 30 pounds (e.g., HP-35670A). In addition, various multichannel computer interface configurations are also commercially available. These devices will be discussed in greater detail in chapter 8 of this text.

Regardless of the specific configuration, or the user interface, all of the DSA's share a common approach in signal processing. Basically, an analog input signal is digitized, and subjected to a time domain windowing function. The frequency data could be considered as processed through a series of digital band-pass filters that are sequentially arranged in what is often referred to as a *picket fence* pattern. A graphical representation of these sequential band-pass filters is shown in Fig. 7-10.

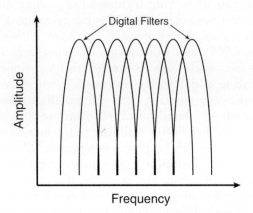

Fig. 7–10 Overlapping
Digital Filters In Dynamic
Signal Analyzer

The analysis frequency bandwidth (span) of the DSA is divided by the number of digital filters to determine the frequency resolution. For instance, if a 0 to 100 Hz display contains 100 digital filters, the resolution is 1.0 Hz. Alternatively, it may be stated that the filter spacing is 1.0 Hz (i.e., 100 Hz/100 filters = 1.0 Hz/filter). The quantity of digital filters is generally referred to as the *number of lines or bins* for the FFT. In this case, a 100 line display produces a low resolution plot, whereas an 800 line spectrum would be considered as a high resolution display. Thus, a 0 to 100 Hz span spectrum with 800 digital filters has a resolution of 100 Hz/800 filters = 0.125 Hz/filter.

The time necessary to acquire a block of data is determined by computing the period of the particular frequency handled by the filter. For a resolution of 0.125 Hz/filter, the sample time is determined from equation (2-1) as:

$$Sample\ Time = Period = \frac{1}{Frequency} = \frac{1\ Cycle}{0.125\ Cycle/Second} = 8.0\ Seconds$$

Obviously, if the sample time of the 100 line filter is subjected to this equation, the result would be a 1.0 second sample period. Thus, the data sampling time is directly related to the number of filters. A 200 line display requires twice as long to sample as a 100 line display. Similarly, an 800 line plot will take eight times as long as a 100 line spectrum for the same frequency span.

Naturally, the frequency bandwidth selected will influence the sample time. If a 6.25 Hz span is selected along with a 400 line display, the resolution is 6.25

Hz/400 lines = 0.015625 Hz/line. The sample time is the reciprocal of this value, or 64.0 seconds to acquire a block of data. If high frequency data was to be examined, the sample time would be significantly reduced. For example, assume that the number of lines remained at 400, and the span was increased to 25,600 Hz. The resolution would be 25,600 Hz/400 lines = 64 Hz/line. Once again, the sample time is the reciprocal of this value, or 0.015625 seconds to acquire a block of data. In all cases, low frequency bandwidths demand long sample periods, and high frequency spans have short sample times.

A convenient way to directly associate frequency span with the length of the time record and the FFT resolution is presented in Table 7-1. The frequency span is listed in the first column. The associated time record length, and resolution are shown in the second and third columns respectively. This data is for a 400 line spectrum, and the presented values must be adjusted for any other frequency span range set on the DSA.

Table 7–1 Measurement Speed Versus Time Record Length And Resolution[a]

Frequency Span (Hertz)	Time Record Length (Seconds)	Filter Resolution (Hertz)
102,400	0.0039065	256
51,200	0.0078125	128
25,600	0.015625	64
12,800	0.03125	32
6,400	0.0625	16
3,200	0.125	8
1,600	0.25	4
800	0.5	2
400	1	1
200	2	0.5
100	4	0.25
50	8	0.125
25	16	0.0625
12.5	32	0.03215
6.25	64	0.015625
3.125	128	0.0078125
1.5625	256	0.00390525
0.78125	512	0.001953125

[a]Values Listed Are For A 400 Line Display Only

Viewed in another manner, for 100 lines at a 100 Hz bandwidth, the time record would be 1 second long, and the resolution would be 1 Hz. That is:

100 Hz, at 100 lines, requires a 1 second time record, at 1 Hz resolution

This *100, 100, 1, 1* sequence is easy to remember. It may be quickly scaled to any measurement configuration by multiplying or dividing the values by the appropriate number of lines, and the span of set on the FFT. The values listed in Table 7-1 are consistent with the previous discussion. In all cases, the lower frequency spans display high resolution combined with long sample times. Conversely, the higher frequency spans have very short time records, coupled with increased filter resolution.

Additional time is required to process the data after sampling. For example, processing time may be 12 to 14 milliseconds for a 1,024 block size. This increases to 25 to 29 milliseconds for a 2,048 block size. The processing time for each time record is normally added to the sample time. However, by using overlap processing — the time records are overlapped and the FFT computation is performed more frequently. Overlap processing reduces the measurement time, and it uses the processor more efficiently. In addition, overlap processing will recover some of the data lost in the windowing process. In all cases, the diagnostician must maintain a proper balance between resolution, data acquisition time, duration of the event, and the machinery type under consideration.

The appearance and resolution bandwidth of the FFT filters is not only dependent upon the selected number of filter lines, it is also controlled by the type of *window*. This window is a time domain weighting function (filter) that is applied to the input signal to remove spurious and non-periodic signals. These digital filters do not influence the input time record, but they affect the displayed information. Most DSA or FFT analyzers employ three basic filter shapes that are referred to as the flat top, Hann, and uniform windows. A visual comparison of these three window types is presented in Fig. 7-11.

The **flat top** window (also called a sinusoidal window) provides input signal

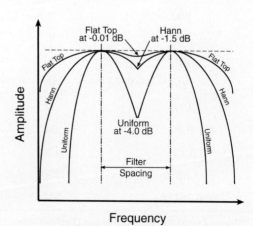

Fig. 7–11 Comparison Of Uniform, Hann, And Flat Top Filter Shapes And Overlap Spacing Characteristics Between Filters

weighting plus a wide digital filter that results in exceptional amplitude accuracy. As shown in Fig. 7-11, the maximum amplitude error between digital filters is only -0.01 dB. This translates to a voltage ratio of 0.9988, which may be expressed as 99.88%. In other words, the component amplitude will be displayed within -0.12% of the true value. Unfortunately, the frequency resolution suffers with a flat top window, but the amplitude accuracy is unsurpassed. An additional view of the flat top window is provided by the time domain and spectrum plots shown in Fig. 7-12. In this example, a 50 Hz (3,000 CPM) sine wave with an

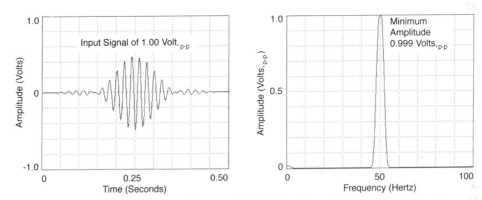

Fig. 7–12 Flat Top Window Characteristics In The Time And Frequency Domains

amplitude of 1.00 Volt,$_{p-p}$ is processed with a flat top window. It is noted that the beginning and end of the time record are heavily attenuated, and the filter is directed at the middle of the sample. This manifests as excellent amplitude accuracy, and the FFT plot in Fig. 7-12 reinforces the previous conclusion that the minimum display amplitude will not be below 0.999 Volts,$_{p-p}$ of the 1.00 Volt,$_{p-p}$ input signal. Again, this high amplitude accuracy is a function of the difference of only -0.01 dB between adjacent filters.

The next filter displayed in Fig. 7-11 is a **Hann** window (also known as a Hanning or Random window). This is the traditional filter shape found on the original Real Time Analyzers. It provides a good compromise between the amplitude accuracy of the flat top window, and the frequency resolution inherent with a uniform window. The Hann window is the most commonly used window for vibration analysis, and random noise measurements.

As shown in Fig. 7-13, the Hann window attenuates the input signal at both ends of the sampled time record, and it forces the signal to appear periodic. The maximum error between Hann filters is typically -1.5 dB. This translates to a voltage ratio of 0.8414, which may be expressed as 84.14%. Thus, a 1.0 Mil,$_{p-p}$ signal would be displayed as 1.0 Mil,$_{p-p}$ if the component frequency was coincident with a filter bin. However, if the frequency was located precisely between two filters, the displayed amplitude would be 0.84 Mils,$_{p-p}$, which is equivalent to an amplitude error of 16%.

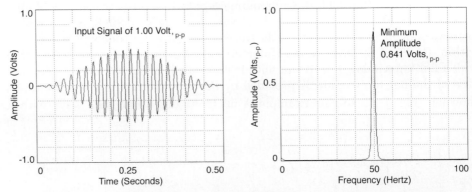

Fig. 7–13 Hann Window Characteristics In The Time And Frequency Domains

For additional explanation, the behavior of a Hann window is depicted in Fig. 7-13, where a 50 Hz (3,000 CPM), 1.00 Volt,$_{p-p}$ sine wave was processed. In this case, the input voltage was subjected to the potential of a 16% amplitude reduction, and the minimum voltage of 0.841 Volts,$_{p-p}$ was noted in the spectrum plot. Again, this potential for an amplitude error with a Hann window is a function of the difference of -1.5 dB between adjacent filters.

The third type of window is called a **uniform** (or a transient) window. This particular window weighs all parts of the sampled time record equally. In essence, the Uniform window does not influence or manipulate the input time record. This is demonstrated in Fig. 7-14 where the 50 Hz (3,000 CPM) sine wave displays equal amplitudes of 1.00 Volt,$_{p-p}$ throughout the entire time record. Some of the technical literature refers to this type of window as a rectangular shape, and other references consider this to be a non-window.

A Uniform window is very useful for examining short duration functions that are basically self-windowing. The force hammer input, and the accelerometer output signals acquired during an impact test are good examples of these

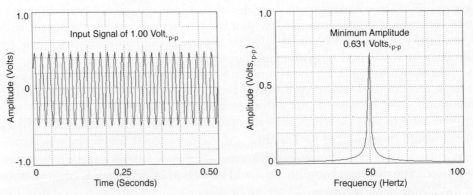

Fig. 7–14 Uniform Window Characteristics In The Time And Frequency Domains

types of functions. For examination of periodic signals, the maximum amplitude uncertainty between Uniform filters is approximately -4 dB as previously shown in Fig. 7-11. This is equivalent to a voltage ratio of 0.6310, or this may be expressed as 63.10%. Hence, if a component falls directly between Uniform filters, the amplitude error could approach 37% as shown in the spectrum plot of Fig. 7-14. Obviously, this is unacceptable for most routine vibration measurements, and one of the other two windows would normally be selected.

For improved handling of transient events, two additional windows known as the force and the exponential windows are in residence within most modern DSAs. The **force** window passes the first portion of the time record, and sets the remainder of the record to some specific value (e.g., zero). This is very useful during hammer impact tests, since this window provides the capability to delete all electronic noise from the force signal past the initial impulse. Truncation in this manner provides a cleaner signal from the impact hammer, and allows for improved clarity of the final frequency response function (FRF).

The **exponential** window attenuates the input response signal at a decaying exponential rate. The actual rate of signal decay is determined by a user defined time constant. Again, this type of window is very useful during hammer impact tests to provide an even exponential decay to the vibration response signal. The validity of applying this type of filter shape was previously demonstrated by equation (2-71) and the companion Fig. 2-14. From this earlier analysis of a simple mechanical system, it was shown that forced vibration will decay in an exponential manner. Hence, this filter shape is quite appropriate. It also should be mentioned that an exponential window does not attenuate the response in the region that contains the best signal to noise ratio, but it does gradually taper off the signal in the time domain where the signal to noise ratio is the lowest. Overall, this window contributes to the elimination of spurious signals, and it improves the accuracy plus clarity of the FRF.

In the overview, it is clear that a variety of electronic filters are available to the machinery diagnostician. Each instrument and each individual filter configuration possesses specific characteristics that may be used to enhance the measured dynamic data, and allow identification and resolution of the mechanical behavior. Conversely, these same filters may totally obliterate the useful data if they are improperly applied. Hence, the diagnostician must become intimately familiar with the filter characteristics on all instruments used for machinery analysis. The reader is also encouraged to study more detailed explanations of FFT signal manipulation in documents such as *The Fundamentals of Signal Analysis*[1], and chapter 14 of the *Shock and Vibration Handbook*[2]. Many other references are also available, but care should be taken to insure that the selected educational materials are consistent with the machinery to be tested, and the selected array of instrumentation.

---

[1] *The Fundamentals of Signal Analysis*, Application Note 243, (Hewlett Packard, printed in USA, 1995).
[2] Cyril M. Harris, *Shock and Vibration Handbook*, Fourth edition, (New York: McGraw-Hill, 1996), chap. 14.

TIME AND ORBITAL DOMAIN

Rotating machinery and associated perturbations usually follow shaft rotation, which always moves in a direction of increasing time. In virtually all cases, careful examination of the time domain behavior of a machine is fundamental to the analysis of that equipment. Since the other dynamic measurements are extracted from the time domain motion, the full understanding of this topic is essential. It is desirable to begin this discussion with an example of how vibratory motion of a machine element is translated into a time domain voltage signal. In this regard, consider Fig. 7-15 of a rotor turning in a clockwise direction, with a vertical proximity probe observing the shaft. A series of eight angular locations are identified with the letters **A** through **H**. These positions are stationary reference locations around the circumference of the rotor (not be confused with the angular coordinate system on the rotating element). To state it another way, locations **A** through **H** represent the *actual* shaft positions as the rotor turns, and follows the circular dotted line path as a function of time.

In this example, presume that the rotor contains a single mass unbalance as indicated by the black dot on the eight shaft positions. Also, assume that this simple machine operates below the rotor natural frequency. It is self-evident that centrifugal force due to the unbalance causes a radial deflection (bow) of the shaft in the direction of the mass unbalance. In essence, the effective rotor *heavy spot* is thrown to the outside of the orbit at all angular locations. This is the same mechanism that was previously discussed in chapter 3 on the Jeffcott rotor. The rotational behavior depicted in Fig. 7-15 is identical to moderate speed operation of the Jeffcott rotor as shown in Fig. 3-30. As the shaft rotates, the heavy spot will pull the rotor into a circular orbit if the stiffness is identical in both vertical and horizontal directions. With assumed identical stiffness values, the resultant circular orbit is depicted by the dotted line in Fig. 7-15.

Clearly, the *heavy spot* (unbalance mass) is identical and coincident with the *high spot* (peak displacement). An actual physical example of this type of

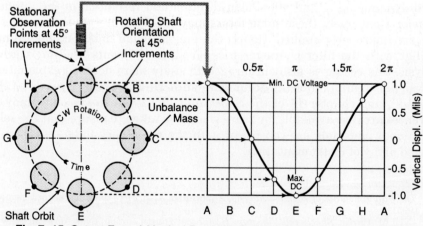

Fig. 7–15 Output From A Vertical Proximity Probe Observing A Rotating Shaft

rotor would be a truck drive shaft. If the rear wheels of the truck are jacked up off the ground, and the drive shaft runs at a moderate speed, the heavy spot may be determined by carefully bringing a piece of chalk up to the rotating shaft. When the chalk makes contact with the shaft, the *high spot* is marked. A balance weight may be attached onto the shaft 180° away from this location. Fig. 7-15 is representative of this physical behavior. However, in this diagram an accurate distance sensing device in the form of a proximity probe is used instead of the piece of chalk. As discussed in chapter 6, the proximity probe converts distance between the probe and the observed surface into a DC plus an AC output voltage.

When the shaft rotation allows the *high spot/heavy spot* to be in a true vertical position, the output DC voltage from a vertical proximity probe will be minimal (smallest gap). This specific condition is defined by point *A* on Fig. 7-15. As the shaft continues to turn in a clockwise direction, reference points *B*, *C*, *D*, etc. expose a different shaft surface to the probe. In each case, the *high spot* continues to move further away from the probe until position *E* is attained. At this point, the shaft has arrived at the farthest distance from the probe tip, and the probe DC gap voltage will be at a maximum.

Continuing in 45° increments through stationary observation points *F*, *G*, and *H*, the shaft returns back to the starting point *A*. The associated time domain plot presented in Fig. 7-15 describes the vertical motion of this rotor with respect to time. In this diagram, time progresses from left to right (same as an oscilloscope); and the vertical displacement is shown to vary between ±1.0 Mil, which is equal to the differential DC voltages divided by the probe sensitivity. In accordance with the convention established at the beginning of this text, the total shaft motion would be expressed as 2.0 Mils,$_{\text{p-p}}$. Furthermore, connecting the eight data points into a continuous curve reveals the repetitive (cosine) nature of this simple example.

The variation in DC voltage is generally considered to be an AC voltage that is proportional to the vibration of the observed surface. Some individuals will argue that this is not a true AC voltage, but an oscillation of a DC level. Although this distinction may be technically rigorous, it is both convenient and appropriate to consider a proximity probe output voltage as consisting of an AC voltage superimposed upon a DC voltage. In all cases, the AC voltage is proportional to the relative vibration of the observed surface. The DC voltage is proportional to the average distance between the probe tip and the observed surface.

Other dynamic transducers such as accelerometers generate an AC voltage that is proportional to the motion of the vibrating surface. These transducers may exhibit an output DC voltage, but that is usually associated with the transducer power supply. Within the typical suite of vibration transducers, the proximity probe is the only sensor that provides static position information via the DC voltage. For additional discussion of specific transducer characteristics, the reader is referred back to chapter 6.

The AC portion of a proximity probe signal is directly proportional to the relative vibration between the probe support element (e.g., bearing housing) and the observed surface (e.g., rotating shaft). The AC portion of the signal produced by a casing mounted vibration transducer (i.e., velocity coil or accelerometer) is proportional to the absolute vibration of the surface to which the transducer is attached (e.g., bearing housing). Within the time domain, the dynamic signal characteristics of the various transducers may appear to be very similar, and the same concepts of time domain observation would apply. Please note that this conclusion does not insinuate that shaft and casing motion are identical. On the contrary, there are normally discernible differences between the two measurements. For instance, a simple machine with residual unbalance and fluid film bearings will display casing signals that lag behind the shaft motion. In addition, the casing vibration amplitudes will normally be attenuated by the bearing housing. On a machine with rolling element bearings, the relative shaft motion will be minimal, and casing vibration will be dominant. Obviously, machines with multiple excitations will exhibit more complicated deviations between shaft and casing vibration measurements.

Although machinery shafts and casings move with distinctly different characteristics, it is common knowledge that all machines vibrate vertically, horizontally, and axially. Typically, the axial vibration is minimal, and the majority of the vibration is observed in the lateral (i.e., radial) planes. This natural vertical and horizontal vibratory motion is typically observed with orthogonally mounted vibration probes. For instance, the inclusion of a horizontal proximity probe into the arrangement discussed in Fig. 7-15 results in the X-Y configuration shown in Fig. 7-16. The additional horizontal probe now provides the capability to view a two-dimensional response of the shaft surface. As discussed earlier in this text, the probes may be in any circumferential location. However, they must be 90° apart to provide an accurate two-dimensional representation of the shaft motion.

Considering the horizontal displacement in Fig. 7-16, it is clear that when the shaft rotation allows the *high spot* to be in a true horizontal position, the output DC voltage from a horizontal proximity probe will be minimal. This condition

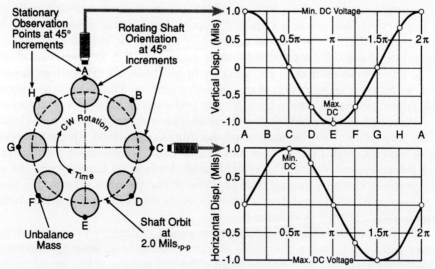

Fig. 7–16 Output From Orthogonal Proximity Probes Observing A Rotating Shaft

is defined by point **C** in Fig. 7-16. As the shaft continues to turn in a clockwise direction, reference points **D**, **E**, and **F**, expose a different shaft surface to the probe. In each case, the *high spot/heavy spot* continues to move further away from the probe until position **G** is attained. At this point, the shaft is at the greatest distance from the probe tip, and the proximity probe DC gap voltage will be at a maximum.

Continuing in 45° increments through stationary observation positions **H**, **A**, and **B**, the shaft returns back to the starting point **C**. Again, time progresses from left to right, and the horizontal displacement varies between ±1.0 Mil, or 2.0 Mils,$_{p-p}$. Connecting the eight data points into a continuous curve reveals the repetitive (sine) nature of this example. If these vertical and horizontal signals were viewed on an oscilloscope, they could be converted into an orbit by selecting the X-Y sweep position, and the display shown in Fig. 7-17 would appear.

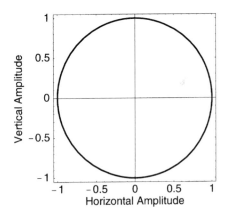

Fig. 7–17 Simple Circular
Orbit From Orthogonal
Proximity Probes

Not surprisingly, this orbital display is identical to the original shaft motion of 2.0 Mils,$_{p-p}$. Since this example was established as a forward and circular motion, the resultant shaft orbit must also be both forward and circular. By visual inspection, it is clear that the resultant orbit is circular. By virtue of the fact that this orbit was developed in a clockwise direction, it is therefore moving with increasing time (i.e., forward). However, on a real machine the determination of the precession of an orbit is somewhat more complicated.

Specifically, the time and orbital domain signals are superimposed with a pulse signal originating from a once-per-rev Keyphasor®. As previously discussed in chapter 6, the Keyphasor® timing pulse will produce a *blank-bright* pattern for a negative going signal pulse such as a notch or a drilled hole in the shaft (e.g., Fig. 6-5). If the Keyphasor® probe is looking at a section of key stock or other projection, the resultant trigger pulse will be a positive going signal, and the oscilloscope blanking will appear as *bright-blank* sequence (e.g., Fig. 6-6).

In order to explain the interaction of the Keyphasor® pulse with the time domain traces, it is useful to expand upon Fig. 7-16 by adding a Keyphasor® probe to the diagram. Assume that a notch is milled into the shaft at an axial location that is several inches away from the radial probe measurement plane.

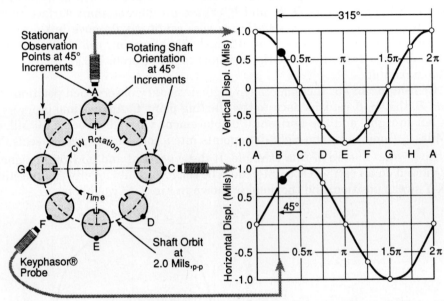

Fig. 7–18 Time Domain Output Signal Outputs From Orthogonal Proximity Probes And Radial Keyphasor® Probe Observing A Rotating Shaft

Further assume that this shaft notch is located diametrically opposite (180° away) from the identified *heavy/high spot*. Physically, the Keyphasor® probe is added to the lower left hand corner of the angular reference frame as depicted in Fig. 7-18. Note that the timing probe is positioned at the **F** location, but the only portion of the rotational cycle that exposes the notch to the Keyphasor® probe occurs when the shaft is at the **B** position. Hence, the Keyphasor® probe will generate a negative going pulse as the shaft rolls past position **B**. If this timing pulse is superimposed upon the vertical and horizontal vibration signals, a *blank-bright* sequence will appear at the **B** location. This situation is graphically shown in Fig. 7-18.

Based upon the standard phase convention presented in chapter 6, the phase angles for vertical and horizontal probes are identified in Fig. 7-18. In each case, the peak of the vibration signal is used as the zero point, and the angle is determined by moving backward in time to the trigger point of the Keyphasor® pulse. In this specific case, each time division is equal to $\pi/4$, or 45° (derived from 1 revolution = 2π radians = 360°). It is clear that the difference between the peak of the horizontal vibration signal and the KØ is one division, or 45°. The vertical probe has seven divisions from the peak amplitude to the KØ pulse. Thus, the vertical phase angle is 315° (= 7 Divisions x 45°/Division).

The physical relationship between these phase angles and the rotating system is illustrated by the drawings in Figs. 7-19 and 7-20. Specifically, Fig. 7-19 describes the relationship between the vertical probe, the Keyphasor® probe, the rotating notch, and the *heavy/high spot*. When the leading edge of the notch is located under the KeyØ probe, the negative trigger pulse is initiated. At this exact point, the vertical probe 315° phase angle may be used to locate the *heavy/ high spot* on the rotor. Since shaft rotation and time are moving in a clockwise direction, the phase lag (backward time) must be in a counterclockwise direction as established in Fig. 2-3. Furthermore, since zero degrees (0°) is always located at the probe, the phase angles are always measured in counter rotation from the probe. Hence, moving 315° counterclockwise from the vertical probe locates the *heavy/high spot* in the upper right hand quadrant of Fig. 7-19.

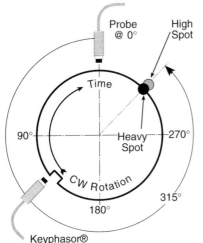

Fig. 7–19 Phase Angle For True Vertical Proximity Probe to *Heavy/High Spot*

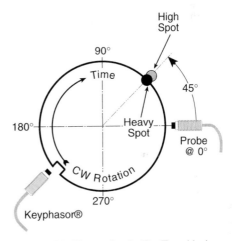

Fig. 7–20 Phase Angle For True Horizontal Proximity Probe to *Heavy/High Spot*

The diagram in Fig. 7-20 displays an identical relationship for the horizontal probe. For this transducer, the 45° phase angle is counter rotation from the horizontal probe (again 0° reference for each probe). The same logic applies, and the same *heavy/high spot* is identified on the rotor. Naturally, this will only be precisely the same point when the orbit is forward and circular. This condition would be visually apparent, and the phase angles would exhibit a 90° difference.

In cases when the specific mechanics of the Keyphasor® installation are unknown, the pulse signal can always be viewed on an oscilloscope to determine if the trigger is negative or positive going. If the blanking characteristics of the oscilloscope are unknown, then the superposition of the Keyphasor® signal upon a time base wave form will identify the actual sequence. In all cases, oscilloscopes sweep from left to right, and the identification of *blank-bright* or *bright-blank* with respect to time may be visually determined on an analog scope.

As applied to a shaft orbit, knowledge of the timing pulse sequence will allow correct identification of the orbit precession. Again, the real key to this identification is the fact that the measured parameters are moving forward in time. Thus, when the time sequence is identified, the orbit precession is also defined. For instance, consider Fig. 7-21 that shows an array of four circular

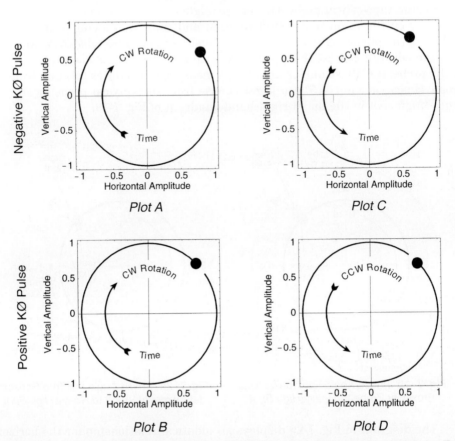

Fig. 7–21 Variation Of Keyphasor® Blank-Bright Sequence With Rotation And Pulse Type

orbits combined with negative and positive going Keyphasor® pulses for both clockwise and counterclockwise shaft rotations.

Plots A and *C* in Fig. 7-21 exhibit the influence of a negative going Keyphasor® trigger pulse. In *Plot A*, the *blank* spot is followed by the *bright* spot in a clockwise direction. This sequential behavior is fully descriptive of a clockwise shaft precession, which in this simple example is the same as the shaft rotation. It should be mentioned that some complex machinery instabilities display a nonsynchronous frequency that has a precession that is opposite to shaft rotation. Without the Time-Orbital-Keyphasor® sequencing concepts presented herein, these reverse precession malfunctions could not be properly diagnosed.

Plot C in Fig. 7-21 reveals a reversal of the *blank-bright* sequence, which must be interpreted as a change in rotation to counterclockwise. Once more, the first event is the *blank spot* (negative slope), followed in a counterclockwise direction by the *bright spot* (positive KØ slope). Hence, the precession (rotation) for the orbit shown in *Plot C* must be counterclockwise.

The two orbits shown in *Plots B* and *D* of Fig. 7-21 represent the shaft rotational directions viewed with a positive going Keyphasor® pulse. As shown in Fig. 6-6, this type of trigger pulse will display a *bright spot* followed by a *blank spot*. Again the same logic applies, and the orbit in *Plot B* may be easily identified as clockwise. Clearly, the *bright-blank* sequence identifies *Plot D* as a shaft rotating in a counterclockwise direction.

As a final comment on this four orbit array in Fig. 7-21, note that the diagonal orbits appear to be identical. That is, the clockwise rotating shaft with a negative Keyphasor® pulse (*Plot A*) looks just like the counterclockwise rotating shaft with a positive pulse (*Plot D*). Similarly, the counterclockwise rotating shaft with a negative pulse (*Plot C*) is identical to the clockwise rotating shaft with a positive Keyphasor® (*Plot B*). This type of visual similarity, with totally different mechanical implications is common throughout this machinery business. Additional examples of similar appearances combined with different mechanical characteristics will be presented in the remainder of this chapter. However, the machinery diagnostician should not take things for granted, and must always verify that the output information is totally consistent with the physical machinery configuration.

In all cases, determination of the specific trigger point for the phase angle measurements is dependent upon the trigger slope (positive or negative), and the direction of the trigger pulse (positive or negative going). This actual trigger point for the Keyphasor® probe is especially important during activities such as field balancing. Hence, the relationship between the rotor trigger point and the Keyphasor® probe will be discussed in greater detail at the beginning of the rotor balancing chapter 11.

The example under discussion has consisted of a forward, circular orbit, that occurs at a single frequency. The circular orbit implies equal amplitudes, plus a 90° phase differential between vertical and horizontal signals. The described response is representative of a machine operating with a single excitation (e.g., unbalance), and with an equal support stiffness in both the vertical and horizontal planes. Unfortunately, process machinery seldom behaves in such an idealistic manner. As stated in chapter 6, a dynamic signal is characterized by the following basic parameters:

○ **Amplitude** (Magnitude or Severity)
○ **Frequency** (Rate of Occurrence)
○ **Timing** (Phase Relationship)
○ **Shape** (Frequency Content)

These fundamental characteristics were discussed for a single dynamic signal. It is understandable that the combination of two dynamic signals into an orbit may significantly complicate the resultant pattern due to variations between the signals. Hence, it is desirable to examine the interaction between variations in amplitude, frequency content, and phase difference between signals. For instance, if the previous horizontal response was reduced by a factor of two, the vertical orbit displayed in Fig. 7-22 would appear. This 2:1 ratio in vertical versus horizontal machine response could be indicative of a 2:1 stiffness ratio.

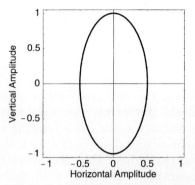

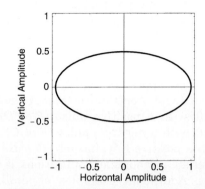

Fig. 7–22 Vertical Orbit With 2:1 Ratio Between Vertical & Horizontal Amplitudes

Fig. 7–23 Horizontal Orbit With 2:1 Ratio Between Horizontal & Vertical Amplitudes

A horizontal preload (e.g., misalignment) that restricts the horizontal motion would also produce a predominantly vertical orbit. In either case, the resultant orbits would appear to be similar, but the mechanics responsible for the deviation in horizontal amplitudes would be considerably different.

If the amplitude ratio is reversed, such that the horizontal vibration is twice the size of the vertical response, the orbit shown in Fig. 7-23 emerges. This type of orbital pattern is quite common for large turbines equipped with elliptical bearings. In these units, the horizontal clearance is often twice the vertical clearance, and the associated differential in bearing stiffness allows the horizontal vibration to exceed the vertical vibration. Again, a vertical preload from a cocked bearing, or a misaligned shaft, could also restrict the vertical motion, and produce a horizontal ellipse. In all cases, the resultant vibration or motion at a bearing is dependent upon the applied force(s) and the associated stiffness. This general relationship between the vibration *Response*, the applied *Force*, and the *Restraint* (stiffness) was previously stated in equation (3-31) as:

$$Response = \frac{Force}{Restraint} \qquad (7\text{-}1)$$

The smaller the restraint, the larger the vibration response for a given unit force input. Similarly, increased force with a constant restraint will result in an increased vibration amplitude. However, for a two-dimensional mechanical system, there will be cross-coupling between horizontal and vertical directions. Thus, an applied force in one direction may result in motion in a perpendicular plane. This mechanical reality does complicate the interpretation of the vibration response data, but the orbital presentation provides quantification of actual journal motion within a bearing.

Clearly, vibration amplitudes in the time and orbital domains will be influenced by a combination of the system forces and restraints. It should also be recognized that the system parameters are often vector quantities. These vectors may appear as constant amplitudes, and mechanical changes may influence or

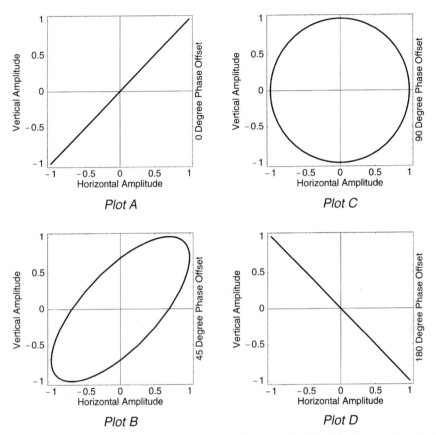

Fig. 7–24 Variation Of Orbits With Changes In Phase — Holding Vertical And Horizontal Amplitudes and Frequencies Constant

cause variations in the relative timing between signals. This will manifest as phase angle changes between the measured vertical and horizontal vibration response vectors. The time domain data will be shifted, and the orbits may be subjected to significant changes. In order to fully appreciate the potential variation in orbital patterns due to phase angle changes, consider the array of four orbits shown in Fig. 7-24. In each case, the amplitudes and the frequencies are identical for both the vertical and the horizontal transducers. The only difference between signals is the time domain phase differential. The orbit presented in *Plot C* is identical to previous examples where a cosine function is plotted against a sine function (90° difference). The elliptical orbit displayed in *Plot B* was constructed with a 45° phase difference between the vertical and horizontal signals. If the differential signal timing was changed to 135°, the elliptical orbit would lean to the left instead of to the right.

The same type of characteristic is displayed by the straight line orbits. The display in *Plot A* of Fig. 7-24 has zero phase difference between channels, and the orbit shown in *Plot D* has a 180° offset. In each case the closed orbital loop has collapsed into a straight line. When the line leans to the right the signals are directly inphase, and when the line leans to left, the signals are exactly out of phase. For this case of identical frequencies and amplitudes between channels — the orbits will unfold from a straight line into a circle as a series of ellipses as the phase angle rolls by 90°.

The next example in Fig. 7-25 is complicated by the fact that the horizontal frequency occurs at precisely twice the vertical frequency. In the case of a heavy radial preload (e.g., severe misalignment), the vertical probe signal might be completely dominated by 1X running speed motion, and the horizontal probe signal might exhibit a major frequency component at twice running speed (2X). Considering this 2:1 ratio of frequencies, the orbits in Fig. 7-25 are generated as the phase angle differential is varied between 0° and 90°. Three of the orbits for this 2:1 frequency ratio are various *Figure 8* shapes, which is consistent with a vertical preload. This concept of preloads will be discussed in more detail in chapter 9. Note that the 45° phase offset in *Plot C* of Fig. 7-25 has resulted in

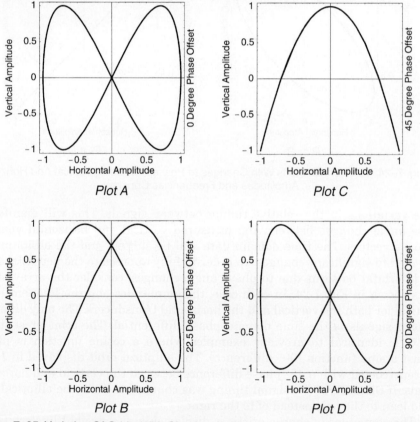

Fig. 7–25 Variation Of Orbits With Changes In Phase — Holding Vertical And Horizontal Amplitudes Constant and Setting A 2:1 Ratio Between Horizontal and Vertical Frequencies

another *line orbit* that resembles a parabola. Again, this is a closed orbit similar to the previously discussed straight line orbits. It appears as a parabola strictly due to the time domain relationship of the two signals. As the differential phase increases past 45° a series of mirror images are generated. For instance at a 135° phase difference between signals, the parabolic orbit becomes inverted, and the apex points downward.

Frequency ratios of 2:1 are occasionally observed on process machines. But pure frequency ratios of 3:1 between orthogonal probes viewing the same shaft surface are seldom encountered. However, for academic purposes, the four orbits presented in Fig. 7-26 were produced with a horizontal frequency equal to three times the vertical frequency. *Plot C* with a phase difference of 30° describes a *three lobed* shape with two crossover points. This symmetrical orbit becomes distorted as the phase angle varies. In fact, the *Lazy S* patterns shown in *Plots A* and *D* are the equivalent of the 1:1 ratio straight line, and the 2:1 ratio parabola previously discussed. Further extensions in frequency ratio between vertical and horizontal probes would increase the number of crossover points, and the com-

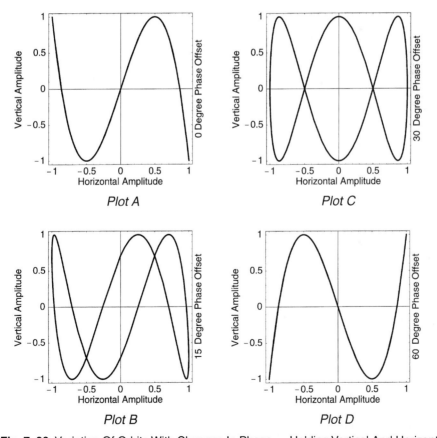

Fig. 7–26 Variation Of Orbits With Changes In Phase — Holding Vertical And Horizontal Amplitudes Constant and Setting A 3:1 Ratio Between Horizontal and Vertical Frequencies

plexity of the closed orbits. However, higher order frequency ratios between orthogonal probes are rarely encountered on real machines. The more common variety of frequency variation in a shaft vibration signal is the appearance of a frequency component that appears in both vertical and horizontal channels. For example, Fig. 7-27 displays sub rotative speed vibration components set at 50% of rotative speed. These half speed components were superimposed upon both horizontal and vertical synchronous signals.

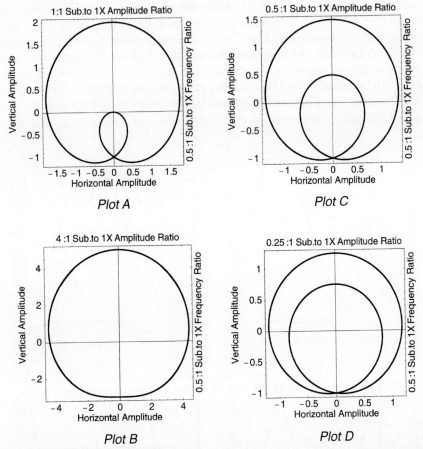

Fig. 7–27 Variation Of Orbits With Subsynchronous Frequency At 50% Of Rotative Speed Combined With Changes In Amplitude Ratio Between Subsynchronous & Rotative Speed

Plot A in Fig. 7-27 was generated with a 1:1 amplitude ratio between the synchronous, and the subsynchronous components. This type of orbit with an inside loop would be representative of a half speed excitation that is equal in magnitude to the rotative speed. If the amplitude of the subsynchronous component increased, the single inside loop would fade away, and a generally circular orbit would eventually appear. For instance, the orbit in *Plot B* of Fig. 7-27 shows the influence of increasing the 50% component size to four times the amplitude of the fundamental rotational speed. A slight flat spot appears at the bottom of this orbit, however further increases of the subsynchronous amplitude will produce an increasingly circular appearing orbit.

If the half speed component decreased in amplitude, and the rotational speed motion remained constant, overall vibration amplitudes would drop. Clearly, the relative size of the inside loop would expand with respect to the size of the outside loop. An example of this condition is shown in *Plot D*, where the magnitude of the half speed frequency was decreased to be only one quarter (25%) of the amplitude of the running speed component. An intermediate condition of a half amplitude, half frequency component is presented as *Plot C* in Fig. 7-27. Hence, the appearance and relationship between *inside* and *outside* loops is really dependent upon the amplitude ratio between the constituents.

During orbital analysis a Keyphasor® pulse should always be incorporated. This Z-axis input is invaluable for determining precession of the overall orbit plus each of the filtered components. The Keyphasor® will also help establish frequency ratios. For instance, if a pair of X-Y probes display an orbit with two fixed Keyphasor® dots, the subsynchronous excitation occurs exactly at 50% of rotative speed. However, if the dots are moving around the orbit, the subsynchronous frequency is not locked in at exactly one half of running speed. Under this condition, the subsynchronous frequency could be 55%, 49%, 43%, or any non-integer fraction of rotative speed. This specific behavior is significant, and it will be examined in greater detail in chapter 9 of this text.

Additional combinations of subsynchronous amplitude and frequency ratios are presented in Fig. 7-28. *Plot A* depicts a condition of equal synchronous to subsynchronous vibration amplitudes (i.e., 1:1). However, in this case, the low frequency component is set at 75% of the rotative speed. Note that multiple loops are generated, and direct interpretation of this behavior might be very difficult. In addition to the need for a synchronous Keyphasor® pulse, the machinery diagnostician should also employ a Dynamic Signal Analyzer (DSA) to assist in specific component identification.

Plot B in Fig. 7-28 displays a 50% subsynchronous frequency combined with a 2:1 amplitude ratio between the subsynchronous and the rotative speed frequency. By direct observation of *Plot B* on an oscilloscope, without the benefit of a Keyphasor® pulse, the viewer might conclude that only one frequency was present in the vibration signals. This type of erroneous conclusion may be avoided by careful examination of the data using orbital, time, and frequency domain analysis with a once-per-revolution trigger pulse. It should be recognized that the examination of vibration signals with various data formats provides the diagnostician with better visibility of the machinery behavior, and reduces the possibility of missing one or more key elements in the response characteristics.

Plots C and *D* in Fig. 7-28 describe the orbits resultant from increasing the amplitude of the sub-harmonic component, plus decreasing the frequency ratio. In *Plot C* the subsynchronous frequency was set at one quarter (25%) of the rotative speed frequency. The sub to 1X amplitude ration was maintained at 2:1 to be consistent with *Plot B* in this same Fig. 7-28. Note that with the same amplitude ratio, the apparently simple orbit of *Plot B* has evolved into a complex orbit with

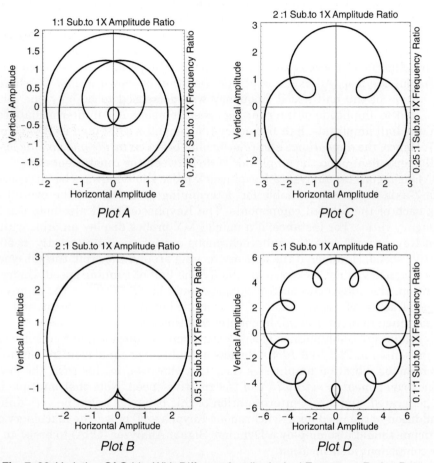

Fig. 7–28 Variation Of Orbits With Different Amplitude And Frequency Ratios Between The Subsynchronous Component And The Rotative Speed Component

three defined inside loops (*Plot C*). Furthermore, if the subsynchronous excitation is reduced to one tenth (10%) of the running speed, and the amplitude ratio is adjusted to 5:1, *Plot D* in Fig. 7-28 evolves. The nine inside loops appear well defined, and there might be a tendency to establish hard and fixed rules for the number of loops versus the subsynchronous frequency. Specifically, it has been touted that the number of inside loops plus one is equal to the ratio between the subsynchronous and synchronous vibration frequency. Although this is correct for *Plots C* and *D*, this type of general observation can lead to some significantly wrong conclusions. The validity of this statement will be substantiated when reviewing of the final group of orbits in Fig. 7-29.

This last set of calculated orbits in Fig. 7-29 addresses the appearance of supersynchronous vibration components combined with the fundamental running speed response. In general, low frequency subsynchronous vibration components have a tendency to exhibit amplitudes that are larger than running speed motion. In some instances, the subsynchronous activity may occupy the entire bearing clearance. Conversely, frequency components that occur above shaft rotative speed tend to display amplitudes that are smaller than the fundamental 1X running speed vibration levels. For discussion and demonstration purposes, the orbits in Fig. 7-29 were constructed with supersynchronous amplitudes that are a fraction of the running speed 1X vibration amplitude.

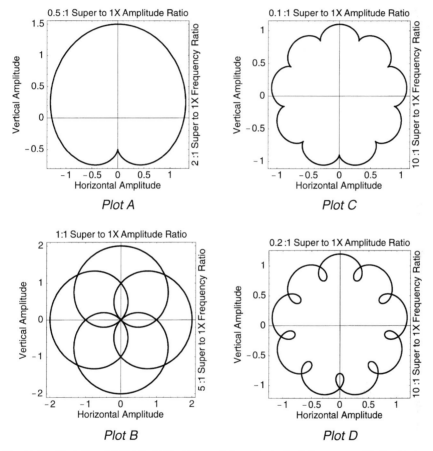

Fig. 7–29 Variation Of Orbits With Different Amplitude And Frequency Ratios Between The Supersynchronous Component And The Rotative Speed Component

The orbit in *Plot A* of Fig. 7-29 exhibits the influence of a frequency at twice rotative speed (2X), with an amplitude set at one half of the fundamental. Note that the orbit shows signs of this potential preload frequency distribution by indentation at the bottom of the orbit. Similar patterns will emerge when orbits with a 50% frequency are compared against orbits with a 2X frequency. Again, the machinery diagnostician must be properly equipped with a Keyphasor® signal plus an oscilloscope and a frequency analyzer to be absolutely sure of the proper frequency and amplitude relationships of the individual components within each dynamic signal.

Plot B in Fig. 7-29 displays a supersynchronous frequency of five times rotative speed, combined with an equal amplitude ratio. The symmetrical *rosette* shape is interesting, but difficult to interpret. Again, a Keyphasor® pulse plus a DSA would be mandatory for proper analysis. This type of behavior may also be a candidate for eliminating the running speed vibration from both signals with a band-reject filter set at rotational speed. Another approach for analysis of this type of orbit would be the application of a band-pass filter set at five times rotative speed. In either approach, the 1X signal must also be examined, and the interaction between frequencies must be documented.

The orbit display *Plot C* of Fig. 7-29 describes a traditional pattern for a machine with a synchronous high frequency excitation. In this case, the high frequency motion was set at ten times rotative speed, and the amplitude was reduced to one tenth of the fundamental running speed vibration. This combination produces a nice scalloped pattern, with a series of nine external loops. It is easy to be fooled into believing that external loops are indicative of high frequency components, and internal loops are representative of low frequency components. In actuality, neither conclusion is totally correct.

The appearance of outside versus inside loops on an orbit is strictly dependent upon the amplitude ratios, and the frequency ratios of the components within each dynamic signal. To demonstrate this point, the 10X frequency in *Plot C* was replotted with an amplitude increase from 0.1 to 0.2. The resultant data is shown in *Plot D* of Fig. 7-29. Note the significant difference between the two orbital plots presented with a 10X component. In fact, the plot with the 10% supersynchronous excitation in *Plot D* of Fig. 7-29 is visually identical to the plot with a 10X subsynchronous component shown in *Plot D* of Fig. 7-28. Hence, a casual examination of the two orbits (without KeyØ®) does not tell the diagnostician whether they are dealing with a high amplitude subsynchronous problem, or a low amplitude supersynchronous vibration problem.

Clearly, the frequency content of the dynamic signals, the ratio between the dominant component amplitudes, and the phase relationship between signals, plus the timing (phase) between individual components will determine the time domain characteristics of the vibration signals. These specific time domain characteristics will govern the shape and pattern of the final orbit(s). In virtually all cases, the complexity of the dynamic signals will dictate the amount and kind of diagnostic instrumentation necessary to properly document, analyze, and understand the measured vibration characteristics.

In addition, it is highly desirable for the machinery diagnostician to fully comprehend and understand the sometimes subtle relationships between the fundamental dynamic signal characteristics of amplitude, frequency, and phase. Although the preceding pages have attempted to describe several different situations, there are a myriad of potential combinations to be explored. Often, it is difficult to simulate the necessary interaction over a wide range of variables with devices such as function generators or rotor kits. However, it is possible to develop mathematical relationships to examine a wide range of variables. This exploration is highly recommended, and it may avoid costly mistakes due to the misinterpretation of the machinery orbital data.

Historically, mathematical simulation of periodic signals has been a difficult task requiring careful equation structure, combined with extensive computer code for input, display, and plotting of the results. However, since the evolution of desktop computers, this task has been significantly simplified. For instance, the calculated orbits within this chapter were produced with a program entitled Mathematica®[3]. This program provides a highly flexible software package for symbolic mathematical plus graphical computations. Mathematica® runs on a variety of platforms with a common Kernel, and various front end interface programs for each specific computer system. Since the program Kernel is the same for all machines, this allows interchangeability of code and statements. It is recommended that the diagnostician acquire this type of computational capability to enhance the ability for self-training. Other math programs provide similar functions, and the final program selection should be based upon the available operating system and specific software requirements.

TIME AND FREQUENCY DOMAIN

Time domain analysis is representative of the actual *sequence of events* occurring on process machinery. As discussed in the previous section, time and orbital domain analysis depict the fundamental history of the machinery. However, there are situations when the complexities of the dynamic signals exceed the data processing capabilities of an oscilloscope. In these cases, frequency domain analysis is required for detailed dissection of the signals.

Over the past twenty-five years significant improvements have been made in the instruments used for frequency analysis. The original swept or tuned filter analyzers, and the initial real time analyzers were strictly analog devices that primarily displayed amplitude as a function of frequency. These instruments have been replaced by analyzers that use digital signal processing techniques, and offer enhanced data acquisition and manipulation.

These current instruments are commonly referred to as Dynamic Signal Analyzers (DSA), or Fast Fourier Transform (FFT) Analyzers. In either case, the basic operation of the unit remains the same. DSAs usually incorporate additional functions such as signal sources for network gain and phase measurements, internal power supplies for driving ICP® transducers, programing capability, plus a variety of signal processing, manipulation, and display options. Typically, these types of instruments operate across a frequency range that extends from less than 1 Hz (60 CPM) to an upper end that exceeds 100 KHz (6,000,000 CPM). Certainly, the vast majority of meaningful machinery vibration response measurements fall well within this frequency domain.

The Fast Fourier Transform (FFT) refers to a fast and efficient method to

---

[3] Stephen Wolfram, *The Mathematica® Book*, 3rd edition, software version 3.0 (Champaign, Illinois: Wolfram Media/Cambridge University Press, 1996).

calculate the Discrete Finite Fourier Transform. This is the mathematical algorithm used for transforming amplitude versus time data into the amplitude versus frequency data. In practice, the input analog time domain signal is converted into a digital equivalent. The appropriate windowing is applied, the user selected number of samples are acquired, and the resultant frequency spectrum of the time record is displayed.

Advanced instruments such as the Hewlett Packard HP-35670A offer a variety of additional data processing and display options. However, the fundamental FFT concept converts the time domain dynamic signals into frequency domain spectra. It is also significant to note that the sampled and digitized signals may be displayed and massaged in the time domain. Thus, a four channel unit such as the HP-35670A may be used as a digital oscilloscope for time and orbital domain analysis, as well as for frequency analysis.

The key to understanding frequency domain analysis is to recognize that machines move and vibrate as a function of time, and frequency domain analysis is another way of observing the machinery time record. In fact, barring any data processing errors, the time and frequency domains may be considered as complementary and interchangeable ways of looking at the same data. Due to this close relationship between the time and frequency domain, it is essential for the diagnostician to understand how common signals appear in both formats. The ability to visualize how dynamic signals that are viewed in one format will appear in the other format can save considerable time during data processing and analysis. To assist in this appreciation of format translation, the following discussion of four common signal types is presented for consideration.

The easiest dynamic signal to convert is a simple sine or cosine wave. As discussed earlier in this text, this type of signal occurs at a single frequency, with an easily definable amplitude. For instance, Fig. 7-30 is a time domain plot of two complete cycles of a 100 Hz sine wave that has an amplitude of 1.0 Volt,$_{o-p}$.

In many cases, dynamic signals may be electronically produced, and analyzed on an FFT. For example, a HP-33120A function generator was set to a sine wave output with a frequency of 100 Hz (6,000 CPM). The voltage level was

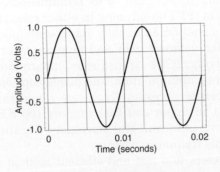

Fig. 7–30 Calculated Time Domain Plot Of A 100 Hz Sine Wave

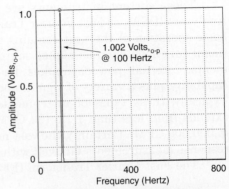

Fig. 7–31 Measured Frequency Domain Plot Of A 100 Hz Sine Wave

adjusted to nominally 1.00 Volt,$_{o-p}$, which is equal to 2.00 Volts,$_{p-p}$. This AC signal was directed to an HP-35670A DSA, and the data processed in a 0 to 800 Hz (48,000 CPM) frequency span. Resolution was set to 800 lines, and a flat top window was used to maximize the amplitude accuracy. The spectrum plot shown in Fig. 7-31 depicts the results of this FFT analysis of the sine wave.

It is understandable that the analysis bandwidth is dominated by a single component occurring at a frequency of 100 Hz (6,000 CPM), and displaying an amplitude of 1.002 Volt,$_{o-p}$. These characteristic values are identical to the input signal frequency and amplitude. From a mathematical standpoint, the time domain signal displayed in the spectrum may be evaluated for the instantaneous voltage (Y_{sin}) at any point in the cycle with the following expression:

$$\boxed{Y_{sin} = A \times \sin(\omega t)} \tag{7-2}$$

where A = Maximum or Peak Voltage (Volts,$_{o-p}$)
          ω = Frequency (Cycles/Second or Hertz)
          t = Time (Seconds)

This is the same general equation that was used for describing a simple periodic motion in chapter 2. The calculated time base plot in Fig. 7-30 displays the time history of this function for two complete cycles. Since the time required for one cycle is 0.01 seconds (Fig. 7-30), that is equivalent to a frequency of 100 cycles per second. The spectrum plot (Fig. 7-31) of the same function reveals the maximum voltage, plus the same frequency. Thus, the characteristic parameters for this periodic motion may be calculated or measured, and plotted in two distinct but interrelated formats.

Although a sine wave is a simple example, it is common knowledge that periodic functions may be expressed as a series of sines and cosines. This concept was originally proposed by the French mathematician and physicist Baron Jean Baptiste Joseph Fourier (1768-1830). His fundamental theories have formed the foundation for wave analysis, and his name has carried through to the Fourier series for basic periodic functions, the Fourier Transform, and the previously mentioned Fast Fourier Transform (FFT).

An improved understanding of the time and frequency domain relationship may be gained by an examination of the Fourier concepts. Certainly the previously discussed sine wave falls into this category, but it is a periodic motion that represents the simplest case. For additional complexity, consider a Fourier series of a triangular wave. This type of information may be obtained from reference books such as *Marks' Handbook*[4], or the *CRC Standard Math Tables*[5]. Based on these references, a convenient formula describing a Fourier series for a triangular wave may be expressed in the following manner:

---

[4] Eugene A. Avallone and Theodore Baumeister III, *Marks' Standard Handbook for Mechanical Engineers*, Tenth Edition, (New York: McGraw-Hill, 1996), pp.2-36.
[5] Daniel Zwillinger and others, *CRC Standard Mathematical Tables and Formulae*, 30th edition (Boca Raton, Florida: CRC Press Inc., 1996), pp 49.

$$Y_{tri} = \left(\frac{8 \times A}{\pi^2}\right) \sum_{n = 1, 3, 5, 7....} \frac{\cos(n \times \omega t)}{n^2} \qquad (7\text{-}3)$$

In this equation, n is equal to the harmonic order. That is, when $n=1$, the frequency component under consideration is the fundamental frequency. When $n=3$, the third harmonic is the associated component, etc. If the maximum voltage A is set equal to 1.0 Volt,$_{\text{o-p}}$, equation (7-3) may be expanded as follows:

$$Y_{tri} = \left(\frac{8 \times 1}{\pi^2}\right)\left\{\frac{\cos(1 \times \omega t)}{1^2} + \frac{\cos(3 \times \omega t)}{3^2} + \frac{\cos(5 \times \omega t)}{5^2} + \frac{\cos(7 \times \omega t)}{7^2} + ...\right\}$$

This expression for a triangular wave may be simplified slightly as:

$$Y_{tri} = \left(\frac{8}{\pi^2}\right)\left\{\cos(\omega t) + \frac{\cos(3\omega t)}{9} + \frac{\cos(5\omega t)}{25} + \frac{\cos(7\omega t)}{49} + ...\right\}$$

The magnitude coefficients for each of the harmonics may now be computed. For instance, the first four components are calculated to be:

$$1st_{tri} = \left(\frac{8}{\pi^2}\right) \times (1) = 0.811 \text{ Volts,}_{\text{o-p}}$$

$$3rd_{tri} = \left(\frac{8}{\pi^2}\right) \times \left(\frac{1}{9}\right) = 0.090 \text{ Volts,}_{\text{o-p}}$$

$$5th_{tri} = \left(\frac{8}{\pi^2}\right) \times \left(\frac{1}{25}\right) = 0.032 \text{ Volts,}_{\text{o-p}}$$

$$7th_{tri} = \left(\frac{8}{\pi^2}\right) \times \left(\frac{1}{49}\right) = 0.016 \text{ Volts,}_{\text{o-p}}$$

The triangular wave amplitude was initially specified as 1.00 Volt,$_{\text{o-p}}$, and each component must carry the same units. The general equation may now be combined with the individual coefficients into the following cosine series:

$$Y_{tri} = (0.811 \times \cos \omega t) + (0.090 \times \cos 3\omega t) + (0.032 \times \cos 5\omega t) + (0.016 \times \cos 7\omega t) + ...$$

The above series of cosine terms may now be plotted for two cycles as shown in Fig. 7-32. Although the above series listed only the first four cosine terms, the time domain plot was generated using the first eight terms. Note that the time domain curve is in fact triangular shaped, with a peak amplitude of 1.00 Volt. In this case, a frequency of 100 Hz, with a period of 0.01 seconds was used for the fundamental. However, the same time domain plot would be generated if a non-dimensional ωt term was used, and the curve plotted between 0 and 4π (2 cycles)

As before, an electronic signal was produced with a HP-33120A function generator. This device was set to a triangular wave output of 1.00 Volt,$_{\text{o-p}}$ at a fre-

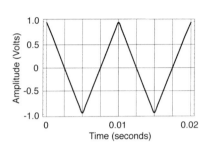

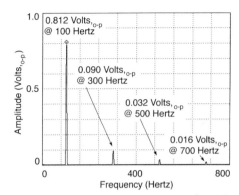

Fig. 7–32 Calculated Time Plot Of A 100 Hz Triangular Wave With 8 Components

Fig. 7–33 Measured Frequency Domain Plot Of A 100 Hz Triangular Wave

quency of 100 Hz. This signal was directed to an HP-35670A, and the data processed in a 0 to 800 Hz bandwidth. Resolution was set to 800 lines, and a flat top window was used to maximize amplitude accuracy. Fig. 7-33 documents the results of this FFT analysis of the triangular wave signal. It is meaningful to extract the amplitudes at each harmonic, and compare the calculated versus the measured peak voltages in Table 7-2:

Table 7–2 Comparison Of Calculated Versus Measured Component Amplitudes For A Triangular Wave With A Frequency Of 100 Hz, And An Amplitude of 1.00 Volt,$_{o-p}$

Harmonic Order	Frequency (Hertz)	Calculated (Volts,$_{o-p}$)	Measured (Volts,$_{o-p}$)
First	100	0.811	0.812
Third	300	0.090	0.090
Fifth	500	0.032	0.032
Seventh	700	0.016	0.016

Excellent agreement is noted between the calculated cosine terms, and the measured FFT amplitudes. The consistency of this data is attributed to the outstanding signal stability of the HP-33120A function generator, plus the signal processing accuracy of the HP-35670A Dynamic Signal Analyzer. Although the higher order harmonics have small amplitudes, the extended dynamic range of the HP-35670A allowed for an accurate measurement of very low voltage levels.

As another example, consider the Fourier series of a square wave. Accessing the previously referenced *Marks' Handbook*, or the *CRC Standard Math Tables*, a Fourier series for a square wave may be expressed with the following common expression:

$$Y_{sqr} = \left(\frac{4 \times A}{\pi}\right) \sum_{n = 1, 3, 5, 7....} \frac{\sin(n \times \omega t)}{n} \qquad \text{(7-4)}$$

If the maximum voltage A is again set equal to 1.0 Volt,$_{\text{o-p}}$ this general square wave equation may be expanded into the following:

$$Y_{sqr} = \left(\frac{4 \times 1}{\pi}\right)\left\{\frac{\sin(1 \times \omega t)}{1} + \frac{\sin(3 \times \omega t)}{3} + \frac{\sin(5 \times \omega t)}{5} + \frac{\sin(7 \times \omega t)}{7} + ...\right\}$$

This expression may now be used to compute the magnitude coefficients for each of the harmonics. For instance, the first four components (1st, 3rd, 5th, and 7th harmonics) are calculated as:

$$1st_{sqr} = \left(\frac{4}{\pi}\right) \times (1) = 1.273 \text{ Volts,}_{\text{o-p}}$$

$$3rd_{sqr} = \left(\frac{4}{\pi}\right) \times \left(\frac{1}{3}\right) = 0.424 \text{ Volts,}_{\text{o-p}}$$

$$5th_{sqr} = \left(\frac{4}{\pi}\right) \times \left(\frac{1}{5}\right) = 0.255 \text{ Volts,}_{\text{o-p}}$$

$$7th_{sqr} = \left(\frac{4}{\pi}\right) \times \left(\frac{1}{7}\right) = 0.182 \text{ Volts,}_{\text{o-p}}$$

This general equation for a square wave may now be combined with the individual magnitude coefficients, and the following sine series results:

$$Y_{sqr} = (1.273 \times \sin \omega t) + (0.424 \times \sin 3\omega t) + (0.255 \times \sin 5\omega t) + (0.182 \times \sin 7\omega t) + ...$$

The square wave series of sine terms will be plotted for two complete cycles that is equivalent to an overall time span of 0.02 seconds for the 100 Hz fundamental frequency. For demonstration purposes, the series used for the plot will consist of only the first four terms. The result of this approach is presented in Fig. 7-34. This rendition is somewhat representative of a square wave, but the tops and bottoms of each curve are ragged edges instead of smooth lines. Clearly, additional higher frequency components are required to improve the simulation. The diagram in Fig. 7-35 represents an extension of this same Fourier sine series to twenty-four coefficients. Note that the time domain curve has been drastically

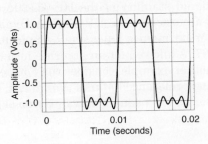

Fig. 7-34 Calculated Time Domain Plot Of A 100 Hz Square Wave Based On The First Four Fourier Components

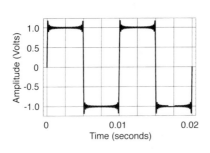

Fig. 7–35 Calculated Time Plot Of A 100 Hz Square Wave With 24 Components

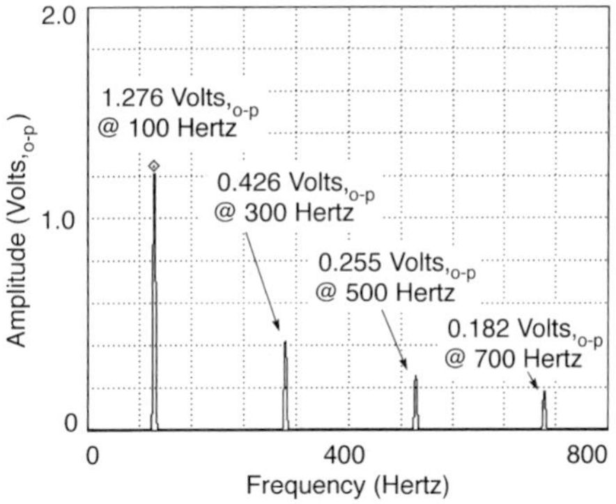

Fig. 7–36 Measured Frequency Domain Presentation Of A 100 Hz Square Wave

improved by the inclusion of the higher frequency components. Clearly, a complex dynamic signal with a structure of significant high frequency components may be substantially distorted if the high frequency components are removed.

An electronic square wave was again produced on an HP-33120A function generator. An output of 1.00 Volt,$_{o-p}$ at 100 Hz was directed to the HP-35670A, and the resultant spectrum plot is displayed in Fig. 7-36. Extracting the amplitudes at each harmonic, and comparing the calculated versus the measured peak voltages, the following Table 7-3 evolves:

Table 7–3 Comparison Of Calculated Versus Measured Component Amplitudes For A Square Wave With A Frequency Of 100 Hz, And An Amplitude of 1.00 Volt,$_{o-p}$

Harmonic Order	Frequency (Hertz)	Calculated (Volts,$_{o-p}$)	Measured (Volts,$_{o-p}$)
First	100	1.273	1.276
Third	300	0.424	0.426
Fifth	500	0.255	0.255
Seventh	700	0.182	0.182

Agreement between calculated sine terms and measured FFT amplitudes are excellent. Minor deviations are attributable to voltage drifts and imprecise settings of the function generator. It should also be noted that the square wave data reveals an amplitude at the fundamental frequency (1.273 Volts,$_{o-p}$) that is larger than the overall unfiltered signal (1.00 Volts,$_{o-p}$). This relationship might seem peculiar to those individuals who believe that: *the unfiltered amplitude must be larger than any of the components*. This casual statement is simply not true as demonstrated by the square wave example. In all cases, the component amplitude relationship is a function of the time domain wave form, and the individual amplitude components and phase relationships required to reconstruct or replicate the original wave form.

As noted, the triangular and square wave examples are both represented by a series of odd numbered harmonics. It is also possible to produce a Fourier series that consists of even and odd harmonics of the fundamental frequency. For instance, a pulse wave will exhibit this type of series. From the *CRC Math Tables*, a Fourier series for a pulse wave may be expressed as follows:

$$Y_{pul} = \left(\frac{2 \times A}{\pi}\right) \sum_{n=1}^{\infty} \frac{(-1)^n \times \sin(n\pi K) \times \cos(n\omega t)}{n} \qquad (7\text{-}5)$$

Within this expression, A represents the peak voltage, the n variable now includes each harmonic (i.e., $n=1, 2, 3, 4, 5...$etc.), and K is used to represent the duty cycle of the pulse. This non-dimensional duty cycle is determined by dividing the pulse width by the period. For example, a pulse with a 100 Hz frequency, and an amplitude of 4.60 Volts$_{,o\text{-}p}$ is shown in Fig. 7-37. Although this signal

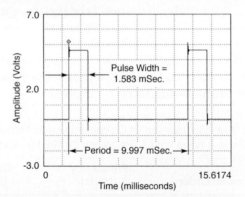

Fig. 7–37 Measured Time Domain Waveform Of A 100 Hz Pulse Signal

originated with the pulse output from a HP-3311A function generator, it is representative of a Keyphasor® from an optical pickup, or a proximity probe observing a key. Fig. 7-37 shows a minor overshoot at the top and the bottom of the signal, which is due to the ringing of the DSA digital filters. The signal period was measured as 9.997 milliseconds. Since frequency is the reciprocal of the period, it is desirable to check this measurement as follows:

$$Frequency = \frac{1}{Period} = \frac{1 \text{ Cycle}}{0.009997 \text{ Seconds}} = 100.03 \text{ Cycles/Second}$$

The period is verified to be consistent with the set frequency of 100 Hz, and it is appropriate to proceed with the following calculation of the duty cycle:

$$K = Duty\ Cycle = \frac{Pulse\ Width}{Period} = \frac{0.001583 \text{ Seconds}}{0.009997 \text{ Seconds}} = 0.158$$

Based upon this calculated duty cycle, and the maximum voltage A of 4.60

Volts,$_\text{p-p}$, the general equation (7-5) for a pulse may be expanded as follows:

$$Y_\text{pul} = \left(\frac{2 \times 4.6}{\pi}\right)\left\{\frac{-1^1 \times \sin(\pi \times 0.158) \times \cos(\omega t)}{1} + \frac{-1^2 \times \sin(2\pi \times 0.158) \times \cos(2\omega t)}{2}\right.$$

$$\frac{-1^3 \times \sin(3\pi \times 0.158) \times \cos(3\omega t)}{3} + \frac{-1^4 \times \sin(4\pi \times 0.158) \times \cos(4\omega t)}{4}$$

$$\left.\frac{-1^5 \times \sin(5\pi \times 0.158) \times \cos(5\omega t)}{5} + \cdots\right\}$$

This complex expression may be simplified slightly, and the number of radians determined for each sine term as follows:

$$Y_\text{pul} = \left(\frac{9.2}{\pi}\right)\left\{-\sin(0.4964) \times \cos(\omega t) + \frac{\sin(0.9927) \times \cos(2\omega t)}{2}\right.$$

$$\left.- \frac{\sin(1.4891) \times \cos(3\omega t)}{3} + \frac{\sin(1.9855) \times \cos(4\omega t)}{4} - \frac{\sin(2.4819) \times \cos(5\omega t)}{5} + \cdots\right\}$$

The magnitude coefficients for each of the first five components (harmonics) may now be calculated in the following manner:

$$1st_\text{pul} = \left(\frac{9.2}{\pi}\right) \times [\sin(0.4964)] = 1.395 \text{ Volts,}_\text{o-p}$$

$$2nd_\text{pul} = \left(\frac{9.2}{\pi}\right) \times \left[\frac{\sin(0.9927)}{2}\right] = 1.226 \text{ Volts,}_\text{o-p}$$

$$3rd_\text{pul} = \left(\frac{9.2}{\pi}\right) \times \left[\frac{\sin(1.4891)}{3}\right] = 0.973 \text{ Volts,}_\text{o-p}$$

$$4th_\text{pul} = \left(\frac{9.2}{\pi}\right) \times \left[\frac{\sin(1.9855)}{4}\right] = 0.670 \text{ Volts,}_\text{o-p}$$

$$5th_\text{pul} = \left(\frac{9.2}{\pi}\right) \times \left[\frac{\sin(2.4819)}{5}\right] = 0.359 \text{ Volts,}_\text{o-p}$$

This general series should now be combined with the individual component magnitude coefficients, and the following cosine series identified:

$$Y_\text{pul} = \{-1.395 \times \cos(\omega t) + 1.226 \times \cos(2\omega t)$$

$$- 0.973 \times \cos(3\omega t) + 0.670 \times \cos(4\omega t) - 0.359 \times \cos(5\omega t) + \cdots\}$$

This series of cosine terms may now be plotted for two cycles ($\omega t = 0$ to 4π) as shown in Fig. 7-38. Due to the complexity of simulating a pulse curve, a total of 50 Fourier coefficients were required for this plot. The minor extraneous peaks on this calculated time domain plot could be improved by extending the number

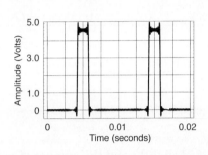

Fig. 7–38 Calculated Time Plot Of A 100 Hz Pulse Signal With 50 Components

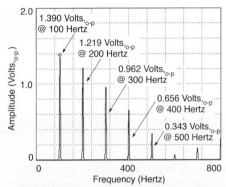

Fig. 7–39 Measured Frequency Domain Presentation Of A 100 Hz Pulse Signal

of coefficients. However, this plot is considered to be a good simulation of the actual measured time domain pulse reviewed in Fig. 7-37. Again, the machinery diagnostician is cautioned against inappropriate filtration on a dynamic signal. If this pulse signal represented a machine Keyphasor®, excessive high-pass filtration would eliminate components that are required to maintain the integrity of the pulse. This could easily result in a distortion of the pulse shape, and the loss of phase continuity for the machinery train. Furthermore, if balancing work was in progress, the alteration of the Keyphasor® signal could result in a change in trigger point, ergo a shift in vector angles. On a large diameter shaft this might not be significant, but on a small diameter shaft, the resultant phase change might prove to be devastating.

Finally, the FFT representation of the HP-3311A function generator pulse output is presented in Fig. 7-39. Again, this is a 4.60 Volt,$_{p-p}$ signal at 100 Hz. This spectrum plot depicts the series of even and odd harmonics of the fundamental. Extracting the amplitudes at each harmonic, and comparing the calculated versus the measured peak voltages, the summary Table 7-4 evolves:

Table 7–4 Comparison Of Calculated Versus Measured Component Amplitudes For A Pulse Signal With A Frequency Of 100 Hz, And An Amplitude of 4.60 Volt,$_{p-p}$

Harmonic Order	Frequency (Hertz)	Calculated (Volts,$_{o-p}$)	Measured (Volts,$_{o-p}$)
First	100	1.395	1.390
Second	200	1.226	1.219
Third	300	0.973	0.962
Fourth	400	0.670	0.656
Fifth	500	0.359	0.343

Agreement between calculated terms, and measured FFT amplitudes is quite acceptable — but the comparative amplitudes are not as close as the previous three examples. Basically, this is attributed to the fact that the other signals were generated from a precision source, whereas the pulse signal was not subjected to rigorous amplitude regulation. Considering the source of the pulse signal, and the minor inaccuracies associated with the determination of the duty cycle, the results shown in Table 7-4 are certainly acceptable.

Other curves such as ramp or a sawtooth patterns will also generate a string of consecutive harmonics. Shaft surface imperfections such as scratches or rust will appear as multiple harmonics on proximity probe output signals. In addition, combinations between various types of signals may also produce a long string of 1X harmonics.

In retrospect, the past four examples of a sine wave, triangular wave, square wave, and the pulse signal produce predictable and consistent results. The interchangeability of time domain to frequency domain data has been repeatability demonstrated. The diagnostician must recognize that FFT analysis of dynamic signals may provide valuable information, and this type of data presentation is a fundamental analysis tool. However, the FFT originates from the machinery vibration time record, and that original time domain signal must always be examined to obtain the full measure of information.

Case History 16: Steam Turbine Exhaust End Bearing Dilemma

When problems occur on major machinery trains, various techniques may be implemented to resolve the difficulty. One popular approach consists of assembling the personnel involved with the machine problem, and discussing the potential origin, plus the corrective solution(s) for the malfunction. This is an effective problem solving forum when the assembled personnel have a good understanding of the equipment, and the abnormal behavior. In some cases the opinions regarding the machinery problem may not coincide, and an acceptable solution cannot be agreed upon. For instance, a dilemma of this type occurred on a turbine compressor set following a routine maintenance inspection.

The physical machinery consisted of a horizontally split centrifugal compressor containing seven impellers, and a rotor weight of 750 pounds. The steam turbine driver contained a three stage rotor that weighed 800 pounds, and produced nominally 5,000 HP at 9,000 RPM. The machines were connected with a fully lubricated gear type coupling. Both machines contained 5 pad tilting pad bearings, and diametrical clearances varied between 5 and 7 Mils. The compressor and the turbine had a good operating history, with only occasional problems at the turbine exhaust end bearing.

During a scheduled plant shutdown, the maintenance personnel elected to physically inspect the condition of the steam turbine exhaust end bearing for evidence of damage or deterioration. This work consisted of a vertical lift check, combined with complete disassembly, and physical inspection of the bearing and turbine journal. It was determined that the diametrical clearance for this turbine exhaust end bearing was 6 Mils. The bearing and journal were both in excellent condition, and the housing was reassembled using the same bearing. No other maintenance work was performed on this machinery train, and operations personnel felt quite confident restarting this equipment. The following startup was performed without any problems, and the machine was successfully placed on-line at a constant speed of 8,955 RPM. Overall vibration amplitudes at each bearing were acceptable, and thrust positions for both rotors had returned to their previous operating position. The only abnormality was a high temperature of 210°F at the turbine exhaust end bearing. This bearing typically operated between 160 and 170°F, and that temperature was measured by a thermocouple embedded in the bottom bearing pad.

As usual, the startup was performed at night. During the warmth of the next day, the bearing temperature increased to approximately 215°F, and concern began to develop about the longevity of this bearing. At this point, various examinations were performed, and the plant personnel used their new spectrum analyzer to produce FFT plots of the proximity probes on this machine. A typical set of spectrum plots across the coupling are shown in Fig. 7-40.

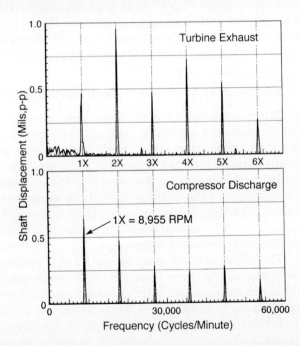

Fig. 7–40 Spectrum Plots from Y-Axis Shaft Proximity Probes Across The Gear Coupling Assembly

This data was acquired from the Y-Axis proximity probes installed at the turbine exhaust bearing, and the coupling or discharge end of the compressor. Note that both FFT diagrams reveal fairly low amplitudes at rotational speed, plus a string of running speed harmonics (i.e., 2X, 3X, 4X, etc.). On the turbine exhaust, the second harmonic of running speed has an amplitude that is twice as large as the fundamental rotational speed motion. This data was of concern to the plant personnel, and two different opinions were soon openly debated.

Some people believed that the series of multiple harmonics were due to mechanical looseness of the turbine exhaust end bearing. Others had the opinion that the measured vibratory behavior was primarily due to misalignment across the gear coupling. The proposed solution was to closely monitor the machinery vibration as the compressor hold down bolts were loosened, and the compressor *allowed* to move into a state of satisfactory hot alignment.

For the uninitiated, it must be mentioned that this type of correction is extraordinarily dangerous to the people as well as the machinery. If one considers the energy contained in a 800 pound rotor rotating at 8,955 RPM, and the potential implications of releasing that energy by unloosening the hold down bolts, the danger in this type of move is obvious.

Following several days of continuous operation, the vibration amplitudes remained constant, and the turbine exhaust end bearing temperature continued to cycle between 210° and 215°F. By virtue of this consistent behavior, the previous two theories were discredited. Clearly, if the turbine exhaust bearing was loose, the behavior would tend to degenerate with time, and that did not occur. The second theory of misalignment was also disproved. Specifically, if heat generation in a bearing was truly due to misalignment — it is logical to believe that the bearing would either fail, or relieve some clearance due to the applied preloads. It is hard to believe that any significant misalignment would appear as constant vibration and elevated temperature without any change. Furthermore, the maintenance inspections performed during the previous shutdown did not disturb the shaft alignment. Clearly, it was necessary to examine the machinery behavior in more detail in order to resolve this problem.

Initially, the vibration signals shown in the frequency domain on Fig. 7-40 were examined in the time domain as displayed in Fig. 7-41. At both measure-

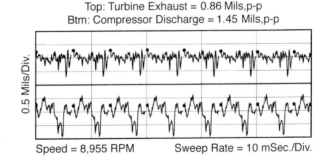

Top: Turbine Exhaust = 0.86 Mils,p-p
Btm: Compressor Discharge = 1.45 Mils,p-p

0.5 Mils/Div.

Fig. 7–41 Time Domain Plots from Y-Axis Shaft Proximity Probes Mounted Across The Gear Coupling Assembly

Speed = 8,955 RPM   Sweep Rate = 10 mSec./Div.

ment planes, it is clear that the time base signals are corrupted by a series of spikes that are indicative of shaft surface imperfections. There is a distinctively different pattern between the turbine exhaust probe signal, and the compressor discharge probe signal. However, both cases are representative of *rough* shaft surfaces below the proximity probes. Comparison of this Y-Axis probe data with the associated X-Axis probes at each measurement location (not shown) reinforces the fact that the vertical spikes are shaft surface scratches that are observed by both probes at each bearing. Based on the previous discussion of the Fourier components of a pulse wave, it is logical to conclude that the majority of the harmonic activity shown in Fig. 7-40 is simply due to the shaft scratches.

Some people will accept this explanation for the running speed harmonics on the shaft vibration signals, and other individuals will not. For additional proof, it is reasonable to obtain a set of casing velocity measurements across the coupling as displayed in Fig. 7-42. This data clearly shows that higher order harmonics do not exist on the bearing housings. This is particularly meaningful on the turbine exhaust bearing. As with many steam turbines of this general size, the exhaust end bearing is a fairly simple unit. The outer shell of the bearing housing retains the tilt pad bearing assembly. There is normally a close relationship between the frequency components measured on the shaft versus the casing. More specifically, if the turbine shaft was really subjected to a strong twice rotational speed component, this frequency component would also appear on the turbine bearing cap vibration data.

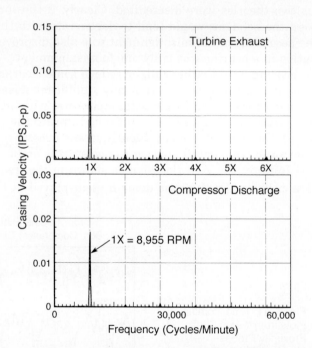

Fig. 7–42 Spectrum Plots From Vertical Casing Velocity Coil Measurements Across The Gear Coupling Assembly

It is very reasonable to conclude that the predominant motion on this machinery train occurs at the rotational speed, and the proximity probe signals are influenced by shaft surface imperfections. The time domain scratches that are clearly visible in Fig. 7-41 are converted into the Fourier components initially presented in Fig. 7-40. There is absolutely no reason to believe that misalignment or mechanical looseness of the turbine exhaust bearing are contributors to the high bearing temperature. However, it is apparent that the bearing pad temperature is higher than normal, and external pyrometer measurements of the turbine exhaust bearing housing also reveals hotter than normal temperatures. The issue then resolves back to the original question of: *How come the exhaust end turbine bearing runs hotter than normal?*

To properly answer this fundamental question, it is necessary to leave the comfort of the air conditioning, and actually examine the operating equipment. This approach goes back to the old days of *go out and look, touch, feel, smell, and listen to the machinery.* In many instances you do not know what you are specifically looking for, and the best advice is to thoroughly examine the machinery for any peculiarities. On this train it was observed that the oil flow leaving the turbine exhaust bearing was minimal. There is an old adage that states: *for every ten drops of oil, only one drop is for lubrication, and the remaining nine drops are for cooling.* On this particular bearing, there was barely enough oil flow for lubrication, and not much left over for cooling.

It was concluded that oil flow to the turbine exhaust bearing was restricted, and this resulted in the elevated bearing temperature. To test this hypothesis, the lube oil supply pressure was increased from 20 to 25 Psig. This change was carefully monitored to insure that there were no detrimental effects to the other machine train bearings. As the oil supply pressure was gradually increased, the bearing temperature dropped. At an oil pressure of 25 Psig, the 215°F bearing temperature was reduced to 203°F. In addition, the oil flowing through the drain sight glass did perceivably increase. The machinery train was successfully operated in this manner for the next six months. At that time, the bearing was opened during a short plant outage. It was discovered that Permatex® was blocking the oil inlet to the turbine exhaust bearing. After this blockage was removed, and the bearing properly assembled, temperatures returned back to normal.

SIGNAL SUMMATION

The previous discussions have generally assumed that the vibratory motion is associated with some fundamental frequency. Typically, this frequency would be the running speed of the machine, and the various harmonic components would track this fundamental frequency. Specifically, if the rotational speed increased, the frequency of the harmonic components would increase proportionally. This type of relationship would apply on supersynchronous excitations such as 2X, 5X, 10X, as well as subsynchronous frequencies that are locked into fractions of the fundamental rotative speed such as X/3, or X/2.

Within the industrial environment, multiple excitations are emitted by different machines, and it is normal to encounter various interactions between frequencies. Excitations are transmitted through fluid streams, through piping systems, and they are also conveyed through foundations and other support structures. This multiplicity of forced and natural resonant excitations are often combined into complex and interesting mechanical movements. Vibration measurements of these machine elements reveal the myriad of excitations, and it then becomes necessary to *sort out* the significant from the inconsequential nonvibratory components. In this regard, the diagnostician must be able to distinguish the common types of signal interactions, and be able to relate these patterns back to the behavior of machine elements.

The three most common types of characteristics encountered are *signal summation*, *amplitude modulation*, and *frequency modulation*. The two types of signal modulation will be discussed in the following sections of this chapter. At this point in the text, the discussion will center on the commonly observed behavior that will be referred to as *signal summation* within this book.

Signal summation can, and does, occur between two or more frequencies that may be closely spaced, or they may be quite divergent in their fundamental frequencies. In general terms, the combination or summation of a series of dynamic signals may be expressed as follows:

$$V_{sum} = F(t)_1 + F(t)_2 + F(t)_3 + ... \tag{7-6}$$

Each separate excitation source is identified as a time variable function such as $F(t)_1$, $F(t)_2$, $F(t)_3$, etc. The time summation of these different excitations is given by V_{sum}. As a specific example, the following equation (7-7) may be used to identify the parameters associated with summing two different frequencies:

$$\boxed{V_{sum} = \{V_1 \times \sin(2\pi F_1 \times t)\} + \{V_2 \times \sin(2\pi F_2 \times t)\}} \tag{7-7}$$

where: V_{sum} = Instantaneous Summation Voltage (Volts)
V_1 = Peak Voltage of Sine Wave #1 (Volts)
F_1 = Frequency of Sine Wave #1 (Hertz)
V_2 = Peak Voltage of Sine Wave #2 (Volts)
F_2 = Frequency of Sine Wave #2 (Hertz)
t = Time (Seconds)

Within this equation it is assumed that two periodic excitations are combined, and each excitation is defined as a simple sine function. These could also be other trigonometric functions, exponential functions, ramps, or a constant amplitude across the frequency domain. However, for convenience during this explanation, a sine function was selected for both signals. The first component was assigned a frequency of F_1, and a peak amplitude of V_1. Similarly, the second part of the signal is defined by a frequency of F_2, and a voltage amplitude of V_2. Since both signals vary with time, the time t is included in equation (7-7).

For demonstration purposes, assume that equation (7-7) represents the vibration characteristics of two adjacent machinery trains. Presume that one machine was running at 100 Hz (6,000 RPM), and producing a vibration signal with a voltage magnitude of 1.0 Volt,$_{o-p}$. Further assume that the second machine was operating at 105 Hz (6,300 RPM), and emitting a vibration signal with a peak voltage of 1.5 Volts,$_{o-p}$. If these values are used to define the sine functions in equation (7-7), and if a time span of 0 to 0.4 seconds was examined, the calculated plot shown in Fig. 7-43 is easily developed to describe the summation of these two sine waves occurring at different frequencies.

To be perfectly clear, this calculated plot in Fig. 7-43 reflects the interaction of the two signals based upon equation (7-7). The signals obviously add together to form the hump in the plot, and they cancel each other to provide the low amplitude portions. Physically, this would be sensed or interpreted as alternating periods of high and low vibration. The frequency differential between the two

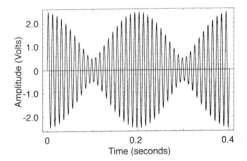

Fig. 7–43 Calculated
Time Domain Plot Of Sig-
nal Summation Between
Two Independent Sine
Waves Of Different Fre-
quencies And Amplitudes

conditions would be a simple beat frequency identified as F_{beat}. This frequency, and the associated period may be stated by the following:

$$\boxed{F_{beat} = F_2 - F_1} \tag{7-8}$$

$$\boxed{Period_{beat} = \frac{1}{F_{beat}}} \tag{7-9}$$

In this example, the differential beat frequency is 5 Hz (300 CPM), and the associated period is 0.2 seconds from equation (7-9). The period of this beat is consistent with low frequency period of 0.2 seconds displayed in Fig. 7-43.

Signal summation is common in virtually all types of machinery systems. It most instances, the summation encompasses multiple frequencies into the final vibration signal. This is normal behavior, and it is one of the common occurrences that the diagnostician must address. One note of caution that should be mentioned, is that this behavior is not an amplitude modulation. Although Fig. 7-43 looks like an amplitude modulated signal, this *is not* amplitude modulation. Some characteristics of summed versus AM signals are similar, but the mechanical implications are quite different. Following the next case history on ID fans, a discussion of amplitude modulated signals will be presented.

Case History 17: Opposed Induced Draft Fans

A classic example of signal summation was displayed by the pair of induced draft fans depicted in Fig. 7-44. These two machinery trains are mounted on top of a primary reformer furnace at an elevation of 240 feet above grade. Each steam turbine driver rotates in a counterclockwise direction, and the pair of speed reducing gear boxes provide clockwise rotation to each fan. These rotational observations are based upon standing at the governor end of either turbine. However, when both machinery trains are viewed from the outboard end of either fan — it is clear that one fan rotates clockwise, and the other fan turns in a counterclockwise direction.

The fans are mounted on a common set of support I-beams that run from east to west across the top of the furnace. Each turbine is under speed control to

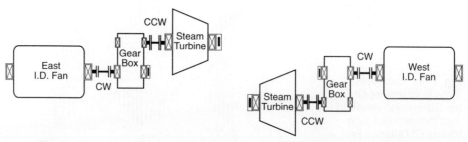

Fig. 7-44 Induced Draft Fans Mounted On Top Of A Primary Reformer Furnace

maintain a specific draft in each respective section of the furnace. Inlet dampers are installed on each fan, but the most of the draft control resides with changes in fan speed (via turbine speed control). The gear box ratio for both trains is constant with 95 bull gear, and 31 pinion teeth. Fan operating speeds typically vary between 1,000 and 1,400 RPM. These fan speeds translate to a turbine operating range of 3,065 to 4,290 RPM.

Historically, both ID fans have experienced multiple problems, and a variety of structural braces and supports have been installed over the years. In some locations, such as the outboard fan pedestals, the braces were supported by additional braces. Both fan rotors are fairly long, and susceptible to bowing. Hence, operations personnel have implemented strict procedures for slow roll and cooling during a normal reformer shutdown sequence.

Unfortunately, a plant emergency forced a trip of the furnace and both fans late one evening. Due to the preoccupation of all available personnel with the crisis situation, the fans were left unattended for several hours. During this time, both rotors experienced substantial shaft bows. In addition, the west fan impeller was damaged, probably during the emergency trip. During the ensuing outage, the furnace tube damage was repaired, but minimal attention was paid to the rotating equipment. However, the extent of the fan shaft bows was quite evident during the next startup. High vibration levels were exhibited by both fans, and the entire superstructure of the furnace seemed to be in sympathetic vibration with the fans.

Operations recognized that full speed and load operation was unattainable, and both east and west units were slowed down to maintain tolerable vibration amplitudes. A vibration analysis of the fans plus the upper structure of the furnace revealed three fundamental frequencies. The two fan running speeds dominated the machinery trains, and the beat frequency between the fan running speeds appeared throughout the furnace superstructure. In fact, the low frequency beat was so strong, it was clearly audible from the front gate.

During this investigation, structural plus bearing housing vibration data was collected on both the east and west machinery trains. For example, the vertical and horizontal time domain motion at the inboard, coupling end bearing of the east fan are presented in Fig. 7-45. In this time domain plot, the maximum vertical amplitude was 6.1 Mils$_{,p-p}$, and the horizontal motion was slightly

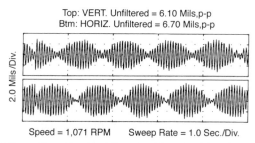

Top: VERT. Unfiltered = 6.10 Mils,p-p
Btm: HORIZ. Unfiltered = 6.70 Mils,p-p

Speed = 1,071 RPM   Sweep Rate = 1.0 Sec./Div.

Fig. 7–45 East Fan - Coupling End Bearing Housing Time Domain Vibration Data

higher at 6.7 Mils,$_{p-p}$. The beat frequency was clearly visible, and it should be mentioned that the other fan bearing housings displayed similar characteristics. A frequency analysis of this data from the east fan is presented in Fig. 7-46.

The vibration measurements were made with casing velocity pickups, and the data was integrated from velocity to casing displacement for both Figs. 7-45 and 7-46. Note that the FFT data also includes the axial casing motion of this east fan coupling end bearing housing. Although the axial vibration appeared at lower amplitudes, it displayed the same frequency distribution as the two lateral transducers.

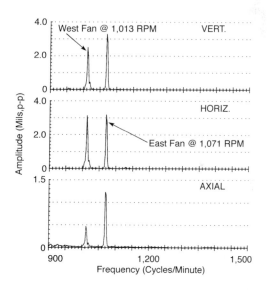

Fig. 7–46 East Fan - Coupling End Bearing Housing Vibration Spectra

There was no apparent vibration at other frequencies, and the spectrum plot in Fig. 7-46 was expanded to provide improved visibility of the dominant running speed components. In this case, a frequency window of 900 to 1,500 CPM was examined. As noted in Fig. 7-46, the east fan speed was 1,071 RPM, and the impeller damaged west fan was operating at a more conservative speed of 1,013 RPM. The beat frequency is computed from equation (7-8) as:

$$F_{beat} = F_2 - F_1 = 1,071 - 1,013 = 58 \text{ CPM}$$

This frequency may now be used to calculate the beat period as follows:

$$Period_{beat} = \frac{1}{F_{beat}} = \frac{1 \text{ Cycle}}{58 \text{ Cycles/Minute}} = 0.0172 \text{ Minutes} \times 60\frac{\text{Sec.}}{\text{Min.}} = 1.03 \text{ Seconds}$$

The period of 1.03 seconds coincides with the low frequency envelope of the time domain plot in Fig. 7-45. There was no question that either a signal summation or an amplitude modulation was occurring between the fan running speeds on top of the furnace. Although this FFT information was definitive in terms of the respective amplitudes and frequencies, it contained minimal information regarding the relative motion between frequencies.

Additional perspective of the fan behavior was provided by examination of this data in the orbital and time domains. For example, Fig. 7-47 displays the orbit time base plots of the east fan inboard bearing housing. This information was filtered precisely at the rotative speed of the east fan. This was accomplished by installing a temporary optical pickup on the east fan shaft, and filtering the data at the Keyphasor® (i.e., running speed) frequency with a DVF. The information shown in Fig. 7-48 was extracted from the same database, but this time the signals were filtered with a temporary optical pickup installed on the west fan.

From the orbital data in Fig. 7-47, the east fan 1X motion was clockwise, and with rotation. This is normal and expected behavior for this type of machine mounted on a compliant support structure. However, the east fan data filtered at the west fan speed reveals a counterclockwise orbit precession in Fig. 7-48. This rotational direction is consistent with the west fan rotational direction when viewed from east to west. Hence, the east fan was excited by running speed vibration from both east and west fans. Similar measurements on the west fan indicated that both fan speeds were likewise driving the west fan structure.

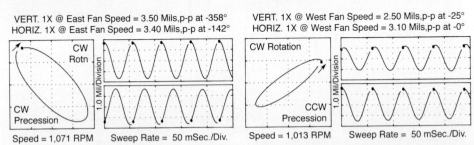

VERT. 1X @ East Fan Speed = 3.50 Mils,p-p at -358°
HORIZ. 1X @ East Fan Speed = 3.40 Mils,p-p at -142°
CW Rotn
CW Precession
1.0 Mil/Division
Speed = 1,071 RPM
Sweep Rate = 50 mSec./Div.

Fig. 7–47 East Fan Coupling End Bearing Housing Vibration At East Fan Speed

VERT. 1X @ West Fan Speed = 2.50 Mils,p-p at -25°
HORIZ. 1X @ West Fan Speed = 3.10 Mils,p-p at -0°
CW Rotation
CCW Precession
1.0 Mil/Division
Speed = 1,013 RPM
Sweep Rate = 50 mSec./Div.

Fig. 7–48 East Fan Coupling End Bearing Housing Vibration At West Fan Frequency

Additional tests were performed by holding one fan speed constant, and varying the other fan speed. In all cases, the beat remained fully active, and the frequency was always equal to the differential between fan operating speeds. It was noted that the vibration severity was largest when the beat frequency was in the vicinity of 60 Cycles/Minute (1 Hz). It was speculated that this frequency might be in the vicinity of a furnace structural resonance.

Based upon this analysis, the east fan rotor was adequately straightened, and this rotor was subjected to a field trim balance. The west fan rotor was damaged beyond a reasonable level of repair, and it was replaced. The new west fan rotor was also field trim balanced. At the conclusion of the field balancing work, casing and shaft vibration amplitudes were all below 1.0 Mil,$_{\text{p-p}}$, and the interaction between fans was virtually eliminated. The nominal 1 Hz beat frequency was also eradicated, and the furnace structure resulted in *better than normal* vibration characteristics. Finally, the multiple braces and supports were removed from each of the outboard fan pedestals.

AMPLITUDE MODULATION

Another form of commonly observed signal patterns is known as *amplitude modulation*, and this is abbreviated as AM. The physics of amplitude modulation varies somewhat from radio broadcasting to machinery interactions. For instance, an AM radio signal is generated by the encoding of a carrier wave by variation of its amplitude in accordance with an input signal. Radio signals usually display high frequency carrier signals, combined with lower frequency modulating signals. Similarly, AM signals on process machines are signals that experience a change in amplitude of one signal due to the amplitude of the second or modulating signal. The machinery AM signals may have large differences between the carrier and the modulating frequencies, or these frequencies may be fairly close together.

Due to the multiple applications of AM, there are different definitions of how two or more signals interact to produce an amplitude modulated signal. For instance, the following equation (7-10) may be used to mathematically describe a common representation of amplitude modulation:

$$V_{am} = V_o \times \{ \sin(2\pi F_m \times t) \} \times \{ \sin(2\pi F_c \times t) \} \qquad \textbf{(7-10)}$$

where: V_{am} = Instantaneous Amplitude Modulated Voltage (Volts)
V_o = Peak Voltage of Wave (Volts)
F_c = Frequency of Carrier Wave (Hertz)
F_m = Frequency of Modulating Wave (Hertz)
t = Time (Seconds)

In many respects, equation (7-10) for amplitude modulation is similar to the previously discussed equation (7-6) for signal summation. Whereas two independent frequencies were specified for the summation case, the AM expression identifies a carrier frequency F_c plus a modulating frequency F_m. In (7-6) the two sine terms were summed in the time domain. For AM, the sine of the carrier and sine of the modulating frequencies are multiplied together as in equation (7-10).

The summation of two components in equation (7-6) resulted in a peak amplitude that was equal to the sum of both independent signals. With an AM signal, the base amplitude is generally considered to be some constant value (e.g., V_o) that is attenuated to various levels throughout the periodic cycle. In certain cases, the similarities between *signal summation* and *amplitude modulation* results in time domain signals that are difficult to distinguish.

For example, equation (7-10) was used to compute the AM signals shown in Figs. 7-49 and 7-50. Within Fig. 7-49, a 5,000 Hz carrier frequency Fc was modulated by a 200 Hz frequency F_m. Note the physical similarities between the true AM signal in Fig. 7-49, and the pair of summed signals previously discussed in Fig. 7-43. Based on this visual similarity, it is no wonder that many cases of simple signal summation are often referred to as amplitude modulation.

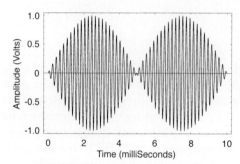 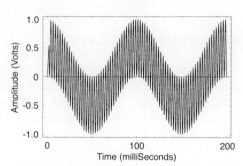

Fig. 7–49 Calculated AM Signal With 5,000 Hz Carrier And 200 Hz Modulator

Fig. 7–50 Calculated AM Signal With 200 Hz Carrier And 190 Hz Modulator

The time domain plot shown in Fig. 7-50 is another common form of an AM signal. This plot was computed with a carrier frequency of 200 Hz, and a modulating frequency of 190 Hz. Although the time domain pattern has been altered, this is still a pure AM signal. Note that the low frequency period for this signal is nominally 100 milliseconds, or 0.1 seconds. This period is equivalent to the beat frequency of 10 Hz, or the differential between the carrier and the modulating frequencies (i.e., 200-190=10 Hz).

For demonstration purposes, an amplitude modulated signal equivalent to the calculated data in Fig. 7-49 was generated with an HP-33120A function generator. The carrier frequency was set at 5,000 Hz (300,000 CPM), and the modulating frequency was adjusted to 200 Hz (12,000 CPM). The synthesized data was directed to HP-35670A, and the processed time base and FFT data are shown in Figs. 7-51 and 7-52. The similarities between the calculated time plot in Fig. 7-49 and the measured data in Fig. 7-51 are self-evident. Within the frequency domain, the wide band FFT revealed a low frequency component at 200 Hz (not shown), and a high frequency component at 5,000 Hz.

For improved visibility, the FFT was zoomed in (translated spectrum) to a frequency range of 4,200 to 5,800 Hz. The data from this 1,600 Hz span is presented in Fig. 7-52. Note that the carrier frequency of 5,000 Hz is bracketed by two strong components at 4,800 and 5,200 Hz. For a true amplitude modulated signal, the side bands at 4,800 and 5,200 Hz represent the sum and difference frequencies between the carrier and the modulating frequency. Stated in another

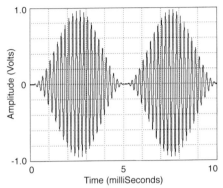

Fig. 7–51 Measured AM Signal With 5,000 Hz Carrier And 200 Hz Modulator

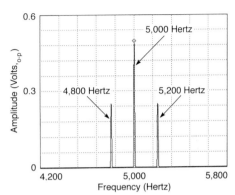

Fig. 7–52 Measured FFT of AM Signal With 5,000 Hz Carrier & 200 Hz Modulator

way, for a true AM signal, the visible frequency components should include the carrier and modulating frequencies, plus sidebands equal to the sum and difference of the two fundamental frequencies.

As another example, the same procedure may be applied to the closely spaced carrier and modulating signals previously computed for Fig. 7-50. If this 200 Hz carrier plus the 190 Hz modulating signal are synthesized and sent to the DSA, the plots shown in Figs. 7-53 and 7-54 may be generated. Once again, the time domain similarity between the calculated signal of Fig. 7-50 and the measured data shown in Fig. 7-53 is quite clear. The spectrum plot in Fig. 7-54 is quite interesting since it encompasses all of the AM frequencies. Specifically, this includes the carrier frequency of 200 Hz, the carrier plus the modulating frequency at 390 Hz, the differential beat frequency at 10 Hz, and a small component at 190 Hz that is the modulating signal.

In most cases, signal summation consists of two independent frequencies that originate from totally different sources. These two frequencies interact in

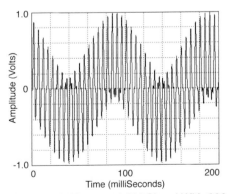

Fig. 7–53 Measured AM Signal With 200 Hz Carrier And 190 Hz Modulator

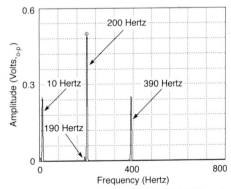

Fig. 7–54 Measured FFT of AM Signal With 200 Hz Carrier And 190 Hz Modulator

such a manner that at times they cancel each other out, and at other times they add together to produce high vibration amplitudes. On these summed signals, the beat frequency is often visible or even audible, and a frequency analysis would exhibit two individual components separated by the beat frequency. However, in an amplitude modulated signal there is often a direct physical link between the carrier and the modulating signal. That is, AM signals are usually associated with interactive excitations on one shaft, or rotating elements that are in direct physical contact (e.g., mating gear sets). The distinguishing or defining frequency characteristics of a true amplitude modulated (AM) signal are the presence of the following frequency components:

○ Carrier frequency
○ Modulating frequency
○ Sum of the carrier and modulating frequencies
○ Difference between carrier and modulating frequencies

The mutual coexistence of these four frequency components is illustrated by the following case history of a centrifugal compressor that displays all of the characteristics of a true amplitude modulation.

Case History 18: Loose and Unbalanced Compressor Wheel

A five stage high pressure centrifugal compressor operated successfully for over 20 years at speeds between 9,000 and 9,600 RPM. The compressor was in refrigeration service, and it experienced minimal mechanical problems throughout its entire operating history. During a scheduled overhaul, the unit was disassembled for mechanical inspection, and the spare rotor installed. The rotor that had been running during the previous 4 years was sent to the shop for a routine set of assembled runout, outer diameter, plus axial position measurements of the wheels with respect to the thrust collar. At the conclusion of these rotor measurements, an over zealous shop supervisor had the rotor un-stacked. At this time they began runout measurements of the bare shaft, keyway inspections, plus the acquisition of bore dimensions on the impellers, thrust disk, and balance drum.

The rotor inspection work was progressing in a timely manner, when the overhaul was completed, and the compressor was restarted with the spare rotor. At a speed of about 3,000 RPM the compressor ingested some type of foreign object that literally destroyed the first stage impeller. The train was not equipped for automatic shutdown, and the compressor operated for several minutes in a seriously distressed condition. Subsequent inspection revealed that the suction end journal bearing was damaged, and the shaft was severely scored.

At this point, the only available option was to split the case, pull the spare rotor, and reinstall the other rotor with the best available set of bearings and seals. Unfortunately, the other rotor was scattered all over the shop, and the responsible shop supervisor was completely embarrassed. In virtually all operating plants, the rotor(s) removed from the machines during a turnaround are not disassembled until the plant has been up and running for a week or more. In this case, the shop supervisor *jumped the gun*, and put the plant startup in jeopardy when he ordered the un-stacking of the compressor rotor.

Options were limited, and the only thing to do was reassemble the old rotor, do a quick check in the shop balancing machine, and stuff this rotor back into the compressor. There was a lot of pressure on the shop personnel, and they ignored the fundamental rotor assembly rule of: *do it right, or do it over*. They rushed the job, and although the residual unbalance of the stacked rotor was different from the incoming inspection report, they went ahead and reinstalled the old rotor back into the compressor casing.

During the ensuing startup, the compressor exhibited fluctuating vibration amplitudes from slow roll up to minimum governor speed. As the unit continued to operate, and temperatures began to stabilize, the fluctuations in radial shaft vibration tended to increase, and then stabilize into clean smooth shaft orbits. As time progressed, the periods of high vibration increased in severity, and the low vibration condition was also degenerating. An example of the shaft vibration just prior to trip of the machine is displayed in Fig. 7-55.

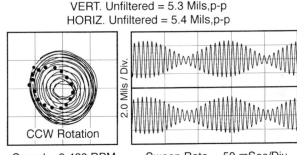

VERT. Unfiltered = 5.3 Mils,p-p
HORIZ. Unfiltered = 5.4 Mils,p-p

Fig. 7–55 Shaft Displacement Orbit And Time Base Plots With Loose And Unbalanced Middle Compressor Wheel

CCW Rotation

Speed = 9,420 RPM

Sweep Rate = 50 mSec/Div.

Note the classical amplitude modulation characteristic of the time domain plot, plus the radial pulsation of the shaft orbit. In essence, the shaft orbit varied from a minimum amplitude of 1.0 to a maximum of 5.4 Mils,$_{\text{p-p}}$. The multiple Keyphasor® dots shown on the orbit indicated the presence of a forward and circular subsynchronous vibration component. Examination of the synchronous, 1X shaft vibration revealed a forward circular precession, with amplitudes that varied between 1.0 and 2.5 Mils,$_{\text{p-p}}$.

Both inboard and outboard compressor bearings revealed similar characteristics. The suction, thrust end, of the compressor was slightly lower than the discharge end bearing. This high vibration condition appeared and subsided at both ends of the compressor in unison. Even the experienced hands who had worked on this machine for many years agreed that something *was really wrong* when the floor grating began to pop out of the retaining channels. Vibration data was recorded, and the unit was finally shutdown to minimize any further damage to the only available compressor rotor assembly.

Following a safe shutdown of the machinery train, the recorded vibration data was examined in greater detail. For instance, the spectrum plot in Fig. 7-56

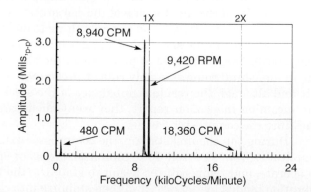

Fig. 7–56 Spectrum Plot of Vertical Shaft Displacement With Loose And Unbalanced Middle Compressor Wheel

documents the frequency behavior sensed by the vertical proximity probe at a time equivalent to the orbit and time base data shown Fig. 7-55. The two dominant frequency components are the 1X running speed vibration at 9,420 RPM and 2.4 Mils,$_{p-p}$, plus a larger component at 8,940 CPM and 3.1 Mils,$_{p-p}$. The differential between these two major components provides a lower sideband at 480 CPM (=9,420-8,940). Similarly, summation of the two major components yields an upper sideband at 18,360 CPM (=9,420+8,940). By the definitions listed in the previous section, it is reasonable to conclude that this observed behavior is a clearly defined amplitude modulation.

Since it is an AM interaction between running speed and the larger component at approximately 95% of running speed, it is reasonable to concentrate on the compressor rotor as the culprit. If this was a *signal summation* condition, the search for potential problems would have to be expanded to a much broader base. However, since this is an amplitude modulation, the efforts were fully directed towards the rotating assembly.

Further examination of the full speed data revealed that the 95% component periodically increased in frequency, and locked into the running speed motion. Under this state, shaft vibration amplitudes subsided, and the compressor appeared fairly normal. Then, for no apparent reason, the subsynchronous component would reappear, and drift back to 5% below running speed with increasing vibration amplitudes. Again, both ends of the compressor exhibited the same type of behavior, with very similar vibration levels. Since this was a reasonably symmetric rotor, this suggested that the malfunction was associated with a problem in the middle of the compressor rotor.

A variation in compressor load would have been beneficial in analyzing this problem, but that option was not achievable. In the final analysis, it was concluded that a midspan compressor impeller was loose. It was speculated that during a free spinning condition, the loose impeller dropped to 95% of operating speed. As it ran under this condition, the residual mass unbalance plus eccentricity of this wheel provided a forward circular forcing function to the rotor. Since this occurred at a different frequency from the shaft running speed, it resulted in two distinct excitations. One excitation occurred at shaft rotational speed, and the other excitation was due to the rotating frequency of the loose wheel. Furthermore, as the loose impeller continued to rotate, friction between the impeller bore and the shaft outer diameter would result in localized heating. This could cause the impeller bore to shrink back onto the shaft, and result in a temporary elimination of the 95% *loose wheel* frequency. As the refrigerant flow cooled off the shaft and impeller, the wheel would eventually re-initiate the entire cycle by loosening up, and slowing down to 95% of operating speed.

After shutdown, and disassembly of the compressor, plus un-stacking of the rotor — it was determined that the middle compressor impeller was indeed loose. The primary reason for this loose wheel was that the impeller key was never reinstalled during the hasty shop rebuild of this rotor. There was indication of this problem on the slow speed shop balance machine, but the responsible individuals choose to ignore this information. Once again, the machinery responds in accordance to the laws of physics, and the associated human beings tend to manage by emotion instead of good engineering practice.

FREQUENCY MODULATION

The final type of common signal interaction commonly encountered around machinery analysis is known as *frequency modulation* (FM). This type of signal was developed between 1925 and 1933 by Edwin Howard Armstrong (1890 to 1954). The original intent of his invention was to eliminate static on radio transmissions. Today, FM signals are used in a variety of different applications. In all cases, these signals originate with a high frequency, constant amplitude carrier signal. The frequency of this carrier wave is varied (or modulated) by the lower frequency modulating wave or signal. A simple form of frequency modulation is presented in the following expression:

$$V_{fm} = V_o \times \cos\{(2\pi F_c \times t) + \sin(2\pi F_m \times t)\} \qquad \text{(7-11)}$$

where:
V_{fm} = Instantaneous Frequency Modulated Voltage (Volts)
V_o = Peak Voltage of Carrier Wave (Volts)
F_c = Frequency of Carrier Wave (Hertz)
F_m = Frequency of Modulating Wave (Hertz)
t = Time (Seconds)

Once again, F_c is defined as the carrier frequency, F_m is the modulating signal, V_o is the carrier voltage, and V_{fm} represents the instantaneous FM voltage over time t. Note that the time dependent sine of the modulating frequency is combined with the time dependent carrier frequency in equation (7-11). The resultant combination is subjected to a cosine trigonometric function. Overall, this type of FM expression appears as a normal cosine function with a constant peak-to-peak amplitude, and a distortion of the frequency with the passage of time. The diagram presented in Fig. 7-57 was computed with a carrier frequency F_c of 5,000 Hz, a modulating frequency F_m of 4,000 Hz, and a base voltage of 1.0

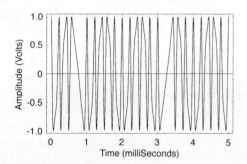

Fig. 7–57 Computed Time
Domain Plot Of FM Signal
Showing Accordion Effect

Volt,$_{\text{o-p}}$. If this type of signal is viewed live on an oscilloscope, the time base wave
will appear to move back and forth horizontally across the screen. This type of
time domain motion appears to oscillate like the bellows of an accordion. Hence,
time base observation of a FM signal is sometimes referred to as an *accordion
effect*. In many instances this is a subtle condition that may be easily missed if
an appropriate sweep rate is not selected on the oscilloscope.

A broad band frequency analysis might also miss an FM signal if the modulating frequency is small compared to the carrier frequency. For example, if a
5,000 Hz carrier is frequency modulated by a 100 Hz component, a *wide* 5,000 Hz
component would be visible in a 0 to 12,800 Hz spectrum. If the DSA frequency
span was translated to a center frequency of 5,000 Hz, and a bandwidth of 800
Hz, the spectrum shown in Fig. 7-58 would appear. In this example, the carrier
frequency appears at the center of the array at 5,000 Hz. In addition, a series of
side bands appear at the carrier frequency, plus and minus multiples of the modulating signal (i.e., $F_c \pm F_m$, $F_c \pm 2F_m$, $F_c \pm 3F_m$, etc.).

This same behavior occurs in Fig. 7-59, where the modulating frequency
has been reduced to 50 Hz. Note that the quantity and size of the sidebands have
been altered. In addition, the magnitude of the carrier frequency at 5,000 Hz has
been reduced. In some cases, the carrier frequency might be reduced to very low

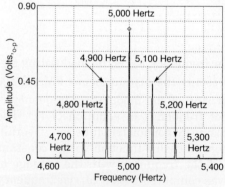

Fig. 7–58 Measured FFT Of FM Signal
With 5,000 Hz Modulated By 100 Hz

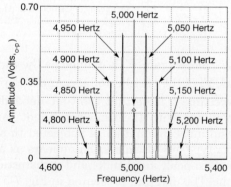

Fig. 7–59 Measured FFT Of FM Signal
With 5,000 Hz Carrier Modulated By 50 Hz

levels. Overall, the distinguishing characteristics of a frequency modulated (FM) signal are summarized as follows:

- ○ Carrier frequency at the center of the component array
- ○ Multiple sidebands at the carrier frequency minus orders of the modulator
- ○ Multiple sidebands at the carrier frequency plus orders of the modulator
- ○ Constant signal amplitude when viewed in the time domain
- ○ No distinct component occurring specifically at the modulating frequency

In the field of machinery analysis, FM signals are encountered in three distinctly different ways. This includes FM tape recorders, measurement of torsional vibration using FM techniques, and mechanical phenomena that generate FM signals. In the first application of FM magnetic tape recorders, a carrier frequency is generated, and the dynamic signals are superimposed upon the carrier. During data reproduction, the high frequency FM signals are demodulated, and the resultant signals provide excellent reproductions of the original dynamic data. FM tape recorders provide high signal to noise ratios, combined with the ability to record data down to zero frequency (DC) of the dynamic signal.

The characteristics of FM recorders are specified by *ISO 3615*. For example, consider a tape transport speed of 15 inches per second (38.1 centimeters/second). A Wide Band Group 1 recorder will have a carrier center frequency 54.0 kHz, and a recording bandwidth extending from DC to 10.0 kHz. This varies with speed, so that a decrease to a tape speed of 7.5 inches per second (19.05 centimeters/second) will cut the center frequency in half to 27.0 kHz. It will also reduce the recording bandwidth by a factor of two to DC to 5.0 kHz. This topic will be discussed in greater detail in the following chapter 8.

The second common utilization for FM signals consists of torsional vibration measurements. In this application, a transducer (e.g., a proximity probe or magnetic pickup) is positioned over a gear or a precision cut slotted wheel. As the machine rotates, the gear tooth passing, or slot passing frequency, acts as a carrier signal. Any torsional vibration of the machinery will result in angular oscillations that frequency modulate the carrier signal. Demodulation of the FM signal provides an electronic signal that is proportional to torsional vibration. This application was previously discussed in chapter 6 of this text.

It should be mentioned that the carrier frequency for torsional measurements generally occurs at a fairly high frequency. For instance, if a 3,600 RPM turbine generator set is equipped with a common 60 tooth wheel, the resultant carrier frequency at synchronous speed will be 216,000 CPM (3,600 Hz). This frequency exceeds the normal operating range for good amplitude accuracy from a proximity probe (e.g., Fig. 6-35). However, it should be recognized that FM measurements vary the frequency of the carrier, and amplitude is not important. As long as the transducer provides a clear pattern of the passing teeth, any changes in timing are detectable, and may therefore be demodulated into torsional displacement data.

The third category of FM signals consists of actual machinery excitations that generate FM signals. This can occur in machines such as gear boxes, electrical machines such as generators and motors, as well as bladed machines such as steam and gas turbines. When multiple excitations are tied into the same fundamental rotational speed frequency on the same rotor, or on mating rotors, the opportunities for generation of FM signals are extensive. In many cases, machinery excitations may appear as amplitude or frequency modulation, or a combination thereof. The machinery diagnostician should be fully aware that AM and FM signals are possible on even simple machines. This basic recognition will save a lot of time identifying sidebands, and allow the diagnostician to examine the potential physical reasons for the AM or FM signals. It should also be mentioned that machines can also exhibit other types of interactions such as phase modulation. These are less common physical occurrences, and they are beyond the current scope of this text.

Case History 19: Gear Box with Excessive Backlash

A simple two element speed reduction gear box consisted of an 80 tooth high speed pinion driving a 220 tooth low speed bull gear. This was a down mesh arrangement with a normal pinion operating speed of 4,228 RPM, and an associated bull gear speed of 1,537 RPM.

Fig. 7-60 documents the normal bull gear spectrum plots obtained from a vertical casing accelerometer at the outboard bearing. The top plot covers a nominal frequency range of 0 to 500,000 CPM. This plot reveals a gear mesh frequency at 338,220 CPM, with an amplitude of 1.3 G's,$_{\text{o-p}}$. The lower plot displays a translated FFT of the frequency domain immediately surrounding the gear mesh frequency. It is clear that the gear mesh frequency is not influenced by any other significant excitation.

The FFT plots in Fig. 7-61 describe the normal pinion characteristics. This acceleration data was obtained in the vertical plane at the blind end pinion bearing cap. Gear mesh frequency occurs at a slightly lower amplitude of 0.7 G's,$_{\text{o-p}}$, and a slight modulation of the mesh by bull gear speed (X_{ls}) is apparent. Since the bull gear is substantially larger than the pinion, it is common to see the influence of the large bull gear element upon the small pinion. This type of activity is visible in the low frequency domains around running speed, and around the identified gear mesh frequency.

These high frequency vibration response characteristics remained quite consistent until the gear box was disassembled for inspection during a routine maintenance overhaul. Unfortunately, the pinion bearings were damaged beyond repair when a 6" 300# weld neck flange was inadvertently dropped on the bearings. This necessitated the installation of a set of spare pinion bearings from the warehouse. Due to time restraints, the spare bearings were installed with no consideration of proper fit, or relative position of the pinion versus the gear.

Subsequent startup and loading of the unit resulted in the generation of a significantly different high frequency *whine* from the box. The vertical casing accelerometer mounted on the bull gear displayed the FFT data in Fig. 7-62. It is noted that speeds are only 0.5% higher than the earlier data set, and that the

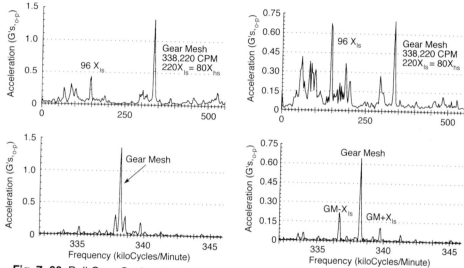

Fig. 7–60 Bull Gear Casing Acceleration With Normal High Frequency Behavior

Fig. 7–61 Pinion Casing Acceleration With Normal High Frequency Behavior

gear mesh amplitude is similar to the previous data. However, the post-overhaul bull gear plots revealed a modulation of the gear mesh (GM) frequency by the low speed bull gear ($\pm X_{ls}$), and the high speed pinion ($\pm X_{hs}$).

These upper and lower side bands of the running speeds were also evident on the pinion data as shown in the Fig. 7-63. Although the relative amplitude of the pinion sidebands appeared to be lower than the bull gear, the absolute values

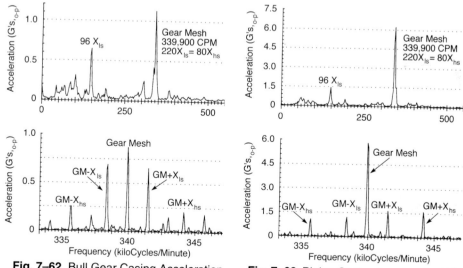

Fig. 7–62 Bull Gear Casing Acceleration With Abnormal High Frequency Behavior

Fig. 7–63 Pinion Casing Acceleration With Abnormal High Frequency Behavior

were actually higher on the pinion. In addition, the gear mesh frequency on the pinion bearing had increased to over 6.0 G's,$_{o-p}$. In the final analysis, a mesh discrepancy was apparent, and the unit was shutdown. Further inspection revealed that the backlash was excessive due to improper lateral position of the pinion bearings. Once again, production pressures are combined with human error and poor judgment to render a good machine inoperable. Fortunately, in this case the correction was easily achievable, with minimal expense and downtime.

BIBLIOGRAPHY

1. Avallone, Eugene A. and Theodore Baumeister III, *Marks' Standard Handbook for Mechanical Engineers*, Tenth Edition, pp. 2-36, 2-37, New York: McGraw-Hill, 1996.

2. Harris, Cyril M., *Shock and Vibration Handbook*, Fourth edition, chap. 14, New York: McGraw-Hill, 1996.

3. *The Fundamentals of Signal Analysis*, Application Note 243, (Hewlett Packard, printed in USA, 1995).

4. Wolfram, Stephen, *The Mathematica® Book*, 3rd edition, software version 3.0 Champaign, Illinois: Wolfram Media/Cambridge University Press, 1996.

5. Zwillinger, Daniel and others, *CRC Standard Mathematical Tables and Formulae*, 30th edition, p. 49, Boca Raton, Florida: CRC Press Inc., 1996.

Data Acquisition and Processing

*I*ndustrial processes operate over a wide range of pressures, temperatures, and flow rates. This variety of operating conditions dictates a similarly large distribution of machinery types and configurations. Although the variables involved are extensive, a common focal point appears when reliability and mechanical integrity are discussed. In virtually all cases, it is economically favorable to evaluate the onstream condition of the machinery. Furthermore, when problems do appear, early detection and proper diagnosis are mandatory. In the pursuit of parameters to access mechanical condition, the analysis of machinery vibration characteristics has consistently proven to be a powerful tool.

Successful vibration analysis requires an intimate familiarity with various types of measurements, application and characteristics of transducers, plus capabilities and limitations of diagnostic instrumentation. Ultimately, the data must be reduced to hard copy for evaluation. The format used for data presentation can enhance the information, yielding a direct identification of the occurring mechanism, or it can submerse the relevant data in a sea of inclusive information. In previous chapters, the basic measurements and transducer characteristics have been discussed. In addition, dynamic signal characteristics and signal manipulation techniques have been reviewed. At this point, it is appropriate to address the integration of a contemporary suite of industrial transducers with the instrumentation required to properly acquire and accurately process high resolution data. Finally, the traditional formats used for documentation of steady state and transient data will be reviewed.

VIBRATION TRANSDUCER SUITE

Since the late 1960s, the eddy current proximity probe has been widely accepted for obtaining relative shaft vibration and position measurements. In most cases, these transducers are used to measure the motion of a rotating shaft with respect to a stationary element such as a bearing. This type of sensor is eminently suitable for machinery protection and trend analysis, due to its capability to measure both vibration and relative changes in position between elements.

Proximity probe calibration is generally consistent with API specifications of 200 mv/Mil, ±5%. The total linear operating range is specified to be a minimum of 80 Mils by the same API[1] specifications. Frequency response for proximity probes are often stated as DC (zero frequency) to 600,000 CPM (10,000 Hz). However, at high frequencies, displacement amplitudes are quite small, and typically fall below the noise floor of the measurement system. For this reason, the most significant probe data occurs between DC and approximately 90,000 CPM (1,500 Hz). Certainly this range is more than adequate for addressing the anticipated vibratory behavior of most large industrial machinery trains.

A pair of mutually perpendicular proximity probes are typically installed at each journal bearing on a machinery train. Normally these probes are mounted at ±45° from the true vertical centerline as shown in Fig. 8-1 of a three bearing turbine generator set. Thrust bearings are protected by dual axial probes as indicated on the turbine. To provide synchronous tracking and filtration capability, a once-per-revolution Keyphasor® probe is usually installed on each train. Since these timing probes are permanently mounted in a fixed location, the resultant phase data is considered as an absolute measurement.

Casing vibration transducers may be installed on a temporary basis for diagnostic measurements, or they may be permanently installed and connected to vibration monitors. Historically, two fundamental types of casing vibration transducers have been applied: the velocity coil, and the piezoelectric accelerometer. Both types of transducers provide relative motion of the casing with respect to free space. These types of sensors are also referred to as seismic, or inertially referenced transducers.

Velocity transducers contain a spring mounted mass enclosed by an outer coil that responds to machine vibration. The motion of the spring mass system is either damped electronically or with an internal viscous fluid. Since this is a mechanically activated system, it is limited in overall frequency response. Typically, a velocity coil will exhibit a low frequency rolloff of 600 CPM (10 Hz), with a high frequency limitation in the vicinity of 90,000 CPM (1,500 Hz). This frequency range is suitable for measurement of casing vibration occurring on medium and high speed machinery. It should be recognized that a velocity coil is generally inappropriate for very low speed measurements. This is due to the attenuation of amplitudes, and phase distortion inherent with the low frequency damping of the spring mass resonance. At high frequencies, this transducer cannot respond mechanically, and the upper frequency limitation appears.

Velocity coils are self-generating devices that do not require an external power source. Output sensitivity for these transducers varies between 500 and 1,080 mv/IPS. Due to these high scale factors, the resultant electronic vibration signals are strong, and easily observed with a variety of readout devices. The major difficulties with this type of transducer are the limitations associated with the sensing mechanism, i.e., the spring mass. This type of pickup is susceptible to spring breakage, and as previously mentioned, it is limited in overall fre-

---

[1] "Vibration, Axial Position, and Bearing Temperature Monitoring Systems — API Standard 670, Third Edition," *American Petroleum Institute*, (Washington, D.C.: American Petroleum Institute, November 1993).

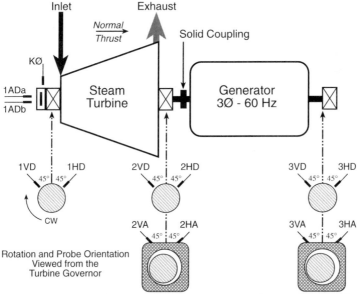

Fig. 8–1 Typical Machinery Arrangement With Shaft And Casing Vibration Transducers

quency response. In addition, most industrial velocity coils are fairly large, and this can restrict installation in many locations.

Due to these velocity coil limitations, accelerometers have emerged as the preferable vibration transducer for seismic measurements. In general, industrial accelerometers are smaller, lighter, more durable, and they cover a significantly larger frequency range than velocity coils. In most accelerometers, the transducer contains an internal mass, a piezoelectric crystal, and an integrated circuit (ICP®). The crystal is generally quartz or ceramic, and the application of force by the internal transducer mass produces an electrical charge. This charge sensitive signal is converted to a voltage signal proportional to acceleration in the ICP® circuit, or in an external charge amplifier. The final conditioned output signal carries the engineering units of millivolts per G of acceleration.

Accelerometers are manufactured in a multitude of configurations. Transducers are available that can successfully operate in cryogenic environments at temperatures below -350°F, and other units are designed for sustained high temperature operation at well over 1,200°F. Some accelerometers are designed for low frequency measurements, with operating ranges of 6 to 1,800 CPM (0.1 to 300 Hz). These transducers are low noise devices with output scale factors that range from 500 to 10,000 mv/G. At the opposite end of the frequency domain, miniature, high frequency accelerometers can reach 1,200,000 to 1,800,000 CPM (20,000 to 30,000 Hz), with typical scale factors ranging from 1.0 to 5.0 mv/G.

The accelerometers used for casing measurements on process machines are less exotic. The normal frequency range for these industrial transducers extends from 600 to 600,000 CPM (10 to 10,000 Hz). The output sensitivity for these accelerometers is generally in the vicinity of 100 mv/G. These probes are not limited by the mechanical movement characteristics of velocity coils, and the top end frequencies are significantly higher. As such, accelerometers are suitable for measuring high frequency machinery excitations such as gear meshing and Blade Passing frequencies.

For demonstration purposes, two sets of mutually perpendicular accelerometers are shown in the machinery diagram in Fig. 8-1. In this example, the bearings on each side of the generator are equipped with X-Y accelerometers mounted at ±45° from the true vertical centerline. These transducers are directly in-line with the shaft sensing proximity probes. Typically, the accelerometers are mounted on the same machine element (e.g., bearing housing) as the shaft displacement probes.

The acceleration signals may be double integrated to yield bearing housing displacement with respect to free space. The shaft proximity probes measure relative motion between the shaft and structure upon which the probes are attached. If the proximity probes are mounted on the bearing housing, then the probes detect relative motion between shaft and the housing.

Performing an electronic summation between the shaft relative and the casing absolute displacement provides the useful measurement of shaft absolute motion. For machines with flexible supports or foundations this can be an extraordinarily important measurement. Conversely, machines with very rigid support structures will display minimal casing motion, and the shaft absolute vibration will be closely approximated by the shaft relative motion.

The vibration transducers are generally designated with some type of logical probe identification scheme. For example, the probes mounted on the machine train shown in Fig. 8-1 are identified with a common three character code. Within this code, the first character refers to a specific bearing. The *1* identifies the turbine governor end, *2* denotes the turbine exhaust, and *3* designates the outboard generator bearing. The second letter refers to probe orientation. Specifically, the letter *V* refers to a vertical or Y-axis probe, *H* specifies a horizontal or X-axis probe, and *A* is used for the axial (thrust) probes. The third character of *D* or *A* identifies the transducer as a displacement proximity probe, or an accelerometer. Thus, a three character identification code is used to uniquely describe the location, orientation, and type of vibration transducer. The only exceptions to this code are the Keyphasor® probe (KeyØ), and the *a* and *b* designations added to the axial probes to define the two channels directed to the thrust monitor.

Other static or dynamic transducers may also be installed on a machinery train to continuously monitor a specific parameter, or measure a particular dynamic characteristic during an investigative test. Additional static devices such as thermocouples, LVDTs, valve position, or any number of process or load measurements may be recorded. The supplementary dynamic transducers include devices such as pressure pulsation pickups, force transducers, strain gauges, and torsional vibration transducers. Many of these probes contain ICP® electronics, and share a common power supply and signal conditioning.

RECORDING INSTRUMENTATION

The machinery transducer systems discussed in the previous section are generally terminated at one or more racks of permanent monitors. Although some racks are field mounted close to the machinery, the majority are installed in a local or a central control room. The monitor racks are normally configured to provide digital or analog outputs to a Distributed Control System (DCS). These monitor output signals are generally used for trending of overall values, plus correlations with process conditions.

The machinery monitoring system may also provide digital outputs to a separate dynamic or transient data acquisition system such as a Bently Nevada Transient Data Manager® (TDM). This type of computer-based system provides improved visibility of vibratory characteristics. Whereas the DCS will trend overall unfiltered vibration amplitudes, this auxiliary system will trend characteristics such as FFT data and vectors at various frequencies (1X, 2X, etc.). This system may also be equipped to capture transient startup or coastdown data. Although these are very useful systems, it should be recognized that these types of systems have limitations in terms of resolution and sampling characteristics.

For situations where an automated data processing system does not exist, or where improved data resolution is required, or in cases where further signal manipulation is anticipated — the use of separate recording instrumentation is mandatory. Fortunately, the tools and techniques to accurately record dynamic data with low noise levels have been available since the advent of the FM tape recorder in the early 1970s. This device allows accurate recording and reproduction of multiple channels of complex dynamic signals in a continuous time record. The recording media is magnetic tape, and configurations ranging from reel-to-reel, cassette, VHS, and DAT formats have been successfully employed.

During the evolution of tape recorders, size and weight have decreased, and capabilities have increased. Although it is tempting to use a tape recorder as a stand-alone instrument, it is much more effective if it is integrated into a complete data acquisition system. For example, consider the diagram in Fig. 8-2 of a typical multichannel analog field data acquisition package.

In Fig. 8-2, the raw transducer signals are terminated at a multichannel switching box. This device allows AC coupling of signals such as proximity

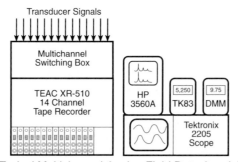

Fig. 8–2 Typical Multichannel Analog Field Data Acquisition System

probes or accelerometers (via internal coupling capacitor). It also provides a way to *switch and compare* each transducer signal with the corresponding tape recorder output signal. This comparison is most effectively performed visually on an oscilloscope. A check of signal integrity is repeated many times during a data acquisition session to insure accuracy of the recorded and reproduced information. Thus, a primary function of the field Tektronix® oscilloscope is to continually verify the validity of the TEAC® tape recorder signals.

Voltage amplifiers may be included as separate units, or the amplifiers may be incorporated as an integral part of the tape recorder. For separate amplifiers, the signal gain is usually indicated by a rotary or toggle switch. Common voltage amplifier settings include gains of x1, x2, x5, x10, x20, x50, and x100. These are often combined with attenuation settings of x0.5, x0.2, and x0.1. For tape recorders that include built-in amplifiers, the determination of signal gain is not necessarily defined directly by the tape recorder. In most cases, the user establishes an input voltage level, plus an output voltage range. The signal gain across the recorder is then determined by equation (8-1).

$$Gain_{recorder} = \frac{Output\ Voltage\ Range}{Input\ Voltage\ Range} \qquad \text{(8-1)}$$

On many recorders the input voltage range covers values such as ±0.1, ±0.2, ±0.5, ±1.0, ±2.0, ±5.0, and ±10.0 volts. Depending on the specific tape recorder, the output voltage range may include a similar set of steps. It may also be fixed at a level of ±1.0 volt, or it may be adjustable with a potentiometer. In any case, the ratio of output to input will determine the signal gain across the tape recorder. There are also occasions where external amplifiers are used in conjunction with the internal recorder voltage amplifiers. In these situations, the overall signal gain is given by equation (8-2).

$$Gain_{overall} = Gain_{external} \times Gain_{recorder} \times \ldots \qquad \text{(8-2)}$$

In the general application, voltage amplifiers are used to amplify low level signals, and reduce the magnitude of large signals. For example, assume that an input signal is amplified by a factor of 5 with the internal (or external) tape recorder amplifier(s). If the direct transducer output signal is viewed on an oscilloscope at 0.2 volts per division, the tape recorder output signal should be checked with an oscilloscope setting of 1.0 volt per division. Conversely, presume that a strong Keyphasor® pulse is recorded on tape at a gain of 0.1. The raw input signal might be viewed at 5 volts per division, and the recorder output signal would be correctly observed at a scope setting of 0.5 volts per division. In either case, the tape recorder gain or attenuation value must be equal to the ratio between the voltage setting on the two oscilloscope channels. Clearly, a visual comparison of dynamic data on the oscilloscope must always be adjusted for any gain or attenuation of the recorded signals.

The internal record and reproduce amplifiers are supplied in two configurations. The most useful type for machinery analysis is the frequency modulated (FM) amplifier. In this type of recording, the signal to be recorded modulates a high frequency carrier. This FM amplifier allows recording of low frequency (DC) voltages, and it provides a flat bandwidth of frequencies that are accurately recorded and reproduced. Specific performance of a typical FM system on a VHS tape recorder are summarized in Table 8-1.

Table 8–1 Typical FM Tape Recording Characteristics In A VHS Format

Tape Speed (Inches/Second)	Tape Speed (Cm/Second)	Bandwidth (Hertz)	Signal to Noise Ratio (dB)	Record Time (Minutes)
30.0	76.2	0 to 20,000	50	5.4
15.0	38.1	0 to 10,000	50	10.8
7.50	19.05	0 to 5,000	50	21.7
3.75	9.52	0 to 2,500	48	43.4
1.875	4.76	0 to 1,250	47	87.0
0.938	2.38	0 to 625	46	173.
0.469	1.19	0 to 313	42	347.

From Table 8-1, it is clear that tape speed and the recording bandwidth are directly related. That is, as tape speed changes by a factor of 2, the bandwidth responds in a similar fashion. For example, at a recording speed of 15 inches per second, an FM channel will have a frequency range extending from 0 (DC) to a top end of 10,000 Hz (600,000 CPM). Doubling the tape speed will increase the bandwidth to 20,000 Hz (1,200,000 CPM). Moving in the opposite direction, cutting the tape speed in half to 7.5 inches per second will reduce the recording frequency range in half to a value of 5,000 Hz (300,000 CPM).

It is noted from Table 8-1 that the signal to noise ratio is defined in terms of decibels (dB). This is just another way of expressing a voltage ratio, and the following equation may be used to convert the voltage ratio to decibels:

$$dB = 20 \times \log\left\{\frac{Voltage_1}{Voltage_2}\right\} \tag{8-3}$$

This expression is computed on a log to the base 10. If the dB value is known, then equation (8-3) may be reconfigured to calculate the voltage ratio as shown in the next expression (8-4).

$$\frac{Voltage_1}{Voltage_2} = 10^{\left(\frac{dB}{20}\right)} \tag{8-4}$$

From Table 8-1, the signal to noise ratio at the higher tape speeds is 50 dB. If the full scale, or reference $Voltage_1$ is set at 1.0 volt, then the tape channel can ideally resolve voltage levels $Voltage_2$ per equation (8-4) as follows:

$$Voltage_2 = \frac{Voltage_1}{10^{\left(\frac{dB}{20}\right)}} = \frac{1.00 \text{ Volt}}{10^{\left(\frac{50}{20}\right)}} = \frac{1.00 \text{ Volt}}{10^{2.5}} = \frac{1.00 \text{ Volt}}{316.2} = 0.0032 \text{ Volts}$$

Thus, the noise floor is nominally 3 millivolts. A properly calibrated amplifier should be able to accurately record and reproduce voltages above this level, and below full scale voltage. In actuality, a conservatively accurate tape recording should not approach the bottom, and it should not exceed the top of the voltage range. One should strive to adjust recording gains so that the minimum signal level is at least ten times the noise floor, and the maximum signal does not exceed the full scale setting. In this example, accurate data could be expected if the recorded signal was between 0.03 and 1.00 volts. If the dynamic signal falls below the noise floor, it will be lost due to the small amplitudes. On the other hand, if the signal significantly exceeds the amplifier full scale voltage setting, the signal will be lost due to overdriving of the amplifier. In either case, the required dynamic data will not be retrievable.

Clearly, tape recording amplifiers must be carefully adjusted to stay above the noise floor, and below the full scale voltage. This is generally easy to do on steady state vibration data where the signals are reasonably constant in amplitude. However, this can become a real challenge on transient data where the machine must pass through a critical speed, and the maximum vibration amplitudes are unknown. In this situation, the machinery diagnostician must provide an educated guess of the potentially largest vibration amplitudes that may be encountered. For instance, if the diametral bearing clearance is 15.0 Mils, it may be assumed that the peak amplitude through the critical speed region would be approximately 50% of the total bearing clearance. This would indicate that the highest shaft vibration amplitude might be in the vicinity of 7.5 Mils,$_{p-p}$. This would be equivalent to 1.5 Volts,$_{p-p}$ if probe sensitivity was 200 millivolts per Mil. Thus, if the voltage amplifier was set at ±1.0 Volt, (or 2.0 Volts,$_{p-p}$) there is good confidence that accurate data could be acquired as the rotor passed through the critical speed range.

Alternatively, the control room operators could be questioned as to the typical maximum vibration amplitudes that are observed during transient speed conditions. This information may be obtained by observations of the vibration monitors, or it may be documented within the DCS or TDM system. It should also be noted that startup versus coastdown vibration amplitudes may be quite different. Hence, the tape recorder amplifier gain settings that were fully acceptable during a startup condition may be totally inappropriate for coastdown data.

From a frequency response standpoint, it was previously stated that displacement and velocity signals are generally limited to frequencies below 1,500 Hz (90,000 CPM). This frequency range is well within the domain of FM amplifiers as shown on Table 8-1. However, if higher frequencies (e.g., accelerometers) must be recorded, Table 8-1 reveals the major weakness of an FM recording sys-

Table 8–2 Typical Direct Tape Recording Frequency Characteristics In A VHS Format

Tape Speed (Inches/Second)	Tape Speed (Cm/Second)	Bandwidth (Hertz)	Signal to Noise Ratio (dB)	Record Time (Minutes)
30.0	76.2	0.1 to 150,000	30	5.4
15.0	38.1	0.1 to 75,000	30	10.8
7.50	19.05	0.1 to 36,000	30	21.7
3.75	9.52	0.1 to 18,000	30	43.4
1.875	4.76	0.1 to 9,000	30	87.0
0.938	2.38	0.1 to 4,500	30	173.
0.469	1.19	0.1 to 2,250	30	347.

tem on VHS tape. That is, the time duration of the recording is severely limited by the high tape speeds, combined with the 246 meter length of the VHS tape cartridge. For instance, at full speed of 30 inches per second, the entire VHS tape will only last for 5.4 minutes.

If it is necessary to record high frequency data for extended periods of time, the FM record and reproduce electronics must be replaced with direct record and reproduce amplifiers. This type of recording system is virtually the same technology that has been used for audio tape recorders for many years. A set of typical characteristics for direct recording on VHS are tabulated in Table 8-2. From this summary, it is clear that the frequency bandwidth of the recording has been substantially increased, but the signal to noise ratio has suffered. At all speeds the direct signal to noise ratio is 30 dB. This is equivalent to a voltage ratio of 31.6. If the full scale voltage is set at 1.0 Volt, then the tape channel can ideally resolve voltage levels of:

$$Voltage_2 = \frac{Voltage_1}{10^{\left(\frac{dB}{20}\right)}} = \frac{1.00 \text{ Volt}}{10^{\left(\frac{30}{20}\right)}} = \frac{1.00 \text{ Volt}}{10^{1.5}} = \frac{1.00 \text{ Volt}}{31.62} = 0.032 \text{ Volts}$$

Thus, the noise floor is nominally 32 millivolts, and a properly calibrated amplifier should be able to record and reproduce voltages above this level. However, it must always be recognized that the direct electronics are 20 dB, or a factor of ten, noisier than FM amplifiers. In actual practice, a tape recorder may be configured with a combination of FM and direct record and reproduce amplifiers. In all cases, the FM cards are used for high resolution, accurate phase coherence between channels, and the inclusion of low frequency characteristics. The direct cards are used for high frequency transducers, and it is understood that voltage accuracy on direct channels is traded for an extended frequency range.

The characteristics presented in Tables 8-1 and 8-2 are typical for analog tape recorders operating with VHS tapes that are 246 meters (807 feet) long. Other reel-to-reel, and cassette recorders provide similar performance, but the total recording time must be adjusted for the actual length of the magnetic tape.

These conventional FM and Direct recorders use stationary record and reproduce heads that contain a defined spacing between tracks. This standardization of tape heads allows interchangeability of tapes between recorders. Various other types of recording configurations are also available. For instance, if FM extra wide band record and reproduce electronics are used, the frequency response ranges in Table 8-1 will be doubled. However, the dynamic range (signal to noise ratio) will be reduced by approximately 3 dB. Again, there are many possible tape recorder configurations, and the machinery diagnostician must be fully aware of the performance of any tape recorder used for serious data acquisition work.

Another approach to acquiring and storing dynamic data consists of using DAT technology. In these types of recorders, an analog to digital converter is used to digitize the data prior to recording. A rotating record head (typical speed of 2,000 RPM) uses a helical scan to record the digital data on small cassettes. During reproduction, the dynamic data is converted from digital to analog format, and directed to the output connectors. DAT recorders typically have a signal to noise ratio of greater than 70 dB. This is 20 dB, or 10 times greater than the previously discussed FM recorders using VHS tape. Tape speed on DAT units is constant at values such as 0.321 inches/second, which provides a two hour recording time on one 60 meter cassette.

Initially, it might appear that DAT recorders have significant technical advantages over traditional FM recording electronics. However, this conclusion is rapidly altered when the recording bandwidth characteristics of DAT units are considered. In this type of magnetic tape recorder, the recording bandwidth is dependent on the number of data channels. For instance, Table 8-3 describes the DAT frequency response characteristics for various channel quantities:

From this tabular summary it is clear that the frequency response characteristics are totally based upon the number of data channels. Hence, 2 channels may be recorded for the entire length of the tape (i.e., 2 hours), with a bandwidth of DC to 20,000 Hz. If 16 channels of data are required, the bandwidth drops to

Table 8–3 Typical DAT Tape Recorder Frequency Response Versus Number Of Channels

Number of Channels	Bandwidth DC to 20 kHz	Bandwidth DC to 10 kHz	Bandwidth DC to 5 kHz	Bandwidth DC to 2.5 kHz
2	2			
4		4		
6		2	4	
8			8	
10		2		8
12			4	8
16				16

DC to 2,500 Hz (still 2 hours recording). If this is an acceptable frequency span, then the DAT recorder is a good choice. However, if more recording controls are mandatory over the full array of available channels, then the diagnostician should give serious consideration to traditional fixed head recorders.

Regardless of the type or configuration of the tape recorder, calibration and operation of the unit must be documented at some point during the data acquisition process (usually at the beginning). This is normally achieved by recording one or more calibration signals on the tape, and logging the specific amplitude and frequency of the applied sine wave. The calibration signal may be internal to the recorder, or it may be obtained from an external function generator. Independent of the source of the calibration signals, the tape output must always match the input calibration signals.

Typically, the frequency of the calibration signal will be set to some reasonable value such as the machine running speed. In other cases, a predetermined frequency at some constant value may be used. In other situations, it may be necessary to cover a wide frequency range with FM and direct amplifiers, and multiple calibration signals should be recorded. For example, consider a machine that operates at 12,000 RPM (200 Hz), and assume that it is driven by a gear box that has a mesh frequency of 540,000 CPM (9,000 Hz). If it is important to record both frequencies, then FM amplifiers would be used for the shaft displacement signals, and direct amplifiers would accommodate the high frequency acceleration signals. A calibration signal at a frequency of 200 Hz would be used for the FM channels, and a 9,000 Hz cal signal would be recorded on the direct channels. In this manner, calibration signals are recorded at the frequencies of interest, and a high level of confidence may be placed in the final data.

In all cases, accurate records of the calibration signals must be maintained. In addition, descriptive information of the various data runs must also be generated. In order to provide the proper level of documentation for accurate processing of the data, it is necessary to consider the other instruments included in the field data acquisition system shown in Fig. 8-2. These additional devices are used to examine the dynamic signals, and to provide reference information for proper reproduction of the tape recording.

For instance, synchronous 1X amplitude and phase information may be obtained with a Digital Vector Filter (either DVF or single channel TK83). This information should be logged for each data channel on each tape run. When viewed in conjunction with the overall, unfiltered amplitudes, and the operating speed, it is possible to perform an initial evaluation of the machinery based upon the tape log. However, the additional importance of this information lies in the ability to support the accuracy of the taped data during post processing. For example, if the hand logged information indicates a 1X vector of 1.6 Mils$_{p-p}$ at 146°, and the data extracted from the tape exhibits 1.57 Mils$_{p-p}$ at 145°, then the values are equivalent. In this case, the diagnostician will have a high level of confidence in the tape recorded data.

The proximity probe gap voltages should always be measured and documented for analysis of lateral and axial positions. This measurement can be easily obtained with a battery operated Digital Multimeter (DMM). It goes without saying that this measurement must be made on the raw or direct transducer signal before AC coupling. The probe gap voltages should initially be acquired at zero speed with the lube oil turned off, and with the lube oil warm and circulating. This establishes the initial starting point for the radial and axial position measurements. Gap voltages should also be obtained at each definitive constant speed point such as slow roll, minimum governor, full speed, and full process load. A DMM is also highly useful for routine items such as checking cable continuity, verifying power supply voltages, and gaping proximity probes.

Observation of frequency domain information during data recording can be quite important, particularly on complex signals. In these situations it is important to use a dual channel DSA (such as the HP3560A) during data acquisition. Modern units are quite compact, and a full set of features are available in a DSA that weighs less than ten pounds. Many field DSAs are also equipped with digital storage. With this capability, they may be used for performing routine surveys, or making multiple structural or piping measurements in an effort to define or locate a particular frequency node or anti-node. A dual channel DSA is also quite handy for performing simple resonance tests with an accelerometer and an impact hammer equipped with a force transducer.

Finally, tape logs must be established and maintained throughout the field tests. A sample log for a 14 channel recorder is shown in Fig. 8-3. This log identifies the machinery, and the magnetic tape number. It also provides necessary calibration information for recording an internal calibration signal. Each data run is specifically identified, and the date, start and stop times, plus tape recorder speed and machine rotation speed are listed. Each of the 14 channels is adequately defined with the pickup location (e.g., *1V, 2H*, etc.), the transducer type (e.g., *D, V,* or *A*), and the overall recording signal gain. Probe gap voltages, plus

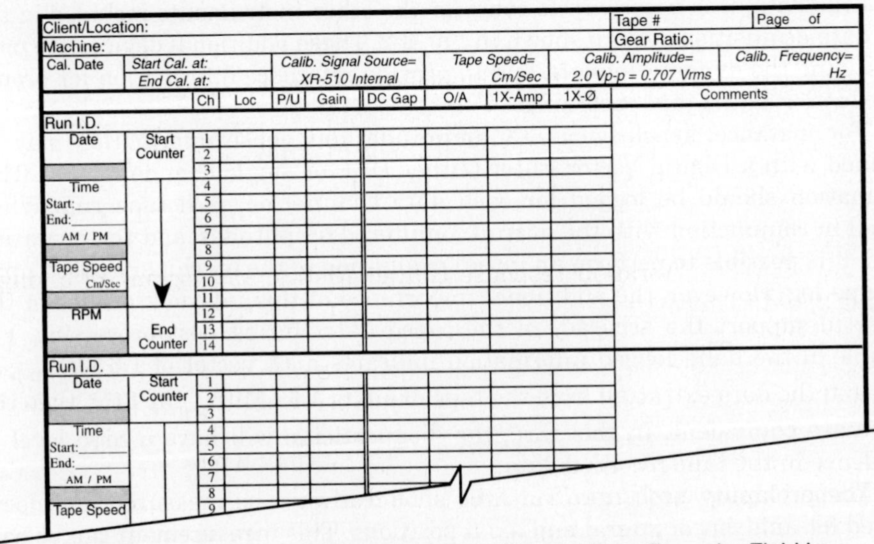

Fig. 8–3 Portion Of A Typical Fourteen Channel Tape Recorder Field Log

overall vibration levels, and synchronous vectors are also tabulated. This type of tape recorder log is very useful when different transducers with different configurations are recorded on consecutive runs.

However, there are many situations when the instrumentation is set up with only one group of transducers, and the system configuration remains constant throughout the entire test. For example, a field balance on a steam turbine might include four radial proximity probes, two casing transducers, two thrust probes, and a Keyphasor®. If the slow roll and full speed vector information was hand logged on the tape recorder logs (like Fig. 8-3), the necessary data would be available, but it would be somewhat difficult to compare. In these situations, a consecutive field vibration log, as shown in Fig. 8-4, might be more suitable.

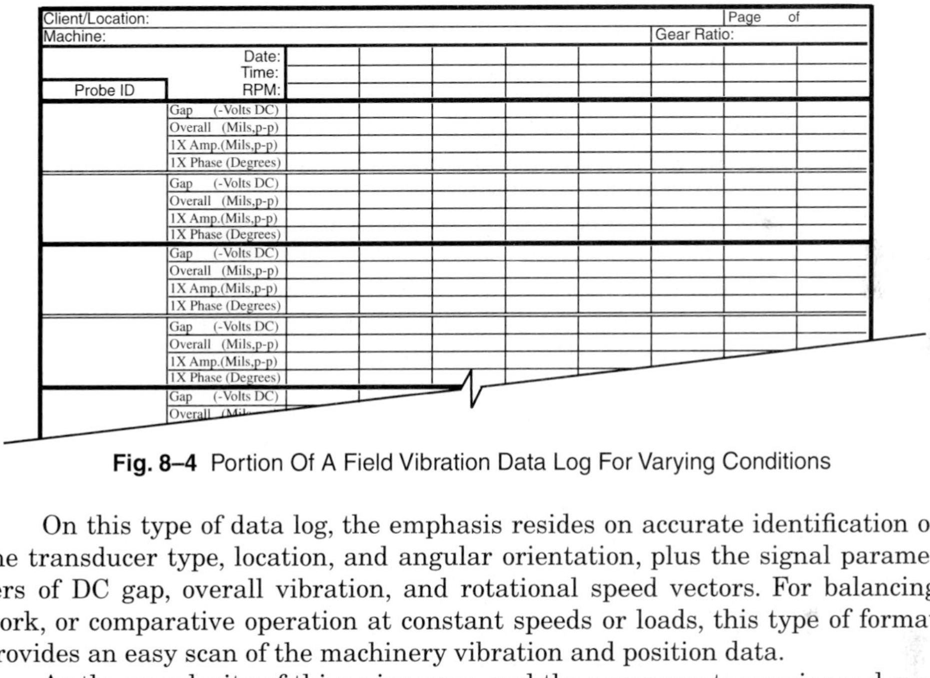

Fig. 8–4 Portion Of A Field Vibration Data Log For Varying Conditions

On this type of data log, the emphasis resides on accurate identification of the transducer type, location, and angular orientation, plus the signal parameters of DC gap, overall vibration, and rotational speed vectors. For balancing work, or comparative operation at constant speeds or loads, this type of format provides an easy scan of the machinery vibration and position data.

As the complexity of things increase, and the pressures to acquire and process a lot of data in a limited amount of time — the time required to fill in numerous hand logs may not be available. In addition, complicated machines often have complex vibration response characteristics, and logging of a single running speed vector may not be adequate to define the behavior. In these cases, computer-based systems may be employed in conjunction with the data recording system previously discussed. For example, Fig. 8-5 describes the addition of a pair of Bently Nevada, 208 Data Acquisition Interface Units (DAIU). These 8 channel units (16 channels total) are wired in parallel with the tape recorder system. The vibration data is acquired and digitized in the DAIU's, and downloaded to a Compaq laptop computer. The data is processed and manipulated into various formats within the computer, and hard copy results are available by parallel

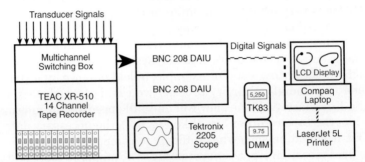

Fig. 8–5 Typical Multichannel Analog Plus Digital Data Acquisition & Processing System

port connection to the HP LaserJet 5L printer.

A variety of flexible options are possible with the data acquisition system described in Fig. 8-5. For instance, if the primary project objective is field balancing, then the main emphasis would be placed on acquisition and processing of the hard copy data. In this instance, the tape recorder system would provide a full analog backup. However, if the fundamental project requirements demanded the acquisition of high resolution data for additional post processing, the tape recording might be assigned as top priority, and the field data processing system used for periodic examination of certain portions of the acquired database. Certainly other combinations are quite probable, and the diagnostician should consider this type of instrument array as complimentary. In all cases, each piece of this instrumentation package should be used to its full capabilities to properly document the behavior of the machinery train.

As a final note, it should be mentioned that utilization of a PC-based data acquisition system can result in the development of numerous data files. These files may become quite large, and it is not unusual to occupy 10 to 12 megabytes of storage with one sample run. Certainly this type of digital data may be stored on the computer hard disk. It is highly desirable to provide a backup set of field data on a separate mass storage media. The use of another hard drive, magneto optical disks, or magnetic media of 100 megabytes or more is a good approach. Procedurally, it is recommended that at the conclusion of each data run, that the diagnostician store the data on the PC hard disk, and then make a backup copy on the external mass storage device. In order to keep track of the various data sets, a computer storage log sheet as described in Fig. 8-6 is recommended.

In addition to the data description, plus the date and time of the acquired field data, a specific file name is assigned to this set of digital information. For obvious reasons, the same file name should be used for the hard disk as well as the backup data file. It is also desirable to include the file size as defined down to the last byte. This gives an indication of the amount of data, and it provides a good checkpoint to verify that the backup is identical in size to the original version stored on hard disk.

The first column in Fig. 8-6 is specified as *Tab Section*. The purpose for this heading is to allow for organization of the hard copy field data. In many cases,

Tab Section	Data Description	Date	Time	File Name	File Size (Bytes)

Fig. 8–6 Portion Of A Computer Data Storage Log For Multiple Sample Runs

the quantity of data plots becomes unmanageable, and a methodical organiza-tion of the data is necessary. In these common situations, one successful approach is to equip a 3-ring binder with a set of divider tabs. As each consecu-tive data set is sampled and processed to hard copy format, the plots are placed in tab sections of the binder, and tab section identification listed in the computer data storage log sheet. This may seem like a minor item, but when cross refer-encing or checking data in the middle of the night, it can be enormously helpful.

DATA PROCESSING INSTRUMENTATION

Following field recording of vibration and other dynamic signals onto mag-netic tape, or digital storage, the next step is to extract the meaningful informa-tion from the data. Historically, most of the devices used for data processing were analog instruments that produced hard copy output with pen plotters, and pho-tographs of instrument CRTs. Today, most instruments are digital devices that supersede their predecessors by exhibiting higher resolution, expanded capabili-ties, and improved interfaces between devices. The current data processing sys-tems are often computer controlled, and they generally employ a combination of instruments from various vendors.

A typical data processing system is shown in Fig. 8-7. This sketch is descriptive of the system that was used to produce most of the data plots con-tained within this text. It is clear that a variety of analog and digital devices have been integrated into this high resolution data processing system. Many of the instruments within this system are both *self-checking* and *self-calibrating*. However, manual system checks are periodically performed using a function gen-erator (HP-33120A) in conjunction with a full 5 $\frac{1}{2}$ digit digital multimeter (HP-3468A), and various frequency measuring devices (e.g., HP-35670A or DVF).

Analog system input is obtained by directing the TEAC XR-510 tape recorder output signals through a switching panel into a rack of voltage amplifi-ers. These step gain amplifiers are used to maintain a suitable and consistent voltage input range for the various instruments. Following voltage amplification

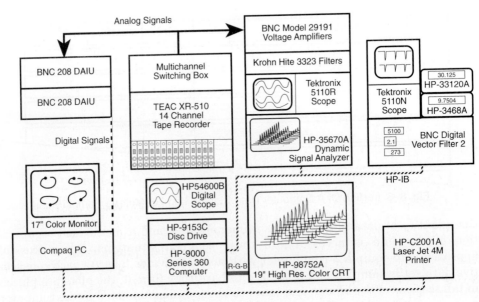

Fig. 8–7 Typical Lab Based Multichannel Data Processing System

or attenuation, the dynamic signals are processed through low-pass and/or high-pass Krohn-Hite filters to allow early removal of any extraneous high or low frequency components. Although most data does not require this type of filtering, it is mandatory in certain data processing situations.

Following the Krohn-Hite filter, the dynamic signals are paralleled into the Digital Vector Filter (DVF), the Dynamic Signal Analyzer (HP-35670A), and the four channel Tektronix (5110R) oscilloscope. Typically, two scope channels are used for the unfiltered X-Y signals, and the other two channels are dedicated to X-Y signals filtered at rotational speed by the DVF. With this arrangement, filtered and unfiltered orbital and time domain signals can be observed on the four channel oscilloscope. This analog data is also compared with the final computer processed plots to insure accurate and consistent results.

A separate two channel Tektronix (5110N) scope is used to monitor the Keyphasor® signal, plus the DVF trigger pulse. Most of the data processing functions are dependent upon a suitable speed trigger. Hence, it is mandatory to continually observe the Keyphasor® pulse, and perform any necessary amplifier adjustments to maintain a clean trigger to the other instruments. In some cases, the machine Keyphasor® signal may be of poor quality due to any number of transducer difficulties. The *cleanup* of an inferior Keyphasor® signal pulse may normally be accomplished by using two stages of voltage amplifiers.

As demonstrated in Plot A of Fig. 8-8, the direct Keyphasor® signal exhibits a substantial amount of surface (baseline) variation, and a trigger pulse that has a height of only 1.0 volt. This Keyphasor® signal shape will not trigger the downstream instruments due to the small pulse height, combined with the potential for false triggers from the erratic upper portion of the curve. Fortunately, the sig-

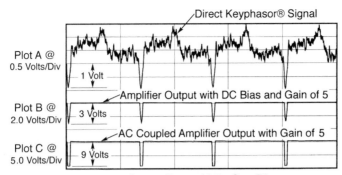

Sweep Rate = 12.5 mSec./Div.

Fig. 8–8 Manipulation Of Machine Keyphasor® Signal With Voltage Amplifiers

nal may be salvaged by applying two voltage amplifiers in series.

The first amplifier is DC coupled, and a DC offset plus a voltage gain are applied to the direct Keyphasor® signal. Specifically, the results of this initial manipulation are shown in Plot B of Fig. 8-8. The DC bias voltage flattened out the erratic upper portion of the initial signal, and the voltage amplification of 5 increased the usable pulse height to approximately 3 volts. Although signal clarity has been significantly improved, the pulse in Plot B is still of minimal height, and the pulse magnitude varies slightly with time. The final *cleanup* of this signal is provided by the second voltage amplifier in the series circuit. This device is AC coupled, and the pulse signal is amplified by 5 to yield the time domain curve shown in Plot C of Fig. 8-8. Note that the final pulse height of 9 volts is combined with a truncated bottom portion of the pulse. This squared off pulse bottom is due to overdriving the amplifier, and effectively *bounding* the trigger pulse.

Returning back to the data processing instrumentation in Fig. 8-7, a separate HP 54600B digital oscilloscope is used for observing data where a rapid sweep rate, or the capabilities of a storage oscilloscope are required. This type of digital scope also provides accurate and rapid on-screen measurements of signal characteristics with the moveable cursors and digital parameter displays.

The HP-35670A Dynamic Signal Analyzer is a four channel FFT spectrum/ network analyzer with a frequency range that extends from nearly DC to slightly over 100 KHz. This unit has a minimum dynamic range that exceeds 80 dB, which allows total extraction of any data that originates from a tape recorder. Although the DSA is primarily a frequency domain analyzer, it is also used to make amplitude and time domain measurements. In fact this instrument performs most of the analog to digital data conversions. The DSA is used for processing Fast Fourier Transform (FFT) information, frequency response functions (FRFs), and time history data. This unit has an adjustable resolution that varies from 100 to 1,600 spectral lines per channel. It incorporates various averagers to obtain statistical accuracy, plus several windows, and a variable zoom to allow precise frequency identification. The DSA is used for sampling both steady state and transient data. Upon confirmation of proper information on the internal

CRT, the data is transferred to the HP 9000 series computer system for final processing and printing.

Most of the plots contained in this text were generated by interfacing the DSA and the DVF with a dedicated HP computer system. As noted in Fig. 8-7, the hardware includes an HP-9000 series workstation computer equipped with a 19" high resolution color CRT, a combination hard and floppy disc drive, plus a LaserJet 4M printer. A dedicated IEEE-488 interface maximizes data transfer rates between the computer and the instruments. A separate HP-IB interface was used for rapid communications between the computer and the hard disk. Finally, a third IEEE-488 interface is employed to transfer data between the computer and laser printer. This system operates under HP Basic, with improved speed provided by a binary compiler. The data processing software evolved from Bently Nevada ADRE® software. The original ADRE® programs have been subjected to numerous revisions by the senior author to provide higher resolution, increased speed, and improved hard copy output. In the overview, the DSA digitized and processed data, and the DVF provided synchronous filtration plus phase and amplitude data. The computer performed sampling, data formatting, and the production of the final hard copy laser plots.

ADRE® for Windows software by Bently Nevada is used during the direct digital processing path shown on the left side of Fig. 8-7. In this data path, the analog signals are connected from the tape recorder output to the Bently Nevada 208 Data Acquisition Interface Units (DAIUs). As previously mentioned, two of these units can simultaneously process 16 data channels of dynamic data. The digital data is passed to a Compaq PC for manipulation, final formatting, printing and storage. This system operates the same in the shop as it does in the field.

In reality, the all digital processing system shares many common qualities with the generally analog system. As time progresses, it is reasonable to assume that more channels and more capabilities will be included in the interface box between the transducer signals and the computer. It is anticipated that PC-based systems will continue to expand in software sophistication and the ability to examine dynamic signals in greater detail with a variety of formats. For instance, the use of Computed Order Tracking® on Hewlett Packard DSAs eliminates the need for analog ratio synthesizers and digital order tracking filters combined with their associated phase accuracy and dynamic range limitations. Most PC-based systems have built-in math and analysis capabilities that allow slow roll removal from orbits, Bode, and polar plots. They also contain curve filling software to allow characterization of frequency response functions (FRFs). This is quite useful in modal analysis, plus the development of polynomial equations to define bearing housing stiffness, as discussed in chapter 4 of this text.

One of the most significant advantages of an all digital sampling system is that the information may only need to be processed one time. This is particularly important on machines such as large turbine generators that require one or two hours to roll from turning gear up to minimum governor. In an all digital system, the PC samples the entire event for all data channels simultaneously. Thus, at the end of the startup, the machinery diagnostician is ready to start examining the sixteen channels of transient data, and committing the necessary information to hard copy format. If the T/G startup required two hours, and the data review and printing required another two hours, the complete startup documentation package would be finished in four hours. However, if a fourteen channel analog tape recorder is employed, and one channel is devoted to the Keyphasor®, the diagnostician is faced with a two hour startup, fourteen hours of dual channel playback, plus another two hours of data examination and printing. In this case, the all digital system is the clear winner with four hours of total data acquisition and processing time compared against eighteen hours for the conventional dual channel processing.

This substantial time savings may lead to the conclusion that analog recording systems are obsolete, and that all future data will be acquired in an all digital format. In actuality, this trend is not completely true. The magnetic tape recorder still provides a true recording of the total dynamic signal, whereas the digital system acquires samples at predetermined incremental speeds or times. If an event occurred between digital samples, the digital system would miss it, and the analog system would capture the event. This ability to do a *cycle by cycle* analysis of the analog signals will not be feasible with all digital systems in the immediate or foreseeable future.

Reliability is also an issue when dealing with the acquisition of unique or non-repeatable data sets such as a machinery startup. In most operating plants the startup or shutdown of a major machinery train is a significant event. The machinery diagnostician has only one chance to acquire this type of necessary transient data. If that opportunity is lost, it probably cannot be recovered. Specifically, if a tape recorder fails during a startup sequence, the section of data recorded before the tape failure will normally be preserved. However, on an all digital system, any type of computer, digital interface, or hard disk malfunction will probably invalidate the entire data file. This is an embarrassing condition for all parties involved, and the desirability of a full backup based upon an analog tape recording of the transient events is self-evident.

It should also be mentioned that backup on magnetic media is only a short-term solution. Magnetic tapes and floppy disks are all subject to damage, and deterioration. These types of magnetic media should only be considered as temporary or interim storage devices. For long-term storage of 20 or more years, the diagnostician should invest in Magneto-Optical or CD-R storage devices.

DATA PRESENTATION FORMATS

Machinery vibration characteristics processed by the previously discussed system are presented on several distinct types of plots. The first format consists of vibration signals in the time domain as shown in Fig. 8-9. This is normally combined with a shaft orbit where mutually perpendicular radial probes are installed, and phase referenced with a Keyphasor® mark.

During field data acquisition, this same information is observed on an oscilloscope. The final hard copy data consists of computer-generated orbits and time base plots that accurately represent the analog signals. Since the radial trans-

VERT. Unfiltered = 0.98 Mils,$_{p-p}$ @ -10.24 Volts DC
HORIZ. Unfiltered = 1.59 Mils,$_{p-p}$ @ -9.72 Volts DC

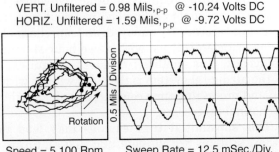

Fig. 8–9 Unfiltered Orbit
And Time Base Plots

Speed = 5,100 Rpm   Sweep Rate = 12.5 mSec./Div.

ducers are often not installed at true vertical and horizontal positions, the orbits must be corrected for this angular deviation. The orbits in Fig. 8-9 are rotated to allow true vertical and horizontal representation. It should also be noted that the computer processed time base plots have not been corrected for transducer orientation. Thus, the time domain plots are representative of the signals as viewed on an oscilloscope.

In order to clearly observe the synchronous 1X vibration of the machinery, the data is also filtered precisely at rotational speed, and another set of orbits and time domain plots are produced. Fig. 8-10 is identical to the previous display, with the inclusion of the 1X running speed filter. On some data sets the influence of shaft scratches and other frequency components substantially reduces the analytical usefulness of the orbital data. In these cases, 1X filtered data is almost mandatory to evaluate the machinery behavior.

VERT. 1X Vector = 0.66 Mils,$_{p-p}$ @ -172°
HORIZ. 1X Vector = 1.27 Mils,$_{p-p}$ @ -23°

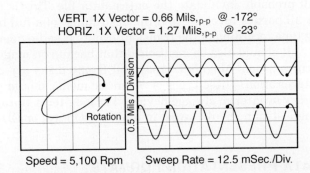

Fig. 8–10 Filtered Orbit
And Time Base Plots

Speed = 5,100 Rpm   Sweep Rate = 12.5 mSec./Div.

Another type of common steady state data consists of a frequency analysis of the vibration signal. This is achieved by the DSA that uses Fourier analysis techniques to separate a time domain signal into discrete frequency components. Information of this type is presented on an X-Y plot where frequency (Cycles/Minute) is plotted along the horizontal axis; with linear vibration amplitude on the vertical axis. The following Fig. 8-11 displays a spectrum plot of a shaft displacement signal. Certainly any dynamic transducer signal such as displacement, casing velocity or acceleration, pressure pulsation, force, strain, torsional

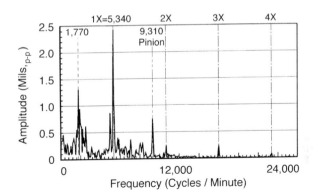

Fig. 8–11 Steady State Spectrum Plot Of Shaft Vibration

displacement or velocity may be processed in this manner.

Statistical accuracy in an FFT plot is obtained by employing RMS averaging. This averaging results in a smoothing of random noise variations, but does not reduce the noise level. Hence, RMS averaging is representative of the actual frequency distribution within each signal. The spectrum data is normally processed with a Hann window filter. This passband is similar to those found in swept frequency spectrum analyzers, and combines a good compromise between amplitude accuracy and frequency resolution across the analysis bandwidth.

As shown in Fig. 8-11, rotational speed (1X) motion generally appears as the highest spectral peak. However, in certain cases such as severe oil whirl, other components may dominate. In order to quantify the frequency components, machine running speed is obtained from the DVF, or from a DSA with a tach input. Once this speed is established, the frequency of the other components are divided by speed to determine their respective harmonic order. Often, an even order will coincide with an internal geometrical configuration, allowing the spectral component to be linked to a machine element. For example, vane passing frequency is the product of the number of impeller vanes times running speed.

In other cases, a particular frequency component is significant towards quantifying the machinery behavior. An example of this behavior would be an excitation from an adjacent unit. Typically, the frequency components are identified on each plot for future documentation and reference. For synchronous components the harmonic order is noted; and for asynchronous peaks, the measured frequency is indicated in Cycles per Minute at the top of each plot. A special designation is generally included when data from a gear box is processed. For example, Fig. 8-11 was obtained from a bull gear, and the associated *Pinion* identifier, plus the respective frequency of 9,310 RPM, are shown on the plot. For high speed pinion signals, the bull gear speed would be similarly displayed.

Another type of steady state data consists of a frequency response function (FRF) between signals. The example presented in Fig. 8-12 describes a FRF between proximity probes mounted at opposite ends of a turbine. This hard copy data consists of three interrelated plots including the Amplitude Ratio (dimensionless Mils/Mils), relative Phase between signals (Degrees), and Coherence (dimensionless) between signals. This type of information is used to examine the amplitude and timing characteristics at specific frequencies. Data validity is verified or ignored based upon the coherence function. In general, coherence must be sufficiently high (e.g., greater than 0.95) to allow confidence in the amplitude and phase data. In Fig. 8-12, only four frequencies display acceptable coherence, and these four frequencies represent the only acceptable data in the entire plot. The machine running speed on Fig. 8-12 is 5,340 RPM, and the amplitude shows about a 2.2:1 ratio between inlet and exhaust ends of this turbine. The phase difference at this frequency is 188° that might seem to be out of phase shaft vibration. However, the radial probes are physically mounted 180° apart, as shown in Fig. 4-7. Hence, the FRF phase value of 188° is really indicative of an 8° phase differential, or an in-phase motion across the gas turbine.

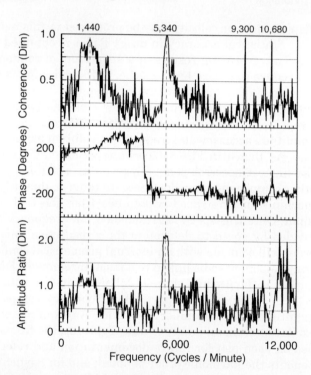

Fig. 8–12 Steady State Frequency Response Function (FRF) Between Proximity Probes At Opposite Ends Of A Gas Turbine

The behavior at twice rotative speed 10,680 CPM, plus the coupled pinion speed at 9,300 RPM may also be examined for appropriate amplitude ratios and differential phase angles. FRF plots are normally performed with RMS averaging, and a Flat Top window to minimize amplitude errors. This type of data presentation is often the only way to accurately relate amplitude and phase characteristics at frequencies other than rotative speed.

The last type of steady state data to be considered is the change in radial rotor position with respect to the stationary bearing. For example, Fig. 8-13 depicts the lateral clearance of a plain circular bearing with two axial oil inlet grooves. In this diagram, the shaft orbit is noted in a normal running position for a counterclockwise rotating machine, and the X-Y proximity probes are also shown in this sketch. Since these probes produce a DC signal proportional to the average gap between the probe tip and the observed surface (shaft), this measurement may be used to locate the journal running position within the bearing.

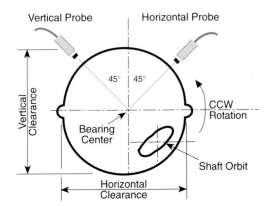

Fig. 8–13 Shaft Centerline Position Shift In A Circular Bearing With Counterclockwise Shaft Rotation

In practice, DC gap voltages from each pair of perpendicular X-Y probes are measured and logged throughout the field tests. This DC voltage data is then converted to a differential basis of *at speed* minus the *zero speed* voltage. The change in shaft centerline with respect to each probe is obtained by division of this differential voltage by the probe sensitivity (e.g., 200 mv/Mil). Finally, a vector summation of the changes with respect to each X-Y probe yields the overall change in shaft radial position. This resultant vector describes the magnitude and direction of the centerline change from the initial starting, or rest point.

Another, and more descriptive way to display this type of radial position information is presented in the Fig. 8-14. This plot exhibits a continuous sample of DC gap voltages that describe the shaft centerline position during a startup

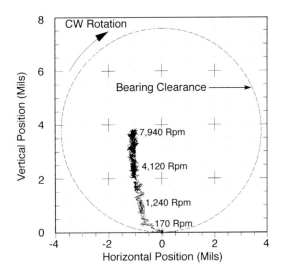

Fig. 8–14 Measured Shaft Centerline Change During Startup Of An Expander With Clockwise Rotation

condition. Again, this data is based upon the changes in static DC proximity probe gap voltages. The starting point is at the bottom of the bearing (zero speed), and changes are plotted with respect to that initial starting point. Information of this gender is used to observe variations in shaft centerline position as a function of speed, time, temperature, or any other appropriate variable. In the specific case of Fig. 8-14, the shaft exhibited a smooth transition from slow roll speed of 170 RPM to full operating speed of 7,940 RPM. Since this expander was equipped with tilt pad bearings, the documented change in journal position was indicative of normal and expected behavior for this clockwise rotating journal.

The previously discussed categories of data are obtained under constant speed, and steady state operating conditions. The next major group of data formats considers the presentation of transient information. Typically, this includes observation of vibratory changes as a function of speed and/or time. This type of information is presented as speed change plots, synchronous rotational speed vector plots, plus frequency spectra, and order tracking from a DSA. In most instances, all four types of data are necessary to provide proper and complete visibility of machinery transient behavior.

The variation of machine rotative speed with respect to time is an often ignored piece of information. As shown in Fig. 8-15, this type of data is a linear plot of shaft rotational speed in RPM on the horizontal axis, versus elapsed time in seconds or minutes on the vertical axis. The data presented in Fig. 8-15 documents the startup of a gas turbine from slow roll to crank speed, plus the final acceleration up to minimum governor at 4,600 RPM. Time history data of this

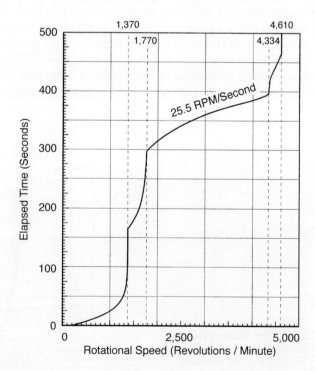

Fig. 8–15 Speed Ramp
Of A Gas Turbine During A
Routine Cold Startup

category allows identification of plateau periods where speed remained relatively constant, plus sections where rapid speed changes occurred. Precise definition of both of these conditions are extremely helpful in understanding the machinery transient characteristics. A comparison of startup versus coastdown behavior will often help explain the differences between the machine vibration differences of a controlled startup versus a coastdown.

Note that these types of data plots also compute the rate of speed change at various portions of the curve (e.g., 25.5 RPM/Second). Thus, it is possible to compare defined speed change rates under different transient conditions or situations. This is also convenient for determination of acceptable speed rate changes, and the settings of proper speed ramps on electronic governors.

Transient vibration data consisting of synchronous rotational speed 1X vectors are initially accommodated by the Bode plot where rotational speed vibration amplitude (1X) and phase angle are plotted as a function of machine speed. The example in Fig. 8-16 depicts a typical data set for a shaft displacement probe. Again, other transducers may be used for examining vibration changes as a function of machine speed. This type of data is appropriate for both startup (speed increasing), and coastdown (speed decreasing) conditions. It is meaningful to note that startup versus coastdown behavior will be different due to the variations of stiffness and damping occurring in the machine bearings. Hence, it is desirable to document and review both the speed increasing as well as the speed decreasing data. A variation of this format considers amplitude and phase versus time. This type of information is useful for evaluating large machines that

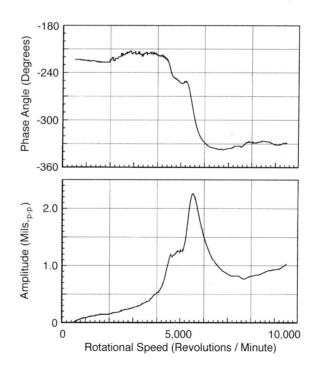

Fig. 8–16 Bode Plot Of A Centrifugal Compressor Startup

require a long time to achieve a full heat soak (e.g., large turbine generator sets).

The Bode is often combined with a polar plot describing the locus of rotational speed vectors during speed changes. Although both of these plots provide the same data array; the Bode provides excellent visibility of changes with respect to speed, and the polar yields improved resolution of phase variations. The diagram presented in Fig. 8-17 is representative of a polar plot for a single

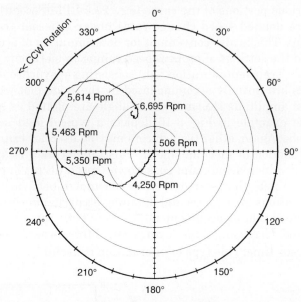

Fig. 8–17 Polar Plot Of A
Centrifugal Compressor
Startup

Plot Radius = 2.5 Mils,$_{p-p}$

radial probe. Note that the zero degree point is always located at the transducer angular position. Thus, a comparison of data from a pair of perpendicular probes mounted at a journal will result in a 90° difference in the zero point on the plot. If a machine has symmetrical bearing stiffness, vertical and horizontal behavior will be identical, and the polar plots from a pair of probes will track together. However, most machines exhibit deviations in stiffness and damping, and this is reflected and visible in the associated polar plots. This type of data is essential for identifying rotor critical speeds, and the influence of secondary system resonances. In most cases, the synchronous 1X vectors are plotted with amplitudes in Mils,$_{p-p}$, and phase angles expressed as Degrees of phase lag. If other transducer signals are plotted in a polar format, the engineering units assigned to the vector magnitude would be appropriately adjusted.

Under machine conditions where significant sub or supersynchronous vibration components are generated, it is necessary to generate a cascade or waterfall plot of individual spectra at incremental operating speeds or times. This type of data presentation provides an excellent overview of the frequency content of the vibration signals as a function of operating speed or time. Fig. 8-18

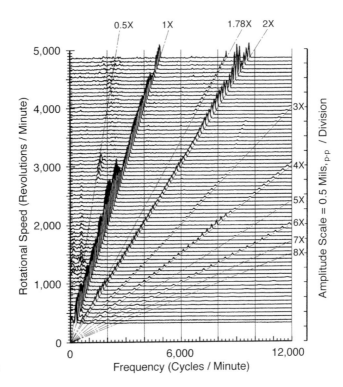

Fig. 8–18 Cascade Plot
Of A Gas Turbine Startup

depicts a typical cascade plot with FFT samples acquired at 60 RPM increments. The harmonic orders for 1X, 2X, 3X, etc. are shown to identify the even order frequency components. In addition, other order lines at 0.5X, and 1.78X are also exhibited. In the case of this machine, the 1.78X represents the vibration frequency of the pinion in the coupled gear box. In all cases, the correct amplitude engineering units are displayed on the right hand axis. Thus, the magnitude of any spectral peak may be compared, and scaled directly with this legend. As noted with the Bode plot, this type of data is generally sampled and processed during both startups and coastdowns. It is also appropriate to process this type of data as a function of time to observe changes that occur with parameters such as load or casing temperature. These types of stacked spectrum plots acquired over time are commonly referred to as waterfall plots, whereas the speed variable plots are generally called cascade plots.

Although cascade or waterfall plots provide excellent visibility of the frequency components, they can miss some information due to the time span between samples. For instance, in Fig. 8-18, the FFT samples were acquired at 60 RPM increments. Although the final cascade plot appeared to be correct, there could be additional unobserved or undocumented activity occurring between sampled blocks of data. To address any potential lapse in information, it is desirable to increase the sampling rate for specific frequency components. In some instances it is also meaningful to observe spectral components as a function of

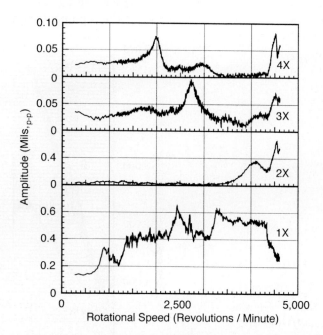

Fig. 8–19 Order Tracking
Of A Gas Turbine Startup

speed or time. To address both of these potential situations, the utilization of
Computed Order Tracking®[2] is highly desirable. An example of this type of information is exhibited in Fig. 8-19, where the first four harmonics of a gas turbine
radial vibration signal are plotted against speed during a normal cold startup.
This type of data allows improved visibility of individual frequency component
behavior over and above the previously discussed cascade or waterfall plot.
Although the same DSA is used for both tasks, the order tracking capability of
the DSA firmware allows substantially more samples for each component, and
thereby provides increased resolution.

Another potentially difficult data processing scenario occurs when the
major excitations are not synchronous. In other words, the largest vibration
amplitudes occur at frequencies that are not integer multiples of running speed.
This type of excitation might be completely missed by an order tracking analysis
of the data. The only positive method to insure that the maximum amplitudes
have been properly identified and documented is to examine the time history
record of each signal. An initial check can be performed by reproducing the taped
data into an oscilloscope. If any high amplitude excursions are detected, then an
additional signal processing technique of transient capture may be employed.

For instance, Fig. 8-20 depicts a transient capture plot of a hydro generator
during load rejection. This data was produced by reproducing the taped data into
the DSA, and configuring the DSA as a high speed digital recorder. Operating in

---

[2] Ron Potter and Mike Gribler, "Computed Order Tracking Obsoletes Older Methods," *SAE
Technical Paper Series — Proceedings of the 1989 Noise and Vibration Conference*, Traverse City,
Michigan (May 1989), pp 63-67.

this mode, the analog data is digitized and stored into the memory of the DSA. The data is not processed during the transient capture, it is just digitized and stored in RAM. In essence, this information replicates a high speed time base recorder, and an initial plot of vibration amplitude versus time is produced. A review of the capture buffer display reveals the areas or events of interest that may be expanded into various formats and examined in greater detail.

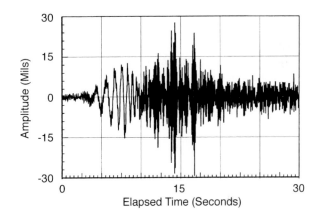

Fig. 8–20 Transient Capture Of A Hydro Turbine Driven Generator During Load Rejection

Once this transient capture data resides within the memory of the DSA, the information may be further dissected, and appropriate post processing applied to the digital data. Depending on the specific data array, the application of cascade, waterfall, or Computed Order Tracking® plots may be necessary. Another option would be to perform a time domain *cycle by cycle* analysis of two or more transient signals to determine relative phase or timing relationships. In any case, the objective is to fully document the recorded dynamic signals, and allow the execution of a meaningful and accurate analysis of the data.

In the final overview, it is quite clear that no single data processing technique is appropriate to all machine configurations, and to all types of mechanical malfunctions. In fact, there are substantial differences in the applicability of the various data processing techniques with different vibration sensors. These differences between shaft displacement proximity probes, casing velocity coils, and casing accelerometers are summarized in Table 8-4 for steady state operation. Clearly, proximity probes are suitable for time, orbital, and frequency domain analysis. Obviously, any of the three transducer outputs may be examined with

Table 8–4 Applicability Of Various Steady State Data Formats To Different Transducers

Steady State Data Type	Shaft Displacement	Casing Velocity	Casing Acceleration
Time Domain	Yes	Sometimes	Sometimes
Orbital Domain	Yes	Sometimes	Seldom
Frequency Domain & FRF	Yes	Yes	Yes

frequency analysis. If the signals are not overly complex, casing velocity may be used for time or orbital domain observation. However, casing acceleration signals are seldom applied and used for orbital analysis. On some types of mechanical events, such as the multiple impacts exhibited by rolling element bearings, casing accelerometers may be successfully used for time domain analysis of the vibration signals.

The applicability of the three different types of vibration transducers to various types of transient data are summarized in Table 8-5. Shaft displacement may be used for any of the categories, but casing velocity or acceleration will not yield position change data. All three transducers may be used with cascade, waterfall, or trend plots, but the signal complexity on the casing pickups may negate some of their effectiveness. The use of Bode and polar plots for casing velocity and acceleration is dependent on the machine type, and the availability of a rotational speed signal for use with the casing mounted transducer.

Table 8–5 Applicability Of Various Transient Data Formats To Different Transducers

Transient Data Type	Shaft Displacement	Casing Velocity	Casing Acceleration
Cascade or Waterfall	Yes	Yes	Yes
Bode or Polar	Yes	Sometimes	Seldom
Position Change	Yes	No	No
Transient Capture or Trend	Yes	Yes	Yes

Overall, these tabular summaries reinforce the previous discussions of using displacement transducers for low frequency measurements, and the need to use FFT processing for the high frequency measurements produced by accelerometers. As with any general rules, exceptions always occur, and the diagnostician must keep an open mind. In all cases, the machinery diagnostician must employ the best combination of transducers, instrumentation, and data processing formats to examine and dissect the information extracted from the data set. Understanding the motion characteristics of the mechanical elements under examination requires looking at the vibratory behavior from many different angles, and applying the best available techniques and instrumentation.

BIBLIOGRAPHY

1. Potter, Ron and Mike Gribler, "Computed Order Tracking Obsoletes Older Methods," *SAE Technical Paper Series — Proceedings of the 1989 Noise and Vibration Conference*, Traverse City, Michigan (May 1989), pp. 63-67.
2. "Vibration, Axial Position, and Bearing Temperature Monitoring Systems - API Standard 670, Third Edition," *American Petroleum Institute*, (Washington, D.C.: American Petroleum Institute, November 1993).

Common Malfunctions

Process machines are subject to many malfunctions that range from internal forcing functions, to self-excited mechanisms, to external forces, plus a myriad of physical phenomena that may impose dynamic loads upon the machinery. Some of the malfunctions (e.g., unbalance) are common to all rotating machines. Other excitations (e.g., gear mesh forces) are unique characteristics for a particular type of mechanical device. These various excitations are normal for all moving elements, and they form a significant portion of the behavioral parameters for the specific piece of machinery.

When the forces responsible for the excitations increase beyond normal or expected limits, this is often detrimental to the integrity of the equipment. This physical change in applied forces is often detectable as a change in the machinery vibration response characteristics. In order to provide the diagnostician with some additional insight into these mechanical relationships, these excitations and the resultant malfunctions will be examined. Sequentially, the common types of machinery malfunctions are reviewed, and specific case histories are presented in this chapter. These common malfunctions are applicable to most rotating machines, and they include forced as well as free vibration mechanisms. In addition, a series of unique excitations associated with specific machine types will be presented and discussed in the following chapter 10.

SYNCHRONOUS RESPONSE

The synchronous, or running speed, or fundamental, or 1X motion of a rotating element is an inherent characteristic of every machine. It should be recognized that all machines function with some level of residual unbalance. All machines must operate with some finite clearance between stationary and rotating elements. Since it is physically impossible to produce a perfectly straight and concentric rotor, another source of synchronous motion is apparent. In addition, all machines are supported by various compliant structures and foundations.

Vibration response measurements on any machine with virtually any transducer will reveal a component at rotational frequency. Not surprisingly, this universally common excitation accounts for the majority of the machinery malfunction mechanisms. Unfortunately, the analysis of 1X vibration is significantly complicated by the fact that many different mechanical malfunctions appear as changes in the rotational motion. Hence, the machinery diagnostician is faced with the real dilemma of observing a single frequency, and attempting to diagnose the origin of an increased vibration amplitude.

The first line of attack resides in reviewing the traditional relationship discussed earlier in this text as:

$$Response = \frac{Force}{Restraint} \qquad \qquad \textbf{(9-1)}$$

It is clear that the vibration response is directly proportional to the applied force when the restraint or stiffness is held constant. As force increases, the resultant vibration response will also increase. This type of relationship is intrinsic with the fundamental concepts associated with activities such as rotor balancing. It is also clear that vibration response is inversely proportional to the restraint or stiffness when the applied force is held constant. In this condition, as the restraint decreases, the resultant vibration response will increase. For instance, a bearing with increasing clearances will typically exhibit a reduction in support stiffness (i.e., restraint). Assuming a constant unbalance force, the rotor will vibrate at a higher response level.

The third possibility for changes in vibration response amplitudes must consider variations in both the force and the restraint. As stated by Donald E. Bently in an issue of *Orbit*[1] magazine:

"Vibration is usually either the result of a bowed rotor or is the result of a force or moment acting on the stiffness of that machine to that force or moment. As such, vibration is actually a ratio and frequently not an end objective measurement in itself. Remember that forces and moments flow through, moments are measured across, and:

$$Dynamic\ Motion\ (Vibration) = \frac{Dynamic\ Forces\ or\ Moments}{Dynamic\ Stiffness}$$

Thus, to best read the behavior of a machine, it is often necessary to know BOTH the numerator and the denominator of this simple relationship ... Obviously, you must make some sort of assumption of stiffness or force in order to have a knowledgeable vibration measurement. We do this regularly and will continue to do so. However, to improve our capabilities of operating machinery, the measurement of observed operating dynamic stiffness will become more important in the future, as either the numerator (Dynamic Forces) may be incorrect, or the denominator (Dynamic Stiffness) may be incorrect."

In many instances it is extraordinarily difficult to quantify the active forces, or the associated restraints (stiffness). This inability to define actual machine parameters often yields to an investigation of changes in response constituents. For instance, the effective rotor support characteristics were discussed in chapter 4. This discussion concluded that for most types of process machinery that

---

[1] Donald E. Bently, "Vibration levels of machinery," *Orbit*, Vol. 13, No. 3 (September 1992), p. 4.

the effective stiffness is related to the oil film stiffness and the overall bearing housing stiffness in equation (4-16). This expression is restated as follows:

$$\frac{1}{K_{eff}} = \frac{1}{K_{oil}} + \frac{1}{K_{hsg}}$$

(9-2)

The bearing housing stiffness includes the support pilings and foundation, the grout and baseplate, the bearings or machinery pedestals, plus the stiffness of the bearing housing itself. Often a visual inspection of the machinery will identify the condition of these mechanical elements. For example, it is quite clear when a bearing housing is loose on a pedestal, or when grout degradation has occurred. If these support elements remain in good condition, then the oil film stiffness characteristics should be examined.

One of the most powerful and commonly available tools for evaluating bearing condition is an examination of the journal position within the bearing. This is performed with radially mounted X-Y proximity probes as discussed in chapter 6 of this text. Specifically, the change in shaft centerline position was determined with the vector example previously displayed in Fig. 6-20. For purposes of completeness, this same diagram is reproduced in the following Fig. 9-1. Within this diagram, the change in probe DC gap voltages may be vectorially summed to determine the overall shift in journal position from an initial *stop* condition to an *operating* position of the shaft within the bearing.

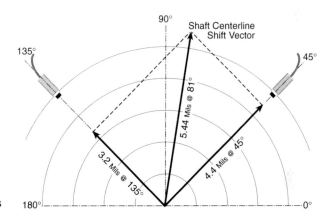

Fig. 9–1 Shaft Centerline Shift Vector As Measured With X-Y Proximity Probes

Substantial changes in radial shaft position are often associated with bearing damage. This is particularly true for a horizontal machine that has experienced damage to the bottom half of the bearing. In these instances, the probe gap voltages will reveal a vertical drop of the shaft into the babbitt. This type of damage often results in a change of the synchronous vibration combined with the position shift. In some cases the 1X shaft vibration will increase, as shown in the induction motor case history 44. In other situations, the running speed vibration will decrease, as illustrated by the refrigeration compressor case history 50. In both cases, the change in rotational speed vibration was associated with a distinct variation in bearing support stiffness. It should also be mentioned that the analysis of both of these case histories depended heavily on the measured changes in shaft centerline position data.

Furthermore, variations in nominal bearing clearance (i.e., too big or too small) represents one of the most common types of mechanical problems on process machinery. These abnormal clearances may be due to poor installation, attrition of the bearings during operation, or the presence of some forcing function that is *hammering out* the bearing clearance. Clearly, there are visual and measurement methods for evaluating changes in rotor support stiffness. If these techniques do not identify any change in restraint (stiffness), then it is reasonable to conclude that the vibratory change is primarily associated with a change in the applied force(s). The remainder of this chapter will be devoted to an examination of common forcing functions, and traditional excitation mechanisms. These malfunctions will also be illustrated with descriptive case histories

MASS UNBALANCE

Mass unbalance represents the most common type of synchronous excitation on rotating machinery. Every rotor consists of a shaft plus a series of integral disks used for turbine wheels, or thrust collars. Turbomachinery rotors may also include a series of slip on elements such as compressor wheels, pump impellers, thrust collars, spacers, coupling hubs, etc. Although each item is typically manufactured to high dimensional tolerances, a residual unbalance is present in each element. It is self-evident that the residual unbalance for a single machine disk may be satisfactory, but the combined effect for a stacked rotor may be completely unacceptable as described by John East[2]. To address this issue, a variety of tools and techniques have evolved to correct mass unbalance problems. Since this is a fundamental problem with all rotating machinery, chapter 11 of this text has been devoted to a detailed explanation of mass unbalance response, and the variety of methods used to determine and correct rotor unbalance.

From a recognition standpoint, mass unbalance will normally produce a transient Bode plot as shown in Fig. 9-2. This calculated plot for a forced unbalance spring-mass-damper system was extracted from Fig. 2-19 of this text. At speeds well below the resonance, the vibration response will vary as the speed squared. The applied centrifugal force may be estimated by equation (9-3):

$$F_{cent} = Mass \times Radius \times \left\{ \frac{RPM}{4,000} \right\}^2 \qquad \textbf{(9-3)}$$

where: F_{cent} = Centrifugal Force Due To Residual Unbalance (Pounds)
         $Mass$ = Effective Mass of Residual Unbalance (Grams)
         $Radius$ = Effective Radius of Residual Unbalance (Inches)
         RPM = Shaft Rotational Speed (Revolutions / Minute)

---

[2] John R. East, "Turbomachinery Balancing Considerations," *Proceedings of the Twentieth Turbomachinery Symposium,* Turbomachinery Laboratory, Texas A&M University, College Station, Texas (September 1991), pp. 209-214.

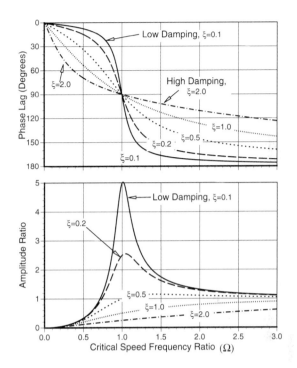

Fig. 9–2 Typical Mass Unbalance Response As Described By A Calculated Bode Plot

Thus, if the speed is doubled, the centrifugal force will be increased by a factor of four. If the mechanical system is totally linear, the observed rotor vibration will also exhibit a fourfold increase in magnitude. For units such as turbines or compressors that operate above a translational critical speed, the shaft rotates about the mass centerline (principal inertia axis). This behavior produces a plateau region of constant 1X amplitude and phase as shown in Fig. 9-2. Thus, moderate changes in speed (within the plateau region) will have minimal effect upon the synchronous vibration vectors.

When a machine is operating at full speed and load, a step change in the running speed vectors (amplitudes and/or phase) may be indicative of a mass unbalance shift. Part of a blade shroud or other minor attachment will manifest as a 1X vector change in accordance with the deflected mode shape. Major unbalance changes such as a blade loss will also produce a running speed vector change. In this case, the balance change will be significant, and concern should be placed on *how are we going to shutdown this machine, and pass through the critical without causing further damage*. In this type of situation, a rapid coastdown is desirable to minimize time within the bandwidth of the critical speed domain. It might be advisable to bring the machinery train down under full load to slow it down as fast as possible.

Typically, a pure mass unbalance problem will appear as a forward and circular shaft orbit. The orbit could also appear elliptical if the machine contains a significant difference in vertical versus horizontal stiffness, or if the rotor is subjected to a shaft preload in addition to the unbalance. Finally, the phase relationship across the rotor will be in accordance with the deflected mode shape, and the location of the probes along the axis. It is easy to be confused by traditional rules that say things like ...*identical phase angles across a machine are representative of mass unbalance.* This statement is only true for a specific set of conditions, for a particular group of machines. Many cases of mass unbalance (e.g., turbine generator case history 39) will exhibit a phase relationship other than a pure inphase motion across the machine.

It is highly recommended that the diagnostician become intimately familiar with the specific topics of shaft mode shapes in chapter 3, dynamic signal characteristics chapter 5, and rotor balancing chapter 11 before attempting to diagnose a mass unbalance problem. Furthermore, the study of 1X synchronous behavior of a rotating system will reveal important information on the specific characteristics of the machine. This type of information will provide significant benefits in malfunction diagnosis of the rotating equipment.

BENT OR BOWED SHAFT

Bent rotors and shaft bows represent another major class of synchronous 1X motion. It was previously mentioned that all machine parts contain some finite amount of residual unbalance. In a similar manner, all assembled horizontal rotors (and some vertical rotors), will exhibit varying degrees of rotor bows. In some cases, such as a light weight pinion with a short bearing span, the midspan deflection will be minimal. In other cases, the rotor will deform due to gravity. That is, the rotor will bend or bow under the influence of its own weight. An example of this type of rotor would be a large gas turbine or steam turbine rotor. These types of rotors will display a change in the gravitational bow by simply lifting the rotor off a set of rollers, and setting it back down.

The shaft bow may be purely a gravitational bow, or it may be a thermally induced bow. In either case, the force associated with the bent shaft is equal to the shaft stiffness times the initial bow radius. This is considerably different from the mass unbalance case, where the initial force is equal to the product of residual unbalance mass, radius, and speed squared. The bent shaft will exhibit a variable speed characteristic that is similar to the diagram in Fig. 9-3 that was extracted from Fig. 2-18. This calculated plot for a forced spring-mass-damper system was based upon the application of a constant force. This is identical to the bent rotor condition where the applied force is equal to the deflection times the shaft spring constant as shown in equation (9-4).

$$F_{bow} = \frac{Deflection \times Shaft\ Stiffness}{1,000} \qquad (9\text{-}4)$$

where: F_{bow} = Applied Force Due to Shaft Bow (Pounds)
    Deflection = Maximum Midspan Deflection Due to Shaft Bow (Mils)
Shaft Stiffness = Lateral Shaft Stiffness (Pounds / Inch)

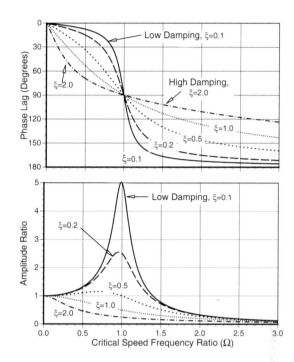

Fig. 9–3 Typical Lateral Response Due To A Shaft Bow As Described By A Calculated Bode Plot

Since shaft stiffness is usually a very large number, a moderate deflection (i.e., bow) will result in a substantial radial force. From Fig. 9-3, it is noted that at speeds well below the resonance, the vibration response is equal to shaft bow. Hence, a symmetrical rotor supported between bearings would display a maximum response at the middle of the rotor. A proximity probe or dial indicator mounted at this location would reveal a runout vector equal to the bow magnitude and location. At speeds well above the resonance, the magnitude of the bow would approach zero as the machine tends to rotate about the mass center, which would be equivalent, or more precisely, coincident with the rotor bow center.

In the vast majority of cases, a thermal or gravitational shaft bow may be *rolled out* by extended operation at slow roll speeds. In most instances the 1X vectors are monitored, and when minimum and constant runout values are achieved, the machinery may be safely started with a minimal shaft bow. In other situations, the severity of the bow is of such a magnitude that the rotor cannot be straightened by slow rolling. Localized heat application may be used to relieve the bow, or the rotor may be suspended vertically and heated in an oven. In other cases the rotor must be completely scrapped (e.g., case history 17).

In all cases, the diagnostician must recognize that rotor bows are inherent with rotating machinery. Furthermore, the shaft bow may consist of complex curves instead of a simple catenary. Also, the synchronous force from any shaft bow will vectorially interact with the synchronous forces due to unbalance or any other 1X forcing function. Hence, the final measured rotational speed vectors probably include contributions from more than one synchronous excitation.

Case History 20: Repetitive Steam Turbine Rotor Bow

A 35,000 horsepower steam turbine was subjected to an extensive overhaul that included replacement of the 11 stage rotor. The maintenance work was performed with minimal difficulties, but major problems were encountered when a routine overspeed check could not get past slow roll conditions. Although initial uncoupled turbine runout vibration levels were quite acceptable, the 1X vectors significantly increased at slow roll. At 1,000 RPM, vibration amplitudes reached 2.5 and 3.1 Mils,$_{p-p}$ at the governor and exhaust ends respectively. This manifestation of high vibration was combined with in-phase deflection of the turbine rotor. Specifically, the governor end horizontal probe had a 261° phase angle, with 265° displayed by the exhaust end horizontal pickup.

This unusual bowed rotor behavior was repeated on multiple runs covering a time period of two days. At this point, historical rotor records were examined, and the following general conclusions and observations were reached:

○ All of the attempted turbine solo runs were aborted due to high rotor vibration. This was indicated by the control room monitoring instrumentation, as well as the physical sensation of excessive deck vibration.

○ During the occurrence of the high turbine vibration, a pure translational (inphase) shaft bow was clearly evident across the rotor.

○ For this particular turbine, shaft vibration amplitudes should remain essentially constant between 300 and 1,400 RPM.

○ Initial appearance of the shaft bow was independent of speed.

○ Initial appearance of the bow was a function of time and temperature. Specifically, the shaft bow appeared following approximately two hours of operation in a warm (>200°F) casing.

○ Cold shaft runout vectors were repeatable, and consideration of any cracked shaft malfunction was discontinued.

○ Rotor inspection records revealed acceptable runout along the length of the turbine. Furthermore, the shop balance was performed to low levels of residual unbalance. Hence, the observed behavior was not associated with a cold rotor bow, or a mass unbalance problem.

○ Records revealed that the spare turbine rotor installed during this overhaul had not operated in the turbine casing for five years. At that point in history, the plant suffered a catastrophic fire, and this spare rotor was essentially *baked* in the casing for several days as the fire burned itself out.

A turbine rotor is machined from a solid forging, and all wheels and thrust collars are integral with the shaft. This type of rotor assembly is heat treated and tempered as part of the manufacturing process. These heat treatments of the alloy forging are performed to obtain specific mechanical properties, and they are implemented by controlled heating and cooling of the rotor. Hence, the turbine rotor is constructed of a steel alloy that was subjected to various heat cycles during fabrication. It is reasonable to believe that such an assembly might be sensitive to the heating and cooling anomalies associated with the previous plant fire.

Based upon these measurements and observations, it was reasoned that the turbine rotor contained a residual stress that was probably inflicted during the fire. This residual stress manifested as a shaft bow whenever the rotor was heated above ambient temperature. Conversely, the shaft bow was not apparent when the rotor was cold. It is logical to assume that the emergency shutdown produced a thermal bow as the hot rotor rested between bearings, on the inter-stage labyrinths. As the fire subsided, the turbine cooled over a period of several days. It was postulated that as the rotor cooled, it returned to a straight condition at ambient temperature, and *locked in* the residual stress from the thermal bow. Since all rotor repairs, inspections, runout checks, and shop balancing were performed with a cold rotor, this type of internal stress would be undetectable.

If this hypothesis was accurate, then correction of the residual thermal shaft bow would require additional heat treatment, combined with continuous rotation of the turbine rotor. Obviously, this type of repair would be difficult to perform in most shop repair facilities. However, the turbine casing provided a means to heat the rotor with inlet steam, plus the ability to turn the rotor at controlled speed. Thus, the opportunity existed to perform an on-line stress relief of the rotor within the actual turbine casing. The ASM International defines stress relieving as: *Heating to a suitable temperature, holding long enough to reduce residual stresses, and then cooling slowly enough to minimize the development of new residual stresses.*

From this common definition, both heating and cooling must be combined to stress relieve the rotor. During original manufacturing of this rotor, the heat treatment temperatures are quite high. It must be recognized that the exhaust casing has much lower temperature limits (circa 300°F). Thus, any field stress relieving of the rotor in the turbine casing must be limited in the heat soak temperature. Based upon the mechanical parameters of the installed turbine system, the following on-line rotor pseudo-stress relieving procedure was developed:

1. Operate the turbine at a slow speed of nominally 300 RPM for approximately 30 minutes with the sealing steam off, and a cool turbine casing. This is the cooling portion of the cycle.

2. Apply shaft sealing steam, and allow turbine speed to increase to approximately 500 RPM with the improved vacuum.

3. Increase speed, and monitor the radial shaft vibration at both journal bearings. Continue to increase speed until the unfiltered radial vibration amplitudes approach a maximum of 4.0 to 5.0 Mils,$_{p-p}$ (based on conservative use of the 12 to 14 Mil diametrical turbine bearing clearance).

4. Operate the turbine at this heat soak condition for approximately 60 minutes. During this time period, the rotational speed and steam flow should be adjusted to maintain a maximum unfiltered radial vibration amplitude between 4.0 and 5.0 Mils,$_{p-p}$, combined with a maximum exhaust casing temperature of 250°F, plus a maximum speed of 1,400 RPM.

5. Following 60 minutes of high speed, high temperature, and high vibration

operation; the turbine speed should be reduced back to 500 RPM. Seal steam should then be removed, and rotational speed returned back to 300 RPM for a repeat of the cooling cycle, Step 1.

6. The previous Steps 1 through 5 should be repeated until the shaft vibration amplitudes remain essentially constant between 300 and 1,400 RPM. When this consistency of slow roll vibration data is achieved, the turbine may be started, and operated normally.

This procedure was implemented, and the first 60 minute hot run was limited to 654 RPM. A total of six additional cold to hot runs were completed, and the results of these consecutive runs are summarized in Fig. 9-4. This diagram consists of 1X radial vibration amplitudes measured at the end of each cold 30 minute run, plus each hot 60 minute run. Note that these 1X filtered amplitudes are slightly less than the unfiltered, overall vibration levels mentioned in the procedure. For simplification, only the horizontal probes at governor and exhaust bearings are shown. The vertical probes exhibited identical characteristics.

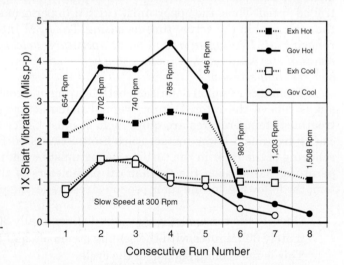

Fig. 9–4 Variation Of 1X Vibration Amplitudes During Multiple Heating and Cooling Cycles

Within Fig. 9-4, the 1X vibration amplitudes at the governor journal are depicted by the solid lines and circular plotting symbols. The exhaust end amplitudes are defined by dotted lines and square plotting symbols. Data points acquired at the end of a cold run at 300 RPM are identified by the open plotting symbols. The 1X vibration amplitudes measured at the end of a hot run are represented by the solid symbols. For each of the hot runs, the rotational speed at the end of the run is listed for each pair of hot data points. From this summary diagram, it is apparent that the maximum attainable speed during each hot run successively increased from run to run. In addition, the hot vibration amplitudes across the turbine tracked up and down in unison.

A significant portion of this plot is noted in the lower right hand corner. Within this region, the measured amplitudes at the conclusion of the seventh run show a close agreement between hot and cold readings. Continued operation at 1,508 RPM resulted in the convergence of hot and cold amplitudes into a common value. Under this condition, the 1X shaft motion at the governor journal was 0.2 Mils,$_{p-p}$, and the exhaust end converged to a 1X amplitude of 1.0 Mil,$_{p-p}$. The difference in vibration magnitudes between ends of the turbine is primarily attributed to the chrome overlay sensed by the proximity probes on the turbine exhaust shaft.

These vector amplitudes should be supplemented by the phase data to gain a better appreciation of the bow characteristics and subsidence. Fig. 9-5 summarizes the running speed phase angle from the same horizontal probes during each of the cold to hot cycles. Note that the cold phase angles are divergent. However, as the shaft warms up during the first five runs, the phase angles across the turbine snap together as the bow becomes active. During the last three runs, the hot phase angles become increasingly coincident with the cold values. At the conclusion of the seven cold to hot pseudo-stress relieving runs, the 1X vector amplitudes and phase angles remained constant. That is, between the cold condition at 300 RPM, and the warm operation at 1,500 RPM, the 1X vectors have not changed. It is concluded that the application of the cold to hot runs allowed a relaxation of the internal rotor stress, and the shaft bow has been relieved.

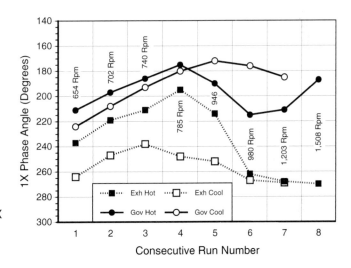

Fig. 9–5 Variation Of 1X Vibration Phase Angles During Multiple Heating and Cooling Cycles

This procedure proved to be quite effective to prepare the rotor for the overspeed trip runs. Following this work, the turbine was coupled to the load compressor, and two days later the machinery train placed on slow roll. It was noted that some of the residual bow activity was reappearing following two days of non-rotation. Four more pseudo stress relieving runs were performed, and the proper turbine slow behavior was re-established. The coupled startup was quite smooth, and the machinery train displayed normal transient vibration characteristics. This bow eliminating procedure is still in use on this turbine, and it is successfully applied after every extended outage on this unit.

ECCENTRICITY

Eccentricity of one machine part with respect to another represents a less common category of rotational speed excitations. Normally, shafts and most rotor elements are ground on-centers. The material center is thereby concentric with the initial center of rotation, and eccentricity is generally not a problem. However, there are occasions when a machine part is bored off-center. Although the majority of the rotating assembly may be straight and concentric, the presence of an eccentric element can impose a significant rotational speed force.

If the eccentric element is a minor part of the rotor assembly, the resultant 1X forces may be insignificant compared to the other active synchronous forces. However, if the eccentric element represents a substantial portion of the rotating assembly, or if it is located at a modally sensitive location (e.g., the coupling hub), then the eccentricity may be a problem. The actual forces associated with an eccentric element may be determined from the following equation (9-5):

$$F_{ecc} = Weight_{element} \times Eccentricity \times \left\{ \frac{RPM}{5,930} \right\}^2 \qquad \textbf{(9-5)}$$

where: F_{ecc} = Radial Force Due to Eccentricity of a Mechanical Element (Pounds)
$Weight_{element}$ = Weight of the Eccentric Mechanical Element (Pounds)
$Eccentricity$ = Radial Eccentricity of Machine Element (Mils)
RPM = Shaft Rotational Speed (Revolutions / Minute)

Large machine elements or high rotational speeds are the most susceptible to high forces due to an eccentric element. In many respects, an eccentric element appears similar to a shaft bow at low rotational speeds. Both mechanisms provide large shaft displacement amplitudes at slow speeds. However, the forces from a bowed rotor may remain constant at all speeds in accordance with equation (9-4). The radial forces from an eccentric element will vary with the speed squared as described by expression (9-5). Naturally this all becomes much more complicated when machines with flexible rotors and multiple mode shapes are discussed. In all cases, eccentric machine elements on a rotor should be avoided, and one source of potential synchronous excitation removed from consideration.

From a detection standpoint, shaft bows and eccentric elements can be determined in the shop with accurate runout checks as described by John East[3]. Once the machine is assembled, runouts can be detected at low speeds with relative shaft sensing proximity probes. Casing velocity coils and accelerometers will probably not detect either mechanism at slow roll speeds. However, the casing vibration transducers will pick-up the influence of a bow or an eccentricity at higher speeds when the radial forces are significant. For instance, case history 21 considers a situation where a pinion coupling hub was bored off-center, and the resultant eccentricity had a considerable influence upon the machinery.

---

[3] John R. East, "Turbomachinery Balancing Considerations," *Proceedings of the Twentieth Turbomachinery Symposium,* Turbomachinery Laboratory, Texas A&M University, College Station, Texas (September 1991), pp. 209-214.

Case History 21: Seven Element Gear Box Coupling Bore

The machinery train in question consists of a lime drying kiln driven by a variable speed synchronous motor through a seven element speed reducing gear box. The drive portion of this train is depicted in Fig. 9-6. The kiln dryer itself consisted of a long cylindrical tube that is mounted on rollers, and it is inclined with respect to grade. Wet lime is loaded into the top end of the kiln, and dry lime is extracted from the lowest elevation. This massive cylindrical kiln rotates slowly, with a maximum speed of 1.8 revolutions per minute.

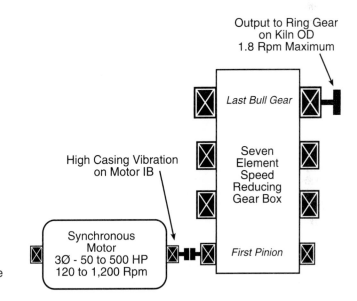

Fig. 9–6 Machinery Arrangement For Large Kiln Drive

The kiln drive system is located on an elevated platform. Initial operation revealed unacceptable vibration amplitudes on the platform, and various structural modifications were implemented. Following these support improvements, the drive train continued to exhibit uncomfortable vibration amplitudes. The machinery was not equipped with any vibration monitoring instrumentation, and initial readings were limited to casing measurements. The large seven element gear box exhibited low vibration amplitudes, and the problem appeared to be confined to the variable speed motor. The outboard motor bearing displayed casing vibration levels of 3.0 to 4.0 Mils,$_{\text{p-p}}$. At the inboard, or coupling end motor bearing, the unfiltered horizontal motion varied between 4.0 to 5.0 Mils,$_{\text{p-p}}$.

It was clear that the kiln and the gear box are massive structures, and any casing vibration would be substantially suppressed by their respective steel structures. Conversely, the motor driver consisted of a fairly light frame, and it was susceptible to excitation from a variety of sources. A complete survey was performed with triaxial casing velocity measurements at each main bearing, plus various locations on the support structure. It was quite evident that the primary excitation frequency occurred at motor running speed. Axial vibration of the motor, gear box, and structure was minimal at all locations. The major vibration appeared to be running speed motion of the motor in a radial direction. The casing orbits presented in Fig. 9-7 depict the unfiltered and 1X filtered behavior at the motor inboard bearing housing. Operating at an average speed of 1,190 RPM, the motor coupling end bearing housing exhibited a forward and elliptical orbit at motor running frequency. As noted in the unfiltered plots, the maximum horizontal motor casing vibration approached 4.0 Mils,$_{p-p}$. Simultaneously, the companion casing orbit on the pinion input bearing housing (not shown) revealed casing displacement amplitudes of less than 0.5 Mils,$_{p-p}$.

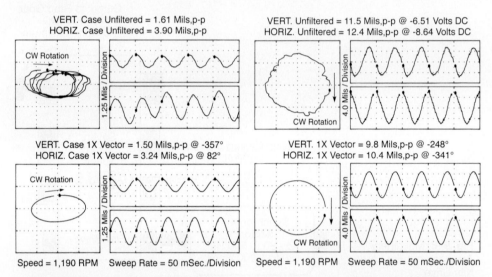

Fig. 9–7 Motor Bearing Housing Casing Vibration At Full Operating Speed

Fig. 9–8 Pinion Relative Shaft Vibration At Full Operating Speed

Further investigation, and visual inspection revealed excessive vibration of the coupling assembly. Since the normal threshold for visual observation of vibrating surfaces is in the vicinity of 10 Mils,$_{p-p}$, it was considered significant that the coupling could be vibrating twice as much as the motor bearing housing.

This observation prompted further investigation into the absolute motion of the coupling. To implement this measurement, a separate stand was constructed from steel plate and angle iron to support a pair of X-Y proximity probes on either side of the coupling. One pair of probes observed the 8.25 inch diameter coupling hub on the motor shaft. The other pair of proximity probes were positioned over the 9.75 inch diameter pinion coupling hub. All four proximity probes were supplemented by velocity coils to allow for potential correction of the shaft vibration signals for motion of the probe supporting structure. As it turned out, this correction was not required, and the direct probe signals from the pinion coupling hub are presented in Fig. 9-8.

This data reveals a distinctive forward circular pattern to the pinion coupling hub displacement. Since the hub outer diameter was not machined or prepared for proximity probes — the hub surface imperfections were visible, and overall amplitudes in the vicinity of 12.0 Mils,$_{p-p}$ were documented. The hub signal filtered at the rotational speed (1X) of 1,190 RPM exhibited an average circular amplitude of 10.0 Mils,$_{p-p}$. It should also be mentioned that the coupling hub on the motor side displayed circular amplitudes of nominally 5.0 Mils,$_{p-p}$. This steady state data at full speed suggested that the high vibration might be originating at the pinion instead of the motor. In support of this preliminary conclusion, it was understood that uncoupled, and unloaded, motor vibration was quite low. That does not necessarily give the motor a clean bill of health, since many motor problems only appear under load. Nevertheless, it did suggest that perhaps the pinion might be the culprit, and the light motor might be just responding to a forced vibration condition.

Additional perspective on this problem was gained by the acquisition and analysis of variable speed information. Specifically, vibration data was recorded during a shutdown of the kiln, and the Bode plot shown in Fig. 9-9 obtained from the X-Y proximity probes positioned over the pinion coupling hub. Note that the synchronous 1X amplitudes and phase angles remained essentially constant from the top speed of 1,190 RPM to the minimum sample point of 186 RPM. This type of behavior is certainly representative of an eccentric mechanical element. In this case, the pinion coupling hub was the primary suspect.

Another perspective of the kiln shutdown was gained from Fig. 9-10, that documents a time history plot of the coastdown. In this diagram, the 1X amplitude and phase are plotted against time from 0 to 60 seconds. In addition, the

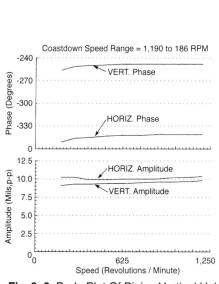

Fig. 9–9 Bode Plot Of Pinion Vertical Hub Vibration During A Typical Coastdown

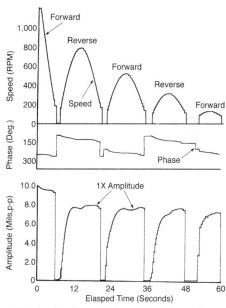

Fig. 9–10 Time History Coastdown From Vertical Prox Probe On Pinion Coupling

rotative speed is included at the top of the same plot. The various humps in this summary plot are due to the fact that the main kiln cylinder carries a tremendous amount of inertia from the rotating kiln, plus the internal lime. During a routine coastdown of this machine, the kiln slows down from full operating speed of 1,190 RPM, and comes to a stop in approximately 6 seconds. The kiln tube (with the moving lime) then begins a reverse rotation, and drives the gear box and motor in a reverse rotation. The input pinion and motor reach a peak speed of 800 RPM before the train starts to slow down. As shown in Fig. 9-10, the unit experiences two more forward, and one more reverse rotation sequence before the train comes to a final stop.

Note that the rotational speed amplitude peaks at about 8.0 Mils,$_{p-p}$ irrespective of a forward or a reverse cycle. When the unit is rotating in a forward direction, the vertical probe phase angle is approximately 250°. During reverse rotation, the 1X phase angle is in the vicinity of 80°. Thus, a nominal 170° reversal in the high spot occurs as the pinion hub rotates in a forward or a reverse direction. This is close enough to 180° to conclude that there was a complete reversal of the phase relationship between forward and reverse rotation. This documented behavior also helps to substantiate the hypothesis of an eccentric coupling hub.

Additional data at various loads provided no other useful information, and it was finally concluded that the coupling hub was bored off center. The physical configuration of the pinion extension did not allow simultaneous dial indicator measurements of the pinion hub versus the pinion shaft. However, when the coupling hub was removed, it was determined that the shaft bore was indeed off center by approximately 8 to 10 Mils. Naturally the coupling supplier was slightly embarrassed, and they provided a concentrically bored coupling assembly in a short period of time. The installation of this correctly bored coupling half on the pinion shaft solved the problem.

SHAFT PRELOADS

Another category of potential malfunctions that are generally applicable to all rotating machinery is the topic of shaft preloads. The presence of various types of unidirectional forces acting upon the rotating mechanical system is a normal and expected characteristic of machinery. Just as residual unbalance, rotor bows, and component eccentricity are inherent with the assembly of rotating elements, the presence of shaft preloads are an unavoidable part of assembled mechanical equipment.

From an initial categorization standpoint, shaft preloads may be divided into two fundamental groups. The first group would address the preloads that originate within the machinery. These internal preloads may be due to any or all of the following common mechanisms:

○ Gravitational Preloads
○ Bearing Preloads
○ Internal Misalignment Preloads
○ Gear Mesh Forces
○ Fluid Preloads

Gravitational preloads on horizontal rotors are responsible for the rotor bow or sag that was discussed earlier in this chapter. Again, this is part of the normal and unavoidable characteristics of process machinery. This is generally not a *design correctable* problem. It is a physical phenomena that must always be considered, and dealt with on a machine by machine basis.

Bearing preloads represent one of the machinery design considerations. As discussed in chapter 4, bearing preloads are typically expressed as a non-dimensional number between 0 and 1. A bearing preload of 0 indicates no bearing load upon the shaft. Conversely, a bearing preload of 1 indicates a shaft to bearing line contact (i.e., maximum preload of 1). The computation of bearing preload is based upon the difference in curvature between the shaft and the individual bearing pad. As expressed by equation (4-6), bearing preload is determined by:

$$Preload = 1 - \left\{ \frac{C_b}{C_p} \right\} \tag{9-6}$$

Where C_b is the bearing clearance which is equal to the bearing radius minus the journal radius. The pad clearance C_p is equal to the pad radius minus the journal radius. On segmented bearings it is common to find that the pad radius is greater than the bearing radius. Thus, the pad clearance C_p is greater than bearing clearance C_b, and their ratio is less than one. From equation (9-6) the bearing preload must therefore be less than 1. Typically, the bearing designers will employ preloads that vary between 0.1 and 0.4.

Clearance reduction at the center of the pad forces the oil to converge into an oil wedge. The operational characteristics of bearing stiffness, damping, and eccentricity position will vary with the preload, and the actual bearing configuration. Obviously, the characteristics of a fixed pad bearing, such as an elliptical bearing, will be different from a five shoe tilting pad bearing. However, in either case an oil wedge will be developed, and that oil wedge will have an associated pressure profile (Fig. 9-20). The direct action of this pressure profile upon the journal is a preload force. Again, this is part of the normal behavior of the rotating machine, but the amount of preload, and the associated journal force is adjustable (within limits) by varying the pad radius and bearing geometry.

The third type of internal machine preload is attributed to **internal misalignment** of elements. This can vary from offset or cocked seals or bearings, to distorted diaphragms or stators, to a variety of rub situations. It is virtually impossible to quantify all of the potential combinations of misaligned internal machine elements. However, the common characteristic that they all share is the generation of a load or force against the shaft. Some of these preloads may relieve themselves during normal operation. For instance, a laby seal rub might occur during initial startup on a machine, and the expanded shaft to seal clearances may never rub again. Other preloads such as a distorted stator will generally remain constant, and will continue to provide a force upon the rotor.

The fourth type of internal preload is associated with **gear mesh forces**. These are significant loads that must always be considered. To demonstrate the magnitude of gear contact forces, the values calculated in case history 24 are repeated. These loads are for a simple pinion - bull gear arrangement. The forces were computed for a transmitted load of 4,000 horsepower, a pinion speed of 5,900 RPM, and a bull gear speed of 1,920 RPM. The significant element weights and forces for this gear box are summarized as follows:

Pinion Weight .... 220 Pounds

Bull Gear Weight .... 1,630 Pounds

Separation Force .... 4,550 Pounds

Tangential Force .... 10,880 Pounds

From this summary, it is clear that the major forces within a gear box are the gear contact forces. The magnitudes of the separation and the tangential forces place the gear weights into the role of a secondary influence. These gear forces are used in the development of gear box bearing coefficients, and in the initial estimation of the journal running position. It is important to consider the journal operating position during alignment of the gear box, and recognize that the bull gear and pinion bearings are subjected to significant radial preloads from the gear forces.

The fifth type of **fluid** preloads is applicable to many types of rotating machines. For instance, the unbalanced radial force in a volute pump is an obvious case of fluid forces acting directly upon a rotor. A less obvious example of fluid forces would be the behavior of a multistage and multilevel turbine during startup. It has been documented that partial steam admission to the first stage nozzles may cause a lifting force on the rotor when the first nozzle segment is located in the bottom half of the turbine casing. This radial force may be sufficient to lift the rotor, and allow the governor end bearing to go unstable. Hence, the vertical shaft preload would work against the stabilizing gravitational force to drive the machine into another type of malfunction. In all cases, the machinery diagnostician must be aware of these types of physical interactions, and must strive to understand and address the fundamental forces behind the observed vibratory motion.

The second major category of shaft preloads considers the array of potential external preloads or forces. For example, the following short list identifies some of the common external shaft or machinery preloads:

○ Coupling Misalignment
○ Locked Coupling
○ Thermal or External Forces

The problem of **coupling misalignment** is common to most types of rotating machinery. Fortunately, the machinery community has devoted considerable time and effort to develop solutions and techniques for execution of correct shaft alignment. In fact, many operating facilities have applied these tools and techniques to develop very successful machinery alignment programs. In these operating plants, misalignment has ceased to be a problem. Nevertheless, when misalignment between machines is present, the shaft preload forces may be substantial, and may result in premature mechanical failure.

A related mechanism to shaft misalignment is the problem of a **locked coupling**. Primarily this occurs on oil lubricated gear type couplings. Since these couplings are designed as flexible members with a tolerance to misalignment, the abnormal condition of locked coupling teeth will violate the intended behavior of the flexible design. A locked coupling may behave similarly to a misaligned coupling. In may cases the coupling is locked due to excessive misalignment. In other cases, the locked coupling may develop during operation due to the accumulation of sludge between the teeth. In either case, the resultant forces on both machinery shafts are unwelcome preloads that may damage machinery components on either side of the locked coupling.

The third category of **external preloads** is associated with the presence of any number of potential external forces or moments on the machinery. This kind of preload could be due to baseplate strain imposed by a degrading grout or foundation. External preloads could also arise from piping strain upon the machine. For instance, the case history 49 describes the effect of a piping moment upon a compressor, and the coupled turbine driver. External preloads may influence the coupling alignment, or they may distort bearing housings, casings, or other mechanical attachments.

In essence, shaft preloads are a normal part of rotating equipment that must be addressed. It is useful to recognize that preloads have different levels of severity. For instance, some preloads such as gravity or fluid based forces may be classified as *soft* preloads that are generally non-destructive. Other preloads, such as misalignment or gear contact forces, may be considered as *hard* preloads that can be damaging to the machinery. A third severity classification for shaft preloads would be the *destabilizing* variety. This type of preload may oppose the normal rotor or bearing forces, and it may act to destabilize the rotor. The severity of these *destabilizing* preloads may also vary from *soft* to *hard*, depending on the final influence upon the process machinery.

Preload detection is predicated upon the recognition of abnormalities in radial vibration. The following three characteristics are used to identify the presence of radial shaft preloads:

○ Normal Orbital Motion
○ Abnormal Shaft Centerline Position
○ Abnormal Shaft versus Casing Motion

Abnormal orbital motion is demonstrated in Fig. 9-11, that depicts an array of four shaft orbits with different levels of radial preloads. The orbit in Sketch A

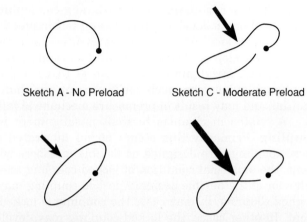

Fig. 9–11 Changes In
Shaft Orbits With Increas-
ing Levels of Radial Shaft
Preloads

displays a normal forward circular pattern with no shaft preload. Application of
a slight preload to the shaft in Sketch B forces an elliptical shape to the shaft
motion. The moderate preload of Sketch C drives the orbit into a *banana* shape.
Finally, the heavy preload case shown in Sketch D results in a *Figure 8* orbit.
This type of analysis is quite appropriate to machines with fluid film bearings,
and shaft sensing proximity probes. In the vast majority of the cases, the direction of the preload is perpendicular to the major axis, and the severity of the preload is dependent on the ellipticity of the orbit.

As a cautionary note, the machinery diagnostician should not confuse the
preloaded orbit with the *normally elliptical orbit*. Machines with significant differences between vertical and horizontal bearing stiffness will naturally display
an elliptical orbit. This type of orbit is strictly a function of the bearing geometry,
and must be considered as normal and proper behavior.

For example, Fig. 9-12 displays a diagram of an elliptical bearing with a
counterclockwise rotating shaft. The normal elliptical orbit is noted in a proper
running position in the lower right hand quadrant. This same diagram also

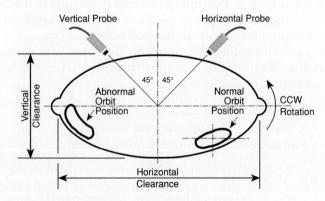

Fig. 9–12 Normal Versus
Abnormal Shaft Centerline
Position In An Elliptical
Bearing Assembly

depicts another elliptical orbit residing in the lower left hand quadrant. The second orbit is obviously in the wrong place for a CCW rotating machine. This improper radial position represents the second method of detection of shaft preloads. That is, the calculated journal centerline position should be in the proper location within the clearance of the bearing. If the shaft centerline position resides at an abnormal location, the possibility of a damaged bearing, or the presence of a radial preload should be suspected.

Three cautionary notes should be added to this type of evaluation. First, the diagnostician must know the proper running position of the journal within the specific bearing before attempting to pass judgment on any field data. For instance, a five shoe tilting pad bearing will display a vertical attitude angle, and normal position for this type of bearing is considerably different from the previously discussed elliptical bearing. If the eccentricity position and the attitude angle are not known, then the diagnostician should consider an FEA analysis of the specific bearing configuration as discussed in chapter 4.

Second, the machinery diagnostician must be working with a properly calibrated proximity probe system to measure the true running position of the journal within the bearing. This includes accurate probe scale factors plus a correct probe orientation diagram. In some instances, it may be necessary to install four radial proximity probes at 90° increments. In this application, the diametrically opposed probes are summed to determine an average shaft position in each orthogonal direction. This is more work, but it does enhance accuracy of the radial position measurement.

Third, the initial proximity probe DC gap voltages must be accurately known to allow a confident calculation of the shaft centerline position. It is difficult to generalize on the precise condition to obtain the *at stop* gap voltages. Normally, this data is obtained prior to startup with warm oil circulating. To be safe, it is recommended that DC voltages be tracked with a computer-based system that will identify time, speed, and other useful information such as oil supply temperature, ambient temperature, etc.

Finally, on machines with accessible bearing housings, it is desirable to acquire X-Y casing vibration response measurements. The casing probes should be placed in the same angular orientation as the shaft sensing proximity probes. In addition, the casing probes should be located as close as possible to the mounting point of the proximity probes. The casing data must be integrated to displacement, and the 1X synchronous vectors compared directly against the runout compensated shaft displacement 1X vectors. Under normal conditions, the casing motion should be smaller than the shaft vibration, and the casing phase angles should lag behind the shaft vibration angles.

For a machine with a radial preload, and a compliant support, the shaft vibration may be suppressed. In this condition, the normal shaft vibration within the bearing is transmitted to the casing and surrounding structure. From a measurement standpoint, the casing 1X vibration amplitudes may exceed the shaft motion, and the shaft to casing phase relationship may appear abnormal. This final criteria is not a totally conclusive test, but it does provide additional insight into the mechanics of the machinery.

RESONANT RESPONSE

Machines and structures all contain natural frequencies that are essentially a function of stiffness and mass. As described in previous chapters, the fundamental relationship may be described by the following expression:

$$Natural\ Frequency \approx \sqrt{\frac{Stiffness}{Mass}}$$

Recall that this expression was developed for a simple spring-mass system, and it basically identified the lowest order resonant frequency. For more complex mechanical systems an entire family of resonant responses must be addressed. For example, consider a turbine compressor set mounted on a mezzanine structure, and connected with a flexible coupling. The potential array of anticipated natural, or resonant frequencies are summarized as follows:

○ Lateral Critical Speeds
  ❏ Turbine Translational (1st)
  ❏ Turbine Pivotal (2nd)
  ❏ Turbine Bending (3rd)
  ❏ Compressor Translational (1st)
  ❏ Compressor Pivotal (2nd)
  ❏ Compressor Bending (3rd)
○ Torsional Critical Speeds
  ❏ Turbine (1st)
  ❏ Turbine (2nd)
  ❏ Compressor (1st)
  ❏ Compressor (2nd)
○ Rotor Element Resonances
  ❏ Coupling Natural Axial
  ❏ Coupling Lateral
  ❏ Turbine Blades
  ❏ Compressor Impellers
○ Acoustic Resonances
  ❏ External Piping Systems (including stubs and branches)
  ❏ Internal Passages Within Casings
○ Structural Resonances
  ❏ Piping Systems
  ❏ Structural Steel Systems
  ❏ Machinery Pedestals
  ❏ Baseplate, Foundation, or Ground Support System

This list of potential resonant frequencies can be intimidating. The range of natural frequencies may vary from 60 CPM (1 Hz) for the foundation and support systems, to 1,800,000 CPM (30,000 Hz) for turbine blade natural frequencies. With this extended range of natural frequencies it is somewhat amazing that process machinery can be designed to operate without violating many of these system natural frequencies. One of the tools used by machinery designers is the Campbell diagram for plotting natural frequencies versus excitations. This was initially introduced in chapter 4. For reference purposes, the Campbell plot from Fig. 5-4 is reproduced in Fig. 9-13. Please note that the natural frequencies (rotor lateral modes) are shown as horizontal lines off the vertical axis, and the various machine excitations are depicted as the slanted lines. These slanted lines are the 1X, 2X, 3X, etc. multiples of rotational speed that the machine will produce as speed increases from slow roll to the normal operating point.

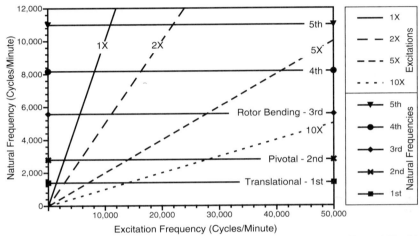

Fig. 9-13 Campbell Diagram Describing Interference Between Several Forced Machinery Excitations And A Partial Group of Turbine Natural Lateral Resonances

An intersection between curves identifies points of potential resonant response. That is, a natural frequency exists, and an excitation source is present at the same frequency. This coincidence results in a stimulation or activation of the resonance. The Campbell plot is an interference diagram that is commonly developed during the design stages of a new machinery train. The actual number of system natural frequencies may vary from 20 to 50 or more. Furthermore, the actual number of machinery excitations may also be significant, and 10 to 30 significant excitations are not unusual.

The general characteristics of each type of resonance must always be considered. For example, a structural resonance has minimal damping, and it will exhibit a sharp response with a high Q amplification factor. Resonances of this type have a narrow bandwidth, and the excitation frequency must be precisely equal to the resonance for any action to occur. On the other hand, rotor lateral resonances include damping from the bearings and seals, and they are active over an appreciable frequency range (bandwidth). These type of rotor resonances may be activated by an excitation force that falls anywhere within the bandwidth of the damped rotor resonance. The resonant frequency may also change with machine speed. For instance, bearing stiffness and damping characteristics vary with rotational speed. In most cases, the bearing parameters that apply as the machine passes through a critical speed are different from the normal operating condition. Hence, the same resonance may appear at different frequencies due to the influence of other associated machine elements (e.g., bearings).

Additionally, the exciting mechanism does not have to occur at rotating speed, or any even order harmonic. The excitation may occur at a subsynchronous (below running speed), or a supersynchronous (above running speed) frequency. For instance, a machine may startup with a locked 43% oil whirl instability. Presumably, the frequency of this instability would increase in direct proportion to the rotational speed. If the machine operated above twice the critical speed, the 43% whirl would eventually coincide with the rotor balance resonance during startup. This coincidence between the rotor resonance and the oil whirl would probably result in a re-excitation of the critical. The oil whirl would evolve into an oil whip, with potentially damaging implications to the machinery. Due to the severity of this problem, additional discussion and explanation will be presented later in this chapter.

As discussed throughout this text, machinery systems exhibit a wide variety of natural resonances. In the vast majority of cases, these resonances remain dormant, and their presence goes undetected. However, when an excitation does appear, or when the mechanical characteristics of the system undergo a change (due to failure or attrition); the idle resonance may become adversely excited. The solution to these occurrences typically resides in identification of the changes in physical machine parameters. With respect to resonance problems, the machinery diagnostician should always examine the mechanical system for evidence of variation of mass, stiffness, or the application of a new force. When the variant is discovered, the solution is close at hand.

Finally, there are groups of natural frequencies that may be discounted from a design standpoint, but they may have to be examined during a detailed machinery analysis. For instance, torsional resonances on a turbine driven compressor would probably not be a major cause for concern. However, if mechanical failures indicated the existence of twisting forces, torsional vibration should be considered. Torsional modes have low damping (high Qs) and they have caused many failures on large turbines.

Conversely, torsional resonances on a reciprocating engine would be of significant interest during initial design, and acceptance testing. However, if the main bearings displayed babbitt failures, the lateral vibration should be evaluated. Thus, a group of potential resonances and/or excitations should not be eliminated from possibility just because they may not apply. The other extreme of performing detailed examinations on mechanisms that have no possible relationship to the immediate problem should also be avoided. As always, the machinery diagnostician must exercise good engineering judgment when selecting or eliminating potential resonances for a machinery problem.

Case History 22: Re-Excitation of Compressor Resonance

Many refrigeration compressors in operating facilities such as ammonia or ethylene plants are configured with multiple side streams plus one or more levels of extraction. These refrigeration compressor systems are designed to maintain specific flow rates and refrigerant temperatures. In many cases, the process requirements are quite stringent, and the compressor is optimized to operate within a very limited performance envelope. These requirements pose some challenging system and machinery design considerations that often results in machines that are difficult to operate at *non-design* conditions.

For instance, consider a large six stage propylene refrigeration compressor that draws in excess of 30,000 horsepower at design operating conditions. The rotor weighs 7,100 pounds, and the normal operating speed for 100% plant load is 3,860 RPM. Obviously this speed will be subjected to control system adjustment dependent on the actual number of operating cracking furnaces, and the associated refrigerant load. A typical startup of this machine is described by the Bode plot shown in Fig. 9-14, plus the polar plot shown in Fig. 9-15. This transient data was acquired from the horizontal proximity probe at the coupling end of the compressor. On this machine, the horizontal vibration is normally larger than the vertical motion, and the discharge bearing vibration is slightly higher than the outboard suction end bearing. Hence, the data presented in Figs. 9-14 and 9-15 represents the highest vibration amplitudes encountered during a normal machinery train startup.

This startup was quite smooth with a clearly defined compressor first critical speed at 2,020 RPM. The observed critical response range extended from

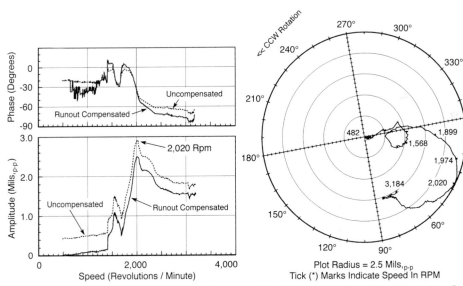

Fig. 9–14 Bode Plot Of Refrigeration Compressor Startup To Minimum Governor

Fig. 9–15 Polar Plot Of Refrigeration Compressor Startup To Minimum Governor

approximately 1,600 to 2,500 RPM. The peak response was somewhat sharp, but that was attributed to the fast ramp programmed into the electronic governor. Hence, this transient response was not indicative of any mechanical abnormality.

A complete set of steady state data was acquired twenty-four hours after startup. The orbit and time domain plots for the discharge end bearing are presented in Fig. 9-16 at an average speed of 3,680 RPM. The compressor orbits are forward, with an elliptical pattern at the discharge, and a circular motion at the suction (not shown). In addition, a subsynchronous instability was visible at the discharge bearing. The dominant direction of this low frequency motion was horizontal, and the frequency oscillated between 1,980 and 2,100 CPM. Extended observation of this subsynchronous component revealed that 2,060 CPM was the major frequency, and peak amplitudes reached 1.5 Mils,$_{p-p}$ at the discharge horizontal. Time averaged behavior is displayed on the FFT plot shown in Fig. 9-17 with a defined peak at 2,060 CPM, and an average horizontal amplitude of 1.1 Mils,$_{p-p}$. It should also be mentioned that subsynchronous vibration amplitudes at the suction bearing generally remained below 0.25 Mils,$_{p-p}$.

This 2,060 CPM frequency is recognized as the first critical speed of the compressor rotor. As previously noted, the startup data displayed a translational balance resonance (first critical) at 2,020 RPM. However, as stiffness characteristics change with speed, journal eccentricity, and temperature, the startup first critical is generally different from the critical response observed at full speed. Hence, the 2,060 CPM component is considered to be a re-excitation of the compressor first critical speed. This phenomena has occurred for over twenty years, and previous studies have correlated the compressor instability with the extrac-

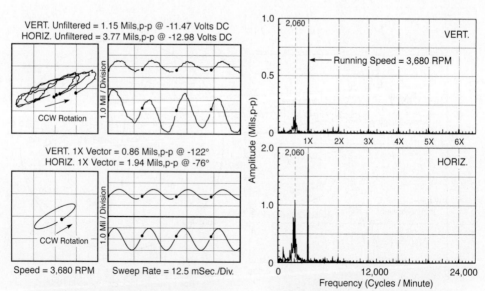

Fig. 9–16 Refrigeration Compressor Inboard Bearing Orbit And Time Base Plots Following Normal Train Startup

Fig. 9–17 Refrigeration Compressor Inboard Bearing Spectrum Plots Following Normal Train Startup

tion sidestream. Historically, startup of the propylene refrigeration system produces flow and pressure fluctuations in the extraction line. These fluid based excitations are transmitted to the discharge end of the compressor rotor via the horizontal extraction nozzle. The observed shaft motion is primarily horizontal, and the largest excitation occurs at the coupling (discharge) end of the rotor. Fluid variations in the extraction stream provides a broadband excitation to the compressor rotor, and the first critical speed is generally excited. In addition, the turbine first critical is also driven due to the close proximity between the compressor and turbine balance resonant speeds. Although the dominant subsynchronous vibration occurs at the compressor discharge bearing, the other machinery train bearings display low level excitations. However, as the refrigeration systems become *lined out*, the extraction stream flow instability diminishes, and excitation of the critical speeds of both rotors normally disappears.

In support of this explanation, the vibration response data acquired with an average speed of 3,860 RPM, and a full process is shown in Figs. 9-18 and 9-19. Although 1X running speed vectors have experienced minor changes, the subsynchronous motion no longer exists. Examination of the turbine data reveals a similar absence of motion at the first critical. Hence, the documented re-excitation of the compressor first critical speed under unstable extraction flows was eliminated by establishing normal loading of the propylene refrigeration system. Again, this behavior is totally consistent with the historical behavior of this machinery train.

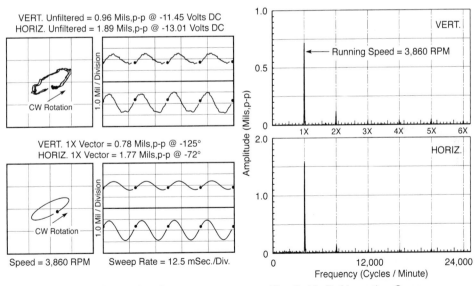

Fig. 9–18 Refrigeration Compressor Inboard Bearing Orbit And Time Base Plots Under Normal Process Load

Fig. 9–19 Refrigeration Compressor Inboard Bearing Spectrum Plots Under Normal Process Load

MACHINERY STABILITY

Throughout this text, the attributes of well balanced and properly aligned machines operating with concentric rotor elements have been repeatedly endorsed. Reductions in shaft preloads are generally associated with reduced forces and extended machinery life. In fact, many machinery problems that appear to be extraordinarily complex are often beat into submission merely by corrections to the basic mechanical parameters of balance, alignment, and element concentricity. There is an added dividend provided by smooth running machines in the area of incrementally improved efficiency. In essence, more of the input energy goes into productive work instead of being wasted on mechanical abnormalities.

However, the uninitiated may be surprised to find that in some situations, these fundamental corrections may result in an inoperable machine. There are many documented instances of properly executed balance or alignment corrections that have resulted in significantly higher vibration response amplitudes. In many of these cases, an examination of the vibratory characteristics has unveiled the presence of a new frequency component. Often this new vibration component occurs at frequencies below rotating speed, and this subsynchronous motion is often associated with machinery instability. Although this general definition of instability is not rigorously correct, it is still used throughout most industrial locations.

The very nature of centrifugal machinery provides the fundamental mechanism for this type of behavior. In all cases, it must be recognized that centrifugal machines consist of rotating cylinders or disks confined within stationary cylinders. If clearances between cylinders are large, there is no possibility for interaction between stationary and rotating parts. For example, a 6 inch diameter shaft rotating within a 20 inch diameter annulus will function in the same manner as it would in free space. However, as clearances decrease, there is increased opportunity for interaction between elements. For instance, if the 6 inch diameter shaft now rotates inside a 6.008 inch diameter bearing; interaction between cylindrical elements now exists across the contained fluid. The fluid might be steam, a process gas, a process liquid, oil in a seal, or oil contained within a bearing. The general type of behavior for a cylinder rotating inside of a stationary cylinder is depicted in Fig. 9-20.

From this diagram, it is anticipated that the rotating element establishes a minimum running clearance to the stationary cylinder. For an oil film bearing, this clearance would normally be identified as the minimum oil film. The active forces across the minimum oil film include the fluid radial force, plus a tangential component. In this simple example, these two forces should be vectorially equal to the shaft load. Thus, the oil film forces are in equilibrium with the shaft load. If Fig. 9-20 was representative of a journal and bearing in a horizontal machine, the shaft load would primarily consist of the shaft weight. Furthermore, the described system would exhibit a minimum oil film in the lower right hand quadrant of the bearing. Due to the counterclockwise rotation, it is intuitive that the shaft would *climb* the lower right hand side of the bearing. Addi-

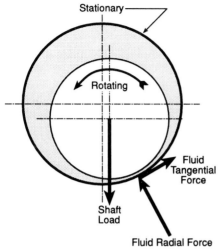

Fig. 9–20 Typical Radial Forces In A Fluid Film Bearing At The Support Point

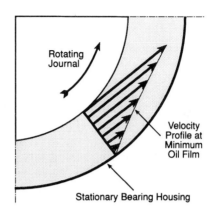

Fig. 9–21 Typical Bearing Oil Velocity Profile At The Point Of Minimum Oil Film

tional shaft loading would tend to drive the shaft further up the bearing wall. Donald E. Bently referred to this action as the *Newton Dogleg Law of Rotating Machinery*[4] where he stated:

"...For every one of the Forward Circular Eigen (Self-Excited) Malfunction Mechanisms of rotating machinery, from simple Oil Whirl and Oil Whip to the complex Aerodynamic Mechanisms, the rotor system responds to radial input force in dogleg style...Thus, if you push a CCW rotating rotor system down towards 6 o'clock position, if that system has any Forward Circular Mechanism, it responds by moving not simply to the 6 o'clock position, but to some position between 6 o'clock and 3 o'clock. This angle α is called the Attitude Angle.

This action is similar to gyroscopic action, but obeys a different set of laws. It is independent of the self-balancing laws, but gets interlocked with both, of course, in the rotor system behavior..."

Clearly, a direct and intimate relationship exists between the rotating system, the fluid film, and the stationary system. It is also obvious that a force balance must exist at the bearings, and that the fluid tangential force strives to move the shaft in a forward direction (i.e., with rotation). Any disruption of the force balance within this pure circular system will allow the minimum oil film to circumnavigate the bearing in a forward manner. The speed of progression is dependent upon the oil film velocity as shown in Fig. 9-21. It is self-evident that the journal rotational speed times the journal radius will provide the shaft surface velocity. This is the maximum oil film velocity. At the stationary bearing, the oil film velocity must be zero due to the non-rotative nature of the bearing. In a

---

[4] Donald E. Bently, "Attitude Angle And The Newton Dogleg Law Of Rotating Machinery" *Bently Nevada Applications Note*, (March 1977).

perfectly linear mechanical system, the average oil film velocity would be 50% of the maximum velocity, or 50% of the rotational frequency. However, in a real world mechanical system, the difference between journal and bearing surface conditions, the bearing geometry, bearing clearances, and loads will result in an average oil film velocity that is somewhat less than a simple 50% arithmetic average. The velocity profile will not be a straight line relationship, and the average oil film velocity will be less than 50%. This manifests as a frequency that appears at something less than 50% of running speed. In practice, ratios between 35% and 49% are commonly observed and documented.

This is the essence of a *self-excited* mechanism such as oil whirl. That is, the physical geometry and loading of the mechanical system allows the establishment of a subsynchronous vibration component in oil lubricated bearing. This motion is forward, circular, and typically appears between 35% and 49% of shaft rotative frequency. The simplest case is known as *oil whirl*, and it is generally detected in machines with plain sleeve bearings.

A visual dichotomy is always encountered when oil whirl orbits are observed on an oscilloscope. Specifically, this is a forward circular mechanism, but the Keyphasor® dots seem to spin backwards against rotation. This gives the appearance of a backwards motion, and a misinterpretation of the data is possible. In actuality, the backwards spinning Keyphasor® dots are perfectly correct for a forward subsynchronous vibratory component. In order to demonstrate this behavior, consider Fig. 9-22 which describes several revolution of a shaft experiencing a forward subsynchronous whirl.

In this diagram of a counterclockwise rotating system, consider the first Keyphasor® dot at position **A**. Assume a negative going Keyphasor® pulse, and the expected *blank-bright* sequence indicates a CCW rotation. As the shaft makes one complete revolution, the predominant subsynchronous component has only completed about 45% of a full cycle ($\approx160°$). At this instant in time, the second KeyØ® dot appears at position **B**. Note that the *blank-bright* sequence is still consistent, and representative of forward CCW motion. As additional cycles are completed, the KeyØ® dot continues to lag further and further behind. Thus,

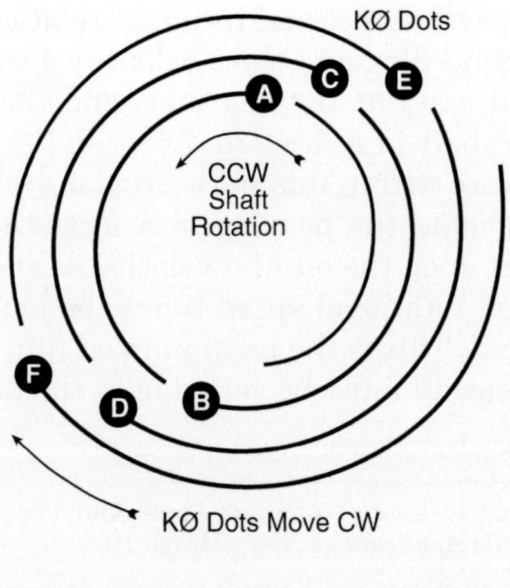

Fig. 9–22 Keyphasor® Dot Precession For Subsynchronous Excitation Occurring At Less Than 50% Of Machine Rotative Speed

a visual observation on an oscilloscope will reveal the dots moving from **A** to **C** to **E**. In the second group, the dots move from **B** to **D** to **F**. Therefore, the dots appear to roll backwards in time, but the *blank-bright* sequence specifically identifies a forward orbit.

This characteristic can be put to good advantage when viewing live subsynchronous data in an orbital display on an oscilloscope. That is, when the dots appear to move in a counter-rotation direction, the subsynchronous component occurs at less than 50% of running speed. Conversely, when the dots appear to move in the same direction as shaft rotation, the subsynchronous component occurs at a frequency that is greater than 50% of rotative speed. The middle condition of a pair of fixed Keyphasor® dots indicates a frequency that is locked onto 50% of running speed.

This is an extraordinarily important concept to remember and apply. It may be extended to three fixed Keyphasor® dots that identify a frequency component at exactly one-third of running speed. Four fixed KeyØ® dots identify a frequency component at exactly one-fourth of running speed, etc. In many instances, the orbital observation of fixed dots is faster, easier, and more accurate than performing an FFT analysis of the data. Finally, the rate of Keyphasor® dot rotation is directly related to the frequency difference between the subsynchronous component and 50% of running speed. If the KeyØ® dots are moving very slowly, the subsynchronous component is quite close to 50% of running speed. Conversely, if the dots are rolling around the orbit at a rapid pace, the subsynchronous component is considerably removed from one half of running speed.

The vibration characteristics of a machine experiencing oil whirl are demonstrated in Fig. 9-23. Operating at 2,832 RPM, the orbit and time base data in the top plot represents the unfiltered signals from a rotor kit with a diametrical

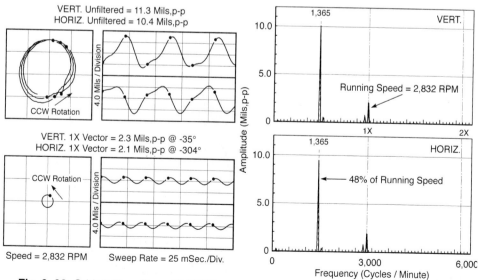

Fig. 9–23 Orbit & Time Base Of Oil Whirl

Fig. 9–24 Spectrum Analysis Of Oil Whirl

bearing clearance of 18 Mils. The orbit-time-base data shown in the bottom of Fig. 9-23 describes the low amplitude running speed (1X) vibratory behavior. For precise frequency identification of the subsynchronous component, the same information is viewed in the frequency domain in Fig. 9-24. From the FFT plots the whirl frequency is 1,365 CPM which is equivalent to 48% of shaft rotative speed. Again, the running speed motion is dwarfed by the violent oil whirl excitation at 48% of running speed.

Under the proper circumstances, oil whirl may turn into oil whip. If the machine operates at a speed of twice the critical, the potential for whip exists. In the previous example of a 48% whirl, the operating speed was about 2,832 RPM. This speed is below the first critical. It should be noted that the bandwidth of the critical speed range for this machine extends from 3,000 to 3,800 RPM. This machine went into whirl at 1,600 RPM, and maintained a steady 48% of running speed whirl during the initial speed ramp. This behavior is shown in Fig. 9-25 describing a cascade plot of vertical vibration spectra versus speed.

In this plot, the 48% Whirl tracked running speed until 6,200 RPM when the Whirl frequency (6,200 RPM x 48% = 2,980 CPM) began to move into the critical speed range. As shown in the cascade plot, the whip frequency remained locked into the critical speed range of 3,000 to 3,800 RPM as speed continued to increase. Under this condition, the increased rotor speed had minimal influence upon the whirl. That is, even at the top speed of 9,800 RPM, the oil whip was still trapped within the natural rotor resonance range. Thus, a resonant response can be obtained from a non-synchronous excitation.

Since this type of oil whip behavior involves the re-excitation of a major rotor resonance, it is reasonable to expect that oil whip may be a potentially dan-

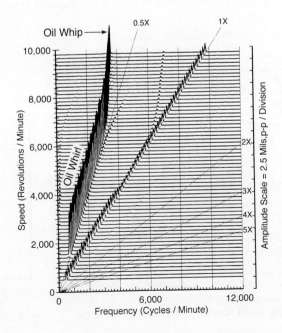

Fig. 9–25 Cascade Plot Revealing The Transition From Oil Whirl To Oil Whip

gerous and destructive mechanism. In numerous field cases, machines have operated successfully for many years with low levels of oil whirl. The whirl existed, it was nondestructive, and it was tolerated. However, few machines have successfully survived any type of extended operation with appreciable levels of oil whip instability.

A closer examination of the orbital and time domain vibration of this oil whip case is presented in the orbit and time base plots of Fig. 9-26. Note that the unfiltered amplitudes are in the vicinity of 9.0 Mils,$_\text{p-p}$, and the rotative speed

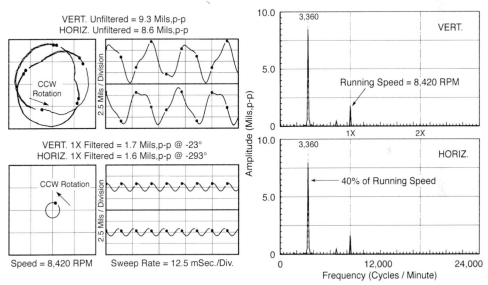

Fig. 9–26 Orbit Time Base Of Oil Whip   **Fig. 9-27** Spectrum Analysis of Oil Whip

was 8,420 RPM. From the orbit plots it is clearly demonstrated that this is a forward and circular mechanism. The companion spectrum plot of the vertical and horizontal probe signals is shown in Fig. 9-27. It reveals the precise oil whip frequency of 3,360 CPM. Under this combination of frequencies, the whip occurs at 40% of rotative speed. Also note that the whip frequency of 3,360 CPM falls directly into the critical speed range of 3,000 to 3,800 RPM previously identified.

Extended operation under this oil whip condition would probably be hazardous to the equipment. Stated in another way, it is generally agreed that startup and shutdown ramps should specifically minimize the time required to pass through rotor critical speeds. Under no conditions shall a machine be allowed to dwell within the bandwidth of the rotor resonance. However, in an oil whip condition, the machine is continually running at operating speed, and the rotor is violently shaking at its natural resonant frequency. It is no wonder that machines with oil whip often experience significant mechanical failures.

Oil whirl and whip serve as an introduction into the broad topic of machinery instability. Process machinery is susceptible to a wide array of instability mechanisms ranging from forced instability, to internal friction, to various types of rubs, and fluid induced aerodynamic instabilities. In some cases the simple self-excited mechanisms evolve into resonant excitations, and in other instances the minor instabilities are either tolerated or ignored. There are many excellent technical papers on this topic. Occasionally, the high level of technology necessary to examine and explain some of these instability mechanisms renders any potential solutions beyond practical approach. The machinery diagnostician is encouraged to seek out articles by Donald E. Bently[5,6], Edgar J. Gunter[7], plus Allaire and Flack.[8] These papers provide practical and understandable explanations for most of the common instability mechanisms.

As mentioned at the beginning of this section, the absence of preloads such as unbalance or misalignment may contribute to instabilities. Some types of whirl or whip may be adequately suppressed by the application of preloads such as intentional misalignment. A hot bearing pedestal may be cooled with water, or a cool pedestal may be heated with a steam hose to provide a minor degree of stabilizing misalignment. Other preload mechanisms may also successfully restrain these types of instability mechanisms.

Design characteristics that contribute to instability have been defined for many years. In fact, machinery designers generally perform many optimization studies during the development or design of a new bearing, seal, or rotor configuration. For instance, consider the stability diagram presented in Fig. 9-28. This diagram displays the stability curves for a plain journal bearing (solid line), and a particular pressure dam bearing (dashed line).

On this type of plot, stability occurs below each line, and instability is predicted for operation above each line. The vertical axis is the stability threshold speed, and the Sommerfeld number is plotted on the horizontal axis. This is the same non-dimensional number introduced in chapter 4 of this text. In Fig. 9-28 the various parameters that form the Sommerfeld number are also listed. Thus, it is possible to examine the general form of the Sommerfeld number for potential clues to stability variations. For example, changes in oil viscosity will directly influence the stability. Changes in viscosity are generally achieved by varying the oil supply temperature. It is clear from the stability diagram that increasing or decreasing the oil viscosity may prove to be beneficial or detrimental. It all depends on the particular operating location within this stability plot.

───────────────

[5] Donald E. Bently, "Forced Subrotative Speed Dynamic Action of Rotating Machinery," *American Society of Mechanical Engineers*, ASME Paper No. 74-Pet-16 (1974).

[6] Donald E. Bently, "Forward Subrotative Speed Resonance Action of Rotating Machinery," *Proceedings of the Fourth Turbomachinery Symposium*, Gas Turbine Laboratories, Texas A&M University, College Station, Texas (October 1975), pp. 103-113.

[7] Edgar J. Gunter, Jr., "Rotor Bearing Stability," *Proceedings of the First Turbomachinery Symposium,* Gas Turbine Laboratories, Texas A&M University, College Station, Texas, (1972), pp. 119-141.

[8] P.E. Allaire and R.D. Flack, "Design Of Journal Bearings For Rotating Machinery," *Proceedings of the Tenth Turbomachinery Symposium*, Gas Turbine Laboratories, Texas A&M University, College Station, Texas (December 1981), pp. 25-45.

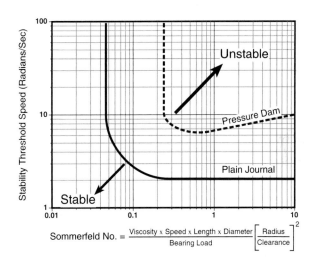

Fig. 9–28 Typical Stability Diagram For Pressure Dam And Plain Journal Bearing Configurations

$$\text{Sommerfeld No.} = \frac{\text{Viscosity} \times \text{Speed} \times \text{Length} \times \text{Diameter}}{\text{Bearing Load}} \left[\frac{\text{Radius}}{\text{Clearance}}\right]^2$$

Fig. 9-28 also shows that changes in speed, load, geometry and bearing clearance will all influence the bearing stability characteristics. Although the machinery diagnostician may not be able to fully compute the specific stability characteristics for a particular machine, the previous array of mechanical parameters does provide some measure of guidance for potential mechanical changes, or variations in operating conditions.

Case History 23: Warehouse Induced Steam Turbine Instability

The 4,700 pound steam turbine rotor shown in Fig. 9-29 produces approximately 15,000 HP at 5,500 RPM. This is a five stage turbine with a double flow exhaust to a surface condenser. The turbine drives a low pressure air compressor, a speed increasing gear box, and a high stage air compressor. This machinery train had a generally successful operating history that was occasionally marred by turbine alignment problems. The most recent difficulties were associated with subsidence of the foundation sub-structure, and the detrimental effect upon alignment within this machinery train.

Prior to a maintenance turnaround, the governor end shaft vibration

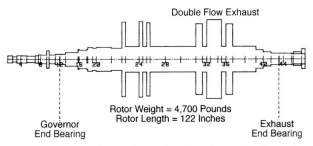

Rotor Weight = 4,700 Pounds
Rotor Length = 122 Inches

Governor End Bearing

Exhaust End Bearing

Fig. 9–29 Five Stage Steam Turbine Rotor Configuration

response data displayed low amplitudes, combined with a flat or preloaded shaft orbit, as displayed in Fig. 9-30. Concurrent with this condition, triaxial casing vibration amplitudes were much higher than anticipated. It appeared that the governor end bearing was heavily preloaded, and substantial energy was transferred through the bearings and into the governor housing. Based on this information, the governor end bearing was opened for inspection during the shutdown. It was discovered that the bearing had been heavily loaded, and it had sustained babbitt damage in the bottom half. A vertical alignment adjustment was made, and the governor end bearing liner was replaced. The exhaust end bearing was also inspected. It was still in good condition, and was reassembled without any replacements.

VERT. Unfiltered = 1.02 Mils,p-p @ -7.25 Volts DC
HORIZ. Unfiltered = 1.23 Mils,p-p @ -7.11 Volts DC

VERT. 1X Vector = 0.65 Mils,p-p @ -35°
HORIZ. 1X Vector = 0.88 Mils,p-p @ -210°

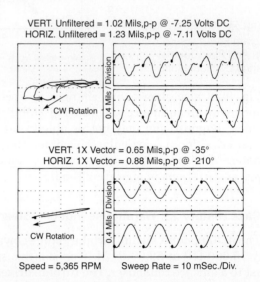

Fig. 9–30 Initial Steady State Data Of Preloaded Steam Turbine Governor End Bearing

Speed = 5,365 RPM   Sweep Rate = 10 mSec./Div.

The ensuing startup of this machine train was uneventful. Unfortunately, after several hours of full speed operation the turbine started to misbehave. The orbits and time base plots in Fig. 9-31 revealed maximum unfiltered amplitudes in the vicinity of 2.5 Mils,$_{p-p}$. A definite reduction in the orbit preload was apparent, along with the appearance of a subsynchronous component at nominally 2,640 CPM (49% of rotative speed). The bottom set of orbit and time domain plots in Fig. 9-31 were filtered at this subsynchronous frequency. Clearly, a forward, horizontally elliptical subsynchronous orbit was indicated. This was considered to be unusual, and further examination of the motion was necessary.

The slow roll data, and the initial DC gap voltages of the probes were somehow lost during the startup. Hence, there was no opportunity to examine any changes in relative shaft lateral position within the journal. There was also a reduction in casing vibration, but the integrity of the entire casing data array became highly questionable when a failed velocity coil spring was discovered.

Spectrum plots of the governor end shaft signals are presented in Fig. 9-32. This FFT data revealed the major subsynchronous component at 49% of rotative

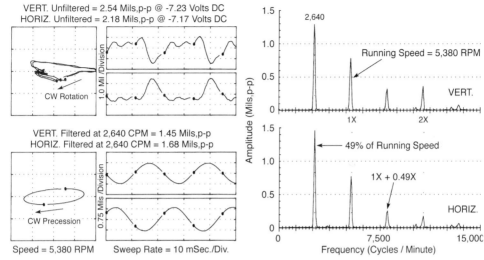

Fig. 9–31 Orbit And Time Base Plots Of Subsynchronous Steam Turbine Instability

Fig. 9–32 Spectrum Plots Of Subsynchronous Steam Turbine Instability

speed, plus the smaller excitation at running speed, and a small upper sideband at 1X+0.49X. As expected, plant operations personnel were reluctant to shutdown this machinery train for any additional mechanical inspections. Hence, the opportunities for comprehending the influence of the turnaround work, plus understanding the current abnormal behavior were rapidly dwindling.

At this point, the bearing stability characteristics were analyzed on a plot similar to the diagram presented in Fig. 9-28. Based on the double axial groove bearing installed in the turbine, it was determined that the stability threshold was approximately 5,500 RPM. This really meant that at speeds approaching 5,500 RPM the bearing would become unstable. Hence, operation at 5,370 RPM was uncomfortably close to this stability margin. Fortunately, operations personnel were agreeable to back off the machine load slightly, and at a speed of 5,200 RPM the subsynchronous response was substantially reduced.

Although this turbine was running in a more desirable manner, there were still many questions to be resolved. A key point was uncovered by one of the mechanics on the job when he pointed out the fact that the turbine used to run with pressure dam bearings instead of axial groove bearings. Additional calculations revealed that pressure dam bearings had a stability threshold in the vicinity of 14,000 RPM. Hence, the use of pressure dam bearings on this turbine would be a major stability improvement over the double axial groove liner installed during the overhaul.

Further investigation revealed that the culprit resided in the plant warehouse. It seems that the turbine was originally supplied with axial groove bearings, but they proved to be unstable at full process rates. The OEM designed a set of pressure dam bearings for this machine, and these modified bearings plus two sets of backup spares allowed many years of successful operation. Following the foundation subsidence problems, the turbine consumed several sets of bearings. As expected, the warehouse reordered spare bearings, but they used the part number for the original axial groove bearings instead of the newer part number for the pressure dam bearings. Thus, the wrong bearings were retained in stock, but no one noticed since the immediate problem was associated with the settling of the foundation.

Eventually, the foundation subsidence was controlled, and the affected baseplates were re-leveled, re-grouted, and the machinery train was properly aligned. The correct pressure dam bearings were installed in the turbine, and the warehouse corrected the erroneous part numbers. Shaft and casing vibration characteristics returned to normal, and the governor end bearing failures ceased. In the final overview, this turbine that required some type of maintenance on a regular basis was transformed into a machine that is only opened for inspection on 8 to 10 year increments.

Case History 24: Pinion Whirl During Coastdown

Most process machinery trains exhibit the highest vibration amplitudes during startup. Occasionally, some machines will vibrate excessively during coastdown. Such a machinery train is depicted in Fig. 9-33. In this train, the steam turbine driver accepts superheated steam, and it exhausts to a low pressure 35 Psi header. The turbine is rated for 9,300 HP at the maximum continuous speed of 11,900 RPM. The turbine operating speed range varies between 8,000 and 11,000 RPM. As shown in Fig. 9-33, the turbine is directly coupled to a two element speed reducing gear box. This is a down mesh box with 41 pinion

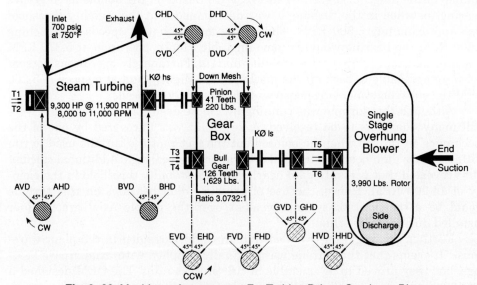

Fig. 9–33 Machinery Arrangement For Turbine Driven Overhung Blower

teeth, and 126 bull gear teeth. The gear set is a double helical arrangement, with a double acting bull gear thrust bearing. The bull gear output coupling drives a single stage overhung blower that contains a 3,990 pound rotor. This air machine accumulates various foreign substances, and it must be cleaned periodically.

Under normal startup conditions, this machinery train seldom exhibits shaft vibration amplitudes in excess of 3.0 Mils,$_{p-p}$. Operating at full speed and load, the runout compensated shaft vibration amplitudes are generally below 1.0 Mil,$_{p-p}$ at all measurement locations. From the train diagram it is clear that all bearings are monitored with two proximity probes. Each journal bearing has X-Y radial probes, and dual axial probes are mounted at each thrust bearing. Hence, the train is well-monitored, and the possibility of undetected high vibration levels on this train is quite remote.

Across the down mesh double helical gear box it is significant to compute the forces associated with this machinery. For example, Fig. 9-34 identifies the bull gear weight as 1,630 pounds, and the pinion weight as 220 pounds. For a reduced load operation of 4,000 HP, the transmitted forces are shown for a pinion speed of 5,900 RPM, and 1,920 RPM on the bull gear.

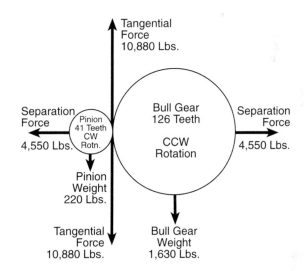

Fig. 9–34 Summary Of Significant Radial Forces Present In Mating Helical Gear Set

Under this reduced load operation, the tangential gear forces are equal to 10,880 pounds, and the companion gear separation forces are equal to 4,550 pounds. These gear contact forces are substantially larger than the weights of the gear elements, and they dominate the force structure on this machine. This relationship between forces is typical for many industrial gear boxes.

It is clear that significant forces are active across the gear teeth, and the bearings respond with large stiffness values. In most startup and operating conditions, the gear elements exhibit minimal shaft relative vibration. However, during coastdown, the normal behavioral pattern is reversed, and the pinion displays high shaft vibration as shown in the partial Bode plot of Fig. 9-35. From

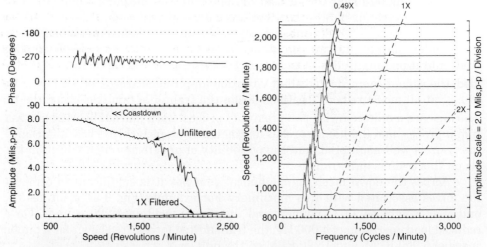

Fig. 9–35 Partial Bode Plot Of Pinion Vibration During Normal Coastdown

Fig. 9–36 Partial Cascade Plot Of Pinion Vibration During Normal Coastdown

full running speed to approximately 2,250 RPM the pinion motion remains below 0.5 Mils,$_{p-p}$. At a speed of 2,250 RPM, the unfiltered pinion vibration levels suddenly begin to climb, and it reaches a maximum amplitude of 8.0 Mils,$_{p-p}$ at approximately 800 RPM. As shown in Fig. 9-35, the synchronous 1X amplitudes remain small, and the 1X phase angle remains essentially constant. The majority of the shaft vibration must occur at a frequency other than rotative speed. The examination of the cascade plot in Fig. 9-36 reveals that the majority of the shaft motion occurs at 49% of running speed. The shaft orbits and time domain plots at 1,000 RPM are shown in Fig. 9-37 at a coastdown speed of 1,000 RPM.

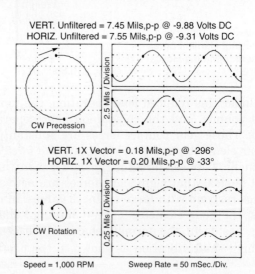

Fig. 9–37 Orbit And Time Base Plots Of Sustained Oil Whirl On Pinion During Coastdown

From this information it is clear that the pinion is experiencing a forward circular oil whirl mechanism at 49% of rotative speed.

This oil whirl instability is almost totally associated with the reduction in gear contact forces as the machinery slows down. The most susceptible element in the machinery train is the 220 pound pinion. It is concluded that, as the unit slows down, the gear forces are gradually diminished, and the unloaded pinion goes into an oil whirl that eventually occupies the full bearing clearance. Other gear boxes also display this same type of coastdown whirl. Since this is an unloaded condition, the transmitted forces between the pinion journals and the bearings are negligible. The whirl will generally remain active until the machine stops rotating. This type of whirl is generally non-destructive, but the diagnostician should always check gap voltages before startup, and at full speed and load to verify that the bearings have not been damaged.

In this particular gear box, the 8.0 Mil,$_{p-p}$ excursions during coastdown have occurred with clock-like regularity. As noted, this type of oil whirl is generally non-destructive, but it cannot be ignored. It could be reduced or eliminated with a pinion bearing redesign. However, the expenditure for this modification could not be justified by the operating company. The unfortunate part of this story is that the unit is equipped with 0 to 5.0 Mil,$_{p-p}$ radial vibration monitors. During coastdown of this train, the operators are accustomed to having the pinion vibration monitors fully pegged at speeds below 2,000 RPM. No one has any idea if the maximum whirl amplitude approached 8.0 Mils,$_{p-p}$ (equivalent to the pinion bearing clearance) — or if the maximum whirl amplitude was 12.0 or 15.0 or 20.0 Mils,$_{p-p}$ (indicative of pinion bearing damage). There are many ways to acquire and present this valuable coastdown information to the control room operators. However, it is anticipated that no action will be taken until the pinion bearings are wiped out during some future coastdown, and the operators attempt to restart the machinery train with damaged pinion bearings. At that time, the repair costs for the gear box, plus the associated production losses, will make the condition monitoring expenditures appear to be insignificant.

MECHANICAL LOOSENESS

The last section addressed instabilities such as oil whirl and whip. The self-excited whirl may be attributable to various sources of *cylinders within cylinders* such as the bearings, seals, balance pistons, or wheels. As previously stated, whirl generally appears at subsynchronous frequencies between 35% and 49% of rotative speed. On the other hand, whip behavior occurs at resonant frequencies that are consistent with the current effective stiffness. For rotors that operate above one or more resonances, the potential whip frequencies can vary over an appreciable frequency range. However, there are other subsynchronous mechanisms that appear on centrifugal machines. Some of these mechanisms produce significant excitations at fractional frequencies of rotational speed. For instance, vibration components at precisely 1/2, 1/3, and 1/4 of running speed may appear.

If rotating speed is changed, the subsynchronous component will change to maintain a constant fractional relationship to the running speed. Thus, the subsynchronous vibration component remains *locked-in*, or fixed at a fraction of rotational speed. It should be clarified that these characteristics are applicable to flexible rotors that operate above one or more shaft critical speeds. For stiff shaft rotors that operate below any shaft critical, the appearance of fractional vibration components with an appreciable amplitude is much less common.

Many of these fractional frequency mechanisms were examined and discussed by Donald E. Bently[9] in 1974. He showed that stiffness variations could result in a change in the effective balance resonance frequency (critical speed) of a machine running at full operating conditions. When this behavior is combined with a force such as rotational speed unbalance in a lightly damped system, this may produce a forced excitation of a system resonance.

Some individuals might argue that stiffness cannot change, and since the mass is constant, the rotor natural frequency must remain at a fixed value. However, it must be recognized that stiffness is not a static quantity. Stiffness does change with variations in parameters such as speed, bearing preload, physical clearances, mechanical fits, applied loads, etc. In chapter 4 of this text it was shown that oil film stiffness varies with machine speed (e.g., Fig. 4-1). In the same chapter, the measurements of housing stiffness (e.g., Figs. 4-18 and 4-20) also reveal variations in the housing stiffness as a function of frequency.

In all cases, the observed critical speed measured during a startup or a coastdown is predominantly governed by the effective stiffness that is active at that specific transient speed condition. To state it another way, the physical parameters that contribute to the transient critical speed do not have the same values at full operating speed and load. Although the effective values of transient versus steady state stiffness may be close to each other, the physical situation has changed, and the full speed critical may vary from the transient critical. This behavior is clearly demonstrated by a comparison of startup versus coastdown vibration response characteristics. The resonance identified during a startup will generally be higher than the coastdown value. In essence, the mechanical system displays an effective relaxation of stiffness during the shutdown. Thus, stiffness will vary, and the rotor resonance(s) will be directly associated with the effective stiffness at any point in time.

Back to the condition of a machine running at full speed. If it is assumed that the rotor is operating above one or more rotor critical speeds, and if there is an appreciable synchronous forcing function such as unbalance — the stage is set for a variety of subsynchronous mechanisms. For example, if a stationary machine part such as a bearing housing becomes loose, the effective support stiffness of the rotor is reduced in accordance with general equation (4-15), or the simpler format of equation (4-16). This combination of a synchronous force combined with a reduced support stiffness allows the rotor resonance to shift into a frequency that is an even multiple, or a fixed fraction of rotating speed.

---

[9] Donald E. Bently, "Forced Subrotative Speed Dynamic Action of Rotating Machinery," *American Society of Mechanical Engineers*, ASME Paper No. 74-Pet-16 (1974).

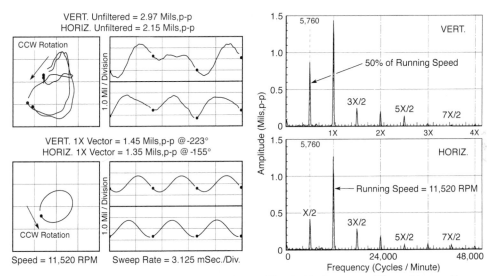

Fig. 9–38 Orbit and Time Base Plots of Loose Bearing with Locked-In Vibration Component At One Half of Running Speed

Fig. 9–39 Spectrum Plots of Loose Bearing With Locked-In Vibration Component At One Half of Running Speed

For demonstration purposes, consider the orbit and time base plots shown in Fig. 9-38, plus the spectrum plots in Fig. 9-39. This data is representative of a unit running at 11,520 RPM with a startup first critical speed of 6,200 RPM. For a machine running at this speed, the nominal 1.4 Mils,$_{p-p}$ of 1X synchronous motion is somewhat higher than desired. This running speed vibration could be due to mass unbalance, excessive bearing clearance, or a combination of the two. In this case, the machine displayed a locked-in 50% component at 5,760 CPM. The FFT plots also displayed interactions between the 1X running speed and the X/2 subsynchronous frequency. This included fractional frequency components at 3X/2, 5X/2, plus a minor response at 7X/2. Overall, the 50% fractional frequency is considered to be the first critical that has been lowered by a reduced support stiffness. Interaction between the 1X and the X/2 frequencies produces the string of fractional speed components.

This data was obtained shortly after startup, and the overall vibration levels were considered to be unacceptable. The unit was shutdown, and it was discovered that the outboard bearing cap bolts were not properly torqued. This loose bearing housing reduced the effective system stiffness, and combined with the unbalance forcing function to generate the behavior displayed in Figs. 9-38 and 9-39. Tightening the bearing cap bolts to the correct torque level corrected this problem, and the 50% of rotative speed component plus the string of fractional components were eliminated.

In the overview, the generation of subsynchronous vibration components that are integer fractions of rotative speed (e.g., X/2, X/3, or X/4) primarily appear in underdamped systems that have a loose stationary mechanical element combined with a high amplitude forcing function such as unbalance. It is possible to display fractions of excitations at other frequencies, but rotational speed unbalance is the most common driving force. When this energy is transferred to the subsynchronous region, the resultant frequency is a re-excitation of a fundamental rotor resonance (with a decreased stiffness). The fractional frequency will generally have a forward precession, and the locked-in subsynchronous component will interact with the other major excitations to produce a string of fractional frequencies.

It should also be mentioned that pulsating torque in a motor or generator may produce similar characteristics. Also, machines with various types of rubs or asymmetrical shaft stiffness (non-circular cross section, or near a keyway) will exhibit some of the same behavior. The diagnostician should also be aware that some cases of mechanical looseness (e.g., loose hold down bolt) may only appear as a high vertical vibration component at running speed on the casing or baseplate. This fundamental synchronous motion may become truncated due the relative movement or hammering of parts, and the truncated 1X sine wave would then exhibit a series of running speed harmonics without the presence of any subsynchronous frequency (e.g., Figs. 7-35 and 7-36).

Case History 25: Loose Steam Turbine Bearing

Another example of this behavior is shown in the steam turbine orbits presented in Fig. 9-40. This data was acquired on the exhaust bearing of an 8 stage turbine that was rated at 5,400 HP at 7,490 RPM. This steam turbine was driving a six stage propane compressor. Due to an operational error, this refrigeration train had been subjected to a period of overspeed operation. This resulted in significant mechanical damage, and it forced a major overhaul of both machines. Following the re-build, the unit had a normal startup, and it displayed higher than desired vibration levels at the turbine exhaust bearing. During the next week, vibration amplitudes gradually increased, and the data shown in Fig. 9-40 represents the shaft vibration 9 days after startup.

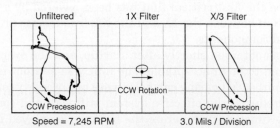

Fig. 9–40 Shaft Orbits Of Loose Steam Turbine Bearing Housing with Locked-In Vibration Component At One Third of Running Speed

The unfiltered orbit at the left side of Fig. 9-40 had vertical amplitudes in excess of 9.0 Mil,$_{p\text{-}p}$. Since the bearing diametrical clearance was only 6.0 Mils, the observed vibration was substantial. By comparison, the 1X rotational speed orbit shown in the middle of Fig. 9-40 seemed to be quite small, but the horizontal shaft motion at 7,245 RPM was 1.7 Mils,$_{p-p}$. Hence, the running speed vibration was appreciable, but it was dwarfed by the subsynchronous component. The orbit on the right side of Fig. 9-40 was filtered at X/3 or 2,415 CPM. Clearly, the 3 stationary Keyphasor® dots are indicative of a locked component at one third of running speed. Furthermore, the precession of this X/3 vibration is counterclockwise. which is forward, and in the direction of shaft rotation.

This shaft vibration data is shown in the frequency domain in Fig. 9-41. Once again, the dominant excitation at one third of running speed appears at 2,415 CPM. In addition, the interaction components between 1X and X/3 are identified at 2X/3, 4X/3, 5X/3, and 7X/3. The same type of behavior is transmitted through to the bearing housing as illustrated in the casing plots of Fig. 9-42.

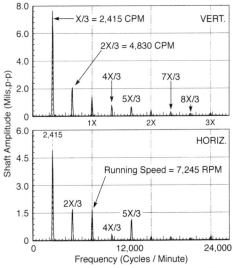 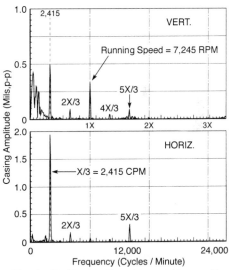

Fig. 9–41 Shaft Spectrums of Loose Turbine Bearing Housing with Locked-In Component At One Third of Running Speed

Fig. 9–42 Casing Spectrums of Loose Turbine Bearing Housing with Locked-In Component At One Third of Running Speed

To verify if this subsynchronous frequency tracked running speed, a 1,000 RPM speed change was performed. At 7,890 RPM the X/3 component had increased to 2,630 CPM, and it was still locked into one third of rotative speed. As the train speed was lowered to 6,870 RPM the subsynchronous frequency dropped to 2,290 CPM which was still exactly one third of running speed. A further speed decrease to 6,840 RPM was sufficient to completely eliminate the subsynchronous vibration component. However, as soon as the X/3 component decayed, the rotational speed vibration increased to 5.6 Mils,$_{p-p}$ horizontally, and 5.2 Mils,$_{p-p}$ vertically. As speed was increased, the rotational speed energy was transformed back into the X/3 subsynchronous component, and it reappeared in an identical manner to the previous Figs. 9-40 through 9-42.

Based on this information, it was concluded that mechanical looseness in the turbine exhaust end bearing housing was the most probable culprit. The unit was operated for another two weeks before it could be shutdown for repair. At that time it was discovered that the exhaust end bearing was improperly fitted into the housing, and the bearing was actually supported by the anti-rotation pins. As an interim measure, stainless steel shims were installed between the bearing and the housing to *fill in* the clearance cavity. This temporary fix proved to be quite effective, and the turbine operated with shaft vibration amplitudes of less than 2.0 Mils,$_{p-p}$ for the duration of the two-year process run.

ROTOR RUBS

The physical contact between rotating elements and stationary machine parts can generate a variety of rub conditions. For example, the following general categories of rubs are encountered on process machinery.

❍ Laby Rubs — Shaft against close clearance aluminum labyrinths.
❍ Intentional Rubs — Rotating labyrinths cutting into abradable seals.
❍ Light Rubs — Short duration rubs due to process or external upsets.
❍ Intermittent Rubs — Due to tight clearances or pseudo bearings.
❍ Heavy Rubs — Due to foreign object ingestion, blade loss, or bearing failure.
❍ Catastrophic Rubs — Heavy radial or axial rub due to broken shaft or failed coupling, also may occur during extensive blade failures.

The first category of **laby rubs** are quite common in machines such as centrifugal compressors with interstage aluminum labyrinths. During an overhaul, these labys are often installed with undersized radial clearances, and the rotor establishes the running clearances during startup and normal operation. These are usually minor rubs that often escape detection due to their minimal severity.

The **intentional rub** category addresses the cases where abradable seals are installed at impeller eyes, between impellers, or around balance drums. These close clearance seals consist of a stationary abradable material combined with rotating labyrinths. Typically, the machine is placed on slow roll, and the rotating labyrinths cut their own running clearance into the abradable material. In some cases the running clearances may be established in less than an hour, and in other situations a full 8 hour shift might be required for the labys to cut the proper clearances. As the machinery is started up, additional rubs may occur as the labys cut further into the abradable material.

Light rubs during normal operation may be due to process upsets where liquids are carried over into a compressor, or a minor surge develops due to downstream process control problems. External influences such as earthquakes, or heavy equipment rolling by the machinery deck may be sufficient to excite the rotor and/or casing and produce a brief contact between the stationary and rotating machine elements. These types of events are also hard to detect unless the diagnostician happens to be viewing vibration signals on an oscilloscope at the time of the rub event.

Intermittent rubs due to tight clearances are the most commonly observed rub. This malfunction will either clear itself, or progress into a heavier rub with associated mechanical damage. The tight clearances may be due to mispositioned stationary machine elements such as seals, or cocked rotating elements such as impellers or thrust collars. These close clearance elements often act as pseudo bearing for the rotor system, and they provide additional lateral restraint. The extra stiffness due to the rubbing element is often sufficient to temporarily raise the critical speed, and allow the rotor to lock into this subsynchronous resonance. In the frequency domain, the intermittent rub looks like a loose bearing housing with integer fractions of rotative speed (e.g., X/2, X/3, or X/4) plus a string of fractional frequencies (e.g. 3X/2, 5X/2, 7X/2, etc.). This mechanism requires an initial driving force such as unbalance, and it is more prevalent in systems with low damping. In many cases, the dominant subsynchronous component will have a backward precession, and the orbits may be unidirectional.

The appearance of **heavy rubs** due to problems such as the ingestion of foreign objects, the breakage or failure of turbine blades or compressor impellers, plus the failure of journal or thrust bearings. These types of rubs are driven by a large 1X rotational speed force, and they typically result in the automatic shutdown of the machinery. For units that are not equipped for auto trip on high vibration, the excessive shaft and casing vibration levels usually convinces the operators to shutdown the equipment. These types of rubs are generally not investigated for very long because the machinery is shutting down.

The final category of **catastrophic rubs** are caused by failures on the rotating assembly. These full radial or axial rubs may be due to broken shafts, failed couplings, cracked gears, multiple blade failures, or any other malfunction that compromises the structural integrity of the rotating assembly. Rubs of this type are fully destructive, and when they occur, the machine is definitely coming down with excessive internal damage. In some situations, various rotor elements may even break through the casing and produce additional destruction.

From the above discussion it is reasonable to conclude that many types of machinery rubs are possible, but very few rubs are of sufficient duration to be properly investigated. The intermittent rubs due to tight clearances are probably the only type of machinery rub that is commonly encountered and documented. For instance, consider the spectrum plots of a shaft rub in Fig. 9-43. This data was obtained on a 5,760 RPM rotor, and it clearly shows a major subsynchronous excitation at 1,440 CPM. This frequency is exactly one fourth of running speed. The two FFT plots also show the interaction between the rotational speed (1X) vibration and the subsynchronous component (X/4) as a string of fractional frequencies (X/2, 3X/4, 5X/4, etc.). For reference purposes, the measured first critical speed of this rotor was approximately 1,300 RPM. It is logical to believe that the subsynchronous component at 1,440 RPM is probably a re-excitation of the first mode with additional lateral stiffness.

Obviously the data in Fig. 9-43 closely resembles the case of a loose stationary element that was discussed in the previous section. This is a difficult situation for the machinery diagnostician since both malfunctions (looseness and rub) look the same in the frequency domain. If you shutdown the equipment and ini-

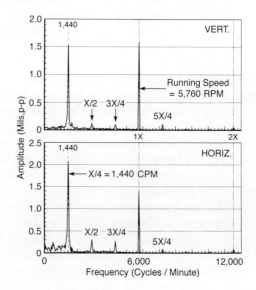

Fig. 9–43 Spectrum Plots of Radial Shaft Rub And Locked-In Frequency Component At One Fourth of Running Speed

tiate an inspection for looseness in one of the stationary elements, and actually discover a shaft rub, this can be quite embarrassing. One way to gather more information on the actual mechanism is to do a complete job of examining the available data. Specifically, if the shaft vibration FFT data shown in Fig. 9-43 is presented in the orbital domain, the data shown in Fig. 9-44 may be examined.

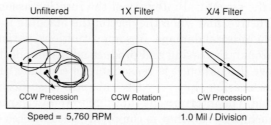

Fig. 9–44 Orbits Of Radial Shaft Rub And Locked-In Frequency Component at One Fourth of Running Speed

The unfiltered orbit on the left side of Fig. 9-44 shows the influence of the one quarter running speed component, plus an overall counterclockwise precession. This is in the same direction as the synchronous 1X orbit showing a counterclockwise rotation in the middle diagram of Fig. 9-44. However, when a band-pass filter is applied to the overall signal, the data may be filtered precisely at the X/4 subsynchronous frequency of 1,440 CPM, as shown on the right side of Fig. 9-44. In this orbit, notice that the precession is clockwise, or in the direction against rotation. This behavior is due to the fact that when the rotor hits the stationary rub point, the shaft is *kicked back against* rotation, which produces the reverse precession orbit at the subsynchronous frequency.

Depending on the type of rub, the subsynchronous frequency might drift around slightly, and it may not stay *locked into* a fraction of running speed. This is simply due to the fact that the rub conditions are continually changing, and this produces variations in the rotor response. In addition, the subsynchronous motion might drift back and forth between forward and reverse precession. Again, this is due to the variations in the rub. In virtually all observed cases, a heavy rub is characterized by a reverse precession at the subsynchronous frequency, plus a string of fractional components at appreciable amplitudes. Finally, the multiple harmonics of running speed and the subsynchronous component normally appear on both the shaft and the casing. However, the precession of the casing subsynchronous motion will not be as reliable of a rub indicator as the precession of the shaft orbit.

CRACKED SHAFT BEHAVIOR

One of the most dangerous problems in a mechanical equipment train is a shaft failure. When this occurs, the transmitted horsepower across the broken rotor is suddenly released. This energy may be consumed or dissipated by the process stream(s), or it may sling parts such as couplings considerable distances from the machinery. There have been reported cases of machine parts traveling half a mile or more from the point of failure. Hence, cracked shafts are potentially hazardous to the physical plant facility, as well as the local personnel.

Machines that are subjected to frequent startups and shutdowns appear to be more susceptible to shaft cracks due to the increased number of cycles through the rotor resonance(s), plus the process heating and cooling. This is particularly true on power generation units that are started and stopped daily, and rotors that are subjected to shaft bending modes. Overhung rotors with large wheels such as vertical pumps also appear to be prime candidates for cracked shafts. Certainly, reciprocating machines with their alternating forces and stresses may likewise experience this type of failure.

Cracks may originate at high stress points such as the square corners of a keyway, or they may occur underneath wheels, impellers, or inside hollow shafts. In some instances, crevices or scratches may be subjected to chemical attack, and these locations may grow into full shaft cracks. In other cases, the rotor may be subjected to stress corrosion cracking. Obviously when lateral shaft bending modes are involved, or stress reversals occur across torsional nodes, there is a potential for this type of failure mechanism. It is virtually impossible to define and categorize all of the plausible cracked shaft failure scenarios due to the potential combinations of lateral and shear forces, bending moments, and torsional forces for every type of machine.

However, there is a common ground for detection of cracked shafts when it is recognized that shaft stiffness must decrease in the presence of a crack. This basic concept allows the detection of shaft growth by measuring the historical changes in synchronous 1X shaft vibration. For instance, Figs. 9-45 and 9-46 document the vibration characteristics at the exhaust end of a steam turbine. This

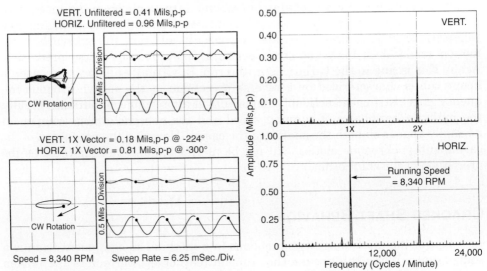

Fig. 9–45 Orbit And Time Base Plots Of Steam Turbine With Solid Shaft

Fig. 9–46 Spectrum Plots Of Steam Turbine With Solid Shaft

data was taken under normal steady conditions with a solid rotor. The presence of the 0.25 Mil$_{p-p}$ component at twice running speed 2X is nothing more than the influence of a surface imperfection below the proximity probe. This same steam turbine with a cracked exhaust end shaft is presented in Figs. 9-47 and 9-48. This 45° crack was due to torsional fatigue, and it was estimated that the crack

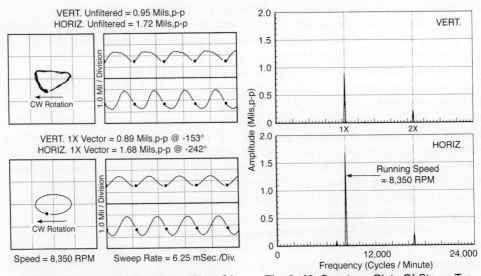

Fig. 9–47 Orbit And Time Base Plots Of Steam Turbine With 10% Shaft Crack

Fig. 9–48 Spectrum Plots Of Steam Turbine With 10% Shaft Crack

extended through 10% of the shaft diameter. Note that the 1X rotational speed vector changed, but there was minimal variation throughout the remainder of the vibration signals. Some might argue that there must be measurable changes in the higher order running speed harmonics such as the 2X, 3X, or 4X. Although these types of harmonic changes occur is some situations, the experience of the senior author of this text is that changes in the running speed 1X vectors represents the primary indication of cracked shafts. Investigators such as Bently and Muszynska[10] have found some useful information at higher order harmonics, but the fundamental mechanism still appears at rotational speed for the vast majority of the documented cracked shafts.

Depending on the type and location of the crack, plus the magnitude of the runout vectors, it is reasonable to expect that 1X vector changes due to shaft cracks may appear as either increasing or decreasing vibration amplitudes. In the example shown in Figs. 9-45 through 9-48 the 1X vectors changed on both of the proximity probes. However, the cracked shaft data shown in 9-47 and 9-48 does not appear to be abnormal. If the initial plots (9-45 and 9-46) were not available for historical comparison, there would be minimal reason to suspect a machinery problem based only on Figs. 9-47 and 9-48. Hence, the careful trending of the 1X vibration vectors is mandatory for proper condition monitoring.

Other symptoms of cracked shafts may appear on many types of machines. For instance, unexplained changes in shaft slow roll vectors may be due to a shaft crack. For any machine, the slow roll runout vectors measured with proximity probes under similar conditions must be repetitive. In addition, any unexpected changes in full speed 1X vectors due to minor process variations, small speed changes, or minor load changes should be viewed with suspicion. Furthermore, startup and shutdown data on major machinery trains should always be acquired and examined. Appreciable changes to balance resonance frequencies (critical speeds) or variations in amplification factors should be questioned.

Some machines are equipped with sufficient radial vibration transducers to be able to detect the general shaft mode shapes of the entire machinery train. This is particularly feasible for large turbine generator sets with hard couplings and multiple locations for radial vibration measurements. For these types of machines it makes sense to plot the runout compensated 1X running speed vectors into a machinery train mode shape using one of the techniques discussed in chapter 3. These train mode shapes should be documented in a *new* or *re-built* condition, and checked periodically with current measurements. Any significant variations in these mode shapes may be indicative of a developing shaft crack.

Finally, abnormal response to balance shots should be carefully examined. It makes sense to compute the full array of balance sensitivity vectors using equation (11-17), and tabulating the results (e.g., Table 11-1). Unusual changes in these presumably constant balance sensitivity vectors may be an early warning of a potential crack in a machine shaft.

---

[10] Donald E. Bently and Agnes Muszynska, "Detection of Rotor Cracks," *Proceedings of the Fifteenth Turbomachinery Symposium*, Turbomachinery Laboratory, Texas A&M University, College Station, Texas (November 1986), pp. 129-139.

Case History 26: Syngas Compressor with Cracked Shaft

Over the years various alliances have been formed between operating companies to share technology and spare parts for similar operating plants. These *spare parts pools* have achieved some good measures of financial and technical success. However, there are occasional problems that develop in this type of environment that are not solved, but are inadvertently passed along from one partner corporation to another. The following case history describes the events associated with one such machinery problem.

The rotor in question was a common spare for a high stage syngas compressor. It was assembled with a completely refurbished set of diaphragms into a complete bundle assembly. This spare bundle was installed in a barrel compressor during a routine maintenance overhaul. The spare rotor dimensions, impeller fits, axial spacing of elements, and all bearing clearance were well within normal specifications. All dimensions had been checked by two independent inspectors, and there was no reason to anticipate any problems with this machine.

Following the conclusion of this maintenance turnaround, the operating personnel experienced the usual array of startup problems. When the syngas train was ready for slow roll, the majority of the problems in the remainder of the plant had been corrected. It certainly appeared that a long month of intensive work was rapidly drawing to an end. The syngas train startup was initially

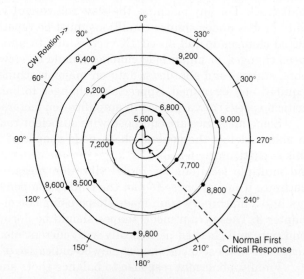

Fig. 9–49 Coastdown Polar Plot Of Syngas Compressor With Subsurface Circumferential Crack In The Shaft

Dots (•) Indicate Speed in RPM
Plot Radius = 6.0 Mils,p-p

uneventful, and all 4 rotors passed through their respective critical speed ranges with no problem. In fact, the maximum vibration level of the high stage compressor through the critical speed was less than 1.0 Mil,$_{p-p}$. At this stage, everyone was starting to smile, and beginning to think about a full nights sleep, plus a Sunday afternoon barbecue. Unfortunately, the festive atmosphere in the control room grew progressively quieter as the syngas train speed increased, and the high stage compressor vibration continued to grow. Normal operating speed for this machinery train was in the vicinity of 10,200 to 10,500 RPM. At an operating speed of 9,800 RPM the shaft vibration was all at 1X running speed, and the runout compensated 1X amplitudes were nominally 5.3 Mils,$_{p-p}$. Since the bearing diametrical clearance was slightly less than 8.0 Mils, it was considered to be unwise to allow any further increases in shaft vibration.

The syngas train was then slowly unloaded, speed was gradually decreased, and the polar plot data presented in Fig. 9-49 was recorded. This data was post processed from the field tape recording. During the actual coastdown, the oscilloscope revealed that the 1X amplitude decreased, and the 1X phase rolled around the orbit as speed and load were reduced. This field observation was fully reflected in the spiraling polar plot shown in Fig. 9-49. That is, the 1X amplitudes decreased with speed, and the phase angle rolled continuously. The phase actually changed by about 360° for every 1,000 RPM of speed drop. This was certainly abnormal behavior, and there was no direct explanation. As the compressor reached the critical speed range, it passed through this balance resonance with low amplitudes, and normal 1X vector response.

After shutdown, the bearings were pulled and inspected, but no significant damage was found. The coupling to the low stage syngas compressor was thoroughly checked, and it was found to be in excellent condition. All four of the radial, and both of the axial proximity probes on the high stage were checked for proper calibration on the actual shaft material. Again, no abnormalities were discovered. All of the compressor hold down bolts were checked, the foundation was checked, all of the large bore piping was examined for correct location and proper spring hanger settings. Again, there was no obvious culprit to blame for the observed behavior of the high stage compressor.

Some of the plant personnel wanted to put it back together, cut the probe wires and *just go and run* the machine. Fortunately, the plant management elected to strip down the compressor and find the root cause of the problem. Although this was a difficult decision, it was certainly the correct thing to do.

Shop disassembly of the unit provided no additional clues at to the origin of the abnormal 1X vector changes with speed and load. All of the impellers, the balance piston, and the thrust collar dimensions were proper, and well within normal assembly tolerances. All of the diaphragm fits and clearances were proper, and axial spacing was correct throughout the entire length of the bundle. All internal passages were clear, and the seals were in good condition. Initially, the only good news from the shop disassembly and inspection was that the mid-span impeller labyrinths had contacted the shaft. This physical observation reinforced the high vibration measured by the shaft proximity probes.

As potential failure mechanisms were systematically eliminated, the probability of a cracked shaft became more and more plausible. To check for cracks in the shaft and impellers a dye penetrant inspection was used. The dye check produced no indication of shaft cracks, and a further test of the shaft was conducted with the additional sensitivity of a Zyglo inspection. Once again, there was no indication of any shaft crack or discontinuity. At this stage, there was no physical evidence of any problem with this high stage barrel compressor rotor.

During a re-examination of all of the shop test data, it became apparent that the visual inspections, dye penetrant checks, and Zyglo tests are all predicated upon the fact that the crack must extent to the surface of the machine element. If the crack was somehow retained below the surface, these inspection techniques would not identify the presence of a crack. On the basis of this conclusion, a final check of the compressor shaft was performed with ultrasonic inspection. Amazingly, the ultrasonic test revealed a circumferential crack below one smooth section of the shaft as illustrated in Fig. 9-50. This crack was approximately 1/4 to 3/8 of an inch below the shaft surface, and it extended over an arc of approximately 60°. The length of the crack could not be accurately determined due to various steps and keyways in the shaft. However, it was clear that this crack was present, and it was significant in size.

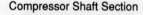

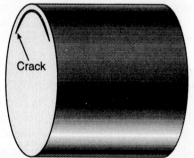

Compressor Shaft Section

Crack

Fig. 9–50 Circumferential Shaft Crack In High Pressure Syngas Compressor

Another spare rotor was pulled from the parts pool. The bundle was then reassembled, and reinstalled in the machine. Although the balance on this spare rotor was not as refined as the previous unit, it did come up and run at full speed without any apparent sensitivity to load. Above 8,000 RPM the 1X vectors were rock solid, and the plant was back in full production within the next 72 hours.

The question that remained was *what caused the subsurface crack in the pool rotor?* In an effort to answer this question, the serial number of this rotor was used to trace the operating history of this shaft. Interestingly enough, it was determined that this shaft had never successfully operated under load. Although it had been installed in several machines over the years, they all experienced high vibration levels during startup or loading. In each case, this rotor was replaced with a functional unit, and the cause of the problem was never determined. The rotor was always stripped and inspected after each failed run, but no problems were detected. Hence, the various members of the spare parts pool that worked with this specific rotor could find no difficulties, and they had it restacked, and returned back to the available pool of spare rotors.

Clearly this subsurface crack had existed for many years. It is speculated that it was an inclusion in the original shaft forging. Unfortunately this unusual shaft crack was not discovered until a considerable amount of time and money had been wasted by several different operating companies.

FOUNDATION CONSIDERATIONS

All machinery trains are supported by some type of foundation. Clearly, the foundation must be rigid enough to maintain alignment of the equipment, and it must be strong enough to accommodate the dynamic or vibratory loads emitted by the machinery. These oscillating loads cover the full range of dynamic forces from simple mass unbalance to complex impact or shock loads. In addition, the supports must be able to handle thermal distortion from the machinery, plus the loads and moments imposed by the piping systems.

In some cases, foundations may be difficult to design, and they may be difficult to analyze. Historically, foundation designs have been based upon empirical solutions, plus general *rules of thumb*. Concepts such as *keep the center of gravity of the entire structure well below grade*, or have *the foundation weight five times the weight of the machinery* are quite nebulous. In recent years, these trial and error designs have been supplemented by computerized structural design programs that can address the static as well as the dynamic loads. Although this approach adds increased sophistication to the analysis, the results of these computer solutions should always be examined by an experienced field engineer to validate the physical appropriateness of the foundation design.

Machinery on airplanes, ships, drilling rigs, trucks, and locomotives are supported by structural steel frames and baseplates. In these applications the foundation is flexible, and considerable computational technology must be employed to provide an adequate support for the equipment. Flexible foundations are also used for machinery in large residential or office buildings. In these installations, the equipment is often isolated from the surrounding structure with spring mounted baseplates, isolation pads, plus shock absorbers. The transmissibility of machine vibration to the surrounding structure, and the eventual impact upon the human occupants are key issues in these types of installations. Again, the dynamic machinery forces must be considered, but the machinery vibration must not be transmitted to the people.

Within the process industries, most of the machinery is installed on rigid steel and concrete foundations. Various configurations are used that vary from foundations built on pilings in swampy regions, to solid monolithic structures built at grade level. There are also a variety of flexible foundations mounted on slender columns. Although the majority of the machinery foundations are solid reinforced concrete structures, the diagnostician must recognize that flexible concrete structures are possible. In many cases, these flexible foundations are elevated units (20 to 100 feet tall) that locate the machinery close to other process equipment such as chillers or upper levels of fractionating columns. These installations should be carefully examined for the potential of structural resonances. It is common to discover one or more structural natural frequencies below machine operating speeds. In these installations, the machinery train is subjected to the structural resonance(s) during every startup and shutdown.

Even on large foundations mounted at grade, structural resonances of the foundation may be a problem. Forced and induced draft fans, plus some turbine generator sets actually have foundation resonances that encroach upon the machinery operating speed domain. It is not uncommon to have a 1,200 CPM natural resonance on an apparently rigid foundation. Intuition can sometimes be deceptive, and a large reinforced concrete structure may have a vertical stiffness of 5,000,000 to 20,000,000 pounds per inch. Combined with the weight of the foundation, this could yield a natural frequency $\sqrt{k/m}$ that is very low, and presumably separated from operating speed. However, when the mechanical system is tested, the intuitive observations may not be supported by the test data.

The field tests performed on a foundation may be quite simple. For example, a deflection profile may be obtained by acquiring vibration measurements at various elevations on the structure. This type of test is normally performed with the machinery train operating, and the vibration data filtered at a specific frequency. Normally, this is the machine running speed, and a synchronous Keyphasor® pulse is typically included to establish rotational speed vectors at each elevation. The use of vectors allows the diagnostician to identify the direction of the vibration component at each elevation. This yields the development of a definitive mode shape plot of the structure. In most conditions, the maximum horizontal motion will exist at the top of the structure, and the foundation vibration at grade will be negligible. However, if the vectors change direction from one elevation to another, this could be indicative of a loose bolted surface or joint (e.g., loose hold down bolt). In another possibility, if the top end motion is excessive, this could be symptomatic of a support structure with insufficient rigidity (i.e., low stiffness). On the other hand, if the support structure is still vibrating at the bottom grade level, the diagnostician should suspect a greater system problem such as discussed in case history 27.

As another approach, a static impact or a controlled shaker test may be performed on the foundation and associated structure. This test may consist of a simple frequency response function (FRF) using an accelerometer plus a force hammer or sledge. Alternatively, it may be a complex modal examination where the entire structure is covered with an accelerometer array. In this type of test the input force may be provided by a battering ram equipped with a force transducer or an electro-mechanical shaker. The expected results from this field test include the natural frequencies of the structure, plus the associated mode shapes. In some instances, it may be necessary to use the modal parameters from these modal tests to *fine tune* a finite element model of the structure to match the actual behavior. Once this step is completed, the FEA model may then be modified with suitable physical changes to correct the machinery problem in question.

In all cases, the diagnostician must remember that any machinery installation must be considered as a complete mechanical system. The behavior of the rotating shaft or reciprocating plunger, the bearing oil film characteristics, the bearing housing support characteristics, plus the baseplate, foundation, and the load bearing soil characteristics all influence the vibratory behavior. These factors must be considered during the design phase of the machinery, and they must be re-examined during troubleshooting of the mechanical equipment.

Case History 27: Floating Induced Draft Fan

The induced draft fan on a large cracking furnace consisted of a dual inlet, radial flow fan rotor that weighed 27,000 pounds. This centrifugal fan rotor was supported between a 226 inch bearing span. The fan was driven by a steam turbine through a speed reducing gear box, and the normal fan operating speed varied between 600 and 780 RPM.

This unit had an uneventful fourteen year operating history that ended one October evening with a fan inboard journal bearing failure. It was noted that the high fan vibration before the failure was reduced by installing a new inboard bearing. Approximately three weeks later, the same bearing failed again, and the fan rotor plus both bearings were replaced. At this time, the fan was in-place trim balanced at full operating speed. In December of the same year, the inboard fan bearing failed again, and the coupling to the bull gear was destroyed. To repair this damage, the bearing was replaced, a new coupling was installed, and the fan rotor was subjected to another field trim balance. Unfortunately, the fan vibration did not remain constant, and two more balance attempts were performed in January of the new year. In essence, a machine that had a good operating history for fourteen years had now experienced three serious mechanical failures in four months. In addition, the replacement rotor apparently would not *hold* a constant balance state.

Although some individuals viewed this behavior as strictly a balancing problem, it was quite clear that some other malfunction was active. Hence, the real failure mechanism must be identified and corrected before worrying about the balancing aspects of this rotor. In an effort to understand the behavior of this machinery, a complete set of casing vibration measurements were made on the fan, gear box, and steam turbine driver. At all measurement locations, the predominant frequency component was the fan rotational speed, and the entire train was preferentially shaking in the horizontal direction. Next a series of three consecutive startup and shutdown runs were performed, and synchronous

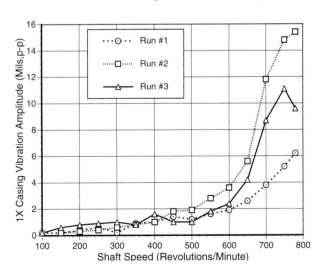

Fig. 9–51 Horizontal 1X Running Speed Casing Vibration Measurements During Three Separate Startups On Induced Draft Fan Bearing Housing

1X fan vibration amplitudes were logged at 50 RPM increments. The horizontal results of this test are presented in Fig. 9-51.

These measurements were acquired with low frequency velocity pickups, and the data was manually corrected for the transducer roll-off characteristics. There were no mechanical changes performed between runs, and a slow roll period of 40 to 50 minutes was included at the beginning of each run to insure a minimal fan rotor bow. From this data it was self-evident that the fan vibration response was quite different during each run. The data from Fig. 9-51 might suggest the possibility of a resonance at 750 RPM, but examination of the corrected phase angles revealed no evidence of a resonance. At this point, there was no direct explanation for the variable behavior of this fan rotor.

The blades were checked for soundness and proper attachment to the back plate and the shrouds. The entire fan impeller assembly was securely attached to the shaft, and there was no angular shifting of the impeller with respect to the shaft. The stationary inlet cones were properly positioned, and there was no visual evidence of any rubs between the inlet cones and the impeller. The pillow block bearings were inspected for damage plus abnormal fits or clearances. There were no problems at either fan bearing, and proper clearances were measured. Furthermore, additional operating tests such as changes in the fan inlet dampers produced minimal changes in the horizontal fan vibration. Overall, the unusual changes depicted in Fig. 9-51 could not be explained.

Although the fan pedestals appeared to be sufficiently rigid for this type of machinery, a vibration profile was obtained on the inboard and outboard structures. These horizontal vibration measurements were made at ten different elevations, and the results are presented on Fig. 9-52. The top measurement elevation (96") was at the centerline of the bearing and shaft. The bottom location (0") was at grade (ground level). The intermediate eight measurements were distributed between the steel support pedestals, and the concrete foundation.

These horizontal vibration measurements were filtered at the fan running

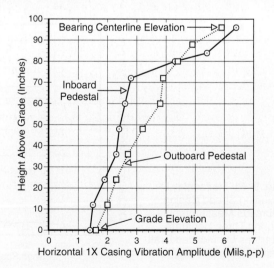

Fig. 9–52 Horizontal 1X Running Speed Casing Vibration Profile On Both Induced Draft Fan Bearing Housings And Support Pedestal Structures

speed of 750 RPM, and all readings were directly inphase. Certainly the amplitudes at the bearing housings were higher than desired, but the main peculiarity of this data set was that the concrete pedestals were vibrating at grade level ($\approx$1.5 Mils,$_{\text{p-p}}$). At this location the pedestal motion should have been very close to zero. It then seemed appropriate to begin digging next to the pedestals and determine the elevation at which the concrete pedestals were not moving.

At approximately two feet below grade, the pedestals were still moving, and warm water was encountered. This was certainly an unusual discovery for a chemical plant in January. Further excavation revealed that there was a small river flowing underneath the entire machinery train. It was eventually determined that this water flow originated with a broken cooling water return line. Operations personnel knew that cooling water makeup rates for the past few months had been much higher than normal, but they had been unable to determine the source of the cooling water loss.

It was also interesting to go back in the records and correlate the fact that the periods of low fan vibration occurred on nights when the ambient temperature was substantially below freezing. Conversely, the highest vibration amplitudes appeared during the warmest days. It was speculated that on the cold nights, the subsurface water would freeze to some extent, and provide support for the fan foundation. During the warmth of the day, the water flow would increase, which essentially reduced the support stiffness, and allowed the foundation to float and vibrate on the subsurface river.

To correct this problem, the broken section of the cooling water return line was located and repaired. However, considerable sub-surface water still remained below the equipment foundation. To remedy this situation, a well-point pumping network was installed around the entire machinery foundation. Following approximately 36 hours of continuous pumping, the area was pumped reasonably dry. At this stage, the machinery foundation mat was excavated, and it was tied into the massive reformer exhaust stack foundation with reinforced concrete. The entire area was then back-filled, the soil compacted, and the unit prepared for operation.

Three runs were made to check the repeatability of the fan vibration. Two of the runs were performed during warm daylight, and the third run was executed shortly after midnight with freezing ambient temperatures. At this time, virtually identical behavior was noted on both the shaft and the casing vibration vectors during each run. Finally, a trim balance was performed by welding a 220 gram weight to the fan center plate. This correction reduced the runout compensated shaft vibration to 0.3 Mils,$_{\text{p-p}}$ at both bearings, and the horizontal housing vibration was less than 0.8 Mils,$_{\text{p-p}}$ at both bearings. The fan continued to perform in this manner regardless of ambient temperature, and a new era of reliable service was initiated on this critical piece of process machinery.

Case History 28: Structural Influence of Insufficient Grout

In some situations, knowledge of the foundation natural frequency and/or mode shape may be sufficient to identify the problem, and point to an obvious solution. For instance, consider the case of an overhung booster compressor that had a history of high casing vibration. This machine operated at a full load speed of 6,500 RPM, and it was driven by a two pole induction motor through a double helical gear box. Interestingly enough, the motor and gear box vibration amplitudes were always quite acceptable, but compressor vibration was much higher than desirable. During the first twenty months of operation, the compressor had suffered two seal failures, and one damaged bearing. Certainly, this was unacceptable behavior for a new machinery installation.

In an effort to determine the true cause of the repeated failures, an extensive overhaul and inspection of the compressor was initiated. The compressor was uncoupled, the case was split, and the rotor, bearings, seals, and coupling were taken to the shop for detailed examination. During this time period, the large bore piping, compressor casing, supports, and foundation were checked for any obvious deficiencies. Externally, the foundation was a solid monolith, and epoxy grout was used between the baseplate and the concrete foundation. As part of the inspection, a simple *hammer test* on the compressor pedestals and the baseplate below the compressor revealed a distinctly hollow sound. This was unusual, and it indicated an incomplete grout pour during the initial machinery installation. The solution to this deficiency resided with filling both pedestals with grout, plus injecting grout between the baseplate and the foundation.

Before implementing this physical correction, it was considered desirable to measure the major mode shapes and natural resonances of this compressor support structure. In previous years, this would have been a complicated project requiring two or more days for setup and testing. However, the use of modern instrumentation plus suitable software reduced this work to a manageable exercise. In all fairness, it should also be mentioned that approximately three hours was devoted to establishing the three-dimensional matrix structure, and proper polarity directions for the tri-axial accelerometer. This setup work was performed offsite, and the actual field test of this structure was performed in less than two hours using an HP-35670 four channel Dynamic Signal Analyzer, plus an impact hammer, and a tri-axial accelerometer.

In this application, the DSA was controlled with Hammer-3D software by David Forrest[11]. This HP I-Basic program runs directly on the HP-35670A, and it eliminates the need for external devices such as a separate computer system. Within Hammer-3D, the test structure geometry and transducer array are defined. The field work consists of acquiring frequency response functions (FRF) on the structure. For this type of test, the impact hammer location remained constant, and the triaxial accelerometer was moved between measurement points. A total of 24 points were identified on this compressor support assembly, and 72 (=

[11] David Forrest, "Hammer-3D Version 2.01," Computer Program in Hewlett Packard Instrument Basic by Seattle Sound and Vibration, inc., Seattle, Washington, 1997.

3 x 24) FRFs were acquired and stored on floppy disk. From a physical dimension standpoint, the pedestals were nominally 36 inches tall, 36 inches in length, and 8 inches wide at the top. The associated baseplate below the pedestals was 36 inches wide, and 58 inches across (pedestal to pedestal). These physical dimensions resulted in a nominal 18 to 20 inch spacing between the 24 measurement points.

After acquisition of the field data, curve fitting was applied to each user defined resonant frequency range on each FRF. The individual modes were then assembled, scaled, and presented as animated mode shapes on the DSA. These are quite realistic modes due to the fact that tri-axial measurements were obtained, and cross-coupling between orthogonal directions is simultaneously displayed. It should be mentioned that this is a complicated procedure, and a manual solution is generally unattainable. Hence, the computational power within the DSA is mandatory for proper and rapid presentation of results.

The animated display on the HP-35670A is quite descriptive, and physically understandable. Unfortunately, reproduction of this dynamic display to a static diagram is often difficult, and proper interpretation may be lost by the examination of a single diagram. Hence, it is generally desirable to examine the minimum and maximum deflections with respect to the stationary structure.

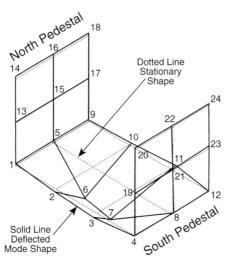
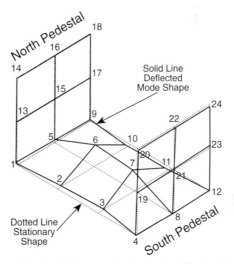

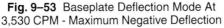

Fig. 9–53 Baseplate Deflection Mode At 3,530 CPM - Maximum Negative Deflection   **Fig. 9–54** Baseplate Deflection Mode At 3,530 CPM - Maximum Positive Deflection

For instance, consider the measured deflection mode shapes presented in Figs. 9-53 and 9-54. Both of these diagrams are associated with a resonant mode at 3,530 CPM. The dotted lines in both drawings depict the stationary structure of the baseplate, plus the north and south compressor support pedestals. The solid lines represent the deflected mode shape of the structure. In Fig. 9-53, the maximum negative deflection is shown, and the companion diagram in Fig. 9-54 displays the maximum positive deflection at 3,530 CPM. Based on these two diagrams, it is clear that a drum mode exists on the compressor baseplate. This is most evident by looking at the relative positions of the center points 6 and 7 in both figures. Surprisingly, the measured frequency for this drum mode occurs at 3,530 CPM, which is uncomfortably close to the normal motor operating speed range of 3,580 to 3,595 RPM. Although both compressors pedestals display minimal motion at this frequency, there still remains an undesirable coincidence between this baseplate natural frequency and the motor running speed.

The second dominant mode measured on this support structure occurs at a frequency of 6,600 CPM. The maximum negative and positive mode shapes at this frequency are presented on Figs. 9-55 and 9-56 respectively. Within this

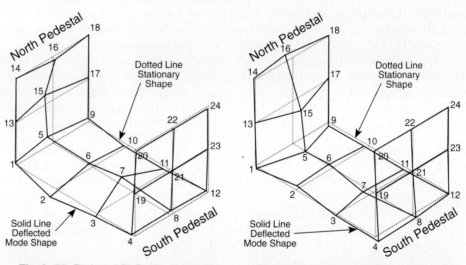

Fig. 9–55 Pedestal Deflection Mode At 6,600 CPM - Maximum Negative Deflection

Fig. 9–56 Pedestal Deflection Mode At 6,600 CPM - Maximum Positive Deflection

mode, the baseplate motion is reduced, but the north pedestal motion is excessive. This is demonstrated by comparing the differential position of point 15 in Figs. 9-55 and 9-56. The motion at the south pedestal is substantially less that the measured deflection of the north pedestal. This is interpreted as a better grout injection in the south versus the north pedestal. Amazingly enough, this mode at 6,600 CPM is close to the compressor operating speed range of 6,500 to 6,530 RPM. Again, this is an undesirable condition for this compressor support.

In the overview, it is clear that these support structure mode shapes are detrimental to the typical requirement for a solid compressor support. The close proximity between these natural frequencies and the excitations due to the machine operating speed range is likewise objectionable. It is quite unusual to encounter a situation where two distinct structural resonances are close to two fundamental machine excitations. However, due to the inconsistencies of the original grout pour, any combination of baseplate and pedestal natural resonant frequencies are physically possible. In this case, the resultant natural frequencies coincided with the machinery operating speeds.

It is concluded that the flexibility of the compressor pedestals and the baseplate are excessive and unacceptable. From a correction standpoint, the previously mentioned solution of filling these voids with epoxy grout remains as the most prudent course of action.

For comparative purposes, after the grout injection was completed, and the compressor reinstalled, a substantial reduction in casing vibration amplitudes was apparent. Historically, unfiltered casing velocity levels varied between 0.18 and 0.22 $IPS_{,o-p}$ on the compressor bearing housing. After the compressor rebuild, and the grout repairs, the maximum casing vibration was 0.03 $IPS_{,o-p}$. The majority of this reduction was due to attenuation of the motor and the compressor rotational speed vibration components.

Prior to this repair, the radial shaft vibration amplitudes on the compressor ranged from 2.2 to 3.0 $Mils_{,p-p}$. After the rebuild and grout repair, all of the centrifugal compressor shaft vibration levels dropped below 1.0 $Mil_{,p-p}$. The success of this simple repair was effectively demonstrated by successful and continuous operation. There were no additional mechanical failures on this machinery train, and vibration amplitudes remained low and constant.

In retrospect, this type of problem is generally encountered on new construction projects. If the field inspectors do not perform a surface *hammer test* after the grout is cured, this type of flaw may go undetected. This could easily result in machinery problems that persist for an extended period of time. Certainly, the initial diagnosis and the physical solution to this insufficient grout problem makes sense to most people. However, some individuals will question the need for performing the structural mode shape measurements discussed in this case history. If these structural tests had not been performed, the solution (additional grout), and the beneficial results (extended run time without failures) would have been the same. However, there would still be an uncertainty as to the root cause of the previous failures. In addition, the potential benefits associated with spending money for injecting epoxy grout to fill in the voids might be hard to sell to management. With the rapid availability of this structural mode shape and frequency data, the cause and effect relationship is understandable, and the repair costs are much easier to justify. Stated in another way, the availability of this information represents the difference between a *shotgun approach*, and a properly *engineered explanation*.

BIBLIOGRAPHY

1. Allaire, P.E., and R.D. Flack, "Design Of Journal Bearings For Rotating Machinery," *Proceedings of the Tenth Turbomachinery Symposium,* Gas Turbine Laboratories, Texas A&M University, College Station, Texas (December 1981), pp. 25-45.

2. Bently, Donald E., "Forced Subrotative Speed Dynamic Action of Rotating Machinery," *American Society of Mechanical Engineers*, ASME Paper No. 74-Pet-16 (1974).

3. Bently, Donald E., "Forward Subrotative Speed Resonance Action of Rotating Machinery," *Proceedings of the Fourth Turbomachinery Symposium,* Gas Turbine Laboratories, Texas A&M University, College Station, Texas (October 1975), pp. 103-113.

4. Bently, Donald E., "Attitude Angle And The Newton Dogleg Law Of Rotating Machinery" *Bently Nevada Applications Note* (March 1977).

5. Bently, Donald E., "Vibration levels of machinery," *Orbit*, Vol. 13, No. 3 (September 1992), p. 4.

6. Bently, Donald E., and Agnes Muszynska, "Detection of Rotor Cracks," *Proceedings of the Fifteenth Turbomachinery Symposium,* Turbomachinery Laboratory, Texas A&M University, College Station, Texas (November 1986), pp. 129-139.

7. East, John R., "Turbomachinery Balancing Considerations," *Proceedings of the Twentieth Turbomachinery Symposium,* Turbomachinery Laboratory, Texas A&M University, College Station, Texas (September 1991), pp. 209-214.

8. Forrest, David "Hammer-3D Version 2.01," Computer Program in Hewlett Packard Instrument Basic by Seattle Sound and Vibration, inc., Seattle, Washington, 1997.

9. Gunter, Jr., Edgar J., "Rotor Bearing Stability," *Proceedings of the First Turbomachinery Symposium,* Gas Turbine Laboratories, Texas A&M University, College Station, Texas (1972), pp. 119-141.

Unique Behavior

*T*he common machinery malfunctions discussed in chapter 9 occur on a wide variety of machines. The typical frequencies observed with those common malfunctions generally occur between one quarter of rotative speed and twice running speed. Many process machines are subjected to additional excitations that impose significant dynamic loads upon the machinery at other frequencies. In chapter 10, the excitations produced within two element and epicyclic gear boxes will be discussed. Common fluid excitations and electrical phenomena will also be examined. Finally, the application of rotating machinery technology to reciprocating compressors will be reviewed. As usual, each of these topics will be highlighted with numeric examples, and actual machinery case histories.

Parallel Shaft - Two Element Gear Boxes

Speed increasing, or speed reducing gear boxes are devices that emit a distinctive set of excitations. Gear box elements move with definable static position changes, and they generate specific frequencies that may be used for mechanical diagnosis. Due to the vast array of gear box configurations, the current discussion will concentrate on the common two element, parallel shaft, single or double helical gears used within the process industries. A review of the complex excitations generated by epicyclic gears is included in following section of this chapter. Gear boxes are complicated machines that have evolved from slow speed water wheels to a vast array of industrial machines. In many respects, gear design, configuration, fabrication, and application is a science unto itself. Due to the complexity of this subject, the reader is encouraged to examine books by authors such as Lester Alban[1], Darle Dudley[2], and M.F. Spotts[3] that go into specific details regarding the mechanics of various types of gear boxes. There are also numerous standards, handbooks, and design guides available from the American

---

[1] Lester E. Alban, *Systematic Analysis of Gear Failures*, (Metals Park, Ohio: American Society for Metals, 1985).
[2] Darle W. Dudley, *Gear Handbook*, (New York: McGraw-Hill Book Company, 1962).
[3] M.F. Spotts, *Design of Machine Elements*, 6th Edition, (Englewood Cliffs, New Jersey: Prentice-Hall, Inc., 1985).

Gear Manufacturers Association (AGMA) in Arlington, Virginia. The gear OEMs also produce some excellent technical references on all aspects of gearing.

The discussion contained in this text is divided into a review of static element shifts, the computation of the major gear contact forces, and the dynamic vibratory characteristics. The position changes or static shifts of gear elements are dependent upon rotation, the driver element, and the applied forces. It is often difficult to maintain a clear perspective of the gear force directions under normal conditions. This issue is clouded by the various potential variations in gear set arrangements. Hence, it is reasonable to examine the expected types of forces and their directions in both the radial and the axial planes.

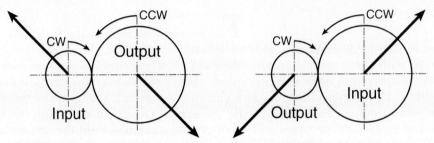

Fig. 10–1 Expected Radial Loading Of Down Mesh Gear Sets - View Towards Input

In order to describe the radial position characteristics of two element gear sets Figs. 10-1 and 10-2 have been constructed. The sketches in Fig. 10-1 describe the expected radial load directions for a down mesh set. The left hand diagram depicts a speed decreasing box where the pinion is the input element, and the bull gear is the reduced speed output. The right hand sketch in Fig. 10-1 describes a speed increasing box where the bull gear is the input, and the pinion is the high speed output. The heavy arrows on each sketch describe the general load direction of the overall forces acting at each respective bearing.

The two sketches presented in Fig. 10-2 depict the expected radial forces for an up mesh gear box. The left hand diagram describes a speed decreasing unit where the bull gear is the input element, and the pinion is the increased speed output. The right hand sketch in Fig. 10-2 shows a speed decreasing gear box where the pinion is the input, and the bull gear provides the slow speed output.

These simple diagrams define the general load directions for each gear ele-

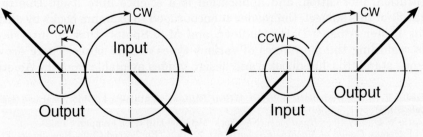

Fig. 10–2 Expected Radial Loading Of Up Mesh Gear Sets - View Towards Input

ment. This is important for the proper location of bearing thermocouples, pressure dams, and the evaluation of radial shaft position shifts as measured by proximity probe DC gap voltages. Many OEMs now provide analytical calculations that predict the vertical and horizontal journal centerline position at full load. These radial positions are unique to each gear box design and should be compared with actual shaft centerline position shifts. Note, this measurement is often difficult to execute due to the tooth engagement between gears at the rest position. Hence, an accurate zero speed starting point (particularly for the pinion) may be difficult to obtain.

Any gear box evaluation should always include a detailed examination of the operating shaft positions at each of the four radial bearings. When available, the measured position should be compared with the radial location calculated by the OEM. An incorrectly positioned bearing will cause significant distress within the gear box. Unless the journal locations are checked for proper running position, the diagnostician may end up chasing a variety of abnormal dynamic characteristics when the real problem is easily identified by the radial journal position data. This also reinforces the argument for installing X-Y radial proximity probes at all gear box bearings. Many facilities tend to install proximity probes only at the input and output bearings, and they often ignore the blind or outboard end bearings. This practice can result in the unavailability of some critical journal position and vibration information.

On a helical gear, the gear tooth contact force is typically resolved into three mutually perpendicular forces. The two radial forces consist of a tangential and a separation force. The tangential force is based upon the transmitted torque and the pitch radius. Calculation of the torque is determined in equation (10-1), followed by the tangential force in equation (10-2):

$$Torque = \frac{33,000 \times HP}{2\pi \times RPM} = \frac{5,252 \times HP}{RPM} \qquad \textbf{(10-1)}$$

$$Force_{Tan} = \frac{12 \times Torque}{R_{pitch}} = \frac{63,024 \times HP}{R_{pitch} \times RPM} \qquad \textbf{(10-2)}$$

where: $Torque$ = Transmitted Torque Across Gear Teeth (Foot-Pounds)
HP = Transmitted Power Across Gear Teeth (Horsepower)
RPM = Gear Element Rotational Speed (Revolutions / Minute)
$Force_{Tan}$ = Tangential Force Across Gear Teeth (Pounds)
R_{pitch} = Pitch Radius of Gear Element (Inches)

In these expressions the speed and pitch radius must be for the same gear element. That is, if the bull gear speed is used to compute the torque, then the bull gear pitch radius must be used to determine the tangential force. Similarly, if the pinion speed is used to calculate the transmitted torque, then the pinion pitch radius must be used to compute the correct tangential force. Note that the transmitted torque is different for the pinion and the bull gear, but the tangential force for both elements must be the same. As another check, the pitch line velocity for both gear elements must also be identical.

The tangential force is the vertical force acting between the gears. Obviously, one gear element is subjected to an upward tangential force, and the mating gear element is subjected to a downward tangential force (necessary to be equal and opposite). Based upon the gear pressure angle, and the helix angle, the gear separation factor may be computed as in equation (10-3). Multiplying the previously calculated tangential force by this non-dimensional gear separation factor provides the gear separation force as shown in equation (10-4):

$$SF = \frac{\tan\Phi}{\cos\Psi} \qquad\qquad (10\text{-}3)$$

$$Force_{sep} = Force_{Tan} \times SF = Force_{Tan} \times \frac{\tan\Phi}{\cos\Psi} \qquad (10\text{-}4)$$

where:　　SF = Separation Factor (Non-Dimensional)
　　　　　Φ = Pressure Angle Measured Perpendicular to the Gear Tooth (Degrees)
　　　　　Ψ = Helix Angle Measured from the Gear Axis (Degrees)
　　$Force_{sep}$ = Separation Force Between Gears (Pounds)

This separation force acts to the right on one gear, and to the left on the mating gear element. Again, a force balance must be achieved in the horizontal plane, and the separation force must be less than the tangential force. For standard gears, the typical pressure angle Φ is either 14.5°, 20°, or 25°. The most common value encountered for the pressure angle is 20°. The helix angle Ψ typically varies between 15° and 35°. Although these angles are similar, it is mandatory for the diagnostician to keep the numbers straight. Finally, the third segment of the overall gear contact force is the axial component. The magnitude of this thrust load is obtained from the following expression:

$$Force_{Thr} = Force_{Tan} \times \tan(\Psi) \qquad\qquad (10\text{-}5)$$

where:　$Force_{Thr}$ = Axial (Thrust) Force Between Gears (Pounds)

As a side note, if the helix angle is 0°, the helical gear equations simplify into spur gear equations. That is, the cosine of 0° is equal to 1, and the separation force is equal to the tangential force times the tangent of the pressure angle. Also, the tangent of 0° is equal to zero, and the thrust load is zero. Obviously, spur gears cannot transmit an axial force.

The axial or thrust loads on a double helical (herringbone) gear are theoretically balanced by the two sides of the gear. If the gear is machined incorrectly, an axial force will occur on a double helical gear, and this may generate significant axial loads. However, on a single helical gear box, the thrust loads are always present. These axial forces must be accommodated by thrust bearings for each element of a single helical gear set. It is meaningful to understand the normal versus the counter thrust directions for a single helical gear. This helps in setting up the thrust monitors properly (i.e., normal versus counter), and it allows a proper evaluation of measured thrust behavior.

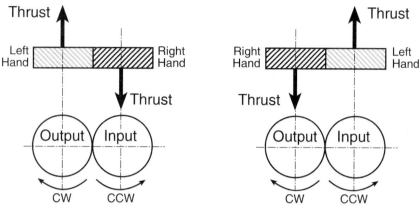

Fig. 10–3 Normal Thrust Direction For Single Helical Down Mesh Gears

The two diagrams presented in Fig. 10-3 describe the normal thrust directions for a single helical gear box equipped with down mesh gears. The drawings in Fig. 10-4 depict the thrust directions for up mesh gears. In each case, the gears are identified as either *right-hand* or *left-hand*. This is a common designation of how the teeth curve away from the mesh line. If the teeth lean or are inclined to the right or the clockwise direction, the element is referred to as a *right-hand* gear. Conversely, if the teeth lean or are inclined to the left, or in a counterclockwise direction, the element is identified as a *left-hand* gear. In any pair of mating helical gears, one element must be *right-handed* and the other gear element must always be *left-handed*.

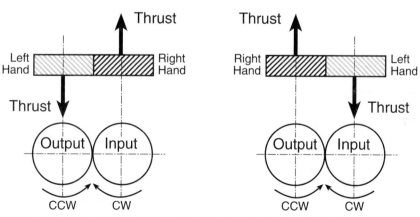

Fig. 10–4 Normal Thrust Direction For Single Helical Up Mesh Gears

As mentioned earlier in this section, gear boxes emit many unique excitations. The actual excitations vary from low to high frequencies. For example, a typical parallel shaft, two element (bull gear and pinion) gear box, will normally produce the following group of discrete frequencies.

○ Bull Gear Rotational Speed — F_{bull}
○ Pinion Rotational Speed — F_{pin}
○ Gear Mesh Frequency — F_{gm}
○ Assembly Phase Passage Frequency — F_{app}
○ Tooth Repeat Frequency — F_{tr}
○ Gear Element Resonant Frequencies
○ Casing Resonant Frequencies

The bull gear and pinion rotational speeds are the actual speeds of each element. These rotative speeds maintain a fixed ratio that is completely dependent on the number of bull gear and pinion teeth. If at all possible, the diagnostician should obtain an exact tooth count on both elements. This must be an exact number (± 0 allowable error). The gear mesh frequency is equal to the speed times the number of teeth as shown in equation (10-6):

$$ F_{gm} = F_{bull} \times T_{bull} = F_{pin} \times T_{pin} \qquad \text{(10-6)} $$

where: F_{gm} = Gear Mesh Frequency (Cycles / Minute)
F_{bull} = Rotational Speed of Bull Gear (Revolutions / Minute)
F_{pin} = Rotational Speed of Pinion (Revolutions / Minute)
T_{bull} = Number of Teeth on Bull Gear
T_{pin} = Number of Teeth on Pinion

The gear mesh frequency must be the same for both the bull gear and the pinion. This commonality also provides a good means to verify the validity of a presumed gear mesh frequency in an FFT plot. In other situations, if the bull gear speed is known, the pinion speed may be determined from (10-7) when the actual number of bull gear and pinion teeth are known. Obviously, the inverse relationship is also applicable.

$$ F_{pin} = F_{bull} \times \frac{T_{bull}}{T_{pin}} \qquad \text{(10-7)} $$

The gear mesh frequency provides general information concerning the gear contact activity and forces. This type of measurement is usually obtained with a high frequency casing mounted accelerometer. Typically, the best data is acquired at the gear box bearing housings, since this is the location where the meshing forces are transmitted to ground. As an example of this type of information, the data presented in case history 19 should be of interest.

The next two excitations of *assembly phase passage frequency* and *tooth repeat frequency* require an understanding of the concept of phase of assembly. This is clearly explained by John Winterton[4] as follows:

"Mathematically, the number of unique assembly phases (N_a) in a given tooth combination is equal to the product of the prime factors common to the num-

---

[4] John G. Winterton, "Component identification of gear-generated spectra," *Orbit*, Vol. 12, No. 2 (June 1991), pp. 11-14.

ber of teeth in the gear and the pinion. The numbers 15 and 9 have the common prime factor of 3. Therefore, three assembly phases exist. The number of assembly phases determines the distribution of wear between the teeth of the gear and pinion..."

Winterton goes on to define the assembly phase passage frequency as shown in equation (10-8):

$$F_{app} = \frac{F_{gm}}{N_a} \qquad (10\text{-}8)$$

where: F_{app} = Assembly Phase Passage Frequency (Cycles / Minute)
         N_a = Number of Assembly Phases (Non-Dimensional Prime Number)

This is followed by the determination of the tooth repeat frequency. This is generally the lowest level excitation within the gear box. Typically it falls below 500 CPM, and sometimes it appears as an amplitude modulation. The tooth repeat frequency may be computed in the following manner:

$$F_{tr} = \frac{F_{gm} \times N_a}{T_{bull} \times T_{pin}} = \frac{F_{bull} \times N_a}{T_{pin}} = \frac{F_{pin} \times N_a}{T_{bull}} \qquad (10\text{-}9)$$

where: F_{tr} = Tooth Repeat Frequency (Cycles / Minute)

Other investigators have a tendency to consider only a true tooth hunting combination where the number of assembly phases N_a is equal to one. This is proper and normal for precision high speed gears. However, equation (10-9) represents the correct computation for all cases.

The natural frequencies of the bull gear and pinion are often located above the normal operating speed range. Thus, examination of the transient Bode plots will generally reveal a stiff shaft response of both gear elements. However, gears do exhibit natural frequencies that generally appear at frequencies above rotational speed. Typically, a bull gear and a pinion will each display a stiff shaft translational followed by a pivotal mode that are both governed by bearing stiffness. At higher frequencies, the shaft stiffness controls the resultant natural frequencies. In a simplistic model, these are *free-free* modes that are dependent on mass and stiffness distribution across each respective gear element. These higher order modes are independent of bearing stiffness, and they are often excited during failure conditions. Typically, discrete frequencies between 60,000 and 180,000 CPM (1,000 and 3,000 Hz) are detectable. These gear element resonant frequencies are often frequency modulated by the running speed of the problem element. For instance, if the bull gear is under distress, a higher level bull gear natural frequency will be modulated by bull gear speed.

It should also be mentioned that most gear boxes display a variety of casing resonant frequencies. The distribution of these frequencies will depend on the gear casing construction. A fabricated box will be lighter than an older cast box. In general, the thinner gear box casings will exhibit higher natural frequencies than heavier and thicker wall casings. Typically, an industrial helical gear box may exhibit multiple casing natural frequencies, and they may appear anywhere between 30,000 to 300,000 CPM (500 to 5,000 Hz). Various attachments to the gear box may also appear as narrow band structural resonances. Items such as unsupported conduit, thermowells, small bore piping, and proximity probe holders may be detectable on the gear box. In one case, long unsupported stingers were used on proximity probes in a large gear box installation. Unfortunately, the natural resonance of the probe stingers was 3,580 CPM, which was excited by the synchronous machine speed of 3,600 RPM.

Parallel shaft gear boxes are also built with multiple gear elements. For example, an intermediate idler gear may be installed between a bull gear and a pinion to obtain a specific speed ratio, or maintain a particular direction of rotation. Some gear boxes contain multiple gears, such as the seven element box discussed in case history 21 in chapter 9. These additional gear elements provide additional rotational speed excitations. If the unit contains direct mesh to mesh contact across the box, the gear mesh frequency will remain constant. However, if the gear box contains any variety of stacked gear arrangements, the unit will emit multiple gear mesh frequencies. These multiple rotational speeds and gear mesh frequencies will often interact in a variety of signal summations, amplitude modulations, and frequency modulations.

Interactions of the multiple frequencies will depend on load, which influences the journal radial positions and the tooth contact between gears. In all cases, the documentation of vibratory data with the box in good condition will be beneficial towards analysis of a variety of potential future malfunctions.

Case History 29: Herringbone Gear Box Tooth Failure

The speed increasing gear box shown is Fig. 10-5 is situated between a low pressure and a high pressure compressor. The normal operating speed for the LP compressor and the bull gear is 4,950 RPM, and the pinion output to the HP compressor runs at 11,585 RPM. A flexible diaphragm coupling is installed between the LP compressor and the bull gear, and another diaphragm coupling is used between the pinion and the HP compressor. The axial stiffness for both couplings are approximately equal, and the gear box has successfully operated in this configuration for many years.

The gear configuration consists of a double helical, or herringbone, arrangement. This type of gear provides a generally balanced axial load between the two sets of gear teeth. Some axial load is inevitable, and a thrust bearing is mounted on the outboard end of the bull gear. Due to the dual mesh interaction, the pinion must follow the bull gear axial position, and a separate pinion thrust bearing is not required. If this unit was a single helical gear, a separate pinion thrust bearing would have been incorporated.

The tooth failure problem on this unit was initiated during a routine *topping off* of the oil reservoir. For whatever reason, excess oil was pumped into the reservoir, and oil backed up the gear box drain line. This reverse flow filled the gear box with lubricant, and oil began spewing from the gear box atmospheric

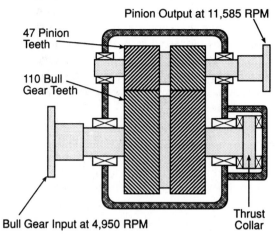

Pinion Output at 11,585 RPM

47 Pinion Teeth

110 Bull Gear Teeth

Bull Gear Input at 4,950 RPM

Thrust Collar

Fig. 10–5 General Arrangement Of Two Element Gear Box With Herringbone Gears

vents. This external oil flow was ignited by a hot steam line, and a fire ensued. The machinery train was tripped and the deluge system activated. These combined actions extinguished the fire with minimal external damage to the machinery. Unfortunately, the train was restarted to a fast slow roll, and allowed to run at 1,500 RPM for approximately 90 minutes. Evidently the gear box was still filled with oil during this abbreviated test run.

Following evaluation of the vibration, mechanical, and process data, the train was shutdown for gear box disassembly and inspection. Upon removal of the top half, it was visibly noted that the gears were in good condition. Following removal of both gear elements it was clear that all four journal bearings were damaged. The inside sections of the four journal bearings that are exposed to the interior of the gear box had melted babbitt, whereas the outside sections of all four bearings retained babbitt. This indicates that the internal gear box temperature was in excess of 500°F to melt part of the bearing babbitt.

Shop examination of the bull gear and pinion revealed that both elements were coated with varnish. This was indicative of burnt or oxidized oil on the surface of the gears. After the varnish was removed, the gears appeared to be in good physical condition with minimal surface wear on the teeth. A dye penetrant inspection did not reveal any cracks or discontinuities, and the shaft journals were considered to be in good condition.

Since the shop inspections revealed no evidence of physical damage to the gear set, the gear elements were reinstalled in the box. Although axial spacing between the bull gear shaft and the LP compressor shaft was maintained, the bull gear coupling hub was mounted 0.25" further on to the shaft than previous installations. The effect of an axially mis-positioned hub on a diaphragm coupling would be the generation of an axial preload on the bull gear. This axial load could force one side of the herringbone gear to carry the majority of the load.

The train was successfully restarted, and machinery behavior appeared to be normal, and consistent with previous vibration data. After six days, a high frequency vibration component around 75,600 CPM (1,260 Hz) was noticed on the gear box. This frequency was approximately 15 times the bull gear speed, and it was intermittently transmitted to both compressors. The unit operated in this manner for approximately one month when a leaking thermowell forced the train down to slow roll speeds for 45 minutes for thermowell replacement. During the subsequent restart, two compressor surges occurred, and this event may have overloaded the gear teeth.

Two days after the restart, shaft vibration amplitudes experienced a series of minor step changes in a gradually increasing trend. The bull gear radial shaft vibration data revealed minor 1X vector changes. However, the largest change occurred on the bull gear axial probes, where the synchronous 1X motion increased from 0.48 to 1.29 Mils,$_{p-p}$. This is certainly abnormal behavior for a double helical gear with normally balanced axial forces.

Simultaneously, the casing vibration amplitudes began to grow in the vicinity of 75,600 CPM (1,260 Hz) with peak levels reaching 12.0 G's,$_{o-p}$. It should also be noted that the previously dormant gear mesh frequency at 544,500 CPM (9,075 Hz) had blossomed into existence, and it was modulated by bull gear rotational speed. Furthermore, audible noise around the gear box had significantly increased. FFT analysis of microphone data recorded on the compressor deck revealed a dominant component at 1,238 Hz (15th harmonic), with sideband modulation at bull gear rotational speed of 4,950 RPM (82.5 Hz).

Considering the available information, the bull gear distress was self-evident, and a controlled shutdown was the only reasonable course of action. Following an orderly shutdown, a visual inspection revealed 12 broken teeth on the coupling side of the bull gear. Additional shop inspection revealed multiple cracked teeth on the bull gear combined with an erratic and accelerated wear pattern on both gear elements.

In retrospect, the gears were probably solution annealed during the period of high internal gear box temperatures. A micro-hardness survey revealed that the broken gear teeth had a *Rockwell C* surface hardness of 20 for the first 2 Mils (0.002 inches) of tooth surface. The hardness then increased with depth to levels consistent with the original gear tooth heat treating. Normal surface hardness for these gears should be 37 on the *Rockwell C* scale. This softening of the gear teeth surfaces represents the root cause of this failure. However, the tooth failure was logically due to a combination of the following events:

1. Initial fire, and the probable surface annealing of the gear teeth.
2. Potential axial load imposed by a mis-positioned bull gear coupling hub.
3. Potential impact loads suffered during compressor surges.
4. Probable high cycle fatigue of the heavily loaded and soft gear teeth.

In all likelihood, the primary damage of softening the gear teeth occurred during the fire. Based upon the metallurgical findings, it is clear that this gear was destined for premature failure. The actual influence of items 2 and 3 in the above list are difficult to quantify. In all probability, these contributors accelerated the failure, but the life span of the bull gear teeth was greatly reduced by the loss of tooth surface hardness.

The symptoms of this failure included minor changes in the shaft radial vibration, significant changes in the bull gear axial vibration, plus substantially increased activity at the gear mesh frequency. These conditions are fully explainable based upon the physical evidence of broken gear teeth. However, the casing excitation at 75,600 CPM (1,260 Hz), and the dominant sound emitted by the gear box at the same general frequency are not immediately obvious.

In an effort to understand the significance of this frequency component, a simple impact test was performed on the failed bull gear resting in the journal bearings (with the pinion removed). The main component encountered during this test occurred at a frequency of 76,800 CPM (1,280 Hz). Clearly, this is quite close to the frequency identified during the failure, and it could be a resonant frequency of the bull gear assembly.

Since additional testing on the bull gear was not a viable option, a 28 station undamped critical speed model for the gear element was developed. The calculated first two modes include a stiff shaft translational response at 8,500 RPM, followed by a stiff shaft pivotal mode at 10,250 RPM. Both resonances have greater than 93% of the strain energy in the bearings, with less than 7% of the strain energy in the shaft. Hence, these first two modes would be influenced by changes in bearing stiffness and damping characteristics. The calculated higher order modes are bending modes of the bull gear rotor. These resonances are completely dependent on shaft stiffness (i.e., bearing stiffness is inconsequential). These higher order resonances were computed for a planar condition of zero speed. This is equivalent to the bull gear sitting at rest in the gear box bearings without a mating pinion. This simplified analysis considers the case of a stationary element without external forces or rotational inertia.

Of particular interest in this simplified analysis was the appearance of a *free-free* mode at a frequency of 73,200 CPM (1,220 Hz). This frequency is close to the 74,280 CPM (1,238 Hz) detected during operation, and the 76,800 CPM (1,280 Hz) measured the stationary impact test on the failed gear. Additional examination of this relationship, and further refinement of the analytical model would be academically interesting. However, this is not a cost-effective exercise, and it is necessary to draw a logical conclusion based upon the available information. In this case, it is reasonable to conclude that the frequency in the vicinity of 75,000 CPM (1,250 Hz) is a natural resonance of the bull gear assembly. This frequency appears during the static impact tests, and it is also excited during operation with failed gear teeth. In this condition, the rotating bull gear is periodically subjected to multiple impacts due to the absence of various gear tooth. It is postulated that these impacts during operation excite this bull gear resonance.

EPICYCLIC GEAR BOXES

Other gear boxes are even more complex due to the internal configuration of the gear elements. One of the most complicated industrial gear boxes is the epicyclic gear train. In these units, a moving axis allows one or more gears to orbit about the central axis of the train. Simple epicyclic gear boxes contain a central *sun gear* that meshes with several *planet gears* that are evenly spaced around the *sun gear*. Both the *sun* and *planet gears* are externally toothed spur gears. The *planet gears* also mesh with an internally toothed *ring gear*. The input and output shafts are coaxial. These shafts may rotate in the same direction, and they may rotate in opposite directions. This is dependent on the actual mechanical configuration of the individual gear box.

Epicyclic gear boxes derive their name from the fact that the planet gears produce epicycloidal curves during rotation. In actuality, there are three general types of epicyclic gear boxes. Perhaps the most common type is the planetary arrangement that consists of a stationary ring gear combined with a rotating sun gear, and moving planet carrier. The star configuration of a epicyclic gear box consists of a stationary planet carrier coupled with a rotating sun gear, and a rotating outer ring gear. The third type, and probably the least common type of arrangement, is the solar gear. This low ratio epicyclic box has a fixed sun gear combined with a moving ring gear, and planet carrier.

Before addressing any specific details on these three epicyclic gear arrangements, it would be beneficial to mention the common characteristics between the three configurations. For instance, on a sun gear input, the tangential tooth load at each planet is derived by an expansion of equation (10-2) into the following:

$$Force_{tan-plt} = \frac{63,024 \times HP}{R_{pitch-sun} \times F_{sun} \times N_p} \qquad \textbf{(10-10)}$$

where: HP = Transmitted Power Across Sun Gear Teeth (Horsepower)
     $R_{pitch-sun}$ = Pitch Radius of Sun Gear (Inches)
     F_{sun} = Sun Gear Rotational Speed (Revolutions / Minute)
     N_p = Number of Planet Gears (Dimensionless)
     $Force_{tan-plt}$ = Tangential Force Across Planet Gear Teeth (Pounds)

The number of external teeth on the sun and planet gears, and the number of internal teeth on the stationary ring gear must maintain a particular tooth ratio to allow assembly. Specifically, the following tooth assembly equations were extracted from Dudley's *Gear Handbook*[5]:

$$T_{ring} = T_{sun} + 2 \times T_{plt} \qquad \textbf{(10-11)}$$

$$\frac{T_{ring} + T_{sun}}{N_p} = Integer \qquad \textbf{(10-12)}$$

---

[5] Darle W. Dudley, *Gear Handbook*, (New York: McGraw-Hill Book Company, 1962), pp. 3-15.

where: T_{sun} = Number of Sun Gear Teeth (Dimensionless)
         T_{plt} = Number of Planet Gear Teeth (Dimensionless)
         T_{ring} = Number of Ring Gear Teeth (Dimensionless)

With three different gear configurations, epicyclic boxes may emit a variety of excitations that vary from low to high frequencies. For example, a generic epicyclic box has the potential to produce the following array of frequencies.

○ Sun Gear Rotational Speed — F_{sun}
○ Planet Gear Rotational Speed — F_{plt}
○ Planet Carrier Rotational Speed — F_{car}
○ Ring Gear Rotational Speed — F_{ring}
○ Planet Pass Frequency — $F_{plt\text{-}pass}$
○ Planet Absolute Frequency — $F_{plt\text{-}abs}$
○ One or More Gear Mesh Frequencies — F_{gm}
○ Gear Element Resonant Frequencies
○ Casing Resonant Frequencies

For a *planetary* gear box the ring gear speed F_{ring} is equal to zero. On a *star* configuration the planet carrier is fixed, and frequency F_{car} is zero. Similarly, the sun gear speed F_{sun} is zero on a *solar* gear. The individual gear mesh frequencies are a bit more complicated, and they will be reviewed in conjunction with each gear box discussion. The gear element natural resonant frequencies, and the casing resonant frequencies listed in the previous summary, originate from the same sources discussed under two element gear boxes.

At this point, it is meaningful to examine the specific frequencies associated with a **planetary** gear box. For instance, consider the typical planetary gear train shown in Fig. 10-6. This is a basic arrangement that may be used as either a speed increasing or a speed decreasing device. For discussion purposes, assume that this gear box is used as a speed increaser. The input shaft is coupled to the planet carrier, and it rotates counterclockwise at a frequency indicated by F_{car}. In this example, three planet gears are attached to the carrier, and each planet mates with the stationary ring gear (F_{ring}=0). As the planet carrier rotates in a counterclockwise direction, the planet gears must turn clockwise at a planet rotational speed of F_{plt}. The center output gear is the sun gear, and it mates with the three planets. The sun gear has a rotational speed of F_{sun} in a counterclockwise direction. In this case, the collinear input and output shafts rotate in the same direction when viewed from one end of the gear box.

The planet spin or rotational frequency F_{plt} is calculated based upon a gear tooth ratio as follows:

$$Planet\ Spin\ Frequency_{planetary} = F_{plt} = F_{car} \times \frac{T_{ring}}{T_{plt}} \qquad \text{(10-13)}$$

The planet passing frequency is determined by multiplying the actual num-

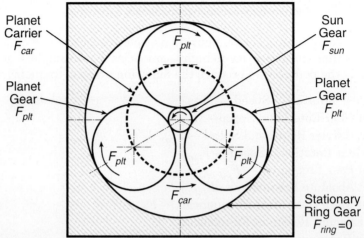

Fig. 10–6 Typical Planetary Configuration Of Epicyclic Gear Box - Stationary Ring Gear

ber of planets N_p by the planet carrier speed F_{car} as in the next equation:

$$Planet\ Pass\ Frequency_{planetary} = F_{plt-pass} = N_p \times F_{car} \qquad \textbf{(10-14)}$$

Since the planets are rotating or spinning on one axis, and that axis is rotating in a circle, the planet absolute frequency is the sum of the planet carrier and the planet spin speed as shown in equation (10-15). This frequency is seldom observed, but it is identified as part of the kinematics of the machine.

$$Planet\ Absolute\ Frequency_{planetary} = F_{plt-abs} = F_{car} + F_{plt} \qquad \textbf{(10-15)}$$

The most difficult calculation is associated with determination of the output sun gear rotational speed F_{sun}. This is not a direct gear ratio due to the element arrangement within the gear box. The calculation may be performed by examining relative speeds and ratios of the various components. In direct terms, the overall gear box ratio may be determined by equation (10-16):

$$Sun\ Gear\ Frequency_{planetary} = F_{sun} = F_{car} \times \left\{ 1 + \frac{T_{ring}}{T_{sun}} \right\} \qquad \textbf{(10-16)}$$

The planet gear mesh frequency is another common excitation for this type of machine. This frequency is easily envisioned as the planet rotational frequency F_{plt} times the number of planet teeth T_{plt}. It may also be computed based upon the number of ring gear teeth T_{ring} and the planet carrier input speed F_{car} as shown in equation (10-17):

$$Planet\ Gear\ Mesh_{planetary} = F_{gm-plt} = F_{plt} \times T_{plt} = F_{car} \times T_{ring} \qquad \textbf{(10-17)}$$

Finally, the high speed sun gear mesh frequency is determined by the product of the sun gear rotational speed F_{sun} and the number of sun gear teeth T_{sun} as shown in equation (10-18).

$$\boxed{Sun\ Gear\ Mesh_{planetary} = F_{gm-sun} = (F_{sun} \times T_{sun})} \qquad \textbf{(10-18)}$$

It should be restated that equations (10-13) through (10-18) are directly applicable *only to a planetary* gear box configuration as shown in Fig. 10-6. For demonstration purposes, these equations will be applied on a planetary gear set containing 214 ring gear teeth, 94 planet gear teeth, and 26 sun gear teeth. Assume that this is a speed increasing box with three planets, and that the input speed is 1,780 RPM. Before starting the gear frequency calculations, the validity of this assembly may be checked with equations (10-11) and (10-12) as follows:

$$T_{ring} = T_{sun} + 2 \times T_{plt} = 26 + 2 \times 94 = 26 + 188 = 214$$

This calculation agrees with the actual ring gear tooth count of 214 teeth. Next, the second assembly equation should be checked as follows:

$$\frac{T_{ring} + T_{sun}}{N_p} = \frac{214 + 26}{3} = \frac{240}{3} = 80 = Integer$$

The final value of 80 is an integer number, and the equation is satisfied. Hence, the basic epicyclic gear assembly equations are satisfied, and it is appropriate to commence the computation of the fundamental excitation frequencies. The planet rotational frequency F_{plt}, also known as the planet spin speed, is determined from equation (10-13) as follows:

$$F_{plt} = F_{car} \times \frac{T_{ring}}{T_{plt}} = 1,780\ \text{RPM} \times \frac{214}{94} = 4,052\ \text{RPM}$$

The planet passing frequency $F_{plt\text{-}pass}$ is computed from equation (10-14):

$$F_{plt-pass} = N_p \times F_{car} = 3 \times 1,780\ \text{RPM} = 5,340\ \text{RPM}$$

Next, the planet absolute frequency $F_{plt\text{-}abs}$ may be determined from equation (10-15) in the following manner:

$$F_{plt-abs} = F_{car} + F_{plt} = 1,780 + 4,052 = 5,832\ \text{RPM}$$

The sun gear rotational frequency F_{sun} may be calculated with (10-16):

$$F_{sun} = F_{car} \times \left\{ 1 + \frac{T_{ring}}{T_{sun}} \right\} = 1,780 \times \left\{ 1 + \frac{214}{26} \right\} = 16,431\ \text{RPM}$$

This final ratio of 9.2308:1 (=1+214/26) may seem excessive, but for a planetary gear box of this general arrangement it is quite common. In fact, overall speed ratios of 12:1 are a common and acceptable practice for planetary boxes.

The planet gear mesh frequency $F_{gm\text{-}plt}$ may be determined from both portions of equation (10-17) to yield the following identical results:

$$F_{gm-plt} = (F_{plt} \times T_{plt}) = (4,052 \times 94) = 380,900 \text{ CPM}$$

or

$$F_{gm-plt} = (F_{car} \times T_{ring}) = (1,780 \times 214) = 380,900 \text{ CPM}$$

Finally, the higher frequency sun gear mesh frequency $F_{gm\text{-}sun}$ is easily calculated from expression (10-18) as follows:

$$F_{gm-sun} = (F_{sun} \times T_{sun}) = (16,431 \times 26) = 427,200 \text{ CPM}$$

From this basic planetary gear train, a total of seven fundamental excitations have been identified. Field vibration measurements on this gear box will generally reveal various interactions and modulations between these frequencies. Due to the potential narrow pulse width of some interactions, the resultant vibration data should always be viewed in both the time and the frequency domain to make sure that all of the significant vibratory motion is detected.

The second type of epicyclic gear box commonly encountered is the **star configuration** as illustrated in Fig. 10-7. For discussion purposes, assume that the input occurs at the center sun gear that rotates at a speed of F_{sun} in a counterclockwise direction. In this type of box, the planets are fixed in stationary bearings, and the planet carrier frequency F_{car} is zero. Although the planets continue to rotate clockwise at a frequency of F_{plt}, there is no translation of the planet bearing centerlines. Also note that the sun gear is not restrained by a bearing, and it essentially floats within the mesh of the planets.

The planet gears in Fig. 10-7 engage an outer ring gear that turns in a clockwise direction at a rotational speed of F_{ring}. The ring gear then connects to the output shaft directly, or it may mate with an outer coupling assembly

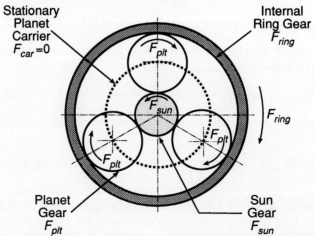

Fig. 10–7 Typical Star Configuration Of Epicyclic Gear Box - Stationary Planet Carrier

through a spline arrangement. Obviously, the speed of the outer coupling assembly must be equal to the rotational speed of the internal ring gear F_{ring}. Normally, the circumferential outer coupling assembly is connected to a common end plate, and this plate is secured to the output shaft, as illustrated in Fig. 10-10 (in case history 30). As shown in this sketch, the input and output shafts rotate in opposite directions for this *star* configuration.

The planet spin or rotational frequency F_{plt} for this star arrangement is calculated from the sun gear as follows:

$$Planet\ Spin\ Frequency_{star} = F_{plt} = F_{sun} \times \frac{T_{sun}}{T_{plt}} \qquad \text{(10-19)}$$

The planet passing frequency is equal to zero, since the planet carrier does not rotate (i.e., F_{car} =0). By the same logic, the planet absolute frequency $F_{plt\text{-}abs}$ is identical to the planet rotational speed F_{plt}. The output ring gear rotational speed F_{ring} may be determined by:

$$Ring\ Gear\ Frequency_{star} = F_{ring} = F_{sun} \times \frac{T_{sun}}{T_{ring}} \qquad \text{(10-20)}$$

The planet gear mesh frequency is constant across all three gears, and it may be computed with equation (10-21):

$$Gear\ Mesh_{star} = F_{gm} = F_{plt} \times T_{plt} = F_{ring} \times T_{ring} = F_{sun} \times T_{sun} \qquad \text{(10-21)}$$

Equations (10-19) through (10-21) only apply to a star arrangement. A set of example calculations for a star configuration are presented in case history 30. Prior to this machinery story, the third type of epicyclic gear box should be reviewed. As previously stated, this is commonly known to as a **solar** configuration. This name stems from the fact that the sun gear remains fixed (i.e., $F_{sun}=0$), and all of the other gears are in motion, as shown in Fig. 10-8. For discussion purposes, assume that the input to this gear train is the clockwise rotating ring gear at a frequency of F_{ring}. The engaged planets are driven in a clockwise direction at a planet spin speed of F_{plt}. The planet gears translate around the fixed sun gear, and they drive the planet carrier in a clockwise direction at a speed of F_{car}. In this case, the input and output shafts rotate in the same direction when viewed from one end of the gear box.

The planet spin or rotational frequency F_{plt} is calculated as follows:

$$Planet\ Spin\ Frequency_{solar} = F_{plt} = F_{ring} \times \frac{T_{ring}}{T_{plt}} \qquad \text{(10-22)}$$

The planet passing frequency is determined by multiplying the actual num-

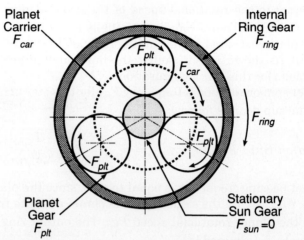

Fig. 10–8 Typical Solar Configuration Of Epicyclic Gear Box - Stationary Sun Gear

ber of planets N_p by the planet carrier speed F_{car} as in the next equation:

$$\boxed{Planet\ Pass\ Frequency_{solar}\ =\ F_{plt-pass}\ =\ N_p \times F_{car}} \qquad \textbf{(10-23)}$$

Since the planets are rotating or spinning on one axis, and that axis is rotating in a circle, the planet absolute frequency is shown in equation (10-24).

$$\boxed{Planet\ Absolute\ Frequency_{solar}\ =\ F_{plt-abs}\ =\ F_{car} + F_{plt}} \qquad \textbf{(10-24)}$$

The determination of the output planet carrier rotational speed F_{car} is presented in the following equation (10-25):

$$\boxed{Carrier\ Frequency_{solar}\ =\ F_{car}\ =\ F_{ring} \times \left\{ 1 + \frac{T_{sun}}{T_{ring}} \right\}} \qquad \textbf{(10-25)}$$

Note that the value resulting from the ratio of T_{sun}/T_{ring} will be considerably less than one. When this value is summed with one, it is clear that the overall gear ratio will be quite low. Hence, a *solar* configuration of an epicyclic gear box is strictly a low speed ratio device. This type of machine exhibits only a single gear mesh frequency that is the planet rotational speed F_{plt} times the number of planet teeth T_{plt}. It may also be computed based upon the number of ring gear teeth T_{ring} and the ring gear speed F_{ring} as shown in the next equation:

$$\boxed{Gear\ Mesh_{solar}\ =\ F_{gm}\ =\ F_{plt} \times T_{plt}\ =\ F_{ring} \times T_{ring}} \qquad \textbf{(10-26)}$$

Once again, it should be noted that equations (10-22) through (10-26) *only apply to a solar* arrangement of an epicyclic gear box. Furthermore, multiple beats and signal modulations are possible on any epicyclic gear box due to the interaction of a large variety of fundamental excitations. There is also the potential for considerable cross-coupling between the lateral and torsional characteristics in these gear boxes. This becomes even more complicated when compound epicyclic gear boxes are examined that contain two planets on the same shaft. In other cases, two epicyclic gear boxes may be used in tandem (i.e., coupled together) to achieve some very high speed ratios. In either case, the array of mechanical excitations becomes quite large, and the potential for interaction with one or more system resonances becomes significant. Specifically, consider the situation described in the following case history.

Case History 30: Star Gear Box Subsynchronous Motion

The machinery train discussed in this case history consists of a 16,500 HP gas turbine driving a synchronous generator through an epicyclic gear box. The arrangement of the gear box and generator, plus the installed proximity probes are shown in Fig. 10-9. The epicyclic gear box was configured in a star arrangement similar to the previously discussed Fig. 10-7. This box contained 3 planet gears with 47 teeth T_{plt} on each gear. The power turbine input speed to the sun gear F_{sun} was 8,568 RPM, and the sun gear contained 25 teeth T_{sun}. The outer ring gear must rotate at 1,800 RPM to drive the generator. This internally toothed ring gear contained 119 teeth. As shown in Fig. 10-10, the ring gear was directly mated to an outer coupling assembly that drove the output gear box shaft. The gear box output shaft was restrained by an outer and an inner journal bearing as shown in Fig. 10-10. The sun gear floated on the planet mesh, and each planet was supported by a fixed journal bearing.

The coupling between the epicyclic gear box and the generator was a close coupled gear coupling. Although technical specifications were not available for this assembly, it was clearly a *hard* coupling with high torsional and lateral stiffness. Hence, any lateral or torsional excitations on the gear box could be easily

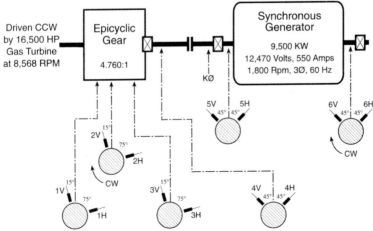

Fig. 10–9 Machinery And Vibration Transducer Arrangement For Star Gear and Generator

transmitted to the generator, and vice versa.

Before addressing the specific problems on this machinery, it would be desirable to check the epicyclic gear box configuration, and identify the anticipated excitation frequencies. As before, the validity of this assembly may be checked with equations (10-11) and (10-12), as follows:

$$T_{ring} = T_{sun} + 2 \times T_{plt} = 25 + 2 \times 47 = 25 + 94 = 119$$

This calculation agrees with the actual ring gear tooth count of 119 teeth. Next, check the second assembly equation as follows:

$$\frac{T_{ring} + T_{sun}}{N_p} = \frac{119 + 25}{3} = \frac{144}{3} = 48 = Integer$$

The value of 48 is an integer number, and the equation is satisfied. Hence, the basic epicyclic gear assembly equations are satisfied, and it is appropriate to compute the fundamental excitation frequencies. The planet rotational or spin frequency F_{plt} is determined from equation (10-19) as follows:

$$F_{plt} = F_{sun} \times \frac{T_{sun}}{T_{plt}} = 8,568 \text{ RPM} \times \frac{25}{47} = 4,557 \text{ RPM}$$

The ring gear rotational frequency F_{ring} may be verified with (10-20):

$$F_{ring} = F_{sun} \times \frac{T_{sun}}{T_{ring}} = 8,568 \text{ RPM} \times \frac{25}{119} = 1,800 \text{ RPM}$$

The gear mesh frequency F_{gm} may be determined from the first portion of equation (10-21) to yield the following:

$$F_{gm} = F_{plt} \times T_{plt} = 8,568 \text{ RPM} \times 25 = 214,200 \text{ CPM}$$

This *star* epicyclic gear configuration provides the proper output speed of 1,800 RPM, and the above calculations also define the planet rotational speed F_{plt}, plus the gear mesh frequency F_{gm}. Under normal machinery behavior the dominant shaft vibration frequency on both the gear box output and the generator should be 1,800 CPM. On the power turbine shaft (sun gear input), the major shaft vibration frequency should be 8,568 CPM. The planet rotational frequency of 4,557 CPM probably would not appear unless one or more planets were in a state of distress. Finally, the major high frequency casing vibration component should logically occur at the gear mesh frequency of 214,200 CPM. Although this is a complex mechanical system, the number of fundamental excitations are limited and definable.

Initial operational tests on this unit revealed normal and acceptable behavior at full load. Vibration levels were low, journal positions were proper, bearing temperatures were normal, and overall skid vibration was acceptable. The transient startup and coastdown Bode plots were likewise normal, and all static plus dynamic measurements pointed towards a normal machinery train. However,

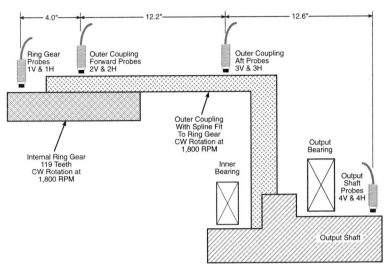

Fig. 10–10 Low Speed Probes Installed In Star Configuration Epicyclic Gear Box

during a reduced load test, an unusual subsynchronous vibration component at 1,240 CPM appeared on the generator. This frequency component at nominally 69% of rotational speed was forward and elliptical. The highest amplitudes occurred at the inboard coupling end bearing. Further investigation revealed that the same frequency appeared on the epicyclic gear box output bearing.

As the examination of the data progressed, it was clear that additional measurement points along the low speed rotor elements would be beneficial. The machinery was originally equipped with X-Y proximity probes mounted at ±45° from vertical center. This included the gear box output (probes 4V & 4H), the generator coupling end bearing (probes 5V & 5H), and the generator outboard or exciter end bearing (probes 6V & 6H). All three sets of original transducers are depicted on the machinery arrangement diagram Fig. 10-9.

In order to obtain more information about this low frequency phenomena, three more sets of X-Y probes were installed through the top cover of the gear box. These supplemental probes were mounted with the vertical or Y-axis probe at 15° to the left of the vertical centerline, and the horizontal or X-axis transducer was positioned at 75° to the right of vertical. The physical location of these additional probes are shown in Figs. 10-9 and 10-10. The first set of probes (1V & 1H) were positioned on the outer diameter of the ring gear. The next two sets of pickups (probes 2V & 2H) were located on the forward part of the outer coupling, and probes 3V & 3H were located on the aft portion of the outer coupling. In essence, probes 3V & 3H were in line with the inner bearing. The location and axial distance between gear box transducer planes is described in Fig. 10-10. This diagram also depicts the outer coupling spline fit to the ring gear, and the general attachment of the outer coupling to the gear box output shaft.

Armed with this full array of transducers observing the low speed rotors, a series of detailed operational tests were conducted in an effort to quantify the characteristics of this subsynchronous vibration component.

A graphical description of this subsynchronous vibration is presented in Fig. 10-11. This steady state data was obtained with the low speed shaft operating at 1,800 RPM, and a 1,600 KW load on the generator. The shaft vibration from each set of X-Y proximity probes was band-pass filtered at the maximum subsynchronous frequency of 1,246 CPM, and the resultant orbits plotted at each measurement station. Since the subsynchronous orbits are all elliptical, a comparison of severity was achieved by identifying the vibration amplitude associated with the major axis of each ellipse. The subsynchronous orbits presented in Fig. 10-11 also include a simulated timing signal at the tuned frequency of 1,246 CPM. This simulated timing signal is not related to a stationary reference point like the shaft once-per-rev Keyphasor® signal. However, the relative phase relationship between all six measurement locations is accurate and consistent.

From Fig. 10-11, it is noted that the subsynchronous motion on the ring gear and the forward end of the outer coupling assembly are in unison. Similar orbit patterns are evident, and the respective amplitudes and phase markers are almost identical. At the aft end of the outer coupling assembly the subsynchronous motion was attenuated. The orbit at the output shaft of the gear box reveals a phase reversal plus a lower amplitude at the subsynchronous frequency. The magnitude of the subsynchronous vibration continues to diminish across the coupling to the generator, with the lowest amplitudes appearing on the outboard end of the generator. In essence, the subsynchronous motion has a conical mode shape with maximum amplitudes at the ring gear and outer coupling; with a nodal point between the outer coupling aft, and the gear box output.

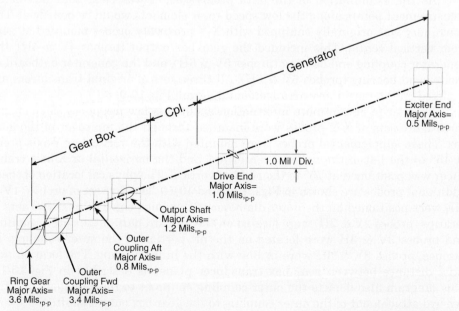

Fig. 10–11 Initial Mechanical Configuration — Subsynchronous Shaft Orbits Filtered At 1,246 CPM With Rotational Speed Of 1,800 RPM And 1,600 KW Load on Generator

Closer examination of the variable speed startup data with cascade plots revealed an independent lateral vibration component that migrates from 1,000 to 1,300 CPM. This behavior was dominant on the gear box, with much lower amplitudes on the generator. In addition, the unfiltered shaft vibration data revealed an amplitude modulation between running speed and the subsynchronous component. This behavior was most visible on the ring gear, and the outer coupling assembly. Finally, when the subsynchronous component was active, casing accelerometers reveal a ±1,240 CPM modulation of the 214,200 CPM gear mesh frequency.

From this data array it is clear that the subsynchronous shaft vibration encountered on the low speed end of this machinery train originates within the epicyclic gear box. It was also determined that the subsynchronous excitation occurs at a frequency that only varies between 1,220 and 1,260 CPM. It was also documented that the largest subsynchronous vibration amplitudes appear during a limited torque range of 5,700 to 6,300 foot-pounds.

This information was substantiated during variable speed generator tests at 1,500 and 1,800 RPM. Specifically, test data at 1,800 RPM and 1,600 KW (6,260 foot pounds) revealed a subsynchronous frequency of 1,240 CPM. By comparison, at a reduced generator speed of 1,500 RPM and a load of 1,212 KW (5,690 foot pounds), the subsynchronous component appeared at essentially the same frequency of 1,220 CPM. Any changes in generator load (up or down) would attenuate, or completely eliminate the subsynchronous vibration.

Load changes, or more specifically torque changes, are directly associated with the repeatable appearance of this subsynchronous component. Since the subsynchronous excitation appears at essentially a constant frequency, and the maximum amplitude occurs within a limited torque range — serious consideration should be directed towards a resonant response in a twisting direction. That is, the excitation of a lower order torsional resonance should be considered as a realistic possibility. This concept was reinforced when the undamped torsional response calculations revealed a first mode at 1,320 RPM. Although some of the mass elastic data was questionable, the coincidence of the calculated torsional critical speed and the measured behavior could not be ignored.

However, the correlation of the measured subsynchronous vibration component with the calculated torsional resonance frequency was an unpopular conclusion. This was complicated by the fact that the subsynchronous component was sensitive to changes in the oil supply temperature, which could be related to changes in damping. In addition, the machinery adversely responded to minor alignment changes across the low speed coupling between the gear box and the generator. It appeared that raising the generator by 10 Mils unloaded the gear output bearing, and allowed it to go unstable. Hence, the popular corporate theory was that the observed behavior was nothing more than a bearing stability problem. Since the major activity occurred within the epicyclic gear box, the parties in charge of the machinery elected to suppress the subsynchronous instability by unbalancing the outer coupling assembly. This change was implemented by adding a 79 gram unbalance weight to the forward axial face of the outer coupling assembly. The influence of this 79 gram unbalance weight on the subsyn-

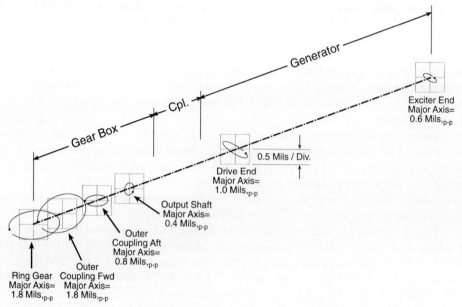

Fig. 10–12 Unbalance Of 79 Grams On Outer Coupling - Subsynchronous Orbits Filtered At 1,248 CPM With Rotational Speed Of 1,800 RPM And 1,600 KW Generator Load

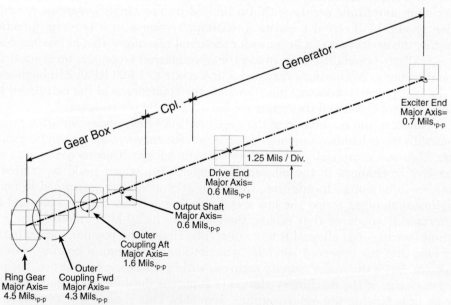

Fig. 10–13 Unbalance Of 79 Grams On Outer Coupling - Synchronous Shaft Orbits Filtered At Rotational Speed Of 1,800 RPM With 1,600 KW Generator Load

chronous motion is shown on Fig. 10-12. Note that the gear box vibration amplitudes at 1,248 CPM on the ring gear and the outer coupling have been attenuated from nominally 3.6 to 1.8 Mils,$_{p-p}$. However, the vibration response over on the generator at this frequency was virtually unaffected. Since the amplitude of this subsynchronous vibration on the generator was the primary concern of the OEM, the addition of the 79 gram unbalance to the gear box outer coupling assembly was not an acceptable solution.

The inappropriateness of this weight addition was further demonstrated by examining the synchronous 1X motion at each of the measurement planes as presented in Fig. 10-13. The previous running speed vibration at 1,800 RPM before installation of the 79 gram unbalance varied between 0.3 and 0.7 Mils,$_{p-p}$ at each in the measurement location. However, after the 79 grams was attached, the 1X amplitudes increased to 4.3 Mils,$_{p-p}$ at the outer coupling, and 4.5 Mils,$_{p-p}$ on the ring gear. Clearly, these increased synchronous amplitudes would be detrimental to the long-term reliability of this gear box.

In the final assessment, the addition of unbalance weights to the gear box does not represent a viable solution to the subsynchronous vibration problem. In fact, it does impose additional dynamic forces upon the gear elements. The proper engineering solution included a modification of the gear box output bearing to cope with the occasional instability due to bearing unloading. In addition, it was necessary to recognize that the subsynchronous motion at nominally 1,240 CPM was logically a torsional natural frequency. This resonance appeared as a lateral vibration due to cross-coupling between torsional and lateral motion across the gear teeth. This torsional resonance was directly related to the stiff gear coupling between the epicyclic gear box and the generator. In all cases, units that contained the stiff gear coupling exhibited this subsynchronous component at reduced load, and identical units that had a torsionally softer flexible disk coupling *did not* experience this subsynchronous lateral response.

PROCESS FLUID EXCITATIONS

Fluid handling machines invariably contain some arrangement of stationary and rotating blades or vanes. This applies to machines handling incompressible fluids such as pumps, hydraulic turbines, and extruders, plus machines that handle compressible fluids such as steam or gas turbines, centrifugal compressors, expanders, and blowers. In many cases, the number of blades or vanes times the shaft rotative speed provides a simple expression for computation of a potential blade passing frequency F_b as shown in the next equation.

$$\boxed{Blade\ Passing\ Frequency = F_b = N_b \times RPM} \qquad \text{(10-27)}$$

where: F_b = Blade Passing Frequency (Cycles / Minute)
N_b = Number of Blades or Vanes (Dimensionless Integer)
RPM = Rotative Speed (Revolutions / Minute)

An application of this concept is presented in Fig. 10-14 that displays a pair of spectrum plots obtained from a large single shaft gas turbine. The top FFT plot in Fig. 10-14 was acquired at the inlet end bearing housing. It displays a frequency array that coincides with virtually all of the axial flow air compressor stages. The sixth stage was not evident and there is no guarantee that some of the other components are attributable to only one stage. In addition, the components that represent several stages (same number of blades on more than one stage) might be due to an excitation from only one stage, or the interaction of several stages to produce one component at a common frequency.

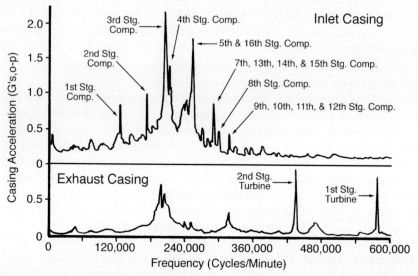

Fig. 10–14 Blade Passing Excitations On A Large Single Shaft Industrial Gas Turbine

The simpler plot at the bottom of Fig. 10-14 was obtained from the exhaust end bearing housing. This diagram is dominated by the hot gas power turbine first and second stage blade passing frequencies. Due to the clear demarcation of blade counts, this data carries more credibility than the complex spectra extracted from the inlet end bearing. However, the diagnostician should still review this data with caution. In most cases, the impedance path between the rotor excitation and a bearing cap acceleration measurement is unknown. It is therefore difficult to correlate these frequency components and amplitudes to specific levels of severity within the machine. At best, the various components may be identified in terms of harmonic order, and potentially associated with specific mechanical elements (e.g., number of turbine blades on a particular stage). From there on, the machinery diagnostician is faced with routine examination of the high frequency spectra, and a trending of the results. Naturally, for this type of program to be effective, the same accelerometer must be mounted in the same location, and the high frequency data acquired and processed in the same manner. Variations of any of these steps would invalidate the accumulated database.

It must also be recognized that high frequency blade passing excitations are often influenced by stationary objects. This interaction between the rotating and the stationary mechanical systems is a nuisance when dealing with compressible fluids, and it forms a mandatory part of the analysis when examining machines that handle incompressible liquids.

For instance, consider Table 10-1 of calculated pump vane passing frequencies. This data is for a centrifugal pump with a vaned diffuser. Typically, the number of diffuser vanes exceed the number of impeller vanes. Due to the close coupled configuration of this type of machinery, there is a definite interrelationship and resultant excitation between stator and rotor parts. Thus, a six vane impeller running inside of a nine vane diffuser will produce a blade passing frequency at four times rotative speed. If the machinery diagnostician is expecting to see a 6X blade passing frequency on the pump, the appearance of a strong 4X component can be most disconcerting.

Table 10–1 Vane Pass Frequency For Various Combinations Of Impeller And Diffuser Vanes

Pump Diffuser Vanes	Number of Pump Impeller Vanes						
	3	**4**	**5**	**6**	**7**	**8**	**9**
4	3	—	5	3	7	2	9
5	6	4	—	6	14	16	9
6	—	4	5	—	7	4	3
7	6	8	15	6	—	8	27
8	9	—	15	9	7	—	9
9	—	8	10	4	28	8	—
10	9	6	—	6	21	16	9
11	12	12	10	12	21	32	45
12	—	—	25	—	35	4	9
13	12	12	25	12	14	40	27
14	15	8	15	6	3	8	27
15	—	16	—	6	14	16	6
16	15	—	15	9	49	—	63
17	18	16	35	18	35	16	18
18	—	10	35	—	35	28	—
19	18	20	20	18	56	56	18
20	21	—	—	21	21	6	81

Table 10-1 was generated with a computer program for computation of vane passing frequencies for diffuser pumps by James E. Corley[6]. This program was implemented over a wide range of impeller and diffuser vane configurations, and the results summarized in Table 10-1. For purposes of clarity, it should be restated that this table and the associated discussion is limited to vaned diffuser type centrifugal pumps. For the more common configuration of volute pumps, the vane pass frequency reverts back to the original definition of speed times the number of impeller vanes stated in equation (10-27).

The previous discussion should not imply that pump excitations are limited to vane passing activity. In reality, a variety of generic excitations as discussed in chapter 9, plus the hydraulic behavior previously mentioned, are possible. Other problems such as cavitation, internal recirculation, flow distribution, plus difficulties associated with mechanical seals and couplings occur. For more detailed information on pump behavior, the reader is encouraged to examine technical papers such as the excellent documents by Nelson and Dufour[7], and Schiavello.[8]

Although the vast majority of process fluid excitations are directly associated with rotating machinery blade or vane passing frequencies, other types of fluid excitations do exist, and they are responsible for some significant mechanical failures. In all cases, it must be recognized that the fluid flow stream (compressible or incompressible) carries a substantial amount of energy, and it may be a significant excitation source. The appearance of pressure fluctuations at the boundary layer of a fluid stream, the problems associated with turbulent flow, the destructive forces associated with vortex shedding frequencies, or any of the acoustic mechanisms are generally formidable engineering problems. In addition, the cross-coupling of fluid excitations into piping systems or support structures may add a new dimension of complexity to an already difficult problem.

The vortex induced vibration problem is particularly interesting, since it encompasses the flow of fluids over stationary objects. This behavior is clearly described by Robert D. Blevins[9] as follows:

"Structures shed vortices in a subsonic flow. The vortex street wakes tend to be very similar regardless of the geometry of the structure. As the vortices are shed from first one side and then the other, surface pressures are imposed upon the structure...The oscillating pressures cause elastic structures to vibrate and generate aeroacoustic sounds...The vibration induced in elastic structures by vortex shedding is of practical importance because of its potentially destructive effect on bridges, stacks, towers, offshore pipelines, and heat exchangers..."

---

[6] James E. Corley, "Tutorial Session on Diagnostics of Pump Vibration Problems," *Proceedings of the Fourth International Pump Symposium*, Turbomachinery Laboratory, Texas A&M University, College Station, Texas (May 1987).

[7] W.E. (Ed) Nelson and J.W. Dufour, "Pump Vibrations," *Proceedings of the Ninth International Pump Users Symposium*, Turbomachinery Laboratory, Texas A&M University, College Station, Texas (March 1992), pp. 137-147.

[8] Bruno Schiavello, "Cavitation and Recirculation Troubleshooting Methodology," *Proceedings of the Tenth International Pump Users Symposium*, Turbomachinery Laboratory, Texas A&M University, College Station, Texas (March 1993), pp. 133-156.

[9] Robert D. Blevins, *Flow-Induced Vibration,* Second Edition, (New York: Van Nostrand Reinhold, 1990), p. 43.

Blevins goes on to explain that vortex shedding from a smooth circular cylinder in a subsonic flow is a function of the Reynolds number. In this context, the Reynolds number N_{Re} is defined in the following manner:

$$N_{Re} = \frac{D \times V}{\nu} \qquad \textbf{(10-28)}$$

where: N_{Re} = Reynolds Number (Dimensionless)
              V = Free Stream Velocity Approaching the Cylinder (Inches / Second)
              D = Cylinder Diameter (Inches)
              ν = Kinematic Viscosity (Inches^2 / Second)

In chapter 4 of this text, equation (4-5) identified the variables used to compute the Reynolds through the minimum oil film of a bearing. An initial comparison between equations (4-5) and (10-28) reveals some differences. However, a closer examination of (4-5) shows that the term "$\omega \times R$" is surface velocity of a rotating shaft with units of inches per second. This is equivalent to the free stream velocity approaching the cylinder V shown in equation (10-28). The oil film height in inches designated by H in equation (4-5) is equivalent to the cylinder diameter D used in (10-28). Finally, the remaining terms are all associated with the moving fluid viscosity. In equation (10-28) the *kinematic* viscosity ν was used, whereas the *absolute* or *dynamic* viscosity μ was applied in (4-5). The two viscosity formats are directly related, as shown in equation (10-29).

$$\nu = \frac{\mu \times G}{\rho} \qquad \textbf{(10-29)}$$

where: μ = Absolute or Dynamic Viscosity (Pounds-Seconds / Inch^2),
              G = Acceleration of Gravity (386.1 Inches / Second^2)
              ρ = Fluid Density (Pounds / Inches^3)

A dimensional analysis of equation (10-29) reveals that the units are correct. Furthermore, both Reynolds number equations (4-5) and (10-28) are equivalent. Whereas (4-5) was used to define the ratio of inertia to viscous forces in a fluid film bearing — expression (10-28) is applied to a fluid stream flowing across a smooth circular cylinder. In this application, the pattern generated by vortices down stream of the cylinder may be predicted based upon the value of the Reynolds number. Specific flow regimes have been identified by various investigators, and the reader is again referenced to the text by Robert Blevins for detailed information. Some of the information on this topic is also available in the *Shock and Vibration Handbook*[10] in the section authored by Blevins. It should also be mentioned that numerous studies have been conducted on vortex shedding, and the associated vortex induced vibration. Hence, this is a well-documented technical field that incorporates many empirical studies and analytical solutions. The references provided within the Blevins text reveal the true breadth of this physi-

---

[10] Cyril M. Harris, *Shock and Vibration Handbook*, Fourth edition, (New York: McGraw-Hill, 1996), pp. 29.1 to 29.19.

cal behavior that stretches across many technical fields.

Of particular importance within the machinery business is the relationship defined by the Strouhal number N_{Str}. This is a non-dimensional number that allows computation of the fundamental or predominant vortex shedding frequency F_s as defined in the following expression:

$$N_{Str} = \frac{F_s \times D}{V} \qquad (10\text{-}30)$$

where: N_{Str} = Strouhal Number (Dimensionless)
               F_s = Vortex Shedding Frequency (Cycles / Second)

The cylinder diameter D, and the constant velocity of the fluid stream V is the same value used in equation (10-28). Thus, for a given cylinder diameter D, and flow velocity V, the vortex shedding frequency F_s may be computed if the Strouhal number N_{Str} is known. Fortunately, there are multiple empirical tests that display a consistent relationship between the parameters specified in equation (10-30). For instance, on pages 48 through 51 of Bevins text, a variety of charts describe the Strouhal number for a circular cylinder, an array of inline and staggered cylinders, plus various other geometric cross sections. From this database, it is clear that the Strouhal number for the vast majority of cases will vary between values of 0.1 and 0.8.

For the simple case of a circular cylinder with a Reynolds number between 500 and 1,000,000, this data suggests that the Strouhal number has a value of nominally 0.2. Many technical references identify this value as 0.22. However, across the range of Reynolds numbers specified, 0.2 represents a more realistic average for the Strouhal number. If value this is substituted into (10-30), the vortex shedding frequency F_s may be computed directly from:

$$F_s = \frac{0.2 \times V}{D} \qquad (10\text{-}31)$$

As a practical example of the application of these vortex shedding concepts, consider the situation of a gas turbine exhaust stack. If the top cylindrical portion of the stack has a diameter of 20 inches, and the environment consisted of standard temperature (60°F) and pressure (14.7 Psia), with wind gusts of 50 miles per hour — it would be desirable to compute the anticipated vortex shedding frequency. From various sources, the absolute viscosity of air under these conditions is 0.018 centipoise. This is equal to 0.00018 poise. From the conversion factors presented in Appendix C of this text, 1 poise is equivalent to 1 dyne-second/centimeter^2. Thus, the absolute or dynamic viscosity μ of air at standard temperature and pressure would be equal to 0.00018 dyne-second/centimeter^2. Converting the metric viscosity units to English units may be accomplished in the following manner:

$$\mu = 0.00018 \text{ Dyne-Sec/Cm}^2 \times 2.248 \times 10^{-6} \text{ Pound/Dyne} \times (2.54 \text{ Cm/Inch})^2$$

$$\mu = 0.00018 \times 2.248 \times 10^6 \times 6.4516 = 2.611 \times 10^{-9} \text{ Pound-Sec/Inch}^2$$

From Table B-2 in the appendix of this text, the density of air at standard temperature and pressure is equal to 0.07632 pounds per foot^3. This density value may be converted to consistent units as follows:

$$\rho = 0.07632 \text{ Pound/Foot}^3 \times (1 \text{ Foot}/12 \text{ Inches})^3 = 4.417 \times 10^{-5} \text{ Pound/Inch}^3$$

Based on these physical properties, the kinematic viscosity of air may now be computed from equation (10-29):

$$\nu = \frac{\mu \times G}{\rho} = \frac{2.611 \times 10^{-9} \text{ Pound-Sec/Inch}^2 \times 386.1 \text{ Inch/Sec}^2}{4.417 \times 10^{-5} \text{ Pound/Inch}^3} = 0.0228 \text{ Inches}^2/\text{Sec.}$$

The peak wind velocity of 50 miles per hour may now be converted into compatible engineering units of inches per second in the following manner:

$$V = 50 \text{ Miles/Hour} \times 5,280 \text{ Feet/Mile} \times 12 \text{ Inches/Foot} \times 1 \text{ Hour}/3,600 \text{ Sec} = 880 \text{ Inch/Sec.}$$

Sufficient information is now available to compute the Reynolds number of the air flow over the cylindrical stack with equation (10-28):

$$N_{Re} = \frac{D \times V}{\nu} = \frac{20 \text{ Inches} \times 880 \text{ Inch/Sec.}}{0.0228 \text{ Inches}^2/\text{Sec.}} = 772,000$$

The Reynolds number of 772,000 falls within the previously specified range of 500 and 1,000,000. This provides confidence in using a Strouhal number of 0.2 for this case of a circular stack. More specifically, these conclusions allow the direct application of equation (10-31) as follows:

$$F_s = \frac{0.2 \times V}{D} = \frac{0.2 \times 880 \text{ Inch/Sec}}{20 \text{ Inches}} = 8.80 \text{ Cycles/Sec} \times 60 \text{ Sec/Min} = 528 \text{ Cycles/Min.}$$

Hence, with a 50 mile per hour wind, the anticipated vortex shedding frequency would be 528 cycles per minute (8.80 Hz). This is an appreciable frequency that could influence the gas turbine, or any of the associated mechanical equipment. Furthermore, this frequency will change with wind speed. If any combination of wind speed and associated vortex shedding frequency coincided with a natural frequency of the stack, the results could be devastating.

The traditional solution to this type of problem resides in the modification of the stack outer diameter to disrupt the vortex shedding, and therefore eliminate (or substantially minimize) the excitation source on the stationary cylinder. Common modifications include helical strakes wrapped around the stack, or a series of external slats or shrouds that are designed to break up the vortices. In some cases, an analytical model using Computational Fluid Dynamic (CFD) software might be sufficient to properly examine the system. In other situations, the development and testing of a scale model in a wind tunnel might be appropriate.

In all cases, the diagnostician must be aware of this vortex shedding phenomena and the potential for induced vibration into the structure.

Another mechanism that periodically appears within fluid handling systems is the acoustic resonance problem. This is the classic *organ pipe* behavior that appears in virtually every physics textbook. The traditional discussion of standing wave theory relates the velocity of sound (i.e., sonic velocity) in the fluid media V_s with the occurring acoustic frequency F_a and the wavelength λ as presented in the following expression:

$$\boxed{V_s = F_a \times \lambda}$$
(10-32)

where: V_s = Sonic or Acoustic Velocity (Feet / Second)
F_a = Acoustic Frequency (Cycles / Second)
λ = Standing Wave Length (Feet)

The velocity of sound V_s will vary according to the media. Solids will generally display the highest values, sonic velocity in liquids will generally be lower, and gases will display even low speeds. For example, Table 10-2 summarizes some common values for the velocity of sound in assorted solids and liquids.

Table 10–2 Typical Values For Sonic Velocity In Various Solids And Liquids

	Material	Velocity of Sound (Feet/Second)
Solids	Lead	4,030
	Brass	11,480
	Copper	11,670
	Iron & Steel	16,410
	Aluminum Alloys	16,740
	Graphite	19,700
Liquids At 60°F and 14.7 Psia	Alcohol	3,810
	Oil - Sp. Gr.=0.9	4,240
	Mercury	4,770
	Fresh Water	4,860
	Glycerin	6,510

The velocity of sound in solids will remain constant over a wide range of conditions. This is due to the fact that the material density ρ_{sol} and modulus of elasticity E remain constant over a wide range of conditions. The sonic velocity in any solid may be computed with the following common expression:

$$V_{s_{sol}} = \sqrt{\frac{G \times E}{144 \times \rho_{sol}}}$$ (10-33)

where: $V_{s\text{-}sol}$ = Velocity of Sound in a Solid (Feet / Second)
G = Acceleration of Gravity (= 386.1 Inches / Second^2)
E = Modulus of Elasticity (Pound / Inch^2)
ρ_{sol} = Solid Material Density (Pounds / Inch^3)

To check the validity of (10-33), the properties from Table B-1 in appendix B may be extracted and inserted into this expression. For example, pure copper has a modulus of elasticity equal to 15,800,000 pounds/inch^2, and a density of 0.323 pounds/inches^3. Equation (10-33) may now be evaluated as follows:

$$V_{s_{sol}} = \sqrt{\frac{386.1 \text{ Inches/Sec}^2 \times 15.8 \times 10^6 \text{ Pounds/Inch}^2}{144 \text{ Inches}^2/\text{Foot}^2 \times 0.323 \text{ Pounds/Inch}^3}} = 11,450 \text{ Feet/Minute}$$

The resultant value of 11,450 feet per minute is comparable to the velocity of sound in copper listed in Table 10-2 of 11,670 feet per minute. The 2% variation between velocities is due to the fact that average values to three significant figures are used for the physical properties in Table B-1. By comparison, experimental results provide the sonic velocity listed in Table 10-2.

The velocity of sound in liquids is determined with an expression equivalent to equation (10-33). The difference between calculating the sonic velocity in solids versus liquids is that Young's modulus of elasticity E is used for solids, and the bulk modulus B is used for liquids. Variations in pressure and temperature of the liquid will be compensated by using the bulk modulus and density at the actual fluid operating conditions in the following expression.

$$V_{s_{liq}} = \sqrt{\frac{G \times B}{144 \times \rho_{liq}}}$$ (10-34)

where: $V_{s\text{-}liq}$ = Velocity of Sound in a Liquid (Feet / Second)
B = Bulk Modulus (Pound / Inch^2)
ρ_{liq} = Liquid Material Density (Pounds / Inch^3)

If a lubricating oil at atmospheric pressure and 60°F has a bulk modulus of 219,000 pounds per inch^2, and a density of 0.0327 pounds per inches^3, the sonic velocity is computed with equation (10-34) in the following manner:

$$V_{s_{liq}} = \sqrt{\frac{386.1 \text{ Inches/Sec}^2 \times 219,000 \text{ Pounds/Inch}^2}{144 \text{ Inches}^2/\text{Foot}^2 \times 0.0327 \text{ Pounds/Inch}^3}} = 4,240 \text{ Feet/Minute}$$

This sonic velocity in oil agrees directly with the value in Table 10-2. Next, the velocity of sound in gases must be addressed. It is well understood that gases are even more sensitive to variations in pressure and temperature. For perfect gases, the sonic velocity may be computed with the following expression:

$$V_{s_{gas}} = \sqrt{\frac{g \times k \times R \times T \times z}{mw}} \qquad \textbf{(10-35)}$$

where: $V_{s\text{-}gas}$ = Velocity of Sound in a Gas (Feet / Second)
        g = Acceleration of Gravity (= 32.17 Feet / Second^2)
        k = Specific Heat Ratio of Cp/Cv (Dimensionless)
        R = Universal Gas Constant (= 1,546 Foot-Pound force/ Pound mole -°R)
        T = Absolute Gas Temperature (°R = °F + 460)
        z = Gas Compressibility Based on Temperature and Pressure (Dimensionless)
     mw = Gas Molecular Weight (Pound / Pound mole)

The compressibility z is typically determined from charts based upon the pseudo-reduced temperature and the pseudo-reduced pressure of the gas. If the constants g and R are included into equation (10-35), the expression may be somewhat simplified as follows:

$$V_{s_{gas}} = 223 \times \sqrt{\frac{k \times T \times z}{mw}} \qquad \textbf{(10-36)}$$

For situations where the specific heat ratio k is unknown, the following equation (10-37) may be used to compute k based upon the molecular weight mw and the specific heat of the gas at constant pressure c_p.

$$k = \frac{1}{1 - \left\{\dfrac{1.986}{mw \times c_p}\right\}} \qquad \textbf{(10-37)}$$

where: c_p = Specific Heat at Constant Pressure (BTU/Pound-°F)
    1.986 = Universal Constant R = 1,546 Foot-Pound force/ Pound mole -°R divided
              by 778.3 BTU/Foot-Pounds (BTU/ Pound mole -°R)

For example, considering a normal composition of air at 60°F (520°R) and 14.7 Psia pressure, the molecular weight mw, and the specific heat at constant pressure c_p, may be obtained directly from Table B-2 located in appendix B of this text. Based on these values, the specific heat ratio k may be calculated with equation (10-37) in the following manner:

$$k = \frac{1}{1 - \left\{\dfrac{\dfrac{1.986 \text{ BTU}}{\text{Pound mole-°R}}}{\dfrac{28.962 \text{ Pound}}{\text{Pound mole}} \times \dfrac{0.2398 \text{ BTU}}{\text{Pound-°F}}}\right\}} = \frac{1}{1 - \left\{\dfrac{1.986}{6.9451}\right\}} = \frac{1}{1 - 0.286} = \frac{1}{0.714} = 1.40$$

This answer of k=1.40 for air at standard temperature and pressure is a normal and expected value. Under these conditions, the pseudo-reduced temperature is 2.18, and the pseudo-reduced pressure is 0.027 — the compressibility factor z is very close to 1.0. The velocity of sound may now be easily determined by substitution of the known parameters into equation (10-36) as follows:

$$V_{s_{air}} = 223 \times \sqrt{\frac{k \times T \times z}{mw}} = 223 \times \sqrt{\frac{1.40 \times 520 \times 1.0}{28.962}} = 223 \times \sqrt{25.136} = 1,118 \text{ Ft/Sec.}$$

This solution is consistent with published values of the velocity of sound in air. For gas streams, the composition and effective mole weight mw, plus the effective specific heat ratio k, the compressibility z, and the absolute temperature T all contribute to the final value of the sonic velocity V_s. Clearly, the general wave equation (10-32) may be subjected to variations due to predictable changes in the velocity of sound in the gas stream. Furthermore, the reader should be aware that sonic velocity in a liquid may be a parabolic shaped curve when velocity is plotted against temperature. In this situation, two entirely different temperatures share the same value for the velocity of sound. For all critical calculations, the machinery diagnostician should obtain the best possible reference for the specific fluid properties, plus the velocity of sound, within the bounds of the actual process conditions.

Returning back to equation (10-32), the concept of a unique frequency F_a relating the acoustic velocity V_a to the wavelength λ is straightforward relationship for standing waves. The frequency F_a is generally identified as the fundamental frequency. Integer multiples of this *fundamental frequency* are often referred to as *overtones* or *harmonics*. A frequency of $2F_a$ would be the first overtone, or the second harmonic. Similarly, a frequency of $3F_a$ would be identified as the second overtone or the third harmonic. The concept of harmonic orders of an acoustical frequency are identical to the multiples or harmonics of any other vibration component or fundamental frequency.

Within a contained system, such as an organ pipe or a length of schedule 40 steel pipe in an oil refinery, it is possible to establish an acoustic resonance. This occurs when an excitation frequency coincides with the physical characteristics of system length and acoustic velocity as described by (10-32). In terms of wave behavior, a good explanation was presented by Schwartz and Nelson[11] as follows:

"...A standing wave is propagated and reflected when a pressure or velocity perturbation encounters a discontinuity in an acoustic system. Resonant conditions will result when the total propagation time up and down the system is such that the reflected wave reaches the excitation source in phase with a subsequent perturbation. Under these conditions the excitation frequency is said to be at one of the natural frequencies of the acoustic system..."

This is exactly the same type of physical behavior exhibited by any other resonant response. That is, the geometry and physical characteristics of the system define a variety of potential resonances. These resonances remain dormant until an excitation frequency coincides with the natural resonant frequency. When that condition occurs, the influenced mechanical element will experience a condition of high vibration.

---

[11] Randal E. Schwartz and Richard M. Nelson, "Acoustic Resonance Phenomena In High Energy Variable Speed Centrifugal Pumps," *Proceedings of the First International Pump Symposium*, Turbomachinery Laboratories, Texas A&M University, College Station, Texas (May 1984), pp. 23-28.

Extension of the concept described in equation (10-32) to integer multiples of the wavelength is generally understandable. However, this same approach applies to particular fractions of the fundamental wavelength. Specifically, passage lengths equal to one half and one quarter of the fundamental wavelength may also support acoustic resonances. For instance, consider the condition of a pipe open at both ends, as shown in Sketches A and B of Fig. 10-15. In this type

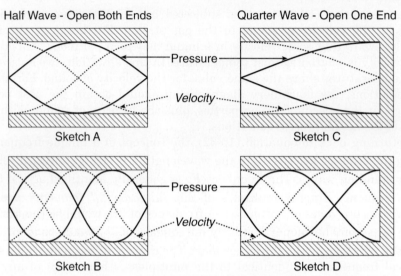

Fig. 10–15 Standing Pressure And Velocity Waves In Pipes

of representation, the end of the pipe is considered to be *open* if the cross-sectional area changes by a factor of two or more. This physically occurs in many types of branched piping systems, pulsation bottles, or internal crossover passages on multistage centrifugal pumps.

Sketch A of Fig. 10-15 depicts a standing half wave condition. As shown in this diagram, the velocity is a maximum at each end of the open pipe (anti-node), and pressure pulsation is at a minimum (node) at both ends. The peak pressure (anti-node) is at the middle of the pipe, and this is coincident with the minimum velocity (node). The first overtone, or second harmonic, of a pipe open at both ends is presented in Sketch B. Again, the velocity is a maximum at the pipe ends, and the pressure is at a minimum. Since this is a half-wave condition, the wave length λ is equal to equal to twice the passage length L_{half} as follows:

$$\lambda = 2 \times L_{half} \tag{10-38}$$

Substituting equation (10-38) into (10-32), and solving for the acoustic natural frequency F_a, the following expression is easily generated:

$$F_a = \frac{V_s}{\lambda} = \frac{V_s}{2 \times L_{half}}$$

However, this equation only covers the condition depicted in Sketch A of Fig. 10-15. The general solution for this half-wave behavior may be extrapolated from the previous expression by including an integer multiplier N_{half} as follows:

$$F_a = N_{half} \times \frac{V_s}{2 \times L_{half}} \qquad \text{(10-39)}$$

where: N_{half} = Harmonic Integer of 1, 2, 3, 4... (Dimensionless)
$\quad\quad\quad L_{half}$ = Physical Passage Length for Half-Wave Resonance (Inches)

It should be mentioned that the equations for a half-wave resonance also apply to a pipe with both ends closed. Although the nodal locations of the pressure and velocity waves become transposed, the physical behavior and the computational equations are the same as a half-wave pipe open at both ends.

The same general approach is applicable to the quarter-wave condition. This is described by a pipe open at one end, and closed at the opposite end as shown in Sketches C and D of Fig. 10-15. Sketch C depicts a standing quarter wave with a maximum velocity (anti-node), and minimum pressure pulsation at the open end. The peak pressure (anti-node) is at the closed end of the pipe, and this is coincident with the minimum velocity (node). The first overtone of a pipe open at one end is presented in Sketch D. Again, the velocity is a maximum at the end, and the pressure is at a minimum. Since this is a quarter-wave condition, the wave length λ is equal to equal to four times the passage length L_{qtr}:

$$\lambda = 4 \times L_{qtr} \qquad \text{(10-40)}$$

Substituting equation (10-40) into (10-32), and solving for the acoustic natural frequency F_a, the following expression is generated:

$$F_a = \frac{V_s}{\lambda} = \frac{V_s}{4 \times L_{qtr}}$$

Again, this expression only covers the condition shown in Sketch C of Fig. 10-15. The general solution for this quarter-wave behavior may be extrapolated from the last equation by including an integer multiplier N_{qtr} as follows:

$$F_a = N_{qtr} \times \frac{V_s}{4 \times L_{qtr}} \qquad \text{(10-41)}$$

where: N_{qtr} = Harmonic Integer of 1, 3, 5, 7... (Dimensionless)
$\quad\quad\quad L_{qtr}$ = Physical Passage Length for Quarter-Wave Resonance (Inches)

A quarter-wave resonance is sometimes referred to as a *stub resonance*. This is an accurate description of a piping element that is connected to a main line, and has a closed end. Various types of piping *dead legs*, *clean-out* nozzles, and nipples fall into this category. Although these are benign looking piping elements, they can harbor quarter-wave resonances that excite the associated piping system. The quarter-wave stub resonance may also adversely influence pressure gauge, and dynamic pressure pulsation measurements. Since the passage length L_{qtr} is often quite short, an acoustic resonant frequency may appear at an undesirable frequency, and corrupt the pressure pulsation data under investigation. One solution to this problem is to effectively extend the passage length with a measurement fixture similar to the diagram previously discussed in Fig. 6-33. This would substantially reduce the acoustic resonant frequency to a range that could be removed via electronic filtration, or just totally ignored.

In most cases, the fluid excitation problem analysis will require a combination of vibration measurements, dynamic pressure pulsation measurements, some level of analytical simulation, plus a recommended fix. Generally, a final round of testing is also necessary to verify the validity of the implemented solution. An example of this type of fluid problem is presented in the following case history of a high pressure boiler feed water pump.

Case History 31: Boiler Feed Water Pump Splitter Vane Failures

A group of three large boiler feed water pumps experienced a series of mechanical failures to their impeller suction splitter vanes. Although the machinery was installed with some questionable procedures, the destructive failures could not be attributed to the marginal installation. The BFW pumps were all steam turbine driven, and all three pumps were connected to common suction and discharge headers. Speed control for each turbine was based upon discharge header pressure.

The internal configuration of these pumps is depicted in Fig. 10-16. A double suction first stage wheel was located at the outboard end of each pump. A short internal crossover connected the discharge of the first stage to the suction of the second stage wheel. A long internal crossover of approximately 70 inches in overall length extended from the discharge of the second stage to the third stage suction. Another short crossover connected the third to the fourth stage, and the final discharge exited the middle of the pump casing at an average pres-

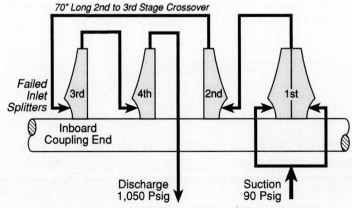

Fig. 10–16 Boiler Feed Water Pump Internal Configuration

sure of nominally 1,050 Psig.

Each pump stage was equipped with a four vane inlet splitter, and each impeller contained seven vanes. The documented failures generally originated at the suction of the third stage as indicated in Fig. 10-16. The splitter vanes at this location were the most susceptible to breakage, and they caused considerable internal pump damage during each failure. In most instances, the fourth stage wheel would be damaged. In some cases, the upstream second stage impeller would also be subjected to varying degrees of damage.

The machinery was subjected to a series of detailed vibration tests, and it was discovered that the major frequency component occurred at seven times rotative speed. This frequency was the impeller vane passing, and it was strongest at the inboard, coupling end of the pump. This location was also coincident with the majority of the splitter vane failures.

Based on this initial data, an additional test was performed, and dynamic pressure pulsation probes were added to the transducer suite. The significant results of this test are presented in the summary plot shown in Fig. 10-17. Within this plot, the amplitude at impeller vane passing frequency (7X) was plotted for a series of controlled operating speeds. The presented information includes the vertical casing velocity at the outboard bearing housing, vertical casing velocity acquired at the coupling or inboard bearing housing, and dynamic pressure pulsation obtained in the 2nd to 3rd stage internal crossover.

It is clear that the maximum 7X vane passing amplitudes occurred at a running speed of 3,410 RPM. It is also evident that velocity levels approaching 0.6 IPS_{o-p}, and pressure pulsation amplitudes in excess of 70 Psi_{p-p}, were intolerable for this machinery. Furthermore, the peaked response on this plot was indicative of a possible resonant behavior. The actual frequency of the peak vane passing response occurred at 23,870 CPM (3,410 CPM times 7 vanes). Based upon the geometry of this system, the excitation frequency, and the high pressure pulsation levels, an acoustic resonance was suspected.

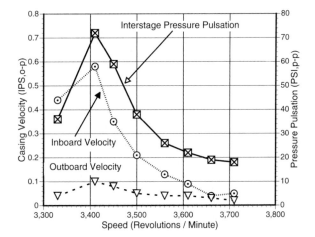

Fig. 10–17 Initial Casing Velocity And 2nd To 3rd Stage Internal Crossover Pressure Pulsation Filtered At 7X Impeller Vane Passing Frequency

The circulating boiler feed water pump operated at a temperature in the vicinity of 310°F. At this temperature, the velocity of sound in water (i.e., sonic velocity) is approximately 4,750 feet/second. Since the vane passing amplitudes increased with speed, it is highly probable that the observed behavior is related to flow. If this is an acoustic related phenomena, the offending member would probably be open at both ends rather than closed at one end. Hence, it makes sense to use the half-wave equation (10-39) for examination of a possible acoustic interaction. The fundamental half-wave frequency would occur when the integer N_{half} was set equal to 1. Based upon the velocity of sound V_a equal to 4,750 feet per second, and the 7X vane passing excitation frequency V_s of 23,870 cycles per minute, equation (10-37) may now be solved for the associated physical passage length as follows:

$$L_{half} = \frac{V_s}{2 \times F_a} = \frac{4,750 \text{ Feet/Sec} \times 12 \text{ Inches/Foot}}{2 \times 23,870 \text{ Cycles/Min} \times 1 \text{ Min}/60 \text{ Sec}} = \frac{57,000}{795.7} = 71.6 \text{ Inches}$$

Recall that the physical passage length of the 2nd to 3rd stage crossover was previously identified as approximately 70 inches. Due to the uncertainties involved, the correlation between the calculated half wave resonance passage length is considered to be in good agreement with the approximate physical length of the 2nd to 3rd stage crossover.

The final assessment concluded that an acoustic resonance in the 2nd to 3rd stage crossover was directly excited by the 7X pump vane passing frequency. Since there was no possibility of changing the internal pump passage length, or changing the pump operating temperature to vary the acoustic velocity — the only remaining viable option was to change the exciting frequency. Hence, the seven vane impellers were replaced with new six vane wheels.

To test the validity of this conclusion, the same field test was repeated with a pump equipped with six vane wheels. The results of this test are presented in Fig. 10-18. Note that the pressure pulsation amplitudes are now in the vicinity of

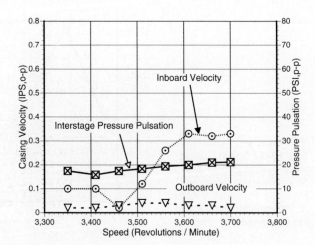

Fig. 10–18 Final Casing Velocity And 2nd To 3rd Stage Internal Crossover Pressure Pulsation Filtered At 6X Impeller Vane Passing Frequency

20 Psi,$_{p-p}$, and the inboard bearing housing velocity amplitudes are slightly above 0.3 IPS,$_{o-p}$ at the highest pump speeds. However, back within the normal operating range of 3,400 to 3,500 RPM, the 6X vane passing amplitudes are in the vicinity of 0.1 IPS,$_{o-p}$, which is quite tolerable for these particular machines. Eventually, all three of the BFW pumps were converted to six vane impellers, and the string of splitter vane failures was completely stopped.

Case History 32: Hydro Turbine Draft Tube Vortex

An entirely different class of machine is examined in the following case history of a vertical hydro turbine driving an 18 megawatt generator at 277 RPM. A general arrangement drawing of this train is presented in Fig. 10-19. The overall height of the rotating assembly is approximately 42 feet, and the overall rotor weight approaches 200,000 pounds. The entire rotor assembly is supported on a thrust bearing mounted at the top of the 26 pole generator. Lateral rotor support is provided by radial bearings at three different elevations. As shown in Fig. 10-19, an upper guide bearing is located at the top of the generator, above the thrust assembly. Directly below the generator, a lower guide bearing is installed. Both the upper and lower generator bearings are tilt pad assemblies that have individual clearance adjustments for each pad. The elevation difference between the upper and lower generator bearings is 127 inches. A turbine guide bearing is located approximately 193 inches below the lower generator bearing. This turbine bearing was a segmented journal bearing that consisted of six fixed pads.

An unusually long 4,200 foot penstock supplied water to this unit. The final water inlet to the spiral case was a straight run of pipe, and the water passed circumferentially through a row of twenty vertical wicket gates as shown in Fig. 10-19. Energy was extracted from the water stream with a Kaplan turbine equipped with six variable pitch blades. A traditional elbow draft tube was installed that consisted of a vertical drop, followed by the elbow, and a final return to an open channel that directed the water downstream.

During original commissioning of this unit, it was determined that steady state operation under load was acceptable, but high vibration amplitudes were encountered during load rejection. In this condition, the load is suddenly removed from the generator (tripped breaker), but the energy of the water flow is still applied to the turbine. This situation results in a significant speed increase of the rotating assembly. For instance, during load rejection from 12 megawatts, the speed will increase from 277 RPM to a runaway speed of well over 400 RPM. The earliest field test data revealed a peak vibration response at a frequency of 660 CPM. This frequency did coincide with a calculated rotor critical speed at the same speed. It was reasoned that although the rotor speed (420 RPM maximum) never reached the resonant frequency (660 RPM), there was sufficient energy imparted to the rotating assembly to excite this major system resonance.

The initial commissioning tests demonstrated that the highest runaway speed was directly related to the generator load (expected result). Similarly, the peak vibration amplitudes steadily increased with load, and maximum runaway

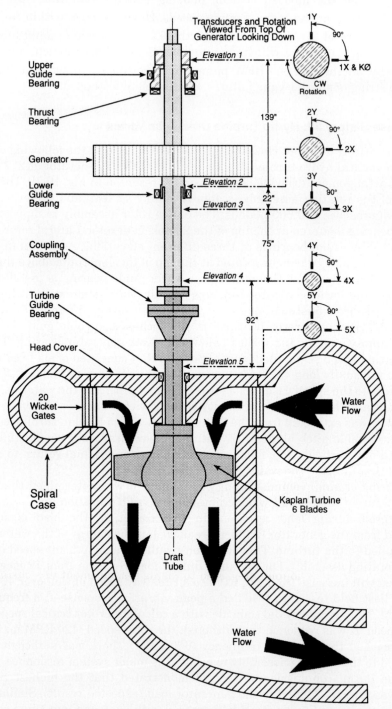

Fig. 10–19 Mechanical Configuration Of Hydro Electric Turbine Generator Set

speed. At the highest loads, the massive structure of the power house was uncomfortably shaken during load rejection. Some relief was obtained by air injection into the head cover, but high vibration amplitudes still occurred after the conclusion of the air injection. Additional tests were conducted with changes in system control variables in an effort to attenuate the response at 660 CPM during load rejection. However, the entire sequence between load rejection, over-speed with high vibration, concluding with coastdown at acceptable vibration levels occurred within 30 seconds. Clearly, the dynamics of the mechanical system overwhelmed the capabilities of the turbine control system.

At the completion of the original field testing, it was generally agreed that operation of the hydro turbine generator should be limited to a maximum load of 6 megawatts (1/3 of rated capacity). Under this steady state load, the vibration levels encountered during load rejection were deemed to be tolerable. It was also suggested that stiffening the lower generator guide bearing support would raise the 660 CPM critical, and reduce the vibration encountered during load rejection. Finally, it was suggested that a fourth lateral bearing be installed above the coupling in order to raise the critical speed. Most parties agreed that the vertical span between the generator lower guide bearing and the turbine guide bearing was abnormally long.

During the next three years more tests were conducted by various individuals, and the unit continued to be limited to 6 megawatts. The external support for the lower guide bearing was stiffened, and alignment plus clearance adjustments were made to the unit. It was acknowledged that the lateral motion of the Kaplan turbine blades was excessive during load rejection from high loads. This was evident by the noise immediately following load rejection, plus the visible radial rubs between the tips of the turbine blades and the casing.

Up to this point, the vibration data was primarily based upon a pair of X-Y proximity probes mounted below the generator lower guide bearing, plus casing readings with various hand held transducers. In an effort to better understand the dynamic behavior of this unit, a series of additional transducers were temporarily installed at five different elevations. These locations are identified in Fig. 10-19 as elevations 1 through 5. Initially, low frequency accelerometers were only installed adjacent to the probes mounted above the lower guide bearing (elevation 2). Following the first test, it was clear that casing motion was appreciable, and all subsequent tests were conducted with low frequency, high sensitivity accelerometers mounted next to each proximity probe. When necessary, the acceleration data was double integrated to casing displacement in a custom analog integration box — and this data was electronically added to the shaft vibration signals. Signal addition of the shaft relative with the casing absolute displacement was achieved with a standard voltage summing amplifier. The resultant output was shaft absolute vibration. Although this data manipulation was not necessary during steady state operation, it did provide an enhanced perspective during some of the transient speed tests.

All of the X-axis probes were aligned in the direction of the downstream water flow, and the Y-axis probes were located 90° away, or perpendicular to the flow. An optical Keyphasor® was installed at the top of the generator in-line with the X-axis vibration transducers. A series of pressure pulsation transducers were mounted in the penstock, head cover, and draft tube for most of the testing. In addition, strain gages were mounted on the lower portion of the exposed shaft. These sensors were powered by batteries, and the dynamic signals transmitted to a receiving antenna for demodulation and final processing.

The generator lower guide bearing pads were replaced due to concerns over an incorrect pad curvature, and the first series of constant load followed by load rejection tests were performed. There was some concern over the proper support of the transducers mounted above the coupling assembly. In order to obtain the best possible data, a rigid uni-strut framework was constructed to support the vibration transducers. The preliminary results of these tests were totally consistent with historical behavior. Vibration amplitudes above the coupling (elevation 4) were the most severe, with typical levels between 40 and 55 Mils,$_{p-p}$ during load rejection from 8 megawatts. An example of this vibratory behavior is presented in the transient capture plot shown in Fig. 10-20.

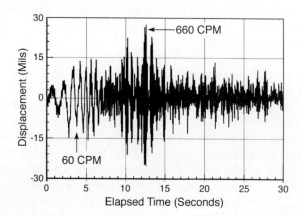

Fig. 10–20 High Speed Transient Capture Of Shaft Vibration At Elevation 4 Above Coupling Assembly Normal Load Rejection From 8 Megawatts

This time domain sample is similar to the type of data that can be obtained with a long sweep rate on an oscilloscope. However this data was digitized and rapidly sampled with a HP-35665A. The DSA allowed accurate expansion of the time domain sample, and it was easily determined that the initial vibration cycles occurred at 60 CPM. This frequency only appeared for a few cycles, and then the signal blossomed into a brief response at 660 CPM that also appeared for only a few seconds, and then decayed away. The maximum shaft vibration response at 660 CPM was 52.7 Mils,$_{p-p}$. This large shaft vibration amplitude was visibly distinguishable, and physically threatening. After observing this shaft vibration and the associated structural motion, most individuals elected to stay out of the turbine pit during the load rejection tests.

It should also be mentioned that the maximum vibration amplitudes were not coincident with rotational speed. Hence, traditional synchronous tracking of the running speed motion would not allow examination of the high amplitude vibration components. In this case, frequency analysis of the data was manda-

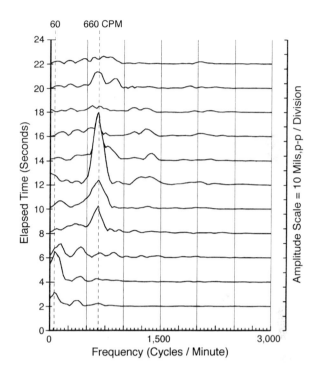

60   660 CPM

Fig. 10–21 Waterfall Plot Extracted From Transient Capture Of Shaft Vibration At Elevation 4 Above Coupling Assembly - Normal Load Rejection From 8 Megawatts

tory. Unfortunately, FFT processing of this transient capture data is very difficult due to the short duration of the specific events, plus the low frequencies involved. In actuality, if the transient FFT data is not properly handled, the significant information may be lost, distorted, or otherwise corrupted. In all cases, the diagnostician should verify that the final processed data in whatever format agrees with the overall time record.

In this specific situation, the transient capture data presented in Fig. 10-20 was post processed in a zero to 3,000 CPM (0 to 50 Hz) span to examine the frequencies of interest. The time record length for a 400 line FFT would be 8 seconds based upon the measurement speed information presented in Table 7-1. However, if the resolution was decreased from 400 to 100 lines, the frequency resolution would suffer, but the time record would decrease from 8 to 2 seconds (=8x100/400). Thus, the FFT data shown in Fig. 10-21 is incremented at 2 second intervals. Furthermore, this data was processed using a flat top filter to enhance the amplitude accuracy. The large component at 660 CPM on the waterfall plot has an amplitude of 47.7 Mils,$_{p-p}$. This value is consistent with the 52.7 Mils,$_{p-p}$ overall level displayed on the time domain plot of Fig. 10-20.

It is clear from the overall database that the largest deflections occurred above the coupling assembly (elevation 4). However, it is meaningful to examine the behavior at the other measurement locations in a consistent manner. It was determined that the motion at all elevations was basically forward and circular. That is, the shaft precession was in the direction of rotation. The amplitudes

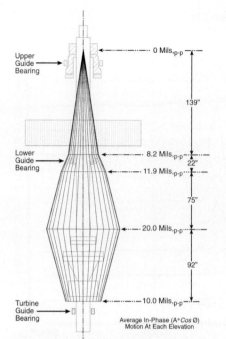

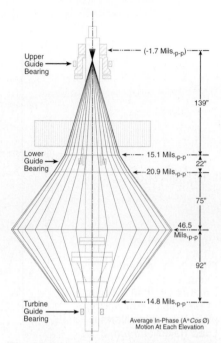

Fig. 10-22 Measured Shaft Mode Shape At 60 CPM Immediately Following Normal Load Rejection From 8 Megawatts

Fig. 10-23 Measured Shaft Mode Shape At 660 CPM At 12 Seconds After Normal Load Rejection From 8 Megawatts

were similar from the X and Y probes at each elevation, and the signals were nominally separated by 90° (i.e., circular). Unfortunately, there were no shaft vibration measurements down in the flooded portion of the turbine shaft, so the analysis had to be based upon data obtained from the dry elevations 1 through 5.

Since the shaft motion was forward and circular at 60 and 660 CPM, it makes sense to average the X and Y data into shaft mode shapes for the exposed portion of the shafts. The resultant mode shape at 60 CPM was obtained immediately following load rejection, and this information is shown in Fig. 10-22. For comparative purposes, the mode shape at 660 CPM obtained approximately 12 seconds after load rejection from 8 megawatts is presented in Fig. 10-23.

The higher frequency mode shape at 660 CPM is associated with the fundamental rotor resonance at 660 RPM that was predicted early in the game. This mode is fundamentally driven by the generator mass, and it is active during all load rejections. As turbine runaway speed increases with higher megawatt loads, the excitation for this 660 CPM resonance becomes greater. It is no wonder that the concrete structure shakes, and windows start breaking when a load rejection occurs from anywhere near design load.

The lower frequency behavior at 60 CPM was initially somewhat of a mystery. It was determined that this frequency really appeared in the general domain of 60 to 90 CPM (1.0 to 1.5 Hz). The dichotomy of this response was that the shaft vibration values shown on the mode shape plot of Fig. 10-22 are not excessive, yet this is the same condition when metal to metal rubbing occurs on the turbine runner. In essence, it was concluded that a shaft mode exists at nominally 60 to 90 CPM that is driven by the heavy Kaplan turbine overhang. This mode has large displacements down at the turbine, and much lower displacement levels up at the dry measurement planes. This conclusion was further substantiated by refinement of the analytical model to reveal a turbine conical mode in the vicinity of 100 RPM.

Throughout the accumulated database it was perfectly clear that high vibration amplitudes at 660 CPM were always preceded by the activity at 60 to 90 CPM. Although this excitation progression was not immediately understood, a variety of mechanical changes were implemented. A stiffer generator shaft was fabricated and installed, and vertical alignment of the entire unit was significantly improved. In addition, the lower guide bearing was set to the correct clearances, and the turbine guide bearing was replaced with a preloaded bearing. All of these changes provided a much smoother running machine at steady state conditions of 277 RPM (at any load up to 18 megawatts).

However, the load rejection behavior from 8 megawatts, and the associated 50+ Mils,$_{p-p}$ of shaft vibration above the coupling remained virtually unchanged. At this stage, folks were getting distressed. A lot of money had been spent on mechanical improvements, and considerably more money had been lost due to the reduced power generation. Fortunately, one of the senior engineers for the operating company theorized that the problem originated with a strong vortex in the draft tube during load rejection. If the vortex swirl frequency was in the vicinity of 60 to 90 CPM, it could easily excite the overhung turbine mode at the same frequency, and that motion could in turn couple to the second mode at 660 CPM. In support of this hypothesis, the draft tube pressure pulsation during load rejection is shown in Fig. 10-24. This time domain transient capture

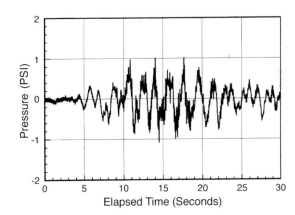

Fig. 10–24 High Speed Transient Capture Of Draft Tube Dynamic Pressure Pulsation Following Normal Load Rejection From 8 Megawatts

revealed the presence of a nominal 60 CPM component throughout the majority of the coastdown. By comparison, the shaft vibration on Fig. 10-20 displays only a few cycles of 60 CPM activity. Hence, if the draft tube pressure pulsation is due to a swirling vortex, the energy transfer to the rotor might be sufficient to excite both rotor criticals including the first mode at nominally 60 CPM, and the second mode at 660 CPM.

In order to test this hypothesis, a high pressure air injection system was devised. This system consisted of two pressure vessels that were tied into the plant air system, and both vessels were pumped up to approximately 100 Psig. A six inch line was then run from the vessels through a flow meter, and into the turbine head cover vacuum breakers. This system was automated to dump a regulated flow of air into the turbine, and potentially break up any vortex in the draft tube. A similar system was tried during initial commissioning of this unit, but it was limited in capacity and associated flow time. The current system was sized for the delivery of high pressure air for approximately one minute.

In order to provide a meaningful test, the hydro generator was subjected to a series of load rejections performed both with and without the air injection. For safety considerations, the maximum allowable shaft vibration above the coupling (elevation 4) was limited to 50 Mils,$_{p-p}$. The summarized vibration data of this final test is presented in Fig. 10-25. At the first test point of 4 megawatts, the air injection had little influence. However, as load was increased, the presence of the air injection became much more significant. In fact, as shown in Fig. 10-25, the load rejection test from 10 megawatts without air displayed a peak amplitude of 50 Mils,$_{p-p}$. By comparison, the automated air injection limited the vibration severity to 32 Mils,$_{p-p}$. At this point, the tests without air injection were terminated due to the previously mentioned vibration limit. However, the tests with air injection continued up to full rated capacity of 18 megawatts.

Although 55 Mils,$_{p-p}$ was reached during this final test, the unit was immediately re-rated for full capacity. In retrospect, the injected air successfully eliminated the draft tube vortex, which minimized the rotor resonant excitation at 60 and 660 CPM. The only side effect was a huge downstream air bubble that relieved itself into the open water channel.

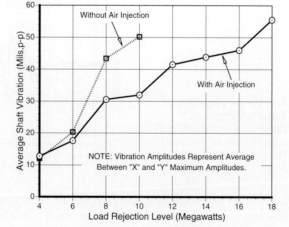

Fig. 10–25 Shaft Vibration Measured Above The Coupling During Load Rejections With And Without Supplemental Air Injection Into The Turbine Head Cover

ELECTRICAL EXCITATIONS

The next category of *machine specific* excitations considers some of the common characteristics encountered on electrical machinery. This type of mechanical equipment includes both motors and generators operating in either a synchronous or an induction configuration. This type of machinery consists of a rotor confined within a stationary stator, and supported by a pair of radial journal bearings. Due to the *self-centering* effect of the magnetic fields between rotor and stator, a thrust bearing is not necessary on horizontal units. However, on vertical motors and generators, a thrust runner will be installed at the top of the unit. The associated thrust bearing assembly will often support the entire weight of the electric machine, plus the coupled unit (e.g. case history 4).

In addition to generic unbalance, eccentricity, resonance, alignment, stability, and assembly problems common to most machines — electric machines are subjected to additional excitations due to the presence of magnetic fields. In many cases, the dilemma encountered during vibration analysis of electric machinery occurs with differentiating, or distinguishing, between mechanical and/or electrical problems. In simple cases, the vibratory evidence clearly points towards a mechanical, or a purely electrical malfunction. In more complex situations, the mechanical and electrical excitations become intertwined, and the diagnostician may be misled into a set of erroneous conclusions.

Before examining the telltale characteristics of electrical machinery malfunctions, it is desirable to define the basic differences between synchronous and induction machines. In most applications, synchronous machines are used for high horsepower, and slow operating speeds. The rotational speed of a synchronous motor or generator is independent of load, and it is strictly governed by the line frequency and the number of poles. The following equation (10-42) is used to determine the synchronous frequency, which is equal to the rotational speed for these machines:

$$F_{sync} = \frac{120 \times F_{line}}{N_p} \qquad\qquad \textbf{(10-42)}$$

where: F_{sync} = Synchronous Speed (Revolutions / Minute)
  F_{line} = Electrical Line Frequency (Cycles / Second or Hertz)
  N_p = Number of Poles (Dimensionless Integer)

This equation is widely publicized as the method to calculate synchronous speed. In reality, this is merely a frequency conversion expression where the line frequency in hertz (cycles/second) is converted to cycles per minute by multiplying by 60. Division of the frequency in cycles per minute by the number of pole pairs yields the synchronous speed. In general practice, the actual number of poles are used rather than the number of pole pairs, and the additional factor of 2 is included. For the normally encountered line frequencies of 60 and 50 Hz, Table 10-3 summarizes the synchronous speed for various numbers of poles:

Large electric machines are built with a large number of poles, and they

Table 10–3 Synchronous Speed As A Function Of Number Of Poles And Line Frequency

Number of Poles	60 Hertz Line Frequency	50 Hertz Line Frequency
2	3,600 RPM	3,000 RPM
4	1,800 RPM	1,500 RPM
6	1,200 RPM	1,000 RPM
8	900 RPM	750 RPM
10	720 RPM	600 RPM
12	600 RPM	500 RPM

run at slow speeds. For instance, a 26 pole generator will run at 277 RPM, whereas a 60 pole unit will only turn at a synchronous speed of 120 RPM. Machines in this category are typically used as generators coupled to hydro turbines (similar to case history 32).

The second major category of electrical machines consists of induction motors and generators. These machines do not operate at synchronous speed, and they change speed in accordance with the load. An induction motor operates close to synchronous speed under a no-load condition. As load is applied to the induction motor, the rotating speed decreases. Similarly, an induction generator operates close to synchronous speed under a no-load condition. As mechanical power is applied to the induction generator, the rotating speed increases as the power output increases.

Induction motors and generators are physically similar, and in some cases identical. For example, in some pump storage facilities, the electric machine is used as a generator during the day to produce electricity. Each night it is used as a motor to pump water back to an elevated reservoir (lake).

The difference between rotating speed and synchronous speed is the slip between the stationary and rotating field. The general equation for determination of slip frequency for an induction machine is presented as follows:

$$F_{slip} = F_{line} \times \left\{ 1 - \frac{F_{rotor}}{F_{sync}} \right\} \tag{10-43}$$

where: $\quad F_{slip}$ = Slip Frequency (Cycles / Minute)
$\qquad \quad F_{line}$ = Electrical Line Frequency (Cycles / Minute))
$\qquad \quad F_{rotor}$ = Shaft Rotating Speed (Revolutions / Minute)

This expression may also be stated in terms of the number of poles by substituting equation (10-42) with consistent units of CPM into (10-43) as follows:

$$F_{slip} = \left\{ \frac{F_{sync} \times N_p}{2} \right\} \times \left\{ 1 - \frac{F_{rotor}}{F_{sync}} \right\} = \left\{ \frac{N_p}{2} \right\} \times \{ F_{sync} - F_{rotor} \} \tag{10-44}$$

This synchronous speed is the frequency applied to the stationary stator coil. The rotative speed is obviously the rotational speed of the rotor. For a two (2) pole machine, the line frequency is equal to the synchronous speed, and equation (10-44) simplifies to the following:

$$F_{slip} = \left\{\frac{2}{2}\right\} \times \{F_{sync} - F_{rotor}\} = \{F_{sync} - F_{rotor}\}$$

As an example, consider a two pole, 60 Hz, induction motor running under load at an actual running speed of 3,580 RPM. The synchronous speed is 60 Hz, or 3,600 CPM, and the slip frequency is easily calculated as:

$$F_{slip} = \{F_{sync} - F_{rotor}\} = 3,600 - 3,580 = 20 \text{ Cycles/Minute}$$

The slip frequency is an important parameter, since it provides one of the tools used to distinguish between various types of malfunctions in electric machines. Unfortunately, the slip frequency is a fairly small value, and it could easily go undetected in many types of routine data acquisition and analysis. For instance, consider the spectrum plot presented in Fig. 10-26. This FFT data was

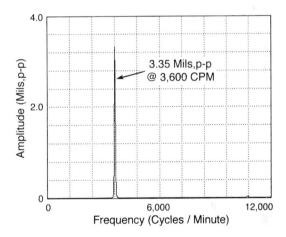

Fig. 10–26 Typical Spectrum Plot Of Electric Motor Shaft Vibration Signal At Frequency Resolution Of 30 CPM Per Filter Line

processed with a respectable 400 line resolution across the 12,000 Cycle/Minute span of this spectrum plot. The frequency resolution for this plot is equal to 12,000 CPM divided by 400 lines, or 30 CPM per line. This FFT resolution does not provide any visibility of the 20 CPM slip frequency. If this plot was the only piece of information available, the diagnostician would be hard pressed to accurately diagnose the origin of the single frequency component. However, if the vibration data is viewed in the time domain, the vibration signal amplitude will appear to pulsate with passing time. In fact, this same signal will display a distinct amplitude modulation as shown in Fig. 10-27.

This time domain plot covers a total of 8.0 seconds, and the vibration amplitude has passed through many high frequency cycles, plus more than 2 cycles of

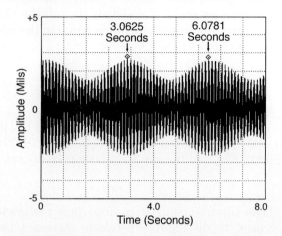

Fig. 10–27 Typical Time Domain Plot Of Electric Motor Shaft Vibration Signal For 8.0 Seconds

a low frequency beat. This low frequency may be determined by measurement of the period, and computation of the associated beat frequency as follows:

$$Period = 6.0781 - 3.0625 = 3.0156 \text{ Seconds}$$

From equation (2-1) the period may be easily determined as follows:

$$Frequency = \frac{1}{Period} = \frac{1 \text{ Cycle}}{3.0156 \text{ Sec.}} \times \frac{60 \text{ Sec.}}{1 \text{ Min.}} = 19.897 \text{ CPM} \approx 20 \text{ CPM}$$

Hence, the beat frequency is equal to the slip frequency of the example induction motor. Although this activity was not evident in the averaged FFT plot shown in Fig. 10-26, it is quite visible in the time domain trace. This type of signal was identified as an amplitude modulation. As discussed in chapter 7, the characteristic beat frequency is the difference between two fundamental frequencies. Thus, further analysis of this signal requires precise identification of the two signals that contribute to the beat frequency.

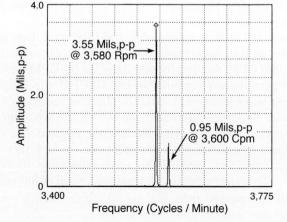

Fig. 10–28 Translated Spectrum Plot Of Electric Motor Shaft Vibration Signal At Resolution Of 0.94 CPM Per Filter Line Displaying A Mechanical Fault

It is clear from the FFT plot in Fig. 10-26 that the signal energy is contained around the rotational frequency. To allow detailed examination of this narrow frequency range, the analysis bandwidth is reduced from 12,000 CPM (200 Hz) to 375 CPM (6.25 Hz) using the frequency translator capabilities of the DSA (zoom transform). The frequency content between 3,400 and 3,775 CPM appears in Fig. 10-28. Note that the majority of the activity (3.55 Mils,$_{p-p}$) occurs at the rotational speed of 3,580 RPM. The other frequency component at the line frequency of 3,600 CPM is only 0.95 Mils,$_{p-p}$. Hence, in this example, the diagnostician would probably concentrate on examination of mechanical problems rather electrical phenomena.

The opposite situation is depicted in the translated spectrum plot in Fig. 10-29. Again the frequency content between 3,400 and 3,775 CPM is examined, and again two major peaks are visible. In this case, the majority of the activity (3.58 Mils,$_{p-p}$) occurs at the synchronous line frequency of 3,600 CPM. The small component is the rotational speed of 3,580 RPM at an amplitude of 0.97 Mils,$_{p-p}$. In this case, the machinery diagnostician would concentrate on examination of electrical rather mechanical problems.

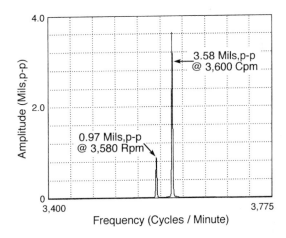

Fig. 10–29 Translated Spectrum Plot Of Electric Motor Shaft Vibration Signal At Resolution Of 0.94 CPM Per Filter Line Displaying An Electrical Fault

It should be mentioned that both of these example plots would appear identical in the original 12,000 CPM span spectrum plot in Fig. 10-26. They would appear to be virtually indistinguishable on an oscilloscope time base trace. However, if the scope is triggered with a Keyphasor® signal, and the majority of the motion becomes locked or frozen on the CRT screen, this would be indicative of a dominant running speed component. Conversely, if the oscilloscope is triggered off line frequency (i.e., 60 Hz), and the majority of the motion becomes locked on the screen, the conclusion of a major component at the electrical synchronous line frequency is correct.

In this example, the beat frequency of 20 CPM is the slip frequency for this motor. In most cases, the slip frequency does not appear as a separate low frequency component. Usually, the effect of the slip, i.e. the beat frequency, is observed on the field vibration data. The same scenario applies to induction generators where a full load slip frequency of 30 or 40 CPM will not be directly visible, but an amplitude modulation with a beat frequency of 30 to 40 CPM will be quite evident in the vibration signals.

Due to the nature of electric machinery, it is often necessary to run additional field tests to determine the origin of a vibration problem. As with most machines, data should be obtained during *initial cold*, plus *normal hot* steady state data under load. The standard transient startup and coastdown characteristics should also be documented. There are mechanisms that occur during transient speed conditions that are not evident during full speed operation. For example, synchronous electric motors display an oscillating torque during startup. This torque oscillation may excite any natural torsional resonance(s) present in the mechanical system. The occurring frequency of this torsional oscillation is easily determined from the following expression:

$$F_{tor} = N_p \times \{F_{sync} - F_{rotor}\}$$ (10-45)

where: F_{tor} = Torsional Oscillation Frequency (Cycles / Minute)

At full operating speed, the shaft rotational speed F_{rotor} is equal to the synchronous speed F_{sync}, and the effective torsional oscillation frequency is zero. However, during startup of a synchronous motor, the torsional excitation frequency decreases as machine speed increases. This oscillating torque is often sufficient to excite the torsional criticals. This is particularly true on a machinery train that includes a gear box coupled to the synchronous motor. The gear box provides a natural mechanism to translate the oscillatory torque into a significant lateral vibration. An example of this behavior is shown in case history 34.

It is also desirable to examine other conditions on electric machines. For instance, a motor may be run solo (i.e., uncoupled), and the shaft vibration data reviewed during *initial cold* versus *final warm* solo operation. Observation of the coastdown behavior of an electric machine, particularly the response due to a termination of input power, may be very useful. If the vibration data is tape recorded in conjunction with a contact closure (e.g., 9 volt battery power source), the actual sequence of events from the trip point may be accurately determined. As discussed in other parts of this text, the field tests should be designed to examine specific mechanisms, or to eliminate potential problems.

Examination of problems such as **mass unbalance** on electric machines will be subjected to the same criteria and characteristics discussed in chapter 11. Obviously, the majority of the shaft vibratory motion should occur at rotative speed, and it should change in accordance with any system balance resonance(s). The 1X vibration response should exhibit minimal change with respect to operating temperature or load. Furthermore, any slip frequency beat should be minimal or nonexistent, and a solo power shut off test should reveal no immediate change as the power is terminated.

It should also be noted that many electric motors are designed as stiff shaft machines. Motors and generators often contain bearings and couplings that are insulated from ground. This isolates the rotating assembly, and minimizes any chance of bearing damage due to electrical currents. However, physical degradation of the insulating material, or increased bearing clearances will reduce the support stiffness of the rotor assembly. In many documented cases, this has resulted in a decrease of the rotor lateral critical speed back into the operating speed range. Once this is recognized, replacement of the damaged bearings or electrical insulators will raise the lateral critical, and restore normal stiff shaft operation to the electric machine.

An induction motor with **broken rotor bars** will exhibit a slip frequency beat to the 1X component when the unit is operated solo. Typically, this beat will cause minor amplitude changes during a solo run, and the ensuing coastdown will exhibit only small variations. In most cases, broken rotor bars generally cannot be detected during a motor solo run. However, operating a motor with broken rotor bars under load will result in a significant increase in vibration amplitudes. Although it is not necessarily a linear relationship, it has been observed that increased load will produce increased vibration. In addition, the signal characteristics will change. The dominant vibration amplitude will still occur at running speed, but upper and lower sidebands, at number of poles times slip frequency, will appear. Thus, for a two pole motor, sidebands at twice slip frequency should appear on both sides of the 1X component. These sidebands will disappear with power removal.

The problem of **high resistance rotor bars** can be confirmed by connecting a clip on current probe to one leg of the motor power line. The current signal is then subjected to a spectrum analysis similar to Fig. 10-29. If high values of the slip frequency side bands (and slip frequency harmonics) around the line frequency are observed, the problem is generally associated with rotor bars. Obviously, the final condition of this behavior is the development of broken rotor bars.

One of the most common problems on electric machines is the mis-positioning of the rotor within the stator to yield an **uneven air gap** between the rotor and stator. On electric machines, a uniform radial air gap around the circumference of the rotor is considered mandatory. If this air gap is distorted by misalignment of the rotor, bearing damage, deformed frame and stator coil, or any other physical mechanism — the vibratory behavior of the electric machine will suffer. Assuming a concentric rotor, the primary characteristic of an uneven air gap is a significant excitation at twice line frequency. Thus, a 60 Hz induction motor would display a major vibration component at 7,200 CPM (irrespective of the number of poles). This excitation would appear under solo operation at no load, plus fully loaded conditions. Since this excitation is due to variations in the magnetic flux between rotor and stator, the twice line frequency vibration component will disappear as soon as the power is cut off (solo or loaded). This dynamic behavior is mimicked by an electric machine with **unbalanced line voltages**. The difference between these malfunctions is usually identified by measuring the voltages and currents for each respective phase.

If an **eccentric rotor** is combined with a round stator that is concentric with the rotor centerline, another type of air gap fault in encountered. In this situation, the rotational speed vibration is high, and it is generally modulated by the slip frequency. In some motors, the modulation occurs at twice the slip frequency, and a discrete frequency component at twice slip speed may be visible. This eccentric rotor behavior appears during unloaded solo operation as well as fully loaded and coupled. The modulation does disappear immediately upon cutting power in either the solo or the loaded condition.

An eccentric rotor should be readily detectable during rotor assembly, and final runout checks. Furthermore, any type of air gap variation between rotor and stator should be identifiable by measuring the assembled machine air gaps. This static measurement is obtained with feeler gauges during final assembly of the electric machine. On small units, it may only be possible to obtain air gap measurements every 90° (4 points). On larger machines, it is desirable to obtain these readings at 8 points (every 45°), or at 12 points (every 30°). These air gap measurements should be taken at both ends of the electric machine, and they should be retained as part of the maintenance documentation.

The question of correct radial rotor positioning, and the acceptable variation of air gap measurements is often difficult to address. It is generally troublesome to make precise air gap measurements due to the physical construction of some machines. Hence, the machinery diagnostician is referred back to the original OEM specifications for guidance in this critical area. Since some specifications change with time, it might also be advisable to contact the OEM directly for an update of current procedures and tolerances.

In years past, the requirement for variable speed electric machines has been satisfied by large DC units. The number of these units are steadily fading due to attrition, and DC electric machines are seldom encountered within the industrial community. In situations where variable speed motors are required due to process considerations or energy conservation, the trend has been towards variable speed AC motors.

From equation (10-42) it is apparent that synchronous speed is dependent on the line frequency, and the number of poles. Since the number of poles is fixed for any particular machine configuration, the only way to vary the motor speed is to change the line frequency. In fact, this is exactly what the newest motor control systems accomplish. These motor control devices are often called frequency converters, or Variable Frequency Drives (VFDs). Functionally, they vary the input frequency to the motor to provide variable speed operation.

These variable speed AC motors add another degree of complexity to the business of analyzing motor vibratory characteristics. Since the motor stator frequency is now a variable parameter instead of a constant 60 Hz, the diagnostician must be even more careful during data acquisition, processing, and analysis. In some cases, the inclusion of motor currents via clamp-on transducers, plus coil temperature distributions may be highly beneficial in identifying and solving a problem on an electric machine. As always, the shaft motion and position should be observed. On flexible supports, the bearing housing motion should be measured, and good engineering judgment applied.

Finally, the views and opinions on electric machines presented in the last few pages are based upon a variety of field experiences by the senior author. Within this chapter there has been no effort to provide any detailed explanations of the electrical forcing functions. This is a complex topic, and the reader is referenced to documents such as the tutorial on motors by James Baumgardner[12]. Within this article, Baumgardner provides a detailed discussion of the vibration characteristics of three phase, squirrel cage, induction motors. The explanation of motor electrical characteristics and forces are both clear and accurate.

The reader is also cautioned against fully believing some of the motor analysis charts and tables that appear in the literature. Some of these guides contain considerable inaccuracies, and they can result in more confusion than positive assistance. As a general rule, the technical information provided by the OEMs and knowledgeable end users will typically be solid and reliable troubleshooting information. Some of the other published sources should be used carefully. Don't get stuck in the groove of trying to find a troubleshooting chart that matches your problem symptoms.

Case History 33: Motor With Unsupported Stator Midspan

Many large motors have long rotors, and the companion stators are equally endowed with physical length. Many of these units are built with support rails that run the full length of the casing, and are designed for support along the entire length of the stator. For example, consider the induction motor driving single stage booster compressor depicted in Fig. 10-30. The 3,000 HP motor was only supported at the four corners. During uncoupled operation, the shaft vibration (from probes 1Y through 2X) was dominated by rotational speed motion at an average value of 1.2 Mils,$_{\text{p-p}}$ for all four probes. The shaft runout varied between 0.2 and 0.3 Mils,$_{\text{p-p}}$, and it was generally in-phase with the synchronous vectors at full speed. Hence, the runout compensated shaft vibration was approximately 1.0 Mil,$_{\text{p-p}}$ at all four measurement points. For a 3,600 RPM motor this was somewhat higher than desirable for an unloaded spin test, but it was accepted by the end user.

Installation of this machinery train was complicated by a poor grouting job. The initial problems were eventually corrected, and a proper full contact epoxy grout was used between the foundation and the sole plate. Examination of the transient startup behavior revealed nothing unusual. However, as load was applied the shaft vibration began to increase, and it eventually reached amplitudes in the vicinity of 3.0 Mils,$_{\text{p-p}}$. Furthermore, the vibration levels did not remain constant, and they continually pulsated back and forth between 3.0 and 0.5 Mils,$_{\text{p-p}}$. As observed on an oscilloscope, the behavior was clearly an amplitude modulation similar to the diagram shown in Fig. 10-27. With a stopwatch, it was determined that the vibration amplitudes were continually changing at a rate of every seven seconds. That is, the time from one peak amplitude, through the vibration decrease, and back to the next peak amplitude was approximately

---

[12] James Baumgardner, "Tutorial Session on Motors," *Proceedings of the Eighteenth Turbomachinery Symposium*, Turbomachinery Laboratory, Texas A&M University System, College Station, Texas (October 1989).

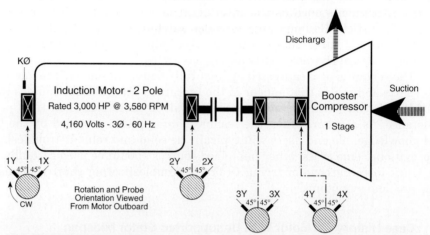

Fig. 10–30 Machinery Arrangement For Motor Driven Single Stage Compressor

seven seconds. Simultaneously, the rotational speed phase angles varied by nominally 100° in a consistent manner with the oscillations in shaft vibration. This behavior occurred at a running speed of 3,591 RPM. From equation (10-42), it was clear that the slip frequency for this two pole motor was 9 cycles per minute. The period of 9 CPM is 6.7 seconds. Hence, the observed amplitude and phase variation period of approximately seven seconds was really the period of the motor slip frequency.

Another perspective of the dynamic motion is presented in the orbital data in Fig. 10-31. This information consists of eight shaft orbits acquired at 1 second intervals. The data was band-pass filtered at the running speed of 3,591 RPM. However, the bandwidth of the filter included the motor line frequency at 3,600

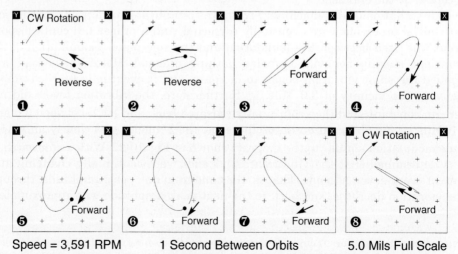

Fig. 10–31 Consecutive Shaft Orbits At Induction Motor Outboard Bearing

CPM (60 Hz). Note that rotation is clockwise as viewed from the outboard end of the motor towards the compressor. For normal machinery behavior, the shaft precession should be in the same direction as shaft rotation. However, orbits ❶ and ❷ in Fig. 10-31 exhibit a distinct reverse precession. This is followed by orbits ❸ through ❽ with a normal clockwise precession. If this data was viewed live on an oscilloscope, the shaft orbit would be in constant motion, and it would consecutively repeat the patterns documented in Fig. 10-31. This dynamic behavior is due to the interaction of two closely spaced frequencies.

The respective amplitudes at 3,591 and 3,600 CPM cannot be properly identified with a 12 CPM band-pass filter. It is necessary to employ the frequency expansion capabilities of a DSA over a suitably small frequency span. In this case, a 6.25 Hz (375 CPM) span was selected between 3,400 and 3,775 CPM. For a 400 line resolution, the time record length was 64 seconds, and the resolution for a 400 line display would be 0.9375 CPM per filter (from Table 7-1). Perform this analysis for 4 averages, the results are summarized in Table 10-4.

Table 10–4 Summary Of Motor Vibration Amplitudes At Running Speed And Line Frequency

Motor Vibration Probe	Amplitude At Running Speed of 3,591 RPM	Amplitude At Line Frequency of 3,600 CPM
Outboard - 1Y	1.64 Mils,$_{p-p}$	1.59 Mils,$_{p-p}$
Outboard - 1X	1.88 Mils,$_{p-p}$	1.68 Mils,$_{p-p}$
Coupling - 2Y	1.17 Mils,$_{p-p}$	1.63 Mils,$_{p-p}$
Coupling - 2X	0.91 Mils,$_{p-p}$	1.67 Mils,$_{p-p}$

The running speed vibration amplitudes at 3,591 RPM are higher than the solo levels of nominally 1.0 Mil,$_{p-p}$. This could be due to a variety of reasons, but the significant result of Table 10-4 is the constant excitation of 1.6 Mils,$_{p-p}$ at the synchronous or line frequency of 3,600 CPM. Based on the previous discussion within this chapter, the diagnostician would certainly suspect an electrically or magnetically induced excitation. Since the 3,600 CPM shaft vibration amplitudes are essentially the same at both ends of the machine, it is reasonable to expect a mechanism that influences the entire rotor in a uniform or evenly distributed manner. The slip frequency behavior of this motor does not directly match any of the traditional symptoms mentioned earlier in this chapter. The documented assembly information on this motor revealed proper bearing clearances, uniform air gap at both ends of the motor, plus a reasonable alignment to the single stage booster compressor. Overall, there was no evidence in the mechanical assembly information to indicate any abnormalities.

A wise mechanical engineer once told the senior author that *"you will never understand a machine unless you go and look at it."* This is certainly an unpopular approach in the modern world of powerful computers, remote information

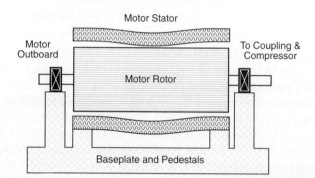

Motor Stator

Motor
Outboard

To Coupling &
Compressor

Motor Rotor

Fig. 10–32 Diagram Of
Induction Motor With
Exaggerated Stator Sag

Baseplate and Pedestals

transfer, and beautiful graphics generated with a variety of analytical and data processing programs. However, even advanced *1's and 0's* technology must always succumb to the realities of the physical installation. In this particular case history, a walk around the unit revealed that the motor was only supported at the four corners. Although a full length support surface under the stator was provided by the OEM, over 90% of this vertical support surface was unused.

Additional field optical measurements revealed that the center of the stator support was about 15 Mils lower than the ends. In essence, the diagram shown in Fig. 10-32 describes the rotor and stator position with only end supports for the stator. In this condition, the air gap was not uniform along the length of the rotor. That is, a straight rotor was running within the confines of a *dropping* stator. The uneven magnetic forces due to this distorted air gap were primarily responsible for the 1.6 Mil,$_{p-p}$ vibration component at 3,600 CPM. Placing a midspan support under the stator relieved the problem, and 60 Hz excitations were reduced to levels of less than 0.2 Mils,$_{p-p}$ at both motor bearings. This midspan stator support also resulted in a reduction of the 1X running speed vibration amplitudes.

As a side note to this problem, recall that the air gap measurements at each end of the motor were acceptable, and within the OEM specifications. As shown in the exaggerated stator deflection diagram of Fig. 10-32, it is quite possible to have reasonably even air gaps around the ends of the rotor combined with an eccentric air gap at the center of the rotor. In this situation, the rotor appears to be properly centered in the stator, but midspan deflection of the stator due to its own weight results in an uneven air gap problem at the center of the rotor.

In virtually all cases, the OEMs recommendations for machine support should be followed. If this information is not supplied in the installation manual, then the OEM should be contacted, and requested to provide their recommended support configuration for the model and size of the machine under consideration. If machine support issues still exist after talking to the OEM, then best engineering judgment should be applied. Remember that few machines have ever experienced problems due to large foundations and rigid supports. However, many machines have been damaged or destroyed due to poor foundations, improper supports, or insufficient supports.

Case History 34: Torsional Excitation From Synchronous Motor

The machinery train shown in Fig. 10-33 consists of a 6 pole, 2,500 HP synchronous motor driving a speed increasing gear box. The pinion output speed is 8,016 RPM, and it is directly coupled to a 5,000 HP hot gas expander turbine. The combined power output from the motor and the expander are used to drive the process air compressor located at the end of the train. Of particular note on this unit is the coupling between the motor and gear box. This 1,200 RPM grease packed gear coupling has a high torsional stiffness of 110,000,000 Inch-Pounds per Radian. This machinery train has been in service for many years, and high vibration amplitudes are generally encountered during startup conditions on the pinion and the adjacent expander bearing.

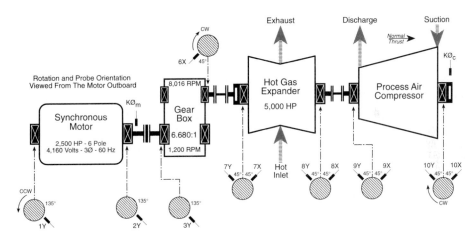

Fig. 10–33 Machinery Arrangement Of Motor, Gear Box, Expander, And Compressor

After the plant is up and running, and a full heat soak has been achieved by all of the cases, shaft vibration levels are generally low and acceptable. However, the initial cold startup generates significant noise in the gear box, plus the high transient vibration levels previously mentioned. Fortunately, the high startup pinion vibration levels only affects bearing 7 (probes 7Y and 7X). Expander bearing 8, and the entire air compressor are generally not influenced by the high pinion vibration amplitudes.

A typical pinion startup is characterized in the transient capture data exhibited in Fig. 10-34. The entire startup from zero to full speed is achieved in slightly less than 23 seconds. It is clear from the time domain trace of Fig. 10-34 that an initial excitation occurred shortly after rolling, and a major response of 6.58 Mils,$_{\text{p-p}}$ occurs at approximately 15 seconds into the ramp-up. Although the overall severity of the pinion vibration is described by Fig. 10-34, this type of information does not provide much visibility as to the origin or frequency composition of the vibration signal.

If the same data is processed in the frequency domain, the cascade plot

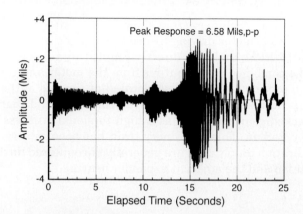

Fig. 10–34 Transient Capture Time Domain Plot Of Pinion Vibration During A Typical Startup

shown in Fig. 10-35 may be produced. This diagram displays a series of spectrum plots at 200 RPM increments between 2,000 and 8,000 RPM. At a pinion speed of 2,000 RPM, the bull gear and motor speed is 300 RPM. At the full pinion speed of 8,016 RPM, the associated motor and bull gear speed is equal to 1,200 RPM. These two end points are connected, and the line labeled as *bull gear and motor speed* on Fig. 10-35. It is clear that throughout the majority of the speed range, the rotational speed vibration amplitudes at motor and bull gear speed are negligible.

The same argument may be applied to the amplitudes associated with the pinion rotating speed. Since the cascade plot shown in Fig. 10-35 is limited in fre-

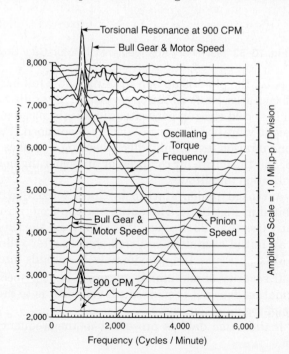

Fig. 10–35 Cascade Plot of Pinion Vibration During A Typical Startup

quency span to 6,000 CPM, the pinion rotational frequency line only extends from 2,000 to 6,000 RPM. This line is labeled as *pinion speed*, and it is clear that vibration amplitudes along this line are minimal.

The major activity throughout the speed domain occurs along a line at 900 CPM (15 Hz). This is the first torsional resonance of the machinery train, and it initially appears between pinion speeds of 2,400 and 4,000 RPM. Logically, this is the initial amplitude *burst* shown in the first few seconds of Fig. 10-34. The next, and largest component on Fig. 10-35 occurs at the intersection of 900 CPM and the line labeled as the *oscillating torque frequency*. The maximum component amplitude at this intersection is 5.0 Mils,$_{p-p}$ that is consistent with the peak value of 6.58 Mils,$_{p-p}$ displayed on Fig. 10-34.

This excitation is the pulsating torque originating from the synchronous motor. The resultant torsional frequency generated conforms to equation (10-45). In this particular case, the number of poles N_p is equal to 6. The synchronous speed of the motor F_{sync} is 1,200 RPM, and the torsional resonant frequency F_{tor} is 900 CPM. If these values are substituted into (10-45), the expression may be solved for the motor rotor speed F_{rotor} that provides a pulsating torque at the torsional resonant frequency of 900 CPM.

$$F_{rotor} = F_{sync} - \frac{F_{tor}}{N_p} = 1,200 \text{ CPM} - \frac{900 \text{ CPM}}{6} = 1,200 - 150 = 1,050 \text{ RPM}$$

Thus, at a motor or bull gear speed of 1,050 RPM, the oscillating torque frequency will be 900 CPM. Multiplying the bull gear speed of 1,050 RPM by the gear box speed ratio of 6.68 yields a pinion speed of 7,014 RPM. Clearly, this is consistent with the data displayed in the cascade plot of Fig. 10-35.

Simultaneous with this high startup vibration on the pinion, the motor and bull gear reveal only minor traces of this torsional motion. It is reasoned that the oscillating torque from the motor is transmitted directly through the torsionally hard coupling to the bull gear. Since there is minimal torsional to lateral cross-coupling in either the motor or the bull gear, the lateral vibration at the torsional frequencies are minimal. However, the tooth contact forces transmitted across the mechanical link between the bull gear and the pinion teeth contain a vector component that translates torsional to lateral motion. Hence, the pinion is forced to vibrate at the oscillating motor torque frequency. This is detrimental for both the pinion and bearings, plus the expander bearing located across the high speed coupling. During one documented startup, new expander bearings with a 6 Mil diametrical clearance experienced a shaft centerline position change of 9 Mils. Hence, the high torsional startup vibration effectively *hammered* the expander bearing, and increased the clearance by 50%.

This situation could be significantly improved by reducing the torsional stiffness of the coupling between the motor and bull gear. This would isolate the oscillating torque behavior to the motor, and would probably influence the effective torsional resonant frequency. However, due to years of successful operation, the end user elected not to correct this problem.

RECIPROCATING MACHINES

Reciprocating machines have been in service for many years prior to the introduction of centrifugal units. In many circles, reciprocating machinery is considered to be less sophisticated, or perhaps less elegant than centrifugal units. In reality, reciprocating machines are more complicated, and contain more individual parts than centrifugal units. Reciprocating machine installations often require customized piping simulation, and pressure pulsation suppression devices such as pulsation bottles or restriction orifices. This technology is seldom required on centrifugal installations. In many cases, reciprocating machines demand larger foundations, and they are less tolerant of supporting structure degradation. In most facilities, the long-term maintenance costs on recips are significantly greater than centrifugals.

In spite of these drawbacks, the total quantity of positive displacement reciprocating engines, pumps, and compressors exceed the number of centrifugal units. From an application standpoint, there are many situations where centrifugal units cannot provide the necessary differential head, or they cannot efficiently operate over the required flow rates demanded by the process. In these cases, reciprocating machines are far superior to their centrifugal counterparts, and the advantages strongly outweigh the obvious disadvantages.

As mentioned throughout this text, the technology required for the measurement and diagnosis of machinery problems on centrifugal units has progressed at a rapid rate during the past three decades. Unfortunately, there has not been similar progress for the analysis of reciprocating machines. In the majority of cases, the reciprocating machinery condition is evaluated based upon process measurements such as suction and discharge temperatures and pressures. These measurements are often supplemented by periodic external measurements such as external valve temperatures, or casing vibration of the frame or crankcase.

Measurement and trending of **valve temperature** has proven to be an effective predictive tool on many reciprocating compressors. In these units the valves are accessible from the outside of the cylinder, and an in-operative valve assembly may be easily replaced. This configuration applies to plate valves, poppet, or channel valves. Some machines are equipped with permanently mounted temperature sensors on each valve, plus an associated scanning recorder. This allows the trending of valve temperatures with time, and provides a good perspective of valve problem development.

Additional perspective of cylinder behavior may be obtained by running a **pressure-volume curve** (PV diagram, also know as an indicator card). The volume is determined by the stroke position, and cylinder pressure is obtained with a dynamic pressure sensor. This provides a good overview of the entire expansion-compression cycle, and problems such as valve chatter and piston ring rattle are discernible. It is also meaningful to observe the time domain pressure pulsation data, and examine the pressure pulsations as a function of both time and stroke position. However, any type of pressure measurement on a cylinder requires a direct physical connection between the interior portion of the cylinder and the externally mounted pressure pickup. In some cases, this is a drilled hole through the cylinder wall, with a tapped external connection for a pressure pickup. This type of installation minimizes the passage length between the cylinder and the transducer, but it does necessitate shutting down the machine for installation and removal of the pressure pickup. In other installations, a nipple and block valve are screwed into the hole that is drilled into the cylinder wall. This approach allows the pressure pickups to be installed and removed from the machine during operating. However, equipping a large machine with ten or twenty small overhung valves may be undesirable from a safety standpoint. These small valves and associated nipples are subject to mechanical damage from people working around the machines, and they are also prone to fatigue failures due to vibration of the cantilevered valves. These type of measurement ports may be acceptable for air or nitrogen compressors, but they should be cautiously applied on reciprocating units handling combustible materials.

On some reciprocating engines, it is possible to purchase spark plugs with pressure pulsation probes that are an integral part of the spark plug. This type of installation is very effective to study detonation or misfiring problems in the engine. In some cases, it is also meaningful to measure casing vibration in conjunction with the pressure pulsation data.

Frame or **crankcase vibration** measurements are common on reciprocating units. These may be periodic measurements obtained with a portable data collector, or they may be permanently installed casing transducers connected to a monitoring system, computer-based DCS, or machinery trending system. Before the appearance of cost-effective piezoelectric transducers, these external frame measurements on recips were generally obtained with velocity coils. Unfortunately, the high vibration levels typically encountered on reciprocating machines results in premature failure of the coil springs in a velocity pickup. Hence, one day the velocity coil has an output of 0.12 IPS$_{o-p}$, and the next day the level has dropped to essentially zero. In this example, the velocity coil has failed, and operations personnel become further convinced that *this stuff doesn't work*.

With the advent of cost-effective piezoelectric accelerometers, and integral transducer electronics to integrate the acceleration signals to velocity, many of the poor measurement reliability issues have been corrected. Hence, solid state vibration transducers may be installed on the external frames or crankcases of reciprocating machines, and they will successfully operate for extended periods of time. However, most end users are still reluctant to use this type of vibration measurement for automatic shutdown of the machinery.

A more meaningful and reliable vibration measurement involves the use of proximity probes. These transducers are mounted on stationary parts of the machine, and they observe the relative motion of moving machine elements. For example, a compressor drive motor such as the 200 RPM synchronous motor shown in Fig. 10-36 (case history 35) may be equipped with X-Y proximity probes to observe the motor shaft vibration relative to the stationary bearing housings. This application is obviously identical to the approach used on any other piece of critical centrifugal machinery. The concept of measuring rotational motion may be extended to the crankshaft main bearings. In some installations, X-Y proximity probes have been installed to measure **crankshaft vibration** relative to the stationary main bearings. The axial clearances around these main bearings are often limited, and the mounting of proximity probes may be quite difficult. Although there are benefits to be gained from this measurement of crankshaft vibration and position, the probe installation is often challenging. In some cases, it is possible to install the probes in the bottom half of the main bearing, and route the extension cables out the side of the crankcase. This location may require boring through the babbitt, and the OEM should be consulted on any potential adverse effect this may have upon the bearing.

From the crankshaft, connecting rods may go directly to the individual pistons, or they may be attached to a crosshead. This is the point in the machine where the rotational motion is converted to reciprocating motion. In the compressor case shown in Fig. 10-36 (case history 35), a main crosshead is mounted at the crankcase for each throw. An auxiliary crosshead is then driven back and fourth by a pair of horizontal drive rods from each main crosshead.

At this point, the primary vibration measurement consists of proximity probes mounted on the seals or cylinders observing the **reciprocating piston rod** or plunger. In a simple installation, one vertical probe is mounted on each rod or plunger. This transducer may be directly above, or directly below, the moving element, but it is important that the probe be in a true vertical direction. By observing and trending DC gap voltages from the proximity probe, it is possible to measure the average horizontal running position of the piston rod. As wear occurs on the piston rings due to normal attrition, the piston rod will move downward, and the vertical proximity probe will detect this event by virtue of a change in probe gap voltage. If the probe is mounted above the piston rod, the DC gap voltage will increase with piston ring wear. Conversely, the gap voltage will decrease if the vertical probe is installed below the piston rod. This is sometimes referred to as the **rod drop** measurement, and it provides actionable maintenance information on this part of the reciprocating machine.

Additional machinery information may be acquired by installing a horizontal probe in conjunction with the true vertical probe as shown in Fig. 10-36. This orthogonal transducer provides information in the horizontal plane, and in many cases this may be more significant than the vertical motion. Furthermore, the use of perpendicular vertical and horizontal proximity probes allows the machinery diagnostician the capability of observing the combined vertical and horizontal motion of the piston rod or plunger as function of stroke. More specifically, on a rotating machine, the probes observe the same circumferential shaft surface as the shaft rotates, and it is customary to examine the shaft orbit. This orbital motion may be examined during a single turn, or for multiple rotations. On a reciprocating machine, the probes observe the same angular position on the rod or plunger as the element reciprocates back and fourth. When observed from the end of the cylinder, the resultant motion is a **Lissajous figure** of the rod centerline during one or multiple strokes. In essence, this is a two-dimensional representation of a three-dimensional event.

The sequential timing of events throughout the machine is achieved by a once-per-cycle **Keyphasor®** probe. This timing transducer may be installed on the motor, crankshaft, or one of the drive rods (as shown in Fig. 10-36). To make the analysis of data as straightforward as possible, it is always desirable to have the trigger point coincident with the top dead center of one of the cylinders within the machine. By positioning this timing probe with respect to a physical event on one cylinder, the entire machine may be phase related.

The electronic tools used to diagnose the behavior of rotating machines are fully applicable to the analysis of the transducer signals generated by reciprocating units. Although the mechanics are different, the processing techniques applied to the electronic signals are virtually identical. From a frequency analysis standpoint, the dominant frequency observed on a reciprocating machine is the fundamental speed of the machine. For instance, on the machine depicted in Fig. 10-36, the motor speed is 200 RPM, and this is directly coupled to the main crankshaft at 200 RPM. The main and auxiliary crossheads, plus the plungers all reciprocate at a frequency of 200 CPM. Since a constant frequency exists throughout the machine, this may complicate any malfunction analysis, since everything is moving at the same frequency. In addition, there are often other cylinders, and other throws that are generating the same frequency, and a variety of interactions plus rotational speed harmonics are often observed.

This measurement of piston rod or plunger motion provides considerable information on the dynamic behavior of the mechanical system. Deflection of the rod or plunger, plus the influence of various preloads, are detectable with these displacement transducers. This dynamic measurement is sometimes referred to as *rod runout*, which is a misnomer. Basically, these machine elements do not have runout, but they do respond to changes in position and dynamic forces. The timing relationships between cylinders, and the position of each individual piston rod or plunger with respect to top dead center and bottom dead center are important parameters. With some malfunctions, the relative position of elements are important. With other mechanical problems, the Lissajous patterns described by the V-H probes provide the necessary clues. In still another class of abnormal behavior, the simple time domain examination of the proximity probe signals will provide useful information. By understanding the characteristic behavior of these dynamic signals, the diagnostician will have additional useful tools for meaningful evaluation of the machinery.

Lower pressure reciprocating machines are equipped with piston rods, pistons, piston rings, plus a variety of seal configurations. These machines may also have manual or automatic valve unloaders, and other appurtenances such as cylinder clearance pockets. These general types of machines may be difficult to diagnose simply due to the mechanical complexity and the associated multiple degrees of freedom of the unit. A somewhat simpler version of the reciprocating machine is the high pressure hyper compressor discussed in case history 35. In this type of machine, a solid tungsten carbide plunger is used as a combined piston rod, piston, and piston rings. This plunger is connected directly to the auxiliary crosshead, and it functions as a rigid member that reciprocates in and out of the cylinder with stationary packing. This type of assembly does simplify some of the system dynamics, but other issues do develop that make the examination of this type of machinery quite interesting.

Case History 35: Hyper Compressor Plunger Failures

The machinery depicted in Fig. 10-36 represents one of four trains installed in a new low density polyethylene plant. The original plant contained a total of four processing lines. Each line consisted of a primary compressor, a high pressure hyper compressor, plus the reactor and associated product handling equipment. The primary compressors in each line are traditionally configured reciprocating machines that boost the ethylene pressure to nominally 4,000 Psi. Depending on product mix, the high pressure compressors provide a reactor inlet pressure that varies from 30,000 to 45,000 Psi. Obviously, it doesn't matter if these pressures are specified as *gauge* or *absolute*, these are very high pressures.

In order to meet these pressure demands, the high stage compressors are a unique variety of machine that was specifically designed for ultra high pressure operation. Although the drive, crankcase, and main crosshead arrangements are fairly typical of most reciprocating machines — a significant change in technology occurs in the auxiliary crossheads, plungers, and cylinders. In this type of machine the piston rod, piston, and piston rings are replaced with a solid tung-

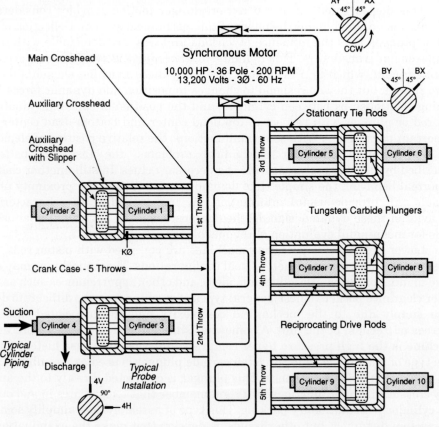

Fig. 10–36 Plan View Of High Pressure Reciprocating Compressor And Motor Driver

sten carbide plunger. Two plungers are attached to each auxiliary crosshead with a patented resilient connection that incorporates a spherical seat joining into a thrust block. These plungers are mated with high pressure cylinders that are mounted on each auxiliary crosshead yoke assembly as shown in Fig. 10-36. The 44.5 inch long plungers have diameters that vary from 2.25 to 3.75 inches, with a total stroke of 15 inches. It is meaningful to note that the tungsten carbide plunger material has a modulus of elasticity of 80,000,000 pounds per inch^2. The tensile strength is equal to 120,000, and the compressive strength is 570,000 pounds per inch^2. This is also a very hard material with *Rockwell C* values in the vicinity of 84. Tungsten carbide will handle enormous compressive loads, but it is much weaker when subjected to tension or bending. In fact, the remnants of plunger failures revealed a brittle fracture of the material.

The suction and discharge piping to each cylinder consists of high pressure tubing. In many respects this piping appears to be a *cannon bore* construction with large outer diameters, thick walls, and small inner diameters. As shown in Fig. 10-36, the suction piping enters at the outboard end of each cylinder. The discharge exits from a connection at the cylinder inboard, close to the stationary yoke assembly. During operation, the cylinders visibly moved back and fourth with each stroke.

The plunger in each cylinder is equipped with several rows of segmented bronze packing that ride directly on the tungsten carbide plungers. The amount of gas leakage past the seals is monitored, and historically this has been used as a partial indicator of packing condition. These units also incorporate plunger coolant circulation around the outer plunger seal assembly.

These unique machines were sequentially started as the construction and commissioning on each line was completed. In the spring of the startup year, all four lines were successfully operating, and the prognosis of continued operation seemed to be excellent. However, in June of that same year a series of three plunger failures occurred within an 18 day period. Total machine operating time to each of the failures varied between 2,400 and 3,000 hours. Fortunately, the first two failures just involved equipment damage. However, the third broken plunger resulted in personnel injuries in addition to fire and mechanical damage.

This type of situation in a chemical plant is difficult for all parties involved. A variety of activities are immediately initiated to determine the root cause of the failures, plus the implementation of suitable corrective and preventative measures. In addition to the complex technical problems, and the personnel protection issues, the plant was faced with a termination of fire insurance on these machines. In the event of a fourth failure, and the possibility of a major fire, the personnel hazard and financial implications were potentially devastating.

In all three mechanical failures, a plunger broke, and high pressure ethylene was released to the atmosphere. One of the candidates that could be responsible for the plunger breakage was misalignment of the plungers to their respective cylinders. Although each plunger was carefully aligned with dial indicators, this only covered the static condition as the machine is barred over. There was no method to measure the plunger position during actual operation of the machine. It was suggested that air gauges be directed at the plungers, but this concept was rejected due to the limited range of this measurement (20 Mils). The only dynamic transducer that made any sense was the proximity probe. Although this sensor would measure the average distance between the probe and the plunger (DC gap), it would also measure the reciprocating motion of the plunger (AC portion of the signal). Initially, it was believed that the reciprocating motion was minimal, and the majority of the information would be derived from changes in probe DC gap voltage with the machine in full operation.

At this point in history, only a handful of people had ever installed proximity probes on reciprocating machines. Most of the industrial contacts that had attempted this measurement were either working on considerably different machines, or they were extremely reluctant to share their experiences. Hence, the installation of proximity probes on tungsten carbide plungers was virtually a new application of an existing technology by the senior author.

The first step in this investigation consisted of determining if proximity probes will work with tungsten carbide. To verify this point, and allow the development of calibration curves, a test fixture was constructed to hold a proximity probe plus a section of a fractured plunger. It was quickly determined that the measurement was quite possible, and a normal calibration curve could be generated from this dense material. On typical compressor shafts made out of 4140 steel, the proximity probes yield a calibration of 200 millivolts per Mil. However, on tungsten carbide plungers, this value increased to 290 millivolts per Mil. Furthermore, the resultant calibration curves were exceptionally clean, the points easily fell into a straight line, and consistency between plungers was exceptional. In retrospect, this was logically due to the high quality of the plungers, and the uniformity of the metallurgy.

Since the proximity probe measurement worked in the shop, it was now necessary to determine if probes could be installed on the compressors. As it turned out, the transducers were mounted on the outer packing flange. One probe was installed in a true vertical direction, and the companion probe was mounted in a true horizontal direction. A *typical probe installation* is shown in Fig. 10-36 with the vertical transducer at the 12 o'clock position, and the horizontal probe mounted at the 3 o'clock location. Due to various interferences with existing hardware, this typical probe installation was occasionally changed by installing the vertical probe below the plunger looking up, and/or mounting the horizontal probe on the left side of the plunger. In all cases, the proximity probes were maintained in a true vertical and a true horizontal orientation.

During the first test of this instrumentation, only six probes were installed on three different plungers. Temporary Proximitors® were connected to the probes, and a jury-rigged power supply and patch panel were used to drive the probes. The test equipment at this stage consisted of an oscilloscope, an oscilloscope camera, a digital voltmeter, and a four channel FM tape recorder. The ensuing startup was quite remarkable since it was clear that DC probe gap was useful information, but the AC or dynamic motion of the plungers was substantially more than anticipated. It was also amazing to watch a startup of this 200 stroke per minute machine, and observe the plunger static plus dynamic motion on a DC coupled oscilloscope. During the first few strokes of the drive rods, the plunger would rise in the cylinder. This was followed by an increase in the vertical motion, and a general decrease in the horizontal excursions.

Following this simple beginning, vertical and horizontal probes were installed on all of the tungsten carbide plungers on this machinery train. The Proximitors® were mounted in suitable enclosures, and the wiring was properly encased in conduit. Routine surveillance then began, and various problems were correlated to changes in plunger position and dynamic motion. Additional measurements were made on the high pressure compressors, including deflection and position changes of the drive rods between the main and auxiliary crossheads. Casing vibration measurements were also acquired along the length of the cylinders, plus various locations on the crank case. Of all the measurements, the proximity probe signals of the relative plunger motion proved to be the most meaningful, and representative of mechanical condition.

During this investigation, it was evident at an early stage that timing between events must be quantified. This was necessary for understanding characteristics of the machine, plus sorting-out of the relative motion of each respective plunger. Since the majority of the signals occurred at 200 CPM, it was reasonable to establish a 200 CPM trigger pulse that could be used to relate the reciprocating to the stationary system. This would be in the same manner that a Keyphasor® probe would be used on a centrifugal machine. In this case, a hose clamp was attached to one of the drive rods, and a proximity probe mounted on the yoke housing (Fig. 10-36). The head of the hose clamp was positioned beneath the probe, and the axial position of the hose clamp was coincident with top dead center (TDC) of the plunger in #1 cylinder. Hence, when #1 plunger was at TDC, the Keyphasor® probe would fire. This proved to be an enormously useful measurement, and highly beneficial towards understanding of the machinery.

The analytical instrumentation was also expanded to include a multi channel brush recorder, tunable filters, and a vector filter. With this additional instrumentation it became possible to dissect the probe signals with even more detail. For instance, the plunger Lissajous patterns shown in Fig. 10-37 were obtained

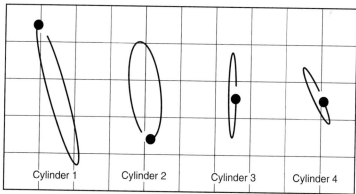

Cylinder 1   Cylinder 2   Cylinder 3   Cylinder 4

Filtered at Reciprocating Frequency of 200 CPM   1.0 Mil per Division

Fig. 10–37 Direct Plunger Lissajous Patterns For Cylinders 1 Through 4

from the first two throws, and the associated cylinders 1 through 4. This data was filtered at 200 CPM, and the previously referenced reciprocating Keyphasor® mark was superimposed to indicate that cylinder 1 was at TDC. At the same time cylinder 2 was at the opposite end of the stroke, and it was at bottom dead center (BDC). Note that the Keyphasor® mark at the top of the pattern for cylinder 1, and at the bottom of the Lissajous for cylinder 3.

It is convenient to refer to plunger axial position in terms of degrees where one stroke equals 360°. Thus, TDC is located at 0° or 360°, and BDC occurs halfway through the stroke, or 180°. Using this approach it is possible to chart peak plunger motion in terms of a digital vector filter phase angle. Table 10-5 provides a typical summary of reciprocating speed vectors for the entire array of ten cylinders with vertical and horizontal measurements on each plunger. A minor com-

Table 10–5 Summary of Reciprocating Speed Vectors Plus Throw Correction

Cylinder Plunger and Probe Location	Recip. 1X Amplitude	Direct Phase	Phase Correction	Corrected Phase
Cylinder #1 Vertical	3.60 Mils,$_{p-p}$	350°	0°	350°
Cylinder #1 Horizontal	1.15 Mils,$_{p-p}$	180°	0°	180°
Cylinder #2 Vertical	2.43 Mils,$_{p-p}$	185°	±180°	5°
Cylinder #2 Horizontal	0.72 Mils,$_{p-p}$	62°	±180°	242°
Cylinder #3 Vertical	2.25 Mils,$_{p-p}$	264°	+72°	336°
Cylinder #3 Horizontal	0.20 Mils,$_{p-p}$	45°	+72°	117°
Cylinder #4 Vertical	1.55 Mils,$_{p-p}$	101°	-108°	353°
Cylinder #4 Horizontal	0.85 Mils,$_{p-p}$	278°	-108°	170°
Cylinder #5 Vertical	2.07 Mils,$_{p-p}$	8°	-36°	332°
Cylinder #5 Horizontal	0.37 Mils,$_{p-p}$	233°	-36°	197°
Cylinder #6 Vertical	3.10 Mils,$_{p-p}$	191°	+144°	335°
Cylinder #6 Horizontal	2.03 Mils,$_{p-p}$	194°	+144°	338°
Cylinder #7 Vertical	2.98 Mils,$_{p-p}$	303°	+36°	339°
Cylinder #7 Horizontal	1.48 Mils,$_{p-p}$	151°	+36°	187°
Cylinder #8 Vertical	1.60 Mils,$_{p-p}$	140°	-144°	356°
Cylinder #8 Horizontal	1.23 Mils,$_{p-p}$	289°	-144°	145°
Cylinder #9 Vertical	3.83 Mils,$_{p-p}$	239°	+108°	347°
Cylinder #9 Horizontal	1.03 Mils,$_{p-p}$	45°	+108°	153°
Cylinder #10 Vertical	1.15 Mils,$_{p-p}$	36°	-72°	324°
Cylinder #10 Horizontal	1.59 Mils,$_{p-p}$	244°	-72°	172°

plication occurs when the other nine cylinders on the machine are referenced to the same Keyphasor®. It then becomes necessary to correct the direct phase angle by the appropriate crank location and plunger orientation. For the five throw machine shown in Fig. 10-36, the angular position between throws must be 72° (=360°/5). For a four throw machine, this incremental step would be 90°.

For the example under consideration, the measured phase angles for cylinder 2 should be corrected by ±180°. For throw 2, cylinder 3, a +72° correction would be necessary, and a -108° correction would be required for cylinder 4. The phase corrections for the remaining six plungers are shown in Table 10-5. A measured phase angle may be easily corrected by the addition or subtraction of the appropriate angle. This will yield a corrected angle that relates plunger position with respect to TDC of that particular plunger. For example, in Table 10-5, the corrected angles are summarized in the right hand column. Note that all of the vertical probes reveal a 1X phase angle in the direct vicinity of 0°. The actual spread varies from 324° to 5°, but it is clear that the timing mark is essentially coincident with the peak of the vertical motion.

Another way to represent this behavior is to apply the phase corrections to the Lissajous patterns previously displayed in Fig. 10-37. The results of this transform are shown in Fig. 10-38. Note that the timing mark appears at the top

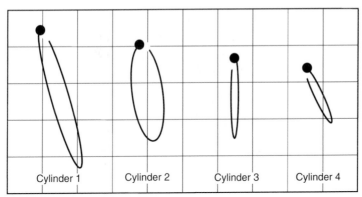

Filtered at Reciprocating Frequency of 200 CPM   1.0 Mil per Division

Fig. 10–38 Plunger Lissajous Patterns For Cylinders 1 Through 4 With Corrected Phase

of each pattern. This is a graphical display of one of the fundamental interrelationships between mechanical components in this machine.

In the overview, it has been documented that the plunger will normally exhibit a 2 to 6 Mil elevation in plunger elevation between zero speed and the normal running position. The plunger will then reciprocate along this new centerline position until acted upon by external forces. As shown in Fig. 10-38, the plunger Lissajous pattern will typically be elliptical, with the vertical motion exceeding the horizontal. In most cases, the vertical displacement at reciprocating frequency varies between 2.0 and 5.0 Mils,$_{p-p}$. In the horizontal plane, the plunger will typically move between 1.0 and 3.0 Mils,$_{p-p}$.

The plunger motion in every observed normal case has been downhill. That is, as the plunger enters the cylinder it drops in elevation. In fact, the plunger will reciprocate at a slight angle with respect to a true horizontal plane. This same motion occurs at both plungers connected to an auxiliary crosshead. It is clear from the data that as one plunger is dropping, the companion plunger on the opposite side of the crosshead is rising. This observation yields the conclusion that a **rocking effect** is present on the auxiliary crosshead. It is generally agreed that this behavior is a function of the following physical occurrences:

1. The *surf boarding effect* of the auxiliary crosshead on the slipper oil film combined with a semi-compliant plunger connection to the crosshead.
2. The differential thermal expansion between the suction and discharge high pressure tubing will tend to elevate the yoke end of the cylinder.
3. Vertical restraint upon the auxiliary crosshead is less than the horizontal.
4. Based on the direction of crankshaft rotation, the main crossheads for throws 1 and 2 may be lifted by the connecting rods. This oscillation at the main crosshead may impart a partial rocking of the auxiliary crosshead.

The rocking motion of the auxiliary crosshead, and the associated plunger motion occurs at the fundamental reciprocating frequency of 200 CPM. This behavior is very consistent on a normal, well-aligned machine. When malfunctions occur, the normal motion will be interrupted. In the final analysis, the precise cause of the 3 plunger failures could not be precisely established. However, the evidence points to changes in plunger alignment during operation. This can be effected by mechanisms such as a loose plunger connection to the crosshead, wear of the crosshead slipper, improperly installed packing, or worn packing. When any of these malfunctions are active, the normal dynamic motion of the plunger is directly influenced. The plunger Lissajous figure will often reveal the preload, and a shift in plunger centerline position is generally observed. If no mechanical damage is caused by the misalignment, then the plunger may be successfully realigned, and normal motion restored. Obviously, when mechanical damage has occurred, realignment will only provide temporary relief. The damage initiated by the misalignment will usually continue to deteriorate, and corrective measures will be required.

For reference purposes, several of the primary mechanical malfunctions detected on this class of machine are summarized as follows.

The detection of **plunger packing failures** (radial or tangential rings) is best described in terms of the restraining effect of the packing. Since the high pressure packing does function with a close clearance to the plunger, the packing must act as one of the lateral plunger restraints (spring). Any increase in packing clearance will decrease the effective packing spring constant, and allow increased plunger motion in the direction of increased clearance. It is usually difficult to separate alignment and packing malfunctions since they often display the same symptoms. It is certainly reasonable for this to occur since alignment and packing difficulties are often coexistent. Therefore, the inception of misalignment may cause packing deterioration, and vice versa. One factor that does occur only with terminal packing failures is a significant increase in *leak gas* rate.

Some packing failures result in the **impregnation of bronze** into the tungsten carbide plungers. These small particles of bronze are highly visible on an oscilloscope time base as sharp spikes originating from the basic plunger sinusoidal waveform. In most observed cases, when the plunger physically contacts the bronze rings, rubbing occurs, and the impregnated bronze will increase as a function of time.

Babbitt loss on the auxiliary crosshead slipper will effect both plungers on a single throw. This mechanism will reveal itself as a significant drop in the centerline position on both attached plungers, combined with a substantial change in plunger dynamic motion. Although the reciprocating amplitudes generally increase when this problem occurs, it is also possible to experience a brief period of reduced motion before amplitudes begin to increase. It should also be noted that this slipper babbitt loss may be due to attrition, or it may be caused by improper lubrication of the slipper.

Loose plunger connection to the auxiliary crosshead is difficult to detect during normal operation, due to the large compressive forces acting between the plunger and the crosshead connection. This problem is usually detected during startup when lube oil is applied to the auxiliary crosshead, and coolant oil is directed to the packing gland. During this condition, a loose connection will show up as a substantial and erratic variation in probe DC gap voltage. This change will appear on one, and in some cases both proximity probes observing the loose plunger. To identify this problem, the probe DC gap voltages should be documented prior to the initiation of oil, and they should be rechecked throughout the startup sequence to be sure that large gap voltage changes have not occurred. Occasionally, this problem may also be visible after startup and before loading of the machine. Running essentially unloaded a loose plunger connection will appear as very high peak to peak displacement amplitudes.

The work described in this case history was performed during a 4 month period. The correlation between machinery integrity and the vibration and position data obtained from the plunger proximity probes was self-evident. The next step was to install permanent X-Y vibration monitors for each plunger. These monitors were equipped with a 15.0 Mil,$_{p-p}$ full scale. Alert levels were set at 8.0 Mils,$_{p-p}$, and an automatic trip of the machine was set at 10.0 Mils,$_{p-p}$. This trip was suppressed during startup, but was fully active after the machine was at speed and pressure. For accurate documentation of the probe DC gap voltages, a DC voltmeter and selector switch was also incorporated with the original monitor racks. Most of the initial data was obtained with clip boards and photographs of the oscilloscope traces. In a current rendition, this type of system would be logically handled in a digital monitoring system combined with a computer-based data acquisition and storage system.

The success or benefit of any machinery analysis project is seldom measured on a short-term basis. The real measure of success is the ability to safely operate year after year with no failures, and no surprises. For this particular project, the initial 3 plunger failures appeared in the vicinity of 3,000 hours of run time (about 4 months). In the early days of this plant, the time interval between major overhauls of these machines was planned for approximately every 5,000 hours (about 6 months). However, since the conclusion of this study, and the installation of the machinery protection instrumentation, there have been no additional plunger failures. Furthermore, the time interval between major overhauls of these machines has been extended to over 20,000 hours (over 2 years), with some reported instances of machines operating more than 30,000 hours (greater than 3 years) between overhauls.

BIBLIOGRAPHY

1. Alban, Lester E., *Systematic Analysis of Gear Failures*, Metals Park, Ohio: American Society for Metals, 1985.
2. Baumgardner, James, "Tutorial Session on Motors," *Proceedings of the Eighteenth Turbomachinery Symposium*, Turbomachinery Laboratory, Texas A&M University System, College Station, Texas (October 1989).
3. Blevins, Robert D., *Flow-Induced Vibration*, Second Edition, New York: Van Nostrand Reinhold, 1990.
4. Corley, James E., "Tutorial Session on Diagnostics of Pump Vibration Problems," *Proceedings of the Fourth International Pump Symposium*, Turbomachinery Laboratory, Texas A&M University, College Station, Texas (May 1987).
5. Dudley, Darle W., *Gear Handbook*, New York: McGraw-Hill Book Company, 1962.
6. Harris, Cyril M., *Shock and Vibration Handbook*, Fourth edition, pp. 29.1 to 29.19, New York: McGraw-Hill, 1996.
7. Nelson, W.E. (Ed), and J.W. Dufour, "Pump Vibrations," *Proceedings of the Ninth International Pump Users Symposium*, Turbomachinery Laboratory, Texas A&M University, College Station, Texas (March 1992), pp. 137-147.
8. Schiavello, Bruno, "Cavitation and Recirculation Troubleshooting Methodology," *Proceedings of the Tenth International Pump Users Symposium*, Turbomachinery Laboratory, Texas A&M University, College Station, Texas (March 1993), pp. 133-156.
9. Schwartz, Randal E., and Richard M. Nelson, "Acoustic Resonance Phenomena In High Energy Variable Speed Centrifugal Pumps," *Proceedings of the First International Pump Symposium*, Turbomachinery Laboratories, Texas A&M University, College Station, Texas (May 1984), pp. 23-28.
10. Spotts, M.F., *Design of Machine Elements*, 6th Edition, Englewood Cliffs, New Jersey: Prentice-Hall, Inc., 1985.
11. Winterton, John G., "Component identification of gear-generated spectra," *Orbit*, Vol. 12, No. 2 (June 1991), pp. 11-14.

Rotor Balancing

Mass unbalance in a rotating system often produces excessive synchronous forces that reduces the life span of various mechanical elements. To minimize the detrimental effects of unbalance, turbo machinery rotors are balanced with a variety of methods. Most rotors are successfully balanced in slow speed shop balancing machines. This approach provides good accessibility to all correction planes, and the option of multiple runs to achieve a satisfactory balance. It is generally understood that balancing at slow speeds with the rotor supported by simple bearings or rollers does not duplicate the rotational dynamics of the field installation. Other rotors are shop balanced on high speed balancing machines installed in vacuum pits or evacuated chambers. These units provide an improved simulation of the installed rotor behavior due to the higher speeds, and the use of bearings that more closely resemble the normal machinery running bearings. In these high speed bunkers, the influence of rotor blades and wheels are substantially reduced by operating within a vacuum. In general, this is a desirable running condition for correcting rotor unbalance, and for studying the synchronous behavior of an unruly rotor.

Some machines, such as large steam turbines, often require a field trim balance due to the influence of higher order modes, or the limited sensitivity of the low speed balance techniques. There is also a small group of machines that contain segmented rotors that are assembled concurrently with the stationary diaphragms or casing. In these types of machines, the final rotor assembly is not achieved until most of the stationary machine elements are bolted into place. Machines of this configuration almost always require some type of field trim balance correction.

In the overview, virtually all rotating machinery rotors are balanced in one way or another. As stated in chapter 9, this is a fundamental property of rotating machinery, and it must be considered in any type of mechanical analysis. Furthermore, it is almost mandatory for the machinery diagnostician to fully understand the behavior of mass unbalance, and the implications of unbalance distribution upon the rotor mode shape, and the overall machinery behavior. If the diagnostician never balances a rotor during his or her professional career, they still must understand the unbalance mechanism to be technically knowledgeable and effective in this business.

Rotor balancing is often considered to be a straightforward procedure that is performed in accordance with the instructions provided by the balancing machine manufacturer. Although this is true in many instances of shop balancing, field balancing is considerably more complicated. It must always be recognized that the rotor responds in accordance to the mechanical characteristics of mass, stiffness, and damping. Thus, a rotor subjected to a low speed shop balance does not necessarily guarantee that field operating characteristics will be acceptable. In most instances, simple rotors may be acceptably shop balanced at low speeds. In some cases, complex rotor systems, or units with sophisticated bearing or seal arrangements, may require a field trim balance at full operating speeds, with the rotors installed in the actual machine casing.

In either situation, the synchronous 1X response of the rotor must be understood, and the influence of balance weights must be quantified. Shop balancing machines typically perform the full array of vector calculations with their internal software. However, field balancing requires the integration of various transducers with vector calculations performed at one or more operating speeds. In order to provide an improved understanding of rotor synchronous motion, the influence of higher order modes, and the typical field balance calculation procedure, chapter 11 is presented.

BEFORE BALANCING

There are several considerations that should be addressed prior to the field balancing of any rotor. The fundamental issue concerns whether or not the vibration is caused by mass unbalance or another malfunction. A variety of other mechanisms can produce synchronous rotational speed vibration. For example, the following list identifies problems that initially can look like rotor unbalance:

○ Excessive Bearing Clearance
○ Bent Shaft or Rotor
○ Load or Electrical Influence
○ Gear Pitch Line Runout
○ Misalignment or Other Preload
○ Cracked Shaft
○ Soft Foot
○ Locked Coupling
○ Gyroscopic Effects
○ Compliant Support or Foundation

Thus, the first step in any balancing project is to properly diagnose the root cause of mechanical behavior. The machinery diagnostician must be reasonably confident that the problem is mass unbalance before proceeding. If this step is ignored, then the balancing work may temporarily compensate for some other malfunction; with direct implications for excessive long-term forces acting upon the rotor assembly.

Balancing speed, load, and temperature are very important considerations. The balancing speed should be representative of rotational unbalance at
operating conditions — yet free of excessive phase or amplitude excursions that
could confuse either the measurements, or the balance calculations. This means
that the rotational speed vectors should remain constant within the speed
domain used for balancing. It is highly recommended that balancing speeds be
selected that are significantly removed from any active system resonance. This is
easily identified by examination of the Bode plots, and the selection of a speed
that resides within a plateau region where 1X amplitude and phase remain constant. It should be recognized that in some cases, the field balancing speed may
not be equal to the normal operating speed. Again, this can only be determined
by a knowledgeable examination of the synchronous transient speed data.

In most instances, the transmitted **load** and **operating temperature** are
concurrent considerations. Balancing a cold rotor under no load may produce
quite different results from balancing a fully heat soaked rotor at full process
load. In many cases, the machine will be reasonably insensitive to the effects of
load and temperature. On other units, such as large turbine generator sets, these
effects may be appreciable. In order to understand the specific characteristics of
any machinery train, the synchronous 1X vibration vectors should be tracked
from *full speed no load*, to *full speed full load* operation at a *full heat soak*. This
should be a continuous record that includes process temperatures and load information. If discrete 1X vectors are acquired at the beginning and end of the loading cycle, the diagnostician has no visibility of how the machine changed from
one condition to the other. Hence, the acquisition of a detailed time record (probably computer-based) is of paramount importance. In some cases, the field balance corrections will be specifically directed at reducing the residual unbalance
in the rotor(s). In other situations, the installation of field balance weights may
compensate for a residual bow, or the effects of some load or heat related mechanism. These should be knowledgeable decisions obtained by detailed examination of the synchronous response of the machine during loading.

Mechanical configuration and construction of the rotor must be reviewed to
determine the **mode shape** at operating speed, plus the location and accessibility of potential balance planes. The mode shape must be understood to select
realistic balance correction planes, and to provide guidance in the location of correction weights. As discussed in chapter 3, the mode shape may be determined
by field measurements, by analytical calculations, or a suitable combination of
the two techniques. If a modally insensitive balance plane is selected, the addition of field balance weights will be totally ineffective. In some situations, weight
changes at couplings, or holes drilled on the outer diameter of thrust collars may
never be sufficient for field balancing a machine. In these cases, the field balancing efforts are futile, and the machine should be disassembled for shop balancing
of the rotor (low or high speed) at modally sensitive lateral locations.

Field **weight corrections** are achieved by various methods depending
upon the machine, and the available balance planes. For example, it is common
to add or remove balancing screws, add or remove sliding weights, add washers
to the coupling, weld weights on the rotor, or drill/grind on the rotor element. The

use of balance screws in an OEM balance plane is usually a safe correction. It is good practice to use an anti-seize compound on the screw threads (mandatory for stainless steel weights screwed into a stainless steel balance disk). In some cases, steel weights may not be heavy enough for the required balance correction. In these situations, consider the use of tungsten alloy balance weights that have nearly twice the density of steel weights.

The sliding weights employed on the face of many turbine wheels fit into a circumferential slot, and are secured in place with a setscrew. These weights have a trapezoidal cross section to fit semi-loosely into the trapezoidal slot. The setscrew passes through the balance weight, and into the axial face of the turbine wheel. Tightening the setscrew locks the weight between the angled walls of the slot and the wheel. Normally, each balance slot of this type has only 2 locations for insertion of the weights. Depending on the weights already installed in the slot, it may be easier to install the weights from one side versus the other.

The addition of coupling washers carries disadvantages, such as the loss of the washers during future disassembly, or the mis-positioning of the washers during future re-assemblies. For the most part, addition of coupling washers represents a temporary balance weight correction measure. This may be the most appropriate way to get a machine up and running in the middle of the night; but more permanent corrections should be made to the coupling or rotor assembly during the next overhaul.

The installation of U-Shaped weights is a common practice on units such as induced or forced draft fans. These weights are temporarily attached to the outer diameter of the fan center divider plate, or a shroud band, using an axial setscrew. After verification that the weights are correct, the balance weights are typically welded to the rotor section. In all cases, welding balance weights on the rotor should be performed carefully. On sensitive machines, the weight of the welding rod (minus flux) should be included in the total weight for the balance correction. In addition, the ground connection from the welding machine must be attached to the rotor close to the location of the balance weight. Under no circumstance should the ground wire be connected to the machine casing, bearing housing, or pedestal. This will only direct the welding machine current flow through the bearings, with a strong probability of immediate bearing damage.

Furthermore, the machinery diagnostician must always be aware of the **metallurgy** of the fan rotor, and the balance weights. On simple carbon steel assemblies, virtually any qualified welder will be able to do a good job with commonly available welding equipment. On more exotic metal combinations, the selection of the proper rod, technique, and welding machine must be coupled with a fully qualified welder for that physical configuration.

For drilling or grinding on a machinery rotor, low stress areas must be selected. Mechanical integrity of the rotor should never be compromised for a balance correction. The rotor material density should be known, so that the amount of weight removed can be computed by knowing the volume of material removed. Finally, the location of angles for weight corrections should be the responsibility of the individual performing the balancing work. It is easy to misinterpret an angular orientation and drill the right hole in the wrong place. Mistakes of this type are expensive, and they are totally unnecessary.

Finally, there are individuals who firmly believe that balancing will provide a cure for all of their mechanical problems. The attitude of *let's go ahead and throw in a balance shot* is prevalent in some process industries. Obviously this philosophy will be correct when the problem really is mass unbalance. However, this can be a dangerous approach to apply towards all conditions. Basically, if the problem is unbalance, then go balance the rotor. If the problem is something else, then go figure out the real malfunction.

STANDARDIZED MEASUREMENTS AND CONVENTIONS

Before embarking on any discussion of balancing concepts, it is highly desirable to establish and maintain a common set of measurements and conventions. These standardized rules will be applied throughout this balancing chapter, and they are consistent with the remainder of this text. As expected, vectors are used for the 1X response measurements and calculations. Vectors are described by both a magnitude and a direction. For instance, a car driven 5 miles (magnitude) due West (direction) defines the exact position of the vehicle with respect to the starting point. For vibration measurements, a running speed vector quantity stated as 5.0 Mils,$_{p-p}$ (magnitude) occurring at an angle of 270° (direction) defines the amount, and angular location of the high spot. In any balancing discipline, both quantities are necessary to properly define a vector.

As discussed in chapter 2, circular functions, exponential functions, and inphase-quadrature terms may be used interchangeably. Although conversion from one format to another can be performed, this does unnecessarily complicate the calculations. Within this chapter, vectors will be expressed as a magnitude, with an angle presented in degrees (1/360 unit circle). Angular measurements in radians, grads, or other units will not be used. Vector amplitudes will vary with the specific quantity to be described. For shaft vibration measurements, magnitudes will be presented as Mils (1 Mil equals 0.001 Inches), peak to peak. This will generally be abbreviated as Mils,$_{p-p}$. Casing measurements for velocity will carry magnitude units of Inches per Second, zero to peak, and will be abbreviated as IPS,$_{o-p}$. Finally, casing vibration measurements made with accelerometers will be shown as G's of acceleration (1 G equals the acceleration of gravity), and zero to peak values will be used. Acceleration magnitudes will generally be abbreviated as G's,$_{o-p}$.

All balance weights, trial weights, and calibration weights must be expressed in consistent units within each balance problem. Typically, small rotors will be balanced with weights measured in grams. Large rotors will generally require larger balance weights, and units of ounces (where 1 ounce equals 28.35 grams), or pounds (where 1 pound equals 16 ounces or 453.6 grams) will be used. It is always desirable to both calculate the weight of a correction mass (density times volume) — plus place the correction mass on a calibrated scale and weight it directly. This is certainly a *belt and suspenders* approach, but it does prevent embarrassing mistakes, and the potential installation of the wrong weight at the right location.

In the following pages, balance sensitivity vectors will be calculated. The magnitude units for these vector calculations will consist of weight (or mass) divided by vibration. For instance, units of Grams/Mils,$_{p-p}$ would be used for sensitivity vectors associated with most rotors. Occasionally, these vectors may be inverted to yield units of Mil,$_{p-p}$/Gram. This format sometimes provides an improved physical significance or meaning. However, the diagnostician should always remember that these are vector quantities. If you invert the magnitude, then the angle must also be corrected to be mathematically correct. In addition, the balance equations are totally interlocked to the balance sensitivity vectors. If someone begins to casually invert sensitivity vectors, the equation structure will become completely violated.

In some situations, a radial length may be included to define the radius of the balance weight from the shaft centerline. This allows more flexibility in selection of the final weight and associated radius. For instance, if a rotor balance sensitivity is 50 Gram-Inches/Mil,$_{p-p}$, and the measured vibration amplitude is 2.0 Mils,$_{p-p}$; then the product of these two quantities would be a balance correction of 100 Gram-Inches. This may be satisfied by a correction weight of 20 grams at a radius of 5 inches, or by a weight of 4 grams at a 25 inch radius.

Generally, the magnitude portion of the vector quantities is easily understood and applied. The major difficulty usually resides with the phase measurements and the associated angular reference frame. Part of the confusion is directly related to the function and application of the timing mark, or trigger point. In all cases, the timing signal electronics provide nothing more than an accurate and consistent manner to relate the rotating element back to the stationary machine. Within this text, the majority of the synchronous timing signals will be based upon a proximity probe observing a notch in the shaft, or a projection such as a shaft key. In either case, the resultant signal emitted by the proximity timing probe will be a function of the average gap between the probe and the observed shaft surface.

As discussed in chapter 6, a Keyphasor® probe will produce a negative going pulse when the transducer is positioned over a notch or keyway as shown in Figs. 11-1 and 11-2. In a similar manner, the Keyphasor® proximity probe will generate a positive going pulse when it observes a projection or key as shown in Figs. 11-3 and 11-4. The actual trigger point is a function of the instrument that receives the pulse signal. This device may be a synchronous tracking Digital Vector Filter (DVF), a Dynamic Signal Analyzer (DSA), or an oscilloscope. All of these traditional instruments require the identification of a positive or a negative slope for the trigger, plus the trigger level within that slope. In many cases, the devices are set for an *Auto Trigger* position, which automatically sets the trigger at the halfway point of the selected positive or negative slope.

The physical significance of the trigger point is illustrated by the diagrams in Figs. 11-1 through 11-4. In all eight cases, the actual trigger point is established by the coincidence of the physical shaft step and the proximity probe. For

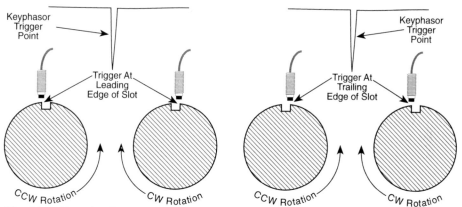

Fig. 11-1 Negative Trigger Slope With Slot   **Fig. 11-2** Positive Trigger Slope With Slot

an instrument set to trigger off a negative slope, the Keyphasor® probe is essentially centered over the leading edge of the notch as shown in Fig. 11-1. For a trigger off a positive slope, the Keyphasor® probe is centered over the trailing edge of the notch as shown in Fig. 11-2.

This positioning between the stationary and the rotating systems is not that critical for machines with large shaft diameters. However, on rotors with small shaft diameters, the establishment of an accurate trigger point is mandatory. For example, on a 2 inch diameter shaft, if the trigger point is off by 1/4 inch, this is equivalent to a 14° error. If this error is encountered during the placement of a balance correction weight, the results would probably be less than desirable. Hence, the establishment of an accurate trigger point is a necessary requirement for successful balancing.

The diagrams presented in Figs. 11-3 and 11-4 describe the trigger condition for a positive going pulse emitted by a timing probe observing a projection or

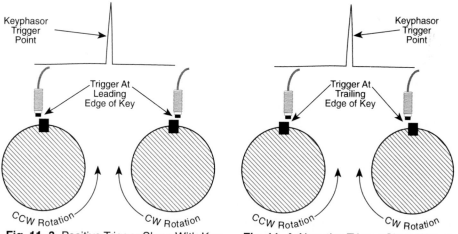

Fig. 11-3 Positive Trigger Slope With Key   **Fig. 11-4** Negative Trigger Slope With Key

other raised surface such as shaft key. Again, the trigger point is established by the coincidence of the shaft step and the proximity probe. For an instrument set to trigger off a positive slope, the Keyphasor® probe is essentially centered over the leading edge of the key as in Fig. 11-3. For a trigger off a negative slope, the KeyØ® probe is centered over the trailing edge of the key as shown in Fig. 11-4. As noted, each trigger point example is illustrated with a clockwise and a counterclockwise example. Typically, the machine rotation is observed from the driver end of the train, and the appropriate Keyphasor® configuration (i.e., notch or projection), is combined with an instrument setup requirement for a positive or a negative trigger slope. This combination of parameters allows the selection of one of the eight previous diagrams as the unique and only trigger point for the machine to be balanced.

In passing, it should also be mentioned that the use of an optical Keyphasor® observing a piece of reflective tape on the shaft will produce a positive going with most optical drivers. Hence, the optical trigger signals will be identical to the drawings shown in Fig. 11-3 and 11-4. Also be advised that reflective tape will not adhere to high speed rotors. Depending on the shaft diameter, a limit of 15,000 to 20,000 RPM is typical for acceptable adhesion of most reflective tapes. For balancing of units at higher speeds that require an optical KeyØ®, the use of reflective paint on the shaft is recommended. Additional contrast enhancement may be obtained by spray painting the shaft with dull black paint or layout bluing. This dark background combined with the reflective paint or tape will yield a strong pulse signal under virtually all conditions.

Regardless of the source of the Keyphasor®, the diagnostician must always check the clarity of the signal pulse on an oscilloscope. A simple time domain observation of this pulse will identify if the voltage levels are sufficient to drive the analytical instruments (typically 3 to 5 volts, peak). Next, the time domain signal will reveal if there are any noise spikes or other electronic *glitches* in the signal. Most of these interferences are due to some problem with the transducer installation, and will have to be corrected back at the timing probe.

There are conditions where baseline noise on the presumably flat part of the trigger curve may be corrected with external voltage amplifiers. In this common manipulation, the direct pulse signal is passed through a DC coupled voltage amplifier, and the bias voltage adjusted (plus or minus) to flatten out the baseline. Next, the signal is passed through an AC coupled voltage amplifier and the signal gain is increased to provide a suitable trigger voltage. Naturally, the outputs of both amplifiers should be observed on an oscilloscope to verify the proper results from both amplifiers (e.g., Fig. 8-8). This same procedure may be used to clean up a signal from a tape recorder. The final objective must be a clean and consistent trigger relationship between the machinery and the electronics.

Once a unique trigger point has been established, the rotor is physically rolled under the KeyØ® probe to satisfy this trigger condition. At this point, an angular coordinate system is established from one of the vibration probes. For example, consider the diagram presented in Fig. 11-5 for vibration probes mounted in a true vertical orientation. In all cases, the angular coordinate system is initiated with 0° at the vibration probe, and the angles always increase

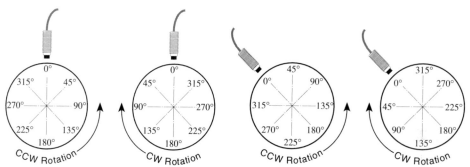

Fig. 11–5 Angular Convention With True Vertical Vibration Probes

Fig. 11–6 Angular Convention With Vertical Probes At 45° Left of Vertical Centerline

against rotation. Fig. 11-5 describes the angular reference system for both a counterclockwise, and a clockwise rotating shaft. Another way to think of this angular coordinate system is to consider the progression of angles as the shaft rotates in a normal direction. Specifically, if one observes the rotor from the perspective of the probe tip, and the shaft is turning in a normal direction, the angles must always increase. This type of logic is mandatory for a proper correlation between the machine, and the resultant polar plots of transient motion, and orbit plots of steady state motion.

If the vibration probes are located at some other physical orientation, the logic remains exactly the same. The 0° position remains fixed at the vibration transducer, and the angles are laid off in a direction that is counter to the shaft rotation. For probes mounted at 45° to the left of a true vertical centerline, the angular coordinate system for a counterclockwise and a clockwise rotating shaft are presented Fig. 11-6.

All vibration vector angles from slow roll to full speed are referenced in this manner. All trial weights, calibration weights, and balance correction weights are referenced in this same manner. The mass unbalance locations are also referenced with this same angular coordinate system. Differential vectors and balance sensitivity vectors also share the exact same angular coordinate system. Although this may seem like a trivial point, it is an enormous advantage to maintain the same angular reference system for all of the vector quantities involved in the field balancing exercise.

A minor variation exists when X-Y probes are installed on a machine. If all of the angles are reference to the Vertical or Y-axis probes, there will be a phase difference between measurements obtained from the X and the Y transducers. If the machine exhibits forward circular orbits, a 90° phase difference will be exhibited at each bearing. This normal phase difference causes only a minor problem. It is recommended that one set of transducers, for example the Y probes, be used as the 0° reference as previously discussed. The vibration vectors measured by the X probes would be directly acquired, and measured angles used in the balancing equations. Since the calibration weights are referenced to the Y probes, the results from the X probes will be self-corrected. This concept will be demonstrated in the case history 37 presented immediately after the development of the single plane balancing equations. Additional explanation of this characteristic will also be provided in case history 38 on a five bearing turbine generator set.

The final concept that must be understood is the relationship between the *Heavy Spot* and the *High Spot*. As discussed in previous chapters, the rotor *Heavy Spot* is the point of effective residual mass unbalance for the rotor (or wheel). As shown in Figs. 11-7 and 11-8, the *Heavy Spot* maintains a fixed angular position with respect to the vibration probe. Regardless of the machine speed (i.e., below or above a critical speed), a constant relationship exists between the vibration probe and the *Heavy Spot*. This is the angular location where weight is removed to balance the rotor (110° in this example). Weight may also be added at 290° in this example to correct for the *Heavy Spot* unbalance at 110°. This is the logic behind the low speed proximity probe balancing rule of:

> *At speeds well below the critical,*
> *remove weight at the phase angle,*
> *or add weight at the phase angle plus 180°*

The *Heavy Spot* and the *High Spot* are coincident at slow rotational speeds that are well below any shaft critical (balance resonance) speeds. Whereas the *Heavy Spot* is indicative of the lumped residual unbalance of the rotor, the *High Spot* is the high point measured by the shaft sensing proximity probe. This is the angular location that maintains the minimum distance between the rotor and the shaft during each revolution.

As rotor speed is increased, the unit passes through a balance resonance (critical speed). For a lightly damped system, the observed *High Spot* passes through a nominal 180° phase shift. This change is due to the rotor performing a self-balancing action where the center of rotation migrates from the original geometric center of the rotor to the mass center, as shown In Fig. 11-8. The proxim-

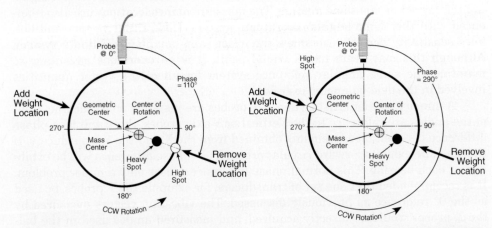

Fig. 11–7 Heavy Spot Versus High Spot At Speeds Well Below The Critical Speed   **Fig. 11–8** Heavy Spot Versus High Spot At Speeds Well Above The Critical Speed

ity probe can only measure distances, and it observes the shift through the critical speed as a change of 180° in the *High Spot*. The probe has no idea of the location of the mass unbalance, it only responds to the change in distance. This behavior is the logic behind the traditional proximity probe balancing rule of:

> *At speeds well above the critical,*
> *add weight at the phase angle,*
> *or remove weight at the phase angle minus 180°*

Clearly, a comparison of Fig. 11-7 with 11-8 reveals that the *Heavy Spot* has remained in the same location at 110°. Removing an equivalent weight at 110° or adding an equivalent weight to 290° will result in a balanced rotor. For additional explanation of this classic behavior through a critical speed, the reader is referenced back to the description of the Jeffcott rotor presented in chapter 3 of this text. In addition, the presence of damping and mechanisms such as combined static and couple unbalance will alter the above general rule.

COMBINED BALANCING TECHNIQUES

The measurements and standardized conventions provide the basis for discussing unbalance corrections. It would be convenient if these concepts could be formed into a set of balancing procedures that are universally applicable to all machines and situations. Unfortunately, such a panacea does not exist, and additional consideration must be given to the actual field balancing techniques. Over the years, various successful techniques have been developed within the machinery community. In 1934, the original influence coefficient vector approach was described by E.L. Thearle[1] of General Electric. Ronald Eshleman[2] of the Illinois Institute of Technology, completed his initial work on methods for balancing flexible rotors in 1962. Balancing with shaft orbits was published in 1971 by Charles Jackson[3] of Monsanto. The combination of modal and influence coefficient techniques was presented in 1976 by Edgar Gunter[4], et. al., University of Virginia. Variable speed polar plot balancing was introduced by Donald Bently[5], Bently Nevada in 1980. Certainly there have been many other contributors to this field; and today there are a variety of balancing techniques available.

The real key to success resides in selecting the techniques most applicable

---

[1] E.L. Thearle, "Dynamic Balancing of Rotating Machinery in the Field," *Transactions of the American Society of Mechanical Engineers*, Vol. 56 (1934), pp. 745-753.

[2] R. Eshelman, "Development of Methods and Equipment for Balancing Flexible Rotors," *Armour Research Foundation, Illinois Institute of Technology*, Final Report NOBS Contract 78753, Chicago, Illinois (May 1962).

[3] Charles Jackson, "Balance Rotors by Orbit Analysis," *Hydrocarbon Processing*, Vol. 50, No. 1 (January 1971).

[4] E.J. Gunter, L.E. Barrett, and P.E. Allaire, "Balancing of Multimass Flexible Rotors," *Proceedings of the Fifth Turbomachinery Symposium*, Gas Turbine Laboratories, Texas A&M University, College Station, Texas (October 1976), pp. 133-147.

[5] Donald E. Bently, "Polar Plotting Applications for Rotating Machinery," *Proceedings of the Vibration Institute Machinery Vibrations IV Seminar*, Cherry Hill, New Jersey (November 1980).

to the machine element requiring balancing — and performing that work in a timely and cost-effective manner. There is an old adage that states: *if your only tool is a hammer, then all of your problems begin to resemble nails*. This is particularly true in the field of onsite rotor balancing. If you only use one specific technique, your options are very limited, and you have no recourse when a machine misbehaves. The balancing techniques used for high speed rotors should integrate the concepts of modal behavior, variable and constant speed vibration measurements, plus balance calibration of the rotor to yield discrete corrections. Although these topics may be considered as separate entities, they are all addressing the same fundamental mass distribution problem. The integrated balancing approach discussed herein attempts to use the available information to provide a logical assessment of field balance corrections.

Initially, a correct understanding of the modal behavior is important for two reasons. First, it helps to identify balance planes with suitable effectiveness upon the residual unbalance. Secondly, it provides direction as to whether the weight correction should be added or removed at a particular phase angle. The mode shape can be determined analytically, or experimentally by vibration measurements. Ideally, the analytical calculations should be substantiated by variable speed field vibration measurements to confirm the presence, and location of system critical speeds.

The next step consists of using vibration response measurements to help identify the lateral and angular location of rotor unbalance. In a case of pure mass unbalance, the runout compensated vibration angles will be indicative of the angular location of the unbalance. In the presence of other forces, amplitude and angular variations will occur. However, the relative vibration amplitudes will help to identify the offending lateral correction planes, and the vector angles provide a good starting point for angular weight locations.

Unless previous balancing information is available, it is usually difficult to anticipate the amount of unbalance. For this reason, many field balancing solutions gravitate towards the *Influence Coefficient* method for calculation of correction weights. Applying this technique, the mechanical system is calibrated with a known weight placed at a known angle. Assuming a reasonably linear system (to be discussed), the response from the calibration or trial weights are used to compute a balance correction that minimizes the measured vibration response amplitudes at the balancing speed.

It should be recognized that the balancing calculations are precise, but they are based upon values that contain different levels of uncertainty. Hence, it is always best to run the calculations with the best possible input measurements, and then make reasonable judgments of the actual corrections to be implemented on the machine. In some cases contradictions will appear in the results, and the individual performing the balancing will have to exercise judgment in selecting corrections that make good mechanical sense.

The following sections in this chapter will address the typical balancing calculations that can be performed. The presented vector balancing equations can be programmed on pocket calculators or personal computers. In fact, operational programs have been available for many years. The use of portable personal computers equipped with spreadsheet programs are ideal for this type of work. It is acknowledged that the calculator programs or computer spreadsheets are only as good as the balancing software. It is always desirable to fully understand the software package, and test it with previously documented balance calculations and/or a mechanical simulation device (e.g. rotor kit), where the integrity and operation of the software can be verified in a noncritical environment.

The final point in any field balance consists of documentation for future reference. In some cases, if a unit requires a field balance, chances are good that periodically this machine may have to be rebalanced. If everything is fully documented, the knowledge gained about the behavior of this particular machine will be useful during the next balance correction. The engineering files should contain all of the technical information, and notes that were generated during the execution of the balancing. This file should be complete enough to allow reconstruction of the entire balancing exercise.

Again, it must be restated that successful field balancing really requires an unbalanced rotor. The mechanical malfunctions listed at the beginning of this chapter will exhibit many symptoms that may be interpreted as unbalance. However, careful examination of the data will often allow a proper identification of the occurring malfunction, and treatment of the actual mechanical problem.

LINEARITY REQUIREMENTS

Traditional balancing calculations generally assume a linear mechanical system. For a system to be considered linear, three basic conditions must be satisfied. First, if a single excitation (i.e., mass unbalance) is applied to a system, a single response (i.e., vibration) can be expected. If the first excitation is removed, and a second excitation applied (i.e., another mass), a second response will result. If both excitations are simultaneously applied, the resultant response will be a superposition of both response functions. Hence, a necessary condition for a system to be considered linear is that the *principle of superposition* applies.

The second requirement for a linear system is that the magnitude or scale factor between the excitation and response is preserved. This characteristic is sometimes referred to as the *property of homogeneity*, and must be satisfied for a system to be linear. The third requirement for a linear system considers the frequency characteristics of dynamic excitations and responses. If the system excitations are periodic functions, then the response characteristics must also be periodic. In addition, the response frequency must be identical to the excitation frequency; and the system *cannot generate new frequencies*.

Most rotating machines behave in a reasonably linear fashion with respect to unbalance. Occasionally, a unit will be encountered that violates one or more of the three described conditions for linearity. When that occurs, the equation array will fail by definition, and a considerably more sophisticated diagnostic and/or analytical approach will be necessary. However, in many instances a direct technique may be used to determine the unbalance in a rotating system.

Case History 36: Complex Rotor Nonlinearities

The machinery discussed in case history 12 will be revisited for this example of nonlinear machinery behavior. Recall that this unit consisted of an overhung hot gas expander wheel, a pair of midspan compressor wheels, and three overhung steam turbine wheels[6] as originally shown in Fig. 5-10. For convenience, this same diagram is duplicated in Fig. 11-9. A series of axial through bolts are used to connect the expander stub shaft through the compressor wheels, and into the turbine stub shaft. In this machine, the rotor must be built concurrently with the inner casing. Specifically, the horizontally split internal bundle is assembled with the compressor wheels, stub shafts, plus bearings, and seals. The end casings are attached, the expander wheel is bolted into position, and the turbine stages are mounted with another set of through bolts.

The eight rotor segments are joined with Curvic® couplings. Although each of the segments are component balanced, any minor shift between elements will produce a synchronous force. Since this unit operates at 18,500 RPM, a slight unbalance or eccentricity will result in excessive shaft vibration. Furthermore, the distribution of operating temperatures noted on Fig. 11-9 reveals the complex thermal effects that must be tolerated by this rotor. The 1,250°F expander inlet is followed by compressor temperatures in excess of 430°F. The steam turbine operates with a 700°F inlet, and a 160°F exhaust.

By any definition, this must be considered as a difficult unit. As discussed in case history 12, the rotor passes through seven resonances between slow roll and normal operating speed. These various damped natural frequencies were summarized in Table 5-4. This rotor normally requires field trim balancing after every overhaul. Previous field balancing activities were successful when a two step correction process was applied. The first step consisted of balancing at a pro-

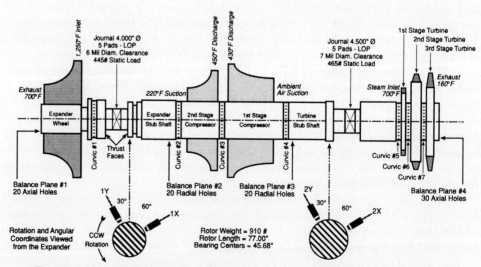

Fig. 11–9 Combined Expander-Air Compressor-Steam Turbine Rotor Configuration

[6] Robert C. Eisenmann, "Some realities of field balancing," *Orbit*, Vol.18, No.2 (June 1997), pp. 12-17.

cess hold point of 14,500 RPM using the outboard balance planes #1 and #4. This was followed by a final trim at 18,500 RPM on the inboard planes #2 and #3 located next to the compressor wheels. It had been repeatedly demonstrated that if the rotor was not adequately balanced at 14,500 RPM, it probably would not run at 18,500 RPM. Hence the plant personnel were committed to performing a field balance at 14,500 as well as 18,500 RPM.

Although the high speed balance at full operating speed was readily achievable, the intermediate speed balance at 14,500 RPM was always difficult. In an effort to improve the understanding of this machinery behavior, the historical balancing records were reviewed, and transient vibration data was examined. In addition, the damped critical speeds plus associated mode shapes were computed as previously discussed in case history 12.

One of the interesting aspects of this machine was the variation in balance sensitivity vectors at 14,500 RPM. As discussed in this chapter, the *balance sensitivity vectors* provide a direct relationship between the rotor mass unbalance vectors and the vibration response vectors. These vectors are determined by installation of known *trial* or *calibration* weights at each of the balance planes, and measuring the resultant shaft vibration response. Suffice it to say, these balance sensitivity vectors must remain reasonably constant in order for the vector balancing calculations to be correct. For this particular rotor, three sets of sensitivity vectors were computed from the available historical information at 14,500 RPM, and the results of these vector calculations are summarized in Table 11-1.

Since this rotor contains two measurement planes, and four balance correction planes, a total of eight balance sensitivity vectors were computed using equation (11-17). The first balance sensitivity vector identified as S_{11} in Table 11-1 defines the vibration response at measurement plane 1, with a calibration weight installed at balance plane 1. Similarly, sensitivity vector S_{12} specifies the vibration response at measurement plane 1, with a weight at balance plane 2,

Table 11–1 Balance Sensitivity Vectors Based On Steady State Data At 14,500 RPM

S Vector	Data Set #1 (Grams/Mil,$_{p-p}$ @ Deg.)	Data Set #2 (Grams/Mil,$_{p-p}$ @ Deg.)	Data Set #3 (Grams/Mil,$_{p-p}$ @ Deg.)
S_{11}	20.3 @ 139°	16.2 @ 179°	*Not Available*
S_{12}	48.1 @ 34°[a]	22.6 @ 309°	76.4 @ 233°
S_{13}	42.6 @ 211°	14.7 @ 177°	24.6 @ 200°
S_{14}	34.1 @ 259°	41.7 @ 289°	16.2 @ 305°
S_{21}	18.3 @ 308°	13.4 @ 345°	*Not Available*
S_{22}	19.1 @ 168°	18.5 @ 142°	18.6 @ 160°
S_{23}	32.3 @ 258°	20.2 @ 269°	31.2 @ 221°
S_{24}	24.2 @ 83°	20.3 @ 147°	14.9 @ 142°

[a]Shaded vectors of questionable accuracy due to small differential vibration vectors with weights.

and so forth throughout the remainder of the tabular summary. The two shaded vectors in Table 11-1 are of questionable accuracy due to the fact that the differential vibration vector was less than 0.1 Mils$_{p-p}$. This small differential vibration is indicative of minimal response to the applied weight, and the validity of the particular balance sensitivity vector is highly questionable. On much larger machines, the validity of the sensitivity vectors would be considered marginal if the differential vibration vectors were less than 0.5 or perhaps 1.0 Mil$_{p-p}$. However, for this small, high speed rotor, a differential shaft vibration value of 0.1 Mils$_{p-p}$ was considered to be an appropriate lower limit.

Examination of the remaining S vectors in Table 11-1 reveals some similarities, but the overall variations are significant. For instance, the magnitude of S_{12} varies from 22.6 to 76.4 Grams per Mil$_{p-p}$, and a 76° angular difference is noted. On S_{24} the amplitudes change from 24.1 to 14.9 Grams per Mil$_{p-p}$, but the angles reveal a 59° spread. At this point, a preliminary conclusion might be reached that this rotor is indeed nonlinear and cannot be field balanced.

Further review showed that the vibration response vectors used for balancing were acquired at a process hold point of 14,500 RPM. Under this condition, the machine speed was held constant, but rotor and casing temperatures were changing as the process stabilized. This could be a major contributor to the spread in sensitivity vectors in Table 11-1. Attempting to balance a machine with these variable coefficients is difficult at best, and many runs are required to attain a barely acceptable balance state.

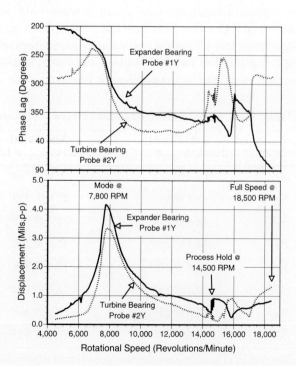

Fig. 11–10 Bode Plot Of Y-Axis Probes During A Typical Machine Startup

Variable speed vibration response vectors were extracted from the historical database, and a typical startup Bode is presented in Fig. 11-10. This data displays the Y-Axis probes from both measurement planes. Both plots are corrected for slow roll runout, and the resultant data is representative of the true dynamic motion of the shaft at each of the two lateral measurement planes. The major resonance occurs at approximately 7,800 RPM, which is consistent with the analytical results discussed in case history 12. It is significant to confirm that the process hold point at 14,500 RPM displays substantial amplitude and phase excursions. This is logically due to the heating of the rotor and casing, plus variations in settle out of the operating system (i.e., pressures, temperatures, flow rates, and molecular weights). Although this process stabilization is a necessary part of the startup, the variations in vibration vectors negates the validity of this information for use as repetitive balance response data.

At speeds above 14,500 RPM, there are additional vector changes, and a desirable plateau in the amplitude and phase curves does not appear. The only consistent area of essentially flat levels occurs in the vicinity of 14,000 RPM. To test the validity of this conclusion, individual vectors at 14,000 RPM were extracted from the historical transient startup files. These displacement vectors, in conjunction with the installed weights, were used to re-compute the balance sensitivity vectors with equation (11-17). The results of these computations are presented in Table 11-2, and are directly comparable to Table 11-1.

By observation and comparison, it is clear that the consistency of S vectors between the three data sets is far superior in the results presented in Table 11-2. This applies to both the magnitude and direction of the computed balance sensitivity vectors. Hence, the repeatability and associated linearity of the balance sensitivity vectors, plus the predictable balance response of the mechanical system was significantly improved by selecting a stable data set for computation of the balance sensitivity vectors.

Table 11–2 Balance Sensitivity Vectors Based On Transient Data At 14,000 RPM

S Vector	Data Set #1 (Grams/Mil,$_{p-p}$ @ Deg.)	Data Set #2 (Grams/Mil,$_{p-p}$ @ Deg.)	Data Set #3 (Grams/Mil,$_{p-p}$ @ Deg.)
S_{11}	24.7 @ 144°	86.2 @ 188°	*Not Available*
S_{12}	125. @ 333°[a]	38.2 @ 228°	42.4 @ 228°
S_{13}	31.8 @ 207°	20.8 @ 184°	22.9 @ 176°
S_{14}	21.8 @ 322°	36.1 @ 321°	30.7 @ 301°
S_{21}	24.9 @ 313°	21.6 @ 311°	*Not Available*
S_{22}	17.3 @ 167°	23.7 @ 175°	18.9 @ 178°
S_{23}	78.7 @ 299°	22.2 @ 221°	27.4 @ 232°
S_{24}	27.6 @ 147°	25.8 @ 166°	28.3 @ 164°

[a]Shaded vectors of questionable accuracy due to small differential vibration vectors with weights.

Single Plane Balance

The simplest form of mass unbalance consists of weight maldistribution in a single plane. This type of unbalance is characterized by an offset of the mass centerline (inertia axis) that is parallel to the geometric centerline of the rotating assembly. This is often called a static unbalance, and in some cases it may be detected by placing a horizontal rotor on knife edges, and allowing gravity to pull the heavy spot down to the bottom of the assembly.

A static unbalance may occur in a thin rotor element such as a turbine disk or a compressor wheel. The mass correction for this condition would probably be very close to the actual plane of the unbalance. A static unbalance may also occur in a long rotor such as a turbine or a generator rotor. In this situation, the correction plane may, or may not, be coincident with the location of the static unbalance. In any balance problem, the following basic questions must be addressed:

○ Lateral Correction Plane To Be Used?
○ Amount of the Balance Weight?
○ Angular Location of the Balance Weight?

The answer to the first question resides in an evaluation of the type of unbalance, combined with the specific rotor configuration, deformed mode shape, and accessible balance planes. The proper weight installed at the wrong plane will not help the rotor. In all cases, the diagnostician must determine the modally effective balance planes, and then narrow the choices down to physically accessible locations. For instance, a generator rotor may display a pure static unbalance that ideally should be corrected at midspan. In reality, the placement of a midspan weight in the generator rotor is not feasible, but the generator unbalance may be corrected by weights installed at the accessible end planes.

The second basic balancing question addresses the amount or size of the balance weight to be used. Ideally, previous information on the specific rotor, or a similar unit, would be available to guide the diagnostician. In field balancing situations where historical data is unavailable, it is customary to install calibration weights that produce centrifugal forces in the vicinity of 5% to 15% of the rotor weight. Machines such as motors or expanders that rapidly accelerate up to speed are candidates for initial weights that produce centrifugal forces in the vicinity of 5% of the rotor weight. A more aggressive approach is often applied towards machines such as steam turbines that may be started up slowly, and the unit tripped if the weight is incorrect. For these types of machines, the installation of an initial weight that produces a centrifugal force equal to 15% of the rotor weight has proven to be an effective starting point.

The third major balancing question of angular location of the weight is often the most difficult to address. Some individuals take the approach of installing the initial balance weight at any angular location, and then computing the vector influence. It must be recognized that this is an ***extremely dangerous practice that can result in serious mechanical damage.*** In virtually all cases, the weight should be installed to reduce the residual unbalance, and lower

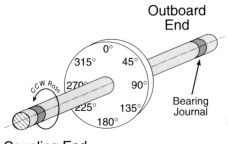

Fig. 11–11 Single Mass
Rotor Kit

the associated vibration amplitudes.

The correct angular location of the initial weight is again dependent upon the type of unbalance, the specific rotor configuration, the deformed mode shape, and the accessible balance planes. For a single plane balance and a simple mode shape, the process is considerably simplified. For demonstration purposes, consider Fig. 11-11 of a one mass rotor kit. The mass is supported between bearings, and proximity probes are mounted inboard of the bearings at both ends of the rotor. Since this is only a single mass system, the dominant viable shaft mode shape is a pure translational mode. At the critical speed of approximately 5,000 RPM the physical orientation of elements, and the maximum shaft deflection is depicted in Fig. 11-12 of an undamped mode shape diagram.

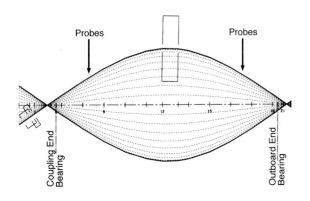

Fig. 11–12 Mode Shape
Of Single Mass Rotor Kit
At Translational First Criti-
cal Speed of 5,000 RPM

It is clear that the maximum deflection occurs at midspan, and that the center mass is in a modally sensitive location. It is also apparent that the proximity probe locations will yield information that is representative of the synchronous 1X response. Therefore, a knowledge of the phase characteristics should provide the information necessary for a logical angular weight placement at the midspan mass.

One of the easiest ways to determine the proper location for a balance weight was proposed by Charles Jackson[7] in his article entitled "Balance Rotors by Orbit Analysis." Quoting directly from this paper, Jackson states that:

"...the orbit represents a graphical picture of the shaft motion pattern. The key-phase mark represents where the shaft is at the very instant the notch passes the probe...Below the first critical, the mark on the orbit represents the location of heavy spot of the shaft relative to that bearing. This point is difficult to see, yet simple once it is understood; i.e., the shaft must be wherever it is because of either external forces or mass imbalance. Limiting this discussion to imbalance, the shaft is displaced by imbalance. The mark shows where the shaft is at that precise instant when the notch passes the probe.

Therefore, if one would stop the machine and turn the shaft until the notch lines up with the probe, the angular position of the shaft is satisfied. Then, laying off the angle from the pattern taken on the CRT gives the heavy spot for correction. Weight can either be subtracted at this point or added at a point 180 degrees diametrically opposite, on the shaft...The orbit diameter will reduce as correction is applied. Should too much weight change be given, the mark will shift across the orbit, indicating the weight added is now the greatest imbalance.

Above the first critical, the rules change. The key-phase mark would have shifted approximately 180 degrees when the shaft mode of motion changed...Therefore, the phase mark will appear opposite the actual heavy spot and the weight addition would be on the key-phase mark position..."

These statements are consistent with the previous discussion presented earlier in this chapter. It is appropriate to apply these techniques to the single mass rotor shown in Fig. 11-11. This rotor kit was run at speeds below and above the first critical. The top portion of Fig. 11-13 displays the 1X filtered orbit and

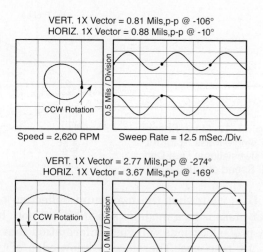

VERT. 1X Vector = 0.81 Mils,p-p @ -106°
HORIZ. 1X Vector = 0.88 Mils,p-p @ -10°

CCW Rotation

0.5 Mils / Division

Speed = 2,620 RPM   Sweep Rate = 12.5 mSec./Div.

VERT. 1X Vector = 2.77 Mils,p-p @ -274°
HORIZ. 1X Vector = 3.67 Mils,p-p @ -169°

Fig. 11–13 Orbit And Time Base Plots For An Unbalanced Single Mass Rotor Kit Running At 2,620 RPM Which Is Below The Critical, And At 8,060 RPM Which Is Above The Rotor First Critical Speed

CCW Rotation

1.0 Mil / Division

Speed = 8,060 RPM   Sweep Rate = 3.125 mSec./Div.

---

[7] Charles Jackson, "Balance Rotors by Orbit Analysis," *Hydrocarbon Processing*, Vol. 50, No. 1 (January 1971).

time domain plots that were extracted from the outboard X-Y probes at 2,620 RPM. This operating speed is below the 5,000 RPM balance resonance (critical speed). The bottom set of 1X filtered orbit and time domain plots in Fig. 11-13 were obtained from the outboard X-Y probes at 8,060 RPM, which is well above the translational critical speed.

Both sets of data reveal forward and reasonably circular orbits at the outboard end of the rotor kit. This data is vectorially corrected for shaft runout, and the presented plots are representative of the true dynamic motion of this single mass rotor. Similar behavior was observed by the coupling end X-Y probes. The coupling end data was not included, since it would be redundant to the outboard plots. However, during an actual field balance, vibration data would always be obtained at both ends of the machine.

For purposes of completeness, both the low and high speed shaft orbits will be evaluated for proper balance weight angular location. The diagram shown in Fig. 11-14 describes the low speed shaft orbit at 2,620 RPM. The vertical probe phase angle documented in Fig. 11-13 was 106°. Thus, moving 106° in a counter rotation direction (i.e., clockwise) from the vertical probe locates the high spot. This high spot is coincident with the heavy spot in this simple example, and the angular location is identical to the Keyphasor® trigger point.

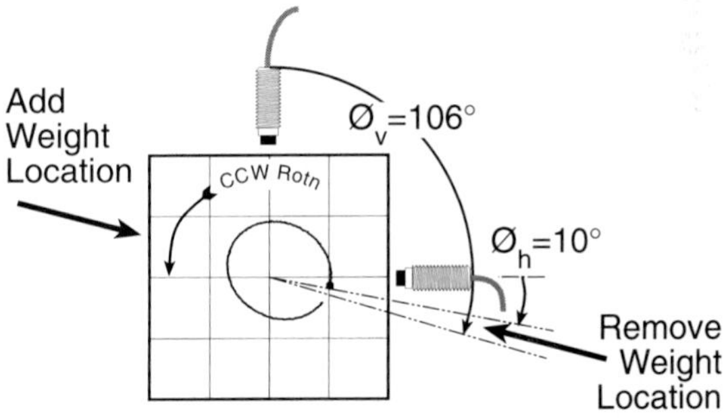

Fig. 11–14 Shaft Orbit And Probe Locations For Rotor Kit Running At 2,620 RPM Which Is Below The Rotor First Critical Speed

The horizontal probe phase angle shown in Fig. 11-13 was 10°. Rotating 10° in a clockwise direction from the horizontal probe in Fig. 11-14 locates the same high spot (i.e., Keyphasor® trigger). Thus, both probes have identified essentially the same angular location, and this point is the high spot, and the KeyØ® trigger point. Since this information was obtained below the first critical speed (translational resonance); the identified point would logically be coincident with the residual heavy spot on the disk. Thus, weight should be removed at the Keyphasor® dot (heavy spot) at nominally 3:30 o'clock angular position. Alternately, weight could be added to the 9:30 o'clock position to correct for the unbalance at the 3:30 o'clock position.

In passing, it should be mentioned that the difference between the angular location identified by the vertical and horizontal phase angles is not exactly the same. In this example, a 6° difference is noted between the two angular locations. This is quite common behavior due to the fact that the orbit is not perfectly circular. In chapter 7, it was shown that a perfectly circular orbit would appear only if the amplitudes in the orthogonal directions were equal, and the phase varied by 90° between the two probes. In the example shown in Fig. 11-14, and in most field balancing situations, the orbit is somewhat elliptical, and the measured phase angles differ from the pure 90° value. Hence, it makes sense to make a weight correction between the two positions, and attempt to satisfy the vertical as well as the horizontal vibration response.

Above the critical speed, the phase should increase by approximately 180°, and the Keyphasor® dot should shift to the other side of the orbit. In fact, this anticipated behavior was displayed by the bottom set of orbit and time base plots presented on Fig. 11-13. This data acquired at 8,060 RPM, which is considerably above the 5,000 RPM critical speed. Extracting the orbit from the high speed data set at the bottom of Fig. 11-13, and including the measured phase angles, the diagram in Fig. 11-15 was generated.

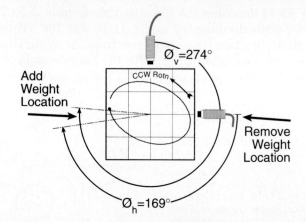

Fig. 11–15 Shaft Orbit And Probe Locations For Rotor Kit Running At 8,060 RPM Which Is Above The Rotor First Critical Speed

The high speed vertical probe phase angle was 274°. Moving 274° in a counter rotation direction (i.e., clockwise) from the vertical probe locates the high spot (coincident with the KeyØ® trigger). Similarly, the horizontal probe phase angle shown in Fig. 11-13 was 169°. Moving 169° in a clockwise direction from the horizontal probe locates basically the same high spot. Thus, both probes have identified essentially the same angular location above the resonance. Since this information was obtained above the first critical speed, the identified point would be opposite to the residual heavy spot. Thus, weight should be added at the Keyphasor® dot around the 9 o'clock position. Alternately, weight could be subtracted at the 3 o'clock position.

In retrospect, the data above the critical speed (Fig. 11-14) is identifying the same general angular location as the data obtained below the critical (Fig. 11-13). The high speed orbit indicates a heavy spot at 3 o'clock, and the low speed orbit reveals a heavy spot that is somewhat lower at 3:30 o'clock. The two values would be identical if both orbits were perfectly round (circular), and a precise 180° phase change occurred through the critical speed range. However, these two ideal conditions seldom occur on real machines, and the discussed data is representative of typical machinery behavior.

In many respects, field balancing consists of a series of compromises. In this case, the average (low to high speed) weight removal should be in the vicinity of 100° clockwise from the vertical probe. However, the available balance holes in this portion of the disk were empty, and there was no opportunity to easily remove additional weight. The next compromise would be to add weight at the light spot at approximately 280° (i.e., 100° plus 180°). It was noted that weights already filled the balance hole at 270°, and the only remaining empty hole was at 292°. A total of 0.5 Grams was installed at this 292° hole, and the resultant response due to this single plane weight addition is presented in Fig. 11-16. As before, the top set of orbit and time base plots at 2,620 RPM depict the shaft vibration below the critical speed. The bottom set of plots in Fig. 11-16 were acquired at 8,060 RPM, which is above the rotor balance resonance frequency.

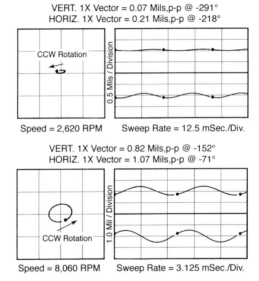

VERT. 1X Vector = 0.07 Mils,p-p @ -291°
HORIZ. 1X Vector = 0.21 Mils,p-p @ -218°

CCW Rotation

0.5 Mils / Division

Speed = 2,620 RPM   Sweep Rate = 12.5 mSec./Div.

VERT. 1X Vector = 0.82 Mils,p-p @ -152°
HORIZ. 1X Vector = 1.07 Mils,p-p @ -71°

Fig. 11–16 Orbit And Time Base Plots For A Balanced Single Mass Rotor Kit Running At 2,620 RPM Which Is Below The Critical, And At 8,060 RPM Which Is Above The Rotor First Critical Speed

CCW Rotation

1.0 Mil / Division

Speed = 8,060 RPM   Sweep Rate = 3.125 mSec./Div.

It is clear that the 0.5 Gram weight at 292° significantly reduced the synchronous 1X unbalance response. It should also be clear that the position of the Keyphasor® dot on the orbit is representative of the high spot. This concept is fundamental to balancing, as well as the analysis and understanding of the behavior of any rotating system.

These concepts might seem to be somewhat different from the automated instrumentation installed on most low speed shop balancing machines. In actuality, the concept is the same, but there other significant differences. For instance, during most shop balancing work, it is inexpensive to make a run, and there is little physical risk to the machinery or the operator. In the case of field balancing, it is often difficult to change weights, and it is generally expensive to make a full speed run. Furthermore, if an incorrect weight is used in a field balance, the results may be hazardous to the machinery, and the health of the operator.

Each field balance shot should be a meaningful move, and it should contribute to the overall database describing the behavior of the machine. Field balancing generally requires the quantification of the basic relationship between the shaft response and the applied force as commonly expressed by:

$$Response = \frac{Force}{Restraint}$$

This general expression has been stated several times in this text due to the fact that it has many specific applications in rotor dynamics. Within the balancing discipline, response is the measured shaft or casing vibration vector. The applied force is represented by the unbalance vector, and the restraint may be thought of as a stiffness vector. In balancing applications, this variable may be considered as a spring-type parameter of a specific unbalance producing a specific deflection or rotor vibration. Another way to view this restraint term is to consider it as the sensitivity of the machine to rotor unbalance. If these balancing terms are substituted for the equivalent values in the previous expression, the following equation (11-1) evolves:

$$Vibration = \frac{Unbalance}{Sensitivity} \qquad \textbf{(11-1)}$$

All variables in (11-1) are vector quantities. Each parameter carries both a magnitude and a direction. If the initial rotor vibration is described by the A vector with amplitude in Mils,$_{p\text{-}p}$, and the unbalance is defined by the U vector with units of Grams, then the balance sensitivity S vector must carry units of Grams per Mil,$_{p\text{-}p}$. Using these designations, equation (11-1) may be rewritten as:

$$\boxed{\vec{A} = \frac{\vec{U}}{\vec{S}}} \qquad \textbf{(11-2)}$$

where: $\vec{A}$ = Initial Vibration Vector (Mils,$_{p\text{-}p}$ at Degrees)

$\vec{U}$ = Mass Unbalance Vector (Grams at Degrees)

$\vec{S}$ = Sensitivity Vector to Unbalance (Grams/Mil,$_{p\text{-}p}$ at Degrees)

This expression may be easily remembered as the $\vec{U} = \vec{S} \times \vec{A}$ equation. In either format, the vibration vector may be measured directly, and the technical problem resolves to one of determining the mass unbalance vector based upon some unknown sensitivity. This sensitivity vector may be experimentally determined by adding a known calibration weight at a known angular location to the rotor, and measuring the vibration response vector. Assume that the calibration weight vector is defined by W, and the resultant rotor vibration is identified as the B vector. If the machine is re-run at the same speed and operating condition, and if the system exhibits linear behavior, then equation (11-2) may be expanded into the following expression:

$$\vec{B} = \frac{\vec{U} + \vec{W}}{\vec{S}} \qquad \qquad \textbf{(11-3)}$$

where: $\vec{B}$ = Vibration Vector with Calibration Weight (Mils,$_{p-p}$ at Degrees)

$\vec{W}$ = Calibration Weight Vector (Grams at Degrees)

Expansion of equation (11-3), and substituting (11-2) yields the following:

$$\vec{B} = \frac{\vec{U} + \vec{W}}{\vec{S}} = \frac{\vec{U}}{\vec{S}} + \frac{\vec{W}}{\vec{S}} = \vec{A} + \frac{\vec{W}}{\vec{S}}$$

or

$$\vec{B} - \vec{A} = \frac{\vec{W}}{\vec{S}}$$

From this expression, the balance sensitivity vector may be computed as:

$$\vec{S} = \left\{ \frac{\vec{W}}{\vec{B} - \vec{A}} \right\} \qquad \qquad \textbf{(11-4)}$$

The unbalance may now be determined from equation (11-2). It should be noted that the measured vibration vector must represent actual dynamic motion of the rotor. Hence, the vector must be corrected for any electrical and/or mechanical runout. This is achieved by vector subtraction of the slow roll from the vibration vector measured at balancing speed. The following equation (11-5) is applicable for all measurements that require a slow roll or runout correction or compensation (e.g., proximity probes). Note that other transducers (e.g., casing accelerometers) do not require this type of vector correction.

$$\vec{A_c} = \vec{A} - \vec{E} \qquad \qquad \textbf{(11-5)}$$

where: $\vec{A_c}$ = Runout Compensated Initial Vibration Vector (Mils,$_{p-p}$ at Degrees)

$\vec{E}$ = Slow Roll Runout Vector (Mils,$_{p-p}$ at Degrees)

Thus, the proper expression for calculation of the mass unbalance is now easily derived from equations (11-2) and (11-5) as follows:

$$\vec{U} = \vec{S} \times \vec{A_c} \qquad \qquad \textbf{(11-6)}$$

As previously noted, the vibration vector amplitudes are measured in Mils,$_{p-p}$, the weight units may be expressed in Grams or Gram-Inches, and the balance sensitivity vectors would carry the units of Mils/Gram or Mils/Gram-Inch respectively. In all cases, the angular orientation is against rotation (i.e., phase lag), from each respective vibration transducer. In addition, the trigger point is established by the coincidence of the physical shaft trigger location, and the center of the Keyphasor® timing probe.

It should be mentioned that the plane of unbalance, the correction plane, and the measurement plane are not defined as coincidental. They may be, and usually are, separate planes in a machine assembly. It is important to recognize that the previous equations are directed towards achieving a minimum value of the vibration vector. That is, the calculations will yield a balance weight, located at the correction plane, that is sized to minimize the vibration response at the measurement plane to the actual mass unbalance distributed in the rotor. In the majority of cases, this is both acceptable and agreeable. However, it is always good practice to consider the shaft mode shape, and verify that the correction weight does not aggravate deflections at other points along the rotor, while reducing vibration at the measurement plane.

This same equation array is applicable to calculations performed both above and below a critical speed. In all cases, the balance computation solves for a zero response amplitude. A similar set of expressions may be developed for multiple measurement and correction planes. It should be restated that the equation set used for the calculations presumes a linear response of the mechanical system to mass unbalance. The presence of significant shaft preloads (due to misalignment, gear contact forces, etc.), thermal effects, fluidic forces, bearing instabilities, and various other mechanisms may render these calculations ineffective. However, for many conditions of rotational mass unbalance, the mechanical system will behave in a reasonably linear fashion.

In many instances, a single plane calculation is not totally adequate, and cross-coupling between two or more correction planes must be considered. This is achieved by expansion of the previously discussed equation set to multiple planes. However, before addressing any additional complexity of the equation structure, it would be advisable to examine the field application of these single plane calculations to the following case history of a forced draft.

Case History 37: Forced Draft Fan Field Balance

A direct application of single plane balancing occurs on simple rotor systems that contain essentially one plane of unbalance. For example, consider the forced draft fan rotor described in Fig. 11-17. This is a fully symmetrical rotor that is nominally 177 inches long, and weights approximately 7,100 pounds. This forced draft fan is driven by a steam turbine through a speed decreasing gear box. Normal speed varies between 1,470 and 1,540 RPM. Unfortunately, the fan translational resonance (critical speed) exists at 1,500 RPM, and the resonance bandwidth is approximately 400 RPM. Thus, the first critical resonance persists between speeds of 1,300 RPM [=1,500-(400/2)] and 1,700 RPM [=1,500+(400/2)]. It is apparent that under normal operating conditions, the fan runs within the bandwidth of the first critical.

The situation is further complicated by the fact that the machine is installed in an undesirable environment that allows ingestion of coke, plus other

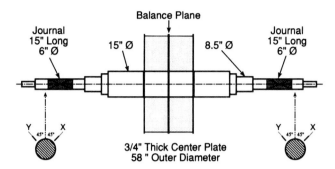

Total Weight = 7,100 Pounds — Overall Length = 177 Inches

Fig. 11-17 Rotor Configuration For Steam Boiler Forced Draft Fan

foreign objects into the fan. Hence, the fan blades are always under attack, and mechanical damage to the blades results in changes to the balance characteristics at the middle of the rotor. Since the fan runs within the translational critical speed domain, and since balance changes due to physical damage continually occur at the midspan, this unit is quite susceptible to rapid changes in vibration. When these undesirable events occur, the only reasonable solution is to correct the change in mass unbalance by field balancing.

The fan is equipped with X-Y proximity probes at each bearing as shown in Fig. 11-17. The probes are installed at ±45° from true vertical, and a Keyphasor® is mounted at the outboard stub end of the fan shaft. It should be mentioned that the bearings are supported on tall pedestals attached to a flexible baseplate. The entire support structure is quite soft, and considerable casing motion occurs. To include this information into the balance calculations, a casing vibration probe is installed on each fan bearing housing. These casing probes are mounted in line with the X-Axis shaft sensing proximity probes. This orientation allows the summation (electronic or by calculation) of the relative shaft signal with the absolute casing vibration signal to obtain absolute shaft vibration.

The fan in question experienced substantial damage during one particularly violent ingestion of a large icicle. Two blades were damaged, and one bearing had a babbitt breakdown. These mechanical problems were corrected, and the unit was restarted. Understandably, the blade repairs were responsible for a major mass unbalance condition, and the startup was terminated at 900 RPM with shaft vibration amplitudes in excess of 7.0 Mils$_{,p-p}$.

Based on previous experience, an 1,190 gram (42 ounces) correction weight was welded to the center plate at an angle of 28°. The machine was restarted, and the correction proved to be effective. This time a desirable balancing speed of 1,650 RPM was achieved. It is usually unwise to try and acquire balance response vibration data close to a critical speed. This is due to the fact that small changes in speed will result in significant changes in the 1X vectors. Hence, when a machine runs in the vicinity of a resonance, it is usually good practice to try and obtain the balance data in the plateau region above the resonance.

In this case, the maximum attainable fan speed was 1,650 RPM, and this speed was used for the remaining balance runs. Response data at both bearings was similar, and for purposes of brevity, only the outboard bearing will be discussed. In this case, the runout compensated X-Y proximity probe data, and the horizontal casing motion will be used for the balancing calculations. The Y-Axis proximity probe was used as the zero degree reference for all of the weights. The initial vibration response vectors at the outboard bearing at 1,650 RPM are summarized in the middle column of Table 11-3. The addition of a 567 gram calibration weight at the fan center plate at an angle of 40° produced the vibration vectors in the right hand column of Table 11-3.

Table 11-3 Forced Draft Fan - Initial Plus Calibration Weight Vibration Vectors

Measurement Location	Initial Vibration (A Vector)	Vibration With Calibration Weight Installed (B Vector)
Outboard Shaft Y-Axis	5.60 Mils,$_{p-p}$ @ 322°	7.54 Mils,$_{p-p}$ @ 226°
Outboard Shaft X-Axis	6.08 Mils,$_{p-p}$ @ 163°	9.82 Mils,$_{p-p}$ @ 81°
Outboard Casing X-Axis	3.85 Mils,$_{p-p}$ @ 144°	5.79 Mils,$_{p-p}$ @ 48°

Unfortunately, vibration amplitudes have increased at all locations, and it is clear that the calibration weight was placed at the wrong angle. It was mistakenly assumed that the calibration weight should go in the same angular location as the 1,190 gram correction weight that was added to allow full speed operation. In actuality, the 567 gram calibration weight should have been placed in the vicinity of 320° instead of the 40° position. The measured amplitudes were high, but they were within the tolerable range for a short duration run. A complete set of information is now available to perform a single plane balance calculation. The first step is to determine the balance sensitivity vectors from equation (11-4). The calculations for the Y-Axis shaft proximity probe are shown as follows:

$$\vec{S} = \left\{ \frac{\vec{W}}{\vec{B} - \vec{A}} \right\} = \left\{ \frac{567 \text{ Grams} \angle 40°}{7.54 \text{ Mils} \angle 226° - 5.60 \text{ Mils} \angle 322°} \right\} = \left\{ \frac{567 \text{ Grams} \angle 40°}{9.85 \text{ Mils} \angle 192°} \right\}$$

$$\vec{S} = 57.56 \text{ Grams/Mil,p-p} \angle 208°$$

This value for the balance sensitivity vector may now be combined with the initial vibration vector to determine the mass unbalance using equation (11-6):

$$\vec{U} = \vec{S} \times \vec{A}_c = \{57.56 \text{ Grams/Mil,p-p} \angle 208°\} \times \{5.60 \text{ Mil,p-p} \angle 322°\} = 322 \text{ Grams} \angle 170°$$

The vector multiplication was performed in accordance with the rules established by equation (2-39). Performing the same calculations for the X-Axis shaft probe and the casing velocity sensor yields Table 11-4. This table summarizes the calculated mass unbalance for this fan as a function of each of the individual vibration measurement probes.

Table 11–4 Forced Draft Fan - Summary Of Calculated Weight Corrections

Measurement Location	Unbalance Weight	Weight To Add
Outboard Shaft Y-Axis	322 Grams @ 170°	322 Grams @ 350°
Outboard Shaft X-Axis	319 Grams @ 156°	319 Grams @ 336°
Outboard Casing X-Axis	300 Grams @ 168°	300 Grams @ 348°

The center column of Table 11-4 identifies the magnitude of the unbalance in grams, plus the location of the unbalance in degrees. If this rotor was to be balanced by weight removal, then approximately 320 grams would be removed at nominally 170° counter rotation from the vertical probe. On the other hand, if balancing will be accomplished by weight addition, then the angular location must be modified by 180° to determine the weight add vectors as shown in the right hand column of Table 11-4. Thus, a weight of approximately 320 grams should be added at an angle of nominally 350° counter rotation from the vertical probe. In actuality, a 342 gram weight was welded onto the center plate at about 350°. This initial correction provided a significant improvement to the 1X synchronous vibration response of this fan. Since additional time was available, a small 30 gram trim correction was installed at 330°. After this trim, the initial and the final response vectors at 1,650 RPM are summarized in Table 11-5.

Table 11–5 Forced Draft Fan - Initial Versus Final Vibration Vectors

Measurement Location	Initial Vibration Vectors	Final Vibration Vectors
Outboard Shaft Y-Axis	5.60 Mils,$_{p\text{-}p}$ @ 322°	0.46 Mils,$_{p\text{-}p}$ @ 104°
Outboard Shaft X-Axis	6.08 Mils,$_{p\text{-}p}$ @ 163°	0.68 Mils,$_{p\text{-}p}$ @ 60°
Outboard Casing X-Axis	3.85 Mils,$_{p\text{-}p}$ @ 144°	0.23 Mils,$_{p\text{-}p}$ @ 246°

The balance corrections identified by these vector calculations were quite close together as demonstrated in Table 11-4. This type of calculated weight distribution makes it easy to select an appropriate and effective correction. Unfortunately, many machines do not produce such a consistent array of required weight corrections. For these units, the diagnostician must carefully examine the vibration response data, and either discount or completely ignore any questionable data. In many cases, field balancing has no unique solution. It is really a compromise of balance weight selection based upon the calculations, plus a further compromise as to what can really be mounted on the machine.

As a final comment, it should be mentioned that the weights, and the vector angles are all referenced to the vertical probe as 0° in accordance with the previous discussions. The horizontal probe calculations are also referenced to this same location, and correct results are obtained. This is due to the self-canceling nature of the calibration and correction weights within the final equation. The validity of this self canceling relationship is demonstrated in case history 38.

TWO PLANE BALANCE

The next level of unbalance complexity consists of weight maldistribution in two separate geometric planes. The simplest form of this type of unbalance is generally referred to as a couple unbalance. A pure couple is characterized by a mass centerline (principal inertia axis) that intersects the geometric centerline of the rotating assembly at the rotor center of gravity. This type of unbalance produces a rocking or pivotal motion in the rotor assembly. Phase angles are separated by 180° across the rotor, and this type of excitation is most detrimental to rotor response through the pivotal balance resonance.

In most process machinery, a pure couple seldom exists by itself. The couple unbalance is usually combined with some type of static unbalance that may, or may not, occur at the same planes as the couple. This weight distribution further complicates the balancing problem, and corrections at more than two planes may be necessary. The term *Quasi-Static* is often used to describe an unbalance condition where the mass centerline (principal inertia axis) intersects the geometric centerline of the rotating assembly away from the rotor center of gravity. This type of unbalance distribution may be sufficient to excite multiple critical speeds, and the machinery diagnostician may experience difficulty in correcting a quasi-static unbalance. The problems associated with this type of unbalance range from recognition of the behavior, to modal separation, and determination of proper angular correction locations. For demonstration purposes, consider Fig. 11-18 illustrating of a two mass rotor kit.

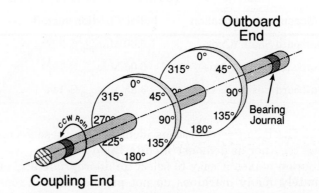

Fig. 11–18 Two Mass Rotor Kit

This is the same device used for discussion of the single plane unbalance, with the addition of a second disk. Again, X-Y proximity probes are mounted inboard of the bearings, and their location allowed detection of synchronous motion. The two masses were positioned at the quarter points of the rotor to enhance the pivotal mode response. The undamped mode shape at the translational first critical speed is depicted in Fig. 11-19 at a speed of 4,800 RPM. This mode shape is virtually identical to the single plane mode shown in Fig. 11-12. In both cases, the maximum deflection occurs at the midspan of the rotor, and modally effective balance weight corrections should be made at the middle of each

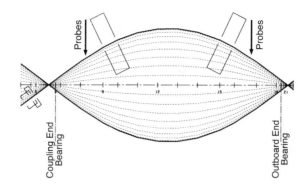

Fig. 11–19 Mode Shape
For Two Mass Rotor Kit At
Translational First Critical
Speed Of 4,800 RPM

rotor for this first translational mode.

The calculated undamped mode shape at the pivotal or second critical speed is displayed in Fig. 11-20. This mode occurs at a measured speed of 8,000 RPM, and is characterized by the midspan nodal point, plus the out-of-phase behavior across the nodal point. From these two mode shape diagrams, it is clear that a static weight correction (same angular location) at both planes would influence the first mode, but would be essentially self-canceling at the second critical. Furthermore, a pure couple correction between the planes would be ineffective at the first mode, but it would have a significant influence upon the pivotal critical.

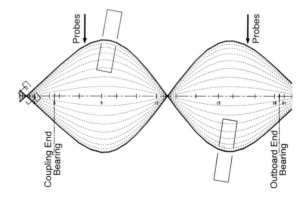

Fig. 11–20 Mode Shape
For Two Mass Rotor Kit At
Pivotal Second Critical
Speed Of 8,000 RPM

The actual transient speed behavior of this two mass rotor kit is depicted in the Bode plot presented in Fig. 11-21. In this data array of 1X amplitude and phase versus rotational speed, the response from the vertical probes at both ends of the machine are documented. It is self-evident that a resonant response is detected by both vertical proximity probes at 4,800 RPM, and 8,000 RPM. The casual observer might identify these resonances as the first and second criticals, and assume that they are the translational and pivotal modes for the rotor. In this case, this assumption is totally correct, but the Bode plot does not directly offer this information. The runout corrected 1X vectors in Fig. 11-21 display the

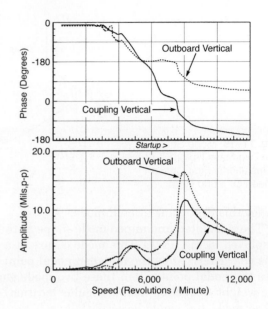

Fig. 11–21 Bode Plot Of
Two Mass Rotor Kit Before
Balancing

same phase through the 4,800 RPM mode, and it is reasonable to conclude that a translational mode is in progress. By the same token, the vertical probe phase angles diverge by 180° through the 8,000 RPM resonance, and a pivotal mode may be deduced. On a more complex machine, reaching these conclusions from a single Bode plot would become much more difficult.

Another approach towards examination of this transient speed vector data was presented by Donald E. Bently [8] in his paper entitled "Polar Plotting Applications for Rotating Machinery." Within this paper, Bently addressed the benefits associated with a polar coordinate presentation of the variable speed vectors. These obvious advantages included assistance in modal separation, plus improved visibility during balancing. Replotting the transient vectors from the Bode plot of Fig. 11-21 into a polar coordinate format produces the plots shown in Figs. 11-22 (coupling end) and 11-23 (outboard end). From these two plots it is clear that both ends of the rotor are moving together, or translating together, as the rotor passes through the first balance resonance (first critical) at 4,800 RPM.

As rotational speed increases, the ends of the rotor begin to move in opposite directions as the unit enters, and then passes through the resonance at 8,000 RPM. From this runout compensated data display it is apparent that the rotor is pivoting through this region, and the conclusion of a pivotal balance resonance is based upon factual evidence. These polar plots also display vector directions for weight additions to correct the pivotal response at 8,000 RPM. These vectors are indicated by thick lines, and they were determined by evaluating the rotor response specifically associated with the pivotal resonance. In most

[8] Donald E. Bently, "Polar Plotting Applications for Rotating Machinery," *Proceedings of the Vibration Institute Machinery Vibrations IV Seminar*, Cherry Hill, New Jersey (November 1980).

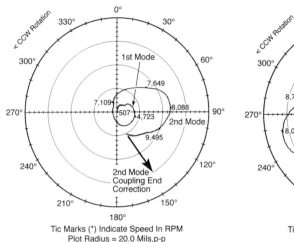

Tic Marks (*) Indicate Speed In RPM
Plot Radius = 20.0 Mils,p-p

Fig. 11–22 Coupling End Polar Plot Of
Two Mass Rotor Kit Before Balancing

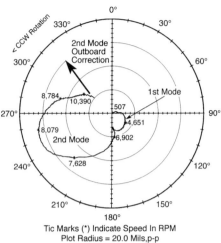

Tic Marks (*) Indicate Speed In RPM
Plot Radius = 20.0 Mils,p-p

Fig. 11–23 Outboard End Polar Plot Of
Two Mass Rotor Kit Before Balancing

cases, the resonances should be viewed individually, and balance corrections or
other analysis performed on each individual resonance loop. The *weight add*
location is determined by drawing a vector from the start of the resonance loop,
to the conclusion of the resonance loop. The direction of this vector will identify
the weight location required to correct the particular mode under examination.
For example, the coupling end polar plot in Fig. 11-22 displays a desired directional correction in the vicinity of 147° (start to end of 2nd mode loop). Simultaneously, the outboard end polar plot in Fig. 11-23 requires a directional
correction in the vicinity of 320° (start to end of 2nd mode loop). The 173° difference between these two weight add locations is indicative of a pivotal resonance,
and the presence of a couple unbalance.

The first mode response will not be corrected in this example, and the balancing effort will be directed to a pure couple correction of the two mass rotor.
Based upon the available balancing holes in both masses, it was determined that
a couple weight set could be easily installed. At the inboard coupling end of the
rotor, a 0.5 gram weight was installed at 135° (147° desired from polar plot). At
the outboard mass, another 0.5 gram weight was also installed. At this location,
the weight was placed at the 315° hole (320° desired from polar plot). At both
ends of the rotor, the weights were installed as close as possible to the desired
angular locations, and a pure 180° couple was maintained between the same
magnitude weights installed at the same radius.

The validity of this approach is demonstrated in Figs. 11-24 and 11-25 that
display the polar plots following this couple balance correction. The initial plots
required a 20 Mil,p-p plot radius to contain the high response through the second
mode. However, following the couple correction, the vibration amplitudes were
significantly reduced. To obtain maximum visibility of the post balance data, the

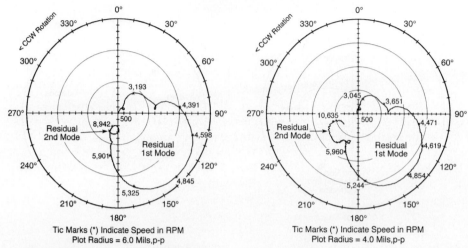

Fig. 11-24 Coupling End Polar Plot Of Two Mass Rotor Kit After Balancing

Fig. 11-25 Outboard End Polar Plot Of Two Mass Rotor Kit After Balancing

coupling end plot in Fig. 11-24 was placed on a 6.0 Mil$_{p-p}$ radius, and the outboard end data in Fig. 11-25 was presented on a 4.0 Mil$_{p-p}$ radius. It should be noted that the vibration response through the second critical has been significantly reduced (almost eliminated), and the motion through the translational first critical has been virtually unaffected by the couple balance shot. This is reasonable and expected behavior based upon the mode shape plots previously discussed for this two mass machine.

The Bode plot of this balanced condition is shown in Fig. 11-26. Again, note the similar behavior through the first mode, and the virtual elimination of the

Fig. 11-26 Bode Plot Of Two Mass Rotor Kit After Balancing 2nd Mode

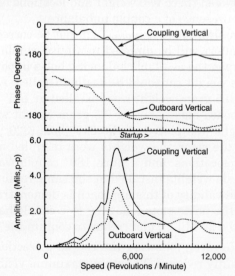

pivotal second mode response. If this Bode plot was viewed on a larger amplitude and phase scale, the uninitiated might not recognize the existence of a pivotal mode on this machine.

Polar plotting is extraordinarily useful due to the direct relationship between the plots and the physical mechanical system. In other sections of this text it is always recommended that the diagnostician maintain a clear and consistent angular relationship between all elements. Thus, all calibration and correction weights, all vibration vectors, all hole locations, and all polar plots should be referenced to zero degrees (0°) at the probe. In all cases, the angular coordinate systems are laid off from the zero degree reference at the probe in a counter rotation direction. Thus, the clockwise rotating system will have angles that progress in a counterclockwise direction. The counterclockwise rotating system will display angles that progress and increase in a clockwise direction. Failure to maintain and enforce this simple angular convention can drive the simplest balance correction into the realm of the unattainable.

Polar plots are also useful for examining closely spaced resonances. This includes split vertical and horizontal criticals, as well as closely coupled but distinctively different modes. Without polar plots the discrimination and relationship between many of these resonant responses may be easily misinterpreted. In addition, secondary resonances such as structural or acoustic resonances may be identified and properly separated from the major rotor balance resonances. Overall, polar plots should be used for improved field balancing, and for final documentation of the results. On very simple machines this may not be necessary; but for complex machinery, the use of polar plots is considered to be mandatory.

In many cases, the balancing problem will be examined in terms of specific vector quantities. Although a complex matrix solution could be constructed for any number of planes, the reality of the field situation typically restricts the allowable corrections to two independent planes. To simplify discussion of the equation set, the measurements will be referenced to bearings 1 and 2 respectively; and weight corrections will be referenced to balance correction planes 1 and 2. Using this nomenclature, the response at each measurement plane is equal to the vector summation of the unbalance response at each balancing plane. This may be expressed by the following traditional two plane vector equations for the initial unbalance response of a linear mechanical system:

$$\overrightarrow{A}_1 = \left\{ \frac{\overrightarrow{U}_1}{\overrightarrow{S}_{11}} \right\} + \left\{ \frac{\overrightarrow{U}_2}{\overrightarrow{S}_{12}} \right\} \qquad \textbf{(11-7)}$$

$$\overrightarrow{A}_2 = \left\{ \frac{\overrightarrow{U}_1}{\overrightarrow{S}_{21}} \right\} + \left\{ \frac{\overrightarrow{U}_2}{\overrightarrow{S}_{22}} \right\} \qquad \textbf{(11-8)}$$

where: $\overrightarrow{A}_1$ = Initial Vibration Vector at Bearing 1 (Mils,$_{\text{p-p}}$ at Degrees)

$\overrightarrow{A}_2$ = Initial Vibration Vector at Bearing 2 (Mils,$_{\text{p-p}}$ at Degrees)

$\overrightarrow{S}_{11}$ = Sensitivity Vector at Bearing 1 to Weight at Plane 1 (Grams/Mil,$_{\text{p-p}}$ at Degrees)

$\overrightarrow{S}_{12}$ = Sensitivity Vector at Bearing 1 to Weight at Plane 2 (Grams/Mil,$_{\text{p-p}}$ at Degrees)

$\overrightarrow{S}_{21}$ = Sensitivity Vector at Bearing 2 to Weight at Plane 1 (Grams/Mil,$_{\text{p-p}}$ at Degrees)

$\overrightarrow{S}_{22}$ = Sensitivity Vector at Bearing 2 to Weight at Plane 2 (Grams/Mil,$_{\text{p-p}}$ at Degrees)

$\overrightarrow{U}_1$ = Mass Unbalance Vector at Plane 1 (Grams at Degrees)

$\overrightarrow{U}_2$ = Mass Unbalance Vector at Plane 2 (Grams at Degrees)

For purposes of clarification, the weights in these equations are expressed in units of grams. Obviously, this is a metric mass unit, and not a weight unit in the English system. The diagnostician may elect to use weights in terms of ounces or pounds. If these units are used consistently, the vector equation structure will not be influenced. However, from a weight measurement standpoint, most operating plants have a laboratory scale, or a triple beam balance that provides accurate weights in units of grams. Hence, grams will be used throughout this text, and the associated examples and case histories.

For a linear system, the addition (or removal) of a calibration weight W_1 at plane 1 should vectorially sum with the existing unbalance U_1. This presumed linear weight summation may be included with the previous equations (11-7) and (11-8) to produce the following new pair of vector equations:

$$\overrightarrow{B}_{11} = \left\{ \frac{\overrightarrow{U}_1 + \overrightarrow{W}_1}{\overrightarrow{S}_{11}} \right\} + \left\{ \frac{\overrightarrow{U}_2}{\overrightarrow{S}_{12}} \right\} \qquad \textbf{(11-9)}$$

$$\overrightarrow{B}_{21} = \left\{ \frac{\overrightarrow{U}_1 + \overrightarrow{W}_1}{\overrightarrow{S}_{21}} \right\} + \left\{ \frac{\overrightarrow{U}_2}{\overrightarrow{S}_{22}} \right\} \qquad \textbf{(11-10)}$$

where: $\overrightarrow{B}_{11}$ = Vibration Vector at Bearing 1 with Weight W_1 at Plane 1 (Mils,$_{\text{p-p}}$ at Degrees)

$\overrightarrow{B}_{21}$ = Vibration Vector at Bearing 2 with Weight W_1 at Plane 1 (Mils,$_{\text{p-p}}$ at Degrees)

$\overrightarrow{W}_1$ = Calibration Weight Vector at Plane 1 (Grams at Degrees)

Removal of the calibration weight W_1 at balance plane 1, plus the addition (or removal) of a calibration weight W_2 at balance plane 2 produces the following pair of vector equations based upon the initial expressions (11-7) and (11-8):

$$\overrightarrow{B_{12}} = \left\{ \frac{\overrightarrow{U_1}}{\overrightarrow{S_{11}}} \right\} + \left\{ \frac{\overrightarrow{U_2} + \overrightarrow{W_2}}{\overrightarrow{S_{12}}} \right\} \qquad (11\text{-}11)$$

$$\overrightarrow{B_{22}} = \left\{ \frac{\overrightarrow{U_1}}{\overrightarrow{S_{21}}} \right\} + \left\{ \frac{\overrightarrow{U_2} + \overrightarrow{W_2}}{\overrightarrow{S_{22}}} \right\} \qquad (11\text{-}12)$$

where: $\overrightarrow{B_{12}}$ = Vibration Vector at Bearing 1 with Weight W_2 at Plane 2 (Mils,$_\text{p-p}$ at Degrees)

   $\overrightarrow{B_{22}}$ = Vibration Vector at Bearing 2 with Weight W_2 at Plane 2 (Mils,$_\text{p-p}$ at Degrees)

   $\overrightarrow{W_2}$ = Calibration Weight Vector at Plane 2 (Grams at Degrees)

In each of these six vector equations, the first subscript defines the measurement plane, and the second subscript describes the correction or balance plane. Hence, this six equation array contains eight known vector quantities, i.e., the six vibration vectors, plus the two calibration weights. The calculation procedure initially solves for the four unknown balance sensitivity vectors, and finally the two mass unbalance vectors are calculated.

The S_{11} vector is determined by first expanding equation (11-9), and then substituting equation (11-7) in the following manner:

$$\overrightarrow{B_{11}} = \left\{ \frac{\overrightarrow{U_1} + \overrightarrow{W_1}}{\overrightarrow{S_{11}}} \right\} + \left\{ \frac{\overrightarrow{U_2}}{\overrightarrow{S_{12}}} \right\} = \left\{ \frac{\overrightarrow{U_1}}{\overrightarrow{S_{11}}} \right\} + \left\{ \frac{\overrightarrow{W_1}}{\overrightarrow{S_{11}}} \right\} + \left\{ \frac{\overrightarrow{U_2}}{\overrightarrow{S_{12}}} \right\} = \overrightarrow{A_1} + \left\{ \frac{\overrightarrow{W_1}}{\overrightarrow{S_{11}}} \right\}$$

This expression may now be solved for the first balance sensitivity vector:

$$\overrightarrow{S_{11}} = \left\{ \frac{\overrightarrow{W_1}}{\overrightarrow{B_{11}} - \overrightarrow{A_1}} \right\} \qquad (11\text{-}13)$$

S_{21} is determined in a similar manner by first expanding equation (11-10), and then substituting equation (11-8) as follows:

$$\overrightarrow{B_{21}} = \left\{ \frac{\overrightarrow{U_1} + \overrightarrow{W_1}}{\overrightarrow{S_{21}}} \right\} + \left\{ \frac{\overrightarrow{U_2}}{\overrightarrow{S_{22}}} \right\} = \left\{ \frac{\overrightarrow{U_1}}{\overrightarrow{S_{21}}} \right\} + \left\{ \frac{\overrightarrow{W_1}}{\overrightarrow{S_{21}}} \right\} + \left\{ \frac{\overrightarrow{U_2}}{\overrightarrow{S_{22}}} \right\} = \overrightarrow{A_2} + \left\{ \frac{\overrightarrow{W_1}}{\overrightarrow{S_{21}}} \right\}$$

This expression may now be solved for the second sensitivity vector:

$$\overrightarrow{S}_{21} = \left\{ \frac{\overrightarrow{W}_1}{\overrightarrow{B}_{21} - \overrightarrow{A}_2} \right\} \tag{11-14}$$

Similarly S_{12} is determined by first expanding equation (11-11), and then substituting equation (11-7) as follows:

$$\overrightarrow{B}_{12} = \left\{ \frac{\overrightarrow{U}_1}{\overrightarrow{S}_{11}} \right\} + \left\{ \frac{\overrightarrow{U}_2 + \overrightarrow{W}_2}{\overrightarrow{S}_{12}} \right\} = \left\{ \frac{\overrightarrow{U}_1}{\overrightarrow{S}_{11}} \right\} + \left\{ \frac{\overrightarrow{U}_2}{\overrightarrow{S}_{12}} \right\} + \left\{ \frac{\overrightarrow{W}_2}{\overrightarrow{S}_{12}} \right\} = \overrightarrow{A}_1 + \left\{ \frac{\overrightarrow{W}_2}{\overrightarrow{S}_{12}} \right\}$$

This expression may now be solved for the third balance sensitivity vector:

$$\overrightarrow{S}_{12} = \left\{ \frac{\overrightarrow{W}_2}{\overrightarrow{B}_{12} - \overrightarrow{A}_1} \right\} \tag{11-15}$$

Finally, S_{22} is determined by first expanding equation (11-12), and then substituting equation (11-8) as follows:

$$\overrightarrow{B}_{22} = \left\{ \frac{\overrightarrow{U}_1}{\overrightarrow{S}_{21}} \right\} + \left\{ \frac{\overrightarrow{U}_2 + \overrightarrow{W}_2}{\overrightarrow{S}_{22}} \right\} = \left\{ \frac{\overrightarrow{U}_1}{\overrightarrow{S}_{21}} \right\} + \left\{ \frac{\overrightarrow{U}_2}{\overrightarrow{S}_{22}} \right\} + \left\{ \frac{\overrightarrow{W}_2}{\overrightarrow{S}_{22}} \right\} = \overrightarrow{A}_2 + \left\{ \frac{\overrightarrow{W}_2}{\overrightarrow{S}_{22}} \right\}$$

This expression may now be solved for the fourth balance sensitivity vector:

$$\overrightarrow{S}_{22} = \left\{ \frac{\overrightarrow{W}_2}{\overrightarrow{B}_{22} - \overrightarrow{A}_2} \right\} \tag{11-16}$$

There is a clear pattern to the development of the balance sensitivity vectors. If these calculations are generalized, the following expression provides a general solution for balance sensitivity vectors for a multilplane solution:

$$\overrightarrow{S}_{mp} = \left\{ \frac{\overrightarrow{W}_p}{\overrightarrow{B}_{mp} - \overrightarrow{A}_m} \right\} \tag{11-17}$$

In equation (11-17), the subscript m specifies the measurement plane, and the subscript p identifies the weight correction plane. This general expression is identical to the previously developed equations (11-13) through (11-16). It may be applied in any balance situation where a weight is installed, vibration data is acquired at speed, and the weight is then removed prior to the installation of the next calibration weight. For example, the balance sensitivity vectors summarized in Tables 11-1 and 11-2 were computed using equation (11-17).

Combining the solutions for the four balance sensitivity vectors within the initial equations (11-7) and (11-8) yields the following result for mass unbalance at both correction planes:

$$\vec{U}_1 = \frac{(\vec{S}_{12} \times \vec{A}_1) - (\vec{S}_{22} \times \vec{A}_2)}{\left(\dfrac{\vec{S}_{12}}{\vec{S}_{11}}\right) - \left(\dfrac{\vec{S}_{22}}{\vec{S}_{21}}\right)} \qquad \textbf{(11-18)}$$

$$\vec{U}_2 = \frac{(\vec{S}_{21} \times \vec{A}_2) - (\vec{S}_{11} \times \vec{A}_1)}{\left(\dfrac{\vec{S}_{21}}{\vec{S}_{22}}\right) - \left(\dfrac{\vec{S}_{11}}{\vec{S}_{12}}\right)} \qquad \textbf{(11-19)}$$

Correcting for proximity probe slow roll runout at both measurement planes, the above two equations should be more properly expressed in terms of runout compensated initial vibration vectors as follows:

$$\vec{U}_1 = \frac{\left(\vec{S}_{12} \times \vec{A}_{1_c}\right) - \left(\vec{S}_{22} \times \vec{A}_{2_c}\right)}{\left(\dfrac{\vec{S}_{12}}{\vec{S}_{11}}\right) - \left(\dfrac{\vec{S}_{22}}{\vec{S}_{21}}\right)} \qquad \textbf{(11-20)}$$

$$\vec{U}_2 = \frac{\left(\vec{S}_{21} \times \vec{A}_{2_c}\right) - \left(\vec{S}_{11} \times \vec{A}_{1_c}\right)}{\left(\dfrac{\vec{S}_{21}}{\vec{S}_{22}}\right) - \left(\dfrac{\vec{S}_{11}}{\vec{S}_{12}}\right)} \qquad \textbf{(11-21)}$$

This final pair of equations (11-20) and (11-21) may be used to calculate a two plane balance correction. The individual expressions for balance sensitivity vectors may also be used separately to compare balance response characteristics between different rotors, or for repetitive calculations on the same rotor.

It is important to recognize that the previous array of balancing equations are predicated upon an explicit application sequence of the calibration weights (W_1 and W_2). Specifically, the first calibration run is performed with weight W_1 mounted at balance plane 1. At the conclusion of this first calibration run, the weight W_1 is removed from the machine, and a calibration weight W_2 is attached at balance plane 2. Following the conclusion of the second calibration run, weight W_2 is also removed from the machine. The unbalance calculations presented in equations (11-20) and (11-21) represent the effective rotor unbalance at each correction plane, irrespective of the balance calibration weights.

The calculated mass unbalance vectors (U_1 and U_2) represent the amount of weight that should be used at each balance correction plane. The angles associated with these unbalance vectors represent the angular location of the mass unbalance. Hence, weight can be removed at the calculated angles, or an equivalent weight may be added at the opposite side of the rotor. That is, if weight must be added, the weight addition angle would be equal to the calculated mass unbalance vector angle plus or minus 180°. Although it is generally desirable to remove weight from a rotor during balancing, there are many situations when it is proper to add balance correction weights.

In some mechanical configurations, balance weights cannot be installed on the machinery, and corrections must be performed by weight removal. In these cases, the weight changes are performed by grinding, or drilling balance holes. In other situations, the applied calibration weights may significantly reduce the vibration amplitudes, and there might be a reluctance to remove weights that provided a positive influence on the machinery. In both of these scenarios, the applied calibration weights become part of the final correction weights. Furthermore, the computation of balance sensitivity vectors must be modified to accommodate this change in weight attachment or removal sequence.

In recent years, balance calculations have been significantly improved by using smaller and faster personal computers. Machines such as the Apple Macintosh®, IBM®, or Compaq provide substantially more capability. All of these computers come in portable laptop configurations (less than 8 pounds) that can easily be carried to a plant site for field balancing work. The utilization of spread sheet programs such as Microsoft® Excel allow rapid data entry combined with almost instantaneous calculations. Hard copy documentation of the balancing data and calculations is easily achieved with a variety of small LaserJet or Ink Jet printers. In essence, these improvements in portable computers have allowed the machinery diagnostician to concentrate on the actual balance problem instead of the intricacies of the multiple vector manipulations.

The same scenario applies to the vibration data acquisition and processing portion of the balancing work. Modern digital systems allow the capture of transient startup and coastdown data, plus steady state information at various loads or heat soak conditions. The integration of these digital instrumentation systems with the laptop computer and the portable printer provides excellent capability for acquiring and printing the Bode and polar plots, plus the constant speed orbit and time base data. Hence, the diagnostician has many tools to acquire a variety of data, perform many complicated calculations, and generate the necessary hard copy documentation in the field. This not only improves the quality of the machinery balancing, it also minimizes the time required to perform the work, and provides improved confidence to the selected balance shots.

With all of the available tools, the machinery diagnostician may have some fundamental questions like *what do I use, and when do I use it?* This is not a casual issue, it is a very serious question regarding the application of the available tools and techniques. Unfortunately, some people try to address field balancing with a *cookbook approach* that can be used on any machine, at any time. These types of *canned* techniques are destined to failure from the beginning. In all cases, the diagnostician must examine the various facets of the machinery problem, and select the appropriate measurements, instruments, and calculations that will solve the problem. This means that you have to know what you're doing instead of blindly following some general procedure.

In an effort to provide some realistic direction, the following three case histories 38, 39, and 40 are presented for consideration. These are three different configurations of generator drives that exhibit three different balance problems. The discussions associated with each of these field case histories are quite detailed, and the specific balancing logic is presented for each unit.

Case History 38: Five Bearing, 120 MW Turbine Generator Set

The five bearing turbine generator set depicted in Fig. 11-27 consists of two turbine cases driving a hydrogen cooled, two pole synchronous generator. The electric generator produces 3 phase, 60 Hz power at 13,800 volts. It is rated at 134,000 KVA, with a power factor of 0.85. The high pressure HP turbine accepts superheated inlet steam at 1,900 Psi, and it exhausts to a reheat section before readmission into the intermediate pressure IP turbine. Exhaust from the IP turbine is directed to a double flow LP turbine that exhausts to a surface condenser at 2 inches of mercury absolute. As shown in Fig. 11-27, the HP and IP rotors are contained in one case, and the inlet steam to the double flow IP is at the center of the machine. This configuration places the large diameter low pressure turbine

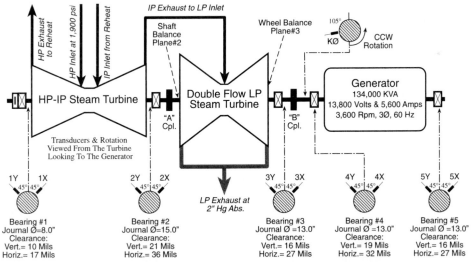

Fig. 11-27 Machinery Arrangement Of Five Bearing Turbine Generator Set

wheels at both exterior ends of the LP casing.

All five journal bearings are elliptical, and their respective vertical and horizontal clearances are shown in Fig. 11-27. The largest clearances are located at the 15 inch diameter journal at bearing #2. In this location, the vertical diametrical clearance of 21 Mils is combined with a 36 Mil horizontal clearance. This is the largest journal diameter in the train, and it must also accommodate some significant thermal changes. The "A" coupling between the HP-IP and the double flow LP turbine, plus the "B" coupling between the LP turbine and the generator are solid couplings. These two couplings must be carefully assembled in conjunction with a sling check at each location.

Prior to the peak generating season, this machinery train was subjected to a major six week outage for a complete maintenance overhaul. During disassembly, it was discovered that the #1 bearing and journal were damaged, and the remaining four bearings had excessive clearances. A set of new buckets were installed in one row of the IP rotor, and the last stage blades on both ends of the LP rotor were replaced with a new and longer blade design. Several new diaphragms were installed, and various other stationary items were repaired or replaced as required. An 8.0 Mil bow was documented on the IP rotor, but it was not corrected. During reassembly, the LP turbine was balanced in-place at a speed of 250 to 300 RPM by OEM personnel.

The rebuilt machinery train experienced a variety of routine mechanical and electrical problems that were sequentially corrected. After eight days of various runs, the turbine generator set was finally placed on-line. The shaft vibration monitors indicated generally acceptable displacement amplitudes. The dominant motion at all measurements points occurred at the shaft rotational speed of 3,600 RPM. Specifically, the shaft 1X running speed orbits at each of the five main bearings are shown on Fig. 11-28 at a constant load of 112 megawatts.

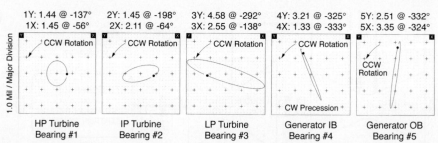

Fig. 11–28 Initial Runout Compensated Orbits At Each Train Bearing At 112 Megawatts

These 1X orbits are all compensated for shaft runout, and they are considered to be representative of the absolute shaft vibration relative to the proximity probes mounted at each main bearing. A normal forward precession combined with low vibration amplitudes are displayed by bearings #1 and #2 on the HP-IP turbine. The Y-Axis probe amplitude on bearing #3 approaches 4.7 Mils,$_{p-p}$, but the horizontal bearing clearance is 27 Mils. Hence, the runout corrected shaft vibration is only 17.4% of the available diametrical clearance. Typically, a shaft vibration of 4.7 Mils,$_{p-p}$ would be a cause for concern on smaller units. However, on a large machine such as this turbine generator set, this shaft vibration level may be higher than desired, but it is still acceptable for operation.

Both generator orbits were predominantly vertical and a reverse precession was noted on the inboard bearing #4. These generator vibration levels were also higher than anticipated, but they were considered to be manageable. It should be mentioned that the radial shaft vibration levels were lower than the values logged before the overhaul. From the standpoint of the panel mounted vibration monitor readings, the machinery train was in good condition. Unfortunately, a significant casing or structural vibration was emitted between the LP turbine and the generator. The severity of this excitation was such that the control room operators were acutely aware of this problem since it produced a physically uncomfortable sensation in the main control room.

This unusual vibratory behavior occurred predominantly at shaft rotational speed, and casing measurements revealed that it was strongest in the axial direction at the #3 bearing. Various load and speed variation tests were performed, and it was repeatedly demonstrated that this casing 1X component varied between 9.0 and 18.0 Mils,$_{p-p}$ axially. For instance, the expanded Bode plot from one cold startup is displayed in Fig. 11-29. This diagram depicts the 1X amplitude and phase from the horizontal #3X and vertical #3Y shaft vibration

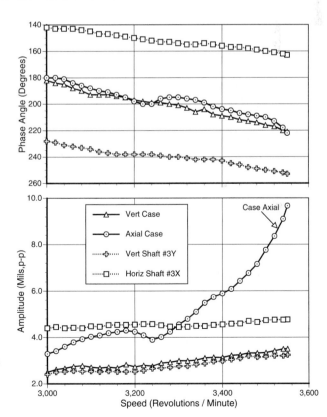

Fig. 11–29 Expanded Bode Plot Of Synchronous 1X Shaft And Casing Vibration At The #3 Bearing On The LP Turbine

probes between 3,000 and 3,550 RPM. It also includes the 1X casing motion measured in the vertical and axial planes at the same bearing. This casing data was acquired on the bearing cap with medium frequency range accelerometers, and double integrated to obtain casing displacement.

The phase data at the top of Fig. 11-29 reveals nothing unusual. It is noted that the phase differential between the X and Y shaft probes remains fixed at about 85°. Furthermore, all four transducers exhibit a nominal 20° change across the plotted frequency range. This startup is typical for this class of machine, and it is not considered to be representative of any type of resonant response.

Examination of the 1X amplitude data presented in the bottom half of Fig. 11-29 reveals moderate changes in synchronous motion for the three radial transducers. However, the axial casing vibration increased from 3.9 Mils,$_{p-p}$ at 3,240 RPM to an amplitude of 9.7 Mils,$_{p-p}$ at 3,550 RPM. Hence, a 10% speed change was accompanied by a 250% increase in axial casing vibration. Although this appears to be a *resonant-like* response, the companion phase data does not support the presence of a casing resonance or rotor critical speed.

Another peculiarity was noted on bearing #3 when oil drain temperatures were checked. The temperatures on the #1 and #2 bearings on the HP-IP turbine consistently ran between 150° and 155°F. The two generator elliptical bearings operated between 140° and 146°F. However, the LP turbine #3 bearing oil drain thermocouple never exceeded 126°F. This thermocouple was replaced, and the reading confirmed with a local dial thermometer. Any way the measurement was made, the oil drain temperature from bearing #3 remained around 126°F.

It was speculated that perhaps bearing #3 was running unloaded. This would account for the lower than average oil drain temperatures, plus the higher than average shaft vibration amplitudes. Unfortunately, this concept was not supported by the shaft centerline position data. For instance, Table 11-6 summarizes the vertical shaft position of the journals within each bearing.

Since this train was equipped with elliptical bearings at all locations, the total shaft centerline position change consisted of a horizontal shift in the direction or rotation, plus a vertical lift. The data presented in Table 11-6 summarizes the vertical clearance at each elliptical bearing, and the measured vertical shaft lift in Mils (based on X-Y proximity probe DC gap voltages). To provide a compar-

Table 11–6 Vertical Position Of Journals With Respect To Each Bearing At 112 Megawatts

Bearing Location	Vertical Bearing Clearance	Vertical Shaft Rise	Percent Journal Rise In Bearing
HP Turbine — Bearing #1	10 Mils	4 Mils	40%
IP Turbine — Bearing #2	21 Mils	8 Mils	38%
LP Turbine — Bearing #3	16 Mils	4 Mils	25%
Generator IB — Bearing #4	19 Mils	7 Mils	37%
Generator OB — Bearing #5	16 Mils	7 Mils	44%

ison of lift versus clearance, the fourth column in Table 11-6 was used for the ratio of these two parameters. Clearly, the journal at bearing #3 is sitting the lowest in its respective bearing, with a total lift equal to only 25% of the available clearance. Hence, the shaft centerline position data indicated a loaded, rather than an unloaded #3 bearing.

Obviously the oil temperature and shaft vibration indicators were in conflict with the journal centerline position data. In an effort to resolve this contradiction, the #3 bearing was elevated by 5 Mils with stainless steel shims. After a restart, and a normal load and temperature stabilization, it was evident that vibration amplitudes had decreased slightly at the #3 bearing. Unfortunately, the shaft vibration had increased at each of the other four bearings. Although the casing axial vibration was temporarily reduced to 6.4 Mils,$_\text{p-p}$, the increases at the other machine train bearings were considered to be unacceptable.

Once again, the oil drain temperature at #3 bearing was about 20°F lower than the other four bearings, and the array of shaft centerline positions remained virtually identical to the behavior described on Table 11-6. Hence, the evidence in support of an unloaded #3 bearing was beginning to dwindle. This issue was finally put to rest when the operating logs from the past few years were examined. Within this database, it was clear that the drain temperature from bearing #3 was always about 20°F lower than the other bearings. Hence, the lower oil drain temperature on #3 bearing could not be associated with an unloaded LP turbine bearing.

Since the generator still displayed flat elliptical orbits, and since the generator vibration had increased with the 5 Mil rise on the #3 bearing — alignment across the "B" coupling was questioned. At this point, the OEM elected to raise the generator by 8 Mils, and see if that helped the situation. After the train was restarted, it was clear that the elevated generator was contributing to the problem. Under this condition, axial casing vibration amplitudes at the #3 bearing exceeded 18.0 Mils,$_\text{p-p}$. Obviously, the project was moving in the wrong direction, and everyone was growing weary of this problem.

External lagging was removed from the LP turbine, and additional vibration readings were acquired on the turbine casing and associated structural elements. This data did not identify any specific mechanical component that could be a contributor or source of the high axial casing motion. Next, the machine was shutdown and allowed to cool. The elevation shims were removed from underneath the #3 bearing (5 Mils), and the generator (8 Mils). This restored these elements back to their original vertical alignments. The LP turbine outer casing was removed, and each strut and cross brace were examined for any indication of a loose structural member within the turbine shell. This inspection extended down into the condenser, and no loose or broken structural members were found.

At this stage, it was clear that there were no contributing structural problems. Furthermore, the original alignment across the "B" coupling was deemed to be proper and acceptable. Since the forcing frequency of the axial casing vibration was at running speed, the attack plan evolved to *do everything possible to minimize the synchronous 1X running speed vibration*. Although there was no evidence of abnormal axial vibration of the HP-IP rotor, it was clear that radial vibration of the LP turbine could be improved. Hence, when all else fails, go back to the basics, and try to eliminate the 1X driving force. This includes any eccentricity, bow, or mass unbalance in the rotor system.

Initially, the "B" coupling between the LP turbine and the generator was disassembled. The #3 bearing was removed, and a sling (or horizontal swing) check was performed at the #3 bearing location. An overhead scale was used to establish the proper vertical load on the free shaft, and dial indicator readings were acquired at 45° increments. The documented runout during initial assembly was 6.0 Mils, and the current maximum runout was found to be in excess of 30 Mils. In addition, these readings were not repeatable. At this point, it was clear that other problems existed, and the #1 and #2 bearings were opened for inspection. The turbine #1 bearing was in good condition, but IP bearing #2 had a section gouged out of the babbitt. The loose babbitt in #2 bearing probably contributed to the erratic sling checks. There was no doubt that this bearing would have experienced an early failure if the loose babbitt had not been detected.

Nevertheless, the damaged #2 bearing was scraped, cleaned, and reassembled. This bearing repair resulted in consistent swing checks at the #3 bearing. The two turbine rotors were realigned at the "A" coupling, and the final coupled sling check runout at the #3 bearing was nominally 5.0 Mils. The #3 bearing was then re-installed, and the "B" coupling was reassembled to factory tolerances. The 8.0 Mil bow on the IP rotor was still present, but there was no realistic opportunity to correct this bow in the available time frame.

The assembled turbine generator set was then restarted to obtain a fresh set of baseline data. The next step was to trim balance the LP turbine rotor running between bearings #2 and #3. This was initially hampered by a midnight startup combined with an additional bow on the HP-IP rotor. Problems like this happen when long hours are combined with production pressures to get the machinery on-line. In most situations, it is best for the key personnel to get some sleep, and try running the machine during the daylight hours. In this specific case, the controlled restart on the following morning proved to be successful, and a new set of baseline or initial reference data was obtained at speed and no load.

Although the temptation exists to perform several runs followed by simultaneous shots on multiple balance planes, it is usually wise to begin with something more realistic like a single plane balance correction. For this machine, the largest radial vibration amplitudes occurred at the #3 bearing, and it makes sense to perform the first correction at this location. As noted in Fig. 11-27, a balance plane exists on the outer face of the exhaust wheel adjacent to the #3 bearing. A diagram of this balance plane with additional physical details is presented in Fig. 11-30. This location is a view from the governor end of the HP turbine, and it is immediately apparent that the 0° reference point is located at the Keyphasor® probe. This transducer is positioned 15° below the horizontal centerline on the left side of the machine. Since the Y-Axis probe is 45° above this centerline, the total angle between transducers is 60° as noted. The X-Axis proximity probe is another 90° against rotation, and it is shown at 150° on Fig. 11-30.

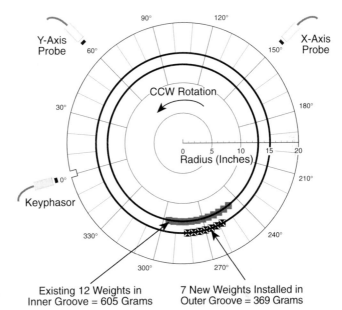

Fig. 11–30 Weight Configuration And Angular Reference Defined on Wheel Balance Plane #3

Existing 12 Weights in Inner Groove = 605 Grams

7 New Weights Installed in Outer Groove = 369 Grams

Inner Groove Radius = 13.04", and One Weight = 4.4°
Outer Groove Radius = 14.92", and One Weight = 3.8°

As discussed throughout this text, the vibration phase measurements are always referenced to the physical location of the vibration probes. Ideally, all phase measurements, weight angles, plots, and computed vectors are referenced to the same radial probe position. From a practical standpoint this is not always possible, and a slight modification in the thought process is necessary. Specifically, if we consider a single plane unbalance condition as described by equation (11-2), and if the vector sensitivity from a calibration run is incorporated from equation (11-4), the following result is obtained.

$$\vec{U} = \vec{S} \times \vec{A} = \left\{ \frac{\vec{W}}{\vec{B} - \vec{A}} \right\} \times \vec{A} = \frac{\vec{W} \times \vec{A}}{\vec{B} - \vec{A}} = \frac{\vec{W}}{(\vec{B}/\vec{A}) - 1}$$

Division of the B vector by the A vector in the denominator of the above expression requires a subtraction of angles (i.e., β-α). Clearly, wherever the reference point is located, the subtraction of vibration vector angles negates any reference to a specific or fixed angular location. Hence, the angle of the calculated unbalance U in the above equation is totally dependent on the angular reference used for the calibration weight W. If the shaft keyway is used as the 0° reference point for the calibration weight (as per Fig. 11-30), then the calculated unbalance will be referenced to the same angular reference point. This obvious simplification makes the field work a lot easier, since it can be correctly stated that:

keyway is 0°, and all angles are counter-rotation from the keyway

This makes life much easier for everyone on the jobsite, and it provides a common and easily understandable angular reference scheme. Furthermore, if you're looking from the turbine to the generator, or the generator to the turbine, the angular location will come out properly (since it is stated as counter-rotation). It is also a good idea to use a bright colored metal marking pen to identify the balancing weight angles on the wheel or shaft. Some balance planes have to be accessed through the condenser inlet piping. This is a hot and miserable environment to work in, and it is easy to lose your sense of direction and perspective. Thus, any type of pre-planned or identified angular layout will be extraordinarily useful in getting the weights installed at the correct angles.

Getting back to the trim balance of the LP turbine, the axial face of the LP wheel adjacent to #3 bearing was equipped with two trapezoidal shaped grooves for similarly shaped balance weights. As shown in Fig. 11-30, the inner groove has a 13.04 inch radius, and it contained 12 weights totaling 605 grams. These weights were installed during the low speed field balance by OEM personnel. The circumferential length of the inner groove was 81.93 inches (= 2 x π x 13.04). Since the balance weights were approximately one inch long, each weight covered an arc of about 4.4° (= 360/82). The outer balancing groove, with a radius of 14.92 inches was used during this field balance. The total length of this outer groove was 93.75 inches (= 2 x π x 14.92). For one inch long balance weights, each weight would cover an arc of about 3.8° (= 360/94).

The angle of the initial calibration weight was based upon the 1X vectors at the #3 bearing. Specifically, the center orbit on Fig. 11-28 showed a Y-Axis vector of 4.58 Mils,$_{p-p}$ at an angle of 292°. Since this turbine runs above the first critical, the phase angle would be representative of the high spot, and this would be the location for adding weight. Since the Y probe is 60° counter from this physical keyway reference, the actual weight installation angle should be 352° (=292°+60°). At the same time, the X-Axis vector was 2.55 Mils,$_{p-p}$ at 138°. Since the X probe is 150° away from the keyway, the weight installation angle should be 288° (=138°+150°). Thus, the Y probe calls for a weight addition at 352°, and the X probe wants weight added at 288°. Both transducers are calling for a weight in the lower left-hand quadrant of Fig. 11-30. Clearly, the angular difference is due to the ellipticity of the shaft orbit. For purposes of simplicity during this initial installation, the weights were mounted at 0° to straddle the keyway.

The amount of weight to add was determined by applying a form of equation (11-64) to this 22,000 pound turbine rotor in the following manner:

$$Cal.\ Weight\ =\ Rotor\ Weight \times \left\{ \frac{291}{RPM} \right\}^2\ =\ 22000 \times \left\{ \frac{291}{3600} \right\}^2\ =\ 144\ \text{Oz-In.}$$

As previously stated, the radius for the outer balance groove was 14.92 inches, and that would require a weight addition of 9.65 ounces (=144/14.92). Since one ounce weighs 28.35 grams, the initial weight addition should be in vicinity of 274 grams (=9.65 x 28.35). The available balance weights were about 55 grams each, and this 274 gram shot would need at least five weights. This amount of weight would also cover a 19° arc (=5 x 3.8). Obviously, the larger the arc, the more offsetting the weights become. Hence, a vector summation of five weights at 3.8° increments would yield an effective weight that must be lighter than the simple arithmetic weight sum. To adjust for this weight spread, a total of six weights were used. This provided a first calibration run using 310 grams installed in this outer groove, straddling the 0° keyway.

At full speed of 3,600 RPM the 1X vibration levels at the #3 bearing were reduced by 0.7 Mils,$_{p-p}$, and the synchronous amplitudes at #2 bearing increased by approximately 0.4 Mils,$_{p-p}$. Fortunately, the 1X vectors at HP bearing #1, plus both generator bearings were not appreciably influenced by this initial weight at balance plane #3. Due to the vibration increase at bearing #2, it was evident that a simultaneous weight correction would be necessary at the opposite end of the LP turbine (close to #2 bearing).

The second calibration run was performed by adding 160 grams on the LP shaft balance plane adjacent to the "A" coupling. At this location, two circumferential balance weight grooves are cut into the shaft. The forward groove by the "A" coupling contained 9 weights totaling 448 grams. These weights were also installed during the low speed field balance by OEM personnel. The aft groove closest to the LP wheel was used for the installation of the 160 grams at 295° clockwise from the keyway. The previous six weights of 310 grams at 0° mounted on the opposite end of the LP rotor (plane #3) remained in place. The vibration response with these weights resulted in decreased vibration levels at both #2 and #3 bearings. In addition, the vectors at the #1 bearing increased slightly, and the shaft vibration at generator bearings #4 and #5 remained fairly constant.

Based on these three runs, a series of two plane balance calculations were performed. The four balance sensitivity vectors (S_{11}, S_{12}, S_{21}, and S_{22}) were computed with equations (11-26) through (11-29). The mass unbalance calculations were performed with two plane equations (11-20) and (11-21). The total *weight add* results of these balance calculations are presented in Table 11-7. It was easily agreed upon that the LP rotor required a static shot at nominally 270° at both ends of the rotor. The magnitude of the correction was subjected to additional debate. It is apparent that large weight corrections are difficult to perform due the self-canceling nature of weights distributed over a large portion of the circumference. Furthermore, it was demonstrated the additional weights placed in

Table 11–7 Weight Additions Based Upon Two Plane Balance Calculations — Plus Vector Average Weight Additions, And Summary of Final Balance Weights Installed

Balance Weight Origin	Shaft Balance Plane #2	Wheel Balance Plane #3
Based on Y-Axis Probes	1,118. Grams @ 245°	648. Grams @ 278°
Based on X-Axis Probes	1,289. Grams @ 270°	826. Grams @ 260°
Vector Average Correction	1,175. Grams @ 259°	728. Grams @ 268°
Final Weights Installed	535. Grams @ 270°	369. Grams @ 270°

the LP rotor around 270° would have a detrimental influence upon the HP-IP rotor, and the vibration at #1 bearing. General experience with this class of machinery has shown that weight corrections in the order of 40% to 50% of the calculated values are appropriately conservative. Hence, the final weights shown in Table 11-7 were physically installed in this LP turbine rotor. It should be mentioned that this weight correction was essentially a static shot that was consistent with the weights previously installed during the low speed balance of this rotor. This type of correction was not unusual, since the low speed balance often underestimates the required weight due to lower sensitivity of the rotor at 250 to 300 RPM versus the actual machine at 3,600 RPM. Overall, the correction weight angles were reasonable, and the weight magnitudes made good sense.

On the balance plane #2 adjacent to the "A" coupling and the #2 bearing, the installed four calibration weights of 160 grams were supplemented by nine additional sliding weights with a mass of 375 grams. These thirteen weights with a combined mass of 535 Grams were positioned around 270°. On balance correction plane #3 (axial face of the aft LP wheel) adjacent to the #3 bearing, the installed six calibration weights of 310 grams were supplemented by one additional sliding weight with a mass of 59 grams. The seven weights with a combined mass of 369 Grams were positioned around 270° as shown in Fig. 11-30.

Additional calculations were performed to predict the vibration response at LP turbine bearings #2 and #3. The results of these calculations are discussed in this chapter under the Response Prediction heading. In addition, the two plane balance calculations were extended to include results between the #1 and #2 bearings, plus the #1 and #3 bearings. Although these computations may not be completely linear, they do provide some indication of the potential vibration severity on the HP-IP rotor due to weight corrections on the LP rotor. These calculations reinforced the fact that aggressive weight additions on the LP turbine could adversely influence the vibratory behavior on the HP rotor #1 bearing.

After the final weight corrections were executed, the machinery train was restarted. Examination of this startup data revealed acceptable vibration amplitudes at all measurement locations. The train was allowed to heat soak, and load was gradually applied. The final runout compensated shaft orbits at each bearing are shown in Fig. 11-31 at a constant load of 115 megawatts.

This data is directly comparable to the initial orbits shown in Fig. 11-28.

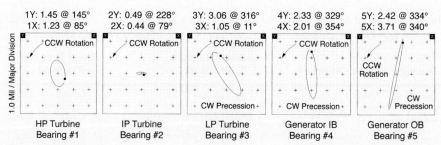

| 1Y: 1.45 @ 145° | 2Y: 0.49 @ 228° | 3Y: 3.06 @ 316° | 4Y: 2.33 @ 329° | 5Y: 2.42 @ 334° |
| 1X: 1.23 @ 85° | 2X: 0.44 @ 79° | 3X: 1.05 @ 11° | 4X: 2.01 @ 354° | 5X: 3.71 @ 340° |

| HP Turbine Bearing #1 | IP Turbine Bearing #2 | LP Turbine Bearing #3 | Generator IB Bearing #4 | Generator OB Bearing #5 |

Fig. 11–31 Final Runout Compensated Orbits At Each Train Bearing At 115 Megawatts

Within the final orbits presented in Fig. 11-31, it is clear that the HP turbine bearing #1 was not adversely influenced by the balance weights. The vibration at the IP bearing #2 was substantially reduced, and a significant improvement was noted at the LP turbine bearing #3. There was also a slight reduction in vertical response on the generator bearings. It is peculiar to note that a reverse precession now appears on bearings #3, #4, and #5. Unfortunately, the analysis and explanation of this behavior does fall under the category of a *whole different story.*

The initial versus the final 1X vibration vectors are presented in a tabular format in Table 11-8 for bearings #2 and #3. One should never lose track of the

Table **11–8** Comparison Of Initial Versus Final 1X Vibration Vectors At Full Load

Vibration Transducer	Initial Condition 112 Megawatts	Final Condition 115 Megawatts
IP Turbine — Brg #2 — Shaft #2Y	1.45 Mils,$_{p-p}$ @ 198°	0.49 Mils,$_{p-p}$ @ 228°
IP Turbine — Brg #2 — Shaft #2X	2.11 Mils,$_{p-p}$ @ 64°	0.44 Mils,$_{p-p}$ @ 79°
LP Turbine — Brg #3 — Shaft #3Y	4.58 Mils,$_{p-p}$ @ 292°	3.06 Mils,$_{p-p}$ @ 316°
LP Turbine — Brg #3 — Shaft #3X	2.55 Mils,$_{p-p}$ @ 138°	1.05 Mils,$_{p-p}$ @ 11°
LP Turbine — Brg #3 — Casing Vert.	3.58 Mils,$_{p-p}$ @ 224°	0.73 Mils,$_{p-p}$ @ 231°
LP Turbine — Brg #3 — Casing Axial	9.18 Mils,$_{p-p}$ @ 226°	0.96 Mils,$_{p-p}$ @ 193°

original project objective. The main objective of this engineering project was to reduce the axial casing vibration on the LP turbine to acceptable levels. At the conclusion of field balancing activities, the vertical and axial casing vibration amplitudes at the #3 bearing were significantly attenuated. Further reductions occurred as the unit was loaded and allowed to heat soak. In Table 11-8, it is clear that following the various corrections and field trim balancing of the LP rotor, the 1X casing axial amplitudes were reduced to nominally 1.0 Mil,$_{p-p}$. In addition, transmitted vibration to the structure, the turbine deck, and the control room were all greatly reduced. It is speculated that the source of the axial casing vibration on the LP turbine was due to an axial wobble of the large last stage turbine wheels. This wobble was logically induced by excessive radial rotor deflections that are dependent on the rotor mode shape at operating speed. Although this hypothesis is difficult to prove with the available information, it certainly does satisfy the observed machinery behavior.

From another perspective, the final shaft and casing vibration amplitudes were very acceptable across the entire load envelope. Transient startup behavior was quite satisfactory, and the turbine generator set remained in constant operation. Other problems, such as excessive hydrogen leakage on the generator seals and stability problems with the electronic speed control system, still had to be resolved, but the two coupled steam turbines were in excellent condition.

WEIGHT SEQUENCE VARIATION

The previous discussion addressed a two plane balance based upon a rigid set of rules regarding the installation and removal of calibration weights. The developed equation array requires the installation of a calibration weight W_1 at the first balance plane, a data acquisition run on the machine, followed by a shutdown, and removal of the W_1 calibration weight. Sequentially, the second calibration weight W_2 is added to the second balance plane, and another data run is made on the machine. Based on the initial vibration response data, plus the two calibration runs, equations (11-20) and (11-21) may be used to compute the balance corrections.

In actual field balancing situations, these rules are often modified due to other influences and considerations (e.g. drilled balance holes). Typically, there are two realistic variations in the sequence of the calibration weights installed for the two calibration runs. First, consider the case where a calibration weight W_1 is mounted at balance plane 1 for the first run. This weight is allowed to remain in position, and another weight W_2 is attached at balance plane 2 for the second calibration run. In this condition, the vibration response during the second calibration run is dependent on the combined effect of both weights W_1 and W_2. Under this condition, the four balance sensitivity vectors must be computed in accordance with the differential vectors for each specific weight change. Thus, the balance sensitivity vectors for this condition are properly defined by the following expressions (11-22) through (11-25):

$$\overrightarrow{S_{11}} = \left\{ \frac{\overrightarrow{W_1}}{\overrightarrow{B_{11}} - \overrightarrow{A_1}} \right\} \tag{11-22}$$

$$\overrightarrow{S_{21}} = \left\{ \frac{\overrightarrow{W_1}}{\overrightarrow{B_{21}} - \overrightarrow{A_2}} \right\} \tag{11-23}$$

$$\overrightarrow{S_{12}} = \left\{ \frac{\overrightarrow{W_2}}{\overrightarrow{B_{12}} - \overrightarrow{B_{11}}} \right\} \tag{11-24}$$

$$\overrightarrow{S_{22}} = \left\{ \frac{\overrightarrow{W_2}}{\overrightarrow{B_{22}} - \overrightarrow{B_{21}}} \right\} \tag{11-25}$$

These resultant balance sensitivity vectors are then used to compute the mass unbalance in each plane as previously specified in equations (11-20) and (11-21). At this point the two calibration weights may be removed, and the final balance correction weights may be installed.

The other viable option is to leave the calibration weights in place (particularly with drilled holes), and perform a final correction that is vectorially equal to the calculated mass unbalance minus the calibration weight (added or removed) at each balance plane. The same final weight correction can also be achieved by running a trim balance calculation using the last set of vibration response vectors as the residual unbalance vectors. The results of this computational approach will provide the final incremental weight change vectors. Either approach is acceptable, and the actual field situation dictates the appropriate procedure.

The second potential variation on the basic procedure case consists of a calibration weight W_2 mounted at balance plane 2 for the first run. This weight is allowed to remain in place, and a new weight W_1 is attached at balance plane 1 for the second calibration run. In this condition, the vibration response is again dependent on the combined influence of both weights W_1 and W_2. The four balance sensitivity vectors for this condition are calculated in accordance with the following four equations (11-26) through (11-29):

$$\overrightarrow{S_{11}} = \left\{ \frac{\overrightarrow{W_1}}{\overrightarrow{B_{11}} - \overrightarrow{B_{12}}} \right\} \qquad \text{(11-26)}$$

$$\overrightarrow{S_{21}} = \left\{ \frac{\overrightarrow{W_1}}{\overrightarrow{B_{21}} - \overrightarrow{B_{22}}} \right\} \qquad \text{(11-27)}$$

$$\overrightarrow{S_{12}} = \left\{ \frac{\overrightarrow{W_2}}{\overrightarrow{B_{12}} - \overrightarrow{A_1}} \right\} \qquad \text{(11-28)}$$

$$\overrightarrow{S_{22}} = \left\{ \frac{\overrightarrow{W_2}}{\overrightarrow{B_{22}} - \overrightarrow{A_2}} \right\} \qquad \text{(11-29)}$$

Following the computation of these four balance sensitivity vectors, the balancing equations (11-20) and (11-21) are used to solve for the effective rotor unbalance at each of the two correction planes. The comments previously provided on the final addition (or removal) of weights to correct for the unbalance vectors are also appropriate to this calibration weight sequence.

Case History 39: Three Bearing Turbine Generator at 3,600 RPM

This case history addresses a 3,600 RPM turbine generator set. This is a 7.5 megawatt (10,000 HP) unit that was commissioned in the 1950s. Two identical trains were installed, and both units display similar characteristics, and virtually identical mechanical difficulties. Although some unique features are associated with each train, they both share a common historical trait of synchronous vibration problems. In recent years, the majority of the problems have been traced to generator unbalance, plus high eccentricity of the solid coupling between the turbine and generator. Although successful field balance corrections have been performed on both units, the logic behind some of the balance corrections had not been completely understood.

The machinery arrangement sketch presented in Fig. 11-32 describes the general arrangement of one unit. It is noted that this is a three bearing train, and the generator inboard must be supported by the turbine exhaust bearing. The 14 stage turbine driver is an extraction unit, with a surface condenser directly below the exhaust flange. The synchronous generator has collector rings mounted at the outboard end and a separate exciter. For many years this train included a direct shaft-driven exciter. However, this was replaced with a separate solid state unit to improve train reliability. The entire machinery unit is mounted on a mezzanine deck, and the bearing supports are reasonably compliant. This is particularly true at the generator outboard bearing, and various hydraulic and screw jacks have been applied over the years to increase the vertical support stiffness at this location.

The machine is equipped with X-Y proximity probes at each journal bear-

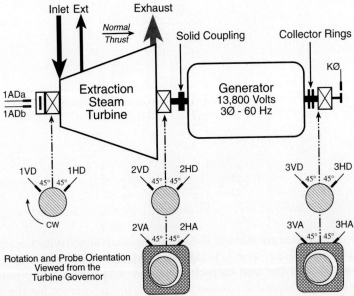

Fig. 11–32 Machinery And Vibration Transducer Arrangement On Turbine Generator Set

ing. As noted in Fig. 11-32, the probes are oriented at ±45° from the true vertical centerline. Two axial probes are mounted at the governor end of the turbine, and a once-per-rev Keyphasor® is installed at the generator outboard. To provide vibratory information on the flexible bearing housings, X-Y casing accelerometers are mounted on the #2 and #3 bearing caps. All of the transducers are connected to a permanent monitoring system. The proximity probes read out in displacement units of Mils,$_{p-p}$, and the accelerometers are integrated to IPS,$_{o-p}$. The data presented in this case history will include another level of integration to convert the casing velocity to Mils,$_{p-p}$ of casing displacement.

Rotor configuration for this machinery train is illustrated in Fig. 11-33. Overall length of the coupled rotors is 25 feet and 5 inches (305 inches). The turbine rotor weighs approximately 9,380 pounds, and the generator rotating assembly weighs 12,080 pounds for a combined rotor weight of 21,460 pounds. The turbine rotor contains three internal radial balance rings, and the generator was constructed with two axial balance rings. The physical locations of the generator balance planes are noted on the rotor drawing.

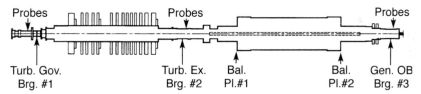

Fig. 11–33 Rotor And Proximity Probe Configuration On Turbine Generator Set

The calculated undamped first critical speed for this rotor system is 1,210 RPM. This is a translatory mode that exhibits the largest amplitudes in the vicinity of the coupling. Generally, the first critical is not a problem for these T/G Sets. The major problem usually occurs at the second or third critical speeds. The calculated undamped mode shapes for both of these resonances are shown in Fig. 11-34. Vibration measurement planes at the #2 and the #3 bearings at both criticals are separated by a nodal point. Hence, a phase reversal would be expected between bearings as the machine passes through these critical speeds. Naturally, the system mass unbalance distribution will dictate the severity of the response through each mode, and the damping would control the phase shift. These characteristics are directly attributable to the mode shapes through the second and third criticals, and this complexity has often confused the field balancing logic.

Due to these complex modes, and the history of balance problems, the operating company elected to send out both the turbine and the generator rotors for high speed balancing whenever a major overhaul was performed. T/G Set #1 was subjected to a major overhaul, and the high speed shop balance of both rotors proved to be effective. Following the correction of several control problems, this unit was restarted and placed on line with minimal problems, and low vibration levels. A year later, a similar overhaul was performed on #2 unit. The turbine

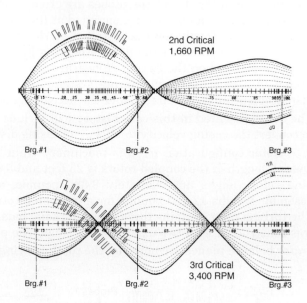

2nd Critical
1,660 RPM

Brg.#1   Brg.#2   Brg.#3

Fig. 11–34 Calculated
Undamped Shaft Mode
Shapes For Turbine Gen-
erator Set - Second and
Third Critical Speeds

3rd Critical
3,400 RPM

Brg.#1   Brg.#2   Brg.#3

rotor was refurbished with some rows of new buckets, and it was successfully
shop balanced at 3,600 RPM. Similarly, the generator rotor was subjected to various repairs at another shop, and it was also shop balanced at the full operating
speed of 3,600 RPM. This work was performed with a temporary shaft stub end
that was bolted to the coupling half, and used as the inboard journal.

Both rotors were reinstalled in their respective cases. The machine was
realigned, and the unit prepared for operation. The initial startup revealed high
vibration amplitudes at the first critical speed of 1,250 RPM. Shaft and casing
vibration levels were unacceptable as the unit entered the second critical, and
this run was terminated at 1,470 RPM with shaft vibration amplitudes
approaching 25 Mils,$_{p-p}$ at the generator outboard #3 bearing.

Following shutdown, it was determined that the governor speed ramp was
set at a low rate. This contributed to the high vibration due to extended operating time at the criticals. The ramp rate was increased, and during the next run
the full speed of 3,600 RPM was achieved. During this startup, the highest vibration amplitude of 15.6 Mils,$_{p-p}$ occurred at the #3 bearing at the second critical
speed of 1,650 RPM. At full speed, the shaft vibration data at the #2 and #3 bearings are shown in Figs. 11-35 and 11-36. Based upon the high vibration amplitudes encountered at the #3 bearing, a physical bearing inspection at this
location was considered necessary. The disassembled #3 bearing revealed babbitt
damage, and expanded clearances. It is reasonable to conclude that the majority
of this damage occurred during the initial aborted run. This damaged bearing
was replaced, and it was checked for proper clearances.

A review of the initial vibration data revealed acceptable amplitudes at the
governor end of the turbine. However, the response at the turbine exhaust #2

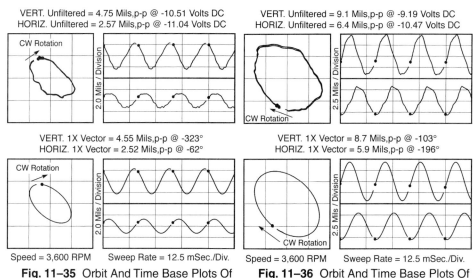

VERT. Unfiltered = 4.75 Mils,p-p @ -10.51 Volts DC
HORIZ. Unfiltered = 2.57 Mils,p-p @ -11.04 Volts DC

VERT. Unfiltered = 9.1 Mils,p-p @ -9.19 Volts DC
HORIZ. Unfiltered = 6.4 Mils,p-p @ -10.47 Volts DC

VERT. 1X Vector = 4.55 Mils,p-p @ -323°
HORIZ. 1X Vector = 2.52 Mils,p-p @ -62°

VERT. 1X Vector = 8.7 Mils,p-p @ -103°
HORIZ. 1X Vector = 5.9 Mils,p-p @ -196°

Speed = 3,600 RPM   Sweep Rate = 12.5 mSec./Div.

Speed = 3,600 RPM   Sweep Rate = 12.5 mSec./Div.

Fig. 11–35 Orbit And Time Base Plots Of Initial Shaft Vibration At #2 Bearing

Fig. 11–36 Orbit And Time Base Plots Of Initial Shaft Vibration At #3 Bearing

bearing, and the generator outboard #3 bearing were unacceptable. In all cases, the vibratory characteristics were dominated by rotational speed 1X motion, and the largest amplitudes appeared on the generator. Based on the available data, and previous experience with this machinery, a balance correction for the generator outboard was computed. At this time a 216 gram, heavy metal weight was installed at 96° at the generator outboard. The physical location of this weight, plus all other balance weights, are shown on the mechanical documentation diagrams presented in Figs. 11-37 and 11-38.

The T/G set was restarted, and full speed data with the 216 gram weight at the generator outboard revealed a significant improvement in vibration amplitudes. Next, a 265 gram, heavy metal weight was installed at 63° at the generator coupling end. The unit was restarted, and full speed data revealed a further reduction in shaft and casing vibration amplitudes. Based upon the installation of these two weights, a complete set of two plane balance calculations were performed. Table 11-9 summarizes the computations based upon shaft proximity probes. A two plane balance was performed between the two vertical probes, and a separate two plane balance was performed between the horizontal probes mounted on each end of the generator. In addition, the casing accelerometers were integrated to displacement, and duplicate calculations based upon the casing motion were performed. The summarized results of these calculations are shown in Table 11-10 for the vertical and horizontal accelerometers. The casing data uses a zero slow roll vector, and all vibratory and weight parameters are consistent with the shaft calculations presented in Table 11-9.

Due to the positive improvement obtained from the original two weights, there was a reluctance to remove these weights. Hence, these weights remained

Table 11–9 Generator Two Plane Balance Calculations Based On Shaft Proximity Probes

	Vertical Shaft Probes						Horizontal Shaft Probes				
	Probe #2VD			Probe #3VD			Probe #2HD			Probe #3HD	
	Mag.	@ Angle		Mag.	@ Angle		Mag.	@ Angle		Mag.	@ Angle
Input						*Input*					
E1=	0.95	@ 212	E2=	1.55	@ 68	E1=	0.90	@ 316	E2=	1.60	@ 152
A1=	4.55	@ 323	A2=	8.70	@ 103	A1=	2.52	@ 62	A2=	5.90	@ 196
B11=	1.50	@ 344	B21=	2.12	@ 106	B11=	0.77	@ 81	B21=	2.08	@ 167
B12=	5.02	@ 309	B22=	7.36	@ 111	B12=	3.07	@ 53	B22=	4.88	@ 183
W1=	265	@ 63	W2=	216	@ 96	W1=	265	@ 63	W2=	216	@ 96
Calculated						*Calculated*					
A1c=	4.97	@ 333	A2c=	7.48	@ 110	A1c=	2.90	@ 79	A2c=	4.88	@ 209
B11c=	2.25	@ 2	B21c=	1.31	@ 153	B11c=	1.48	@ 111	B21c=	0.68	@ 205
B12c=	5.22	@ 319	B22c=	6.32	@ 121	B12c=	3.30	@ 69	B22c=	3.60	@ 196
C11=	3.89	@ 116	C21=	5.25	@ 293	C11=	2.42	@ 224	C21=	2.94	@ 14
C12=	1.26	@ 248	C22=	1.74	@ 247	C12=	0.70	@ 19	C22=	1.59	@ 60
W1e=	265	@ 63	W2e=	216	@ 96	W1e=	265	@ 63	W2e=	216	@ 96
S11=	68.12	@ 307	S21=	50.48	@ 130	S11=	109.5	@ 199	S21=	90.14	@ 49
S12=	171.4	@ 208	S22=	124.1	@ 209	S12=	308.6	@ 77	S22=	135.9	@ 36
Output						*Output*					
U1=	334	@ 259	U2=	310	@ 259	U1=	307	@ 264	U2=	210	@ 231
WA1=	334	@ 79	WA2=	310	@ 79	WA1=	307	@ 84	WA2=	210	@ 51

in place, and the balance sensitivity vectors for each of the four data sets were computed in accordance with equations (11-26) through (11-29). The two plane balance calculations were then performed with equations (11-20) and (11-21). In both cases, the calculations were executed on a Microsoft® Excel spreadsheet.

For purposes of explanation, the slow roll vectors in Table 11-9 are identi-

Table 11–10 Generator Two Plane Balance Calculations Based On Casing Accelerometers

	Vertical Casing Probes						Horizontal Casing Probes				
	Probe #2VA>D			Probe #3VA>D			Probe #2HA>D			Probe #3HA>D	
	Mag.	@ Angle		Mag.	@ Angle		Mag.	@ Angle		Mag.	@ Angle
Input						*Input*					
E1=	0.00	@ 0	E2=	0.00	@ 0	E1=	0.00	@ 0	E2=	0.00	@ 0
A1=	2.33	@ 322	A2=	6.52	@ 107	A1=	1.07	@ 327	A2=	5.93	@ 115
B11=	1.04	@ 346	B21=	0.84	@ 123	B11=	0.23	@ 342	B21=	0.82	@ 137
B12=	2.55	@ 310	B22=	2.79	@ 92	B12=	1.14	@ 320	B22=	2.63	@ 120
W1=	265	@ 63	W2=	216	@ 96	W1=	265	@ 63	W2=	216	@ 96
Calculated						*Calculated*					
A1c=	2.33	@ 322	A2c=	6.52	@ 107	A1c=	1.07	@ 327	A2c=	5.93	@ 115
B11c=	1.04	@ 346	B21c=	0.84	@ 123	B11c=	0.23	@ 342	B21c=	0.82	@ 137
B12c=	2.55	@ 310	B22c=	2.79	@ 92	B12c=	1.14	@ 320	B22c=	2.63	@ 120
C11=	1.81	@ 110	C21=	2.11	@ 260	C11=	0.93	@ 135	C21=	1.86	@ 293
C12=	0.56	@ 249	C22=	3.89	@ 298	C12=	0.15	@ 261	C22=	3.32	@ 291
W1e=	265	@ 63	W2e=	216	@ 96	W1e=	265	@ 63	W2e=	216	@ 96
S11=	146.4	@ 313	S21=	125.6	@ 163	S11=	285.0	@ 288	S21=	142.5	@ 130
S12=	385.7	@ 207	S22=	55.5	@ 158	S12=	1,440.	@ 195	S22=	65.1	@ 165
Output						*Output*					
U1=	364	@ 263	U2=	203	@ 271	U1=	328	@ 248	U2=	236	@ 278
WA1=	364	@ 83	WA2=	203	@ 91	WA1=	328	@ 68	WA2=	236	@ 98

fied by E_1 and E_2. The initial vibration at each plane is specified by A_1 and A_2. The measured vibration response with calibration weights installed are identified by the four vectors $B_{11}, B_{12}, B_{21},$ and B_{22}. All eight of these vector quantities carry the units of Mils,$_{\text{p-p}}$ at Degrees. The two calibration weights W_1 and W_2 have engineering units of Grams at Degrees. These vector quantities provide the *input* portion of the spreadsheet. In the *calculated* section of the spreadsheet the slow roll runout is subtracted from each of the balancing speed vibration vectors to provide visibility of the actual shaft motion. Next, a series of four new vectors identified as $C_{11}, C_{12}, C_{21},$ and C_{22} are presented. These are the differential vibration vectors that represent the change in vibration due to the installation of the calibration weights. If these differential vectors are very small, the associated balance sensitivity vector will be quite large. This is indicative of a condition where the calibration weight was undersized, and it was insufficient to produce a measurable response. The other possibility is that the installed calibration weight at a specific balance plane has minimal effect upon a particular measurement plane. In either case, the diagnostician must have visibility of the magnitude of these differential C vectors.

The four balance sensitivity vectors $S_{11}, S_{12}, S_{21},$ and S_{22} are listed at the bottom of calculated data sections on Table 11-9. These vectors have engineering units of Grams per Mil,$_{\text{p-p}}$ at Degrees. The smaller the magnitude of this number, the more sensitive the location will be to weight addition. Conversely, a large sensitivity vector magnitude reveals an insensitive combination as discussed in the previous paragraph. Finally, the *output* section of the spreadsheet summarized the calculated unbalance at each plane by U_1 and U_2. For situations where weight will be added to the machine, the unbalance vectors are adjusted by 180°, and WA_1 and WA_2 are used to identify the two weight add vectors.

At this point it is meaningful to summarize the results of the four sets of balancing calculations. The magnitude and location of the calculated mass unbalance was extracted from Table 11-9 for the proximity probes, and the companion Table 11-10 for the casing accelerometers. These calculated mass unbalance results are presented in the following Table 11-11:

A good comparison exists between the vertical versus the horizontal calculations. In addition, the shaft and casing computations are in general agreement. From this summation, it may be concluded that the coupling end unbalance is

Table 11–11 Summary Of Calculated Generator Unbalance Vectors

Transducers	Calculated Unbalance At Coupling End	Calculated Unbalance At Outboard End
Shaft Vertical	334 Grams @ 259°	310 Grams @ 259°
Shaft Horizontal	307 Grams @ 264°	210 Grams @ 231°
Casing Vertical	364 Grams @ 263°	203 Grams @ 271°
Casing Horizontal	328 Grams @ 248°	236 Grams @ 278°

approximately 340 grams at an angle of nominally 260°. Similarly, the effective mass unbalance at the outboard plane will have a magnitude of approximately 240 grams at a nominal angle of 260°.

The unbalance should be correctable by a weight addition of 340 grams at 80° at the coupling plane, plus 240 grams at 80° at the outboard end. However, the inboard coupling end balance plane already contains the calibration weight of 265 grams at 63°. The outboard plane has a 216 gram calibration weight at an angle of 96°. Rather than remove these existing calibration weights, and install new correction weights, it makes sense to supplement the existing weights with smaller additional weights to equal the desired correction vectors. To be perfectly clear on this point, the weight vectors are summarized in Table 11-12. The first row in this summary specifies the desired correction weight vectors. The second row of this table presents the magnitude and angular location of the existing calibration weights in both planes. The vector difference between the desired and existing weights is listed as the additional weight vector in the bottom row of Table 11-12. From this vector subtraction, an additional weight of 116 grams at 122° should be installed at the coupling end plane, and 68 grams should be mounted at 19° on the generator outboard correction plane.

Table 11-12 Balance Weight Corrections Based Upon The Vector Difference Between Desired Correction Weights And Existing Calibration Weights Installed In The Generator

Weight	Coupling Correction	Outboard Correction
Desired Weight Correction	340 Grams @ 80°	240 Grams @ 80°
Existing Calibration Weights	265 Grams @ 63°	216 Grams @ 96°
Additional Weight (Vector Difference)	116 Grams @ 122°	68 Grams @ 19°

This *weight add* determination may also be performed by running another set of balancing calculations. In this approach, the last set of vibration response vectors (with both calibration weights installed) are combined with the previously calculated set of four balance sensitivity vectors. The resultant balance calculation are generally referred to as a *trim* balance. That is, the balance calculations are applied to the last set of vibration response measurements, and a final trim balance for the unit is selected. In some instances, the computed trim corrections are insignificant, and the balancing exercise is thereby completed. In the case of this generator, the resultant trim corrections were meaningful. For comparative purposes, Table 11-13 summarizes the vector weight difference from Table 11-12, combined with the average correction calculated from the final set of vibration response vectors. These two sets of potential weight corrections are listed with the actual set of final balance weights that were physically installed at both ends of the this generator.

As previously stated in this chapter, field balancing is often composed of a series of compromise decisions. As shown on Figs. 11-37 and 11-38, the generator only contains eleven axial balance holes at each end of the rotor. Each threaded

Table 11–13 Summary Of Potential And Actual Balance Weight Corrections

Weight	Coupling Correction	Outboard Correction
Additional Weight (Vector Difference)	116 Grams @ 122°	68 Grams @ 19°
Additional Weight (Trim Calculation)	108 Grams @ 102°	90 Grams @ 15°
Actual Weight Installed	94 Grams @ 96°	85 Grams @ 31°

hole is one inch in diameter, and one and a quarter inches deep. Hence, there are definite physical limits to the location and the maximum size of the balance weights. For this type of weight correction, it makes sense to have a stick of all-thread available to use for the fabrication of balance weights. Once weight magnitudes are selected, the equivalent length may be determined, and that length cut off of the section of all-thread. The balance weight should have the threads dressed, and a screw driver slot milled at one end. Before installation, the balance weight should be accurately weighed to verify that the final weight is of the correct mass.

With respect to the current generator balance problem, it was clear from the first two rows of Table 11-13 that the coupling end plane requires another weight of approximately 110 grams at an average angle of 110°. The outboard balance plane needs about 80 grams in the vicinity of 20°. These potential vector weight additions were compared against the available balance weights, plus the empty balance holes. Following this reconciliation, a 94 gram weight was installed at 96° on the coupling end, and an 85 gram weight was installed at 31° on outboard balance plane. These corrections were considered to be acceptably close to the calculated weight requirements, and they were physically achievable on the actual machine.

The validity of these final weight corrections was tested during the next startup. The synchronous 1X vibration amplitudes were again reduced during the startup through the three critical speeds, plus the normal operating speed condition at 3,600 RPM. In the initial startup, transient vibration amplitudes exceeded 20.0 Mils,$_{p\text{-}p}$. After the final balance correction, the maximum vibration was 3.6 Mils,$_{p\text{-}p}$ through the sensitive second critical speed region. At full operating speed of 3,600 RPM the final vibration amplitudes were below 1.0 Mil,$_{p\text{-}p}$ at all measurement locations. This included the runout compensated shaft vibration amplitudes, plus casing displacement. Furthermore, this low vibratory behavior was retained as the unit reached full load and heat soak.

The last issue to be addressed is the final documentation of the balancing project. Certainly the detailed spreadsheets of the balancing calculations are important. Equally significant are the mechanical descriptions of the weights installed (or removed), plus the details of the balance correction planes. As an example of this type of information, Figs. 11-37 and 11-38 were produced.

In retrospect it should be noted that the two plane balance correction on this generator ended up as a simple static shot. This is self-evident by comparing the effective weight add vectors in Figs. 11-37 and 11-38. The effective weight

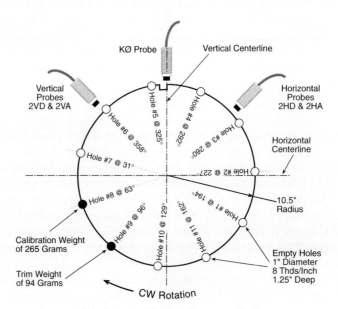

View from the Turbine Governor End
Correction Weights Vectorially = 348 Grams at 71°

Fig. 11–37 Balance Correction Plane At Inboard Coupling End Of Generator

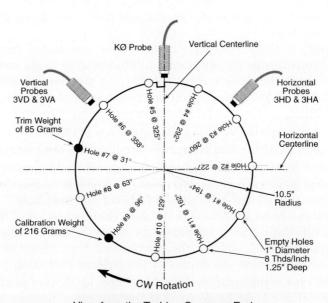

View from the Turbine Governor End
Correction Weights Vectorially = 263 Grams at 79°

Fig. 11–38 Balance Correction Plane At Outboard End Of Generator

angle at the coupling end was 71°, and the outboard correction occurred at 79°. This static correction might seem to be incorrect when viewed against the apparent pivotal response displayed on the orbits in Figs. 11-35 and 11-36. However, the undamped mode shapes in Fig. 11-34 did reveal the presence of a nodal point between the vibration probes at the #2 turbine exhaust, and the #3 generator outboard bearing. Hence, the documented vibratory behavior, the analytical mode shapes, and the effect of the balance corrections are in unison.

Case History 40: Balancing A 36,330 RPM Pinion Assembly

The following example considers a high speed pinion that is driven by a cryogenic expander turbine at a normal speed of 36,330 RPM. As shown in Fig. 11-39, the pinion mates with a low speed bull gear that is coupled to an induction generator that runs at 3,622 RPM. This machinery train extracts energy from the gas stream, and converts this energy into electrical power. This energy extraction also lowers the temperature of the gas passing through the expander.

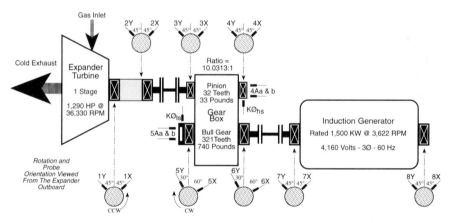

Fig. 11–39 Machinery & Transducer Arrangement On Expander Turbine Driven Generator

From an instrumentation standpoint, the train is well-equipped with X-Y proximity probes at each journal bearing. In addition, dual thrust probes are mounted at the outboard end of the bull gear, and the pinion. To provide synchronous tracking, high and low speed Keyphasor® probes are installed at the blind end of both gear elements. This transducer array was easily accommodated at the outboard end of the bull gear. However, the outboard, or blind end of the pinion, became crowded with five proximity probes installed at the end of a two-inch diameter shaft. To provide a proper installation of probes, with adequate spacing between transducers to eliminate any cross talk between probes, the OEM included a four-inch diameter stub end for the pinion shaft. This instrumentation hub is shown at the right end of the pinion drawing in Fig. 11-40.

This drawing also describes the general pinion configuration. It is noted that the assembly is a 32 tooth, double helical gear. The overall pinion length is

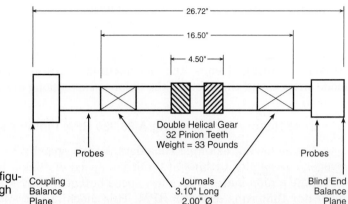

Fig. 11–40 Rotor Configuration On Gear Box High Speed Pinion

26.72 inches, and it runs with an up mesh. Total weight of the pinion assembly was approximately 33 pounds, and the rotor is supported on tilt pad bearings with 13.40 inches between centers. Physically, the pinion was in good mechanical condition. The assembly was slow speed balanced on a flexible pedestal balance machine to residual unbalance levels below 0.1 gram inches. Dial indicator measurements on the pinion revealed a straight assembly with mechanical runout varying between 0.10 and 0.15 Mils, T.I.R.

Analytical calculations revealed the presence of two critical speeds that the pinion must transcend during startup. As shown in Fig. 11-41, the first critical speed occurs at approximately 18,510 RPM. This is an overhung mode that is primarily associated with the coupling hub. The calculated second critical speed appears at 23,380 RPM, as shown in Fig. 11-42. This mode is likewise an overhung mode, and it is primarily driven by the four-inch diameter instrument hub on the outboard end of the pinion. It should be mentioned that the cantilevered masses at the pinion ends constitute 40% of the total weight. The coupling half accounts for 23% of the assembly weight, and the instrument hub provides 17% of the total assembly weight. Hence, the analytical mode shapes describing a pivotal action at each end plane are considered to be realistic. In addition, the sequence of critical speeds makes sense with respect to the weight distribution, plus the greater overhang of the coupling hub.

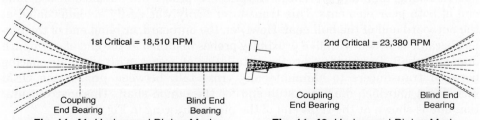

Fig. 11–41 Undamped Pinion Mode Shape At 1st Critical Speed = 18,510 RPM

Fig. 11–42 Undamped Pinion Mode Shape At 2nd Critical Speed =23,380 RPM

The radial vibration measurement planes are located sufficiently close to the planes of maximum motion to provide good representative data of pinion vibration. Furthermore, the only available balance correction planes are at the pinion outboard ends. There is no opportunity to add weights to this assembly. Thus, any pinion balance corrections must be performed by drilling holes into the coupling hub, or by drilling holes into the blind end instrumentation hub.

Initial condition of the pinion assembly is presented in Fig. 11-43 that displays Bode plots of the runout compensated Y-Axis probes at each end of the rotor. From this data, it is apparent that the pinion exhibits a flexible shaft response with a resonance speed range between 16,000 and 28,000 RPM. This range agrees with the previously calculated undamped critical speeds. At full operating speed, the amplitude and phase appear within a reasonably stable plateau region. Thus, any changes in running speed will have minimal effect upon the synchronous 1X vibration vector amplitudes or phase angles. Hence, it was decided to use 36,000 RPM as a suitable balancing speed for this pinion.

It is important to develop an overview of the balance state. From the previous information, it is apparent that the pinion exhibits a flexible shaft response, with the majority of the 1X vibration occurring at the blind outboard end. The pinion is straight, with no evidence of secondary resonances, or gyroscopic effects. Hence, the characteristics are indicative of mass unbalance that can be addressed by weight corrections at the modally effective end planes.

The initial runout compensated 1X shaft orbits and time base plots for both ends of the pinion are shown in Fig. 11-44. The coupling end vectors indicate a weight add angle of 288° with respect to the vertical probe, and 214° from the horizontal probe. Adding 90° to the horizontal probe angle yields 304°, which is compatible with the vertical angle of 288°. Since the pinion is operating above the critical speed, the indicated probe angles are representative of the high spot, which is about 180° from the unbalance. Subtracting 180° from the runout com-

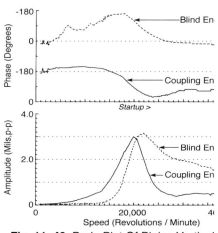

Fig. 11–43 Bode Plot Of Pinion Vertical Probes Before Balancing

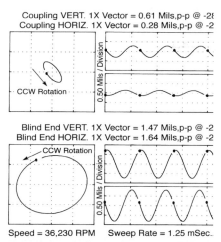

Fig. 11–44 Orbit and Time Plots Of Pinion Synchronous Vibration Before Balancing

pensated angles yields an unbalance location (weight removal) of 108° to 124°.

The outboard end shows a similar response through the resonance, but a higher vibration at full operating speed. The vertical outboard probe shows a phase of 24°, with 286° exhibited by the horizontal probe. Again, adding 90° to the horizontal angle yields 16° high spot indication with respect to the vertical probe position. Thus, the vertical and horizontal probes are identifying high spots of 24° and 16°, respectively. The unbalance location would be 180° removed from the high spot, and the weight removal angle would be 204° to 196°.

As previously mentioned, the weight changes had to be drilled holes, and each weight change was a final correction. Although the outboard end displayed the highest vibration levels, the coupling end was eminently more accessible. Hence, the first correction was made on the coupling end. In accordance with the previous discussion of phase angles, the coupling correction was made by drilling an 1/8" diameter hole at 120°. The selected depth was 5/32", which resulted in a weight removal of 0.22 grams. At the 2.0 inch radius, and the balance speed of 36,000 RPM, this was equivalent to a centrifugal force of 36 pounds. Since the pinion weighed only 33 pounds, this might seem like an excessive correction.

On slower speed machinery, it is common to use initial weights that produce a centrifugal force equal to 5 to 15% of the rotor weight. However, on high speed pinions this traditional *rule of thumb* does not work very well, and extra weight is normally required. It is presumed that the additional weight requirements are due to the higher stiffness of most pinion assemblies, plus the large gear contact forces. As always, if the machine has a variable speed driver, the initial weight correction may be more aggressive, and the unit carefully observed during a controlled startup. Conversely, if a fixed speed driver (e.g. motor) is employed, then the initial weight should be channeled towards a more conservative direction. Fortunately, this unit was equipped with a variable speed driver, and the 0.22 gram correction was made at 120° on the coupling.

The results of this initial correction are presented in Table 11-14 that summarizes the runout compensated vibration response vectors, and the computed single plane mass unbalance calculations from equations (11-4) and (11-6):

Table 11–14 Initial Single Plane Balance Correction On Pinion Coupling Hub

Transducer	Shaft Vibration	Calculated Unbalance
Coupling - Probe 3Y	0.16 Mils,$_\text{p-p}$ @ 270°	0.29 Grams @ 114°
Coupling - Probe 3X	0.04 Mils,$_\text{p-p}$ @ 10°	0.19 Grams @ 123°
Outboard - Probe 4Y	1.43 Mils,$_\text{p-p}$ @ 6°	0.74 Grams @ 42°
Outboard - Probe 4X	1.53 Mils,$_\text{p-p}$ @ 275°	1.08 Grams @ 227°

This initial hole drilled at the hub was quite effective in reducing the coupling end vibration vectors. In fact, the calculated initial unbalance vectors from both coupling probes (0.29 Grams @ 114°, and 0.19 Grams @ 123°) are quite close to the selected correction of 0.22 Grams @ 120°. Hence, this initial correction was adequate to properly balance the coupling end of the pinion.

At the pinion outboard, vibration vectors appeared to be unaffected by the coupling end weight correction. However, the following Table 11-15 summarizes the vibration vector changes due to the coupling end weight removal:

Table 11–15 Differential Vibration Vectors Due To Correction On Pinion Coupling Hub

Transducer	Vibration Change
Coupling - Probe 3Y	0.45 Mils,$_{p-p}$ @ 115°
Coupling - Probe 3X	0.31 Mils,$_{p-p}$ @ 31°
Outboard - Probe 4Y	0.47 Mils,$_{p-p}$ @ 280°
Outboard - Probe 4X	0.33 Mils,$_{p-p}$ @ 173°

From this viewpoint, it is clear that the 0.22 gram correction at the coupling produced a net amplitude response at the outboard bearing that matched the change at the coupling end. This leads to the conclusion that cross effects between balance planes cannot be ignored. The next obvious step would be to make a correction at the outboard instrumentation hub, and then compute a two plane balance solution.

As per the previous discussion, the correction angle for weight removal of the initial outboard unbalance should be between 196° and 204°. The first correction at the coupling end rolled these initial angles slightly, and the new outboard high spot was 6° on the 4Y probe, and 275° with respect to the 4X probe (from Table 11-14). Adding 90° to the 4X probe angle yields a 5° high spot that matches the 6° high spot indicated by the 4Y probe. Hence, the weight at the outboard should be installed at 5° to 6°, or removed at 185°. This angular location was combined with a 1/8" diameter hole, 5/32" deep, for a total weight removed of 0.22 grams at the outboard instrumentation hub. The results of this second balance correction are summarized in Table 11-16.

Table 11–16 Results Of Balance Correction On Pinion Outboard Instrumentation Hub

Transducer	Shaft Vibration	Vibration Change
Coupling - Probe 3Y	0.19 Mils,$_{p-p}$ @ 284°	0.05 Mils,$_{p-p}$ @ 342°
Coupling - Probe 3X	0.02 Mils,$_{p-p}$ @ 0°	0.03 Mils,$_{p-p}$ @ 196°
Outboard - Probe 4Y	1.04 Mils,$_{p-p}$ @ 14°	0.43 Mils,$_{p-p}$ @ 166°
Outboard - Probe 4X	1.15 Mils,$_{p-p}$ @ 283°	0.43 Mils,$_{p-p}$ @ 74°

Following this run with the correction at the outboard blind end plane, it is noted that a good response was evident on the outboard, and minimal change was noted on the coupling end. Now the vectors from both runs are combined into a two plane balance calculation, using equations (11-22) through (11-25) for determination of the balance sensitivity vectors. Equations (11-20) and (11-21) are used for the pinion unbalance calculations. These calculations are based upon the original vibration vectors, plus the required trim from the last set of vibration vectors. The spreadsheet describing the complete array of vectors, and intermediate results for both sets of probes are presented in Table 11-17.

Table 11–17 Pinion Two Plane Balance Calculations Based On Shaft Proximity Probes

	Vertical Probe Calculations					Horizontal Probe Calculations				
	Probe #3Y			Probe #4Y			Probe #3X			Probe #4X
	Mag.	@ Angle		Mag.	@ Angle		Mag.	@ Angle		Mag.	@ Angle
Input						*Input*					
E1=	0.27	@ 324	E2=	0.10	@ 208	E1=	0.24	@ 229	E2=	0.08	@ 116
A1=	0.84	@ 299	A2=	1.37	@ 24	A1=	0.52	@ 221	A2=	1.56	@ 286
B11=	0.39	@ 304	B21=	1.34	@ 4	B11=	0.21	@ 236	B21=	1.46	@ 274
B12=	0.43	@ 308	B22=	0.94	@ 12	B12=	0.23	@ 232	B22=	1.07	@ 282
W1=	-0.220	@ 120	W2=	-0.220	@ 185	W1=	-0.220	@ 120	W2=	-0.220	@ 185
Calculated						*Calculated*					
A1c=	0.61	@ 288	A2c=	1.47	@ 24	A1c=	0.28	@ 214	A2c=	1.64	@ 286
B11c=	0.16	@ 270	B21c=	1.43	@ 6	B11c=	0.04	@ 10	B21c=	1.53	@ 275
B12c=	0.19	@ 284	B22c=	1.04	@ 14	B12c=	0.02	@ 360	B22c=	1.15	@ 283
C11=	0.45	@ 115	C21=	0.47	@ 280	C11=	0.32	@ 31	C21=	0.33	@ 173
C12=	0.05	@ 342	C22=	0.43	@ 166	C12=	0.03	@ 196	C22=	0.43	@ 74
W1e=	0.22	@ 300	W2e=	0.22	@ 5	W1e=	0.22	@ 300	W2e=	0.22	@ 5
S11=	0.489	@ 185	S21=	0.468	@ 20	S11=	0.688	@ 269	S21=	0.667	@ 127
S12=	4.400	@ 23	S22=	0.512	@ 199	S12=	7.333	@ 169	S22=	0.512	@ 291
Output						*Output*					
U1=	0.34	@ 127	U2=	0.80	@ 195	U1=	0.26	@ 123	U2=	0.80	@ 204
WA1=	0.34	@ 307	WA2=	0.80	@ 15	WA1=	0.26	@ 303	WA2=	0.80	@ 24
Trim						*Trim*					
A1=	0.43	@ 308	A2=	0.94	@ 12	A1=	0.23	@ 232	A2=	1.07	@ 282
A1c=	0.19	@ 284	A2c=	1.04	@ 14	A1c=	0.02	@ 360	A2c=	1.15	@ 283
U1=	0.13	@ 138	U2=	0.58	@ 199	U1=	0.04	@ 143	U2=	0.59	@ 211
WA1=	0.13	@ 318	WA2=	0.58	@ 19	WA1=	0.04	@ 323	WA2=	0.59	@ 31

The trim calculations shown in Table 11-17 are identical to the unbalance calculations from the standpoint of equation structure. That is, the same equations (11-20) and (11-21) are used for both sets of unbalance calculations. However, the unbalance values U_1 and U_2 shown in the output section of Table 11-17 are based upon the original vibration vectors (A_1 and A_2). Under the trim section, the U_1 and U_2 unbalance vectors are based upon the final vibration amplitudes recorded after drilling the hole at the outboard hub (i.e., B_{12} and B_{22}). Thus, the ending point for one problem becomes the starting point for the next set of balance calculations.

It should also be noted that holes were drilled for the two calibration weights. This was accommodated by the spreadsheet by entering a negative weight (i.e., -0.220 grams) into the *Input* section of the balance spreadsheet. To maintain consistency throughout the balance calculations, this was converted to a positive weight by adding 180° to the location of the calibration weights, and listing this as an equivalent weight W_{1e} and W_{2e} in the calculation portion of the spreadsheet. Thus, -0.22 grams at 120° was converted to +0.22 grams at 300°, and the outboard end -0.22 grams at 185° was converted to +0.22 grams at 5°.

For direct comparative purposes, the computed initial unbalance and trim vectors are extracted from the spreadsheet in Table 11-17, and are summarized in Table 11-18:

Table 11–18 Summary Of Two Plane Calculations For Original Plus Trim Unbalance Weights

Transducer	Original Unbalance	Residual Trim
Vertical - Coupling 3Y	0.34 Grams @ 127°	0.13 Grams @ 138°
Vertical - Outboard 4Y	0.80 Grams @ 195°	0.58 Grams @ 199°
Horizontal - Coupling 3X	0.26 Grams @ 123°	0.04 Grams @ 143°
Horizontal - Outboard 4X	0.80 Grams @ 204°	0.59 Grams @ 211°

From this data, the vertical Y-Axis, and the horizontal X-Axis probes are providing consistent results both in terms of the initial pinion unbalance, and the required trim correction. The coupling end changes are minimal, but then the coupling end response to the outboard weight change was very small. In the general assessment, the two plane balance calculations called for an additional outboard weight removal of 0.58 grams @ 199° based upon the response from the vertical probes, and a comparable weight removal of 0.59 grams @ 211° based upon the horizontal probes.

The actual action taken at the outboard balance correction plane was to drill another 1/8" diameter hole, 3/8" of an inch deep, for a total weight removal of 0.54 grams. This hole was drilled at an angle of 205°. At this time, a correction was not made at the coupling end of the pinion due to the low residual trim corrections computed for this location. At the balancing speed of 36,000 RPM, the resultant runout compensated vibration vectors, and the vectorial vibration change for all four radial probes due to the outboard weight removal are summarized in Table 11-19 as follows:

Table 11–19 Results Of Second Balance Correction On Pinion Outboard Hub

Transducer	Shaft Vibration	Vibration Change
Coupling - Probe 3Y	0.37 Mils,$_{p-p}$ @ 301°	0.20 Mils,$_{p-p}$ @ 314°
Coupling - Probe 3X	0.23 Mils,$_{p-p}$ @ 235°	0.24 Mils,$_{p-p}$ @ 232°
Outboard - Probe 4Y	0.35 Mils,$_{p-p}$ @ 102°	1.08 Mils,$_{p-p}$ @ 174°
Outboard - Probe 4X	0.30 Mils,$_{p-p}$ @ 328°	0.96 Mils,$_{p-p}$ @ 60°

From Table 11-19 it is clear that the runout compensated shaft vibration vectors are well under control. The weight change at the outboard end also produced a measurable response on the coupling plane. Overall, the outboard plane was significantly improved, but the coupling end was slightly degraded. Rerunning the two plane calculations resulted in the final set of residual trim corrections, as summarized in Table 11-20:

Table 11–20 Calculated Final Two Plane Trim Correction on Pinion

Transducer	Residual Trim
Vertical - Coupling 3Y	0.18 Grams @ 127°
Vertical - Outboard 4Y	0.03 Grams @ 173°
Horizontal - Coupling 3X	0.17 Grams @ 143°
Horizontal - Outboard 4X	0.12 Grams @ 25°

These calculations indicate a minor correction at the outboard end, and a slight correction at the coupling side. Due to the time required to disassemble the outboard housing, it was decided to make a single correction at the coupling plane. The current runout compensated vectors called for a high spot from the coupling Y-Axis probe of 301°, and 235° from the X probe. Again, adding 90° to the X-Axis probe response angle yields 325° which is compatible with the Y-Axis probe angle. The weight removal location would be 180° away; thus, the two coupling probes show a weight removal angle of 121°, and 145°, for the Y and X probes respectively. These angles correspond directly to the results obtained from the two plane calculations summarized in Table 11-20.

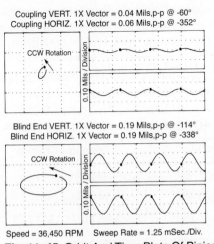

Coupling VERT. 1X Vector = 0.04 Mils,p-p @ -60°
Coupling HORIZ. 1X Vector = 0.06 Mils,p-p @ -352°

CCW Rotation

0.10 Mils / Division

Blind End VERT. 1X Vector = 0.19 Mils,p-p @ -114°
Blind End HORIZ. 1X Vector = 0.19 Mils,p-p @ -338°

CCW Rotation

0.10 Mils / Division

Speed = 36,450 RPM   Sweep Rate = 1.25 mSec./Div.

Fig. 11–45 Orbit And Time Plots Of Pinion Synchronous Vibration After Balancing

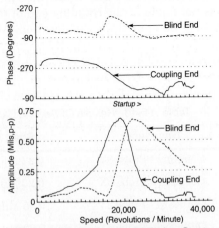

Phase (Degrees)

Blind End
Coupling End
Startup >

Amplitude (Mils,p-p)

Blind End
Coupling End

Speed (Revolutions / Minute)

Fig. 11–46 Bode Plot Of Pinion Startup After Two Plane Balancing

Now, recall that the first correction was a hole drilled at 120°. Based on the positive results of this initial correction, it was decided to extend the depth of the 120° hole to 5/16". This extra hole depth resulted in an additional 0.24 grams removed at the coupling. The results of this last correction were quite positive, as shown in Fig. 11-45 that displays the orbit, and time base plots obtained at 36,450 RPM. These plots are certainly indicative of very acceptable shaft relative motion at full speed. For direct comparison purposes, the initial and final full speed vectors are summarized in Table 11-21:

Table 11–21 Comparison Of Initial Versus Final Runout Compensated Shaft Vibration

Transducer	Initial Vibration	Final Vibration
Coupling - Probe 3Y	0.61 Mils,$_{p-p}$ @ 288°	0.04 Mils,$_{p-p}$ @ 60°
Coupling - Probe 3X	0.28 Mils,$_{p-p}$ @ 214°	0.06 Mils,$_{p-p}$ @ 352°
Outboard - Probe 4Y	1.47 Mils,$_{p-p}$ @ 24°	0.19 Mils,$_{p-p}$ @ 114°
Outboard - Probe 4X	1.64 Mils,$_{p-p}$ @ 286°	0.19 Mils,$_{p-p}$ @ 338°

The balance corrections provided a significant improvement in the rotational speed vectors at full operating speed. Runout compensated vectors after the four balance runs are quite acceptable, and the balancing activities were concluded. The total field time required for execution of this 36,000 RPM pinion balance was approximately 5 hours from initial data acquisition to final startup. The improvement in 1X response is also apparent in the Bode plot shown in Fig. 11-46. Note that the response across the entire speed domain is attenuated. In particular, the peak amplitudes exhibited during transition through the critical speeds have been decreased from maximum levels of 3.2 Mils,$_{p-p}$ to current peaks of 0.7 Mils,$_{p-p}$. Also note that the definition between the first and second critical speeds has improved. This is probably indicative of reduced modal coupling due to the improved balance state.

The pinion thrust probes responded to improvements in the pinion balance. For example, Table 11-22 compares the 1X axial vectors before and after balancing: The thrust probes measure position and vibration of the outboard hub. It is reasonable for radial motion to manifest as an axial wobble (e.g., as on mode shape plots). Hence, as the balance state is improved, the axial motion should decrease. This anticipated behavior is supported by the vectors in Table 11-22.

Table 11–22 Comparison Of Initial Versus Final Pinion Axial Shaft Vibration

Transducer	Initial Vibration	Final Vibration
Axial - Probe 4Aa	0.46 Mils,$_{p-p}$ @ 22°	0.08 Mils,$_{p-p}$ @ 190°
Axial - Probe 4Ab	0.47 Mils,$_{p-p}$ @ 214°	0.10 Mils,$_{p-p}$ @ 328°

THREE PLANE BALANCE

As previously mentioned, there are unbalance arrays that consist of couple and static combinations that may require unusual weight corrections. These types of dynamic unbalance problems may be solved in two planes, or they may be more conveniently addressed in three correction planes. For instance, consider Fig. 11-47 describing a three mass rotor kit. This is the same device used for discussion of the two plane unbalance, with the addition of a third disk. Again, X-Y proximity probes are mounted inboard of the bearings, and their location allows proper detection of synchronous motion. The two outboard masses are positioned at the quarter points of the rotor to enhance the pivotal mode response. The new mass was placed at the midspan to drive the translational (first) mode.

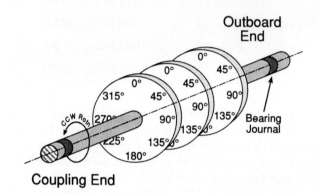

Fig. 11–47 Three Mass Rotor Kit

The undamped mode shape at the translational first critical speed is depicted in Fig. 11-48 at the critical speed of 4,200 RPM. From this translational mode shape it is clear that the midspan mass is the most modally effective location for correcting vibration response amplitudes through the first critical speed. Certainly the two outboard masses will provide a positive contribution in correcting the translational mode, but the major influence occurs at the midspan.

The calculated undamped mode shape at the pivotal or second critical speed is displayed in Fig. 11-49. This mode occurs at a nominal speed of 8,000 RPM,

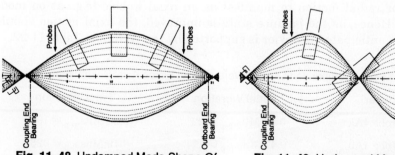

Fig. 11–48 Undamped Mode Shape Of Three Mass Rotor Kit At 4,200 RPM

Fig. 11–49 Undamped Mode Shape Of Three Mass Rotor Kit At 8,000 RPM

and is characterized by the midspan nodal point, plus the out-of-phase behavior across the nodal point. It is clear that the midspan mass will be ineffective for pivotal mode corrections, since it is very close to a shaft nodal point. Thus, a couple correction between the outboard planes would be required to control the vibratory characteristics at the pivotal resonance.

The initial behavior of this three mass rotor kit is shown in the Bode plot in Fig. 11-50. Within this transient data, the previously mentioned first critical at 4,200 RPM, and the pivotal second mode at 8,000 RPM are clearly visible. In addition, two lower frequency resonances at 2,000 and 3,000 RPM are also apparent. Further analysis reveals that the 2,000 RPM response is a structural support resonance for the rotor kit. It was not highly visible in either the one or two plane examples, due to the lower rotor masses and lower excitation forces in this speed domain. The polar plots in Figs. 11-51 and 11-52 show this 2,000 RPM resonance as a small inside loop occurring just above the plot origin (i.e., just above slow roll speed).

The 3,000 RPM peak is diagnosed as a horizontal translational resonance that is coupled to the vertical translational resonance at 4,200 RPM. This is the classic *split critical* condition where a machine passes through a soft horizontal resonance, followed by a harder vertical resonance. The polar plots in Figs. 11-51 and 11-52 exhibit identical behavior through this speed range. That is, the rotor translates through one critical (horizontal), and then it translates through the vertical critical. Additional verification was provided by the horizontal proximity probes (not shown). These orthogonal transducers confirmed the split critical hypothesis by displaying much higher amplitudes at 3,000 versus 4,200 RPM.

The casual observer might identify these resonances as four different shaft critical speeds. This would result in substantial levels of grief and confusion if one attempted to modally correct four criticals on a rotor system that really contained only a split first, combined with a pivotal second mode. In all cases, it should be recognized that a proper understanding of the motion characteristics

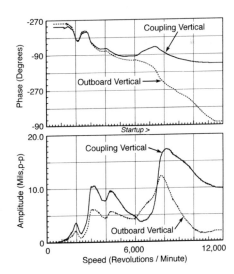

Fig. 11–50 Bode Plot Of Vertical Probes On Three Mass Rotor Kit Before Balancing

of the mechanical system will allow the most rapid and cost-effective field balance solution.

The initial vertical polar plots that are equivalent to the Bode plot in Fig. 11-50 are presented in Figs. 11-51 and 11-52. The double loop for the split first critical is clearly evident on both the coupling and the outboard plots. Constructing a directional vector from the plot origin to the end of the first mode response results in a nominal 93° angle at the coupling end, combined with a 118° angle at the outboard. These 1st mode midspan correction vector directions are shown on these plots with the heavy vector lines.

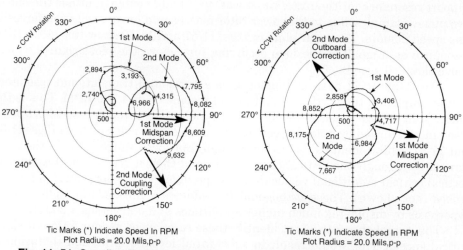

Fig. 11–51 Coupling End Polar Plot Of Three Mass Rotor Kit Before Balancing

Fig. 11–52 Outboard End Polar Plot Of Three Mass Rotor Kit Before Balancing

The polar plots also display vector directions for weight additions to correct the pivotal response at 8,000 RPM. Again, these vectors were determined by evaluating the rotor response specifically associated with the pivotal resonance. The weight add locations were determined by drawing a vector from the start of the pivotal resonance loop, to the end of the loop. Specifically, the coupling end plot displays a desired directional correction in the vicinity of 143° (start to end of 2nd mode). Simultaneously, the outboard end plot requires a directional correction in the vicinity of 322° (start to end of 2nd mode loop). A 179° difference exists between these directions which helps substantiate the couple nature of the pivotal mode and the associated unbalance.

Not surprisingly, these weight add directions are almost identical to the two plane example previously discussed. In fact, this should be the case, since the same disks and the same balance weight distributions were used for the two plane and three plane balancing examples. It is meaningful to note that the addition of the heavy midspan mass had minimal influence upon the pivotal mode. Thus, the physical measurements confirm and reinforce the theoretical calculations and the validity of the presented mode shapes.

As before, it was determined that a couple weight set could be installed to correct the pivotal second mode. At the inboard coupling end of the rotor, a 0.5 gram weight was placed at 135° (143° desired from the polar plot). At the outboard mass, another 0.5 gram weight was also installed. At this location, the weight was placed at the 315° hole (322° desired from the polar plot).

The midspan weight should be located between 93° and 118° as previously discussed. The heavy midspan wheel already contained weight at the 112° hole, and the 90° hole was empty. Hence, a 1.2 gram weight was installed at the midspan disk at 90°. A larger balance weight was used at the midspan, due to the significantly heavier weight of the new disk installed in the middle of the rotor.

The results of this three plane correction for two modes are displayed in Figs. 11-53 and 11-54. Note that the initial data required a 20 Mil$_{p-p}$ plot radius to contain the high vibration amplitudes. However, following the three plane correction, vibration levels were substantially reduced, and both plots were easily accommodated by a 2.5 Mil$_{p-p}$ radius. The residual first mode and the residual unbalance for the second mode are quite evident on these two polar plots.

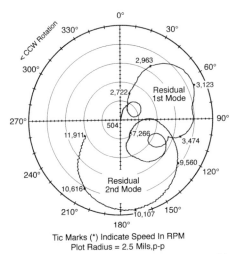

Tic Marks (*) Indicate Speed In RPM
Plot Radius = 2.5 Mils,p-p

Fig. 11–53 Coupling End Polar Plot Of Three Mass Rotor Kit After Balancing

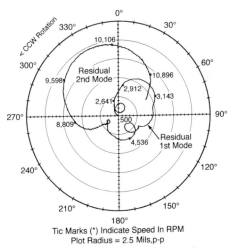

Tic Marks (*) Indicate Speed In RPM
Plot Radius = 2.5 Mils,p-p

Fig. 11–54 Outboard End Polar Plot Of Three Mass Rotor Kit After Balancing

Overall, the polar plotting techniques provided the necessary visibility to allow a three plane balance correction based upon two measurement planes. It is also quite clear that the final vibration vector from one mode (e.g. the first mode), becomes the initial runout vector for the next mode (e.g., the second mode). Hence, even uncoupled modes may be adversely influenced by the residual unbalance from an earlier mode.

The equations and techniques used for the single plane and the two plane balance may also be applied to more complicated mechanical systems. For example, a three plane unbalance problem on a linear mechanical system may be described by the following equations that address the vibration response measured at three independent locations due to unbalance at three separate planes:

$$\vec{A_1} = \left\{ \frac{\vec{U_1}}{\vec{S_{11}}} \right\} + \left\{ \frac{\vec{U_2}}{\vec{S_{12}}} \right\} + \left\{ \frac{\vec{U_3}}{\vec{S_{13}}} \right\} \qquad (11\text{-}30)$$

$$\vec{A_2} = \left\{ \frac{\vec{U_1}}{\vec{S_{21}}} \right\} + \left\{ \frac{\vec{U_2}}{\vec{S_{22}}} \right\} + \left\{ \frac{\vec{U_3}}{\vec{S_{23}}} \right\} \qquad (11\text{-}31)$$

$$\vec{A_3} = \left\{ \frac{\vec{U_1}}{\vec{S_{31}}} \right\} + \left\{ \frac{\vec{U_2}}{\vec{S_{32}}} \right\} + \left\{ \frac{\vec{U_3}}{\vec{S_{33}}} \right\} \qquad (11\text{-}32)$$

where: $\vec{A_1}$ = Initial Vibration Vector at Bearing 1 (Mils,$_\text{p-p}$ at Degrees)

$\vec{A_2}$ = Initial Vibration Vector at Bearing 2 (Mils,$_\text{p-p}$ at Degrees)

$\vec{A_3}$ = Initial Vibration Vector at Bearing 3 (Mils,$_\text{p-p}$ at Degrees)

$\vec{S_{11}}$ = Sensitivity Vector at Bearing 1 to Weight at Plane 1 (Grams/Mil,$_\text{p-p}$ at Degrees)

$\vec{S_{12}}$ = Sensitivity Vector at Bearing 1 to Weight at Plane 2 (Grams/Mil,$_\text{p-p}$ at Degrees)

$\vec{S_{13}}$ = Sensitivity Vector at Bearing 1 to Weight at Plane 3 (Grams/Mil,$_\text{p-p}$ at Degrees)

$\vec{S_{21}}$ = Sensitivity Vector at Bearing 2 to Weight at Plane 1 (Grams/Mil,$_\text{p-p}$ at Degrees)

$\vec{S_{22}}$ = Sensitivity Vector at Bearing 2 to Weight at Plane 2 (Grams/Mil,$_\text{p-p}$ at Degrees)

$\vec{S_{23}}$ = Sensitivity Vector at Bearing 2 to Weight at Plane 3 (Grams/Mil,$_\text{p-p}$ at Degrees)

$\vec{S_{31}}$ = Sensitivity Vector at Bearing 3 to Weight at Plane 1 (Grams/Mil,$_\text{p-p}$ at Degrees)

$\vec{S_{32}}$ = Sensitivity Vector at Bearing 3 to Weight at Plane 2(Grams/Mil,$_\text{p-p}$ at Degrees)

$\vec{S_{33}}$ = Sensitivity Vector at Bearing 3 to Weight at Plane 3 (Grams/Mil,$_\text{p-p}$ at Degrees)

$\vec{U_1}$ = Mass Unbalance Vector at Plane 1 (Grams at Degrees)

$\vec{U_2}$ = Mass Unbalance Vector at Plane 2 (Grams at Degrees)

$\vec{U_3}$ = Mass Unbalance Vector at Plane 3 (Grams at Degrees)

Addition of a calibration weight W_1 at plane 1 results in the following set of three new vector response equations:

$$\overrightarrow{B}_{11} = \left\{ \frac{\overrightarrow{U}_1 + \overrightarrow{W}_1}{\overrightarrow{S}_{11}} \right\} + \left\{ \frac{\overrightarrow{U}_2}{\overrightarrow{S}_{12}} \right\} + \left\{ \frac{\overrightarrow{U}_3}{\overrightarrow{S}_{13}} \right\} \qquad \textbf{(11-33)}$$

$$\overrightarrow{B}_{21} = \left\{ \frac{\overrightarrow{U}_1 + \overrightarrow{W}_1}{\overrightarrow{S}_{21}} \right\} + \left\{ \frac{\overrightarrow{U}_2}{\overrightarrow{S}_{22}} \right\} + \left\{ \frac{\overrightarrow{U}_3}{\overrightarrow{S}_{23}} \right\} \qquad \textbf{(11-34)}$$

$$\overrightarrow{B}_{31} = \left\{ \frac{\overrightarrow{U}_1 + \overrightarrow{W}_1}{\overrightarrow{S}_{31}} \right\} + \left\{ \frac{\overrightarrow{U}_2}{\overrightarrow{S}_{32}} \right\} + \left\{ \frac{\overrightarrow{U}_3}{\overrightarrow{S}_{33}} \right\} \qquad \textbf{(11-35)}$$

where: $\overrightarrow{B}_{11}$ = Vibration Vector at Bearing 1 with Weight W_1 at Plane 1 (Mils,$_{\text{p-p}}$ at Degrees)

$\overrightarrow{B}_{21}$ = Vibration Vector at Bearing 2 with Weight W_1 at Plane 1 (Mils,$_{\text{p-p}}$ at Degrees)

$\overrightarrow{B}_{31}$ = Vibration Vector at Bearing 3 with Weight W_1 at Plane 1 (Mils,$_{\text{p-p}}$ at Degrees)

$\overrightarrow{W}_1$ = Calibration Weight Vector at Plane 1 (Grams at Degrees)

Removal of calibration weight W_1 at balance plane 1; plus the addition of a calibration weight W_2 at plane 2 produces the next set of vector equations:

$$\overrightarrow{B}_{12} = \left\{ \frac{\overrightarrow{U}_1}{\overrightarrow{S}_{11}} \right\} + \left\{ \frac{\overrightarrow{U}_2 + \overrightarrow{W}_2}{\overrightarrow{S}_{12}} \right\} + \left\{ \frac{\overrightarrow{U}_3}{\overrightarrow{S}_{13}} \right\} \qquad \textbf{(11-36)}$$

$$\overrightarrow{B}_{22} = \left\{ \frac{\overrightarrow{U}_1}{\overrightarrow{S}_{21}} \right\} + \left\{ \frac{\overrightarrow{U}_2 + \overrightarrow{W}_2}{\overrightarrow{S}_{22}} \right\} + \left\{ \frac{\overrightarrow{U}_3}{\overrightarrow{S}_{23}} \right\} \qquad \textbf{(11-37)}$$

$$\overrightarrow{B}_{32} = \left\{ \frac{\overrightarrow{U}_1}{\overrightarrow{S}_{31}} \right\} + \left\{ \frac{\overrightarrow{U}_2 + \overrightarrow{W}_2}{\overrightarrow{S}_{32}} \right\} + \left\{ \frac{\overrightarrow{U}_3}{\overrightarrow{S}_{33}} \right\} \qquad \textbf{(11-38)}$$

where: $\overrightarrow{B}_{12}$ = Vibration Vector at Bearing 1 with Weight W_2 at Plane 2 (Mils,$_{p-p}$ at Degrees)

$\overrightarrow{B}_{22}$ = Vibration Vector at Bearing 2 with Weight W_2 at Plane 2 (Mils,$_{p-p}$ at Degrees)

$\overrightarrow{B}_{32}$ = Vibration Vector at Bearing 3 with Weight W_2 at Plane 2 (Mils,$_{p-p}$ at Degrees)

$\overrightarrow{W}_2$ = Calibration Weight Vector at Plane 2 (Grams at Degrees)

Removal of calibration weight W_2 at balance plane 2, followed by the addition of calibration weight W_3 at plane 3 produces the final set of vector equations:

$$\overrightarrow{B}_{13} = \left\{\frac{\overrightarrow{U}_1}{\overrightarrow{S}_{11}}\right\} + \left\{\frac{\overrightarrow{U}_2}{\overrightarrow{S}_{12}}\right\} + \left\{\frac{\overrightarrow{U}_3 + \overrightarrow{W}_3}{\overrightarrow{S}_{13}}\right\} \qquad (11\text{-}39)$$

$$\overrightarrow{B}_{23} = \left\{\frac{\overrightarrow{U}_1}{\overrightarrow{S}_{21}}\right\} + \left\{\frac{\overrightarrow{U}_2}{\overrightarrow{S}_{22}}\right\} + \left\{\frac{\overrightarrow{U}_3 + \overrightarrow{W}_3}{\overrightarrow{S}_{23}}\right\} \qquad (11\text{-}40)$$

$$\overrightarrow{B}_{33} = \left\{\frac{\overrightarrow{U}_1}{\overrightarrow{S}_{31}}\right\} + \left\{\frac{\overrightarrow{U}_2}{\overrightarrow{S}_{32}}\right\} + \left\{\frac{\overrightarrow{U}_3 + \overrightarrow{W}_3}{\overrightarrow{S}_{33}}\right\} \qquad (11\text{-}41)$$

where: $\overrightarrow{B}_{13}$ = Vibration Vector at Bearing 1 with Weight W_3 at Plane 3 (Mils,$_{p-p}$ at Degrees)

$\overrightarrow{B}_{23}$ = Vibration Vector at Bearing 2 with Weight W_3 at Plane 3 (Mils,$_{p-p}$ at Degrees)

$\overrightarrow{B}_{33}$ = Vibration Vector at Bearing 3 with Weight W_3 at Plane 3 (Mils,$_{p-p}$ at Degrees)

$\overrightarrow{W}_3$ = Calibration Weight Vector at Plane 3 (Grams at Degrees)

In each of these equations, the first subscript defines the measurement plane, and the second subscript describes the correction or balance plane. Hence, this twelve equation array contains fifteen known vector quantities, i.e., the twelve measured vibration response vectors, A_1 through B_{33}, plus the three calibration weight vectors W_1 through W_3. The calculation procedure initially solves for the nine unknown balance sensitivity vectors, and finally the three unknown mass unbalance vectors, U_1, U_2, and U_3.

As derived during the two plane balance, the balance sensitivity vectors are determined by dividing the appropriate calibration weight vector by the differential vibration amplitude vectors. Rather than deriving each individual expression, it can be shown that the balance sensitivity vectors for a three plane balance are computed with the following expressions:

$$\vec{S}_{11} = \left\{ \frac{\vec{W}_1}{\vec{B}_{11} - \vec{A}_1} \right\} \tag{11-42}$$

$$\vec{S}_{12} = \left\{ \frac{\vec{W}_2}{\vec{B}_{12} - \vec{A}_1} \right\} \tag{11-43}$$

$$\vec{S}_{13} = \left\{ \frac{\vec{W}_3}{\vec{B}_{13} - \vec{A}_1} \right\} \tag{11-44}$$

$$\vec{S}_{21} = \left\{ \frac{\vec{W}_1}{\vec{B}_{21} - \vec{A}_2} \right\} \tag{11-45}$$

$$\vec{S}_{22} = \left\{ \frac{\vec{W}_2}{\vec{B}_{22} - \vec{A}_2} \right\} \tag{11-46}$$

$$\vec{S}_{23} = \left\{ \frac{\vec{W}_3}{\vec{B}_{23} - \vec{A}_2} \right\} \tag{11-47}$$

$$\vec{S}_{31} = \left\{ \frac{\vec{W}_1}{\vec{B}_{31} - \vec{A}_3} \right\} \tag{11-48}$$

$$\vec{S}_{32} = \left\{ \frac{\vec{W}_2}{\vec{B}_{32} - \vec{A}_3} \right\} \tag{11-49}$$

$$\vec{S}_{33} = \left\{ \frac{\vec{W}_3}{\vec{B}_{33} - \vec{A}_3} \right\} \tag{11-50}$$

Equations (11-42) through (11-50) may be derived from equation (11-30) through (11-41), or they may be expanded directly from equation (11-17). In either case, the differential vibration vectors should always be examined to insure that a realistic change was imposed by the calibration weight. If the installed calibration weight does not produce a significant change in the vibration vector(s), the balance calculations may end up based upon *noise level* data. Hence, it is mandatory to review the vectors to verify that the calibration weight(s) have produced an acceptable and measurable change in the vibration response vector(s). This fundamental concept is just as important in a single plane balance as it is in a three plane problem.

Once the vibration data has been acquired, and the validity of the vectors checked — the nine sensitivity vectors in equations (11-42) through (11-50) may be calculated. These vectors are then combined with the initial vibration response vectors to compute the mass unbalance at each correction plane. One solution of the three initial expressions (11-30) to (11-32) is presented in the following equations (11-51), (11-52), and (11-53).

$$\vec{U}_1 = \frac{\vec{A}_1\left\{\dfrac{1}{\vec{S}_{22}\vec{S}_{33}} - \dfrac{1}{\vec{S}_{23}\vec{S}_{32}}\right\} + \vec{A}_2\left\{\dfrac{1}{\vec{S}_{13}\vec{S}_{32}} - \dfrac{1}{\vec{S}_{12}\vec{S}_{33}}\right\} + \vec{A}_3\left\{\dfrac{1}{\vec{S}_{12}\vec{S}_{23}} - \dfrac{1}{\vec{S}_{13}\vec{S}_{22}}\right\}}{\left\{\dfrac{1}{\vec{S}_{11}\vec{S}_{22}\vec{S}_{33}} - \dfrac{1}{\vec{S}_{12}\vec{S}_{21}\vec{S}_{33}} - \dfrac{1}{\vec{S}_{11}\vec{S}_{23}\vec{S}_{32}} + \dfrac{1}{\vec{S}_{13}\vec{S}_{21}\vec{S}_{32}} + \dfrac{1}{\vec{S}_{12}\vec{S}_{23}\vec{S}_{31}} - \dfrac{1}{\vec{S}_{13}\vec{S}_{22}\vec{S}_{31}}\right\}} \quad \textbf{(11-51)}$$

$$\vec{U}_2 = \frac{\vec{A}_1\left\{\dfrac{1}{\vec{S}_{21}\vec{S}_{33}} - \dfrac{1}{\vec{S}_{23}\vec{S}_{31}}\right\} + \vec{A}_2\left\{\dfrac{1}{\vec{S}_{13}\vec{S}_{31}} - \dfrac{1}{\vec{S}_{11}\vec{S}_{33}}\right\} + \vec{A}_3\left\{\dfrac{1}{\vec{S}_{11}\vec{S}_{23}} - \dfrac{1}{\vec{S}_{13}\vec{S}_{21}}\right\}}{\left\{\dfrac{1}{\vec{S}_{12}\vec{S}_{21}\vec{S}_{33}} - \dfrac{1}{\vec{S}_{11}\vec{S}_{22}\vec{S}_{33}} - \dfrac{1}{\vec{S}_{11}\vec{S}_{23}\vec{S}_{32}} + \dfrac{1}{\vec{S}_{13}\vec{S}_{21}\vec{S}_{32}} + \dfrac{1}{\vec{S}_{12}\vec{S}_{23}\vec{S}_{31}} - \dfrac{1}{\vec{S}_{13}\vec{S}_{31}\vec{S}_{22}}\right\}} \quad \textbf{(11-52)}$$

$$\vec{U}_3 = \frac{\vec{A}_1\left\{\dfrac{1}{\vec{S}_{31}\vec{S}_{22}} - \dfrac{1}{\vec{S}_{21}\vec{S}_{32}}\right\} + \vec{A}_2\left\{\dfrac{1}{\vec{S}_{11}\vec{S}_{32}} - \dfrac{1}{\vec{S}_{12}\vec{S}_{31}}\right\} + \vec{A}_3\left\{\dfrac{1}{\vec{S}_{12}\vec{S}_{21}} - \dfrac{1}{\vec{S}_{11}\vec{S}_{22}}\right\}}{\left\{\dfrac{1}{\vec{S}_{12}\vec{S}_{21}\vec{S}_{33}} - \dfrac{1}{\vec{S}_{11}\vec{S}_{22}\vec{S}_{33}} + \dfrac{1}{\vec{S}_{11}\vec{S}_{23}\vec{S}_{32}} - \dfrac{1}{\vec{S}_{13}\vec{S}_{21}\vec{S}_{32}} - \dfrac{1}{\vec{S}_{12}\vec{S}_{23}\vec{S}_{31}} + \dfrac{1}{\vec{S}_{13}\vec{S}_{22}\vec{S}_{31}}\right\}} \quad \textbf{(11-53)}$$

These equations may be structured in different ways, and still obtain the correct results. A manual solution to the above three plane equations is prohibitive due to the complexity of the manipulations, plus the time required for performing these calculations. Once again, these equations have been programmed on a Microsoft® Excel spreadsheet, and a representative example is presented in Table 11-23. The vibration data set was obtained from a three mass rotor, with probes mounted close to each disk. In a manner similar to the other balancing examples in this chapter, this spreadsheet describes the entire array of traditional calculations. This includes runout subtraction, vector changes, sensitivity vectors, plus the unbalance calculations. In addition, the supplemental calculations of response prediction and trim calculations are included.

Table 11–23 Three Plane Balance Calculations Based On Shaft Proximity Probes

	Vertical Probe Calculations								
	Probe #1V			Probe #2V			Probe #3V		
	Mag.	@ Angle		Mag.	@ Angle		Mag.	@ Angle	
Input									
E1=	0.35	@ 126	E2=	0.44	@ 218	E3=	0.22	@ 67	
A1=	2.22	@ 264	A2=	2.61	@ 246	A3=	2.17	@ 161	
B11=	0.56	@ 113	B21=	1.05	@ 224	B31=	3.14	@ 209	
B12=	0.96	@ 287	B22=	2.39	@ 248	B32=	2.66	@ 116	
B13=	3.14	@ 224	B23=	2.16	@ 260	B33=	0.22	@ 171	
W1=	4.780	@ 270	W2=	5.860	@ 248	W3=	4.780	@ 180	
Calculated									
A1c=	2.49	@ 269	A2c=	2.23	@ 251	A3c=	2.20	@ 167	
B11c=	0.23	@ 93	B21c=	0.61	@ 228	B31c=	3.32	@ 211	
B12c=	1.30	@ 292	B22c=	2.02	@ 254	B32c=	2.52	@ 120	
B13c=	3.21	@ 230	B23c=	1.86	@ 269	B33c=	0.35	@ 209	
C11=	2.72	@ 90	C21=	1.68	@ 80	C31=	2.33	@ 253	
C12=	1.39	@ 68	C22=	0.24	@ 45	C32=	1.90	@ 62	
C13=	2.03	@ 179	C23=	0.73	@ 21	C33=	1.95	@ 340	
W1e=	4.78	@ 270	W2e=	5.86	@ 248	W3e=	4.78	@ 180	
S11=	1.757	@ 180	S21=	2.845	@ 190	S31=	2.052	@ 17	
S12=	4.216	@ 180	S22=	24.42	@ 203	S32=	3.084	@ 186	
S13=	2.355	@ 1	S23=	6.548	@ 159	S33=	2.451	@ 200	
Output									
U1=	4.01	@ 114	U2=	5.77	@ 40	U3=	3.33	@ 45	
WA1=	4.01	@ 294	WA2=	5.77	@ 220	WA3=	3.33	@ 225	
Predict									
Z1=	4.20	@ 270	Z2=	5.80	@ 225	Z3=	3.25	@ 225	
C11z=	2.39	@ 90	C21z=	1.48	@ 80	C31z=	2.05	@ 253	
C12z=	1.37	@ 45	C22z=	0.23	@ 22	C32z=	1.88	@ 39	
C13z=	1.38	@ 224	C23z=	0.50	@ 66	C33z=	1.33	@ 25	
A1c=	0.09	@ 247	A2c=	0.12	@ 259	A3c=	0.31	@ 106	
A1=	0.31	@ 140	A2=	0.54	@ 226	A3=	0.50	@ 90	
Trim									
A1=	0.40	@ 122	A2=	0.45	@ 227	A3=	0.38	@ 85	
A1c=	0.06	@ 96	A2c=	0.07	@ 304	A3c=	0.18	@ 107	
U1=	0.15	@ 136	U2=	0.44	@ 290	U3=	0.13	@ 117	
WA1=	0.15	@ 316	WA2=	0.44	@ 110	WA3=	0.13	@ 297	

It is apparent that these three plane calculations are quite complicated, and they offer many opportunities for computational errors during the execution of the vector manipulations. The casual form offered by the single plane balance calculations is missing, and the equation simplification possible with the two plane balance is also unattainable. It is also clear that balance calculations for additional planes (e.g., 4 or more planes) can become extraordinarily complicated. Due to this calculation complexity, it is generally advisable to perform these calculations with computer assistance. Since circa 1980, several dedicated programs have been available for these types of balancing calculations. Generally, these multiplane balance programs are configured as matrix solutions.

The matrix solutions can be easily configured as a single plane or a multiple plane problem. The number of balance planes is typically equal to the number of independent vibration measurement planes n. Generally, matrix solutions can accommodate 10 or 20 planes. Although measurement accuracy and true influence of each correction plane upon each measurement plane becomes questionable with more than three planes, the vector calculations can be significantly improved by using a matrix of the following configuration:

$$
\begin{bmatrix} \vec{U}_1 \\ \vec{U}_2 \\ \vec{U}_3 \\ \dots \\ \vec{U}_n \end{bmatrix} = \begin{bmatrix} \vec{S}_{11} & \vec{S}_{12} & \vec{S}_{13} & \dots & \vec{S}_{1n} \\ \vec{S}_{21} & \vec{S}_{22} & \vec{S}_{23} & \dots & \vec{S}_{2n} \\ \vec{S}_{31} & \vec{S}_{32} & \vec{S}_{33} & \dots & \vec{S}_{3n} \\ \dots & \dots & \dots & \dots & \dots \\ \vec{S}_{n1} & \vec{S}_{n2} & \vec{S}_{n3} & \dots & \vec{S}_{nn} \end{bmatrix} \times \begin{bmatrix} \vec{A}_1 \\ \vec{A}_2 \\ \vec{A}_3 \\ \dots \\ \vec{A}_n \end{bmatrix} \qquad \textbf{(11-54)}
$$

The form of this matrix is the same as the original equation (11-6). Basically, identical statements are made in both expressions. That is, mass unbalance is equal to vibration times the balance sensitivity vectors for each plane. Again, solution of single plane problems are quite easy, and two or three plane problems may be conveniently handled on a spreadsheet. When balance calculations are performed for four or more planes, the spreadsheet complexity becomes significant, and matrix solutions become the only reasonable approach. In any of these cases, the fundamental concepts of linearity apply, and the machinery diagnostician must be meticulous in the acquisition of transient and steady state vibration data. Furthermore, the mode shapes must be understood, and the most effective balance planes fully utilized.

STATIC COUPLE CORRECTIONS

The previously discussed three plane balance requires three balance correction planes in addition to three separate and independent vibration measurement planes. When dealing with a single machine case, there are very few machines that will allow meaningful shaft vibration measurements at more than two locations. Even on machines that may have three or more accessible rotor balance planes, the techniques described in the last section cannot be applied due to a lack of suitable vibration measurement planes.

One solution to this dilemma resides in a minor restructure of the logic associated with the weight corrections. Specifically, two balance planes may be defined and associated with a weight couple, and a third plane may be used for a static correction. Other combinations of balancing weights may also be used, but at this point simply consider a situation of two measurement planes combined with three correction planes. Presume that the machine contains a static unbalance U, combined with a couple or moment unbalance M. The previous general two plane equations (11-7) and (11-8) may be modified to represent this particular situation as follows:

$$\overrightarrow{A}_1 = \left\{ \frac{\overrightarrow{U}}{\overrightarrow{S}_{1u}} \right\} + \left\{ \frac{\overrightarrow{M}}{\overrightarrow{S}_{1m}} \right\} \qquad \textbf{(11-55)}$$

$$\overrightarrow{A}_2 = \left\{ \frac{\overrightarrow{U}}{\overrightarrow{S}_{2u}} \right\} + \left\{ \frac{\overrightarrow{M}}{\overrightarrow{S}_{2m}} \right\} \qquad \textbf{(11-56)}$$

In these two expressions, the sensitivity vectors are associated with the static unbalance and the moment unbalance, The calibration run for the static unbalance U is performed exactly the same as previously described. The calibration run for the moment unbalance M is executed with a coupled pair of weights. That is, two identical weights are installed on the rotor at two different balance planes, and the weights are mounted 180° apart. This combined angular and axial separation of the pair of weights produces a balance couple, or moment upon the rotor.

Generally, one of the couple weights is used as the reference for the moment calculations. It is recognized that the other couple weight is the same size (and probably mounted at the same radius from the rotor centerline). Also, the second couple weight is always located 180° away from the reference couple weight. The sketch in Fig. 11-55 is representative of typical weight placement on a rotor. The static correction at the rotor midspan is located at approximately 330°. The couple correction at the coupling end balance plane is shown at nominally 300°, with an angle of 120° for the second couple weight at the outboard end balance plane.

The static-couple balance calculations are performed exactly the same as a standard two plane balance. The static calibration and correction weights are applied at the same balance plane. Typically, this would be close to the midspan

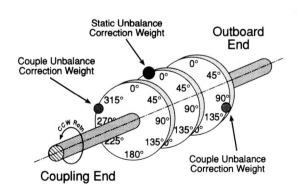

Fig. 11–55 Combined Static And Couple Unbalance On A Three Mass Rotor Kit

of the rotor. The couple or moment calibration and correction weights are applied at two separate balance planes that are often close to the ends of the rotor. The final correction moment weights are scaled from the calibration moment weight in terms of magnitude and angular location.

Using this approach, it is possible to perform a three plane balance using two measurement planes. It is also meaningful to recognize that the midspan (static) correction weight is the balance weight that will provide the greatest influence upon the first (translational) critical speed. Similarly, the outboard (couple) correction weights are the balance weights that will provide the greatest influence upon the second (pivotal) critical speed. Hence, as with all balancing projects, the actual deflected shape of the rotor must be considered in conjunction with any applied weights.

MULTIPLE SPEED CALCULATIONS

In all of the previously discussed cases, the balance sensitivity vectors describe the rotor response to unbalance at specified correction planes, at one operating speed. For balancing flexible rotors that operate above several critical speeds, it may be necessary to identify several distinct balance correction speeds. For example, consider a rotor that operates above the second critical speed. If difficulty is encountered in balancing this rotating assembly, then balance data unique to each critical speed may be acquired to allow additional visibility of modal response. For instance, the following five speeds may be identified:

- ○ Slow Roll
- ○ Below First Critical Speed
- ○ Between First and Second Critical Speed
- ○ Above Second Critical
- ○ At Full Operating Speed

Obviously the first speed point is used to identify shaft runout characteristics at slow roll speeds (e.g., 500 RPM). As previously discussed, these runout vectors are vectorially subtracted from each of the balancing speed vibration vectors in order to determine the actual rotor vibration response vectors. In this example, vibration data would be acquired at each speed, during the initial run, and during each of the calibration weight runs. This data would then be used to calculate unbalance correction vectors at each of the four speed points. If this was a two plane balance, then the resultant information would consist of two plane weight corrections for each speed point.

If the static-couple technique discussed in the last section is applied, then the resultant data would consist of corrections for three balance planes. In either case, the final array of information could be visually examined, and a reasonable weight set selected. Sometimes one or more points may be violated in order to achieve acceptable vibration levels at operating speed. In many instances the modes are uncoupled. Thus, a midspan correction would not influence the second critical, and a couple correction at the end planes would have minimal effect upon the first critical.

Another way of evaluating this multiple speed data would be to apply a Least Squares Balancing calculation in the manner defined by E.J. Gunter and A. P. Palazzolo[9] at the University of Virginia. Their technique provides the lowest overall response across the entire set of variable speed data points, but it does not necessarily guarantee the lowest possible response at any specific operating speed. Again, human judgment must be applied to select the most reasonable set of balance correction weights, to achieve the lowest possible vibration amplitudes, in the time available for field balancing of the rotor.

RESPONSE PREDICTION

Evaluation of the suitability of a particular set of balance weights can be achieved by installing the weights into the machine, and running the machine up to normal operating speed. If the weight selection is correct, then the synchronous vibration amplitudes will decrease. However, if the weight selection is incorrect due to any number of reasons, then the results can vary from embarrassing to catastrophic. In addition, operating and fueling costs for large units generally prohibit unnecessary runs on the machinery.

Hence, it is often desirable to evaluate the influence of balance weights upon a rotor before installation of the weights and running the machine. Since the mechanical systems under consideration are presumed to be linear, the most direct way to perform this evaluation consists of using the calculated balance sensitivity vectors to estimate the vibration response. For instance, consider a single plane balance that is performed in accordance with equations (11-4) and (11-6). Assume that a correction weight Z is to be installed on the rotor. Although the mass unbalance vector U has been computed, the correction weight and angle often vary from the ideal or calculated unbalance. The predicted vibration change due to the installation of this correction weight is calculated by a version of equation (11-2) as follows:

$$\overrightarrow{A}_z = \frac{\overrightarrow{Z}}{\overrightarrow{S}} \qquad \textbf{(11-57)}$$

where: $\overrightarrow{A}_z$ = Predicted Vibration Vector Due To Weight Z (Mils,$_{\text{p-p}}$ at Degrees)

            $\overrightarrow{S}$ = Sensitivity Vector To Unbalance (Grams/Mils,$_{\text{p-p}}$ at Degrees)

            $\overrightarrow{Z}$ = Weight Vector To Be Installed On Machine (Grams at Degrees)

[9] Alan B. Palazzolo and Edgar J. Gunter, "Multimass Flexible Rotor Balancing By The Least Squares Error Method," *Rotor Dynamics Course Notes, Part IV*, (University of Virginia, Charlottesville, Virginia, 1981), pp 92-123.

The predicted change in rotor vibration must now be summed with the initial runout compensated vibration to determine the new anticipated vibration vector as shown in equation (11-58):

$$\overrightarrow{A_{C\text{-}New}} = \overrightarrow{A_z} + \overrightarrow{A_c} \qquad\qquad (11\text{-}58)$$

where: $\overrightarrow{A_c}$ = Previous Runout Compensated Shaft Vibration (Mils,$_{p\text{-}p}$ at Degrees)

$\overrightarrow{A_{C\text{-}New}}$ = Predicted Runout Compensated Shaft Vibration Vector (Mils,$_{p\text{-}p}$ at Degrees)

The predicted runout compensated vibration response vector $A_{C\text{-}New}$ may now be summed with the shaft runout vector E to obtain the estimated direct synchronous vibration vector as presented in the expression (11-59):

$$\overrightarrow{A_C} = \overrightarrow{A_{C\text{-}New}} + \overrightarrow{E} \qquad\qquad (11\text{-}59)$$

This procedure for calculating the effect of a balance weight, equation (11-57), combining this weight effect with the initial runout compensated shaft vibration to determine the new shaft vibration vector, equation (11-58), plus the addition of the slow roll runout, equation (11-59), will prove to be quite accurate for a linear mechanical system. Experience on a variety of machines has shown that the vector magnitude is generally within ±0.5 Mils,$_{p\text{-}p}$, and the predicted angle is within ±30° of the final measured value. If this type of correlation does not exist, then consideration should be given to bearing configuration, system nonlinearities, shaft preloads, thermal effects, fluidic forces, bearing instability, or other mechanisms that could influence the synchronous response of the rotor.

To demonstrate some typical results of these response prediction calculations, consider the information shown in Table 11-24. This data was extracted from case history 38, which describes a two plane balance of a large double flow low pressure turbine rotor. The predicted shaft vibration vectors are shown in column 2 of this table, and the actual measured values with the balance weights installed are listed in the center column 3. The differential vibration vector amplitudes are presented in column 4, and the difference between the measured and the predicted phase angles are listed in column 5.

Table 11–24 Comparison Of Predicted Versus Measured Vibration Response Vectors

Shaft Vibration Transducer	Predicted 1X Vibration Response Vectors	Measured 1X Vibration Response Vector	Differential Vector Amplitude	Differential Phase Angle
Brg #2 - #2Y	0.48 Mils,$_{p\text{-}p}$ @ 226°	0.95 Mils,$_{p\text{-}p}$ @ 208°	0.51 Mils,$_{p\text{-}p}$	18°
Brg #2 - #2X	0.76 Mils,$_{p\text{-}p}$ @ 55°	0.91 Mils,$_{p\text{-}p}$ @ 57°	0.15 Mils,$_{p\text{-}p}$	2°
Brg #3 - #3Y	1.50 Mils,$_{p\text{-}p}$ @ 225°	1.47 Mils,$_{p\text{-}p}$ @ 255°	0.77 Mils,$_{p\text{-}p}$	30°
Brg #3 - #3X	2.16 Mils,$_{p\text{-}p}$ @ 121°	2.31 Mils,$_{p\text{-}p}$ @ 129°	0.35 Mils,$_{p\text{-}p}$	8°

Note that the predictions from the two X-Axis probes are much closer than the results from the Y-Axis transducers. Since this machine rotates counterclockwise in elliptical bearings, the shaft orbits will tend to be elliptical. Often, the major orbit axis will be closer to the measurement direction of the X probes (45° to the right of vertical). Simultaneously, the minor orbit axis will tend to be aligned closer to the measurement direction of the Y probes (45° left of vertical). For this specific mechanical configuration, a balance weight change will generally be more apparent on the X probes rather than the Y probes. This basic characteristic of the orbit shape combined with the location of the X-Y proximity probes often yield greater accuracy for the X probe balance sensitivity vectors. In addition, the response measured by the #3Y probe is also influenced by the reverse precession on the generator rotor that is briefly discussed in case history 38. This additional influence contributes to the inaccuracy of the Y probe measurements for this specific example. Overall, the behavior documented in Table 11-24 is quite typical. Again, it must be recognized that the *predict* calculations are fully dependent on the machine geometry, external influences, shaft preloads, and the measurement accuracy of the input data.

This response prediction technique is applicable to any situation where the balance sensitivity vectors are known, or can be calculated from available data. This procedure can be extended to two, three, or multiplane balance problems. In each case, the weight added (or removed) at each plane is vectorially multiplied by the appropriate sensitivity vectors, and the individual vibration response vectors summed at each measurement plane. An example of a three plane predict is shown in the spreadsheet in Table 11-23. In this case, three different weights provided three different vibration response vectors that had to be vectorially summed at each bearing. This predicted weight response was then summed with the last shaft vibration vector to determine the predicted runout compensated shaft vibration. For direct comparison with field instrumentation readings, the slow roll runout may be vectorially added to each runout compensated result to determine the final anticipated vibration vector at each measurement plane.

This tool provides a realistic means for evaluating the effect of different weight combinations on a rotor. For example, the results due to the application of various correction weights to a series of multiple speed calculations can be quantified and compared. Although these *predict* calculations may not be totally accurate due to minor nonlinearities in the mechanical system, they will generally discriminate between an acceptable and an unacceptable weight correction. Thus, reasonable weight correction decisions may be based upon documented *predict* calculations rather than *gut feel* speculation.

TRIM CALCULATIONS

Following the installation of balance correction weights on a rotor, the machine may still exhibit some residual vibration. In many cases this may be due to an inability to match the required balance corrections with field weights. In other situations, nonlinearities in the mechanical system, the influence of external forces, or a variation in the balance sensitivity vectors may be responsible. On machines with fluid film bearings, it has been shown that journal centerline position within the bearing changes with the applied loads. Thus, as rotor balance is improved, the unbalance force at each bearing will be reduced, and changes in eccentricity position, oil film thickness, and bearing stiffness and damping coefficients will occur. These changes in bearing characteristics may appear as alterations of the balance sensitivity vectors. Hence, the balance sensitivity vectors that were established under one balance condition have changed in magnitude or angle as the balance state changes.

Regardless of the source of the variations, there is often a need to perform a final trim balance calculation, combined with a final weight adjustment. The execution of this *trim balance* calculation is virtually identical to the original balance calculations. For example, a two plane balance solution was previously presented in equations (11-20) and (11-21). In order to perform a trim calculation, the initial vibration vectors (A_{1c} and A_{2c}) are simply replaced by the current vibration response vectors. The resulting calculation will identify the final trim balance weights. This approach reuses the same balance sensitivity vectors. Although some variations to the S vectors have probably occurred, this final trim calculation will generally bring the machinery vibration amplitudes into a desirable and acceptable range.

Specific examples of trim balance calculations are presented and discussed as part of the turbine generator case history 39. In addition, the high speed pinion balancing case history 40 also applies this concept of trim computations. This is a highly useful technique that may be employed in conjunction with the response prediction calculations, and the multiple speed calculations. On large machines that incur significant startup and shutdown costs, the machinery diagnostician should use all of the available computational tools at his or her disposal to provide a mechanically suitable and cost-effective field balance in the minimum number of runs.

The diagnostician should also be fully aware that multiple or repetitive trim calculations are indicative of abnormal machinery behavior. If the machine calls for a new trim shot during successive startups, this suggests that some type of mechanical malfunction may be active. In this situation, the diagnostician should start looking for other malfunctions, such as a loose impeller, entrained fluids, loose thrust collar, progressive bearing damage, or a cracked shaft.

BALANCING FORCE CALCULATIONS

It is always important to consider the physical forces applied to a rotor by the addition or removal of balance weights. It can be shown that the centrifugal force ($mr\omega^2$) from a balance weight may be calculated by:

$$Force_{pounds} = M_{grams} \times R_{inches} \times \left\{ \frac{RPM}{4,000} \right\}^2 \qquad \textbf{(11-60)}$$

where: F = Centrifugal Force (Pounds)
M = Mass of Correction Weight (Grams)
R = Correction Weight Radius (Inches)
RPM = Rotational Speed (Revolutions / Minute)

In this expression, the units for a balance weight correction in grams are combined with the balance weight radius measured in inches, and the machine speed in revolutions per minute. The resultant centrifugal force is a radial force that the balance weight applies to the rotor. This lateral force occurs at the location of the balance weight correction plane.

This same expression for centrifugal force may be expressed in other units. For instance, if the balance weight is measured in ounces, and the weight radius remains in inches, the common unbalance units of *ounce-inches* may be used. To calculate the centrifugal force F in pounds at any speed, the following equation (11-61) may be applied:

$$Force_{pounds} = W_{ounces} \times R_{inches} \times \left\{ \frac{RPM}{750} \right\}^2 \qquad \textbf{(11-61)}$$

where: W_{ounces} = Weight of Correction Weight (Ounces)

For large machines, the balance weight may be expressed in pounds, with the weight radius remaining in inches. For this condition, the unbalance units of *pound-inches* may be used to calculate the centrifugal force F in pounds, at any speed with equation (11-62):

$$Force_{pounds} = W_{pounds} \times R_{inches} \times \left\{ \frac{RPM}{188} \right\}^2 \qquad \textbf{(11-62)}$$

where: W_{pounds} = Weight of Correction Weight (Pounds)

The magnitude of this centrifugal force is a significant consideration during field balancing. Specifically, if the applied force is very small, such as with a small calibration weight, the machine response will be minimal, and the resultant vibration change vectors and the associated balance sensitivity vectors will be understated. In this condition, the validity of the balancing calculations will probably be compromised.

At the other extreme, if the centrifugal force from the calibration or balance weight is large, then the rotor may be damaged due to excessive radial forces. In essence, the addition of a small weight will be useless, and a large weight may be dangerous. Hence, it is necessary to evaluate the appropriateness of the weight selection. A reasonable method to quantify the severity of the balance or correction weight centrifugal force is to compare this force against the rotor weight.

For example, if a 100 gram weight is installed at a 10 inch radius, on a machine that runs at 5,000 RPM; the centrifugal force is computed from equation 39 to be 1,562 Pounds. If this balance weight is placed on a 10,000 pound rotor, the centrifugal force is only 16% of the total rotor weight. However, if this same weight is mounted on a 500 pound rotor; the resultant centrifugal force is more than three times the total rotor weight at 312%. This type of centrifugal force to rotor weight ratio is normally considered to be excessive.

In most field balancing situations, it is customary to install initial calibration weights that produce centrifugal forces in the vicinity of 5% to 15% of the rotor weight. The conservative 5% value is generally applied to machines that rapidly accelerate up to speed. Units such as electric motors, or high speed expanders fall under this category. If the previous equation (11-61) is equated to 5% of the rotor weight, the initial calibration weight (in ounce-inches) may be computed with the following equation (11-63):

$$ 5\%\ Cal.\ Weight_{Ounce\ Inches}\ =\ Rotor\ Weight_{pounds} \times \left\{ \frac{168}{RPM} \right\}^2 \qquad \textbf{(11-63)} $$

The more aggressive 15% force value is generally reserved for machines such as steam turbines, or machines driven by turbines, that may be started up slowly, and easily tripped if the weight is incorrect. Again, if the previous equation (11-61) is equated to 15% of the rotor weight in pounds, the initial calibration weight (in ounce-inches) may be computed with equation (11-64):

$$ 15\%\ Cal.\ Weight_{Ounce\ Inches}\ =\ Rotor\ Weight_{pounds} \times \left\{ \frac{291}{RPM} \right\}^2 \qquad \textbf{(11-64)} $$

Forces from balance correction weights are often in the range of 10% to 50% of rotor weight. Certainly larger weights may be successfully installed, but good engineering judgment and caution must be applied. The diagnostician must also be aware of the fact that large eccentric rotor elements, or shaft bows, will produce substantial radial forces on a rotor assembly. For example, if equation (11-62) is modified to consider a rotor element with a weight measured in pounds, and an eccentricity of the mass center measured in Mils, the resultant force at a given speed may be determined by the following equation (11-65):

$$ Force_{pounds}\ =\ Element_{pounds} \times Eccentricity_{mils} \times \left\{ \frac{RPM}{5,930} \right\}^2 \qquad \textbf{(11-65)} $$

If a single disk or wheel is properly balanced, and that element is installed eccentrically, equation (11-65) will identify the effective force based upon the element weight, eccentricity, and speed. Similarly, if a balanced rotor has a significant shaft bow, the midspan eccentricity will be indicative of the displacement of the mass centerline. Thus, the rotor weight, bow radius, and speed may be used to determine the effective force upon the rotating assembly.

It must always be recognized that field trim balancing of a rotor does have limitations. If the required balance correction weights are excessive, then there may be internal damage to the rotor, and disassembly of the machine and inspection of the rotor may be warranted. In other situations, if the balance weights cannot be placed at modally effective locations, then the calculated correction weights may be abnormally large, and they may generate excessive radial forces. To assist in defining reasonable calibration or correction weights placed at modally effective locations, Table 11-25 is offered for consideration:

Table 11–25 Typical Balance Sensitivity Magnitudes For Various Rotor Configurations That Have Been Field Balanced With Modally Effective Weights At Available Balance Planes

Machine Type	Rotor Weight (Pounds)	Rotor Speed (RPM)	Sensitivity (Gram-Inches/Mil)
High Speed Pinion	35	36,000	1.0 to 4.0
Turbo-Expander	900	14,000	20 to 100
Two Pole Motor	3,000	3,600	50 to 250
Two Pole Generator	10,000	3,600	500 to 2,000
Steam Turbine	18,000	3,600	3,500 to 15,000
Single Shaft Gas Turbine	21,000	5,200	150 to 600[a]
Induced & Forced Draft Fans	6,000	1,200	1,000 to 5,000

[a]Based on weights located on external coupling hubs.

Table 11-25 represents a general overview of balance sensitivity magnitudes obtained during field balancing on various types of machines. The installed balance weights are located at modally effective planes, or at the only available locations (e.g., generator end planes). This table is useful for a quick *sanity* check on the magnitude of a calibration weight. For instance, if a two pole motor is to be balanced, a reasonable sensitivity value from Table 11-25 would be 120 Gram-Inches per Mil. For a 4.0 inch weight correction radius, and a desired 1.0 Mil,$_{p-p}$ response, the calibration weight should be approximately 30 grams (=1 x 120 / 4).

In the overview, it must also be acknowledged that field balancing is generally not a cure for a damaged or distressed rotor. Field balancing is ideally suited to provide a final trim to a rotor running in the actual operating mechanical configuration (case, bearings, seals, and coupling). Field balancing is definitely *not* a *Magic Elixir* to be casually applied whenever vibration levels increase.

BALANCE WEIGHT SPLITTING

Machines such as large industrial fans are typically balanced by welding weights directly to the wheel. Other rotors, such as high speed pinions, are often balanced by drilling holes in the vertical faces on each side of the gears. For these types of balance corrections, the weight additions or removals can be performed at any required angular location. However, there are many machines, such as steam turbines and generators, that are equipped with radial or axial balance rings that contain a series of circumferential drilled and tapped balance holes.

For machines equipped with balance rings, the angular spacing between holes may vary from 5° on large diameter rotors, to a coarse set of holes at 30° on small rotors. It is rare for a calculated balance correction weight to be coincident with one of the existing holes. Hence, it is often necessary to split a required balance correction weight into two or more holes. The fundamental requirement for this weight split is for the vector summation of all the installed weights to be equal to the desired balance correction weight vector. For an example of this type of situation, consider the vector diagram in Fig. 11-56.

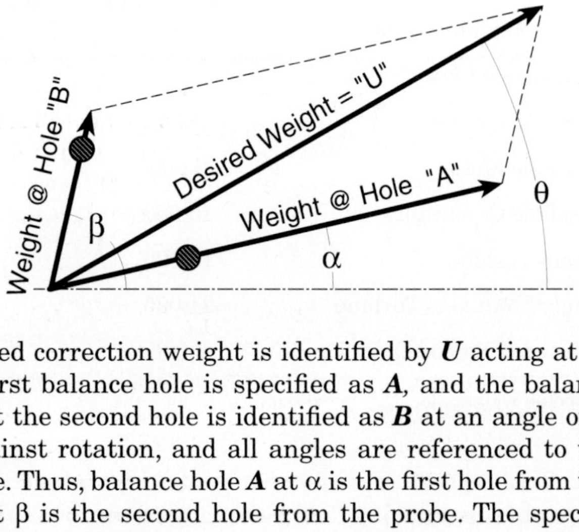

Fig. 11–56 Balance Weight Splitting Between Two Adjacent Holes

In this sketch, the desired correction weight is identified by U acting at an angle θ. The weight at the first balance hole is specified as A, and the balance hole angle is α. The weight at the second hole is identified as B at an angle of β. All angles are measured against rotation, and all angles are referenced to the location of the vibration probe. Thus, balance hole A at α is the first hole from the probe, and balance hole B at β is the second hole from the probe. The specific relationship between these three weight vectors can be stated as:

$$\boxed{\vec{U} = \vec{A} + \vec{B}}$$

(11-66)

This vector equation may be expanded into a summation of vertical and horizontal forces as shown in the following two expressions:

$$U \times \sin\theta = A \times \sin\alpha + B \times \sin\beta$$
$$U \times \cos\theta = A \times \cos\alpha + B \times \cos\beta$$

This pair of expressions may now be solved for the weights A and B with the following equations (11-67) and (11-68):

$$A = U \times \left\{ \frac{\cos\theta \times \sin\beta - \sin\theta \times \cos\beta}{\cos\alpha \times \sin\beta - \sin\alpha \times \cos\beta} \right\} \qquad \text{(11-67)}$$

$$B = U \times \left\{ \frac{\sin\theta \times \cos\beta\alpha - \cos\theta \times \sin\alpha}{\cos\alpha \times \sin\beta - \sin\alpha \times \cos\beta} \right\} \qquad \text{(11-68)}$$

As an example of vector hole splitting, assume that a balance calculation concludes with a required correction of 50 grams at an angle of 50°. Further assume that the available balance holes are located at 30° and 60°. The required weight for the first hole at 30° may be calculated using equation (11-67).

$$A = 50 \left\{ \frac{\cos 50° \times \sin 60° - \sin 50° \times \cos 60°}{\cos 30° \times \sin 60° - \sin 30° \times \cos 60°} \right\} = 50 \left\{ \frac{0.643 \times 0.866 - 0.766 \times 0.500}{0.866 \times 0.866 - 0.500 \times 0.500} \right\}$$

$$A = 50 \left\{ \frac{0.557 - 0.383}{0.750 - 0.250} \right\} = 50 \left\{ \frac{0.174}{0.500} \right\} = 17.4 \text{ Grams}$$

Similarly, the weight required at the 60° hole is determined by (11-68):

$$B = 50 \left\{ \frac{\sin 50° \times \cos 30° - \cos 50° \times \sin 30°}{\cos 30° \times \sin 60° - \sin 30° \times \cos 60°} \right\} = 50 \left\{ \frac{0.766 \times 0.866 - 0.643 \times 0.500}{0.866 \times 0.866 - 0.500 \times 0.500} \right\}$$

$$B = 50 \left\{ \frac{0.663 - 0.321}{0.750 - 0.250} \right\} = 50 \left\{ \frac{0.342}{0.500} \right\} = 34.2 \text{ Grams}$$

Hence, this required 50 gram balance weight at 50° may be exactly duplicated by installing:

$$\text{Hole A} = 17.4 \text{ Grams} \angle 30°$$

$$\text{Hole B} = 34.2 \text{ Grams} \angle 60°$$

The validity of these calculations may be verified by performing a vector summation of the results. In this case, the sum of the above two vectors are equal to 50 grams at 50°. This concept may be expanded to several holes by simply extending the equation array. Obviously, the effectiveness of hole splitting diminishes as the angular spread increases. When it is necessary to install a large amount of weight in a balance ring, there may not be sufficient holes available to accommodate the required weight. In these situations the diagnostician should consider the use of heavy weights fabricated from tungsten based alloys. These weights are nearly twice as heavy as equivalent steel weights, and they may provide the necessary mass when a large correction is required.

WEIGHT REMOVAL

As previously mentioned, some machines require weight removal to implement a proper rotor balance. This weight may be removed by grinding acceptable surfaces, or by drilling holes into the rotor. Although grinding operations are common for many shop balancing procedures, they are difficult to quantify during field balancing. Any type of field grinding should always be approached with great care. The two main areas of concern are excessive weight removal that exceeds the desired balance change, and the potential for weakening the machine parts subjected to the grinding wheel. Major mechanical failures have been traced to excessive grinding on critical machine surfaces.

Fortunately, weight removal by drilling holes is much more controllable and definable. For instance, Table 11-26 is offered for reference purposes. This tabular summary documents the weight associated with standard diameter holes (inches) drilled to various depths (inches). These weights are based upon a 118° drill bit tip, and steel density of 0.283 Pounds per Inch^3.

Table 11-26 Hole Weights In Grams For 118° Drill Bit In Steel

Diam.	1/8"	3/16"	1/4"	5/16"	3/8"	7/16"	1/2"	9/16"	5/8"	11/16"	3/4"	1"
Depth												
3/32"	0.108	0.199	0.275	-	-	-	-	-	-	-	-	-
1/8"	0.158	0.310	0.472	0.615	0.708	-	-	-	-	-	-	-
5/32"	0.207	0.421	0.669	0.922	1.151	1.325	1.414	-	-	-	-	-
3/16"	0.256	0.532	0.866	1.230	1.594	1.928	2.202	2.388	-	-	-	-
7/32"	0.305	0.642	1.063	1.538	2.037	2.531	2.990	3.385	3.686	3.863	-	-
1/4"	0.354	0.753	1.260	1.846	2.480	3.134	3.778	4.382	4.917	5.353	-	-
9/32"	0.404	0.864	1.457	2.153	2.923	3.737	4.566	5.379	6.148	6.842	7.433	-
5/16"	0.453	0.975	1.654	2.461	3.367	4.341	5.354	6.376	7.379	8.332	9.206	11.32
11/32"	0.502	1.086	1.851	2.769	3.810	4.944	6.142	7.374	8.610	9.821	10.98	14.47
3/8"	0.551	1.196	2.048	3.077	4.253	5.547	6.929	8.371	9.841	11.31	12.75	17.62
13/32"	0.601	1.307	2.245	3.384	4.696	6.150	7.717	9.368	11.07	12.80	14.52	20.77
7/16"	0.650	1.418	2.442	3.692	5.139	6.753	8.505	10.36	12.30	14.29	16.30	23.92
15/32"	0.699	1.529	2.639	4.000	5.582	7.357	9.293	11.36	13.53	15.78	18.07	27.07
1/2"	0.748	1.640	2.836	4.308	6.026	7.960	10.08	12.36	14.77	17.27	19.84	30.22
17/32"	0.798	1.750	3.033	4.615	6.469	8.563	10.87	13.36	16.00	18.76	21.61	33.38
9/16"	0.847	1.861	3.230	4.923	6.912	9.166	11.66	14.35	17.23	20.25	23.39	36.53
19/32"	0.896	1.972	3.427	5.231	7.355	9.769	12.44	15.35	18.46	21.74	25.16	39.68
5/8"	0.945	2.083	3.624	5.539	7.798	10.37	13.23	16.35	19.69	23.23	26.93	42.83
21/32"	0.995	2.193	3.821	5.846	8.241	10.98	14.02	17.34	20.92	24.72	28.70	45.98
11/16"	1.044	2.304	4.018	6.154	8.685	11.58	14.81	18.34	22.15	26.21	30.48	49.13
23/32"	1.093	2.415	4.215	6.462	9.128	12.18	15.60	19.34	23.38	27.70	32.25	52.28
3/4"	1.142	2.526	4.412	6.770	9.571	12.79	16.38	20.34	24.61	29.19	34.02	55.44
25/32"	1.192	2.637	4.608	7.077	10.01	13.39	17.17	21.33	25.84	30.67	35.80	58.59
13/16"	1.241	2.747	4.805	7.385	10.46	13.99	17.96	22.33	27.08	32.16	37.57	61.74
27/32"	1.290	2.858	5.002	7.693	10.90	14.59	18.75	23.33	28.31	33.65	39.34	64.89
7/8"	1.339	2.969	5.199	8.001	11.34	15.20	19.54	24.32	29.54	35.14	41.11	68.04
29/32"	1.389	3.080	5.396	8.308	11.79	15.80	20.32	25.32	30.77	36.63	42.89	71.19
15/16"	1.438	3.191	5.593	8.616	12.23	16.40	21.11	26.32	32.00	38.12	44.66	74.34
31/32"	1.487	3.301	5.790	8.924	12.67	17.01	21.90	27.32	33.23	39.61	46.43	77.50
1"	1.536	3.412	5.987	9.232	13.12	17.61	22.69	28.31	34.46	41.10	48.20	80.65

The tabulated weights may be scaled up or down to accommodate materials of different density by using the following equation (11-69):

$$Weight\ Removed_{grams}\ =\ Table\ Value \times \left\{ \frac{Material\ Density}{0.283} \right\} \quad \textbf{(11-69)}$$

Densities for many common metals are presented in Appendix B of this text. For instance, if a 1/2" diameter hole is drilled to a depth of 5/8" in steel, the weight removed from Table 11-26 is found to be 13.23 grams. If the drilled material was aluminum, the density from Appendix B is equal to 0.0975 Pounds per Inch^3, and the amount of weight removed is computed by:

$$Aluminum\ Weight\ Removed_{grams}\ =\ 13.23\ Grams \times \left\{ \frac{0.0975}{0.283} \right\} = 4.56\ Grams$$

If the material subjected to drilling was heavier than steel, for example brass with a density of 0.308 Pounds per Inch^3, the weight removed by drilling the same sized hole may be determined by:

$$Brass\ Weight\ Removed_{grams}\ =\ 13.23\ Grams \times \left\{ \frac{0.308}{0.283} \right\} = 14.40\ Grams$$

Hence, using the combination of Table 11-26 and the actual material density, the diagnostician may accurately estimate the weight removed by drilling holes in any material. In addition, different hole depths may be extrapolated from Table 11-26. For instance, if a hole depth of 1.25 inches was selected, the weight for a 1.0" may be combined with a 1/4" hole to approximate the weight of metal removed. Note, there will be some variation with the actual metal removed due to the inclusion of two drill tip cone weights from the table.

SHOP BALANCING

The vast majority of the rotating elements for process machines are shop balanced with slow speed balancing machines. As discussed in previous portions of this text, this is a generally acceptable procedure if all rotating components are dimensionally correct, and all components are properly balanced before and during assembly. In addition, rotor flexibility and the number of balance resonances (critical speeds) to be transcended, and the associated shaft mode shapes at each resonance should be considered. The general characteristics of these slow speed shop balancing machines will be discussed in this section.

There are also two families of high speed balancing machines that are operational in some shops. This includes the vertical high speed balance pits used for small expander and compressor impellers. These machines may run in excess of 60,000 RPM, and they are generally employed for balancing individual components. Vertical balancing machines in this category are typically used by OEMs for component balancing, and they are generally not used for assembled rotors.

A few large horizontal machines are in existence that can accommodate turbomachinery rotors up to 40,000 pounds in weight, and they can operate at speeds in excess of 20,000 RPM. This second category of high speed balancing machines are installed in vacuum chambers to minimize aerodynamic (windage) forces from turbine blades and compressor impeller vanes. These installations include complete lube oil supply systems, sophisticated speed control systems, vacuum pumps, variable stiffness pedestals, plus extensive vibration data acquisition and processing instrumentation. These facilities are expensive to build and maintain, but when a rotor demands a high speed shop balance, no other type of balancing machine is acceptable. Most of these machines are quite unique, and many units approach the complexity of a formal shop test stand. Due to the many intricacies of these units, a detailed examination of high speed shop balancing machines exceeds the current scope of this text.

The discussion contained herein is directed at the common types of shop balancing machines. These are slow speed balancing machines that operate between 200 and 1,000 RPM. In these units, the rotor to be balanced is placed between a pair of rollers, and the rotor is driven by a direct belt drive mechanism. Vibration pickups are mounted by each set of rollers (bearings), and an optical trigger probe is employed to measure rotor speed and phase. Machines in this category are further subdivided as soft bearing or hard bearing machines.

A rendition of a typical **soft bearing** shop balancing machine is presented in Fig. 11-52. Soft bearing balancing machines are designed with rollers that are mounted on either flexible supports or sliding carriages that react to the unbalance forces. The horizontal motion of the carriage assembly is measured with a velocity or a displacement transducer as shown in the Fig. 11-57. An electronic data collection system is used to record the vibration response of the calibration weight, and calculate the weight correction. Machines of this type are very flexible, and the horizontal stiffness is quite low to allow maximum measurable motion. As such, the first natural frequency for the balancing machine is very low, and the rotors are balanced at speeds above the first critical of the balance machine. Due to this soft horizontal stiffness, an unbalanced rotor plus the sup-

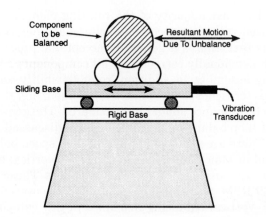

Fig. 11–57 Soft Bearing
Shop Balancing Machine

port rollers vibrate freely in the horizontal direction. From an analogy stand-point, a soft bearing balancing machine is like a velocity coil. That is, it operates above the fundamental natural frequency of the spring-mass-damper system.

Soft bearing machines generally readout in displacement, and must be cali-brated by placing known weights at defined angles at each of the selected bal-ance correction planes. In many ways this is similar to field balancing with calibration weights. However, in the case of the shop soft bearing machine, the detailed balancing calculations are performed by the *electronics box* attached to the measurement instrumentation. It should be noted that rotors of different weights with the same amount of unbalance will vibrate differently on a soft bearing machine. This is due to the influence of the rotor weight upon the reso-nance, and the resultant behavior of the rotor above the balancing machine reso-nant speed. These machines may also be somewhat dangerous in the hands of an inexperienced operator. If a rotor contains a large unbalance (original or induced by the operator), the horizontal forces may be sufficient to drive the spinning rotor out of the balance machine.

The more common type of shop balancing unit is known as a **hard bearing** balancing machine. An example of this machine is presented in Fig. 11-58. This type of balancing machine consists of rigid rollers that are mounted on vertical support pedestals. This type of balancing machine operates well below the natu-ral frequency of the combined rotor and support system. Using a vibration trans-ducer analogy, a soft bearing machine runs above the resonance like a velocity pickup — and a hard bearing balancing machine operates below the system reso-nance like an accelerometer. In many cases, a soft bearing machine with flexible pedestals may physically look like a hard bearing machine with rigid supports. To be safe, the diagnostician should always inquire about the specific character-istics of any machine used for balancing machinery rotors.

Hard bearing balancing machines come in a variety of sizes, and they can handle a wide range of rotor weights and configurations. The mechanical motion

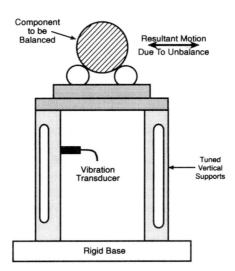

Fig. 11–58 Hard Bearing Shop Balancing Machine

produced by a rotor on a hard bearing machine is generally proportional to the centrifugal force in the rotor to be balanced. This is an advantage over the soft bearing units. Specifically, on a hard bearing machine, the vibration signal is not influenced by rotor weight, or the bearing and pedestal mass. Hence, a hard bearing machine typically displays a constant relationship between measured vibration and residual rotor unbalance.

Hard bearing machines are also equipped with various types of electronic packages and built-in computers to process the measurements and display vector outputs of unbalance in units of grams, ounces, gram-inches, or ounce-inches, plus the appropriate phase angle. Some machines incorporate a digital display, and other units have a desirable polar coordinate display screen for each balance plane. Modern units are also combined with various types of printers for hard copy documentation, plus digital storage of the rotor balance response characteristics. This is quite important for production balancing of many similar rotors. Overall, the hard bearing machine is easier and faster to operate. However, certain rotating machinery may require a soft bearing machine because of the operating conditions and/or sensitivity of the unit.

Before balancing is attempted on *stacked rotors*, the individual rotor elements, such as the shaft, balance piston, thrust collar, and all of the impellers, should be component balanced. The shaft runout must always be checked and any discrepancies addressed. In addition, the journals must be measured and verified as both *round* and *straight*. Failure to start with a straight shaft and proper journals will compromise the entire balance project by doing nothing but balancing the rotor bow and/or eccentricity. On large shafts or solid turbine rotors it is often necessary to roll the shafts for several hours *in the balancing machine* before attempting any balance corrections. This also applies on built-up rotors such as large gas turbines with precision axial fits between the wheels.

On bare shafts with keyways, it is mandatory that all keyways are filled with half keys. On rotating elements such as impellers, thrust collars, and balance pistons, a precision shop mandrel must be used to component balance each of these elements. The mandrel itself must be a perfectly straight shaft that is balanced to the lowest possible levels of residual unbalance. The mandrel must not contribute to any unbalance or eccentricity of the machine element to be balanced. A light interference fit should be used between the rotor element and the mandrel. There can be no looseness or wobble of the component to be balanced on the mandrel. In all cases, accurate mandrels will help insure a properly balanced rotor assembly, and sloppy mandrels will result in unnecessary grief.

After all components are balanced to within the specified tolerances, it is time to assemble the rotor. In some cases, the components are stacked on the rotor, and the assembled rotor is subjected to a final trim balance. On more sensitive rotors, the balancing continues concurrently with the rotor assembly. That is, the rotor assembly begins by mounting the middle one or two wheels with the proper keys, and interference fit to the shaft. After these initial wheel(s) are mounted, the shaft is rechecked for runout, and the wheels are checked for radial runout plus axial wobble. These readings become part of the job documentation, and they must be retained for comparison as the rotor balance and assembly progresses. If runouts are acceptable, then the balance of the shaft plus the initial wheel(s) are checked. At this stage, only minor corrections should be necessary. If the balance machine calls for major weight corrections, then something is wrong, and the only realistic option is to stop and find the origin of the problem.

Following the mounting of the initial wheels, the rotor is *built-up* or *stacked* by adding wheels in pairs. One wheel is added from each end of the rotor, and any necessary two plane balance corrections are performed on the recently added wheels. This procedure is continued until the entire rotor assembly is completed, and all runouts and critical radial and axial dimensions are within specification. All rotating components should be mounted and indexed to some repeatable point of reference. This insures that all of the rotor components are reassembled to the same reference point during each rebuild of the rotor.

Some rotors can only be stacked from one direction. On these units the wheels are added one at a time, and typically a single plane balance would be performed on the last installed wheel. On solid or integral rotors such as steam turbines, and segmented rotors such as gas turbines, the ability to *stack balance* does not exist. On these rotor configurations, the trim corrections performed on the shop balancing machine should only be performed after a suitable slow roll period to insure that any shaft bows are minimal.

When balancing an assembled rotor with multiple elements, weight removal or addition must be carefully evaluated. When a static type unbalance is found, weight corrections should be made at the center planes to minimize deflection through the translational critical. If a couple unbalance is encountered, weight corrections should be made to the modally effective outboard balance planes. Preferably, weight changes should be made by removing material from a mechanically sound location. In most cases, it is advisable to clay the rotor before any permanent weight changes are made to the rotor. It is highly desirable to refrain from using field balancing holes for shop balancing. When a rotor leaves the shop balancing machine, the accessible field balancing holes should either be totally empty, or totally filled with full-length balance weights. Either approach will allow the machinery diagnostician the ability to fully utilize the effectiveness of the field balancing planes. Finally, when balance weights are welded to a rotor, they should be located were they will not be a personnel hazard, and they must not affect rotor performance or provide any potential interference with any stationary parts of the assembled machine.

The issue of acceptable balance tolerances is an integral part of any discussion of shop balancing. The various standards and specifications that are applied to this work are somewhat overwhelming. Numerous standards organizations and various types of OEMs have different views of what constitutes an acceptable residual unbalance. These variations are understandable, since machines such as gear boxes tolerate higher levels of unbalance due to the fact that the transmitted gear forces are considerably larger than any mass unbalance forces. In another example, centrifugal pump rotors are usually restrained by impeller wear rings, and the pumped fluid provides significant damping to the rotor. Hence, many pump rotors operate successfully with higher residual unbalance levels. These existing standards vary from prescribing a particular fraction of the acceleration of gravity (e.g., 0.1 G's) to some combination of rotor weight and speed. Many industries have adapted the API standard of $4W/N$ as the acceptable shop balance tolerance. This expression is defined as follows:

$$U_{oz-in} = \frac{4 \times W}{N}$$ (11-70)

where: U_{oz-in} = Maximum Residual Unbalance Per Plane (Ounce-Inches)
         W = Journal Static Weight (Pounds)
         N = Maximum Continuous Machine Speed (Revolutions / Minute)

For instance, if a 1,500 pound rotor has a maximum speed of 5,500 RPM, the acceptable residual unbalance per journal would be calculated as follows:

$$U_{oz-in} = \frac{4 \times W}{N} = \frac{4 \times (1,500/2)}{5,500} = \frac{4 \times 750}{5,500} = 0.545 \text{ Ounce-Inches}$$

This level of residual unbalance is generally achievable with the typical shop balancing machines. In fact, some end users have reduced this tolerance to $2W/N$ for critical machines, or during component balancing to reduce the accumulated residual unbalance. Equation (11-70) requires a knowledge of the static rotor weight per journal. If the rotor is symmetrical, the total rotor weight may be divided in half to determine the weight per journal. For a non-uniform rotor, the journal weights must be measured or calculated as in case history 3. This is particularly important on rotors such as large steam turbines that have more weight supported by the exhaust bearing versus the governor end bearing. The maximum continuous operating speed of the machine must be used to calculate the balance tolerance in (11-70). Please note, this *is not the balancing machine speed*, it is the maximum continuous speed of the rotor installed in the field. For those that prefer balance units of grams, equation (11-70) may be restated in equation (11-71) where the residual unbalance is in gram inches:

$$U_{gm-in} = \frac{113.4 \times W}{N}$$ (11-71)

The journal static weight W is still expressed in pounds, and the maximum speed N is still stated in revolutions per minute. The only difference between equations (11-70) and (11-71) is that the conversion factor of 28.35 grams per ounce has been included in equation (11-71).

Once a rotor has met the balance specifications, it is desirable to perform a residual unbalance check to verify that the rotor is balanced correctly at each plane. This not only proves that the rotor is balanced, but it will verify that the machine is working properly. A residual unbalance check is accomplished by taking a known amount of weight at a known radius, and a known angle, and verifying that the balance machine properly tracks the weight. For example, if the $4W/N$ balance tolerance is 0.50 ounce-inches, this is equivalent to 14.18 gram-inches. For a 10 inch correction radius, then the tolerable unbalance would be 1.4 grams. After the component (or rotor) has been balanced, assume that the balancing machine displays a 1.2 gram unbalance at 90°, and a radius of 10 inches. This residual unbalance can be verified by placing a 10 gram weight sequentially at 12 locations (30° apart) at the 10 inch radius. The balance machine reading (grams at degrees) are recorded at each position, and the twelve measured vectors are plotted in a polar coordinate format. Fig. 11-59 displays the results obtained by adding 10 grams at 0°, and then moving this weight at 30° increments. Note that the measured balance machine angles are close to 30° increments, but some deviations appear. This is due to slight differences in actual placement of the clay test weight. If a rotor was equipped with 12 evenly spaced balance holes, the angular variation would be minimal, and adjacent readings would generally be separated by 30°.

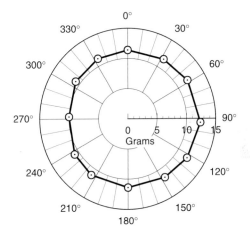

Fig. 11–59 Twelve Point Residual Unbalance Check On Shop Balance

If the final polar plot does not reveal a reasonably circular pattern, then the offending point(s) should be rerun. The residual unbalance is simply the center of the data circle. To determine the residual unbalance, take the highest amplitude reading, subtract the lowest reading, and divide the difference by two as in the following equation (11-72):

$$Residual\ Unbalance = \left\{ \frac{Maximum - Minimum}{2} \right\} \qquad \textbf{(11-72)}$$

With respect to the data shown in Fig. 11-59, the residual unbalance is easily determined from equation (11-72) as follows:

$$Residual\ Unbalance = \left\{ \frac{Max - Min}{2} \right\} = \left\{ \frac{12.4 - 10.0}{2} \right\} = 1.2\ \text{Grams}$$

This residual unbalance should be equal to the final display on the balancing machine. If these residuals are equal, the balancing machine is functioning properly. If the 10.0 inch balance weight radius is included, the following applies

$$Residual = 1.2 \text{ Grams} \times 10.0 \text{ Inches} = 12.0 \text{ Gm-In} \times \frac{1 \text{ Oz}}{28.35 \text{ Gm}} = 0.42 \text{ Ounce-Inches}$$

If the balancing machine is set to read in gram-inches instead of grams, then this final step is not required. In some cases it is also desirable to determine if the balancing machine is too sensitive, or not sensitive enough for the component. Hence, the performance of a sensitivity test may be desirable, and the manufacturer of the balancing machine should provide specific test instructions.

There are certain types of machines, such as three bearing units, highly flexible rotors, and/or rotors that operate above several critical speeds that cannot be successfully shop balanced. Machines of this type should be shop balanced to the lowest possible levels, and final trim balance corrections performed in the field when the rotor is installed in the casing. This approach is also necessary for rotors that display unusual synchronous 1X variations as a function of temperature or load. If the machine will faithfully duplicate a particular 1X response (e.g., hot to cold, and cold to hot), there is a good possibility that this type of repetitive behavior can be attenuated with a properly sized and positioned balance weight correction.

As stated earlier in this chapter, mass unbalance corrections are very useful for extending the life span of a piece of mechanical equipment by reducing the level of transmitted forces. However, like any corrective measure, shop or field balancing should always be placed in the context of *what are we really balancing out, and where did it come from*. In some cases, the application of rotor balancing may mask a more serious problem (e.g., cracked shaft). In all cases, the machinery diagnostician should thoroughly review the situation, and strive to maintain equilibrium between sound engineering judgment and economic responsibility.

BIBLIOGRAPHY

1. Bently, Donald E., "Polar Plotting Applications for Rotating Machinery," *Proceedings of the Vibration Institute Machinery Vibrations IV Seminar*, Cherry Hill, New Jersey (November 1980).

2. Eisenmann, Robert C., "Some realities of field balancing," *Orbit*, Vol. 18, No. 2 (June 1997), pp. 12-17.

3. Eshelman, R., "Development of Methods and Equipment for Balancing Flexible Rotors," *Armour Research Foundation, Illinois Institute of Technology*, Final Report NOBS Contract 78753, Chicago, Illinois (May 1962).

4. Gunter, E.J., L.E. Barrett, and P.E. Allaire, "Balancing of Multimass Flexible Rotors," *Proceedings of the Fifth Turbomachinery Symposium*, Gas Turbine Laboratories, Texas A&M University, College Station, Texas (October 1976), pp. 133-147.

5. Jackson, Charles, "Balance Rotors by Orbit Analysis," *Hydrocarbon Processing*, Vol. 50, No. 1 (January 1971).

6. Palazzolo, Alan B., and Edgar J. Gunter, "Multimass Flexible Rotor Balancing By The Least Squares Error Method," *Rotor Dynamics Course Notes, Part IV*, (University of Virginia, Charlottesville, Virginia, 1981), pp. 92-123.

7. Thearle, E.L.,"Dynamic Balancing of Rotating Machinery in the Field," *Transactions of the American Society of Mechanical Engineers*, Vol. 56 (1934), pp. 745-753.

Machinery Alignment

*M*isalignment problems have plagued machines ever since the need evolved to transmit torque from one mechanical device to another. Early industrial machines did not have extensive alignment problems due to low speeds, low horsepower, and compliant connections. However, as machinery sophistication developed over time, so have the requirements for improved alignment. Modern machinery trains usually consist of primary drivers directly coupled to the driven equipment. The induction motor driving the process pump, the gas turbine driving the feed gas compressors, and the steam turbines driving the generator all share a common characteristic. All of these machinery trains require alignment of components, internals, and coupled shafts for safe and reliable operation.

To meet the needs for improved machinery alignment, a number of technical approaches have been devised, and many good references are available to the machinery diagnostician. One of the earliest techniques to employ electronic instrumentation for this task was the 1968 paper by Charles Jackson[1]. More recently, John Piotrowski[2] authored the latest revision of his book that includes the details of laser tools for improved alignment measurements and accuracy. These techniques have been further refined with the use of portable computers that interface directly with the measurement system. These devices allow consistent recording of the moves, plus automated calculations and graphing.

It should be self-evident that proper alignment is critical to the life of a machine, and the consequences of misalignment can be seen through the train. Coupling wear or failure, bearing failures, bent rotors or crankshafts, plus bearing housing damage are all common results of poor alignment. The extent of the damage is directly related to the magnitude of the misalignment. For example, a general purpose motor driven pump with a slight shaft misalignment might experience premature seal failure, bearing damage, or coupling wear. In this case, the marginal shaft alignment decreases the time between failures, and increases the annual repair costs. However, a severe misalignment between this pump and motor could be potentially destructive to the plant surroundings, and

---

[1] Jackson, Charles. "Shaft Alignment Using Proximity Probes," *ASME Paper* 68-PET-25, Dallas, Texas (September 1968).

[2] Piotrowski, John, *Shaft Alignment Handbook* - 2nd ed., (New York: Marcel Dekker, Inc., 1995).

the attendant personnel. In fact, there have been many documented cases of couplings that have traveled for several thousand feet after a failure. In this scenario, the primary concerns include personnel safety, plus extensive repair costs.

In the overview, alignment consists of three distinct categories that are identified as shaft, bore, and position alignment. Shaft alignment is the most common form of alignment performed on machinery. There are many procedures, practices, and tools available to obtain a precision shaft alignment. Bore alignment addresses the position of the internal machine components with relation to fixed items such as the main bearings. Bore alignment is used for tasks such as locating diaphragms with respect to bearing centerlines. Position alignment is primarily reserved for machine location or elevation. It is also commonly used to measure and correct for thermal growth of the machinery. Within this chapter, the fundamentals of machinery position, bore, and shaft alignment will be discussed, and descriptive case histories will be presented.

PRE-ALIGNMENT CONSIDERATIONS

Prior to embarking on any alignment project, the diagnostician must evaluate the machinery installation, and select the method, tools, and procedures to be applied. Since each machinery installation differs is size, speed, power, location, and function, it is necessary to integrate all of the alignment variables in a cohesive plan before commencing the actual work. The fundamental items to be addressed are summarized as follows:

- ○ Machine arrangement, type, bearing configuration, and viewing position.
- ○ Coupling type, condition, runout, speed, and transmitted torque.
- ○ Potential thermal growth or shrinkage.
- ○ Potential pipe strain.
- ○ Condition of foundation, baseplate, sole plate, and anchor bolts.
- ○ Location and condition of leveling bolts and jack bolts.
- ○ Shim selection, and soft foot checks.
- ○ Obstructions to alignment work.
- ○ Machinery alignment offsets and tolerances.

Machine arrangement often dictates the alignment process and method. On any train, it is necessary to identify the fixed plus the moveable machines. The fixed equipment is the unit that will not be moved during the alignment work. Conversely, the moveable machines will be moved to obtain the correct alignment. For instance, in a typical pump-motor application, the pump remains fixed, and the motor is the moveable machine. In a turbine driven compressor train, the turbine remains fixed, and the compressor is the moveable unit. If a gear box is included between the turbine and compressor, the gear box becomes the fixed machine, and the turbine and the compressor both become moveable. The criteria for determination of which machine is fixed or moveable is basically a decision defining the *moveability* of the various machines. A motor typically has no external forces or pipe strain, and it may be easily moved. Turbines are sensitive to external forces, and they usually become the fixed units. A gear box in a machine train almost always becomes the fixed equipment. If the gear box is moved, the alignment moves required farther down the machine train may not be possible. Certainly there are exceptions, such as reciprocating compressors that are driven through a gear. In this arrangement, the compressor becomes the fixed equipment, and the gear box plus driver become the moveable units.

Machine types and specific **bearing configurations** can greatly influence the alignment process due to special requirements or considerations. Since most alignment techniques are performed at zero speed, it is necessary to anticipate and accommodate the machinery behavior between zero speed and full operating conditions. For example, large machines with sleeve bearings will experience a shift in shaft centerline position as the journal progresses from a rest position to full running speed. The shaft will rise on the oil film in the direction of rotation, and this centerline change should be considered in the cold alignment offset. As another example, tilting pad bearings can cause problems during shaft alignment. Machines with load between pads (LBP) bearings retain the shaft during alignment. However, machines equipped with load on pad (LOP) bearings can cause problems if the shafts are turned in different directions during alignment sweeps. This may cause the bottom pad to pivot, and this could cause the shaft to shift horizontally, which would corrupt the alignment data.

Special conditions exist for heavy rotors such as industrial gas turbines, or long generator rotors. These types of assemblies will sag due to gravity when the shaft is not turning. Once the machine is running, centrifugal and gyroscopic forces will straighten the rotor. However, the static gravitational sag, or bow, will cause the shaft ends to deflect upward outboard of the journal bearings. It is necessary to know this shaft deflection to set the machine at the proper *at rest* location. If the catenary curve describing the static rotor is not known, then it may be computed with the analytical techniques described in chapter 5 of this text.

Overhung machines have a similar problem, due to the fact that the overhung wheel often pulls the rotor down at zero speed. In essence, the shaft pivots across the wheel end bearing, and this action forces the coupling end journal to the top of the available bearing clearance (or vice-versa depending on specific rotor geometry). To compensate for this motion, it is common practice to push the shaft back down into the bearing with a pair of rollers positioned on top of the coupling end journal. Gear boxes may also provide difficult situations during the alignment process. As shown in Figs. 10-1 and 10-2, the direction of the gear contact forces are opposite for each gear. One element tries to sit down in the bearing, and the other gear wants to climb to the top of its respective bearing. Again, this actual running position should be taken into account during alignment.

Although radial offsets generally command most of the attention during machinery alignment, the diagnostician must also consider the axial position of the respective rotors. For instance, gear boxes with double helical elements will typically have only one thrust bearing on the bull gear. The pinion will center itself in the helix, and this axial running position must identified in order to set the proper axial coupling spacing for the high speed pinion. Similarly, the magnetic center for motors must be identified so that the motor rotor may be properly located in the running condition. This knowledge will allow the correct spacing to be established between the motor and the driven coupling hub. Although axial spacing is important on any machine, it is essential to maintain the proper axial dimensions on units equipped with diaphragm couplings.

The **viewing position** of the machinery to be aligned is critical to any alignment work. The fundamental concepts of *up-down*, *left-right*, and *fore-aft* become totally meaningless unless a definitive observation or reference point is used. This observation point or direction must be established at the start of the alignment work, and that point is maintained throughout the entire project. Furthermore, the final alignment documentation must reflect this viewing direction. It is also desirable to reference the compass points for additional clarification. This viewing position often varies between centrifugal and reciprocating machines. Hence, complete documentation of the machinery layout is the only way to maintain historical continuity.

Coupling types will have a significant influence on alignment, and it is important to understand how a particular coupling works. Many technical papers have been published on the various coupling types. In addition, excellent overviews on the entire topic are presented in the 1986 book by Jon Mancuso[3] plus the 1994 text by Mike Calistrat[4]. As mentioned in the previous paragraph, part of shaft alignment is the proper setting of coupling spacing, and/or shaft end gap. This type of information, plus the associated tolerances, are generally specified on the certified coupling drawing. Coupling types will also govern the configuration of the dial indicator alignment bracket(s). Also, it is important to determine how the shafts will be rotated to take the alignment readings. On flange type gear couplings, the flange bolts work well to turn the shafts, but due to their size they often require the use of larger brackets.

Furthermore, in planning the alignment job, consideration should be given to the handling and intermediate storage of the coupling parts. For instance, a gear coupling may have two dozen coupling bolts to be removed, and saved for the reassembly. However, some diaphragm couplings with intricate spool piece designs fall into the category of *1,000 bolt* couplings. These units require a lot of time to disassemble, and even more time to put back together. Coupling bolts are usually body fit bolts that are matched and balanced. If someone happens to lose a coupling bolt or nut, the entire project becomes unavoidably delayed.

The physical **condition** of the coupling can greatly affect the alignment. A worn or damaged coupling may produce erratic readings, or have high runouts. It is also easy to misdiagnose vibration data as misalignment when the real problem is a damaged coupling. In all cases, the coupling should be thoroughly inspected prior to beginning any alignment job. Check the gear teeth, shim packs, grid members, bolts, or whatever components exist in the coupling assem-

---

[3] Jon R. Mancuso, *Couplings and Joints - Design, Selection, and Application*, (New York: Marcel Dekker, Inc., 1986).

[4] Michael M. Calistrat, *Flexible Couplings, Their Design Selection and Use*, (Houston: Caroline Publishing, 1994).

bly. Elongated bolt holes, signs of excessive heat, or any evidence of damaged or excessive wear should be sufficient information to have the coupling committed to the dumpster. This simple act will save considerable time and effort during alignment, plus it will improve overall machinery reliability.

Shaft or **coupling hub runout** can influence alignment readings, and it can reflect the machine condition, or the presence of other mechanical problems. Runouts should always be checked prior to the alignment process. Shaft runout may reveal internal mechanical problems such as a cracked or distorted shaft. In these situations, the problem should be resolved before continuing with the alignment. If the shaft runout originates due to a poorly fabricated shaft or rough surface finish, the surface may be cleaned up or polished. Coupling hub runout can be indicative of problems such as a hub bored off center, tapered or cocked hub, loose shaft fit, or any number of other reasons. Some of these problems must be corrected, and others may be dealt with in the alignment process. The critical factor is that the shaft and coupling runouts must be checked, recorded, and their origin specifically understood.

From equation (2-98) it was shown that **torque** is a function of both speed and horsepower. To state it another way, the transmitted horsepower is the product of speed and torque. Naturally a constant must be applied to obtain the correct engineering units, but the fundamental relationship persists. That is, increases in speed or torque will result in an increase of the transmitted horsepower. In many ways, this formulates part of a difficult engineering problem of accommodating machinery power increases, while maintaining mechanical integrity of the entire coupling assembly.

Truly, the coupling in many modern machines represents a separate rotor system that is often isolated via flexible connections to the adjacent machines. As such, the diagnostician must be aware of the potential for coupling problems associated with mass unbalance, eccentricity, excessive clearances, a variety of preloads, plus the possibility for resonant behavior on axially compliant couplings. In the majority of large or critical machines, the couplings are precision elements that are designed to meet a specific set of criteria. Any attempt to increase the speed or load rating of these units should be made with proper engineering study and surveillance. Although some operating facilities seem to thrive on continually changing coupling types and configurations — there is considerable wisdom in staying with a coupling design that has a proven track record for continuous and reliable service.

Most machines experience temperature changes between the cold condition during which the alignment is performed, and the loaded condition at full process rates. It is common knowledge that hot machines such as steam turbines expand, and cold machines such as cryogenic pumps shrink. Typically, the cold alignment condition is adjusted to compensate for the anticipated change in physical dimensions. In some cases, the **thermal growth** or **contraction** is measured as discussed later in this chapter. In other instances, particularly on new installations, the amount of thermal growth must be estimated during alignment. Fortunately, industrial metals have a reasonably uniform coefficient of thermal expansion. This material constant may be combined with the length of the machine element and the differential temperature change, to determine the dimensional change. The specific relationship is expressed as follows:

$$\Delta L = \Delta T \times L \times C$$ (12-1)

Where ΔL = Anticipated Thermal Growth or Contraction (Mils)
ΔT = Average Change in Temperature (°F)
L = Length or Height of Heat Affected Area (Inches)
C = Coefficient of Thermal Expansion (Mils / Inch-°F)

For example, consider a gear box between an electric motor and a centrifugal compressor. The gear box centerline height is 47.5 inches, and the average skin temperature is 125°F. If the alignment was performed at an ambient temperature of 75°F, the thermal growth may be estimated from equation (12-1). If the gear box is a cast steel housing, the coefficient of thermal expansion may be obtained from Table B-1 in the appendix of this text. For this material, the coefficient of thermal expansion is 0.0063 Mils/Inch-°F. Combining this value with the previously mentioned physical parameters, the thermal growth may be computed with equation (12-1) as follows:

$$\Delta L = \Delta T \times L \times C = \{125 - 75°F\} \times 47.5 \text{ Inches} \times 0.0063 \text{ Mils/Inch°F} = 15.0 \text{ Mils}$$

Thus, the anticipated gear box growth would be 15 Mils vertically between ambient and operating temperature. Another way to obtain the same result would be to use the chart presented in Fig. 12-1. In this plot, equation (12-1) was

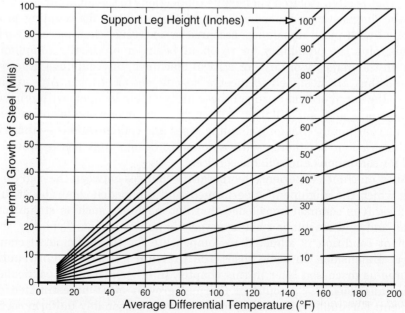

Fig. 12-1 Thermal Growth Of Cast Steel Machine Elements

solved for various combinations of differential temperatures (10 to 200°F), combined with pedestal or support leg heights ranging from 10 to 100 inches. As indicated, this chart is for cast steel, and it is sometimes easier to use than the equation. For this current example, an average differential temperature of 50°F may be located on the bottom axis. A vertical line up to the centerline height of approximately 47.5 inches, and a horizontal line to the left axis reveals a thermal growth of 15 Mils. Naturally this chart would change for other materials with different coefficients of thermal expansion.

It should also be mentioned that the *average differential temperature* should be taken literally. That is, at operating conditions, a temperature profile will exist along the length of the support leg, or casing. A series of surface temperature measurements should be made along the length of the element, and a realistic average operating temperature determined. Under normal alignment conditions, the entire pedestal will be at the same temperature, which is probably equal to ambient temperature. Hence, it is not necessary to run a temperature profile under this condition. For clarity, the differential temperature ΔT in equation (12-1) should be more precisely defined as follows:

$$\Delta T = (T_{avg\text{-}run}) - (T_{align}) \tag{12-2}$$

where: $T_{avg\text{-}run}$ = Average Pedestal Temperature During Normal Running Operation (°F)
T_{align} = Ambient Pedestal Temperature During Machinery Alignment (°F)

Please note that the average support temperature during normal operating conditions $T_{avg\text{-}run}$ is not necessarily the same as the temperature at the middle of the support. Steam leaks or other materials that blow directly onto the pedestal may significantly influence the average pedestal temperature.

The same expressions may be used for the computation of the shrinkage of machines that operate in cold service. For instance, the 316 SS supports on a cryogenic expander are 24 inches tall, and the average support temperature during operation is -45°F. This may be difficult to determine since the pedestals are completely covered with ice under this condition. Obviously, the temperature that must be used is the average metal temperature, and not the ice temperature. If the alignment was performed at an ambient temperature of +75°F, the differential temperature ΔT may be determined from equation (12-2) as:

$$\Delta T = (T_{avg\text{-}run}) - (T_{align}) = (-45°F) - (75°F) = -120°F$$

As stated, the supports are 316 SS, and from Table B-1, the coefficient of thermal expansion is 0.0092 Mils/Inch-°F. If this value is combined with the differential temperature, and the pedestal height, the thermal shrinkage may be calculated with equation (12-1) as follows:

$$\Delta L = \Delta T \times L \times C = \{-120°F\} \times 24 \text{ Inches} \times 0.0092 \text{ Mils/Inch-°F} = -26.5 \text{ Mils}$$

It is sometimes hard to believe that metal can shrink that much when it gets cold. Nevertheless, it should be realized that the differential temperatures for cryogenic services are often quite large, and this results in substantial contraction of the machinery dimensions. Of course, this is the same problem that must be addressed in starting up cryogenic equipment to insure that the stationary parts do not contract and seize the rotating elements.

Both hot and cold services have the potential for generating **pipe strain** on the machinery. This is a three-dimensional problem that potentially combines forces and moments on the machinery flanges. These piping loads often vary from ambient temperature to normal operating conditions. Clearly, pipe strain can make it impossible to achieve a precision alignment, and it can cause serious damage to the associated equipment. In most facilities, the original piping systems are designed by the Engineering and Construction firm (E&C). These piping designs must fit in a specified space, and provide the best possible flow characteristics, combined with acceptable pipe strain. This is a complex design effort, and computer simulations are used to model the piping system dynamics. For reciprocating machines, the problem is even more complex, since the acoustic and pressure pulsations characteristics must also be evaluated and optimized. The final piping configuration usually includes a variety of pipe hangers, supports, expansion loops, expansion joints, plus various spring loaded cans. In some cases, dampers or snubbers are required to dissipate the piping system energy.

It should also be mentioned that spring cans are a common and effective method to help relieve piping strain when they are operational. However, these spring loaded cans are a piece of hardware that can malfunction. It is important to periodically inspect these cans, and insure that the pipe is really supported by the spring. In new plants, the factory installed shipping stops may not be removed, and the effectiveness of the can is substantially reduced. In addition, long-term corrosion or internal dirt buildup can cause the spring can to lock up, and again minimize the effectiveness of the can.

If possible, machines should be aligned with the large bore piping detached. Prior to attaching this process piping to the machine, each flange should be checked for aligned vertical and horizontal flange position, parallel flange faces, and proper bolt hole orientation. In general, angular face alignment, and concentric hole orientation provide greater concerns than minor vertical or horizontal offsets. Tolerances for pipe alignment are typically available from the E&C, or the end user engineering specifications. However, the real test of a successful flange mating between the piping system and the machinery is obtained by monitoring the coupling position as the process flanges are bolted in place. This activity is monitored by mounting vertical and horizontal dial indicators on an adjacent machine, and observing the shaft or coupling hub motion as the flanges are connected. Any pipe strain that produces position changes at the coupling of greater that 2.0 Mils should be corrected.

In all cases, the machinery **foundation** must be visually inspected for any cracked, broken, or missing concrete or grout. The primary function of a foundation is to support the gravity and the dynamic loads imposed by the machinery system. A weak or damaged foundation will not act as a rigid member, and unacceptable movement may occur. Furthermore, oil and chemical contamination can weaken or corrode a foundation with minimal evidence of surface destruction. In some cases it is desirable to check the concrete or grout integrity with a chipping hammer, or perhaps a core sample is required in other situations.

Due to the superior compressive and tensile strength of epoxy grout, most machinery installations use this material between the concrete foundations and the baseplates. Epoxy grouts also have a greater resistance to water, oil, and chemical damage. However, these materials are sensitive to stress risers, and proper installation requires all sharp corners or edges to be radiused. There are many excellent epoxy grouts available, and the machinery diagnostician should carefully compare the physical and mechanical properties, plus the advantages and limitations of each grout to select the proper material for each job.

Prior to any alignment job, it is highly desirable to inspect **anchor bolts** for damage. Broken or loose bolts must be replaced, and proper torque values should be maintained. Many times anchor bolts break above or at the surface of the foundation. This is usually caused from corrosion, over torquing, misaligned holes, or loose equipment. Anchor bolts also break below the surface, and this may not be discovered until the bolt is tightened or the machine fails. Several inspection techniques exist to test anchor bolts. Ultrasonic testing (UT) is the most common form of inspection to determine the distance from the top of the bolt to the first inclusion. This could be the bottom of the bolt, a bend, or a crack. If the original bolt configuration is not known, the UT data could be useless.

Proper design and installation of machine **baseplates** or **sole plates** is also imperative to a good running machine, and a proper alignment. Initially, the diagnostician should inspect for corrosion or erosion that may weaken the baseplate, or even separate the plate from the grout. Hollow cavities or low spots are locations for water or corrosive chemicals to collect and damage the baseplate. A common practice of *sounding* a baseplate by tapping the surface with a hammer can reveal voids or separations. The sound or chime produced will vary between solid areas where the baseplate and grout are in direct contact versus spots where voids or separations exist.

Machine or baseplate **leveling bolts** (sometimes called jack bolts) are important considerations during initial installation of a baseplate. They allow the baseplate to be precision leveled quickly and accurately. Often they are installed on top of a steel plate that rests directly on the concrete foundation. These bolts must be numerous and sufficiently strong to support the baseplate. They should be sealed prior to grouting, and they should be removed after the grout has set. Retained leveling bolts can be detrimental to the installation. It is possible to distort or separate a baseplate from the grout if leveling bolts are still in place during final anchor bolt torquing.

In new installations, the baseplates or sole plates must be checked for levelness and flatness. In most plants, the only tool available to check a baseplate is a precision machinist level. Readings can be used for coplanar comparisons as well as slope measurements. In most applications a perfectly level installation is desired, but not mandatory. However, the flatness of the baseplate or sole plate is critical. If the machine mounting surfaces are not perfectly flat, the internal bore alignment, as well as the shaft alignment, is in jeopardy. When addressing flatness concerns beyond a simple coplanar comparison, it is often necessary to employ optical or laser measurements.

Unlike the leveling bolts used to set baseplates, **jack bolts** are used to move machines around on a fully installed baseplate. Standards for horizontal, vertical, and axial jack bolts should be rigidly mounted on the baseplate to allow quick and easy machine movements. Temporary screws or clamps work well with smaller machines, but they become ineffective on large, heavy machines. After the alignment is complete, the jackbolts may be removed from the machine. In most plants, the jack bolts are allowed to remain in-place for the next alignment job. In these situations, the jack bolts should be backed off at least 0.25 inches from the machine foot. This will allow the machine to grow, and not be restrained by a close fitting jack bolt. The concept that the jack bolts keep the machine in line during operation is incorrect. In actuality, a tight jack bolt can provide a detrimental restraining force on the machine.

Shims should always be thoroughly inspected on any alignment job. Corroded, damaged, or separated shims must be replaced. Although some individuals tend to reuse old shims for convenience, the replacement of questionable shims will save a great deal of time during the alignment process. There are many companies that manufacture and sell precut shims in various configurations. These shims are usually sold in sets that range in thickness from 1.5 to 250 Mils. Shim stock material can also be purchased in rolls, and cut to match the dimensions of the machine foot. Laminated (or glued together) shims are also available in standard fractional thickness with 1.5 to 3.0 Mil thick laminations.

The selection of proper shim materials is a very important consideration. Carbon steel shims should not be used due the effective thickness growth as carbon steel begins to rust. Stainless steel or brass shims should be used whenever possible. Many times carbon steel shims are used on custom fit applications where the surface grinders are equipped with magnetic tables. Obviously, stainless steel or brass cannot be retained by a magnetic table, and only carbon steel shims can be ground. Nevertheless, stainless steel is the preferred shim stock material, even though it is more expensive and harder to cut than brass. In applications where brass will not be chemically attacked or subjected to severe environments, brass shims may be used. Brass is often used for large shims due to their ease of use and general availability.

During initial installation of a machinery train, the individual machines should be set on very thick shims. In small applications, a shim thickness of 0.125 inches is acceptable. For larger machines, the initial shim thickness should be at least 0.375 to 0.500 inches to provide an adequate range for vertical alignment moves. On very hot machines, the initial shim thickness may be even larger. Rust, paint, dirt and other foreign objects can cause the machine to move unpredictably, or allow the machine to move while in service. To insure a good alignment, the machine foot, sole plates, and shims must all be clean and dry.

The number of shims installed under a machine foot should typically be limited to 5 shims. If more than 5 shims are required to align a piece of equipment, thinner shims should be swapped out for thicker ones. Contamination and shim springiness become problems when excessive shims are used. The size of the shim should be large enough to support the machine, and allow the hold down bolts to pull down evenly. Precut shims are sufficient for smaller general purpose equipment, but large machines may require custom fit shims. When installing shims, the shim area should be a minimum of 75% of the machine foot. An example of an improperly supported machine was presented in case history 33, which describes the problems encountered on a 3,000 HP induction motor.

In many machines, one or more support feet may be higher or lower than the others, or one or more support feet may be bent in relation to the others. This condition is commonly referred to as **soft foot**, and it can cause many problems during alignment and operation. Soft foot may be checked with a feeler gauge, or a dial indicator. Ideally, this is checked before starting alignment by removing all shims from under each machine foot, and setting the machine flat on the baseplate. Feeler gauges should then be used to measure any clearance between each machine support foot and the baseplate. Clearances of less than 2.0 Mils are generally acceptable. Next, tighten down all hold down bolts, and then loosen each foot consecutively, and measure any clearance with the feeler gauges. This simple test is then repeated for each of the remaining machine feet. If clearances exceed 2.0 Mils in either feeler gauge check, the soft foot must either be corrected by machining or shimming.

Once shims have been installed, the feeler gauge method may become difficult. The second way to check soft foot is by using a dial indicator. To check the soft foot with this method tighten down all hold down bolts, and position a magnetic base and dial indicator to measure deflection of the machine foot with respect to the base. The indicator should be located so that it will not interfere with loosing or tightening of the foot bolts. After the indicator is zeroed, loosen the foot bolt, and read the total movement shown by the dial indicator. If the indicator reads less than 2.0 or 3.0 Mils the foot is acceptable. Repeat the process for each of the machine feet and shim as required.

During the alignment process it is necessary to set up measurement equipment, rotate shafts, tighten and loosen bolts, etc. Physical **obstructions** may seriously hamper this work. In position alignment, equipment placement is very important for accessibility to all measurements points. Walls, beams, piping, and machine components can all interfere with the line of sight. In any type of optical alignment, the placement of the tripod, and the targets may also interfere with the tightening of bolts, or the installation of other machine components. During shaft alignment, coupling guards, oil piping, or typical bearing instrumentation may interfere with the indicator brackets. These items can damage equipment or greatly lengthen the alignment process. In all cases, the diagnostician should *think through* the alignment process, and attempt to *eliminate* or *work around* as many of the potential physical obstructions as possible.

Once the mechanical items are resolved, the proper cold **alignment offsets** must be included to allow the machines to move into an aligned position during normal operation. The available techniques for measuring the thermal growth (or contraction) are discussed in the last part of this chapter. After these alignment offsets have been established, allowable target values should be identified. These target values may be listed as acceptable variations to centerline offsets, indicator readings, or elevation readings.

In a perfect world, the final alignment numbers would be identical to the

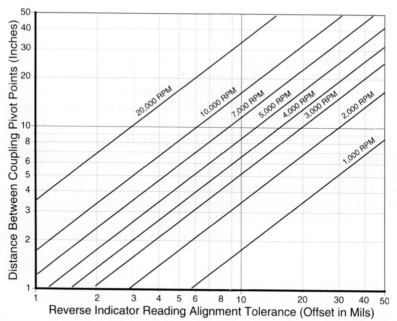

Fig. 12–2 Typical Tolerance Chart Relating Coupling Span, Speed, And Allowable Offset

desired offsets. However, this is seldom attained, and **alignment tolerances** are employed to define an acceptable alignment range. For instance, Fig. 12-2 displays a typical plot of coupling length versus alignment tolerance for various machine speeds. From this diagram it is clear that longer couplings have greater misalignment tolerances. Furthermore, low speed machines are more tolerant to misalignment than high speed units. Both of these conclusions make good practical sense, and no one will question the need for better alignment on higher speed machines with shorter couplings. The actual alignment tolerances will vary in accordance with the type of coupling, the machinery configuration, the applicable engineering specifications, the sophistication and accuracy of the measurements, plus the experience level of the people performing the alignment work. In virtually all cases, the more experience and knowledge contained within the alignment crew, the better the final alignment job.

Clearly, there are a variety of tools and techniques available to perform the position, bore, and shaft alignment tasks. Excellent references are available from the vendors of optical and laser alignment equipment. Due to the variations between lasers, and the continuing evolution of these products, it is difficult to supply up-to-date references in this text. However, the diagnostician should have no problem finding current information at any trade show. The final selection of the best approach for machinery alignment generally comes down to the experience of the individuals performing the work, plus the available hardware, and the specific tasks to be performed. In order to provide the diagnostician with an overview of the options, the following discussions are offered for consideration.

OPTICAL POSITION ALIGNMENT

Position alignment is a three-dimensional problem of locating machinery with respect to stationary bench marks. This includes the initial surveying work to establish the proper location for a foundation, the accurate positioning of a baseplate, as well as optimization of the casing position. This type of alignment work is necessary during new construction, and it is also required during various types of machinery maintenance. Optical tooling is the most common type of measurement equipment used for position alignment. However, laser tooling is also used, and that approach will be discussed in the next section. Although much of the position alignment work is performed by land surveyors, the tools and concepts discussed in this text are directed at machinery alignment.

Optical alignment tooling has been used for many years, and it has been successfully adapted to position, bore, and shaft alignment. Obviously, special tooling is often required to perform a specific task on a particular machine. Fortunately, a wide variety of tools are readily available for almost every use. The primary instrument for optical alignment is the **optical telescope**. This device is a telescope that consists of a tube containing an objective lens, focusing lens, reticle with cross lines or similar pattern, plus an eyepiece. The focusing lens is a moveable element located between the objective lens and the reticle. By moving this lens, images can be accurately focused on the reticle. The objective lens inverts the image, and the reticle is inverted as well. The eyepiece inverts and magnifies the image so the viewed image is correct to the human eye.

Fig. 12–3 Tilting Level
K&E Paragon® Series

An example of an optical telescope is the Keuffel & Esser (K&E) **tilting level** shown in Fig. 12-3. This device is used to establish a single plane in space perpendicular to gravity (i.e., level). This instrument is ideal for setting elevations and checking levels and flatness. Several companies provide optical levels, and the specifications vary between vendors, but the overall accuracies are similar. The tilting level is a telescope mounted on a four screw leveling system that is attached to a tripod. The four leveling screws in the tilt axis are coincident with the azimuth axis, and this eliminates errors during the leveling process. This telescope is equipped with an optical micrometer accurate to thousandths of an inch up to 100 feet. The magnification automatically varies from 20x at a near distance to 30x at infinity. Two positioning or setup levels are integral with this instrument. The first device is a circular vial level that has a sensitivity of 10 minutes per 2 millimeters of movement. The second device is a coincidence type level with a sensitivity of 20 seconds per 2 millimeters of movement. Combined with an optical magnification, this allows leveling to within 1 second of arc.

More sophistication and capability in the optical telescope is obtained with a **jig transit**, such as the K&E unit depicted in Fig. 12-4. This is a versatile instrument that consists of a telescope mounted on a base that allows rotation around both the elevation and azimuth axes. Rotation around the elevation axis allows the user to view a vertical plane, and establish plumb lines. The azimuth center is hollow, and the telescope line of sight is centered above this yoke to allow the capability of viewing vertically downward. Like the tilting level, a jig transit is mounted on a four screw leveling system to the tripod plate. The four leveling screws in the tilt axis are coincident with the azimuth axis, and this eliminates errors during the leveling process. Elevation leveling is accomplished

Fig. 12–4 Jig Transit
K&E Paragon® Series

with two speed tangent screws to allow precision leveling. For setup, the jig transit is equipped with one circular vial level and one coincidence type level, much like a tilting level. However, on a jig transit, the circular vial level is installed on the transit base, and the coincidence level is mounted on the telescope.

Jig transits incorporate a variety of special features. Many units have a mirror on the telescope axle to be used for collimation activities. Jig transit squares have a hollow telescope axle with a semi-reflective mirror on one end. The telescope square has a telescope installed in the hollow horizontal axle. This cross axis telescope may be fitted with an auto-collimation unit. This enables an optical reference to be kept from a mirror, while using the line of sight telescope.

In all cases, the optical telescope must be properly supported on a rigid stand or tripod. Many commercial types of supports are available. This includes rigid stationary units, plus a variety of portable stands with adjustable legs. A quality field stand must be able to be located on loose and uneven surfaces, and the entire assembly should have mechanical locks to prevent movement during alignment readings. In some instances, a portable stand cannot be used due to physical limitations such as an unstable deck or insufficient space for a tripod. In these cases, a customized stand or support plate may be purchased or fabricated to suit the particular configuration.

The position of any machine element in space may be quantified by mounting precision **optical scales** on the object. Differential optical measurements between these scales and fixed bench marks will allow an accurate determination of the three-dimensional location of a machine element. If differential optical measurements are made between locations, the absolute (bench mark) reference does not exist, but the user still retains the ability to check for level or flatness of the machine surface. Obviously, there are many ways to use an optical telescope in conjunction with a precision optical scale.

Fig. 12–5 Portion of Ten Inch Long K&E Optical Bar Scale With 20 Divisions Per Inch

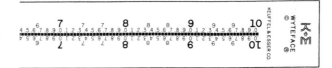

In order to extract the required accuracy from the optical measurements, the scales must be accurately engraved, and easy to read. For example, Fig. 12-5 depicts the end of a 10 inch long K&E optical scale. The graduated lines are black to provide good visibility against the Wyteface® background. In this diagram, the graduated optical scale where the center of each pair of lines to the next is 0.050 ±0.001 inches. The scales are numbered in inches, and appropriate subdivisions. Other K&E scales are provided with four different patterns to allow for various sight distances. These typical optical scale separations and their associated sight distances are summarized in Table 12-1.

To accommodate various physical installations, K&E provides optical scales in standard lengths of 3, 10, 20, and 40 inches. The scales are held in position on the machinery with a variety of methods. For instance, Starrett® produces a weighted scale holder that is useful on nonmagnetic surfaces. A variety of other supports are available that use magnetic bases. In some cases a scale mounting base may be attached with epoxy cement. On critical machinery applications where precise repetition of the mounting point is mandatory, a dowel hole and surface spot-face are simultaneously produced with a special drill bit. A stainless steel dowel pin is then inserted into the hole, and the target holder is mounted

Table 12–1 Optical Scale Line Separations Versus Sight Distance

Scale Separation	Sight Distance
4.0 Mils	Up To 7 Feet
10.0 Mils	7 to 20 Feet
25.0 Mils	20 to 50 Feet
60.0 Mils	50 to 130 Feet

on the dowel pin. This type of installation requires more setup time for the first set of readings, but repeatability is excellent, and future comparative reading can be acquired much faster.

Following selection of the optical telescope, support stand, and the appropriate scales or targets, the instrument must be properly setup. Detailed instructions are provided in the instruction manuals for each device. In general terms, the setup begins by focusing the image for the individual viewer. If the eyepiece is not properly focused **parallax** may occur, and this causes the image and the reticle to appear in two separate planes. The instrument must be **leveled** with the four screw leveling system mentioned earlier. The circular vial level is the primary indicator for this task. For telescopes used for leveling, a **peg adjustment** may be used to check the instrument calibration, and identify any necessary corrections. For situations where the transit must reference two points of interest, the instrument must be **bucked-in** to both points. The two points of interest may be bench marks that are established by the observer, or specific locations on a machine. In other situations, readings have to be taken on a vertical face to check **square** and **plumb**. The jig transit may be used to establish vertical plumb lines, or check an entire surface for plumb.

Some telescopes are equipped with an **auto-collimation** eyepiece (semi-transparent mirror and a light). When this type of telescope is pointed at a mirror, the observer can see the reticle, and its reflected image. By moving the transit until the cross lines from reticle and the reflected image converge, the instrument will be collimated. This adjustment of the line of sight is equivalent to the observation of parallel light rays that are focused on the reticle. If an auto-collimation eyepiece is not available, **auto-reflection** may be used. The procedure consists of mounting a printed target on the front end of the telescope, and adjusting the telescope or mirror until the reticle and the reflection of the target coincide. Auto-reflection is not as accurate as auto-collimation.

In position alignment, it is often necessary to measure the **squareness** of an object. This implies that one plane is perpendicular to another. This can be accomplished by auto-collimation from a second instrument to the axle mirror of the primary unit. In another method, a pentaprism can be mounted on the front of the telescope, and rotated to sweep a perpendicular plane. Alternately, a jig transit telescope square may be set with the cross axis telescope. The cross axis telescope can be auto-collimated or auto-reflected. Finally, the telescope can be set to individual bench marks and bucked-in, as previously mentioned.

There are certainly many setup steps required for the proper use of optical equipment. Although it may not be immediately apparent, optical readings are often based upon small changes measured over long distances. For instance, Table 12-2 is a conversion chart for changing angular arc readings measured in seconds into arc lengths in Mils. From this table, it is noted that a 60 second arc (= 1 minute = 0.017 degrees), has an arc length of 349 Mils for a 100 foot span. Thus, a very small change in an angle will represent a significant change in linear position of an object. For this reason, optical alignment measurements must

Table 12–2 Conversion Of Seconds Of An Arc To Arc Length

Arc Length (Seconds)	Equivalent Arc Length in Mils					
	10 Foot Span	20 Foot Span	30 Foot Span	40 Foot Span	50 Foot Span	100 Foot Span
1	0.6	1.2	1.7	2.3	2.9	5.8
2	1.2	2.3	3.5	4.6	5.8	11.6
5	2.9	5.8	8.7	11.6	14.5	29.1
10	5.8	11.6	17.4	23.3	29.1	58.2
20	11.6	23.3	34.9	46.5	58.2	116.4
60	34.9	69.8	104.7	139.6	174.5	349.1

be carefully performed to provide meaningful results. Optical measurements should always be rechecked and sometimes rechecked. On critical machinery, it is highly desirable to have two people obtain independent measurements to verify the accuracy of the data.

As applied to position alignment, optical measurements are primarily directed towards obtaining accurate optical scale readings in three dimensions. These readings may be combined with suitable bench marks to accurately determine the physical position of a point in space. This is necessary for tasks such as locating foundations, positioning baseplates, or positioning machinery cases. The numbers can also be used to reference flatness or level of an object. This type of measurement is employed to insure that support feet on a baseplate are properly positioned, and correctly machined. For an improved perspective on position alignment, case history 41 provides an overview on a major position alignment project dealing with a high pressure reciprocating compressor.

Case History 41: Hyper Compressor Position Alignment

The machine under consideration represents one of four high pressure ethylene compressors in a low density polyethylene plant. High stage discharge pressure for this machine varies between 28,000 and 35,000 Psi (depending on product). A general arrangement of this reciprocating compressor is described in the plan view of Fig. 12-6. This compressor is driven by a 3,500 HP synchronous motor. This 30 pole rotor is mounted directly on the compressor crankshaft, and it runs at a synchronous speed of 240 RPM. As shown in Fig. 12-6, the crankshaft is supported on five main bearings, with the #1 bearing at the oil pump (outboard) end of the crankcase, and the #5 bearing supporting the motor rotor. The compressor is a four throw unit, with auxiliary crossheads that drive two cylinders on each throw (eight cylinders total). Each auxiliary crosshead is mounted on a separate concrete pedestal, and the motor stator is on separate sole plate.

Visual inspection revealed that the auxiliary crosshead and the motor foundations were in good condition. However, the main foundation under the crankcase had horizontal cracks at 5 to 7 feet down from the grout cap. Oil and chemicals had impregnated the foundation cracks, which showed the foundation *winking* during operation. In essence, the top of the main foundation was severed from the support base. Several foundation repairs had been tried over the years, none of which were successful. The machine was continually plagued with numerous mechanical problems. For instance, main bearing failures occurred at a rate of one failure every two years. Since there are four identical compressors with the same problems, the average failure repetition rate was one failure every six months. This is certainly an unacceptable main bearing failure rate, and the separated foundation was considered to be one of the obvious culprits.

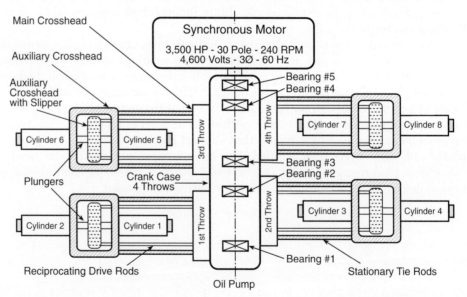

Fig. 12–6 Machinery Arrangement Plan For Hyper Reciprocating Compressor

Clearly, the main foundation under the crankcase required extensive repairs, and that task dictated the removal of the compressor crankcase, and main crossheads. To properly repair this foundation, the existing concrete must be chipped out to an elevation below the cracked and oil soaked sections. Fortunately, the motor and auxiliary crosshead pedestals do not require any repairs, and they remained intact. From a timing standpoint, the machinery train was scheduled for a major overhaul, and all machine elements were to be removed for complete refurbishing by the OEM. Not surprisingly, when the compressor was shutdown for this overhaul, it was discovered that the main crankcase #2 bearing had failed. This failure was consistent with previous bearing history.

On high pressure mechanical equipment of this type, it is customary to reference or position the various machine elements with respect to the common crankcase and associated crankshaft. During the type of extensive overhaul, it is imperative to remove the crankcase, rebuild the foundation, and reinstall the crankcase in exactly the same location. If the crankcase in not reinstalled in the same place, the motor, auxiliary crossheads, piping, and other appurtenances will not line up properly. This can cause undue difficulty, and seriously delay the completion of the maintenance turnaround on this unit.

Since the main foundation will be removed, the compressor anchor bolts must also be replaced. Prior to pouring a new foundation, new anchor bolts will have to be positioned to match the compressor footing plan, plus they must be at the correct elevation. Obviously, this type of repair project requires accurate position alignment measurements prior to crankcase removal, plus additional position alignment during reassembly.

The fundamental task during disassembly of this unit was to document the actual positions of the machine elements with respect to a defined set of three-dimensional bench marks. Permanent X-Y-Z bench marks did not exist, and it was necessary to establish these fixed reference points during disassembly. Since the compressor elements are normally referenced to the crankshaft, the crankshaft bore centerline was considered as the primary control parameter. Hence, it is necessary to document the crankshaft bore to the new bench marks.

For purposes of consistent definition, the Y-axis will be defined as horizontal, and collinear with the crankshaft axial centerline (i.e., fore and aft). Horizontal motion (i.e., left and right) will be considered as the X-Axis. Changes in elevation (i.e., up and down) will be defined as the Z-axis. This X-Y-Z coordinate system will have an origin at the crankcase outboard looking towards the motor. The readings are further identified with a traditional plus or minus (±) direction. Note that this convention is different from the viewing position on centrifugal machinery, and it is also different from the specialized plunger labeling applied in case history 35. The current type of coordinate system is commonly employed on reciprocating machinery alignment.

The installed Z-axis bench mark consisted of a 1/4 inch bolt on a structural steel supporting beam. This bolt was assigned an arbitrary elevation of 100 inches. To determine crank shaft centerline elevation, slipper guide targets were set and centered (±0.25 Mils) in the #1 and #5 bearing bores. Using the precision ground outer diameter of the target, a 40 inch optical scale was mounted on the top of each target, and read with a tilting level. The scale reading was corrected by 1.125 inches to determine the centerline elevation of both end bearings. In addition, the elevations at eight locations on the crankcase top gasket surface were obtained with the precision tilting level.

The Y-axis centerline was referenced from the motor face of the crankcase. A jig transit was bucked-in to this face with a 3.200 inch offset. The transit was then sighted on the motor sole plates on each side of the crankcase, and reference lines were scribed on the sole plates to establish the bench mark. Scale readings were then taken from the two front anchor bolts to establish the Y-axis control for the crankcase and the anchor bolts. This was followed by elevation readings of the anchor bolts and flanges. For reinstallation purposes, elevation readings were also obtained from each of the dog house (main crosshead) anchor bolts and flanges.

Establishment of the X-axis bench marks was more difficult. Initially, an alignment telescope was positioned in front of the #5 bearing on *cone mounts*, and this telescope was bucked-in to the bore of bearings #1 and #5. An aluminum plate was attached to the concrete wall behind the crankcase, and lines were scribed on the plate to reference the crankshaft centerline. This bench mark provided a reference point for one end of the X-axis. Next, a jig transit was positioned 87.5 inches from the #5 bearing. It was centered on the #1 and #5 target centerlines, plus the scribe lines on the wall mounted aluminum bench mark. The transit was then sighted down through the hollow center to the concrete floor, and this location was marked. The transit was then sighted to a floor location 75 inches behind the transit, and the floor was again marked.

Holes were drilled into the concrete floor, and brass plugs were epoxied in place at each location. Both plugs were recessed below the floor surface to minimize damage. After the epoxy hardened, the jig transit was reset above the plug at 87.5 inches from the #5 bearing. The transit was again centered on the targets at bearings #1 and #5, plus the scribe lines on the aluminum wall bench mark. The transit was again sighted down through the hollow center to the brass plug, and the plug was center punched to the center of the telescope crosshairs. The same procedure was repeated for the second brass plug 75 inches behind the transit. These two floor mounted brass bench marks combined with the scribed aluminum plate to uniquely define the X axis.

Based on these bench marks, the exact crankcase location was identified and recorded. The locations of the main bench marks in the immediate vicinity of the crankshaft are shown in Fig. 12-7. Following these initial position measurements, the machinery was removed, and sent to the OEM for rejuvenation. At this point, the old foundation was removed past the cracks and contamination. New rebar was installed in the retained foundation to connect the existing with the new concrete. In addition, a precision aluminum anchor bolt template was fabricated based upon the OEM anchor bolt locations. This template was optically positioned, and new full length anchor bolts were installed. After the anchor bolts were tightened, and the template secured, a final set of optical position readings was taken to verify that nothing was moved during the securing process.

The concrete was successfully poured, and the new foundation was prepared to accept the crank case for setting and grouting. The rebuilt crankcase was set and leveled with the jack bolts to the top gasket surface. All 26 anchor bolts lined up perfectly with the crankcase holes. Slipper guide targets were once again set and centered in the #1 and #5 bearing bores. To reference the previously established bench marks, a jig transit was centered with the two brass floor plugs, and the scribe lines on the aluminum wall plate. A second transit was bucked-in to the motor sole plate scribe lines, and a tilting level was set to measure the crankcase elevation verses the vertical bench mark. Following this setup of optical equipment, the crankcase was moved back to the original position. The bore alignment was checked (case history 42), and then all 26 anchor bolts were torqued to an initial value of 50 foot-pounds to hold the machine during grouting.

After the crankcase was grouted, and the epoxy cured, the optical equipment was used to obtain position readings before and after the anchor bolts were fully torqued. The final results of this work are depicted in Fig. 12-7. As noted, the #1 bearing position was 5.0 Mils to the North, and 2.0 Mils lower than the original position. The #5 bearing was 16.0 Mils North, and 20.0 Mils lower than initially found. The crankcase face position was 3.0 Mils West of the original location, and the compressor top cover elevations varied from 1.0 to 6.0 Mils from their initial positions. The final offsets at the motor #5 bearing were larger than desired, but the remaining locations were judged to be quite acceptable. Additional bore alignment work was performed on this compressor, and this is discussed in case history 42. Overall, the efforts to accurately measure the crankcase position prior to removal, and during installation, allowed a *clean* reinstallation of this machinery. The final success of this work is evident from consecutive years of smooth operation, plus the absence of bearing failures.

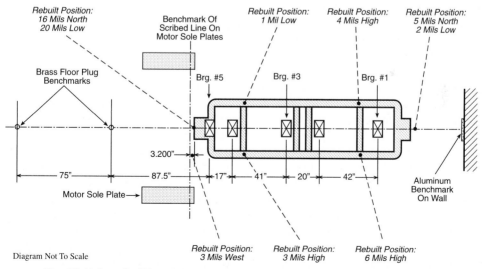

Fig. 12–7 Installed Bench Marks Plus Final Position Of Compressor Crankcase

LASER POSITION ALIGNMENT

For many years, optical alignment was the only versatile method for positioning process machinery. Since 1975, laser alignment equipment has been developed, and utilized in the machinery business. The accuracy of the laser has been demonstrated in many calibration and inspection labs. As technology and instrumentation has improved, the industrial laser has evolved as a rugged and versatile device. The greatest advantage of a laser is the high degree of accuracy over a long distance. Some surveying lasers have effective ranges of over 10 miles with an accuracy of ±0.2 inches. With respect to machinery alignment, the laser alignment methods are based upon the same line-of-sight principles as the optical telescope. However, the laser beam represents a straight line in space that can be positioned and accurately measured.

Lasers offers several advantages over optical tooling. For instance, laser alignment systems for machinery applications typically have a range up to 150 feet, with an accuracy of ±1.0 Mil. The laser beam is not subject to interpretation like optics, and the readout displays the target offset directly. Since the readout is a continuous display that tracks the laser beam, any movements are immediately displayed by the readout system. This allows components to be moved and recorded with the laser. Furthermore, the laser beam is not affected by physical sag like many mechanical measurement methods.

The laser does have several drawbacks over optics. A laser is very sensitive to moisture, and high humidity, water vapor, steam, and heat hazes can all influence the laser beam. In many cases the beam may be bent due to water vapor. The removal of the human factor from the data interpretation requires that the operator be experienced, and able to recognize good versus questionable data. From a commercial standpoint, lasers are expensive to purchase and repair. Furthermore, most lasers and accessories are not interchangeable between vendors, and the user is committed to purchase everything from one supplier.

In virtually all cases, a calibrated laser used in its proper application with qualified personnel will provide results comparable to that of optics. The operation and procedures of the laser are very similar to optics, and the intended results are the same when measuring for flat, level, square, or plumb. Laser tooling consists of six major components. Specifically, this includes the laser head, detector, beam directors, beam splitters, power supply, and readout. The stands and holders in some cases are identical to the optical equipment. In other situations, custom holders and fixtures are needed to locate and set the laser tooling.

A **laser head** is the device that generates and directs the laser beam. For machinery applications, the laser head is often a rectangular box with the laser beam emitted from one end. Lasers are also available in tri-axial units, where beams are directed out of two locations 90° apart. The physical size, shape, and construction of the laser differs for each manufacture. Fortunately, some vendors of alignment lasers build units in the shape of long tubes with the same diameter as an optical telescope. This allows much of the traditional optical tooling to be utilized with the alignment laser. Currently, most lasers are visible beam systems, and this eliminates the need for a beam finder. The effective working range and accuracy of the laser depends on the specific unit, and commercial systems typically identify an operating range between 50 and 300 feet.

The **detector** is the target used in laser alignment. For machinery alignment, the detector typically contains four cell units to cover the horizontal and vertical axes. The detector has a finite range over which the beam can be sensed. If the measurement range is not compatible with the detector, it may be difficult to retain the beam in the detector. This is especially true during large changes during rough alignment moves. The physical size and operation of the detector dictates the mounting. Typically, detectors are mounted on magnetic bases or angle plates. As noted in the optical instruments, special holders for bores and irregular surfaces may be fabricated if the vendor does not offer such a mount.

Since a laser beam operates in a straight line, it is sometimes necessary to use a **beam director**. This is a mirror and prism device that deflects the beam by 90°. This allows a vertical plane to be established, and it also provides a way to deflect the beam around stationary obstructions. **Beam splitters** allow the laser beam to be split into two beams with a 90° separation. With the proper placement of beam splitters and deflectors, a total of 8 axes can be measured with one laser head.

A **power supply** is used to drive the laser, detector, and readout. This device converts AC or DC power to the voltage and frequency required to operated the laser. Amplifiers, if needed, are typically located in the power supply, and many systems combine the power supply and readout into one unit.

The laser **readout** displays the current position of the laser beam on the detector. Machinery alignment readouts generally show a two axis digital display with readings in the vertical and horizontal directions. The values are typically labeled as plus or minus to indicate the direction of offset. Some systems have the ability to display offsets as position change vectors, which is quite desirable in some applications. Current systems incorporate computers to record, display, and manipulate the readings. This provides the operator with quick and accurate results. In addition, the software allows the results to be logged and saved. In this situation, the computer does reduce the chance for a human recording error. It should be recognized that computers have a tendency to isolate the diagnostician from the machine, and provide a set of concise alignment moves. This is fine when everything works properly, but if measurement or mechanical problems appear, the computer solution may not provide sufficient visibility to identify and correct the error.

It should also be noted that the methods behind laser alignment are the same as optical alignment. Measuring flat, level, square, and plumb are all possible with proper placement and beam manipulation. The setting, leveling, and zeroing of each laser will vary depending on the vendor. For instance, some units have internal automatic levels, others have manual levels, and some lasers are not equipped with levels. Regardless of the specific laser, the important point to remember is that the principles and procedures for laser position alignment are similar to that of optical alignment. In all cases, if the basic concepts are understood, the diagnostician has two viable instruments that may be applied to a field machinery problem.

OPTICAL AND LASER BORE ALIGNMENT

Bore alignment is the process of measuring and repositioning bore center lines in relation to a fixed point. This type of alignment is performed on a daily basis in many machine shops, where machine quills or boring bars are used to indicate internal machine bores and faces. With respect to process machinery, bore alignment is also performed with optical tooling, laser instruments, and wire alignment equipment. These three methods can be applied in the shop and in the field on virtually any type of equipment.

The optical tooling used in bore alignment is similar to devices used for position alignment. Telescopes, targets, and supports are required, and many of the tools and procedures used for position alignment are directly applicable to bore alignment. Optical tooling for bore alignment is generally expensive, and specific fixtures are often required for each individual machine. These measurements are subject to environmental conditions such as vibration and heat hazes, plus the resultant data is subject to interpretation. As with any technical topic, the understanding and interpretation of results improves with experience, and a sound working knowledge of the equipment and procedures.

The heart of an optical bore alignment system is the **alignment telescope** that is used to set a single line of sight. The K&E telescope shown in Fig. 12-8 is similar to that of a transit telescope. However, this model offers specific features required for bore alignment. The telescope barrel has been hard chromed, and ground to a diameter of $2.2498 \, ^{+0.0000}_{-0.0003}$ inches. This provides a precision surface to mount the scope on a cone mount, or a sphere and cup mount. The telescope can be focused from zero to infinity with a magnification range from 4X to 46X. The alignment scope is also equipped with two micrometers for horizontal and vertical movement. The micrometer numbers are colored black and red on opposite sides of zero to indicate the direction of displacement. Alignment telescopes can be fitted with auto-collimation eyepieces as well as right angle eyepieces. Attach-

Fig. 12–8 K&E Alignment Telescope Supported On Cone Mounts For Borescope Measurements

ments such as optical squares, auto-reflection targets, and angle reading attachments are also available.

The telescope **cone mount** shown in Fig 12-8 is a four cone mounting system that allows precise adjustment of the telescope in the vertical and horizontal directions. The mount consists of front and back twin cone assemblies with top retaining strips. The support cones are precision devices with threaded shafts that screw into the base. The individual cones are identified as 1 through 4. As viewed from the eyepiece end of the telescope, cones 1 and 3 are on the left side, and cones 2 and 4 are mounted on the right side. Furthermore, cones 1 and 2 are at the objective lens end of the telescope, whereas cones 3 and 4 at located back at the eyepiece end. If all four cones are initially set to the same height, and in the center of the total threaded length, the telescope mount will have the maximum adjustment range. For reference purposes, the movement characteristics of the telescope with specific cone rotations are identified in Table 12-3.

Table 12–3 Adjustment Of Support Cones For Specific Optical Telescope Movements

Operation	CW Cone Rotation	CCW Cone Rotation
Raise	—	1,2,3,4
Lower	1,2,3,4	—
Move Right	2,4	1,3
Move Left	1,3	2,4
Aim Up	3,4	1,2
Aim Down	1,2	3,4
Aim Right	2,3	1,4
Aim Left	1,4	2,3

As discussed in the position alignment section, optical targets are used to establish a line of sight with the telescope. For bore measurements, the **targets** are also used to determine offset values with respect to the line of sight. Targets used for bore alignment typically contain paired lines in both the horizontal and vertical directions. The construction varies from glass or plastic printed targets, to wire strung targets. These targets are precision ground on the outer diameter, and the crosshairs are centered to the outer diameter within 1.0 Mil. For bore alignment, two distinct types of target holders exist. Specifically, there are three-legged adjustable holder that will accommodate both glass and wire targets. The leg extensions allow the holder to be used on bore diameters ranging from 5.5 to 21.5 inches. The targets are centered in the bore with a Y centering device. This works well for rough centering, but can cause some difficulties for final setting. Another target holder configuration is shown in Fig. 12-9. Bayshore Surveying in Deer Park, Texas, produces this target holder, which adapts to most applications.

Fig. 12–9 Bore Target
Holder For Slipper Guide
By Bayshore Surveying

The holder accepts all standard glass, and wire targets. The device is designed with a solid support base, and two vertical rails. The cross rail assembly is adjustable on the vertical rails, and the target assembly is mounted on the cross rail with position adjustment screws. Thus, two position adjustment screws control the vertical and horizontal movement. The system is extremely versatile, and can easily be modified to fit in special locations.

For accurate bore measurements, the targets must be located in the center of each bore. To accomplish this task, Bayshore Surveying manufactures a **bore sweeper**. This device is a brass fixture that clamps to the outside of the target, and it supports an adjustable arm with a dial indicator for sweeping the bore. The bore may be indicated, and the target centered within 0.25 Mils. Ideally, the bore sweeper is set to indicate the bore in the same plane as the crosshairs. Also, if the bore is elliptical, the indicator readings between horizontal and vertical will not be equal. However, the target will be centered in the bore as long as the opposing readings are equal (i.e., left equals right, and top equals bottom).

As a cautionary note, an improperly set target can produce significant errors that may not be obvious. Before the equipment is installed, all targets and holders must be inspected for damage. Glass targets should be clean, and wire targets should be straight and free of kinked or broken wires. In addition, the targets should be calibrated on a routine cycle by a qualified shop.

After the targets have been positioned in each bore, it is necessary to set the alignment telescope. The telescope line of sight must be set to two of the target locations (i.e., two points determine a line). Each individual machine will dictate which two bores should be used as the zero points. Many times centrifugal machines will set zero on the two bearing bores while the seal fits are checked. In the case of an engine or reciprocating compressor, the two most outer bores are typically set as zero. The procedure for setting the alignment telescope is dependent upon the type of scope and mount (e.g., Fig. 12-8). The physical setup may require the use of a portable tripod, or the telescope assembly may be bolted directly to the end cover of the machine. In either case, once the telescope has been centered on the targets, the offsets at each point of interest (i.e., other bores) may be directly read, recorded, and physically adjusted as required.

The laser is well suited for bore alignment applications, and this has been common practice for many years. The instrumentation is the same basic equipment previously discussed in the laser position alignment section of this chapter. Once again, the bore alignment procedure for lasers is based upon the optical method. The specific procedures are unique to the laser system supplied by each vendor. It should be mentioned that most laser system targets are not indicated to the bore like the optical targets. In laser bore measurements, the detectors are set to the laser head alignment, and the detectors are then rotated 180° and zeroed. The detectors are then moved from location to location for the individual offset readings. Like any optical device, heat hazes and steam leaks will influence the laser beam, and they should be removed from the line of sight.

WIRE BORE ALIGNMENT

Wire alignment has been utilized for many years to check engine, compressor, and cylinder bores. This technique is adaptable to any situation where wire can be strung, centered, and read. The wire alignment equipment is simple, compact, and readily available at a reasonable cost. The approach is quite flexible for checking various machine components. Once a wire is strung and set in a cylinder bore, any gland, shoulder, or surface can be measured with the same setup.

Like any measurement system, wire alignment has limitations. Even though the wire is tensioned, it still sags due to gravity. For spans over 35 feet (420 inches), the sag becomes considerable. The sag must be determined and compensated for in every wire alignment. The required sag calculations are simple, but due to the repetitive nature, unconscious errors can occur. In all cases, the wire must be straight, and free from kinks or corrosion. Wire vibration due to inadvertent *twanging* of the wire, or induced vibration from adjacent machines will produce measurement errors. The wire is also sensitive to thermal changes, and mounting brackets will grow faster than the machinery case when exposed to sunlight. In addition, wire alignment requires direct physical access to all measurement areas. Small bores, long and closed spans, and tight locations can make the process difficult, and in some cases impossible.

Wire alignment requires the use of a wire, anchors, oscillator, micrometer, and a tensioner. In most commercially available wire alignment kits, the primary components are the same, with some variance in wire diameters and the tensioner. For consistency, the following discussion will be based upon the hardware contained in a standard Dresser-Rand wire alignment kit.

In the alignment kit, a **tensioner** is used to preload the wire. The two common means of tensioning the wire are by using a spring or a dead weight. The weight method offers the advantage of constant load; however, the weight is heavy and often cumbersome to hang in the field. The spring design used by Dresser-Rand provides a compact package that accommodates most applications.

The drawback to the spring tensioner is that it must be periodically calibrated to insure proper wire tension.

The piano **wire** used by Dresser-Rand has a diameter of 18.0 ±0.3 Mils. During setup, the wire must be inspected to insure that it is both clean and straight. If any physical problems are discovered with the wire, it should be discarded and replaced. The wire is supported between two **anchors**, one fixed and one adjustable. The anchors are designed to retain the wire and allow position adjustment of the wire within the bore. The anchor and wire must be electrically insulated from the machine casing. The fixed anchor is designed to clamp and retain, and it has four adjustment screws to allow positioning of the clamp in the vertical and horizontal directions. The adjustable (or tension) anchor clamps the wire, provides vertical and horizontal adjustment, and it contains the calibrated spring used to tension the wire. For the following discussions, the Dresser-Rand tensioner is presumed to be calibrated to 60 pounds of tension.

Physical alignment measurements are obtained between the wire and the bore with standard rod style inside **micrometers** such as the Starrett® 124. This micrometer configuration allows variable length rods to be inserted in the head to reach various diameters. These straight extension rods can be fabricated from 1/4 or 5/32 inch drill rod. If different diameters are to be measured, it is convenient to use several micrometers, and have one mic setup for each diameter.

The inside micrometer is used to measure the distance between the stationary bore, and the tensioned wire. Obviously, the micrometer will have a solid surface on the bore. The trick to wire alignment is to measure the distance from the bore to the wire *without* deflecting the wire. One way to accomplish this task is by connecting a low power electrical **oscillator** between the machine case and the insulated wire (usually with alligator clips). During the measurement process, the micrometer completes the circuit between the wire and ground, and an audible tone is generated through a **headset** connected to the oscillator. The oscillator may be supplied with volume and frequency controls to allow adjustment of the sound to a level compatible with the surroundings. The same information may be obtained by connecting a digital multimeter (DMM) between the wire and the machine case. In this technique, when the micrometer stem approaches the wire, the circuit is completed, and the continuity checker on the DMM emits an audible tone. Using a DMM eliminates the oscillator and the headset, but the resultant tone may not be loud enough in a noisy environment.

Regardless of the method used to identify circuit completion, the wire must be stretched between fixed and adjustable mounting points. The wire must be centered at each reference bore (*points of zero center*), and the sag at each measurement point determined. To establish the gravitational wire sag, it is necessary to obtain accurate span dimensions between the reference points of center, and the distance from the center of the wire to each measurement point. Obviously, the wire will deflect as a catenary curve, and the maximum deflection will occur at the midspan. For the Dresser-Rand system employing 18 Mil piano wire with a 60 pound tension, the wire will sag in as shown in Fig. 12-10 (rendered from referenced Dresser-Rand service manual[5]). This diagram covers an overall range of 400 inches (±200 inches from the center of the sagging wire). If a wire is

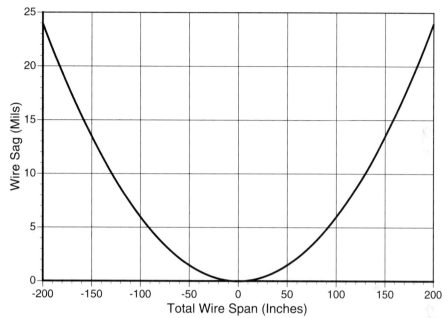

Fig. 12–10 Sag of 18 Mil Piano Wire With 60 Pound Tension - Original by Dresser Rand

stretched for the total length of Fig. 12-10, the center or midspan deflection will be 24 Mils. This is an appreciable sag, and this physical characteristic restricts wire alignment to moderately short spans.

As an example of wire sag, consider the bearing spacing shown in Fig. 12-7 in case history 41. On this machine, the overall span between the end bearings is 120 inches (=42+20+41+17). Thus, the range on Fig. 12-10 would be ± 60 inches (=120/2). For determination of wire sag at the #2 bearing, the axial location of this bearing from #1 bearing is 42 inches. The above curve can now be read at the location of the #2 bearing of 18 inches (=60-42). The wire sag at 18 inches is about 0.1 Mil, and the value at 60 inches (#1 bearing) is about 2.1 Mils. Thus the differential wire sag between the center point at #1 bearing and the measurement point at #2 bearing is approximately 2.0 Mils (=2.1-0.1). This 2.0 Mil correction would be applied to all of the vertical wire alignment data at this bearing. The sag can also be calculated from the following Dresser-Rand equation.

$$Y_{bet} = 0.0006 \times D \times \{S - D\} \qquad \text{(12-3)}$$

where:
Y_{bet} = Wire Sag Between Centers at Distance D from the Point of Center (Mils)
D = Axial Distance from the Point of Center to the Measurement Point (Inches)
S = Total Axial Span of Wire Between Points of Center (Inches)
0.0006 = Sag Constant for 18 Mil Diameter Wire with 60 Pound Tension (Mils / Inch^2)

---

[5] *Gas Engine and Compressor Field Service Manual* - Section 16 - Drawing R19307A, (Painted Post/Corning, New York: Dresser-Rand, 1986), pp. 1-12.

If the data from the previous example is inserted into equation (12-3), the span S would be 120 inches, and the distance from the zero point at #1 bearing to the #2 bearing D would be 42 inches. The sag is calculated as follows:

$$Y_{bet} = 0.0006 \times D \times \{S - D\} = 0.0006 \times 42 \times \{120 - 42\} = 1.96 \approx 2.0 \text{ Mils}$$

This is the same result previously obtained with the graphical solution from Fig. 12-10. In all cases, the calculated sag values should be compared to the graphical approach, and the answers should match within 0.25 Mils. Sag outside of the points of zero center may also be obtained by plotting the machine dimensions on Fig. 12-10. The graph is read the same as before, with the exception that the sag is now opposite to the direction of gravity. This sag (actually a rise above the zero point) can be calculated for the locations outside the points of center with the following Dresser-Rand equation:

$$Y_{out} = 0.0006 \times \left\{ d^2 - \left(\frac{S}{2}\right)^2 \right\} \tag{12-4}$$

where: Y_{out} = Wire Sag Outside of Centers at Distance d from the Center of Span (Mils)
                d = Axial Distance from Center of Span to the Measurement Point (inches)

Note that the reference points of center always have zero sag. The wire sag is determined with respect to these zero sag points. For locations inside the points of zero center, the sag value must be added to the bottom vertical reading. Conversely, for locations outside of the points of center, the sag must be subtracted from the bottom vertical readings. Horizontal readings are not influenced by sag, and do not require correction. Furthermore, if the micrometer readings are obtained with calibrated stems, the distance readings are dimensionally accurate values. However, if the readings are taken with uncalibrated drill rods in the micrometer, the resultant values are not actual lengths, and they are only suitable for differential comparisons (i.e., right versus left).

Once the wire has been centered, and the sags determined, the bore readings may be taken. As always, physical orientation must be clearly identified. Most turbomachinery trains are viewed from the driver to the driven, which clearly establishes left and right. Engines and reciprocating machines are typically viewed from the oil pump end. This may be changed, but the direction of view must be clearly identified on the alignment documentation.

After all data has been recorded and verified, the wire sag values must be applied to each measurement location. The vertical offset of a full bore is one half of the difference between the corrected top and bottom readings. Naturally, the direction of offset is governed by the larger number. Thus, if the top value is greater than the bottom value, the bore is higher than the established centers. In the case of a half bore, the bottom readings are corrected as before, and the bore offset is the difference between the bottom corrected value and the bottom reading of the center points (presumably zero). In the horizontal direction, the bore offset is one half of the difference between the left and right readings. The final bore alignment results may be presented in a graphical or a tabular format.

Case History 42: Hyper Compressor Bore Alignment

The removal and replacement of the high pressure reciprocating compressor previously discussed in case history 41 required extensive position alignment. In addition, it is mandatory to obtain proper bore alignment of the five main bearings. Since bore alignment addresses the offset between main bearings, this type of alignment is critical to the life and successful operation of the machine. For the hyper compressor from case history 41, the bores should be aligned to a tolerance of 1.0 Mil step between each of the main bearings.

The crankcase bore alignment was checked prior to the removal of the crankcase to establish a pre-overhaul base line. The initial bore alignment revealed steps between bearings that varied from 1.0 and 1.5 Mils. In addition, the bores were out of round (elliptical) and oversized. As previously mentioned, during disassembly the #2 main bearing was found to be broken.

A machine of this size and construction may easily be twisted or distorted during installation. Once the machine has been grouted in place, major alignment movements are impossible. Due to the importance of the bore alignment, both wire and optical alignment measurements were performed. Slipper guide targets were set into all five bearing bores. The targets were centered to within 0.25 Mil, and the alignment telescope set on cone mounts at the #5 bearing. The telescope was bucked-in to the #1 and #5 targets, and the offsets at bearings #2, #3, and #4 were read. These readings were taken prior to grout, and they were recorded and plotted. The optical equipment was removed, and the wire alignment equipment was installed. The wire readings were found to be within 0.5 Mils of the optical results.

After the crankcase was grouted, and allowed to cure, the crankcase was ready for the bore alignment checks. Optical tooling was again installed and the measurements after grouting were compared to the pre-grout values. There was no significant distortion of the crankcase due to the grouting, and optical equipment was then used to check the bore while the frame anchor bolts were hydrau-

Table 12–4 Summary Of Final Wire Bore Alignment Measurements And Vertical Sag Corrections — All Measurements Shown With Consistent Units Of Mils.

| Parameter | Bearing Number | | | | |
	1	2	3	4	5
Bottom Vertical	72.0	70.5	70.0	70.5	72.0
Wire Sag	0	2.0	2.0	1.0	0
Corrected Vertical	72.0	72.5	72.0	71.5	72.0
Vertical Offset	0	0.5 Low	0	0.5 High	0
Right Horizontal	72.0	71.5	72.5	72.5	72.0
Left Horizontal	72.0	72.5	72.5	72.0	72.0
Horizontal Offset	0	0.5 Left	0	0.25 Right	0

lically torqued to 1,400 foot pounds. The anchor bolt torquing was performed in several steps, and the bore alignment was continually monitored. Once all twenty-six frame anchor bolts were torqued to their final values, a set of alignment readings were obtained and recorded. The optical equipment was then removed, and the wire alignment equipment was strung and read. The final set of wire alignment readings are shown in Table 12-4. The top four rows of this summary table display the vertical wire bore readings. Since this was a half bore measurement, the bottom vertical readings were obtained, and corrected for sag. The corrected vertical readings were then compared with the end points, and the resultant differentials listed as the vertical offsets. The horizontal offsets in Table 12-4 were obtained by one half of the difference between the horizontal readings. Again, wire sag will not appreciably influence these values.

Finally, the results between the optical alignment and the wire alignment measurements are presented in Table 12-5. It is noted that these offset measurements agree within 0.5 Mils at all locations. The tolerance of 1.0 Mil maximum per step between bearings was met, and there was good confidence that the main bearing bores were properly aligned. Clearly, both techniques provide accurate dependable results for bore alignment. The optical alignment is quicker and easier to use during the setting, adjusting, and torquing. The wire may have been used during some of those operations, but would have been considerably slower. Certainly, laser measurements could also be employed for this work, and similar results would be anticipated. Finally, it should be restated that the overhaul on this high pressure reciprocating compressor was quite successful, with extended run times, and the elimination of the periodic bearing failures.

Table 12–5 Final Bore Position At Five Main Bearings On High Pressure Reciprocating Compressor Crankcase — Comparison Between Optical And Wire Alignment Techniques

| Bearing Number | Measured Bore Offset (Mils) | | | |
| | Optical Measurements | | Wire Measurements | |
	Vertical (Elevation)	Horizontal (Plan)	Vertical (Elevation)	Horizontal (Plan)
1	0	0	0	0
2	0.5 Low	0	0.5 Low	0.5 Left
3	0.5 Low	0.5 Right	0	0
4	0.5 High	0.5 Right	0.5 High	0.25 Right
5	0	0	0	0

SHAFT ALIGNMENT CONCEPTS

Shaft alignment is the most common form of machinery alignment practiced and discussed in industry. The general topics of discussion range from alignment related failures, shaft alignment methods, plus modern tools and instruments available. Study after study has ranked shaft misalignment as the primary cause of rotating equipment failures. Hence, shaft alignment is commonly performed on virtually all rotating equipment from small pumps to massive turbine generator sets. Shaft alignment is the final step to aligning a piece of rotating equipment. As discussed in this chapter, machinery alignment begins with the position of baseplates, sole plates, and the initial location of machinery casings. As part of this installation, the support members are checked to verify proper level and flatness. Next, bore alignment is employed to position all critical elements within a machine casing. This may be achieved by sweeping the machine internals with a bearing mounted mandrel, or by one of the measurement techniques discussed in the previous section.

Many of the techniques and instruments used for position and bore alignment are directly applicable to shaft alignment. However, before addressing the specific tools and procedures for shaft alignment, it is meaningful to understand the principles of shaft alignment. In essence, shaft alignment is the placement of two shafts in a coaxial position. Ideally, in a *perfect* alignment, the centerline of one shaft would be coincident with the centerline of the other shaft. In reality this type of *perfect* shaft alignment is seldom, if ever achieved. Due to the influence of variable thermal conditions, manufacturing tolerances, assembly clearances, runouts, and different operating loads — a *perfect* alignment is often considered to be an *acceptable* alignment that falls within the misalignment tolerance of the coupling. Hence, shaft alignment really is the final positioning of the machinery in a cold condition to compensate for the machinery full load condition, and allow for a hot running alignment that is within the coupling misalignment tolerance.

Interestingly enough, machines such as gas turbines may have a large cold offset to allow for thermal growth as the turbine reaches full operating temperature. In many units it is common to encounter cold offsets of 250 Mils or more. With this type of cold offset across a coupling, it is clear that the machinery train starts up with a significant misalignment. The unit then moves into an acceptable alignment state as the machines reach thermal equilibrium. In more than one case, a coupling has been subjected to serious distress due to extended operation at presumably *safe* slow roll speeds.

To appreciate the requirements and problems associated with shaft alignment, it is reasonable to begin with a basic understanding of the three fundamental types of shaft misalignment. This includes a simple parallel offset between shafts, and angular offset of one shaft to another, plus a combination of the two offsets into a parallel plus an angular offset. For demonstration purposes, consider the example of parallel misalignment shown in Fig. 12-11. In this diagram, two shafts are shown to be perfectly parallel. That is, the centerline of one shaft is parallel to the centerline of the other shaft. The misalignment occurs

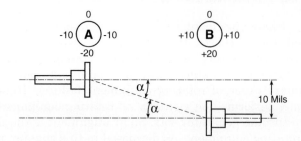

Fig. 12–11 Parallel Shaft
Misalignment

when the centerline of the **B** shaft is dropped by 10 Mils below the centerline of the **A** shaft. If reverse dial indicators are mounted across these two shafts, the resultant sweep readings are shown at the top of Fig. 12-11. Reverse dial indicators will be discussed later in this chapter, but the immediate comment on Fig. 12-11 might be: *if the shafts are displaced vertically by 10 Mils, how come the indicators show a 20 Mil change?* Initially, this might be a difficult concept to grasp. However, the dial indicator readings are directly explainable.

Specifically, assume that a pair of dial indicators are mounted on the **A** coupling, and they are indicating on the **B** coupling. Assume that both shafts are perfectly aligned, and that one indicator is mounted on top at the 12 o'clock position, and the other dial indicator is mounted on the bottom at 6 o'clock. Further assume that both indicators are set to zero with the shafts perfectly aligned. If shaft **B** is lowered by 10 Mils as shown in Fig. 12-11, the top dial indicator would show a reading of -10 Mils indicating that the stem was moving away from the indicator. Simultaneously, the bottom indicator would show a reading of +10 Mils indicating that the stem was collapsed, or moving into the indicator. Therefore, the total change sensed by both indicators would be +20 Mils {=+10-(-10)}. Looking at this move another way, if the top indicator that is reading -10 was reset to zero, and the **A** shaft rotated by 180°, the resultant reading at the 6 o'clock position would be +20 Mils. This simple example identifies a fundamental rule in shaft alignment as expressed by the following equation:

$$Vert_o = \left| \frac{Bottom - Top}{2} \right| \qquad\qquad \textbf{(12-5)}$$

where: $Vert_o$ = Vertical Offset (Mils)
         $Bottom$ = Indicator Reading at Bottom of Sweep (Mils)
         Top = Indicator Reading at Top of Sweep, Usually set to Zero (Mils)

The same argument may be applied to the horizontal offset as follows:

$$Horiz_o = \frac{Right - Left}{2} \qquad\qquad \textbf{(12-6)}$$

where: $Horiz_o$ = Horizontal Offset (Mils)
        $Right$ = Indicator Reading at Right Side of Sweep (Mils)
        $Left$ = Indicator Reading at Left Side of Sweep (Mils)

The next common type of misalignment is the angular condition where the centerline of the **B** shaft intersects the center line of the **A** shaft at the coupling. This condition is shown in Fig. 12-12, and the associated reverse indicator readings are presented at the top of the diagram. Since the shaft centerlines intersect at the **A** coupling hub, the dial indicator on the **B** coupling does not detect any change during a sweep. However, the indicator mounted on the **A** coupling reveals that the **B** coupling centerline is low by 10 Mils (=TIR/2). The only way that both shafts can coincide at the **A** coupling, and simultaneously be 10 Mils low at the **B** coupling is the angular misalignment condition of Fig. 12-12.

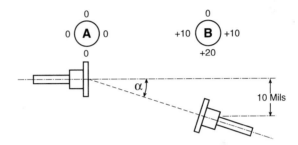

Fig. 12–12 Angular Shaft Misalignment

The third type of shaft misalignment is the combination of angular plus parallel misalignment as depicted in Fig. 12-13. In this condition the shaft centerlines cross at any point other then the coupling faces. This is a much more common situation where adjustments in offset and angularity are required. For comparative purposes, the reverse indicator readings across the span are again shown at the top of Fig. 12-13. A casual observation of these readings might lend one to conclude that the offset is a simple parallel offset as previously shown in Fig. 12-11. However, the key element in this combination of parallel and angular offsets is the fact that both indicators show a value of +20 Mils. Whereas on Fig.12-11, one indicator read +20 Mils and the other indicator showed -20 Mils. This subtle distinction of polarity on the dial indicator reading makes a significant difference in the type of misalignment to be addressed.

Once again the need to properly document plus and minus signs and maintain a fixed viewing position of the alignment job is reinforced. In all cases, if the machinery diagnostician starts changing viewing directions, right becomes left, and vice versa. Basically, the entire alignment project crumples into disarray.

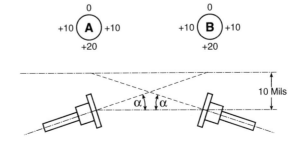

Fig. 12–13 Combination Parallel And Angular Shaft Misalignment

Obviously the previous three cases represent the most simple types of shaft misalignment. Process machines seldom display misalignment in only one plane. In most instances, the diagnostician has to deal with a complex offset in both the vertical and the horizontal directions. In an effort to minimize these offsets, it often makes sense to place the machines in a *rough aligned* position. This can be accomplished using a straight edge or precision scale, and generally will bring the alignment close enough to begin the precision alignment.

Most shaft alignment techniques use dial indicators and indicator brackets. These brackets are attached to a machine shaft or coupling hub. A dial indicator is then mounted on the end of the bracket, and it is positioned to indicate the mating hub or shaft. Brackets are constructed of everything from solid re-bar to hollow aluminum tubes. Unfortunately, as rigidity increases so does the bracket weight. The diagnostician should strive for brackets that provide a good compromise between high rigidity and low weight. In all cases, the indicator bracket will sag due to the weight of the bracket plus the indicator. For short distances the sag is minimal and repeatable. For long coupling spans the bracket sag can become both excessive and unpredictable. Regardless of the span and bracket size, the sag must be checked and documented prior to use. The sag value must be recorded and used to correct the alignment readings. To check the indicator bracket sag the following procedure may be used.

1. Determine where brackets will be mounted on the equipment to be aligned.
2. Measure and record the length between alignment planes (bracket span).
3. Attach indicator(s) and bracket to a lightweight rigid pipe (mandrel) at the same position as they will be mounted on the machine.
4. Zero the radial indicator, and lightly tap the indicator face to check for stability. If the needle does not return to zero, check and retighten all clamps. If the indicator does not re-zero, replace the indicator and/or bracket.
5. Pick up the assembly and hold it in front of your body.
6. Verify that the radial indicator is still reading zero. If the indicator has moved the mandrel is flexing, and should be replaced with a rigid unit.
7. Rotate the assembly to the 6 o'clock position, and hold it over your head so the bracket and indicator(s) are hanging towards the ground.
8. Read the radial indicator and record this value.
9. Next, hold the assembly at the 3 or 9 o'clock position and read the radial indicator. The value should be one half of the sag at the 6 o'clock position.
10. Return the assembly back to the 12 o'clock position and re-check zero.

In specialty situations with large or heavy brackets, it may not be possible to physically lift the assembly. In this case, the assembly may be set in V-blocks, rollers, or the mandrel may be positioned in a lathe for the sag test. Regardless of the physical setup, the indicated bottom value is twice the bracket sag. This is due to the fact that the bracket sags both at the top vertical 12 o'clock position, as well as the bottom vertical 6 o'clock position. Hence, the overall bracket sag may be obtained by dividing the bottom reading by 2. This relationship is an obvious extension of equation (12-5), and it may be stated as:

$$Bracket\ Sag = \frac{Bottom\ Reading}{2} \qquad \textbf{(12-7)}$$

This expression requires that the indicator is zeroed at the top 12 o'clock position, and the *bottom reading* is obtained with the entire assembly (mandrel, bracket, and indicator) inverted by 180°. This bracket sag must always be a negative number. If the dial indicator displays a positive value, the mandrel is probably more flexible than the indicator bracket and it should be replaced.

It should also be mentioned that a bracket counterweight may be applied to *balance out* some of the overhung weight associated with the dial indicator(s). The position and size of the counterweight may be optimized during the bracket sag test. Generally, if bracket sag is less than 20 to 25 Mils, an indicator counterweight is not necessary. As always, the repetitiveness of the data may carry more significance than the actual magnitude — and the diagnostician must be aware of the specific measurement situation, plus the associated *trade-offs*. From a utilization standpoint, the bracket sag correction will be demonstrated during the following discussions about rim and face, plus the reverse indicator sections.

RIM AND FACE SHAFT ALIGNMENT

Rim and face shaft alignment is one of the oldest and most common forms of aligning two shafts. This method measures the shaft offset by indicating one shaft with respect to the rim and face of a mating shaft. A typical machinery configuration subjected to this type of alignment is shown in Fig. 12-14. The rim dial indicator readings determine vertical and horizontal offsets, and the face readings identify angular misalignment. The face readings may be obtained with a dial indicator. Alternatively, inside micrometers may be used to measure the distance between coupling hubs. On couplings with close axial clearances, taper or feeler gauges may be used to measure the gap between faces. Although this method has lost some appeal as newer techniques were developed, rim and face alignment is still used in the following applications:

❍ Trains where one shaft cannot be rotated during the alignment process.
❍ Machines with coupling hubs that are axially close to each other.
❍ Machines that have large diameter couplings (i.e., the coupling diameter is much greater than the coupling span).
❍ Small general purpose machines that are typically less than 5 HP.

Rim and face readings provide a good visualization of the relative positions of two shafts. However, this technique has several disadvantages, and many applications require reverse indicator measurements (to be discussed in the next section). With respect to the current topic, the diagnostician generally has to contend with the following limitations associated with rim and face alignment.

❒ Coupling hub runout will induce an error into the alignment readings. If runout is excessive or inconsistent, it can be very difficult to compensate.

❒ Machines with sleeve bearings can have a face reading error due to rotor axial float. It is necessary to locate the rotor in a fixed axial position for each sweep. On machines with a large axial floats, such as motors with sleeve bearings, some type of axial *stops* must be employed.

❒ Generally, the coupling spool piece must be removed.

❒ Provides marginal accuracy on units with small diameter couplings and/or long spans (i.e., coupling diameter is much less than the coupling span).

❒ Any indicator face sag can be difficult to accurately compensate, and this may adversely influence the angularity corrections.

❒ On long spans, the indicator brackets can become very complex, and flexible.

Along with the standard pre-alignment considerations discussed earlier in this chapter, rim and face alignment requires several additional steps. Specifically, the coupling spool piece must be disconnected, so that one rotor will remain stationary while the other rotor is rotated. The coupling spool also interferes with the face readings, and in most cases it should be removed.

Rim and face coupling hub runouts must be documented on the stationary unit. If the shaft in the stationary machine can be turned, it is a simple matter to mount a dial indicator on any fixed mechanical element, and measure the coupling hub runout (every 90°). However, if the rotor of the stationary machine cannot be rotated, the radial hub runout can be determined by a dial indicator mounted on the shaft that can be turned. In this situation, the rotatable shaft is turned in 90° increments, and readings are obtained from the stationary coupling hub. The dial indicator is then repositioned to the shaft of the stationary machine, and a second set of radial readings are obtained. Keep in mind that the bracket sag may change with this move, and must be rechecked. The difference between the shaft and hub readings will be the radial coupling hub runout (assuming minimal shaft runout). Excessive radial runout (greater than 2.0 Mils) may be indicative of other problems with the machinery, and this should be investigated and corrected prior to any alignment activities. Similarly, excessive face runout of greater than 0.5 Mils should also be queried and rectified.

The tools required to perform rim and face shaft alignment include the dial indicator and some form of indicator bracket. The dial indicator is a precision mechanical distance measurement device consisting of a face, plunger, and body. The circular face has a graduated scale with an indicator needle. For shaft alignment applications, the device will typically read in graduations of 1.0 Mil per division. Indicators are also available in 0.5 and 0.1 Mil steps. The graduated face contains a zero point with circular scales progressing in a positive clockwise direction, and a negative counterclockwise direction. As the plunger is depressed, the needle moves in a positive clockwise direction. Similarly, as the plunger is extended, the needle moves in a negative or counterclockwise direction. In actual use, the plunger is placed in direct contact with the surface to be measured, and the indicator face is rotated to the zero point. Subsequent measurements may then be identified with *plus* and *minus* signs to define the direction of plunger travel. This polarity information identifies the convergence or divergence of the indicator with respect to the measured surface.

An inclinometer is a shaft mounted level that displays degrees of rotation. This device is used for determination of true horizontal and vertical locations. In many cases, a pair of perpendicular bubble levels attached to a magnetic base may be mounted on the coupling face. By observation of the two bubbles, the shaft may be rotated in fairly concise 90° increments.

A strap wrench or chain wrench is typically used to rotate the moveable shaft during alignment. With very heavy rotors, a specialized *cheater bar* with pins to fit into the coupling holes may be constructed. As a precautionary note, the shaft must never be turned by twisting the dial indicator bracket. Application of torque at the bracket will generally invalidate any useful alignment data. In addition to the standard millwright hand tools and measurement devices, the diagnostician should address the tool requirements for actually moving the machinery. These tools will vary between jobs, and options such as slings and chainfalls, hydraulic jacks, pry bars, jack bolts, or jack screws are acceptable methods of moving equipment. Hammers and wedges should not be used to move equipment, and hydraulic jacks should be very carefully applied. More than one machine has been damaged due to excessive force from a hydraulic jack.

In order to convert the rim and face measurements to physical moves of the machinery, similar triangles are established between the dial indicator readings and the machinery arrangement. Specifically, Fig. 12-14 describes a typical setup of rim and face dial indicators, plus the required support dimensions. The coupling diameter observed by the rim indicator is identified as the *A* dimension. The distance from the face indicator plane to the center of the *near foot* is identified as the *B* dimension. Similarly, the *C* dimension defines the distance between the *far foot* and the face indicator plane. Based upon these distances, and the rim

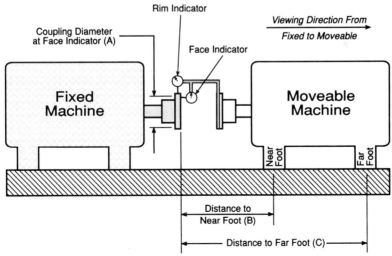

Fig. 12–14 General Arrangement For Rim And Face Shaft Alignment

and face dial indicator readings, the two shafts may be aligned. Since this is such an important activity, the specific details and requirements to check, document, and correct the machinery alignment are presented in the following procedure.

1. The *rough* shaft alignment should be verified with a straightedge, and precision scale. This is a three-dimensional check that covers vertical and horizontal position, plus axial spacing across the coupling. Large errors in any direction should be corrected at this time. In all cases, the expected alignment moves should be well within the travel range of the dial indicators.

2. Measure and record the coupling diameter *A*, the distance from face indicator plane to the moveable near foot *B*, and distance from the face indicator plane to moveable far foot *C*. Check and record the coupling span.

3. Based on the machine dimensions, set the indicator brackets on a mandrel and check bracket sag in accordance with the previously stated procedure.

4. Check and correct for soft foot as previously discussed.

5. Mark the 3, 6, 9 and 12 o'clock positions on both coupling hubs with a paint or ink marker. It is very important to maintain the two shaft orientations, especially when runout compensation is necessary.

6. Mount the indicator bracket and the dial indicator(s) on the shaft of the moveable machine. This shaft must be able to be rotated for the alignment readings. If a face indicator cannot be used, establish an alternate measurement of axial coupling span (e.g., inside micrometer or feeler gauge).

7. Position both rim and face dial indicators at the top of the coupling hub. Make certain the rim indicator is perpendicular to the shaft centerline, and the face indicator is perpendicular to the coupling face. Set both indicators to the middle of their respective travel range, and zero both indicators.

8. If possible, rotate the shaft on the stationary machine, and record coupling hub and face runouts every 90°. If this stationary rotor cannot be turned, then alternate methods for determination of runout should be employed.

9. Turn the moveable shaft through one complete rotation and check for adequate indicator travel. It is always a good idea to follow the indicator around with a mirror to make sure that the plunger is in contact with the shaft surface throughout the sweep, and to confirm the reading polarity. Also check for obstructions that may interfere with the indicator bracket.

10. Reposition the indicators at the top 12 o'clock location, and re-zero. Again rotate the shafts through one complete rotation and verify the indicators return to zero at the 12 o'clock position. If the indicators do not return to zero, check the setup for any possible looseness or movement, and repeat.

11. Rotate the moveable shaft around to 3, 6, and 9 o'clock positions, and record the dial indicator readings and direction (polarity) at each location. Orientation is typically viewed from the fixed to the moveable, and this must be stated with the recorded indicator readings for future reference. Fig. 12-15 describes the traditional method used for documenting this type of dial

indicator information. The zeroes at the 12 o'clock position represent the manually set zero points. The rim readings are typically shown on the outer diameter of the circle, and the face readings are presented inside the circle. It is always a good idea to obtain several sweeps, and present a realistic average dial indicator reading at the 3, 6, and 9 o'clock positions.

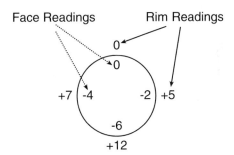

Fig. 12–15 Direct Indicator Readings For Rim And Face Measurements

12. Check the validity of the averaged dial indicator readings. In all cases, the following expression must be satisfied:

$$Top + Bottom = Right + Left \qquad \text{(12-8)}$$

Since the top readings are set to zero, this equation may be simplified as:

$$\boxed{Bottom = Right + Left} \qquad \text{(12-9)}$$

For the rim indicator readings shown in Fig. 12-15, it is clear that +12 is equal to +7 added to +5. This is a fundamental validity test, and if it does not agree to within 1.0 or 2.0 Mils, the readings are corrupted. In this situation, the cause of the inequality must be determined and corrected before proceeding with the alignment. In all cases, the plus or minus polarity of the indicator readings must be considered and included into (12-9).

13. The averaged readings from step 11 must now be corrected for bracket sag. As previously stated, bracket sag is a negative number, and it must be subtracted from the dial indicator readings. The bottom vertical reading must have the total measured sag, or twice the actual sag subtracted from the indicator reading. The horizontal readings must have the actual sag, or one half of the measured sag subtracted from the indicator reading. For example, if the indicator bracket used for the measurements shown in Fig. 12-15 displayed a total bottom sag reading of -6.0 Mils, and a pair of side sag readings of -3.0 Mils, the sag corrections would be performed as follows:

$$Left = (+7) - (-3) = +10$$
$$Right = (+5) - (-3) = +8$$
$$Bottom = (+12) - (-6) = +18$$

These corrected rim readings are combined with the face readings, and the results are shown in Fig. 12-16. In most instances face sag is insignificant and it is not considered. Hence, the rim and face readings in Fig. 12-16 are the values that will be used to determine the vertical alignment changes.

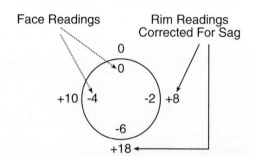

Face Readings   Rim Readings
Corrected For Sag

Fig. 12-16 Face Measurements Combined With Rim Measurements That Are Corrected For Bracket Sag

14. It may be possible to save time by performing vertical and horizontal alignment moves simultaneously. However, machines have a tendency to slide horizontally during vertical adjustments. The most direct approach is to make the vertical moves first and then proceed with the horizontal shifts. For face measurements, the following equations (12-10) and (12-11) may be used to calculate the required correction at each foot of the moveable equipment for angular misalignment.

$$Ang_{nf} = (Face) \times \left\{ \frac{B}{A} \right\} \tag{12-10}$$

$$Ang_{ff} = (Face) \times \left\{ \frac{C}{A} \right\} \tag{12-11}$$

where: Ang_{nf} = Shim Change at Near Foot for Vertical Angular Misalignment (Mils)
Ang_{ff} = Shim Change at Far Foot for Vertical Angular Misalignment (Mils)
$Face$ = Face Dial Indicator Reading At 6 o'clock on The Coupling Face (Mils)
A = Diameter of the Coupling Hub where the Face Reading Was Obtained (Inches)
B = Distance Between the Indicator Plane and the Center of the Near Foot (Inches)
C = Distance Between the Indicator Plane and the Center of the Far Foot (Inches)

These dimensions in equations (12-10) through (12-13) are consistent with the machinery arrangement diagram presented in Fig. 12-15. Intuitively, if the face readings are zero, there is no angular misalignment, and the problem reverts to a simple parallel offset, such as shown in Fig. 12-11. If the bottom face reading is positive, the results from (12-10) and (12-11) will be positive. The plus sign will indicate the requirement to add shims, and raise the moveable machine. Conversely, a negative value is indicative of a need to remove shims and lower the moveable machine.

15. The vertical offset in rim and face alignment is determined by the bottom corrected indicator reading. The bottom corrected reading is divided in half to give the centerline offset as previously described by equation (12-5). If the angular offset from equations (12-10) and (12-11) are combined with the vertical offset, the following two equations may be used to calculate the total vertical correction at each support foot of the moveable machine.

$$Vert_{nf} = (Face) \times \left\{ \frac{B}{A} \right\} - \left\{ \frac{Rim}{2} \right\} \qquad \textbf{(12-12)}$$

$$Vert_{ff} = (Face) \times \left\{ \frac{C}{A} \right\} - \left\{ \frac{Rim}{2} \right\} \qquad \textbf{(12-13)}$$

where: $Vert_{nf}$ = Total Shim Change at Near Foot for Vertical Angular and Offset Misalignment (Mils)
   $Vert_{ff}$ = Total Shim Change at Far Foot for Vertical Angular and Offset Misalignment (Mils)
   Rim = Rim Dial Indicator Reading At 6 o'clock on the Coupling Hub (Mils)

Again, the required movement direction is determined by the sign (±). A positive (+) sign for equations (12-12) and (12-13) indicates that the moveable machine must be raised. A final negative (-) sign requires that the moveable machine be lowered at both support feet.

As an example of these calculations, consider the corrected rim and face readings shown in Fig. 12-16. From these dial indicator readings, the *Face* value was equal to -6 Mils, and the corrected *Rim* reading was +18 Mils. Assume that the swept face diameter *A* was 8 inches, and the distances from the indicator plane to the support feet *B* and *C* were equal to 21 and 40 inches respectively. Based on this information, the required vertical corrections may be computed from equations (12-12) and (12-13) as follows.

$$Vert_{nf} = (-6) \times \left\{ \frac{21}{8} \right\} - \left\{ \frac{18}{2} \right\} = -15.75 - 9 = -24.75 \approx -25 \text{ Mils}$$

$$Vert_{ff} = (-6) \times \left\{ \frac{40}{8} \right\} - \left\{ \frac{18}{2} \right\} = -30 - 9 = -39 \text{ Mils}$$

From these calculations, it is clear that the near foot should be lowered by 25 Mils, and the far foot should be lowered in elevation by 39 Mils.

16. After completion of the vertical elevation changes on both of the moveable machine supports — the rim and face indicators should be re-zeroed, and another sweep made of the coupling. At this time, the vertical angularity and offset should be corrected. However, if residual misalignment still persists, the previous steps 9 through 15 should be repeated.

17. Once the vertical corrections have been completed, the horizontal movements can be made. If the machine was not moved horizontally, the previous sweeps may be used to calculate the required horizontal movements. In essence, the vertical readings identify the vertical misalignment, and the horizontal dial indicator readings are used to identify and correct the horizontal misalignment between shafts. Stated in another way, the vertical dial indicator readings (top and bottom) are used to compute the vertical shim changes at the near and far support feet. During these vertical calculations the horizontal dial indicator readings are not used. Similarly, the horizontal dial indicator readings (left and right) are used to determine the horizontal or sideways moves. For the horizontal changes, the vertical dial indicator readings are not required.

18. In most cases it is desirable to re-sweep the coupling to obtain new dial indicator readings. This will identify the current horizontal position, and check the vertical position. The horizontal shifts are calculated in the same manner as the vertical movements. Specifically, the horizontal alignment changes may be determined with the following equations:

$$Horiz_{nf} = (Face) \times \left\{ \frac{B}{A} \right\} - \left\{ \frac{Rim}{2} \right\} \qquad \text{(12-14)}$$

$$Horiz_{ff} = (Face) \times \left\{ \frac{C}{A} \right\} - \left\{ \frac{Rim}{2} \right\} \qquad \text{(12-15)}$$

where: $Horiz_{nf}$ = Shift at Near Foot for Horizontal Angular and Offset Misalignment (Mils)
$Horiz_{ff}$ = Shift at Far Foot for Horizontal Angular and Offset Misalignment (Mils)

The *Face* and *Rim* numbers used in equations (12-14) and (12-15) are based upon the horizontal dial indicator readings. Some individuals prefer to re-zero the indicators horizontally prior to the horizontal moves. This is perfectly acceptable as long as consistency is applied, and the validity rule of equation (12-8) is satisfied.

Also, if equations (12-14) and (12-15) display positive (+) signs, the moveable machine must be moved to the left. If these two expressions produce negative (-) signs, then the machine must be moved to the right.

19. After calculating the required horizontal moves in the previous step 18, the diagnostician should position horizontal dial indicators at all four corners of the moveable machine. This will allow the horizontal machine movement to be accurately monitored as the machine is shifted into position.

20. Repeat steps 18 and 19 until the desired horizontal alignment has been achieved. Double-check that the coupling gap is still within specification.

21. Once the final alignment condition has been reached, obtain one final set of indicator readings. Correct these readings for sag, and identify this information as the *final* alignment readings.

This procedure and the associated discussion was directed at achieving a straight 0-0-0-0 rim and face alignment between two shafts. In reality, thermal growth and other running conditions must be factored into the final alignment. In most machinery applications, target values are established to accommodate the changes between the cold alignment and the normal running condition. The methods to obtain these offsets are discussed at the end of this chapter. As with any other form of machinery alignment, the previously outlined procedure can be adapted to a variety of specific applications. In all cases, once the fundamental concept is understood, the modification of a standard procedure to address a particular mechanical situation is easily accomplished.

REVERSE INDICATOR SHAFT ALIGNMENT

Reverse indicator shaft alignment is similar to the rim and face technique. This method measures the shaft centerline offsets between two adjacent shafts or couplings. In most cases, the two shafts are rotated together, and simultaneous rim readings are acquired. A typical machinery arrangement subjected to reverse indicator readings is shown in Fig. 12-17. The machine on the left side of this diagram is designated as the fixed or stationary machine, and the machine on the right side is considered as the moveable unit. A dial indicator bracket is attached to the shaft or coupling of the fixed machine, and the indicator is positioned in contact with the shaft or coupling hub of the moveable machine. A second indicator bracket is mounted on the shaft or coupling of the moveable machine, and the attached dial indicator is placed in contact with the shaft or

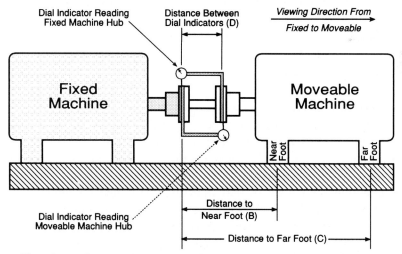

Fig. 12–17 General Arrangement For Reverse Indicator Shaft Alignment

coupling hub of the stationary machine. This setup for shaft alignment has become quite popular due to the following advantages.

○ Reverse indicator provides better dimensional accuracy for small diameter, long span machines (i.e., the coupling diameter is less than coupling span).
○ Reverse indicator does not require removal of the coupling spool piece.
○ In most cases, both rotors are turned together, and coupling hub runout does not influence the dial indicator readings.
○ Axial float does not significantly influence the alignment measurements.
○ Only one indicator is mounted on each indicator bracket and this decrease in overhung weight reduces the dial indicator bracket sag.
○ A graphical display of the relative shaft positions and the final alignment condition are easily generated, and are visually meaningful.

Virtually all mechanical procedures display various benefits combined with some drawbacks. Reverse indicator shaft alignment is no exception and the traditional disadvantages of this technique are summarized as follows:

❐ Reverse indicator shaft alignment requires the rotation of both rotors. When one machine cannot be turned, reverse indicators cannot be used.
❐ Provides marginal accuracy on close coupled machines with large diameter couplings (i.e., the coupling diameter is greater than coupling span).
❐ Cannot be used on small machines with insufficient room to install two indicator brackets and associated dial indicators.
❐ On machines with large diameter couplings and extremely long spans, the indicator brackets become difficult to handle and obtain consistent data.

The tools and equipment used for reverse indicator shaft alignment are the same devices required for rim and face measurements. The setup is fundamentally the same, except for the use of brackets with only one dial indicator, and the fact that the coupling spool piece is generally not removed. From Fig. 12-17, the distance between the indicator plane on the fixed machine to the center of the *near foot* of the moveable machine is again identified as the **B** dimension. Similarly, the distance between the dial indicator plane on the fixed machine to the center of the *far foot* on the moveable machine is once again called the **C** dimension. The axial distance between dial indicator planes will be referred to as the **D** dimension. In most installations, this **D** distance will be greater than the shaft end spacing, but **D** will always be less than the **B** or **C** dimensions.

It should be mentioned that reverse indicator alignment may be performed in several different ways. For instance, one bracket with one dial indicator may be mounted on the fixed machine and used to sweep the coupling hub of the moveable machine. This bracket and indicator may then be removed from the fixed machine, and then *reversed* and attached to the moveable machine. In this location, the indicator is used to sweep the coupling hub of the fixed machine. Moving of brackets and dial indicators does minimize the required hardware to perform the alignment. However, it is time consuming, and prone to error if the

mounting positions are not accurately repeated. In most industrial applications, a pair of dial indicators and associated support brackets are mounted across the coupling. This takes slightly longer during the initial setup, but it is much easier and faster during the actual alignment work. The method discussed in the following procedure locates both indicators at the same angular position on the shafts. Some methods use indicators that are opposed by 180° across the coupling. Although the results are identical, the sign convention can become confusing. In all cases, the sign convention, orientation, and equipment viewing direction must be documented and maintained throughout the alignment work.

Reverse indicator alignment is a common practice in many industries. Due to the widespread use of this technique, the following detailed procedure and discussion is presented for both information and reference purposes.

1. Inspect the coupling to verify proper assembly and correct bolt torque.
2. The *rough* shaft alignment should be verified with a straightedge and precision scale. This is a three-dimensional check that covers vertical and horizontal position, plus axial spacing across the coupling. Large errors in any direction should be corrected at this time. In all cases, the expected alignment moves should be well within the travel range of the dial indicators.
3. Measure and record the distance D between indicator planes, the distance from the indicator plane on the fixed machine to the moveable near foot B, plus the distance from the indicator on the fixed machine to moveable far foot C. Also, measure and record the coupling span.
4. Based on the machine dimensions, set the indicator brackets on a mandrel and check bracket sag in accordance with the previously stated procedure.
5. Check and correct for soft foot as previously discussed.
6. Mount both indicator brackets and the associated dial indicators on both shafts (or couplings), and verify that all brackets and clamps are tight.
7. Position both dial indicators to read their respective hub or shaft surfaces. Make certain that the indicators are perpendicular to the shaft centerline. Set both indicators to the middle of their respective travel range, and zero both indicators.
8. Turn the shaft through one complete rotation and check for adequate indicator travel. It is always a good idea to follow the indicators around with a mirror to make sure that the plungers are in contact with the shaft surface throughout the sweep, and to confirm the reading polarity. Also check for obstructions that may interfere with the indicator bracket.
9. Reposition the indicators at the top 12 o'clock location, and re-zero. Again rotate the shafts through one complete rotation and verify that the indicators return to zero at the 12 o'clock position. If the indicators do not return to zero, check the setup for any possible looseness or movement, and repeat.
10. Rotate both shafts to the 3, 6, and 9 o'clock positions, and record the dial indicator readings and the direction (polarity) at each location. Use an incli-

nometer or dual bubble levels to locate the precise positions at 3, 6, and 9 o'clock. In most instances, the shafts should be stopped precisely at 90° increments. If there is any *overshoot* of the angular position, it is usually best to continue around in the direction of rotation until the correct angular position is obtained. This is normally preferable to stopping and reversing direction (especially with LOP tilt pad bearings). However, there is an exception to this rule when dealing with machines coupled to gear boxes. In this situation, the shafts may get *hung up* in the gear mesh. It is often advisable to rotate slightly past the desired angular positions, and then reverse rotation to allow the gear to *back off* the mesh.

11. Document the reverse indicator data in the manner shown in Fig. 12-18. Again, these dial indicator readings are obtained with a viewing direction that originates at the outboard end of the fixed machine and looks towards the moveable machine. The zeros at the 12 o'clock position on both sweeps represent the manually set zero points. The readings shown on the outside of each circle are the radial or rim readings from each dial indicator. Again, it is always a good idea to obtain several sweeps and present a realistic average dial indicator reading at the 3, 6, and 9 o'clock positions.

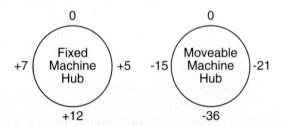

Fig. 12–18 Direct Reverse
Indicator Readings

12. Check the validity of the dial indicator readings with equation (12-9). This expression requires that the sum of the left and right indicator readings are equal to the bottom value. A variance of 1.0 to 2.0 Mils is considered acceptable. However, if these tolerances are exceeded, the dial indicator readings are corrupted. Once again, the source of the error must be determined and corrected before proceeding with the shaft alignment.

13. The averaged readings from step 11 must now be corrected for bracket sag. As previously stated, bracket sag is a negative number and it must be subtracted from the dial indicator readings. The bottom vertical reading must have the total measured sag or twice the actual sag subtracted from the indicator reading. The horizontal readings must have the actual sag or one half of the measured sag subtracted from the indicator reading. For instance, if the indicator brackets used for the readings shown in Fig. 12-18 displayed a total bottom sag reading of -6.0 Mils, and a pair of side sag readings of -3.0 Mils, the sag corrections would be performed as follows:

Fixed Hub Readings:

$Left = (+7) - (-3) = +10$

$Right = (+5) - (-3) = +8$

$Bottom = (+12) - (-6) = +18$

Moveable Hub Readings:

$Left = (-15) - (-3) = -12$

$Right = (-21) - (-3) = -18$

$Bottom = (-36) - (-6) = -30$

These sag corrected readings are shown in Fig. 12-19. These vertical dial indicator readings will be used to determine the vertical shim changes, and the horizontal values will be used to compute the horizontal moves.

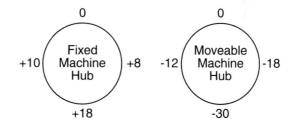

Fig. 12–19 Reverse Dial Indicator Readings Corrected For Bracket Sag

14. Based on the machine geometry (dimensions **B**, **C**, and **D**) and the sag corrected dial indicator readings — the vertical and horizontal alignment moves may now be determined with direct calculations or a graphical solution. These computations may be accomplished with a variety of handheld calculator alignment programs, or the more sophisticated personal computer programs that provide both computed and graphical results. These programs are excellent tools; however, it is necessary to understand the basic concepts before depending on any type of automated system. Thus, the continuation of this reverse indicator procedure will address the specific details for calculating the moves, plus plotting the graphical solution. This work begins by determining the vertical and horizontal centerline offsets using the following set of equations:

$$V_{o_{fix}} = \left\{ \frac{Bottom - Top}{2} \right\}_{fix} = \left\{ \frac{Bottom}{2} \right\}_{fix} \qquad (12\text{-}16)$$

$$V_{o_{mov}} = \left\{ \frac{Bottom - Top}{2} \right\}_{mov} = \left\{ \frac{Bottom}{2} \right\}_{mov} \qquad (12\text{-}17)$$

$$H_{o_{fix}} = \left\{ \frac{Right - Left}{2} \right\}_{fix} \qquad (12\text{-}18)$$

$$H_{o_{mov}} = \left\{ \frac{Right - Left}{2} \right\}_{mov} \qquad (12\text{-}19)$$

where: $V_{o\text{-}fix}$ = Vertical Shaft Offset at Fixed Machine (Mils)
        $V_{o\text{-}mov}$ = Vertical Shaft Offset at Moveable Machine (Mils)
        $H_{o\text{-}fix}$ = Horizontal Shaft Offset at Fixed Machine (Mils)
      $H_{o\text{-}mov}$ = Horizontal Shaft Offset at Moveable Machine (Mils)
      $Bottom$ = Indicator Reading at Bottom of Sweep (Mils)
          Top = Indicator Reading at Top of Sweep, Normally Equal to Zero (Mils)
       $Right$ = Indicator Reading at Right Side of Sweep (Mils)
         $Left$ = Indicator Reading at Left Side of Sweep (Mils)

Obviously, equations (12-16) through (12-19) are nothing more than extensions of the previous shaft offset equations (12-5) and (12-6) applied to each set of dial indicator readings. It is also clear that bracket sag corrected numbers must be used from this point on. Again, it is very important to view the machine from a constant position and maintain the proper sign convention. For example, the sag corrected dial indicator readings from Fig, 12-19 may be used to compute the vertical and horizontal offsets using equations (12-16) through (12-19) as follows:

$$V_{o_{fix}} = \left\{ \frac{Bottom}{2} \right\}_{fix} = \left\{ \frac{18}{2} \right\}_{fix} = +9 \text{ Mils}$$

$$V_{o_{mov}} = \left\{ \frac{Bottom}{2} \right\}_{mov} = \left\{ \frac{-30}{2} \right\}_{mov} = -15 \text{ Mils}$$

$$H_{o_{fix}} = \left\{ \frac{Right - Left}{2} \right\}_{fix} = \left\{ \frac{8 - 10}{2} \right\}_{fix} = -1 \text{ Mil}$$

$$H_{o_{mov}} = \left\{ \frac{Right - Left}{2} \right\}_{mov} = \left\{ \frac{(-18) - (-12)}{2} \right\}_{mov} = -3 \text{ Mils}$$

15. Based on the vertical and horizontal offsets, the following common set of equations may be used to compute the required movements for the near foot and the far foot of the moveable machine.

$$Vert_{nf} = \left\{ (V_{o_{fix}} + V_{o_{mov}}) \times \left(\frac{B}{D} \right) \right\} - \{ V_{o_{fix}} \} \tag{12-20}$$

$$Vert_{ff} = \left\{ (V_{o_{fix}} + V_{o_{mov}}) \times \left(\frac{C}{D} \right) \right\} - \{ V_{o_{fix}} \} \tag{12-21}$$

$$Horiz_{nf} = \left\{ (H_{o_{fix}} + H_{o_{mov}}) \times \left(\frac{B}{D} \right) \right\} - \{ H_{o_{fix}} \} \tag{12-22}$$

$$Horiz_{ff} = \left\{ (H_{o_{fix}} + H_{o_{mov}}) \times \left(\frac{C}{D} \right) \right\} - \{ H_{o_{fix}} \} \tag{12-23}$$

Once again, the **B** and **C** dimensions are the distances between the indicator plane on the fixed machine to the center of the near foot and far foot on the moveable machine as shown in Fig. 12-17. The dimension **D** is the axial distance between indicator planes. As previously noted, the **B**, **C**, and **D** dimensions are all expressed in inches. The final results for vertical shim changes at the near foot $Vert_{nf}$ are computed with equation (12-20), and the shim changes at the far foot $Vert_{ff}$ are determined with (12-21). Again, a plus sign (+) reveals a need to add shims and raise the machine. Conversely, a final negative sign (-) requires a removal of shims to lower the machine. With respect to the horizontal equations (12-22) and (12-23), the horizontal shift is defined by $Horiz_{nf}$ at the near foot, and $Horiz_{ff}$ at the far foot respectively. A positive sign (+) on the horizontal value indicates a need to move the machine to the left. A negative sign (-) requires a shift of the moveable machine to the right. Since all of the dial indicator readings carry the units of Mils, the final alignment shift values will also be in Mils.

As an example of these movement calculations, equations (12-20) through (12-23) will be used to compute the overall vertical and horizontal moves based upon the previously calculated vertical and horizontal offsets. These values will be combined with the same machine geometry used for the rim and face example. Namely, the distance to the near foot **B** will be maintained at 21 inches, and the distance to the far foot **C** will remain at 40 inches. If the spacing between reverse indicator planes **D** is 8 inches, the required movements are calculated as follows.

$$Vert_{nf} = \left\{ (9 - 15) \times \left(\frac{21}{8} \right) \right\} - \{9\} = \frac{(-6) \times 21}{8} - (9) = -24.75 \approx -25 \text{ Mils}$$

$$Vert_{ff} = \left\{ (9 - 15) \times \left(\frac{40}{8} \right) \right\} - \{9\} = \frac{(-6) \times 40}{8} - (9) = -39 \text{ Mils}$$

$$Horiz_{nf} = \left\{ (-1 - 3) \times \left(\frac{21}{8} \right) \right\} - \{-1\} = \frac{(-4) \times 21}{8} - (-1) = -9.5 \text{ Mils}$$

$$Horiz_{ff} = \left\{ (-1 - 3) \times \left(\frac{40}{8} \right) \right\} - \{-1\} = \frac{(-4) \times 40}{8} - (-1) = -19 \text{ Mils}$$

From these calculations and the associated sign convention it is clear that shims must be removed from both feet of the moveable machine. Furthermore, the moveable machine must be shifted to the right. Although these calculations are fairly straightforward, it is easy to make a numerical or a sign mistake. Hence, it is always desirable to perform a graphical solution of the alignment move to check the calculations.

16. A graphical solution of the alignment data yields the required movement at the feet of the moveable machine. This procedure is based upon plotting the calculated offsets at each dial indicator plane, and then extending a straight line to the support feet of the moveable machine. Traditionally, the horizontal axis is used to define the axial locations of both dial indicator planes, plus the center of the two support feet on the moveable machine. Units of inches are generally employed on this axis. The vertical axis displays the offsets in engineering units of Mils. In all cases, the offsets between shafts are calculated for both vertical and horizontal directions using equations (12-16) to (12-19).

17. It is good practice to plot the vertical and the horizontal data separately to avoid confusion, and errors in polarity. The actual graph is generated by plotting the offsets at both dial indicator planes, and extending the straight line out to the moveable support feet. When plotting the point at the stationary plane a positive number (+) is placed above the centerline, and a negative value (-) is plotted below the centerline. When locating the point on the moveable dial indicator plane, a positive number (+) is place below the centerline, and a negative (-) offset is located above the centerline. A line intersecting the two dial indicator offset points is then drawn and extended past the locations of the support feet. The required offsets are the distance from the centerline to the point where the extended shaft centerline intersects the lines associated with the moveable support feet. If the intersection occurs above the centerline, that indicates that the moveable machine is high and it must be lowered. Similarly, if the intersection point of the extended centerline falls below the desired centerline, the moveable machine is low and it must be elevated.

18. For demonstration purposes, the vertical offset data in the current example is plotted on Fig. 12-20. In this case, equation (12-16) revealed a vertical offset at the indicator plane on the fixed machine of +9 Mils. This point is plotted at the 0 inch location on Fig. 12-20. Across the coupling at the dial indicator plane on the moveable machine, equation (12-17) showed an offset of -15 Mils. Since the span between indicator planes was stated as 8 inches, the -15 Mil vertical offset was plotted at 8 inches. Since this value carried a negative sign, it was plotted above the desired centerline. These two measurement points establish the extended shaft centerline through the location of the near foot and the far foot. From Fig. 12-20, it is clear that the near foot location intersects the extended centerline at 25 Mils. This means that the moveable machine is 25 Mils high and shims equal to this distance should be removed. At the far foot, the extended shaft centerline is 39 Mils above the desired centerline. Hence, a shim thickness equal to 39 Mils should be removed at this location to lower the outboard end. These values and directions are identical to the previously calculated moves.

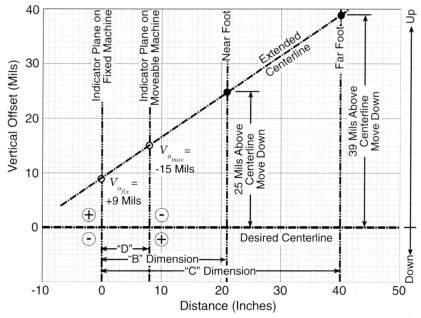

Fig. 12–20 Graphical Solution Of Vertical Alignment Based On Reverse Dial Indicators

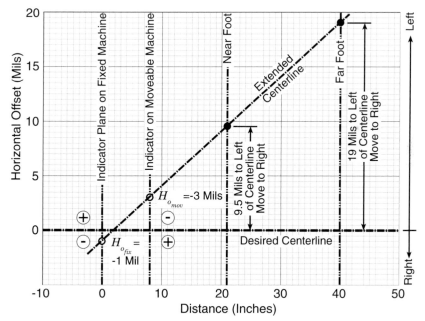

Fig. 12–21 Graphical Solution Of Horizontal Alignment Based On Reverse Dial Indicators

19. The graphical solution may now be repeated for the horizontal data as shown in Fig. 12-21. Equation (12-18) revealed a horizontal offset at the fixed machine of -1 Mil. This point is plotted at the 0 inch position on Fig. 12-21. Across the coupling at the moveable machine indicator plane, equation (12-19) displayed a -3 Mil offset. Again, the span between indicator planes was 8 inches, and the -3 Mil horizontal offset was plotted at 8 inches. Since this value carried a negative sign, it was plotted above the desired centerline. These two horizontal offsets establish the extended shaft centerline up through the location of the near and the far foot. From Fig. 12-21, the near foot location intersects the extended centerline at 9.5 Mils. This means that the moveable machine is 9.5 Mils to the left of the desired centerline, and the machine should be shifted 9.5 Mils to the right. At the far foot, the extended shaft centerline is 19 Mils to the left of the desired centerline, and the outboard end of the moveable machine should be shifted 19 Mils to the right. Again, these values duplicate the previous calculations.

20. As previously discussed, the most direct approach is to make the vertical shim changes first, and then proceed with the horizontal moves.

21. For the horizontal moves, position horizontal dial indicators at all four corners of the moveable machine. This will allow the horizontal machine movement to be accurately monitored as the machine is shifted into position.

22. Repeat steps 10 through 21 until the desired alignment has been achieved. Double-check that the coupling gap is still within specification.

23. Once the final alignment condition has been reached, obtain one final set of indicator readings. Correct these readings for sag, and identify this information as the *final* alignment readings.

The diagnostician is encouraged to use both the calculation and the graphical procedure to cross-check results. In all cases, clearly label the data with consistent designations. On graphical solutions, do not clutter up the page with sloppy or meaningless information. Always separate the vertical and horizontal alignment plots, and use separate pages for the calculation of the cold offsets. It is also a good idea to write the date and time on each piece of acquired and calculated data. This date and time stamping helps keeps things organized and sequential on difficult alignment projects. As with most technical endeavors, *clear and accurate work will be rewarded by excellent results.*

As stated at the end of the section on rim and face alignment, the presented procedures are directed at achieving a perfect alignment between shafts under a static condition. Obviously, as the machinery journals move around in their respective bearings, and process fluids are introduced, and thermal growth (or contraction) initiates, the shaft alignment will change. The variation of shaft position under normal operating conditions versus the cold alignment condition is presented in the text section on hot alignment techniques. This is a very important consideration, but it cannot be properly addressed until the fundamental concepts of rim and face plus reverse indicator alignment are understood.

OPTICS, LASERS, AND WIRES FOR SHAFT ALIGNMENT

Optical tools have many machinery applications including shaft alignment. As discussed earlier in this chapter, optics may be used to measure the relative positions of two shafts, or the differential machinery growth between *at stop* and *normal* operating conditions. This makes optical measurements a strong candidate for measurement of the thermal movement of process machinery. In addition, optics is quite useful for tasks such as setting shaft centerlines. By placing a mirror on a shaft end, and auto-collimating a scope, shaft centerlines may be set and measured. Optical tooling is quite adept for measuring shaft gaps, and face parallelism. Although the potential uses of optical measurements on machinery alignment are extensive, many optical methods are slow, and they are often quite complex. There is also considerable skill required to correctly execute these optical measurements. Unless the machinery diagnostician is specialty trained in all of the particular nuances associated with optical equipment, the application of this technology may be both difficult and time consuming.

Laser alignment has become an extremely popular form of shaft alignment due to the ease and speed of operation. The tools for laser alignment are identical to the devices discussed in the section on laser position alignment. For shaft alignment, specialized mounting brackets are available to allow the laser and detector to be secured to the shaft surfaces. Most systems utilize some sort of chain clamp for this purpose. Magnetic holders are also available to allow units to be mounted on faces where circumferential clamps may not be practical. Although variations exist between systems from different vendors, most laser measurements mimic the operation of shaft mounted dial indicators.

The laser provides a high degree of accuracy, combined with computer-based automation of the measurements. These computer controls have become quite sophisticated, and are capable of determining hot and cold alignment offsets, acceptable alignment tolerances, plus many other functions. The associated computer can generally display the results in numeric and graphical formats. As with any computer-based system, the final results are totally dependent on operator input, plus proper operation of the system. Hence, the diagnostician must have a complete working knowledge of the particular laser system, plus a full understanding of basic alignment principles.

Shaft wire alignment techniques are generally not performed on horizontal machines due to obvious limitations. However, wire alignment is commonly applied on large vertical machines, such as hydro turbines and large motor driven pumps. These types of units are often several stories tall and they employ a series of guide bearings at various elevations. Many of these machines are built with access passages to allow dropping alignment wires from the top to the bottom guide bearing. In this application wire sag is not a problem, but the wires must be secured in a perfectly vertical position to have meaningful distance comparisons between elevations. The measurement tools previously discussed for wire bore alignment are fully applicable to these large vertical units. It should also be mentioned that laser alignment is becoming popular for this application due to reduced setup time and faster processing of the results.

HOT ALIGNMENT TECHNIQUES

As previously mentioned in this chapter, the desired alignment for many applications must include offsets to compensate for thermal expansion (or contraction) of the machinery at normal operating conditions. Typically, the thermal related movements are calculated or measured, and these values are converted into vertical and horizontal offsets at the coupling hubs. These desired offsets are then combined with the current position of the shafts, and appropriate corrections computed or charted.

For instance, if the final desired position of the moveable machine is 10 Mils high and 4 Mils to the right, the equivalent reverse indicator readings are shown in Fig. 12-22. These desired indicator readings do not include bracket sag, and they may be directly converted to vertical and horizontal offsets.

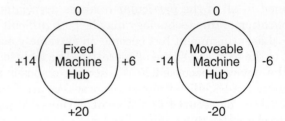

Fig. 12–22 Desired Reverse Indicator Readings For 10 Mil Vertical, And 4 Mil Horizontal Offsets Between Shafts

In the general case, the desired vertical and horizontal offsets may be computed in the same manner as equations (12-5) and (12-6). Expanding these fundamental expressions into the desired vertical and horizontal offsets at each dial indicator plane yields the following four expressions.

$$V_{d_{fix}} = \left\{ \frac{Bottom}{2} \right\}_{d_{fix}} \tag{12-24}$$

$$V_{d_{mov}} = \left\{ \frac{Bottom}{2} \right\}_{d_{mov}} \tag{12-25}$$

$$H_{d_{fix}} = \left\{ \frac{Right - Left}{2} \right\}_{d_{fix}} \tag{12-26}$$

$$H_{d_{mov}} = \left\{ \frac{Right - Left}{2} \right\}_{d_{mov}} \tag{12-27}$$

where: $V_{d\text{-}fix}$ = Desired Vertical Shaft Offset at Fixed Machine (Mils)
  $V_{d\text{-}mov}$ = Desired Vertical Shaft Offset at Moveable Machine (Mils)
  $H_{d\text{-}fix}$ = Desired Horizontal Shaft Offset at Fixed Machine (Mils)
  $H_{d\text{-}mov}$ = Desired Horizontal Shaft Offset at Moveable Machine (Mils)

To verify the validity of the reverse indicator readings shown in Fig. 12-22, equations (12-24) to (12-27) will be used to compute the desired vertical and horizontal offsets in the following manner:

$$V_{d_{fix}} = \left\{\frac{Bottom}{2}\right\}_{d_{fix}} = \left\{\frac{20}{2}\right\}_{d_{fix}} = +10 \text{ Mils}$$

$$V_{d_{mov}} = \left\{\frac{Bottom}{2}\right\}_{d_{mov}} = \left\{\frac{-20}{2}\right\}_{d_{mov}} = -10 \text{ Mils}$$

$$H_{d_{fix}} = \left\{\frac{Right - Left}{2}\right\}_{d_{fix}} = \left\{\frac{6 - 14}{2}\right\}_{d_{fix}} = -4 \text{ Mils}$$

$$H_{d_{mov}} = \left\{\frac{Right - Left}{2}\right\}_{d_{mov}} = \left\{\frac{(-6) - (-14)}{2}\right\}_{d_{mov}} = +4 \text{ Mils}$$

Clearly the desired parallel offsets of 10 Mils vertically and 4 Mils horizontally are displayed in the above calculations. These desired offsets must now be combined with the current or ambient offsets previously described by equations (12-16) through (12-19). These two sets of offset equations may now be integrated into the following general solutions for vertical and horizontal alignment moves at the near foot and the far foot of the moveable machine.

$$Vert_{nf} = \left\{(V_{o_{fix}} - V_{d_{fix}} + V_{o_{mov}} - V_{d_{mov}}) \times \left(\frac{B}{D}\right)\right\} - \{V_{o_{fix}} - V_{d_{fix}}\} \qquad \textbf{(12-28)}$$

$$Vert_{ff} = \left\{(V_{o_{fix}} - V_{d_{fix}} + V_{o_{mov}} - V_{d_{mov}}) \times \left(\frac{C}{D}\right)\right\} - \{V_{o_{fix}} - V_{d_{fix}}\} \qquad \textbf{(12-29)}$$

$$Horiz_{nf} = \left\{(H_{o_{fix}} - H_{d_f} + H_{o_{mov}} - H_{d_{mov}}) \times \left(\frac{B}{D}\right)\right\} - \{H_{o_{fix}} - H_{d_{fix}}\} \qquad \textbf{(12-30)}$$

$$Horiz_{ff} = \left\{(H_{o_{fix}} - H_{d_f} + H_{o_{mov}} - H_{d_{mov}}) \times \left(\frac{C}{D}\right)\right\} - \{H_{o_{fix}} - H_{d_{fix}}\} \qquad \textbf{(12-31)}$$

Obviously, if the desired offsets in equations (12-28) through (12-31) are set equal to zero (i.e., no process related or thermal offset), these expressions would default back to the earlier simplistic set of equations (12-20) through (12-23).

Using the example values from the previous section on reverse indicators, plus the desired offsets from Fig. 12-22, equations (12-28) through (12-31) may now be used to compute the required vertical and horizontal corrections at the near foot and far foot as follows:

$$Vert_{nf} = \left\{ (9 - 10 - 15 - (-10)) \times \left(\frac{21}{8}\right) \right\} - \{9 - 10\} = \frac{(-6) \times 21}{8} - (-1) = -14.8 \text{ Mils}$$

$$Vert_{ff} = \left\{ (9 - 10 - 15 - (-10)) \times \left(\frac{40}{8}\right) \right\} - \{9 - 10\} = \frac{(-6) \times 40}{8} - (-1) = -29 \text{ Mils}$$

$$Horiz_{nf} = \left\{ (-1 - (-4) - 3 - 4) \times \left(\frac{21}{8}\right) \right\} - \{-1 - (-4)\} = \frac{(-4) \times 21}{8} - (3) = -13.5 \text{ Mils}$$

$$Horiz_{ff} = \left\{ (-1 - (-4) - 3 - 4) \times \left(\frac{40}{8}\right) \right\} - \{-1 - (-4)\} = \frac{(-4) \times 40}{8} - (3) = -23 \text{ Mils}$$

Clearly, these alignment calculations are simple to perform, but they are prone to error due to the constant manipulation of plus (+) and minus (-) signs. This may not be a problem at the beginning of an alignment job, but pluses and minuses have a tendency to get crossed up in the middle of the night. The machinery diagnostician must be fully aware of this potential for silly errors, and he or she should implement procedures to prevent, or at least minimize, errors. As a minimum, all calculations should be rechecked at least twice. In an ideal situation, it is best if two people can acquire the data and independently perform the calculations.

Based on the above calculations, approximately 15 Mils of shims must be removed from the near foot, and 29 Mils of shims removed from the far foot of the moveable machine. In addition, the moveable machine must be moved to the right by 13.5 Mils at the near foot, and 23 Mils at the far foot. The same results could be obtained by plotting the reverse indicator data. Case history 43 at the end of this chapter will demonstrate the combination of this calculation technique, plus a graphical solution of a hot pump driven by an induction motor.

The previous example and associated discussion have assumed that the desired offsets due to thermal expansion are known. In actuality, this is a presumptuous posture, since the determination of the thermal growth is often a difficult endeavor. In many instances, the physical alignment moves are easily computed and accomplished. However, the majority of the technical effort is directed at determining the positional changes of the rotating shafts between the initial cold or ambient alignment conditions, and full load operation at normal process temperatures.

Initially, the techniques used to calculate thermal growth (or contraction) presented earlier in this chapter may be used to estimate the changes in machine casing position based on differential temperatures. The issue of determining average operating pedestal temperature may be difficult on complex machines installed in cramped quarters. In some cases, thermography may be used to determine the temperature gradient across the machine support members. This used to be a difficult measurement, but the advent of electronic thermography based upon infrared video cameras and associated digital signal processing has considerably simplified this task, and the associated cost. Equipment of this type may also be used to examine and document the temperature profile of machine casings, couplings, bearings housings, and associated piping and support structures. Hence, tools are available to accurately determine surface temperatures in a wide variety of situations.

The machinery diagnostician should not forget that many machines have a simple temperature gradient on their support members. The use of a contact thermometer can often fully describe the surface temperature profile, and provide the basis for calculation of the thermal growth. Certainly these thermal growth calculations are useful and necessary, but one should not forget the old adage of ...*one good measurement is worth a thousand expert opinions.*

The measurement of process machinery thermal growth may be accomplished in several different ways. Specifically, the optical measurements discussed earlier in this chapter represents one of the first techniques for accurate measurement of machinery position changes. The 1973 article by Al Campbell[6] describes the use of a jig transit combined with reverse dial indicator readings to properly align rotating equipment. In the same year, and at the same symposium, Charles Jackson[7] presented a paper that included optical alignment, plus the use of proximity probes mounted on cold water stands to measure thermal growth. The proximity probes added a new dimension to alignment measurements by providing an electronic output (i.e., probe DC gap voltage) that could be directed to various types of voltage recorders. With this information, machine position may be measured as a function of time, and the machinery movement could be tracked during startup and thermal heat soak. The Jackson cold water stands could be used to measure changes in casing position, or direct shaft position measurements on exposed shaft surfaces.

Since proximity probes are electronic micrometers, their use for alignment growth measurements represents a direct and logical application. However, in many machines, cold water stands supporting proximity probes cannot be used due to space limitations on the baseplate. Even with Invar brackets, it may not be possible to mount probes close to the shaft or a suitable external machinery surface. In these situations, proximity probes and companion measurement surfaces may be mounted on Dodd bars[8]. These devices consist of two sets of bars that span across the coupling and allow differential vertical and horizontal thermal growth measurements. Typically, two sets of X-Y probes are mounted on one bar, and their respective targets are mounted on the second bar. This technique provides a relative growth (or contraction) measurement across the coupling, but it does not allow the determination of absolute position changes as obtained with cold water stands or optics.

---

[6] A. J. Campbell, "Optical Alignment of Turbomachinery," *Proceedings of the Second Turbomachinery Symposium,* Turbomachinery Laboratories, Texas A&M University, College Station, Texas (October 1973), pp 8-12.

[7] Charles Jackson, "Cold and Hot Alignment Techniques of Turbomachinery," *Proceedings of the Second Turbomachinery Symposium,* Turbomachinery Laboratories, Texas A&M University, College Station, Texas (October 1973), pp. 1-7.

[8] V.R. Dodd, *Total Alignment*, (Tulsa, Oklahoma: The Petroleum Publishing Company, 1975).

Another approach to the measurement of machinery growth was developed by Jack Essinger[9]. This technique requires the installation of two tooling balls on the machinery bearing housings, plus mating tooling balls on the baseplate or foundation. Precision distance measurements are then acquired between these tooling balls under ambient alignment conditions, and they are compared against similar measurements with a full heat soak of the machinery. The distances between balls are measured with extended range dial indicators mounted in spring loaded telescoping gauge columns. These devices allow accurate length measurements and the differential machinery growth may be determined with calculator programs or graphical solutions.

Certainly laser alignment systems have evolved to increasing levels of sophistication and capability. In many applications, lasers have replaced the previously discussed techniques for measurement of position changes between the ambient conditions and the fully heat soaked machinery operating at normal process rates. Although this is wonderful technology, the diagnostician must always remember the machinery fundamentals, and be able to recognize when the alignment computer yields bogus or questionable results. In addition, the diagnostician must keep an open mind, and use the best available measurement and alignment technique for each specific application.

Finally, the shaft alignment discussions in this chapter have addressed only a simple two case machinery train, where one unit is fixed or stationary and the other case is moveable. Although this covers a large percentage of the installed machinery, it does not specifically address the multicase trains used for critical service. For these installations, one case is generally designated as the fixed machine and the other cases are sequentially aligned to the fixed unit. It is highly desirable to plot the graphic solutions for all rotors in the machinery train on the same piece of paper. Actually, one piece of graph paper should be used for the vertical solution and another for the horizontal alignment. This overall train view is necessary from a visibility standpoint, and it soon becomes apparent when the unit gets *bolt bound* or *shim limited* due to the consecutive stacking of case to case offsets. In many ways field shaft alignment is similar to field rotor balancing. That is, in most cases there are no ideal or universal solutions, but there are generally more than one acceptable solution. It is always a constant challenge to use the most appropriate techniques to satisfy the engineering requirements in the minimum amount of time.

[9] Jack N. Essinger, "Benchmark Gauges for Hot Alignment of Turbomachinery," *Proceedings of the Ninth Turbomachinery Symposium,* Turbomachinery Laboratories, Texas A&M University, College Station, Texas (October 1980), pp 127-133.

Case History 43: Motor To Hot Process Pump Alignment

Shaft alignment of most general purpose motor driven process pumps is a fairly straightforward exercise. In many instances, the operating conditions do not warrant thermal offsets or any running positions corrections. Hence, the methods to achieve a straight ambient alignment condition discussed earlier in this chapter may be directly applied. However, for machines that require an initial offset to compensate for a specific hot or cold running condition, the job becomes somewhat more complicated.

For demonstration purposes, consider the pump-motor machinery train described in Fig. 12-23. The two pole, 400 HP induction motor drives a multistage boiler feed water pump. The motor is equipped with roller bearings, and it operates at 3,570 RPM under full load. The horizontally split pump supplies boiler feed water to a package boiler. Pump suction is maintained at a constant temperature of 150°F, and the pump discharge pressure is 650 Psig at approximately 250°F. The pump is equipped with sleeve bearings and a fluid film thrust bearing. This machinery arrangement is the same configuration that has been used for examples in the previous discussions of rim and face, plus reverse indicator shaft alignment.

From the pump coupling flange, a distance of 21 inches is noted to the near foot of the motor, and 40 inches to the far foot. In the opposite direction, the pump inboard, or near foot, is 12 inches from the pump coupling flange, and the outboard, or far foot, is located 30 inches from the coupling flange. Once again, the spacing between dial indicator planes is 8 inches. The major area of concern during alignment of this unit is the anticipated thermal growth of the pump during operation. Minimal expansion of the motor is expected, but the elevated pump operating temperatures suggest that thermal alignment offsets might be appropriate. As shown in Fig. 12-23, the inboard pump pedestals experience a 52°F rise, combined with an average 80°F rise at the outboard supports. Due to the rigidity of the pump suction and discharge piping, the pump is considered as

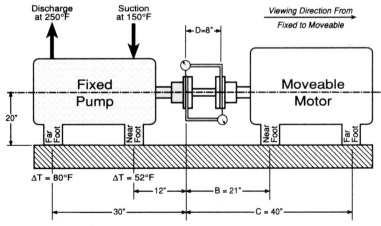

Fig. 12–23 Machine Arrangement Of A Hot Pump Driven By An Induction Motor

the fixed unit, and the motor will be the moveable unit for alignment. Based on the 20 inch height between the baseplate and the pump centerline, the expected thermal growth may be calculated as follows:

$$\Delta L_{pump_{nf}} = \Delta T \times L \times C = \{52°F\} \times 20 \text{ Inches} \times 0.0063 \text{ Mils/Inch°F} = 6.5 \text{ Mils}$$

$$\Delta L_{pump_{ff}} = \Delta T \times L \times C = \{80°F\} \times 20 \text{ Inches} \times 0.0063 \text{ Mils/Inch°F} = 10.1 \text{ Mils}$$

Hence, the thermal growth at the pump near foot should be 6.5 Mils, combined with 10.1 Mils on the pump far foot. These values may be easily confirmed with Fig. 12-1, describing the thermal growth of cast steel elements. These calculations could be supplemented by physical measurements of the actual thermal growth. In either case, the diagnostician has thermal growth information at the support feet of the stationary machine, and these values must be converted to vertical offsets at the dial indicator planes. The easiest way to accomplish this conversion is by plotting the thermal offsets at each pump support leg, and extending this desired pump centerline back across the coupling. For instance, on Fig. 12-27, the 10 Mil growth on the pump outboard was plotted at the pump far foot location of 30 inches from the pump coupling flange. Similarly, the 6.5 Mil thermal growth was located 12 inches from the pump coupling flange. Connecting these two points, and extending the desired pump centerline across the coupling provided the intercepts with both dial indicator planes.

From Fig. 12-27, the desired vertical offset at the pump coupling hub $V_{d\text{-}fix}$ may be read directly from the plot scale as -4 Mils. On the motor coupling hub, the desired vertical offset $V_{d\text{-}mov}$ is similarly determined to be +2.5 Mils. As always, the polarity changes across the coupling, and the sign goes from negative to positive. Assuming no appreciable horizontal offset, these desired vertical offsets may be easily converted to the reverse dial indicator readings in Fig. 12-24. To state it another way, when the pump-motor set is correctly aligned to compensate for the pump thermal growth, a set of ambient reverse indicator readings should be quite similar to the values displayed in Fig. 12-24.

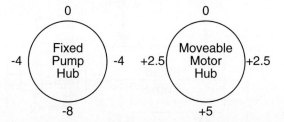

Fig. 12–24 Desired Reverse Indicator Readings To Compensate For Pump Thermal Growth

Since these are true offsets, indicator bracket sag has not been included. However, bracket sag does exist in the initial sweep values depicted in Fig. 12-25. Obviously, a true comparison of reverse indicator readings cannot be addressed until the initial data has been corrected for bracket sag. Although these calculations were performed earlier in this chapter, the exercise will be

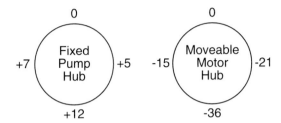

Fig. 12–25 Initial Reverse Indicator Readings Describing Pump To Motor Cold Offset With Bracket Sag Included

repeated for the sake of completeness. For both brackets, the total indicator reading was -6 Mils (= 2 x Sag) with the bracket inverted. Hence, the initial reverse indicator readings shown in Fig. 12-25 may be sag corrected as follows:

Fixed Pump Hub Readings:

$Left = (+7) - (-3) = +10$

$Right = (+5) - (-3) = +8$

$Bottom = (+12) - (-6) = +18$

Moveable Motor Hub Readings:

$Left = (-15) - (-3) = -12$

$Right = (-21) - (-3) = -18$

$Bottom = (-36) - (-6) = -30$

These results may now be displayed as the sag corrected reverse indicator readings shown in Fig. 12-26. This data reveals a minor horizontal offset, combined with a significant vertical deviation from the desired readings previously developed and displayed in Fig. 12-24.

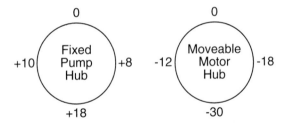

Fig. 12–26 Initial Reverse Indicator Readings Describing Pump To Motor Cold Offset With Bracket Sag Removed

From Fig. 12-26, it is clear that the vertical offset $V_{o\text{-}fix}$ at the pump coupling hub is +9 Mils (=+18/2). In a similar fashion, the vertical offset at the moveable motor coupling hub $V_{o\text{-}mov}$ is equal to -15 Mils (=-30/2). These values may now be combined with the desired offsets of $V_{d\text{-}fix}$ =-4 Mils, and $V_{d\text{-}mov}$ =+2.5 Mils. Based upon these desired and actual vertical offsets, plus the machine train dimensions, the required vertical corrections at the motor feet may be computed with equations (12-28) and (12-29) in the following manner:

$$Vert_{nf} = \left\{ (9 - (-4) - 15 - 2.5) \times \left(\frac{21}{8}\right) \right\} - \{9 - (-4)\} = \frac{(-4.5) \times 21}{8} - 13 = -24.8 \text{ Mils}$$

$$Vert_{ff} = \left\{ (9 - (-4) - 15 - 2.5) \times \left(\frac{40}{8}\right) \right\} - \{9 - (-4)\} = \frac{(-4.5) \times 40}{8} - 13 = -34.5 \text{ Mils}$$

The computed results indicate that approximately 25 Mils of shims must be removed from the inboard (near foot) of the motor. Additionally, 35 Mils of shims must be extracted from the motor far foot. These are clearly shim reductions as defined by the negative sign on both sets of calculations. The same information may be displayed graphically as presented in Fig. 12-27.

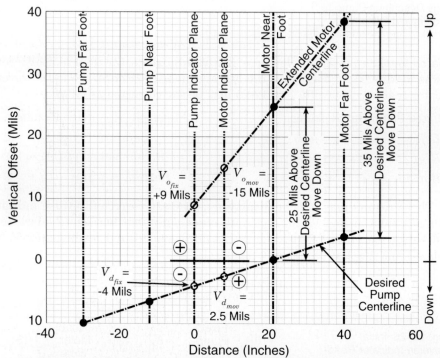

Fig. 12–27 Graphical Solution Of Vertical Alignment Of Induction Motor To Hot Boiler Feed Water Pump With Calculated Thermal Offsets At Pump Support Feet

In this diagram, the previously discussed thermal offsets are shown along with the *desired pump centerline* at the bottom of the plot. This desired pump centerline is extended past the motor support feet, and it is the reference line that the motor shaft attempts to meet. The initial vertical position of the motor shaft is defined by the $V_{o\text{-}fix}$ and $V_{o\text{-}mov}$ points plotted at dial indicator planes. An extension of this line produces the *extended motor centerline*. The actual shim corrections required to bring the *motor centerline* back down to the *desired pump centerline* are determined by the vertical difference between the two centerlines. Hence, a 25 Mil correction is required at the motor near foot, and a 35 Mil change is indicated at the motor outboard (far foot). It is also clear from Fig. 12-27 that the motor shaft is sitting higher than the desired position of the pump shaft. Hence, the motor must be lowered to become properly aligned with the pump. These results are totally consistent with the previous alignment calculations.

Note that the complexity of Fig. 12-27 is significantly greater than the simple offset alignment previously discussed in Fig. 12-20. The inclusion of the thermal offset, plus the requirements to properly identify the new points has substantially complicated the entire graphical solution. If this machinery train consisted of three or four individual cases, and various physical offsets had to be incorporated for each case, the complexity of the alignment diagram would be exponentially increased. It is once more highly recommended that the diagnostician be very meticulous in this work.

The horizontal corrections on this machine were determined in accordance with the previous horizontal alignment plot displayed in Fig. 12-21. As recommended earlier in this chapter, the vertical shim changes are performed first, and then the couplings are re-swept to check the horizontal offsets. The changes in horizontal motor position are achieved by loosening the outboard end motor bolts, and one of the inboard bolts. The motor was then pivoted on the remaining tight inboard bolt to achieve the proper centerline position.

The final aligned position of this induction motor and the hot boiler feed water pump was within 1.5 Mils of the desired indicator readings in Fig. 12-24. The slight remaining offset was considered to be well within the misalignment tolerance of the coupling. After completion of this work, the subsequent startup was quite smooth. The steady state vibration data at full load and heat soak revealed a well-aligned machinery train, with no evidence of pre-loads at any bearing.

In the overview, proper machinery alignment is mandatory for successful long-term operation of the mechanical equipment. The tools and techniques discussed in this chapter are directed at providing a general introduction to this subject, plus specific instructions on performing some basic alignment functions. Some diagnosticians might say...*I'll probably never do a field alignment job, so why bother learning about machinery alignment...* In actuality, so many mechanical problems are directly associated with misalignment and the resultant rotor preloads, this topic cannot be ignored. Furthermore, it must be recognized that other mechanisms besides misalignment may be actively engaged in destroying a mechanical coupling. Poor lubrication, incorrect installation, overloading, plus a variety of corrosion attacks, and fatigue failures may be responsible for coupling difficulties. For descriptions and photographs of a wide range of coupling problems, the reader is encouraged to examine the array of failed couplings presented by Mike Calistrat[10] in his text entitled *Flexible Couplings*.

---

[10] Michael M. Calistrat, *Flexible Couplings, Their Design Selection and Use,* (Houston: Caroline Publishing, 1994), pp. 390-424.

BIBLIOGRAPHY

1. Calistrat, Michael M., *Flexible Couplings, Their Design Selection and Use*, Houston: Caroline Publishing, 1994.
2. Campbell, A. J., "Optical Alignment of Turbomachinery," *Proceedings of the Second Turbomachinery Symposium,* Turbomachinery Laboratories, Texas A&M University, College Station, Texas (October 1973), pp. 8-12.
3. Dodd, V. R., *Total Alignment*, Tulsa, Oklahoma: The Petroleum Publishing Company, 1975.
4. Essinger, Jack N., "Benchmark Gauges for Hot Alignment of Turbomachinery," *Proceedings of the Ninth Turbomachinery Symposium,* Turbomachinery Laboratories, Texas A&M University, College Station, Texas (October 1980), pp. 127-133.
5. *Gas Engine and Compressor Field Service Manual* - Section 16, Drawing R19307A (Painted Post/Corning, New York: Dresser-Rand, 1986), pp. 1-12.
6. Jackson, Charles, "Shaft Alignment Using Proximity Probes," *ASME Paper* 68-PET-25, Dallas, Texas (September 1968).
7. Jackson, Charles, "Cold and Hot Alignment Techniques of Turbomachinery," *Proceedings of the Second Turbomachinery Symposium,* Turbomachinery Laboratories, Texas A&M University, College Station, Texas (October 1973), pp. 1-7.
8. Mancuso, Jon R., *Couplings and Joints - Design, Selection, and Application*, New York: Marcel Dekker, Inc., 1986.
9. Piotrowski, John, *Shaft Alignment Handbook,* 2nd edition, New York: Marcel Dekker, Inc., 1995.

Applied Condition Monitoring

*T*his chapter is based upon a presentation for the *Turbomachinery Symposium* of Texas A&M University. The original document was part of a Short Course[1] entitled "Inspection and Overhaul of Major Turbomachinery" presented in conjunction with personnel from Chevron USA. This course covered a variety of machinery subjects, and the three sections prepared by the senior author have been extracted and modified for this chapter. The first portion provides an introduction to condition monitoring concepts, with examples of diagnosed problems on operating machines. The second section covers calibration and verification checks on instrumentation systems that should be performed during a turnaround. The third section addresses the application of condition monitoring during machinery startup following an overhaul. Again, illustrative examples of machinery case histories have been incorporated.

Since the original presentation of this short course, the instrumentation systems have evolved to more sophisticated levels. However, the basic measurements and approaches associated with condition monitoring remain unchanged. Hence, this document is still considered to be appropriate and representative of the application of condition monitoring to turbomachinery systems.

MAINTENANCE PHILOSOPHIES

Maintenance philosophies concerning major machinery trains are many and varied. Each company, geographic plant location, and individual operating unit are subjected to numerous factors that influence the normal maintenance approach. Items such as product demand, required machine availability, historical behavior, and management attitudes are integrated into an overall doctrine. Generally, maintenance activities may be categorized as either *reaction-based*, *time-based*, or *condition-based* maintenance

It is reasonable to briefly review each category, and identify the respective merits of each approach. Since the most cost-effective approach resides within

---

[1] Robert C. Eisenmann, John East, and Art Jensen, "Short Course 1 - Inspection and Overhaul of Major Turbomachinery," *Proceedings of the Seventeenth Turbomachinery Symposium and Short Courses*, Turbomachinery Laboratory, Texas A&M University, Dallas, Texas (November 1988).

the third group of condition maintenance, additional discussion will be provided. This will include the tools and techniques necessary for evaluating machine condition, plus case histories to illustrate the application of condition monitoring towards operating machinery.

Reaction-based maintenance requires little if any advanced planning. It typically takes the form of continuous operation until a failure occurs. Then it becomes a purely reactionary mode to repair the damage and return the machinery train to operation as soon as possible. Maintenance of this type is expensive, and the potential of an unexpected catastrophic failure is significant. In addition, normal safety considerations may be compromised when a machine is allowed to run until a failure occurs. Reaction maintenance might be appropriate in facilities where the installed machinery is minimal; and the plant is not totally dependent on the reliability of any individual machine. A realistic example would be a unit that contains only small process pumps that are each 100% spared.

Expenditures for instrumentation, documentation, and personnel training are low with reaction maintenance. However, the risks are often significant, and most companies gradually moved away from this type of maintenance during the 1970s and 1980s. Unfortunately, the emphasis on corporate profitability during the 1990s have driven many industrial plants into severely curtailing a variety of preventative and/or predictive maintenance programs. These plants are returning to reaction maintenance. Although the immediate financial implications may be impressive, the long-term prognosis of this approach is frightening.

The concept of **time-based maintenance** evolved due to the economic impact and safety unsuitability of reaction maintenance. Within this category, machinery maintenance is planned and scheduled on a periodic basis. Often an annual cycle is used for shutdown, disassembly, and mechanical inspection of the various machine elements. Although this approach is substantially more cost-effective than simply reacting to failures, two obvious deficiencies are apparent.

First, a mechanical malfunction may occur during the normally scheduled operating time period. This may result in an unexpected machine failure, with a necessary default to reactive maintenance. At this stage, repair costs may become exorbitant, and production losses may be irrecoverable. Second, time-based maintenance often results in the disassembly of perfectly good machinery. This is not only expensive from a labor and downtime position, there is another risk involved. That is, there may be a problem induced during machinery reassembly. Hence, the turbine/compressor train that was in excellent condition prior to a turnaround, cannot be restarted due to an assembly error.

Critical mechanical items such as bearing clearances, seal installation, or coupling hub position may be violated during an inspection, and the machinery rendered inoperative. The solution to avoiding this situation is the direct approach of: *If it's not broken, then don't fix it*. This basic philosophy leads to the evolution of condition maintenance. That is, machine repair is based upon a reasonable knowledge of the mechanical condition of the machinery, and it is not governed by the calendar or a potentially destructive mechanical failure.

Condition-based maintenance is safer, and economically more attractive than either of the reaction or time-based methods. The superiority of this

approach has been demonstrated in various ways. In many plants the maintenance dollar expenditures per installed horsepower have significantly decreased
following the implementation of condition maintenance. This is generally due to
a reduction in direct maintenance, an extension of the time period between overhauls, plus secondary effects such as the reduction in the quantity of warehouse
spare parts. The fundamental concept behind condition maintenance consists of
evaluating the process machinery from many aspects and assessing the current
mechanical condition. Maintenance plans and material requirements are then
driven by the anticipated work scope. This is certainly more advantageous than
either of the two previously mentioned maintenance categories.

Evaluation of machinery condition requires management support and com-
mitment. Resources must be allocated for the devices to determine machinery
behavior via performance and vibration response measurements. In addition,
documentation and historical records must be established and maintained.
Finally, trained personnel are required to implement and sustain the program.
In an effectively run program, hardware and administration costs are more than
offset by the savings incurred from reduced maintenance. The success or failure
of this approach is dependent upon a good knowledge of the condition of the
installed machinery. More specifically, condition monitoring of the machinery is a
necessary prerequisite. The tools and techniques required for implementation of
condition monitoring are discussed in the following sections of this chapter.

CONDITION MONITORING

Detailed observation, plus monitoring of machinery behavior, provides the
database for condition monitoring. As discussed by John Mitchell[2] there are
many technical aspects to consider in any condition monitoring program. However, for this text, the two major characteristics to be addressed are the machinery performance and the vibration response characteristics.

These characteristics typically include continuous monitoring plus periodic
samples obtained with greater detail/resolution. For example, machine performance is constantly monitored with the normal operating instrumentation. Specific performance tests may be periodically conducted to compute operating
points and the efficiency of individual machinery cases. During the course of
these precision measurements, the accuracy of the operating instruments would
also be verified. The frequency of the performance tests is a function of the
machinery service. Obviously, a clean refrigeration compressor might only be
subjected to a performance test on an annual basis. Conversely, a cracked gas
compressor with the potential for internal coke accumulation might be checked
on a monthly or even a weekly basis.

Similarly, the vibratory behavior of a major machinery train should be continually monitored and protected with an automated system. Overall radial

---

[2] John S. Mitchell, *Introduction to Machinery Analysis and Monitoring*, second edition (Tulsa,
OK: Pennwell Publishing Company, 1993), pp. 291-345.

vibration levels plus thrust position are used for automatic shutdown. Additionally, detailed vibration response data may be acquired on a periodic basis. These dynamic vibration signals are viewed in a variety of formats, and compared with previous results under similar operating conditions. During the course of this routine data acquisition, accuracy of the monitoring instrumentation may also be verified. Some computer systems allow digital storage of the dynamic data, and detailed trending of all normal vibration parameters. However, it is comforting to routinely commit this data to a hard copy paper format for historical comparison and future reference purposes.

Other types of information, such as lube oil analysis, thermography, and current analysis on electric machinery, might be incorporated into the condition monitoring program. In all cases, the diagnostician must exercise common sense when including additional technologies into the program. For instance, if a routine examination of an oil sample from a process pump revealed 10.0 parts per million of a metallic bearing material, there would be cause for concern. However, on large turbomachinery, the main reservoir may contain several thousand gallons of oil. If a routine oil analysis detected 1.0 part per million of bearing metal, it would already be too late. In reality, by the time the oil analysis detected the presence of trace quantities of the bearing metal, the bearing was already destroyed. In this case, the small quantity of bearing metal is virtually undetectable within the large volume of the lube and seal oil system. Hence, many technologies are available in the marketplace, but the diagnostician must carefully select only those measurements that are related to the direct determination of the machinery condition.

The machinery database should be optimized to quantify or characterize normal behavior and highlight abnormal characteristics. Generally, the diagnostician should establish realistic limits that define normal variation of the measured parameters. For example, a rotational speed vector may exhibit minor changes in amplitude and phase during routine machinery operation. It is necessary to define an appropriate window or envelope of normal behavior for each parameter, or group of associated parameters. When these various windows of normal behavior are exceeded, the diagnostician must then analyze the machine to determine the probable cause. In this manner, maintenance options may be reasonably discussed, and a suitable balance maintained between the economics of continued operation versus various levels of corrective maintenance.

MACHINERY PERFORMANCE

A performance measurement and evaluation program will vary with the sophistication of the machinery and the available process data. In an ideal situation, the process controls are interfaced through a Distributed Control System (DCS). All necessary pressures, temperatures, flow rates, and molecular weights are measured and scanned as part of the normal control scheme. Hopefully, a direct shaft torque measurement is included, and this data is also directly available. In this type of system all of the necessary parameters are logged by the computer. In addition, the performance calculations for head, flow, and efficiency may also be stored within the DCS. Respective data points can be compared with the OEM performance test curves, and deviations from expected conditions identified. The calculated data points may also be trended as a function of time.

With an analog control system, the available process information may not have suitable resolution. In this case, a separate set of data may be obtained with calibrated, high resolution gauges. This includes pressures, temperatures, and flow rates. Fluid samples may be extracted and processed to determine fluid properties. For condensing turbines the condensate flow may be measured, and for electrical equipment the voltage and current may be obtained. Although these indirect measurements are not as accurate as a direct shaft torque reading, they do differentiate between poor versus acceptable performance.

The machinery test data may be used to compute the performance parameters, and this data may be compared against the performance curves, and trended with time. Detailed test procedures, measurements, and calculations are published by the ASME. Standards such as the Power Test Code for Compressors and Exhausters[3] (PTC-10), and the Performance Test Code for Steam Turbines[4] (PTC-6) are readily available. It should be recognized that a strict ASME performance test is often difficult to conduct in a field environment. In many cases, the necessary hardware provisions are not available to obtain the required precision. In addition, measurement of items such as transmitted torque across a coupling is impossible if the proper transducers are not installed. However, for many machines, an overall polytropic or isentropic efficiency may be sufficient for trending purposes. Since the basic requirement is to consistently trend and identify any degradation in performance, a less sophisticated test may be acceptable.

In addition to the ASME standards, other references on the performance characteristics of process machinery are available. This includes mechanical engineering handbooks, OEM literature, and various textbooks. The diagnostician should design a performance determination system that is suitable and compatible with the installed machinery. Factors to consider for performance measurements are summarized in the following list:

○ Machine Types
○ Expected Degradation Mode(s)
○ Measurements Available
○ Accuracy of Measurements
○ Process or Test Gauges
○ Method of Performance Calculation
○ Method of Trending Results
○ Performance Limits

---

[3] "ASME Power Test Codes - Test Code for Compressors and Exhausters, PTC-10," *The American Society of Mechanical Engineers*, (New York: The American Society of Mechanical Engineers, 1965, reaffirmed 1986).

[4] "ASME Performance Test Codes - Code on Steam Turbines, PTC-6," *The American Society of Mechanical Engineers*, (New York: The American Society of Mechanical Engineers, 1976, reaffirmed 1982).

Considering these items will allow the implementation of a usable system that will provide visibility of reduced performance on a machinery train. This type of information is highly beneficial by itself, and it becomes even more meaningful when coupled with the detailed machinery vibration response data.

VIBRATION RESPONSE DATA

It has been recognized for many years that the mechanical integrity of a machinery train can be evaluated by a detailed analysis of the vibratory motion. This is achieved by using a variety of dynamic transducers that provide information on both the casing vibration and the relative shaft motion plus position.

From an evolutionary standpoint, the original vibration transducers were seismic devices. As discussed in chapter 6, these transducers measure the vibration of the attached surface with respect to an inertial reference frame (free space). They can be mounted directly on bearing housings, machine casings, foundations, piping, or other stationary elements. For measurements in the lower frequency domains (600 to 90,000 CPM), velocity coils may be used. Accelerometers are generally employed to obtain high frequency characteristics (typical maximum of 1,200,000 CPM). These external vibration transducers provide considerable information about the motion of the surface on which they are mounted. However, many industrial machines display minimal external casing motion. For instance, a high speed rotor in a barrel compressor may experience an attenuation of 100:1 when shaft and casing measurements are compared. This is certainly reasonable when one considers a 600 pound rotor contained within a 12,000 pound casing. Any casing vibration due to shaft motion is attenuated by the enormous mass of the casing, plus the viscous damping inherent with oil film bearings and seals. It should also be mentioned that the mechanical impedance between a rotor excitation and an external measurement point is not constant. Impedance variations with speed and journal position are normal, and should be expected. In short, there is no such thing as a *constant amplitude ratio between shaft and casing vibration*.

By the early 1970s shaft sensing eddy current displacement probes evolved into common usage. These proximity probes are noncontacting devices that allow measurement of rotor motion relative to the probe mounting surface. Typically, the probes are attached to the bearing housings. Thus, the shaft displacement measurement provides the relative motion between the rotor and the bearing housing. This type of measurement is very useful for diagnosis of machinery malfunctions. In addition, since displacement probes are not susceptible to external (not machine related) excitations, they are excellent candidates for continuous monitoring and trending of rotor vibration.

Today, most large machinery trains are equipped with shaft displacement probes. In many cases, the probes are connected to permanent machinery protection systems. These systems include radial vibration monitors, thrust position monitors, tachometers, differential expansion indicators, plus other machine related information such as bearing temperatures from embedded thermocouples or resistance temperature detectors (RTDs).

Although definitive process measurements may vary considerably between plants and machinery trains, the shaft displacement transducer suite will be reasonably similar. That is, radial bearings should contain a pair of orthogonal probes, and thrust bearings will usually be equipped with a pair of axial probes. Standard configurations are addressed in documents such as the American Petroleum Institute (API) standard for Vibration, Axial Position, and Bearing Temperature Monitoring Systems[5]. In general, the API standards depict the minimum instrumentation requirements for a piece of machinery. The actual installed instruments may be increased as a function of unique machinery, or the need for additional diagnostic capabilities.

As discussed earlier in this text, the coplanar radial probes are typically located above the shaft at ±45° from the true vertical centerline. The probes mounted to the left of top dead center are referred to as the Y-Axis or vertical probes. The transducers mounted to the right of the vertical centerline are called the X-Axis or horizontal probes. For a machine with a single thrust collar, two axial probes may be used to observe the end of the shaft. In other arrangements, one axial probe might be inserted between the thrust pads to observe the thrust collar, and a second axial probe might be mounted at the end of the turbine shaft. Finally, a once-per-revolution timing probe is installed. This transducer is used to phase reference the machinery train, and provide a tachometer input.

Typically, all the monitors from one machinery train will be located in one rack of monitors. The location of monitors in the rack should follow the location of transducers installed on the equipment. That is, the first monitor position should be reserved for the turbine thrust, followed by the governor end radial vibration monitor, and so forth throughout the machinery train. On large units, the monitors may occupy two or more racks. In this situation, the left side of the second rack begins where the right side of the first rack ends.

As previously mentioned, these types of systems are often armed for automatic shutdown. Certainly this is totally dependent on the company and their maintenance philosophy. In organizations that fully support the concepts of machinery protection and condition monitoring, the units will be fully armed for radial vibration, and thrust position shutdown. Organizations that are not fully convinced of the merits of these devices will probably not tie the monitors into automatic trip circuits.

Shaft displacement probes are suitable for measurement of frequencies between zero and approximately 90,000 CPM. At higher frequencies, displacement amplitudes are quite small, and respective changes would probably go unnoticed. However, the low frequency capability allows the probe to measure actual changes in the average rotor position. Typically, this is applied for the thrust position measurements. However, it should be noted that both radial and axial displacement probes do measure relative position (DC gap) as well as the

---

[5] "Vibration, Axial Position, and Bearing Temperature Monitoring Systems — API Standard 670, Third Edition," *American Petroleum Institute*, (Washington, D.C.: American Petroleum Institute, November 1993).

shaft dynamic motion (vibration). Hence, it is highly desirable to alarm, or even shutdown, on excessive changes in radial shaft position as indicated by variations in probe DC gap voltages. As demonstrated throughout this book, the average radial or axial rotor position is significant information, and should always be incorporated in an assessment of machinery vibration.

Output signals from shaft sensing proximity probes are typically presented in units of Mils (0.001 Inches). For vibration readings, the actual measurement is a peak-to-peak value that represents the total motion within the bearing. Position measurements (axial or radial) are normally identified as a plus or minus position change from some defined reference location. Since the probes are observing rotating shafts that may be subjected to a variety of forces and frequencies, the output signals are often complex waveforms. In order to consistently characterize these dynamic signals, the data is typically discussed in terms of the following parameters:

- Overall Vibration Amplitude
- Running Speed (1X) Vibration Amplitude
- Running Speed (1X) Vibration Phase Angle
- Frequency Content
- Relative Position Change

The overall vibration amplitude is the unfiltered peak to peak motion measured by the probe. Running speed (1X) vibration amplitude is the motion filtered precisely at rotational speed. Similarly, the phase angle at (1X) running speed is the phase relationship between the rotational speed vibration signal, and the timing probe. Although instrumentation manufacturers vary somewhat in their phase circuitry, the most logical phase convention consists of the angular relationship between the leading edge of the timing notch and the next positive vibration peak. As discussed in previous chapters, this is expressed as a phase lag, with engineering units of degrees. The frequency content of a signal describes the mixture of frequencies and their respective amplitudes. Generally, the steady state vibration data is presented in the following hard copy formats:

- Unfiltered Orbit and Time Base
- Running Speed (1X) Filtered Orbit and Time Base
- Spectrum Plot
- Shaft Position Plot

The unfiltered orbit and time base may be observed directly on an oscilloscope connected to the probe output signals. The running speed (1X) filtered data is viewed in a similar manner, with the addition of a narrow band-pass filter on each signal. Typically, this is accomplished with a synchronous tracking filter that adjusts the filter center frequency to match the exact running speed of the machine. The oscilloscope screen may be photographed with a camera to provide permanent documentation of the orbital and time domain information. Various computer-based systems also perform the same basic operation, and many examples of this type of data are presented throughout this text.

Spectrum analysis is used to determine the frequency content of the signal. This task is handled as a Fast Fourier Transform (FFT) with either a dedicated instrument, or with a software routine and an appropriate analog to digital conversion. Data of this gender provides visibility of the frequencies present, and their respective amplitudes. One more, examples of this type of data are shown throughout this text. Naturally the shaft position plot is a summary of changes in probe DC gap voltages (e.g., zero speed versus full operating load). This type of information may be manually calculated and plotted. However, automated diagnostic and monitoring systems perform this task quite easily and accurately.

Although computer-based systems are in common use, it is still advisable to provide analog backup for checking and verification of the output data. The software required to capture and present vibration data is inherently complex. In most cases, a simple oscilloscope check of the raw waveforms helps to maintain sanity and provide credibility to the resultant database.

BEARING TEMPERATURE DATA

Correlation of information to support a common conclusion is always desirable. Hence, measurement and observation of bearing temperatures via embedded Thermocouples or RTDs are highly useful, if not mandatory, for most large process machines. The traditional bearing temperature measurements recommended in the previously referenced API 670 are summarized as follows:

○ Both Machine Journal Bearings
○ Active Thrust Shoes
○ Inactive Thrust Shoes

The location and quantity of temperature sensors is dependent upon the specific bearing configuration. In most instances, a total of six temperature probes are installed on each machinery case. This would include temperature sensors on each of two active and two inactive thrust pads. In addition, a single temperature pickup would be installed on each journal bearing. Thus, a quantity of six temperature sensors would be normal for many installations. The number of radial bearing temperature sensors would be increased if the bearings are long (e.g., L/D ratio of greater than 0.5). For a long journal bearing, two sensors might be required to properly detect the maximum bearing temperatures. Specific temperature transducer arrangements and recommended locations are clearly illustrated in API-670[6].

The mounting locations of the bearing temperature pickups are preferentially positioned to be close to the load zone on each bearing. This location allows the measurement of the maximum bearing temperature. It must be mentioned that the thermocouple is installed only into the base metal of the bearing. Typi-

---

[6] "Vibration, Axial Position, and Bearing Temperature Monitoring Systems — API Standard 670, Third Edition," *American Petroleum Institute*, (Washington, D.C.: American Petroleum Institute, November 1993), pp 23 and 24.

cally, the tip of the thermocouple remains 30 to 50 Mils (0.030 to 0.050 inches) from the bearing babbitt. Under no circumstances should the thermocouple be allowed to penetrate the babbitt. This type of installation would result in an early, and totally unnecessary, bearing failure.

Furthermore, the angular location of the temperature sensors must be carefully controlled to be in the load zone, as determined by shaft rotation and specific bearing configuration. If the thermocouple is on the wrong side of a tilting pad pivot point, the measured temperature will be less than the desired maximum bearing temperature. In addition, it should be noted that units such as overhung machines will often have a temperature sensor mounted in the bottom half of the coupling end bearing and the top half of the wheel end bearing. This is correct placement of the thermocouples, since the shaft on the overhung rotor will generally ride in the bottom half of the coupling end bearing and the top half of the inboard wheel end bearing. When in doubt, the actual load zone of the shaft within the bearing may be determined by measuring the shift in shaft centerline position with DC gap voltages from proximity probes.

Other machines such as large vertical units often have fully flooded bearings that contain one or more sensors installed in each bearing pad. Again, the specific machine configuration dictates the quantity and location of bearing temperature probes. Redundant or spare temperature pickups are often installed to backup the primary sensors. This is especially true for trains that are required to operate over long intervals between overhauls. In the 1960s the common practice was to perform disassembly and inspection of critical machines on a yearly cycle. In the 1990s, the expected run time between major overhauls often exceeds five years. Hence, redundant temperature sensors make good sense. It is also desirable to install and monitor additional temperature measurements such as the lube and seal oil supply, plus the ambient temperature. The solution to some mechanical malfunctions is often dependent upon proper correlation with oil or atmospheric temperatures.

DATA TRENDING

It is highly recommended that two distinct types of documentation be employed to chart machinery attrition versus operating time. As previously mentioned, the combination of continuous trend plots, plus a periodic detailed examination of machinery behavior, provides the most useful base for machinery evaluation. In most cases, the data should consist of information regarding both vibratory behavior plus performance data. The mechanical information is normally obtained by interfacing the machinery protection system with a data trending system. In older plants, the vibration and thrust position monitor recorder outputs were connected to analog strip chart recorders. Process measurements were recorded in a similar fashion with strip chart, circular chart, or multipoint recorders. This type of instrumentation provided a good deal of information on the machinery, and many problems have been identified and solved with this type of analog data.

Improved resolution, accuracy, and convenience may be obtained from the digital storage of trend data. For modern process plants containing a Distributed Control System (DCS) the process operating history is easily stored and retrieved. In many cases, process and vibration data are archived on nonmagnetic media such as magneto optical disks that have a storage life of twenty or more years. Hence, the immediate operating conditions, as well as all of the historical information, may be accessed and examined. As a cautionary note, many digital data storage systems employ averaging algorithms to compress the measured data. This compressed data may be misleading, since a transient event may be automatically averaged into digital oblivion.

The second important category of trend data consists of periodically acquiring detailed vibration and performance data at constant load conditions. As a minimum, the vibration information should be processed in both the orbital/time and frequency domains. In addition, relative shaft positions and bearing temperatures should be obtained for all radial and thrust bearings. The minimum performance data includes all applicable measurements of process pressures, temperatures, flow rates, loads, and molecular weights. It is important to obtain these field measurements in a consistent and repeatable manner. The resultant hard copy data may then be compared from one sample period to the next. In a typical situation, data is obtained during the initial startup of a new or rebuilt train. Following temperature and load stabilization (nominally 12 to 48 hours), a complete set of baseline data is acquired. This baseline information documents the initial vibratory and performance characteristics of the machinery, and it provides the reference point for future comparison.

This type of detailed machinery evaluation should be periodically repeated. The actual time between samples is a function of the type of machinery, severity of service, gross variations in load and/or speed, and the development of any specific problems. For example, a machinery train in good operating condition may be examined on a monthly or quarterly schedule. However, if it appears that a particular mechanical element is showing signs of distress, it would be appropriate to increase the frequency of the samples to weekly or even daily. This higher sample rate would be maintained until the problem is resolved. It should also be recognized that during the time periods between detailed data sets, the machinery monitoring system is fully operational, and continuous protection is available via the automated trip system.

In all cases, the trend data should be directed at providing reliable documentation of the machinery historical behavior. In many cases, significant changes can be recognized and addressed before a mechanical failure occurs. Often, the degrading or failing mechanical part can be identified. This provides the ability to make a reasonable judgment call regarding continued operation versus the shutdown and repair of a specific item. In some situations it is possible to take a brief outage and repair or replace the offending piece of hardware. To illustrate these points in greater detail, the following three field case histories are offered for consideration and demonstration of the application of on-line machinery condition monitoring.

Case History 44: Four Pole Induction Motor Bearing Failure

A small catalytic cracking unit contains the main air blower machinery train described in Fig. 13-1. Within this arrangement a 3,500 horsepower, four pole, induction motor drives a five stage centrifugal air compressor through a down mesh, double helical gear box. The vibration transducers installed on this train are directed to a rack of control room monitors. Unfortunately, the monitors for this machinery train are mounted about eighteen inches from the floor, and it is reasonably difficult to obtain readings from these *knee high monitors*. Due to the location of these monitors, it is understandable why the control room operators pay minimal attention to the vibration amplitudes.

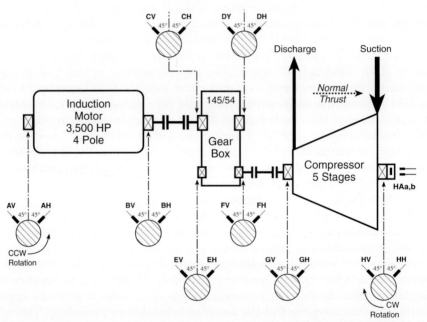

Fig. 13–1 Machinery And Transducer Arrangement For Motor Driven Air Compressor

This particular machinery had been running for an extended time period with no evidence of any abnormality. Motor shaft vibration response characteristics under typical operating conditions are shown in Fig. 13-2. This diagram documents the motion at the inboard, or coupling end, motor bearing. The top orbit and time base plots are unfiltered, and the bottom set of traces are filtered at the running speed of 1,790 RPM. From this data it is clear that the majority of the motion occurs at motor rotational speed. Although the running speed amplitude is approximately 2.5 Mils,$_{p-p}$, the runout compensated levels are in the vicinity of 1.5 Mils,$_{p-p}$. These amplitudes are historically typical, and mechanically reasonable for this class of motor. In addition, proximity probe DC gap voltages revealed proper journal position within the sleeve bearing.

Late one Saturday night, a process upset occurred, and the coupled com-

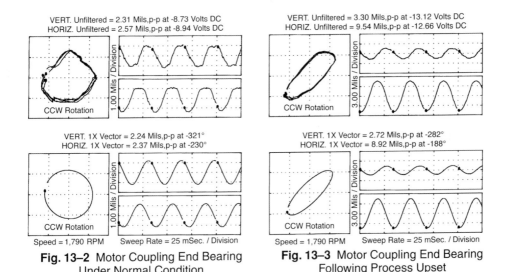

VERT. Unfiltered = 2.31 Mils,p-p at -8.73 Volts DC
HORIZ. Unfiltered = 2.57 Mils,p-p at -8.94 Volts DC

CCW Rotation

VERT. 1X Vector = 2.24 Mils,p-p at -321°
HORIZ. 1X Vector = 2.37 Mils,p-p at -230°

CCW Rotation

Speed = 1,790 RPM   Sweep Rate = 25 mSec. / Division

Fig. 13–2 Motor Coupling End Bearing
Under Normal Condition

VERT. Unfiltered = 3.30 Mils,p-p at -13.12 Volts DC
HORIZ. Unfiltered = 9.54 Mils,p-p at -12.66 Volts DC

CCW Rotation

VERT. 1X Vector = 2.72 Mils,p-p at -282°
HORIZ. 1X Vector = 8.92 Mils,p-p at -188°

CCW Rotation

Speed = 1,790 RPM   Sweep Rate = 25 mSec. / Division

Fig. 13–3 Motor Coupling End Bearing
Following Process Upset

pressor surged for an undefined period of time. Vibration and temperature alarms were acknowledged, but the machine was not shutdown (no automatic trip system). After the process was back under control, it was noted that the motor coupling end bearing vibration reading was off scale on the 5.0 Mil,$_{p-p}$ vibration monitor. In addition, the bearing temperature strip chart showed a ten minute excursion to 210°F, followed by a leveling out to an unusually low journal bearing temperature of 70°F. Following this upset, the meter on the motor inboard vibration monitor remained pegged. E&I personnel discovered that the vertical channel vibration amplitude was considerably less than the horizontal reading. At this time, the motor inboard horizontal probe was disconnected from the monitor to allow visibility of the vertical probe amplitude. It was noted by control room personnel that the peak meter reading had changed from 2.6 to about 3.3 Mils,$_{p-p}$. It was erroneously concluded that the shaft vibration had only increased by 0.7 Mils,$_{p-p}$, and that the horizontal probe had probably failed.

In actuality, the motor coupling end horizontal probe was still operational, and emitting a proper signal. After reconnecting this transducer, the vibration response characteristics at the motor inboard following the upset are displayed on Fig. 13-3. Note that the vertical probe reveals a slight vector shift, but the horizontal transducer has a rotational speed amplitude of 8.9 Mils,$_{p-p}$. It should also be mentioned that the shaft orbits shown in both Figs. 13-2 and 13-3 have been rotated 45° to allow a presentation of the orbits in a true vertical and horizontal orientation. When the location of proximity probes at ±45° from the vertical centerline are considered, it is clear that the vertical probe observes the minor axis of the elliptical orbit. Simultaneously, the position of the installed horizontal transducer is directly in line with the major axis of the shaft orbit.

This coincidence of the major and minor axes of the elliptical orbit with the physical location of the proximity probes reinforces the need for orthogonal (90°) transducers. In this case, if the machinery assessment was based totally upon the vibration amplitude measured by the vertical probe, the conclusion would probably favor continued operation. However, when the motion sensed by the companion horizontal probe is considered, the extent of the machinery distress is clearly evident in the data shown in Fig. 13-3.

From another perspective, the measured DC gap voltages on both proximity probes revealed a substantial increase. Specifically, the left-hand probe had a voltage change from -8.73 to -13.12 volts DC. The differential value of 4.39 volts DC is equivalent to a distance of 22.0 Mils at a probe sensitivity of 200 mv/Mil. Similarly, the right-hand probe exhibited a change of -8.94 to -12.66 volts DC. This differential voltage of 3.72 is equivalent to a radial position shift of 18.6 Mils. Since the probes were mounted at ±45° from vertical, the total shaft center-line position change may be determined by a vector summation of the two position shift vector amplitudes as follows:

$$Shaft\ Centerline\ Change = \sqrt{(22.0)^2 + (18.6)^2} = \sqrt{830.0} = 28.8 \text{ Mils}$$

Since the gap voltages increased, the distance between the probe tips and the observed shaft surface must have increased. If it is assumed that the proximity probes remained in a fixed position, the measured changes in gap voltage must be indicative of the shaft riding lower in the bearing. The only way that this condition can occur is if the bottom half of the bearing has been damaged.

If that prognosis proved to be correct, then the expanded bearing clearances will provide less restraint upon the journal (lower stiffness). Assuming a constant driving force (e.g., rotor unbalance), synchronous vibration levels should increase. Also, it was discovered that the bearing temperature sensor was improperly located in the top of the bearing assembly. In this position, the sensor would observe a lower than normal temperature since the load zone was physically on the opposite side of the bearing from the installed thermocouple.

Finally, one last piece of evidence should be considered in this investigation. Recall that the operating speed of this induction motor was measured at 1,790 RPM. For a normal USA line frequency of 60 Hz, the synchronous speed for a four pole motor is 1,800 RPM. The slip frequency for a four pole induction motor is computed from equation (10-44) in the following manner:

$$Slip = \left(\frac{Poles}{2}\right) \times (Synchronous\ Speed - Rotor\ Speed)$$

$$Slip = \left(\frac{4}{2}\right) \times (1,800 - 1,790) = 2 \times 10 = 20 \text{ CPM}$$

The extended time domain traces of the motor and the bull gear revealed an amplitude modulation occurring every 1.5 seconds. The frequency of this motion is determined from equation (2-1) as follows:

$$Frequency = \frac{1}{Period} = \frac{1 \; Cycle}{1.5 \; Seconds}$$

$$Frequency = 0.667 \frac{Cycles}{Second} \times 60 \frac{Seconds}{Minute} = 40 \; CPM$$

This amplitude modulation at 40 CPM occurs at twice the motor slip frequency of 20 CPM. It is reasoned that the bull gear was responding to this motor excitation. The vibration severity of this low frequency modulation is summarized in Table 13-1. This data suggests that the motor excitation at twice slip frequency was transmitted to the bull gear in a much stronger manner after the process upset. One explanation for this behavior would be an ineffective motor inboard bearing, and the requirement for the gear bearings to carry more load through a reduced flexibility coupling.

Table 13–1 Variation Of Low Frequency Amplitude Modulation

Measurement Location	Initial Shaft Vibration	Shaft Vibration After Upset
Motor — Coupling End Bearing	2.0 Mils,$_{p-p}$	3.0 Mils,$_{p-p}$
Bull Gear — Coupling End Bearing	0.9 Mils,$_{p-p}$	2.0 Mils,$_{p-p}$
Bull Gear — Outboard End Bearing	≈ 0	1.1 Mils,$_{p-p}$

Based on this overall data, the unit was shutdown in an orderly manner for replacement of the inboard motor bearing. To minimize production loss, the plant was placed in recycle. This plan allowed the cracking furnace to remain in hot standby, and permit a plant restart with minimal delay.

Prior to any disassembly, a lift check was performed at the motor inboard. The dial indicator revealed a total diametrical clearance of 31 Mils. This value is consistent with the previously calculated changes in probe gap voltages (28.8 Mils). Since the correct bearing clearance should be between 8 to 10 Mils, it was clear that all measurements pointed towards excessive bearing clearance.

Following disassembly, inspection revealed severe babbitt damage to the lower half of the inboard bearing liner. Fortunately, the motor journal was not marred, the bearing liner was replaced, and clearances checked. The new bearing provided correct diametrical clearances, but air gap measurements (rotor to stator) revealed that the motor rotor was not concentric with the stator. These variations in air gap exceeded the OEM specifications, and were logically responsible for the previously discussed modulation at twice slip frequency. Due to a lack of available resources, this problem was not corrected at this point in time. The train was restarted, and the plant suffered minimal impact due to the brevity of this machine train shutdown.

After a 24 hour thermal soak, the steady state vibration and temperature data returned to the originally documented levels. This was considered to be indicative of normal behavior for this four pole induction motor.

Case History 45: Cracked Gas Compressor Intermittent Instability

The four case machinery train described in Fig. 13-4 operates in a large ethylene plant. This equipment is used to compress cracked furnace gas prior to drying and fractionation. The machinery train consists of a steam turbine driving a low pressure compressor that is coupled to the bull gear of a down mesh, double helical gear box. The pinion output shaft is coupled to a high stage compressor that operates between 10,700 and 11,500 RPM. The suction end of this high stage is positioned at the coupling end. The final discharge nozzle and the thrust assembly are located at the outboard end of the high pressure compressor.

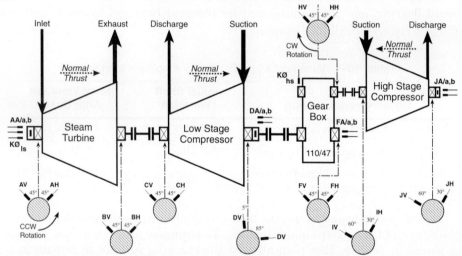

Fig. 13–4 Machinery And Transducer Arrangement For Cracked Gas Compressor Train

This compressor contains eight impellers and a large balance piston. The rotor was supported with five pad, tilting pad journal bearings. Both radial bearings are contained within a housing configuration that includes oil control rings on each end. These control rings were designed by the OEM to float within a circumferential groove cut into each respective bearing housing. The inner diameter of the four control rings contains a babbitt coating and the diametrical clearances of the control rings are greater than the adjacent tilting pad bearing assemblies. The high stage compressor was also equipped with a traditional double acting Kingsbury type thrust bearing. A single thrust disk was keyed to the shaft, and restrained axially by the stationary thrust bearing assembly. The coupling between the pinion and the high pressure compressor was a gear type, and the hubs were keyed to their respective shafts.

Since this type of service is susceptible to internal coke buildup on the wheels and diaphragms, performance data is acquired on a regular basis. In some instances, injection or wash oil flow rates are increased to maintain head and efficiency across each of the cracked gas compressors.

Following eight months of successful operation, it was noted that efficiency

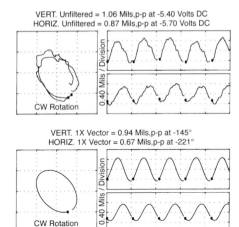

Fig. 13–5 Orbit And Time Base Of High Stage Compressor Discharge Bearing Under Normal Operating Conditions

was decreasing across the high stage compressor. It was anticipated that coke buildup was responsible, and injection oil rates were increased. Shortly thereafter, the outboard radial bearing began to display intermittent vibration alarm conditions. During periods of normal vibration levels, the average shaft motion was in the vicinity of 1.0 Mil,$_{p-p}$ as shown in Fig. 13-5. During this normal condition dominant motion occurred at rotational speed, and the shaft exhibited a slightly elliptical orbit, with a normal forward precession (i.e., with rotation).

Initially, the vibration alarm events occurred sporadically. Eventually, the alarm conditions occurred with reasonable regularity early every morning. It was observed that vibration levels at this bearing would begin to increase at

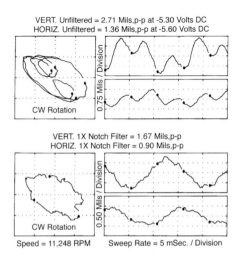

Fig. 13–6 Orbit &Time Base Of Discharge Bearing With Subsynchronous Motion

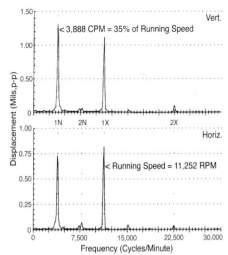

Fig. 13–7 Spectrum Plots Of Discharge Bearing With Subsynchronous Motion

approximately 4:00 AM. On most days, by 5:00 to 5:30 AM the first alert setpoint of 2.5 Mils,$_{p-p}$ was exceeded. Typically, by 7:00 AM the vibration levels were decreasing. By 9:00 AM, the high stage compressor was back to normal behavior, and it generally remained that way until the next morning. From a management standpoint, this behavior was viewed as a non-problem. That conclusion was attributed to the fact that by the time the morning meetings were concluded, the high vibration condition had subsided for the remainder of the day.

During one of the early morning excursions, dynamic shaft vibration was observed and documented on magnetic tape. The orbital and time domain information during this period is presented on Fig. 13-6. The upper data set in this diagram displays the unfiltered vibration characteristics. The bottom orbit and time base traces documents the shaft motion at all frequencies except rotational speed (i.e., 1X notch filter). From this data, it is evident that the major vibratory activity occurs at a frequency located below running speed. This subsynchronous motion is forward (with rotation) with a slightly elliptical shape.

Precise frequency identification of the subsynchronous component is made in the spectrum plot of the vertical and horizontal probes, as shown in Fig. 13-7. This data reveals that the subsynchronous component occurs at a frequency of 3,888 CPM, with a shaft rotative speed of 11,252 RPM. From a percentage standpoint, this low frequency motion is at 35% of running speed. It is noteworthy that this is the initiation frequency, and that continued operation resulted in a gradual increase in the subsynchronous frequency. In fact the final frequency was at 42% of rotative speed (4,720 CPM). Shortly after attaining the 42% condition, the subsynchronous component decayed on schedule at 9:00 AM.

It was also noted that bearing temperatures were lower prior to the appearance of the subsynchronous excitation. Lube oil supply temperatures were logged at 98°F at the beginning of the cycle, and 106°F at the end of the daily excursion. Based upon the acquired data, the subsynchronous vibration was characterized in the following manner:

1. Forward Precession
2. Nearly Circular Motion
3. Frequency Varies from 35% to 42% of Speed
4. Frequency Appears Below 1st Critical
5. Sensitive to Oil Supply Temperature
6. Repeatable, but Not Self-Sustaining

From this evidence, the self-excited vibration was considered to be an oil whirl. Another theory that was suggested by one of the plant engineers was an aerodynamic whirl between the rotor and diaphragms as coke buildup occurs. However, the machine had no change in balance, and the potential for coke accumulation was discounted. In reality, the triggering mechanism was oil supply temperature, and a problem within the outboard bearing was suspected.

It is often stated that tilt pad bearings cannot exhibit oil whirl. This statement is generically true since vertical to horizontal cross-coupling does not exist in a tilt pad configuration (i.e., $K_{xy}=K_{yx}=0$). Without cross-coupling, the rotating oil wedge cannot be sustained, and the shaft cannot whirl. However, the oil control rings on either side of the tilt pads provide an annular surface that could sustain a self-excited whirl. This is possible if the clearance of the tilt pad assembly expanded, and the journal became supported on the control rings. Under normal conditions, the diametral bearing clearance should be 5 to 7 Mils. At clearances approaching 20 Mils, the control rings could act as bearing surfaces. It was concluded that this was the most plausible explanation for the whirl.

As mentioned earlier, the lube oil supply temperature was directly related to the appearance of the whirl. As the reservoir cooled down each evening, a minimum oil temperature of 98°F was observed. As the sun came up each morning, the reservoir temperature increased and the whirl disappeared. Since this behavior was the apparent triggering mechanism, it also provided the control tool. Specifically, the reservoir heating coils were used to maintain a minimum oil supply temperature of 110°F throughout the night.

This approach was successful in controlling the whirl, and preventing a reoccurrence. From a process standpoint, the ethylene plant remained in full operation for the next few weeks until a short shutdown could be scheduled. A one shift outage was planned to coincide with high product storage inventory. The ensuing inspection revealed a lift check at this outboard compressor bearing in excess of 0.014 Inches (i.e., twice the maximum allowable clearance). Although the babbitted tilt pads were not damaged, the pad seats were worn, and overall clearance was excessive. A replacement bearing assembly was installed with a 0.006 Inch overall diametral clearance. In addition, the oil control rings were modified with axially milled slots every 60° to prevent these rings from allowing any future whirl mechanisms. The train was successfully restarted, and the plant experienced minimum downtime. Again, the application of condition monitoring allowed the selection of the most favorable options to maximize production rates, and minimize maintenance and material costs.

Case History 46: High Stage Compressor Loose Thrust Collar

This problem addresses a four case machinery train operating in ammonia plant syngas service. The train consists of a condensing steam turbine coupled to a high pressure topping turbine, plus two barrel compressors. The low pressure compressor is coupled to the topping turbine, and it drives the high pressure barrel. Normal train operating speeds vary from 10,400 to 10,800 RPM.

The high stage compressor is oriented with the suction nozzle, and thrust assembly at the outboard end of the machine. The two radial bearings are five pad, tilting pad assemblies. A double acting Kingsbury type thrust bearing is installed around the single thrust disk. This thrust disk is keyed into position, and secured with an outboard thrust lock nut.

Following an overhaul, this train operated at full load for several days. Unexpectedly, radial vibration levels increased at the thrust (suction) end of the high stage. The vibration monitor drifted up and down over a 2.4 Mil$_{,p-p}$ span for several hours before the abnormality disappeared. Although axial position remained reasonably constant, the axial 1X vibration revealed a 0.9 Mil,$_{p-p}$ swing. The high stage discharge bearing exhibited similar characteristics, but at lower amplitude fluctuations of 0.8 Mils,$_{p-p}$. The low stage compressor was slightly influenced and both turbines remained unchanged.

Orbits at the high stage suction end bearing revealed a significant phase change in addition to the variation in synchronous vibration. This is depicted on Fig. 13-8 that shows the decay and initiation of this mechanism. This computer-generated four hour trend plot has a horizontal scale in minutes. The top diagram depicts 1X phase from the X-Axis probe at the compressor suction, with 1X amplitude on the bottom trend plot from the same probe.

It is noted that 1X amplitudes vary from a minimum of 0.5, to a maximum of 2.8 Mils,$_{p-p}$. Simultaneously, the phase lag angle changed from 90° to 7°. In essence, the rotational speed vectors are constantly changing while this mechanism is active. Operating under constant process loads with only minor speed changes, the fluctuating 1X vectors would come and go in a random manner.

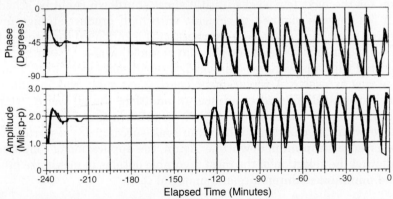

Fig. 13–8 High Stage Syngas Suction Bearing - Four Hour Trend Plot Of 1X Vector

Thrust and radial bearing temperatures remained constant, and no particular event could be identified to initiate or stop this unusual behavior. The time duration of this fluctuating vector also varied. In some instances the mechanism would be active for twenty minutes, and at other times the fluctuation would persist for five to six hours. The only consistency was the amount of cycling once the mechanism was active.

Since the observed motion was primarily a change in 1X vectors, it is reasonable to develop a summary vector diagram of the measured response characteristics. Fig. 13-9 shows such a plot for the suction end horizontal probe. Three running speed vectors were plotted to develop this diagram. First, the normal steady state vector of 1.7 Mils,$_{p-p}$ at 45° was drawn. Next, the maximum amplitude vector of 2.8 Mils,$_{p-p}$ at an angle of 32° was drawn. Finally, the minimum amplitude vector was included at 0.5 Mils,$_{p-p}$, at 60°. It should be mentioned that the respective phase angles associated with the maximum and minimum amplitudes were read off the trend plot in Fig. 13-8. Hence, some minor phase errors are inherent with this estimation of the charted phase angles.

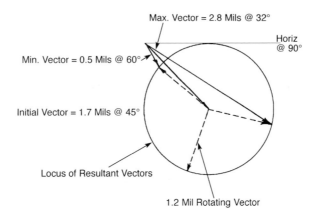

Max. Vector = 2.8 Mils @ 32°

Horiz @ 90°

Min. Vector = 0.5 Mils @ 60°

Initial Vector = 1.7 Mils @ 45°

Locus of Resultant Vectors

1.2 Mil Rotating Vector

Fig. 13–9 High Stage Syngas Compressor Suction Bearing Vector Summary Diagram

Nevertheless, the resultant plot reveals that the initial vector is positioned between the minimum and maximum vectors. Drawing a circle from the head of the initial vector through the heads of the identified minimum and maximum vectors reveals a reasonably constant amplitude of 1.2 Mils,$_{p-p}$. This is interpreted as a 1.2 Mil,$_{p-p}$ rotating vector that either adds or subtracts from the initial steady state vector. Actually, the circle represents the locus of resultant vectors (i.e., vector summation between the initial vector of 1.7 Mils,$_{p-p}$ at 45° plus and minus the 1.2 Mil rotating vector). It is also noted that the phase angle exhibits a minimum phase of 0° and a maximum phase angle of 92° that coincides with the trend plot on Fig. 13-8.

It was concluded that the measured behavior was driven by a 1.2 Mil,$_{p-p}$ rotating vector. Since the compressor performance remained constant, the gas flow characteristics were not influenced. Thus, it is reasonable to eliminate the high stage compressor impellers as a potential problem source. Consideration must then be given to some other rotating element. Obviously, the thrust disk at the suction end of the high stage compressor is a potential candidate. This is even more reasonable when it is recalled that axial vibration was also fluctuating by 0.9 Mils,$_{p-p}$. Hence, any wobble of the thrust disk would result in a change in axial vibration, plus a balance and/or gyroscopic response on the lateral (radial) vibration characteristics. It was also discovered that the V groove on the outer diameter of the thrust collar had been used for balance corrections.

Shutdown and inspection of this thrust assembly revealed a loose thrust collar and a worn thrust lock nut. The thrust assembly was rebuilt, and the train restarted within a single shift. Again, condition monitoring identified a specific problem area, and repairs were executed in an orderly and cost-effective manner.

Clearly, condition monitoring can reduce maintenance labor and materials. It is also reasonable to expect increased production stream factors due to more efficient and effective maintenance. Furthermore, proper machinery analysis provides benefits for turnaround planning. In situations where a process unit is shutdown for pressure vessel inspections, heat exchanger cleaning or re-tubing, piping replacement, etc. — condition monitoring can identify the machinery items that require attention, as well as those inspections that are not necessary. Overall, this is an important maintenance planning tool, and these techniques can extend the time period between major overhauls. Today, there are many machinery trains that are not subjected to annual overhauls. Some of these machines have operated for five to ten years without a major disassembly. In an ideal situation, a machine might continue to display parameters indicative of good mechanical condition. Over the years, the performance of a machine may degenerate due to widening interstage labyrinth clearances. Based on condition monitoring, the machine would be opened and overhauled for a specific reason (i.e., to restore the efficiency) rather than the old and expensive approach of: *while we're down, why don't we open it up, and look inside?*

PRE-STARTUP INSPECTION AND TESTING

After a machinery overhaul, many of the systems have been inoperative for a period of time, or they have been disassembled. A successful startup, followed by a trouble free extended run is dependent on the integrity of all devices. Thus, it is necessary to methodically check each of the following machinery systems:

○ Process Alarms and Shutdowns
○ Lube and Seal Oil System
○ Speed Control System
○ Surge Control System
○ API 670 Machinery Protection System

It is the intent of this section to review each category, present some guidelines for system verification, and identify some of the potential pitfalls. For the sake of consistency in addressing this topic, a generalized machinery train will be discussed. Fig. 13-10 depicts a typical steam turbine driving a multistage centrifugal compressor. The turbine is not equipped with extraction or injection nozzles, and the compressor contains a single suction and a single discharge nozzle. To facilitate discussion of the API 670 instrumentation, all traditional shaft sensing displacement probes are shown on this sketch.

Each of the five subsystems contains measurement functions, display and monitoring capabilities, plus a myriad of alarms and shutdown features. In addition, the first four systems are engaged in active control functions. It is difficult to rank the importance of one system over another, since malfunctions in any one of the categories can cause a false shutdown. In all cases, it is desirable to check each system independently, and develop a suitable testing and verification plan.

Before addressing the individual subsystems, it should be recognized that the only safe way to stop the turbine and compressor set is to cut off the steam

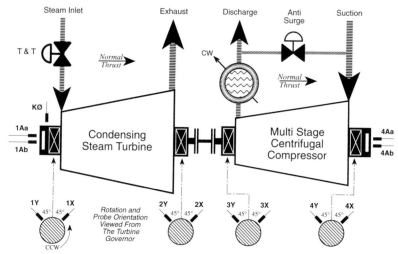

Fig. 13–10 Typical Steam Turbine Driven Centrifugal Compressor Machinery Train

supply. The ability to do this rapidly is provided by the **trip and throttle valve** (T&T). This valve is used as a hand throttle valve for admitting steam to the turbine when bringing it up to minimum governor. The other, and perhaps the most important function, is to rapidly terminate the steam inlet to the turbine during a trip condition (typically in less than 0.3 seconds). The closure of the T&T valve can generally be accomplished by any one of the following conditions:

○ Manual Trip
○ Turbine Overspeed Trip
○ Low Governor Oil Pressure Trip
○ Opening the Solenoid Dump Valve

Irrespective of the type of trip, the result is the same; that is, the virtually instantaneous closure of the T&T valve. This valve contains a hydraulically operated mechanism that is activated by governor oil pressure. Thus, any of the above four actions will relieve this pressure, and the valve will close. It must be manually reset before it can be reopened. If governor oil pressure is not available, the valve remains closed. The various trips discussed in the following sections have one thing in common. That is, all of the trip signals have their final termination point at this solenoid operated dump valve (SOV).

Furthermore, it is very important to consider the overall system time delay between trip initiation and the closure of the T&T valve. Although the T&T valve closes rapidly after the oil is relieved, the time delay in getting the signal from the sensor to the trip solenoid may be intolerable. Based on the particular instruments installed, there may be a group of three or five second delays in series. In some cases, system time delays of over thirty seconds have been measured. Generally, overall system delays in excess of five seconds are unacceptable.

Many of the newer steam turbine control systems include sophisticated electronic governor controls. In some installations this may be a dedicated Process Loop Controller (PLC) within the Distributed Control System (DCS), or it may be a completely separate free standing console dedicated to controlling machine speed. A separate machine console is even more likely for complex drivers such as gas turbines that must incorporate a variety of firing controls and safety backup systems into the control scheme. In any configuration, the turbine control system must be capable of rapidly shutting down the turbine in case of emergency — and this capability should be thoroughly check and physically verified during each maintenance turnaround.

In addition to the normal process control loops, many devices are installed with the specific purpose of preventing damage to the machinery trains. These **process related alarms and shutdowns** typically provide two independent setpoints. The first point is an alarm that may annunciate locally, or in conjunction with a parallel control room alarm. If action is not taken to correct the impending problem, the second setpoint is usually an automated trip of the machinery train. In other cases, the second setpoint may not be an instrument, but it may be a relief valve or a rupture disk that protects the machinery from an overpressure condition. Process problems can cause severe damage to gas handling equipment. This includes the ingestion of liquids, plus excessive high or low compressor casing temperatures. The following list represents typical process instruments that are wired for alarm and machinery train shutdown:

○ Compressor Suction Drum Level
 ❏ Low Level Alarm
 ❏ High Level Alarm
 ❏ High Level Shutdown
○ Compressor Suction Pressure
 ❏ High Pressure Alarm
 ❏ High Pressure Relief and Shutdown
○ Compressor Suction Temperature
 ❏ High Temperature Alarm
 ❏ High Temperature Shutdown
○ Compressor Discharge Pressure
 ❏ High Pressure Alarm
 ❏ High Pressure Relief and Shutdown
○ Compressor Discharge Temperature
 ❏ High Temperature Alarm
 ❏ High Temperature Shutdown
○ Condensate Hotwell Level
 ❏ High Level Alarm
 ❏ Spare Hotwell Pump On Alarm
 ❏ High Level Shutdown

Prior to startup of a machinery train, the level transmitters should be checked by building a suitable high or low level within the respective suction drum or turbine hotwell. This level can usually be verified by direct observation in the attached gauge glass. For the shutdown points, it is mandatory to verify that the T&T valve does shut. It is convenient to just check the transmitter output. However, one is never really sure of a trip until the T&T valve slams shut.

With a high level in the hotwell, the spare condensate pump should be allowed to auto-start, and verification should be made that the alarm does function. In some installations the main condensate pump may be steam turbine driven, and the spare unit motor driven. In other plants, both the main and the spare condensate pumps are motor driven. In all cases, each unit that is equipped for automatic startup based on high condensate level should be tested for proper operation with a high water level in the hotwell.

The compressor temperature alarms and shutdowns are usually simulated electronically at the thermocouple junction. Again, it must be verified that the T&T valve actually does close at the proper trip temperatures. Pressure transmitters may be tested with dead weight or other hydraulic pressure devices. Rupture disks cannot be field tested, but relief valves should be removed and bench tested on a scheduled basis.

Due to the complexity of the turbine trip signals, it is necessary to have an updated *ladder diagram* of all trip relays and any associated interposing relays. This relay sequencing and logic generally cannot be deduced from the field wiring. An updated logic or ladder drawing is mandatory for proper verification of the system. In addition, the logic diagram does not do any good locked away in somebody's office or file cabinet. In all cases, the logic diagram should be available in the control room with twenty-four hour a day accessibility. In many instances, it is highly desirable to mount a current copy of this diagram on the wall or inside the cabinet door of the trip instrumentation (e.g., relay cabinet).

The **lube and seal oil system** generally provides three basic functions for the process machinery. First, it supplies lubricating oil to the machine train bearings. For the example train shown in Fig. 13-10, this would include the four journal bearings, the two double acting thrust bearings, and the gear coupling. Second, the oil system delivers governor oil to the T&T valve. Third, the oil console provides seal oil to the overhead seal oil tank to maintain a nominal fifteen foot head above the compressor shaft seals. Some operating companies prefer to divide these functions into separate lube oil and seal oil consoles for each train. Other end users elect to combine these functions for several machinery trains into one large lube and seal oil system.

Regardless of the specific console configuration, the required system checks are in accordance with normal maintenance practices. This includes traditional level, pressure, and temperature tests to insure that the various alarms and trips are functioning properly. All control valves should be stroked and packing checked. Filters should be changed, and the oil reservoir should be verified for cleanliness. Oil coolers should also be checked to insure that both the shell and the tube side are clean and suitable for continued service. Also, the level floats in the sour oil traps should be verified for proper condition and operation. Sometimes it is easy to ignore the oil console due to the built-in redundancy of spare pumps, coolers, filters, etc. However, a thorough flushing and system operational check is quite comforting, and it may pay big dividends in terms of extended unit operation. For reference purposes, the normal alarms and shutdowns associated with a typical lube and seal oil system are summarized as follows:

- ○ Lube Oil Pressure
  - ❏ Low Pressure Alarm
  - ❏ Start Auxiliary Oil Pump Alarm
  - ❏ Low Pressure Shutdown
- ○ Seal Oil Pressure
  - ❏ Low Pressure Alarm
  - ❏ Start Auxiliary Seal Oil Pump Alarm
  - ❏ Low Pressure Shutdown
- ○ Seal Oil Level in Overhead Tank
  - ❏ High Level Alarm
  - ❏ Low Level Alarm
  - ❏ Start Auxiliary Seal Oil Pump Alarm
  - ❏ Low Level Shutdown
- ○ Lube & Seal Oil Reservoir
  - ❏ Low Level Alarm
  - ❏ Low Temperature Alarm
- ○ Governor Oil Pressure
  - ❏ Low Pressure Alarm
  - ❏ Low Pressure Shutdown

The turbine **speed control system** is designed in accordance with the requirements of the particular process. In many instances, speed is regulated by compressor suction or discharge pressure. The turbine governor accepts this input signal, and regulates the angular position of the rack (camshaft), which controls the sequential opening and closing of the individual steam inlet valves.

Mechanical linkages should be checked for proper positions and pin arrangements. It is not unusual to miss a linkage by one hole, and not be able to run the train above minimum governor speed. The older oil relay governors are being replaced by electronic governors. These units have excellent control capabilities, but they do require accurate speed sensing. It is customary to use a magnetic probe observing a multiple tooth gear. These sensors can be damaged during an overhaul and not noticed until the turbine is ready to run. If possible, install one or more redundant probes, and check the transducer outputs with a portable test rig, or during initial slow roll.

The last or final speed control device is the mechanical turbine overspeed trip. The setting and repeatability of this device should be verified during the turbine solo run. Generally, three overspeed trips are made to establish confidence in the mechanism. It is usually desirable to measure the actual trip speed with a digital tachometer set to capture the maximum RPM. Since the actual overspeed trip occurs in a very short period of time, the assistance provided by the peak speed sensing capability of a digital tach is highly desirable.

Surge control systems are installed on many centrifugal compressors. These machines are susceptible to a low flow, high head phenomena known as surge. On a standard head capacity curve, the surge point is located at the left-hand edge of the curve. When various operating speeds are considered, a family of curves are produced, and the surge line forms the left-hand, or low flow, limit for these curves. If the compressor is operated past this surge line, the system back pressure exceeds the compressor discharge pressure. At this point, flow is momentarily reversed through the machine. This action lowers the system back pressure, and forward flow is re-established. Unless some flow or pressure changes are made, this cycle will continue until some mechanical element fails and aborts the cycle. Some common techniques for surge control are as follows:

○ Vent to Atmosphere
○ Vent to Flare
○ Inlet Control or Variable Guide Vanes
○ Speed Control or Reduction
○ Bypass to Lower Pressure (e.g., Suction)

The atmospheric vent is generally unacceptable for anything but air compressors. A vent to flare is a waste of resources and is generally frowned upon by the local residents. Inlet control or variable guide vanes are expensive from an initial procurement standpoint and from the costs associated with periodic maintenance. Speed reduction is a reasonable candidate, assuming that the control and mechanical systems can respond within an effective time frame. The last option of bypassing gas to a lower pressure is the technique generally applied on machinery trains such as the example in Fig. 13-10. In this simple system, gas from the compressor discharge is returned to the suction. Flow control opens the anti-surge valve, and higher flow rates are maintained through the machine. Thus, the compressor operates at a comfortable distance from the surge line. In most instances, cooling is required on the final discharge stream to prevent excessive heat buildup during operation of this kick back loop.

In this traditional control scheme, a flow or anti-surge controller is used to activate the bypass valve. Often these devices are field calibrated at several speeds. Specifically, the compressor is briefly placed in surge, and the associated flow rates and speeds are documented. The surge occurrence is verified by differential pressure measurements and shaft vibration data. Following the acquisition of three to five data points the controller is set to open the anti-surge valve well in advance of the empirically determined surge curve.

Following an overhaul, the surge controller can be calibrated, and verified electronically to duplicate the previous test data. In addition, the valve stroke and response time can be checked. Due to the variety of system configurations, and instruments, the specific test details should be developed for each individual system. As more complicated machine trains are considered with multiple cases, and multiple side streams per case, the surge control problem increases exponentially. The resultant anti-surge system may be a combination of two or more of the previously discussed options. Again, the calibration and checkout procedures are often unique to each system.

The highest concentration of alarm and shutdown points typically resides within a fully armed **API 670 machinery protection system**. For the example shown in Fig. 13-10, a total of thirteen displacement proximity probes are installed. Based on this array of transducers, the following list summarizes the potential alarms and shutdowns on the steam turbine driver:

- ○ Turbine Speed Alarm
  - ❏ Low Slow Roll Speed Alarm
  - ❏ High Operating Speed Alarm
  - ❏ Electronic Overspeed Shutdown
- ○ Turbine Thrust Position
  - ❏ Active Thrust Alarm
  - ❏ Inactive Thrust Alarm
  - ❏ Active Thrust Shutdown
  - ❏ Inactive Thrust Shutdown
- ○ Turbine Inlet Bearing
  - ❏ Radial Vibration Alarm
  - ❏ Radial Vibration Shutdown
- ○ Turbine Exhaust Bearing
  - ❏ Radial Vibration Alarm
  - ❏ Radial Vibration Shutdown

A similar set of alarms and trips may be employed on the centrifugal compressor as summarized in the next list:

- ○ Compressor Discharge Bearing
  - ❏ Radial Vibration Alarm
  - ❏ Radial Vibration Shutdown
- ○ Compressor Suction Bearing
  - ❏ Radial Vibration Alarm
  - ❏ Radial Vibration Shutdown
- ○ Compressor Thrust Position
  - ❏ Active Thrust Alarm
  - ❏ Inactive Thrust Alarm
  - ❏ Active Thrust Shutdown
  - ❏ Inactive Thrust Shutdown

The four journal and two thrust bearings are equipped with embedded thermocouples or RTDs. In addition, the ambient and lube oil supply temperature may also be monitored with permanently installed temperature sensors. These transducers are often terminated in the same rack as the digital tachometers, plus the vibration and thrust position monitors. A typical array of temperature alarms are described in the following list:

○ Turbine Thrust Bearing Temperature
   ❑ Active High Temperature Alarm
   ❑ Inactive High Temperature Alarm
○ Turbine Inlet Bearing Temperature
   ❑ High Temperature Alarm
○ Turbine Exhaust Bearing Temperature
   ❑ High Temperature Alarm
○ Compressor Discharge Bearing Temperature
   ❑ High Temperature Alarm
○ Compressor Suction Bearing Temperature
   ❑ High Temperature Alarm
○ Compressor Thrust Bearing Temperature
   ❑ Active High Temperature Alarm
   ❑ Inactive High Temperature Alarm
○ Auxiliary Temperature Measurements
   ❑ Low Lube Oil Temperature Alarm
   ❑ High Lube Oil Temperature Alarm
   ❑ Low and/or High Ambient Temperature Alarm

Due to the quantity of transducers involved, and the various possibilities for automatically tripping the machinery, considerable care must be exercised in system checkout and verification. It is presumed that the original system installation was documented with the following minimum descriptive information:

○ Probe Calibration Data
○ Probe Installation Diagrams
○ Machine Train Wiring Diagram
○ Plot Plan of Cable Runs
○ Machine to Monitor Wiring Diagram
○ Monitor Arrangement Diagram
○ Monitor Back Plane Wiring Diagram
○ Shutdown Logic and Hardware Diagram

Calibration should be verified for all the thrust probes, and a reasonable sampling of the radial probes. Transducer systems that do not meet the minimum API requirements for a linear range of 80 Mils, or sensitivity of 200 mv/Mil ±5% should be discarded. Whenever possible, the calibration checks should be run on the shaft material. Thrust probes should be installed and gapped in accordance with predetermined position values. In all cases, the thrust probes shall be installed with concurrent agreement between the distance measured by probe gap voltages, thrust monitor readings, and axial dial indicator readings. The three values must agree as the rotor is moved between active and inactive thrust shoes. It is also a good idea to rotate the rotor 180° and repeat the thrust position measurements. There should be little if any difference between the 0° and the 180° thrust position readings.

Continuity between each probe and the assigned monitor location shall be verified. In addition, a dynamic AC vibration signal should be generated at each radial probe location, and the respective vibration monitor checked for proper values, and minimal signal noise. Where required, calibration shall be performed to meet normal tolerances. All vibration transducers shall be checked for proper field isolation, and single point grounding at the intrinsic safety barriers or the monitor rack (whichever is appropriate). Also, the probe installation hardware shall be reviewed for suitability, durability, and functional rigidity.

The associated trip circuitry shall be physically as well as functionally checked. Particular attention shall be paid to relay connections. The standard nomenclature of NC (Normally Closed) and NO (Normally Open) can be quite confusing unless the intended relay state (energized or de-energized) is also defined. The functional checks shall include the generation of AC and DC signals to active alarms and trips on each channel. All alert (alarm) and danger (shut-down) setpoints on each monitor shall be verified. A shutdown (trip) indication is not completed until the T&T valve is closed. Also, any voting logic between transducers should be checked for proper and consistent operation.

All probe to pigtail connectors should be securely wrapped in a suitable insulating material to prevent any possibility of stray ground loops between the probes and monitors. Insulating materials such as Scotch® 70 Self-Fusing Silicone Rubber Electrical Tape is far superior to fragile materials such as teflon tape. Finally, all changes in hardware, configurations, or calibration shall be permanently documented in the system file. Ideally, this would occur before startup of the machinery train.

From these discussions, it is apparent that the five auxiliary systems are critical to the operation of the main machinery train. If these systems are ignored, the best possible situation would be a series of nuisance trips. In the worst possible situation, the protective trips would malfunction during a real emergency condition. The required automated trip would not occur, and a catastrophic machine failure might be the final result.

STARTUP INSPECTION AND TESTING

At the completion of an overhaul, and the conclusion of the inspections and system tests, the moment of truth eventually arrives. That is, the time has come to startup the unit. With motor or turbine driven equipment, the first phase is to run the driver solo. Even though the driver has no load on it, the *spin up* does verify that the unit is capable of attaining operating speed. For turbine drives, the solo run provides an opportunity to check overspeed trips. In addition, various types of data can be obtained, and operational checks may be performed on the auxiliary systems. For motor drives, this solo run provides the opportunity to mark the axial magnetic center of the rotor running within the stationary field.

The solo run is followed by the second phase that consists of a coupled train startup. The load carrying capability of the driver is tested for the first time, and the driven piece of machinery is subjected to an initial trial run. More behavioral data can be collected during this startup and an initial assessment made of the machinery condition. If the equipment was subjected to an overhaul this startup vibration data is vital for future documentation, and for providing the transient speed base for future condition monitoring.

Assuming that the machinery has successfully passed through the first two phases, the final test is imposed. This third phase addresses the thermal heat soak, combined with process and load stabilization. The time duration of this last startup phase is difficult to quantify due to the number of potential problems that may be encountered. In very general terms, a twenty-four hour period might be considered a reasonable norm. At the end of this third phase, a good evaluation of the machinery condition and behavior should be possible. Hence, a new data point of performance and vibration response would be established for continued steady state condition monitoring.

It is the intent of the following sections to review the vibration measurements made during variable speed and load transients. The different data acquisition and documentation requirements will be discussed, and each of the three startup phases will be reviewed in the form of a case history. Performance measurements as discussed earlier in this chapter would not be applied until constant speed and load has been attained. Hence, the majority of the following discussion will center around the transient vibration measurements.

Various discussions of diagnostic vibration hardware have been presented throughout this text. Clearly, the basic tools required for constant speed, full load, data observation and acquisition include an oscilloscope, Digital Vector Filter (DVF), Digital Signal Analyzer (DSA), plotter, and a multimeter. Supplemental filters, meters, interface devices, and various levels of computer-assisted data handling are useful additions. However, this basic hardware provides 95% of the data documentation capability required for steady state conditions. Using these tools, the information may be formatted as orbit and time base data, spectrum plots, plus axial and radial rotor position plots.

As noted in previous chapters, the oscilloscope may be used to document unfiltered shaft orbits and time base information. Inserting a tracking filter between the transducer signals and the oscilloscope provides the capability for 1X filtered data. The spectrum analyzer allows an examination of the frequency content of the signals. The multimeter is used to measure probe gap voltages. From this DC voltage data, shaft position plots may be generated.

Generally, this class of instrumentation will only handle one or two data channels at a time. This is acceptable when machine conditions are constant. However, during a startup there are many transducers to simultaneously examine. Even the simple turbine compressor set shown in Fig. 13-10 requires 13 data channels. In these situations, it is impossible to properly observe all channels during the startup sequence. Assistance is required to handle this task, and the multichannel tape recorder or direct digital data storage is the proper device for this type of transient information.

From a data integrity standpoint, the tape or digital recorder should be considered as just another tool. It does not replace other instruments, it just provides multichannel recording and playback capabilities. In some situations this can be extremely helpful. However, if the entire database is committed to a single recorder, and the recording device malfunctions, the data may be lost forever. The safest approach is to use a combination digital and analog data acquisition system (e.g., Fig. 8-5). The recording may be supplemented by on-line hard copy data, or appropriate log sheets. This provides some level of backup, and it also improves visibility of machine behavior since the diagnostician is looking at the dynamic signals. That is certainly more productive than sitting and watching the reels spin. In addition, the hard copy data can be compared against the reproduced recorder signals and calibration verified. Besides, if recording redundancy is combined with on-line hard copy data, the recorders will probably never fail.

It should also be mentioned that during a startup, the instrumentation array can be used in a variety of manners. For example, the oscilloscope may be used to observe output signals from monitors by moving the leads from jack to jack. The scope may also be used to verify signals in and out of the tape recorder to verify proper recorder operation. The DVF tracking filter may be used to log 1X vectors from different probes at different *hold* points, or it may be dedicated to plotting data continuously from one specific transducer. Similarly, the DSA spectrum analyzer may be used for examining inputs from live transducers.

It is almost mandatory to produce hard copy documentation of the slow roll characteristics from each pair of proximity probes. Since these transducers detect shaft surface imperfections, changes in permeability of the observed surface, localized magnetism, etc., it is necessary to define the slow roll, or runout characteristics. Ideally this would be a combination of orbit/time base plots combined with logs of rotational speed vector amplitudes and angles.

Also, DC gap voltages from all probes should be logged at identifiable conditions. For example, the log should include gap voltages *at stop* before oil circulation and with oil circulating. The next point would be during slow roll on the machinery. Then there may be two or three plateau regions on the way up to speed. Certainly, DC gaps at minimum governor and full speed data should be obtained, and a final set acquired under full load and full heat soak.

The recorded information may be reproduced into a variety of analog instruments or a computed-based data processing system (e.g., Fig. 8-7). If the data was initially stored in a digital format, it may be subjected to additional manipulation without the need for analog instruments. Once again, synchronous vectors are accommodated by the Bode plot, where rotational speed vibration amplitude and phase angle are simultaneously plotted as a function of machine speed. The Bode is combined with a polar plot describing the locus of rotational speed vectors during variable speed operation. Although both of these plots provide the same basic data array, the Bode provides excellent visibility of changes with respect to speed, and the polar plot yields improved resolution of phase variation. Data of this type is essential for identifying rotor critical speeds and the influence of various resonances. Under machine conditions where significant subsynchronous or supersynchronous vibration components are generated, it is desirable to generate a cascade plot of individual spectra at incremental operating speeds. This type of data presentation provides an excellent overview of the frequency content of the vibration signals as a function of speed. Spectrum data may also be processed as a function of time in a waterfall plot. In addition, individual frequency components or orders may be tracked with respect to speed or time, as required.

The reader is referenced back to chapter 8 for a detailed discussion of the types of analog and digital instrumentation systems, and the traditional data presentation formats. For situations where high resolution, rapid data processing, significant post processing requirements, or extensive data manipulation is anticipated, a combination between analog and digital systems may be mandatory. Based on this brief review of transient data formats, it is reasonable to discuss three case histories and observe actual machinery vibration response characteristics under various normal and abnormal conditions.

Case History 47: Turbine Solo Operation with Tapered Journal

This case study addresses a twenty-year-old steam turbine solo run. In this example, a four stage turbine was subjected to a thorough overhaul. The stationary diaphragms were replaced, a spare rotor was installed, and new seals and bearings were installed. This unit had a good operating history, and minimal if any problems were anticipated. During the turbine solo, the machine was operated from a slow roll of 550 to an overspeed trip at 9,510 RPM. The response data from the governor end horizontal probe is presented in Fig. 13-11. The bottom portion of this Bode plot displays rotational speed (1X) amplitude, and the top plot exhibits phase angle as a function of speed. An unusually large runout vector of 2.32 Mils,$_{p-p}$ at 67° was documented. This vector was subtracted from the sampled data, and dual plots of the runout corrected (or compensated), and the uncompensated (or direct) traces are displayed in Fig. 13-11.

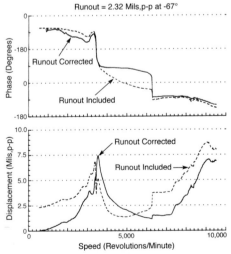

Fig. 13–11 Turbine Solo Run - Lateral Response At Governor End Radial Bearing

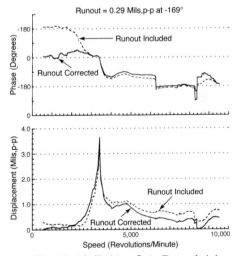

Fig. 13–12 Turbine Solo Run - Axial Response At Governor End Thrust Bearing

Prior to startup, the governor end diametrical bearing clearance was measured to be approximately 6.5 Mils. That value was exceeded as the turbine passed through the translational first critical at a speed of 3,500 RPM with an amplitude of 7.5 Mils,$_{p-p}$. At operating speeds above 7,000 RPM, a steadily degenerating situation was quite evident as 1X radial amplitudes continued to increase. Observation of rotor axial vibration during a solo run is normally uneventful. However, the axial response characteristics of this turbine were quite unusual. The direct and runout compensated Bodes are shown on Fig. 13-12. It is evident that the sharp peak at 3.8 Mils,$_{p-p}$ occurs at 3,500 RPM, which is coincident with the 7.5 Mil,$_{p-p}$ peak observed on the lateral Bode plot.

In most cases, it is normal to detect a minor level of lateral to axial cross-coupling in a rotor system. However, the observed ratio of almost 2:1 is unusual. In addition, the high amplification factor of the axial response was judged to be abnormal. Fortunately, the turbine was tripped by the overspeed trip assembly and was not restarted. As always, some individuals wanted to couple up the turbine to the compressor on the basis that the compressor would *calm down* the turbine. This is nothing but wishful thinking, and the machinery diagnostician should always remember that *you don't go on to the next step until the current step or test has been successfully completed.*

In this case, the turbine was shutdown following this single uncoupled (solo) run. A subsequent bearing inspection revealed a wiped governor end radial bearing, and a polished exhaust end bearing. The rotor was pulled for further inspection, and it was determined that the governor end journal was tapered. In fact, the diameter difference across the width of the bearing was in excess of 3.0 Mils. It was postulated that the bearing could not develop a proper oil wedge across the axial length of the bearing. This journal taper plus the lack of a proper radial support evidently resulted in the abnormal lateral and axial behavior.

This was a case where a machining oversight could have manifested into a major turbine failure. Fortunately, the peculiarities of the vibration data called attention to the mechanical problem during the uncoupled solo run of the turbine, and a suitable correction was achieved by re-grinding the turbine journal.

Case History 48: Coupled Turbine Generator Startup

The next case history considers a steam turbine driven generator set. In this train, the turbine had a very acceptable solo run, but the generator had a slightly rougher time during the coupled startup. The cascade plot presented in Fig. 13-13 was acquired from the horizontal probe mounted at the exciter (outboard) end of the generator. Note that the individual spectra are reasonably clean, with the dominant motion occurring at rotational speed (1X). The minor harmonics of this fundamental frequency are primarily attributable to shaft surface imperfections rather than actual rotor vibration. Since the rotor surface observed by the proximity probe was external to the bearing housing, and exposed to the atmosphere, this type of signal noise should be expected.

Further visibility of the horizontal 1X motion is obtained from the Bode plot

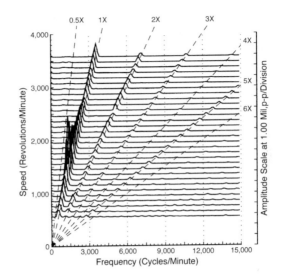

Fig. 13-13 Cascade Plot
Of Generator Startup On
Outboard Exciter End
Journal Bearing

shown in Fig. 13-14, and the polar plot of Fig. 13-15. Notice that the Bode displays two distinct amplitude peaks and associated phase excursions between 1,000 and 2,000 RPM. In this case, the observed response may be representative of a split critical. That is, there may be a discernible difference in the vertical versus horizontal rotor support stiffness. Since the horizontal stiffness is generally lower, the peak at 1,350 RPM might be the horizontal first critical speed, and the response at 1,650 RPM might be representative of a vertical first critical. Thus, the 300 RPM difference between peaks might indicate a split critical.

The same type of behavior could also be due to two closely spaced reso-

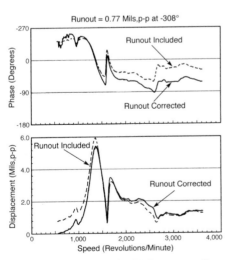

Fig. 13-14 Bode Plot Of Generator Startup On Outboard Exciter End Bearing

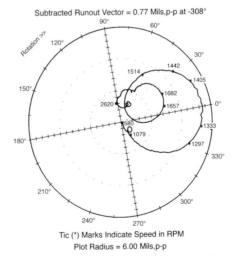

Fig. 13-15 Polar Plot Of Generator Startup On Outboard Exciter End Bearing

nances that are representative of two independent modes. In some cases, there may be a modal coupling between modes — and the motion from one resonance may excite or drive the other resonance. This type of modal coupling usually appears on higher order criticals (e.g., 2nd and 3rd).

The additional phase angle resolution offered by the polar plot (Fig. 13-15) reveals a slightly different perspective of the same vector data. Note that the large polar loop represents the horizontal resonance at 1,350 RPM, followed by the smaller inner loop that is the critical at 1,650 RPM. Since the polar loops are coincident, this information begins to look like a pair of rotor resonances. However, another interpretation of this data might be a major rotor resonance followed by a minor secondary or structural resonance. In essence, examination of the Bode and polar plots from one probe results in three possibilities for the peaks at 1,350 and 1,650 RPM. As discussed, this could be attributed to:

○ Split Horizontal and Vertical Critical
○ Independent but Closely Coupled Shaft Resonant Modes
○ Shaft Critical followed by Secondary or Structural Resonance

The available data is insufficient to provide any additional clarification of the origin of the two resonances. However, when the Bode and polar plots for the other three radial probes on the generator are examined, it becomes clear that the behavior is due to a split (horizontal then vertical) translational critical speed. The presented plots (Figs. 13-14 and 13-15) are representative of the general behavior. However, examination of the overall machinery response from the X-Y probes at both the coupling and the exciter end is necessary to fully identify the modal behavior of this generator. This type of data is certainly beneficial for condition monitoring, and it must be considered as necessary information for evaluation of any future problems on this generator.

It should also be mentioned that this turbine generator set was installed on an adequate foundation. The casing motion was minimal, and shaft relative proximity probes were sufficient to fully describe the machinery response characteristics. However, if the foundation was softer (i.e., flexible or compliant), the casing motion would have to be measured and considered. The inclusion of casing absolute with the shaft relative measurements allows the determination of shaft absolute vibration as previously discussed in chapter 6. This does complicate the data, but sometimes the ability to accurately evaluate the behavior of a machine is directly related to the thoroughness of the database.

On this generator, it is clear that the unit was fairly active through the resonant region. A trim balance correction would improve these synchronous characteristics through the criticals. However, the 1X vectors at full speed were all in the vicinity of 1.0 Mil,$_{\text{p-p}}$. These observed shaft displacement amplitudes are quite acceptable when compared against the 18 Mil diametrical bearing clearance of both generator bearings. Hence, the unit remained on-line, and the generator was producing full rated power in less than four hours.

Case History 49: Heat Soak and Load Stabilization

The final phase in most machinery startups consists of a heat soak (twelve to forty-eight hours) combined with process and load stabilization. Small machines with thin cases and lower operating temperatures may stabilize in twelve hours or less. Conversely, larger units with thick machinery cases, and higher operating temperatures may easily require up to forty-eight hours to achieve an acceptable operating equilibrium. During this condition, machine speeds are fairly constant. Speed changes are generally performed a few RPM at a time, and the rapid data sampling rates associated with the startup conditions are no longer required. In many cases, the machinery behavior may be adequately examined and documented by reverting back to steady state data acquisition and processing techniques.

For instance, consider a steam turbine driven compressor following a routine startup. After attaining an initial speed of 5,054 RPM, the orbit and time domain plots across the coupling are presented in Figs. 13-16 and 13-17. This data was consistent with previous measurements, and it was considered to be indicative of normal mechanical behavior. As noted in Fig. 13-16, the turbine exhaust orbit was elliptical and predominantly horizontal. The maximum unfiltered horizontal probe amplitude was 1.85 Mils$_{p-p}$. This value was identical with previous startups. The predominantly flat orbit was indicative of a vertical preload, and historically this had been attributed to cold misalignment across the turbine. From previous startups, the turbine exhaust end orbit would always evolve into a circular pattern as the front of the turbine warmed up to a normal operating elevation.

Across the flexible coupling, the compressor inboard bearing had vibration amplitudes of slightly less than 1.0 Mil$_{p-p}$. The motion was forward, circular, and

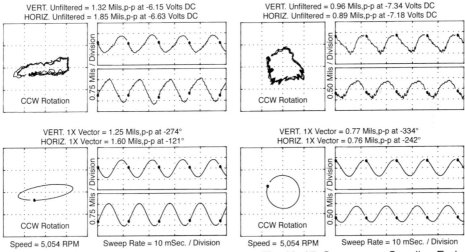

Fig. 13–16 Turbine Coupling End Bearing Immediately After Train Startup

Fig. 13–17 Compressor Coupling End Bearing Immediately After Train Startup

quite stable. This documented orbital and time domain data was directly comparable with previous information. As the turbine achieved a uniform heat soak,
the lower temperature compressor casing exhibited minimal thermal growth. In
fact, as time progressed, the compressor coupling end vibration normally did not
change appreciably from the patterns shown in Fig. 13-17.

Many machinery startups are achieved after long hours of correcting
numerous instrumentation and system problems. Usually, when the equipment
is up and running on the governor, the problems are generally over. However, on
this particular machinery train, symptoms of substantial coupling misalignment
appeared 25 hours after startup. The turbine exhaust shaft vibration increased
somewhat to maximum amplitudes of 3.4 Mils,$_{p-p}$ as shown on Fig. 13-18. Concurrently, the horizontal probe on the coupling end of the compressor increased
from slightly less than 1.0 to 6.1 Mils,$_{p-p}$ as documented in Fig. 13-19.

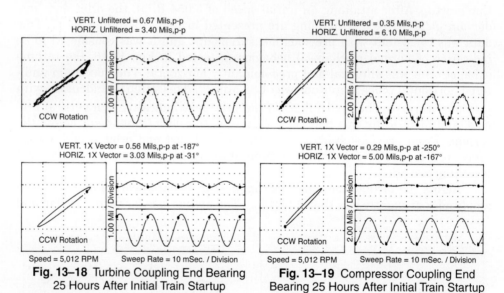

Fig. 13–18 Turbine Coupling End Bearing
25 Hours After Initial Train Startup

Fig. 13–19 Compressor Coupling End
Bearing 25 Hours After Initial Train Startup

Operating at a slightly lower speed of 5,010 RPM, both rotors exhibited
severely pre-loaded (flat) orbits. In addition, the two orbits are essentially out-of-
phase with each other. This behavior is most graphically illustrated by a direct
comparison of the 1X filtered orbits. The turbine exhaust displays a Keyphasor®
mark at the upper right-hand corner of the orbit. At the same time, the compressor inboard reveals a Keyphasor® dot at the lower left side of the orbit. This
behavior is representative of a pivotal motion across the coupling, and it is probably associated with a locked up coupling.

In this abnormal situation, the culprit was traced to the large diameter
compressor suction line. A pipe shoe somehow worked its way off a support, and
this section of the suction line dropped several inches. This overhung weight provided a tremendous torque on the compressor casing, and produced the observed
misalignment. Fortunately, the problem was detected rapidly, and the machinery

train was shutdown in an orderly fashion.

Mechanical inspection revealed that the gear coupling and the compressor inboard bearing required replacement. Both turbine bearings and the compressor outboard bearing were not damaged. The thrust bearing on each case had experienced some elevated temperatures, but the shoes were found to be in good shape. Following completion of these repairs, the unit was successfully restarted, and placed on-line. In the final analysis, it was concluded that scheduled condition monitoring performed one day after startup averted any further damage to the machinery or the associated facility.

Based on these discussions, and the case histories, it is apparent that condition monitoring is a valuable and viable tool during startup situations. The fundamental logic is similar to the observation of machinery characteristics at constant speed and load. Specifically, machinery in normal condition does respond in a fairly repeatable and consistent manner. Mechanical problems or developing failures will alter these typical response characteristics. Since these variations are detectable, they are also identifiable, and therefore correctable.

BIBLIOGRAPHY

1. "ASME Performance Test Codes — Code on Steam Turbines, PTC-6," *The American Society of Mechanical Engineers*, (New York: The American Society of Mechanical Engineers, 1976, reaffirmed 1982).

2. "ASME Power Test Codes — Test Code for Compressors and Exhausters, PTC-10," *The American Society of Mechanical Engineers*, (New York: The American Society of Mechanical Engineers, 1965, reaffirmed 1986).

3. Eisenmann, Robert C., John East, and Art Jensen "Short Course 1 - Inspection and Overhaul of Major Turbomachinery." *Proceedings of the Seventeenth Turbomachinery Symposium and Short Courses*, Turbomachinery Laboratory, Texas A&M University, Dallas, Texas (November 1988).

4. Mitchell, John S., *Introduction to Machinery Analysis and Monitoring*, second edition, pp. 291-345, Tulsa, OK: Pennwell Publishing Company, 1993.

5. "Vibration, Axial Position, and Bearing Temperature Monitoring Systems - API Standard 670, Third Edition," *American Petroleum Institute*, (Washington, D.C.: American Petroleum Institute, November 1993).

Machinery Diagnostic Methodology

*T*he following chapter is based upon a mini course prepared by the senior author for an annual meeting of the Vibration Institute[1]. Although years have elapsed since this presentation and the tools and instrumentation systems have evolved to more sophisticated levels, the basic problem-solving approach towards machinery malfunctions remains unchanged. Thus, the systematic thought process contained herein is still considered to be quite appropriate. Over the years, this approach has proven to be a successful and effective methodology for solving machinery problems.

This chapter will review these field-proven methods for the diagnosis of machinery problems. The correct identification and proper diagnosis of most malfunctions requires the correlation of parameters such as mechanical construction, process influence, maintenance history, and vibratory behavior. Although machinery diagnosis has occasionally been associated with mysticism, it really should be considered as an engineering project. This type of project demands a lot of hard work, a belief in physical laws, plus a methodical problem solving approach. Requirements also exist for a certain level of skill, diagnostic tools, and relevant experience. For the purposes of this chapter, the emphasis will be placed upon the problem-solving approach, that is, the *Diagnostic Methodology*. Although the specific approach will vary from problem to problem, there are similarities that can be categorized into the following seven major areas:

1. **Diagnostic Objectives**
2. **Mechanical Inspection**
3. **Test Plan Development**
4. **Data Acquisition and Processing**
5. **Data Interpretation**
6. **Conclusions and Recommendations**
7. **Corrective Action Plan**

Each of the above categories will be addressed, and the methodology associated with transforming the unknown mechanical failures into the realm of docu-

---

[1] Robert C. Eisenmann, "Machinery Diagnostic Methodology." *Vibration Institute - Mini Course Notes - Machinery Vibration Monitoring and Analysis Meeting*, New Orleans, Louisiana (May 1985).

mented information will be discussed. These concepts will be illustrated with a series of field case histories dealing with startup problems encountered on three different machinery trains. Although these problems occurred on new machines during initial operation, the same techniques are applicable to existing mechanical equipment and condition monitoring programs.

DIAGNOSTIC OBJECTIVES

The need to establish diagnostic objectives is often overlooked, due to the fact that these objectives are obvious to all parties involved. However, there is an old bit of philosophy that says: *the first step to getting anything done is to write it down*. This statement may seem trivial, but it does represent a basic truth. In the realm of machinery analysis, it is easy to lose sight of the real objectives on complex projects that last a long time. It is not uncommon for individuals to become wrapped up studying some unrelated frequency that has absolutely nothing to do with the real mechanical fault.

The form or format of the statement of objectives can be very simple, or it may be quite formal. On some minor problems this may consist of a note that you carry around in your wallet or a line item on the weekly *To Do List*. For a complex problem with high visibility, significant financial impact, and multiple party involvement, the objective statement may evolve into a formal daily status review meeting. In any case, the idea is to maintain a continuous reminder of the real project or analysis objective(s), and to continually move in the direction of satisfying the objectives, and solving the problem.

MECHANICAL INSPECTION

On-site mechanical inspection is a significant part of the database for any machinery problem. This includes a *hands on* examination of the machinery, and the actual installation. Familiarization with the operating and maintenance history is highly desirable, if not mandatory. On a new installation, refer to the OEM shop testing records and customer witness reports. In recent years, the tendency throughout many industries has been to rely upon remote analysis via FAX machines and transmission of data. Although this is a cost-effective approach for routine mechanical problems, it is not advisable for difficult and/or complex problems. Hence, direct inspection of the machine in distress still provides the most information about the equipment.

The mechanical construction of the machinery must be examined. It is virtually impossible to diagnose machine behavior without knowing what lies beneath the outer casing. This should include a review of assembly clearances and tolerances, bearing, seal, and coupling configuration, plus overall rotor assembly. Ideally, there is an opportunity to work with the millwrights, and be right there to observe (or participate) in the actual physical measurements. As an alternate, there is often a warehouse with identical spare parts. Examination of the stored spare parts also provides a perspective on the storage condition of the parts. Questions arise as to how the rotors are stored to prevent shaft bows, and how the machine elements are protected from environmental degradation.

On problems where failed machine parts are available, they should be thoroughly scrutinized to help determine the failure mode(s). Sometimes metallurgical examination of these items can provide positive benefits and data. Also, look at the shop and/or field balancing techniques and procedures. How good are the records? How good is the work? The same applies to alignment data. Are indicator readings corrected for bracket sag? Are soft foot checks made? What type of shim stock is normally used?

From an external standpoint, the associated lube and seal oil system should be reviewed to determine any associated peculiarities that could adversely influence the mechanical behavior of the machinery. Sometimes this can identify problems such as a damaged oil pump, and other times it may uncover a collapsed float in a seal oil drain pot. The machine foundation and support structure should be examined, and the grout between the baseplate and the foundation should be visually inspected. The associated piping system should be reviewed, and particular attention paid to the location of sliding and fixed pipe supports, plus spring hangers and dampers. The location and general condition of the main process block and check valves should be checked, and the fundamental machine process control technique should be identified. For instance, machine speed may be regulated by a pressure controller sensing discharge pressure. If the pressure sensor or the field controller misbehaves, the result might be an erratic speed control from the governor.

The bottom line for this section is simply *you should become intimately familiar with the mechanics of the machinery and the associated systems*. Failure to perform this initial familiarization with the installation can easily result in a misdirected analysis effort.

TEST PLAN DEVELOPMENT

Based upon the diagnostic objectives, and the mechanical inspection, the development of a realistic test plan would be the next logical step. This test plan should be directed at learning about the behavior of the machine and the defined problem area(s). The actual test may take many different shapes and formats, depending on the type of problem(s) encountered. The testing may include variable speed runs, variations of process load or operating conditions, changes in oil supply temperature or pressure, or any other parameter that can be altered in a controlled and repeatable manner. The testing may include mechanical changes such as variations of mass unbalance, machine alignment, baseplate attachment, piping support, or physical machine element changes.

In other cases the testing may center on structural, or discrete resonance testing. The items to be tested may include the specific machine components, a spare rotor assembly, or a variety of structural measurements. The options are many and varied, and it should be recognized that not every field test will directly identify a problem. In some situations, a test is conducted to verify that a particular problem is not occurring. For example, differential pressure across a compressor may be increased just to prove that the machine is not sensitive to load changes.

For a test plan to be effective it should include a description of the physical parameters that will be varied. Next, an outline of the specific test procedure must be developed. The procedure needs to be agreed upon by all involved parties before testing begins. In addition, the procedure should go into specific detail as to the measurements that will be made, and some statement regarding the expected results or knowledge to be gained from the test. If possible, the normal or expected response of the unit should be stated, and if the field test provides different results, the deviation may be immediately recognized. From a safety standpoint, appropriate limits should be established for the critical values (e.g., shaft vibration or bearing temperature). In addition, general agreement should be reached beforehand that the field test would be aborted if any of the predetermined safety limits are exceeded.

Finally, during the investigation of a complex problem it is possible to conduct tests that appear to provide minimal results. It is important to maintain overall perspective and recognize that there is something to be learned from virtually every test. Always maintain the perspective that some tests help to eliminate potential malfunctions and other tests help identify the real problem(s).

DATA ACQUISITION AND PROCESSING

The acquisition of field data during a planned test, or on a routine basis, requires the assembly of the proper transducers and test instrumentation. The transducer suite must be compatible with the machinery and the measurements to be made. The test instrumentation must likewise fit into the requirements of the test program. Prior to initiation of testing, it is also necessary to check continuity of all wiring, perform system calibration checks, and include sufficient means for post-test verification of the data. In some cases, the transducers may be subjected to another calibration check after the test is completed. For example, if the operating environment for some of the transducers approaches the temperature limit for the pickups, then a post test calibration would be advisable to verify that the pickup was not permanently damaged during the test. In some instances, the test results may be rendered invalid due to the failure or malfunction of an important measurement.

It should be recognized that the data acquired is really a function of the specific machine, the associated problem, and the individual test plan. Typically, field data acquisition combines a variety of dynamic and static information. Measurements of displacement and velocity are used to examine the shaft and casing motion in the frequency domain surrounding running speed. High frequency casing acceleration measurements are used to observe excitations such as blade passing or gear meshing. Specialized measurements, such as pressure pulsation or torsional vibration, are applied as required. These dynamic measurements are usually recorded on multichannel instrumentation grade magnetic tape recorders and/or directed to a digital signal processing system (e.g., Fig. 8-5).

Analog outputs proportional to process variables such as temperature, pressure, or flow rate may also be included. During many situations this type of data is hand logged, or obtained from strip chart recorders. In other cases, where time correlation is important, the analog parameters may be simultaneously tape recorded with the dynamic vibration signals, and suitable time stamping provided. On modern plants with Distributed Control Systems (DCS), a variety of graphic and tabular process data presentation options are available.

The recorded test data should be supplemented by hand logged information and applicable on-line data reduction. This provides immediate documentation of significant information, and direct verification of the information recorded on magnetic tape or digitally processed. If this step is bypassed, then the credibility of the data suffers; and the final solution to the machinery problem may be unnecessarily delayed.

Detailed data processing follows the completion of field data acquisition. For digitally sampled data the information is often processed and available concurrent with the completion of the field test. This hard copy data is reviewed, and additional formats processed as necessary. For tape recorded information, the taped data is reproduced into additional diagnostic instruments to observe both the steady state and the transient behavior of the machinery. During this phase, the data is processed to hard copy format. The accuracy of the data recording system must be checked, and all calibration signals must be reviewed. Comparisons should be made between the final data plots and the field logged data (e.g., the 1X vectors acquired during testing should be duplicated on the final hard copy plots).

As a minimum requirement, the tape recorded signals should be viewed on an oscilloscope. In most cases, the signals are processed through a Digital Vector Filter (DVF) to view the synchronous amplitude and phase, plus a Dynamic Signal Analyzer (DSA) to observe the frequency content of the signals. The DSA also performs special operations such as statistical averaging, frequency expansion, frequency response functions, and coherence. Normally, the signals are simultaneously documented in the time, orbital, and frequency domain. When applicable, radial and/or axial position data is also calculated and charted.

Often the data is further summarized or correlated into appropriate graphic or tabular formats. This allows for a concise summary of the pertinent information, and usually provides more data visibility during the interpretation stage. Various tools and techniques are available for data reduction. Some are simple, others are sophisticated. In general, the data processing system must be reliable, provide sufficient resolution of the data, and integrate with the field data acquisition system to produce accurate and repeatable results.

Although the diagnostician is generally concerned with the complexities of acquiring and accurately processing the dynamic vibration data, the associated steady state measurements should not be ignored. This includes the operating process information of pressures, temperatures, and flow rates. Load information such as measured shaft horsepower or motor currents may also be meaningful during the evaluation of the test data. Other sources of useful information include derived data such as lab results for molecular weights, or a heat and material balance conducted with a Distributed Control System (DCS). As a cautionary note, it should be mentioned that many DCSs retain detailed process information for a limited amount of time. It is often necessary to extract the desired process data within 24 hours of the test conclusion to retain full resolution of the sampled information. After this time period, the data is averaged and stored with various compression algorithms, and the sample resolution available during the field test is lost forever. If detailed process information is required during a field test, it is highly desirable to request this data prior to starting the test rather than as an afterthought.

During the examination of test data such as summary or trend plots, various nonlinearities may appear. Sometimes these variations are due to actual mechanical events, and sometimes the deviations are due to test or measurement errors. One way to address these variations would be to run a curve fitting routine on the experimental data, and use the resultant polynomial equation for further analysis. This is also a useful technique for extrapolating a data set to a region that resides outside of the measured information. Depending on the complexity of the information, a second or third degree polynomial may be acceptable. Occasionally a fifth or sixth degree polynomial curve fit is required to accommodate significant changes in curve pattern. An example of this type of information is presented in chapter 4, where frequency response measurements are used to measure bearing housing support stiffness. In this case, a sixth degree polynomial curve fit was performed to develop a characteristic equation for housing stiffness as a function of speed.

Analytical computations may also be performed to compare the results between a computer model and the response characteristics of the real machine. This type of verification between analytics and measurements may be highly beneficial to verify the accuracy of a particular model. Once this step is performed, the analytical model may be used to examine a variety of changes in machine characteristics. These changes could be due to different mechanical elements, or the influence of different operating speeds or conditions that could not be conveniently examined during a machinery field test. Hence, the machinery diagnostician should keep an open perspective of the potential data processing and analysis techniques that may be applied. In many instances, if one approach is blocked, another technique may provide the necessary information to solve the machinery problem.

DATA INTERPRETATION

Data interpretation is the formative phase of any machinery diagnostic project. This activity includes a summary and correlation of all pertinent data acquired during the project. This includes the mechanical configuration, process and maintenance history, field testing data, plus any supportive calculations or analytical models. All of this information constitutes various parts of the puzzle. Often the problem cannot be solved unless enough puzzle pieces are available to develop a logical and consistent overview. The diagnostician is cautioned against taking a superficial approach during this phase of the project. In many cases, a quick review of the data plots does not solve the problem. Normally, each set of test conditions must be compared, the behavior of the machine must be examined, and the results committed to some type of summary log. Although this may be a brutally difficult exercise, it is the only way to properly quantify the field test results.

Actual interpretation of the test data is strongly dependent on the type of machinery and the operating conditions. The 30,000 horsepower gas turbine behaves quite differently from a similarly rated steam turbine. Hence, the data must be interpreted in accordance with the physical characteristics of the particular machine type and the operating environment. One approach is to view the data in terms of normal behavior for a particular machine type, and then look for the abnormalities in response characteristics.

It should also be mentioned that the diagnostician may want to use some of the evolving computer tools for acquisition, processing, and initial interpretation of the data. The general availability of computer programs using artificial intelligence are generally termed *Expert Systems* for the diagnosis of machinery problems. These programs may provide significant time and accuracy advantages for the diagnostician. The most powerful programs acquire, correlate, and compare transient and steady state data. Based upon these results, a series of potential malfunctions are identified.

Less sophisticated diagnostic programs rely on an *interview process* where the user provides answers to a series of questions, and the program responds with a list of potential malfunctions. In all cases, the machinery diagnostician should recognize that the depth of problems addressed by these software programs are limited to the knowledge and experience of the people that established the initial database, and the rules for problem identification. Hence, if the machinery problem is fairly common (e.g., the malfunctions discussed in chapter 9), expert systems will produce good results. Conversely, if the problem falls into a unique category (e.g., as presented in chapter 10), these programs may not be able to identify the true origin of the malfunction. Although computers are wonderful data processing devices, they still pale in deductive reasoning powers when compared with the human brain.

CONCLUSIONS AND RECOMMENDATIONS

The final conclusions and recommendations constitute two separate, but concurrent thought processes. The development of the conclusions is considered to be a summary of the knowledge gained by executing the activities discussed during the previous six sections. The conclusions are really a summary statement of *what's wrong with the machine*. In many situations, the conclusions may also summarize *what's right with the machine*. In some cases the conclusions may be direct and precise. On other problems, the final conclusions may be less definitive due to the nature of the malfunction and/or the lack of accurate or sufficient data. It is possible to reach a conclusion that the test data is insufficient for solving the problem, and a retest with additional parameters or measurements is warranted. Certainly this may be expensive, and professionally embarrassing. However, it is always much better to be honest about the test results, rather than attempt to hide inclusive test results in a mountain of technical *dither and trivia*.

The recommendations, on the other hand, are specific statements of what can be done to fix the machine. This may be as simple as: *replace the outboard bearing*. Conversely, the recommendations may be complicated, and require a series of phases or steps presented in a logical sequence. Specifically, the first phase of the recommendations may consist of actions that should be executed in an immediate time frame to keep the machine on-line. The second phase may consist of action items for the next scheduled turnaround. The third phase may include long-term items that require additional investigation, analytical simulation, or redesign for improved safety, reliability, or availability. In all cases, the diagnostician should include a healthy dose of *reality* in the development and presentation of the final recommendations.

CORRECTIVE ACTION PLAN

The corrective action plan is the final stage of any machinery malfunction scenario. During this stage the recommendations are discussed, economic influence is introduced, and a final action plan is generated. This action plan may be different from the recommendations due to factors such as plant production requirements, turnaround schedules, or situations where the machine modification cost exceeds the repair costs during a fixed time period. Like many other situations, the final action plan is a compromise between the parties involved. The best possible corrective action plan is one that maintains a reasonable balance between sound engineering judgment and proper financial considerations. It should always be recognized that repairs or modifications to large process machinery must compete with other projects within the corporation requesting funding. In some cases, the economic decision dictates that the mechanical problem be tolerated. In other extreme conditions, the loss or failure of one machinery train may result in the corporate bankruptcy.

Case History 50: Steam Turbine Electrostatic Voltage Discharge

On many machinery problems there is direct physical evidence to help identify the origin of the malfunction. This evidence is often combined with measurements, such as increasing vibration amplitudes, to identify the problem onset. However, there are machinery malfunctions that exhibit minimal physical symptoms, and sometimes result in the reduction of vibration levels with time. The following case history describes such a failure that occurred on the machinery train presented in Fig. 14-1. This unit consists of a six stage, horizontally split refrigeration compressor driven by a 28,600 HP condensing steam turbine. Typical operating speeds vary between 3,700 and 3,900 RPM. The turbine accepts superheated steam at 600 Psi, and exhausts directly to a close coupled surface condenser. The turbine was originally equipped with oil dam sleeve bearings, and the compressor rotor was supported by tilt pad journal bearings.

Approximately two months after original commissioning and a successful plant startup, the compressor inboard bearing failed. Fortunately, the shaft was not damaged and a spare set of pads was installed. During the next two months of operation, the vertical vibration at this coupling end bearing decreased from an initial starting level of 0.7 Mils,$_{p-p}$ to a minimal amplitude of 0.3 Mils,$_{p-p}$. At the time, several individuals commented on the fact that the compressor vibration was improving and the machine was *healing* itself on-line. This proved to be *fuzzy* thinking, but it was a popular theory for a short period of time.

Eventually, shaft vibration amplitudes began a steady growth from 0.3 up to 0.7 Mils,$_{p-p}$ during a three week period. Since this was back to the original amplitude, little attention was paid to the change. Compressor shaft vibration continued to gradually increase and one night it jumped up to 3.4 Mils,$_{p-p}$. The inboard vertical probe gap voltage revealed an overall 4.9 volt change. This DC

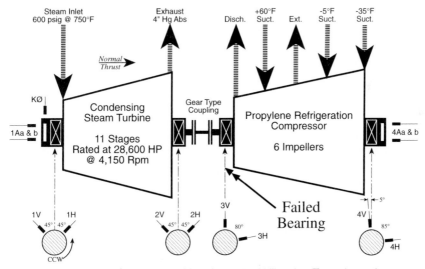

Fig. 14–1 Propylene Compressor Machinery And Vibration Transducer Arrangement

gap change was equivalent to 24.5 Mils of vertical shaft position drop (i.e., 4.9 Volts x 5 Mils/Volts = 24.5 Mils). Since the babbitt thickness on the bearing pads was nominally 15 to 20 Mils thick, there was concern that substantial damage had occurred at the compressor inboard. The next morning the train was shut-down to inspect and probably replace that coupling end bearing. It was discovered that the bearing damage extended to the point that the journal was severely scored. Hence, the compressor case had to be split and the spare rotor installed.

In retrospect, the shaft damage should have been anticipated, since the total vertical shaft drop (24.5 Mils) exceeded the available bearing pad babbitt thickness (15 to 20 Mils). Thus, the steel shaft was riding on the steel bearing pad, and any breakdown of the minimum oil film would result in steel to steel contact (i.e., shaft to bearing backing). However, at that point in time, minimal attention was given to the trend plots of gap voltage versus time.

The third failure on this machine was monitored very carefully. Another vertical proximity probe was installed directly opposite the existing transducer for confirmation of the vertical shaft position. During the next twenty-four days, the vertical vibration amplitude decreased from 0.64 to 0.26 Mils,$_{p-p}$. Concurrently, the top vertical probe DC gap increased from -7.0 to -9.3 volts, as shown in the Fig. 14-2. This voltage change was equivalent to an 11.5 Mil vertical drop of the compressor shaft. This change was also reflected by the new probe mounted

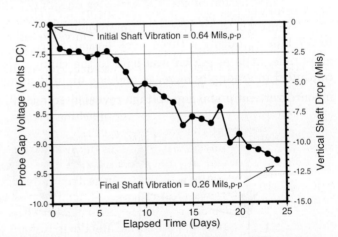

Fig. 14–2 Compressor Coupling End Bearing Vertical Shaft Position Change With Time

on the bottom of the bearing assembly. Based on the previous failure, the train was shutdown, and inspection revealed that the babbitt was almost totally missing from the bottom shoe. The journal exhibited a frosty-satin-gray appearance, much like the failed bottom shoe. Further inspection showed similar characteristics on both turbine and compressor thrust bearings. A slight indication of frosting was also noted on the turbine exhaust end bearing, and the bronze governor drive gear mounted at the turbine outboard.

Fortunately, one of the OEM reps had previously encountered similar failures on electrical machinery. He reported that the discharge of shaft voltages could cause the failures, and recommended the installation of rotor grounding brushes. It was postulated that the increasing vertical probe voltage really meant that the journal was sinking into the bearing, as represented in Fig. 14-3. Based on this physical evidence, and the OEM recommendation, insulated brushes were installed on the outboard shaft end of the turbine and the compressor. Both brushes were to be grounded directly to the machine baseplate.

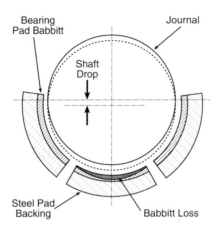

Fig. 14–3 Compressor Coupling End Bearing Vertical Shaft Position Change

Reexamination of the historical log sheets revealed that all three failures exhibited identical characteristics of decreasing shaft vibration amplitude, and increasing vertical probe gap voltage. Only during the second failure did the shaft vibration increase, and this was attributed to the fact that the nominal 15 Mil bearing babbitt thickness was exceeded. Hence, there was a good deal of confidence that all three failures were due to the same mechanism.

The ensuing startup with the installed ground brushes was viewed with optimism. Shortly after startup, the machinery was operating smoothly, but it was discovered that the compressor brush had not been connected to ground. During the act of physically attaching the loose wire to the baseplate ground, a substantial electrical arc was encountered. The surrounding atmosphere was filled with propylene fumes, and it was fortunate that ignition did not occur. After the machinery was on-line, the voltage potential between rotor and ground was measured. Upon disconnecting the brush lead from the baseplate, an electrical spark was again experienced. Readings taken with a series voltmeter revealed levels in excess of 20 volts. Hence, there was no question that the machinery was producing an electrical voltage.

A more detailed analysis was made following eighteen months of successful operation. The significant plots from this analysis are displayed in Fig. 14-4. The top time base and spectrum plots document the compressor shaft vibration characteristics measured by the vertical probe at the inboard bearing. The major frequency component occurs at running speed of 3,750 RPM, with sub-harmonic shaft motion at 48% and 63% of speed. Shaft voltage signal characteristics are presented on the bottom half of Fig. 14-4. It is of particular interest to note that the spectral content of the shaft voltage is virtually identical to the vibration signal. This similarity even extends to the appearance of the 48% and 63% of run-

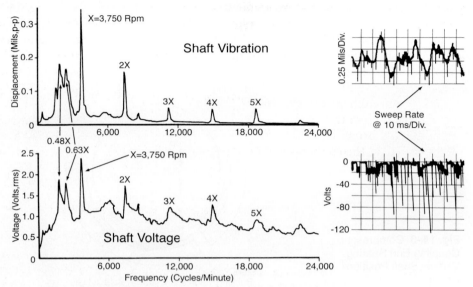

Fig. 14–4 Vertical Shaft Vibration And Shaft Voltage From Grounding Brush

ning speed components. It is reasoned that as the shaft orbits, the electrical discharge will respond to the dynamics of the minimum oil film. Thus, similar frequency content should be expected from both the displacement vibration probe and the electrical discharge signal. It should also be mentioned that the time domain shaft voltage data is presented at a sweep rate of 10 milliseconds per division. If this scale was expanded to 2 milliseconds per division, the long vertical spikes (120 volts), can be observed to dissipate in approximately 3 milliseconds (0.003 seconds). This voltage change is equivalent to a discharge rate of 40,000 volts/second!

Generation of internal shaft voltages is apparently limited to condensing turbines where the last stages are subjected to saturated steam. The author has never encountered this problem on a back pressure turbine. It is generally hypothesized that the brushing effect of water droplets (condensate) across the blades develops an electrostatic charge that builds up on the rotor. This charge periodically discharges to ground through the point of least resistance. When the rotor voltage dissipates across an oil film bearing, microscopic pits are produced in the babbitt. This results in the continual removal of metal from the bearing surface. Trouble may be noticed in periods ranging from a few days to several years, depending on the magnitude of the current flow. This phenomenon shows up as a frosted surface in the loaded zone of the bearing where the oil film is at a minimum thickness. This type of failure also appears on thrust bearings, seals, and any other type of mechanical device in close contact with the shaft. The voltages generated in the turbine rotor are transmitted throughout a train of mechanical equipment. They can pass across couplings and through gear boxes. Hence, the eventual failure might occur at a bearing far removed from the original source of the voltage. It has been documented that failures due to this electrostatic voltage may be transmitted through three couplings and associated compressor bodies.

The brush design originally installed on this refrigeration train is shown in Fig. 14-5. This preliminary design places the grounding brush at a low velocity point on the shaft in order to minimize brush wear. The brush is insulated from the case in order to direct the shaft voltage to ground. Next, the brush ground wire passes through a meter to verify brush operation. This holder can be changed during operation and it is sealed with RTV to minimize oil leaks. One axially mounted brush was installed on the turbine governor end, and a second brush mounted on the compressor outboard shaft end.

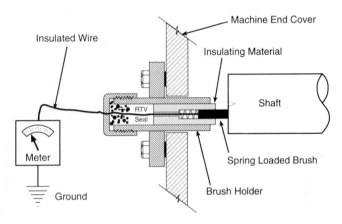

Fig. 14–5 Original Rotor Ground Brush Installation

Since this original brush installation by the author, several more sophisticated designs have evolved. Today, there are grounding brushes commercially available from several sources. However, the fundamental mechanism of dissipation of electrostatic shaft voltages remains the same.

It must be mentioned that this particular problem has been around for many years. One of the earliest complete references to this specific malfunction was an ASME paper by two General Electric engineers: J.M. Gruber, and E.F. Hansen[2]. Their paper represented untold man-hours of laboratory research, field measurements, and in-depth literature surveillance. Gruber and Hansen dealt primarily with large turbine generator sets, and they addressed the destructive effects of shaft voltages upon bearings. Categorically they identified five distinct types of shaft voltages: *"the electromagnetic or 60-cycle a-c voltage, a ground-detector 120-cycle a-c voltage, the ignitron excitation voltage, the high-frequency-exciter ripple voltage, and the electrostatic d-c voltage"*. They reviewed each of the categories and then went into more detail on electrostatic voltage. They stated:

"The electrostatic shaft voltage has been found to have several reasonably well-pronounced characteristics as follows:

---

[2] J.M. Gruber and E.F. Hansen, "Electrostatic Shaft Voltage on Steam Turbine Rotors," *Transactions of the American Society of Mechanical Engineers*, Paper Number 58-SA-5, (1958).

1. The voltage between shaft and bed plate is direct current. This means that the polarity does not reverse periodically.

2. The magnitude is not usually constant and in some cases falls repeatedly to low values after which it climbs back up to higher values. This means that the voltage contains both a-c and d-c components even though the polarity does not reverse.

3. The maximum magnitude observed by oscilloscope was about 250 volts peak.

4. The rate of rise of shaft voltage was often in the range of 200 volts per 1/60 sec. or 12,000 volts per second.

5. The voltage decay when falling to zero is less than 0.1 millisecond.

6. The minimum magnitude observed was a few tenths of a volt.

7. Typical magnitudes were between 30 and 100 volts peak value.

8. The shaft polarity was positive on many turbines and negative on fewer turbines.

9. The potential at any instant is essentially the same anywhere along the turbine or generator shaft. The shaft voltage appears between shaft and bedplate which is grounded.

10. The maximum current observed in a resistance circuit connected between shaft and ground, regardless of how small the magnitude of resistance, was approximately 1 milliamp."

Sound engineering conclusions are timeless. This technical summary by Gruber and Hansen is as applicable now as it was in 1958. Certainly this list could be modified to reflect some different measurements, or different machines, but the fundamental concepts, descriptions and characterizations of the phenomena are still the same. It is comforting to note that physical principles remain constant, and that our understanding of many of these physical principles has a tendency to grow with improved technology, measurements, and communication.

Finally, the role of the seven-step problem solving approach introduced at the beginning of this chapter should be reviewed. In this case, the diagnostic objectives were clearly directed at stopping the coupling end compressor bearing failures. Secondary objectives included an understanding of the failure mechanism, plus the issue of long-term reliability of the refrigeration train. Due to the consistent and repetitive nature of the failures, the identification and statement of the diagnostic objectives were quite straightforward.

Mechanical inspection provided no significant information during the initial failures, since the physical evidence was destroyed by the occurring failure mechanism. However, one of the key elements in the final detection and analysis of this problem evolved from the observation of the *frosted* bearings by an OEM service representative. If that individual had not encountered similar failures on electric machinery, the series of failures could have continued for several more iterations.

The test plan development for investigation of these failures was confined to traditional process variations, combined with steady state analysis of shaft and casing vibration response characteristics. This detailed system and vibration analysis provided little insight into the actual failure mechanism, and the test plan evolved into a simple monitoring activity. Hence, during the final two failures the data acquisition program consisted of daily measurements of shaft vibration and position.

The position data was obtained by measurement of the proximity probe DC gap voltages in both radial and axial directions. At the time of these failures, the machinery train was equipped with a full array of X-Y and thrust proximity probes. However, Proximitors® and monitors were not installed, and the measurements were obtained with a portable data collection box. The information from these daily readings was subjected to the simple data processing technique of manual trend plots. Although these plots did not directly solve the problem, they certainly did identify the existence of the abnormality (e.g., Fig. 14-2).

Initial data interpretation of this malfunction was subjected to various degrees of speculation. Although the *shaft sinking into the bearing description* was acceptable to most of the engineering personnel, the fundamental cause for this behavior remained a mystery. The full interpretation of this trend data was finally achieved by the previously mentioned OEM representative who had experienced electrostatic discharge problems on condensing steam turbines within the power generation industry. The conclusions and recommendations were to install a rotor grounding system to remove the electrostatic shaft voltage. The corrective action plan consisted of a temporary installation to verify the hypothesis, followed by an improved design for long-term operation and reliability.

The appropriateness of this grounding brush solution was demonstrated by several years of good operation of this machinery train. Although there were some problems with the compressor bearing design, the shaft voltage problem appeared to be under control. The only failure that occurred was due to complacency by the local personnel. They did not replace a worn out compressor grounding brush, and the inboard compressor bearing began to migrate down into the bottom pad. Many of the local management personnel had changed, and there was a general insensitivity towards this problem. Fortunately, one of the mechanical engineers involved with the original problem diagnosis and solution was still employed by the operating company. He reviewed the data and convinced management of the reoccurrence of this problem. Reluctantly, they shutdown the machinery train, and discovered that the compressor coupling end bearing was failing in the manner previously described. This bearing was replaced, both grounding brushes were replaced, and the train was returned to normal operation.

Approximately eight years after the original plant startup a new wrinkle into the shaft voltage problem was introduced. By this time, various modifications had been made to the original OEM bearings — and consultants concluded that the compressor was experiencing electromagnetic problems. In this type of malfunction, a rotating magnetic field produces shaft voltages that might exceed the current carrying capability of the ground brushes. One of the popular opinions expressed was that the electromagnetic problems originated from a wheel to diaphragm rub experienced during the OEM shop testing of the compressor. Of course, none of the founders of that opinion were associated with this machine during the shop test. However, speculation was somehow converted into an engineering conclusion. The reality of the situation was that a wheel rub occurred during a shop performance test by the OEM on a different first stage wheel design. This test was conducted with OEM compressor internals that were not part of the contract. Furthermore, after the occurrence of the rub, the contract casing was thoroughly checked for residual magnetism, and none was found.

A machine shutdown and disassembly following nine years of operation revealed residual magnetic fields on some elements of the compressor, turbine, and piping. Since these localized magnetic fields did not originate during the OEM shop tests, another source should be considered. For instance, improper welding and grounding practices on the compressor desk would be more suspect than any possible influence from a ten-year-old shop test rub.

In the final analysis, there are at least three distinct lessons to be learned from this case history. First, electrostatic shaft voltages are possible on machine trains driven by condensing steam turbines. These electrostatic shaft voltages are controllable by isolation and/or grounding of the rotating assemblies. Secondly, electromagnetic shaft voltages can appear when rotating magnetic fields are generated by magnetized machine elements. These electromagnetic shaft voltages can only be eliminated by degaussing the magnetized parts. Thirdly, during the investigation of any machinery malfunction, if the data becomes corrupted by idle speculation and/or conjecture, the final results will not be accurate, and the conclusions will often be misleading.

Case History 51: Barrel Compressor Fluidic Excitation

During the design of machinery for a new chemical plant or refinery, the OEM is faced with providing a cost-effective offering to satisfy the process requirements of the end user. Fluid parameters such as flow rates, temperatures, pressures, and gas compositions are used to define an operating envelope for the machinery. In most instances the fluid properties are well defined (e.g., air, propylene, ammonia, etc.). However, there are times when a new plant process may result in a fluid stream with properties that are not fully quantified. Although pilot plants and process computer simulations provide a wealth of knowledge, the final stream compositions may not be known until the full-scale plant is built and placed in operation. As a case in point, consider the following example of a prototype urea plant and the main centrifugal machinery train handling a mixture of carbon monoxide, carbon dioxide, plus trace quantities of hydrocarbons.

The machinery train in question consists of an extraction steam turbine driving two barrel compressors. The first two process stages are contained within the low stage compressor. The third and fourth process stages are handled within the high stage compressor. Each compression stage was equipped with interstage cooling and knockout drums. Configuration of the high stage rotor is presented in Fig. 14-6. The ten impellers (five per compression stage) are located back to back to minimize thrust load. This compressor was equipped with five pad, tilt pad radial bearings, and a traditional thrust arrangement. The rotor exhibited a first critical of 5,100 RPM, combined with a design speed of 9,290 RPM.

During initial field startup with process gas, the high stage compressor

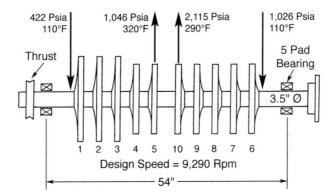

Fig. 14–6 Ten Stage Rotor Configuration For High Stage CO$_2$ Barrel Compressor

wrecked. The general opinion was that the system was under damped, and the OEM designed and retrofitted a set of squeeze film damper bearings. This modification consists of a machined annulus between the casing and the bearing housing that is filled with lube oil. The ends of the annulus are generally sealed with O-Rings, and it allows the bearing housing to float within the confines of the O-Rings and the oil film. The additional damping provided by this change proved successful, and the unit was easily brought up to normal operating speeds.

It appeared that the majority of the problems were over, and the remainder

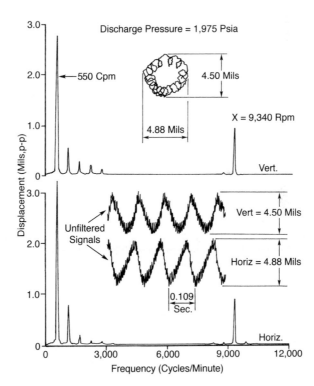

Fig. 14–7 Initial Coupling End Shaft Vibration On High Stage Barrel Compressor

Chapter-14

of the plant startup could proceed. However, during loading of the machine, vibration levels increased dramatically. At discharge pressures around 1,500 Psi, a low frequency excitation appeared. As discharge pressure was increased, the magnitude of the low frequency vibration would likewise increase. Typical shaft response for this unit is shown in Fig. 14-7. Rotational speed vibration amplitudes were approximately 1.0 Mil,$_{p-p}$, and the majority of the motion occurred at a frequency of 550 CPM. This motion was strongest on the coupling end bearing and somewhat lower on the outboard bearing. The shaft orbit was nearly circular at 550 CPM, and a series of inside loops at running frequency were noted. Under this set of operating conditions for every one cycle of the low frequency excitation, the rotor made almost seventeen revolutions.

This low frequency vibration was audible on the compressor deck and the piping. In fact, the unit sounded more like a reciprocating engine than a centrifugal compressor. One of the folks on the job claimed that it sounded like an armadillo running loose in the piping. A test program was established to run the compressor during various conditions to characterize the vibration response. The results of these initial tests are summarized in the following list:

1. The low frequency vibration was present at discharge pressures above 1,500 Psi, and did not exist at lower discharge pressures.
2. The low frequency vibration initially appeared at 660 CPM. As discharge pressure and temperature increased, it shifted to a frequency of 550 CPM.
3. This frequency was measurable on the structure and the associated piping.
4. The support structure reduced this motion and the foundation was quiet.
5. During startup and shutdown, the rotor system was well-behaved, and passed through its first critical in a normal and repeatable fashion.
6. There were no reciprocating pumps or compressors in the immediate area that could influence this centrifugal compressor.

This initial test program confirmed the presence of this low frequency vibration, and tied down the fact that it was strongly related to discharge pressure. These tests did not provide any substantial clues as to the origin of the low frequency vibration. One suspicion was that there might be an excitation originating from the fourth stage aftercooler. During one of many shutdowns, the exchanger head was pulled, and a bundle of corrosion coupons approximately twelve inches long were discovered. Well, that was the proverbial armadillo in the piping system. The coupons were removed and the machine restarted. At 1,500 Psi the low frequency component reappeared, and no further mention was made to the piping armadillo. Throughout the initial tests, the compressor fourth stage discharge temperature was considerably lower than the process design predicted. It was clear that the final after cooler was not necessary. Eventually, this heat exchanger was completely removed and replaced with a spool piece. Again, there was no influence on the low frequency motion.

Following the death of the armadillo theory, one of the process engineers suggested filtering the shaft vibration signals to eliminate the low frequency vibration component. This approach began to gain popular approval and many individuals considered this electronic filter approach to be fully acceptable. The logic behind this approach revolved around the concept that the majority of the vibration occurred at low frequencies, and *everyone knows that low frequencies are less destructive than high frequencies.* Not only is the application of this concept wrong, the argument is further compromised by the relative amplitudes. Specifically, the total shaft motion was very close to the overall clearance of the bearing assembly. Hence, any minor excitation could easily drive the shaft into metal to metal contact with the bearing pads, with significant damage to the entire machine. Fortunately, the plant manager was a reasonable and experienced engineer who recognized that the problem must be solved by eliminating the excitation, and not by masking the critical measurements.

After several other blind alleys, a test plan was established to make a series of pressure pulsation measurements throughout the high stage compressor piping system. From external vibration measurements on the outer bore of the piping it was known that the low frequency component was present. However, it was not totally defined whether the excitation was transmitted through the fluid stream or through the structure. To answer this question pressure pulsation measurements were employed. Fig. 14-8 shows the approximate location

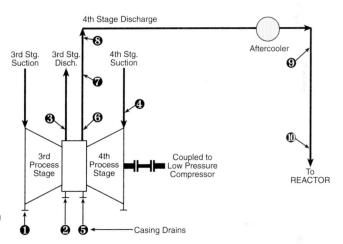

Fig. 14–8 Physical Location Of Pressure Pulsation Measurement Points

of test points ❶ through ❿ corresponding to the casing drain at the third stage suction through the fourth stage discharge. During this test, the compressor was shaking at a low frequency of 550 CPM. The third stage measurements did not display this low frequency component, and neither did the suction of the fourth stage. However, the fourth stage discharge casing drain ❺, and the fourth stage discharge piping ❻ through ❿ revealed a clearly defined pressure pulsation at the elusive frequency of 550 CPM. Plotting dynamic pressure pulsation amplitudes at this frequency versus distance from the fourth stage discharge flange produced the summarized data shown in Fig. 14-9.

From this test result, it was concluded that the exciting force was internal

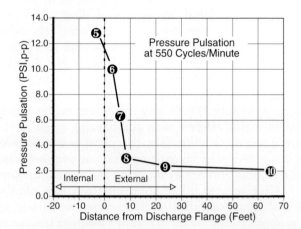

Fig. 14–9 Dynamic Pressure Pulsation Versus Discharge Pipe Length

to the compressor, and probably located between the fourth stage suction and discharge flange. The next step was to visit the OEM that designed and manufactured the compressor. The ensuing meeting was very frank and open. After many hours, it was agreed that the rotating assembly could not generate the type of frequency characteristics measured. Since everything appeared normal on the rotor, the next step was to begin examining the entire drawing set for anything that was different from traditional design. Again this appeared to be getting nowhere, when someone mentioned the balance port at both discharge nozzles.

The balance port configuration consisted of a rectangular passage that connected one side of the discharge nozzle with the opposite side of the nozzle, as shown in Fig. 14-10. This passage was an integral part of the barrel and was included as a means to pressure balance each discharge nozzle. Both the third and the fourth stage nozzles were similarly fabricated. From this meeting, it was decided to plug both ends of each balance port, hand grind the surface in contour with the nozzle, and drill a weep hole in one side. Following return to the plant

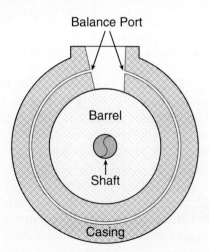

Fig. 14–10 Cross Section Through High Stage Compressor Discharge Nozzle

site, it was confirmed that a balance port did exist in each discharge. By running a piece of wire through the hole on one side, it was verified that the balance port was a circumferential passage that connected one side of the discharge nozzle with the opposite side of the same nozzle. Following the agreed upon action plan, the plugs were installed, and the machinery train restarted.

During the ensuing startup, the compressor came up smoothly through the first critical, and reached normal operating speed without any difficulty. As load was gradually increased, the compressor reached 1,500, and then 1,600 Psi, without the low frequency excitation. Finally, the machine was fully loaded, and the previously encountered low frequency excitation at 550 or 660 CPM did not appear. The shaft and casing vibration, plus the pressure pulsation measurements were void of the previous low frequency component. Specifically, Fig. 14-11 is representative of the final condition of this machine at a speed of 9,340 RPM, operating at a discharge pressure of 2,150 Psi.

The running speed motion had decreased to 0.13 Mils,$_{p-p}$ vertically, and 0.09 Mils,$_{p-p}$ horizontally. This was due to refinement of rotor balance during the course of this project. There are two minor components at 900 and 2,580 CPM that appeared at amplitudes of 0.04, and 0.06 Mils,$_{p-p}$ respectively. These components were considered to be nondestructive, and were later found to be associated with the stiffness of the O-rings installed with the squeeze film dampers (based on OEM comments and recommendations). However, the previously dominant low frequency excitation at 550 or 660 CPM was no longer visible at any operating condition. Although the exact origin of the low frequency was not totally defined, plugging of the balance ports certainly eliminated the excitation.

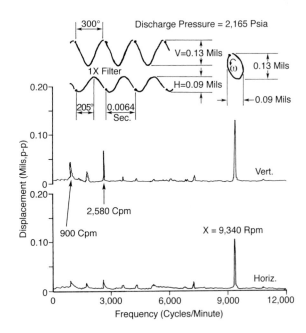

Fig. 14–11 Final Coupling End Shaft Vibration

As part of the overall testing program, X-Y shaft proximity probes were installed on the floating squeeze film bearing housing to observe shaft motion relative to the squeeze film bearing housing. In addition, standard X-Y shaft probes were mounted on the case and provided a measurement of shaft motion with respect to the casing. The 1X running speed vectors from both sets of probes are presented on Fig. 14-12. This vector plot displays relative shaft motion with respect to the case and with respect to the bearing. Clearly, the vector difference between the two relative measurements must be the motion of the squeeze film bearing housing with respect to the case. Since relative motion existed, it was concluded that the squeeze film housing was moving and providing some measure of additional damping.

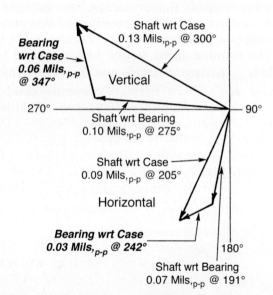

Fig. 14–12 Relative
Motion Of Coupling End
Bearing Housing

In retrospect, this problem was temporarily considered to be in the realm of a semi-magical solution. That is, the low frequency vibration and pressure pulsation in the fourth stage discharge were eliminated. However, it was difficult to verify the nature of the forcing function. This was primarily due to the confusion that existed as to the physical properties of the process gas at high pressures.

Sometime after the conclusion of this project, it was determined that the balance port passage acted as an acoustic resonator. Based upon the passage length, and the velocity of sound at specific operating conditions, the fundamental acoustic resonant frequency was one half of the measured low frequency excitation. It was concluded that the occurring frequency was in all probability the first overtone of the fundamental acoustic resonant frequency. In addition, the shift from 660 to 550 CPM was attributed to the fact that the 660 CPM was produced in the third stage discharge, and the 550 CPM excitation was generated at the fourth stage nozzle. Hence, during loading, the acoustic resonance shifted nozzles as a function of process fluid properties.

Again, the role of the seven-step problem solving approach introduced at the beginning of this chapter should be examined. In this case, the diagnostic objectives were primarily directed at reducing and/or stopping the low frequency compressor vibration. Secondary objectives included an understanding of the excitation mechanism, plus the client acceptance of this new machinery train. Mechanical inspection provided no useful data during the initial phases of investigation. However, once the design drawings revealed the potential presence of the circumferential balance port at the discharge nozzle, field inspection verified the existence of this cavity. Hence, on-site mechanical inspection played a role in the identification of this problem.

The test plan development for investigation of this low frequency motion included a full array of speed and load tests combined with measurement of shaft vibration response characteristics. This detailed analysis identified the direct relationship between the appearance of the low frequency vibration component, and specific operating conditions of pressure and temperature. The test plan was then expanded to include structural and piping vibration measurements, plus pressure pulsation measurements throughout the system.

Field data acquisition was conducted with a traditional set of diagnostic and recording instrumentation. The internal shaft sensing proximity probes mounted on the compressors were recorded simultaneously with the casing, structural, and piping vibration transducers. These external measurements were made with velocity coils attached with high strength magnets. The pressure pulsation measurements acquired around the high pressure compressor and the process piping were made with piezoelectric transducers. These pressure probes were attached to existing drain or vent valves. There was minimal opportunity to obtain additional test points due to the piping metallurgy and the reluctance to add additional pipe connections.

At high stage discharge pressures below 1,500 Psi, the mechanical system was void of the low frequency vibration component. Hence, extended comparisons between data sets was not required, and the data processing efforts were concentrated on a detailed documentation of the dynamic measurements during a condition when the low frequency perturbation was active. Since the subsynchronous motion was present only at full loads, the database was limited to steady state formats of orbits, time base, and spectrum plots. This data was generally summarized on traditional plots such as Fig. 14-7.

Data interpretation revealed a clear association between the low frequency shaft vibration, and the low frequency pressure pulsation. This excitation appeared to originate within the fourth stage of the high stage compressor. Meetings with the OEM revealed nothing unusual within the compressor bundle, and the discharge nozzle balance port was suspected. The project conclusions centered around the potential existence of an acoustic resonator, and the recommendation was to plug the balance port, and rerun the previous field tests. This conclusion proved to be correct, and the corrective action plan consisted of simply restating that the circumferential discharge nozzle balance port should not be exposed to the process fluid. Although the full technical explanation for this behavior lagged the field correction by several months, the plan was field proven to be the proper solution.

Case History 52: High Speed Pinion Instability

The final case history in this book deals with a four poster high speed air compressor. This type of machine is also referred to as an integral gear, or a multiple casing compressor. In this particular unit, a central bull gear is driven by a directly coupled motor. The bull gear in turn drives four high speed pinions that are mounted at 90° increments around the bull gear. Each pinion operates in a separate volute casing, and intercooling plus liquid knockout is provided between each stage. The unit under consideration was a new design based upon good experience with similar, but smaller machines. Whereas the previous case history dealt with unknown fluid properties, this current problem is associated with the capacity upgrade of a successful design. Although the dimensions were conservatively expanded, the upgraded unit experienced problems. For the purpose of this discussion, the behavior of the fourth stage pinion assembly will be considered. The rotor configuration for this assembly is shown in Fig. 14-13. As expected, the rotor operates in a pivotal manner as illustrated in the mode shape at the bottom of this diagram.

The motor driven bull gear for this machine runs at 1,200 RPM, and the fourth stage operates at 17,770 RPM. Following initial startup of the unit, a significant amount of subsynchronous motion was detected. The vibratory behavior under typical operating conditions is presented on Fig. 14-14. The two FFT plots represent averaged data on the bottom, and peak hold data on the top plot. Both sets of information were obtained during an identical time period of eight sec-

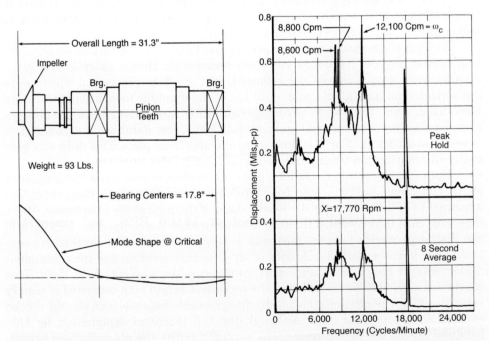

Fig. 14–13 Overhung Pinion Configuration   **Fig. 14–14** Initial Steady State Vibration

onds. From this initial information, it is apparent that the subsynchronous vibration covers a broad band (≈6,000 to 15,000 CPM), with identifiable components at 8,600 and 8,800 CPM, plus another discrete frequency at 12,100 CPM.

In passing, it should be mentioned that averaged data can successfully hide short duration peaks of any frequency. Hence, this type of information should always be compared against peak response in the time and the frequency domains. This step will verify that a realistic sample has been averaged, and that the proper amplitudes are presented in the frequency domain plot.

Further investigation revealed that the subsynchronous vibration varied as a function of compressor discharge pressure. As shown in Fig. 14-15, the running speed (1X) vibration remained fairly constant with increasing pressure. However, the subsynchronous bands increased as the unit was loaded up to normal operating conditions.

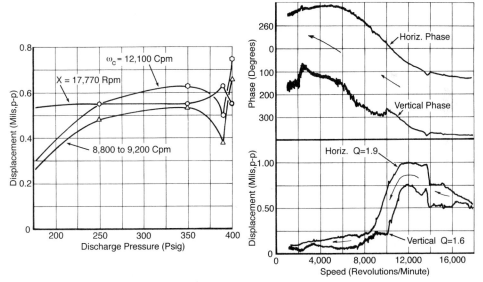

Fig. 14–15 Initial Shaft Vibration Response Versus Discharge Pressure

Fig. 14–16 Initial Bode Plot Of Coastdown Behavior Before Modification

The variable speed coastdown of the initial fourth stage pinion configuration is presented in Fig. 14-16. This Bode plot displays normal response characteristics, with a pivotal critical speed at 12,000 RPM, and reasonable amplification factors. This resonant speed coincides with the measured steady state frequency at 12,100 CPM observed during full speed operation. It was concluded that the major subsynchronous component at 12,000 CPM was a re-excitation of the pinion balance resonance. The OEM then performed a re-audit on the rotor design and developed a series of modifications that included stiffening the overhung assembly to raise the pinion natural frequency. Physically this was accomplished by selectively increasing the outer diameters of the overhung portion of the pinion. Following fabrication and installation of the new fourth stage pinion, the compressor exhibited the full load vibration data in Fig. 14-17.

The shaft vibratory characteristics following modification were considerably improved, with peak amplitudes of 0.3 Mils$_{p-p}$ at 10,600 CPM, and 0.25 Mils$_{p-p}$ at 13,900 RPM. Rotational speed motion had increased somewhat from 0.56 to 0.67 Mils$_{p-p}$. In addition, rotor sensitivity to subsynchronous motion as a function of discharge pressure had decreased, as exhibited on the Fig. 14-18.

The pinion coastdown following modification is presented on Fig. 14-19. This Bode plot displays a measured pivotal resonance at 13,600 RPM, which is 13% higher than the initial design. The amplification factor has apparently increased from Fig. 14-14. The OEM anticipated a lower amplification factor due to a computed increase in damping for the modified fourth stage pinion. Superficially, the transient vibration response data indicated a slight increase in the amplification factor, which would be associated with decreased damping. However, the amplification factor calculation is somewhat compromised by the fact that a gear driven rotor (pinion) is involved, and minor changes in the herringbone gear contact may result in appreciable changes to the shaft motion. In addition, the vibration amplitudes are quite small, and any variation may appear as a large change in the final amplification factor. When all factors are considered and appropriately weighed, the final variable speed behavior of the modified fourth stage pinion is acceptable.

In order to complete the project, the fourth stage pinion was field balanced,

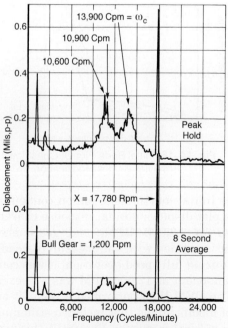

Fig. 14–17 Steady State Shaft Vibration After Pinion Modification

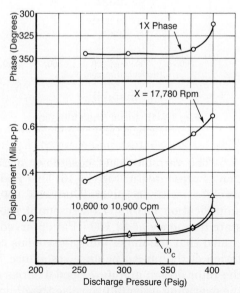

Fig. 14–18 Shaft Vibration Versus Discharge Pressure After Pinion Modification

and running speed amplitudes were reduced to slightly more than 0.2 Mils,$_{p-p}$. As expected, this balance improvement had minimal effect on the broad band subsynchronous motion, as shown in Fig. 14-20. Finally, comparison of Fig. 14-20 with the initial data shown in the first steady state plot of Fig. 14-14 reveals the substantial improvement obtained by the rotor redesign and trim balance. Although minor subsynchronous activity persists, the maximum amplitudes are quite small. At this stage, the machine was given a clean bill of health by the OEM and it was considered to be acceptable to the end user.

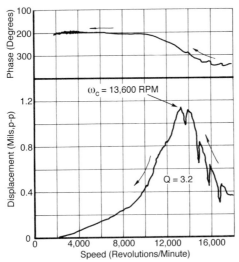

Fig. 14–19 Normal Bode Plot Of Coastdown After Pinion Modification

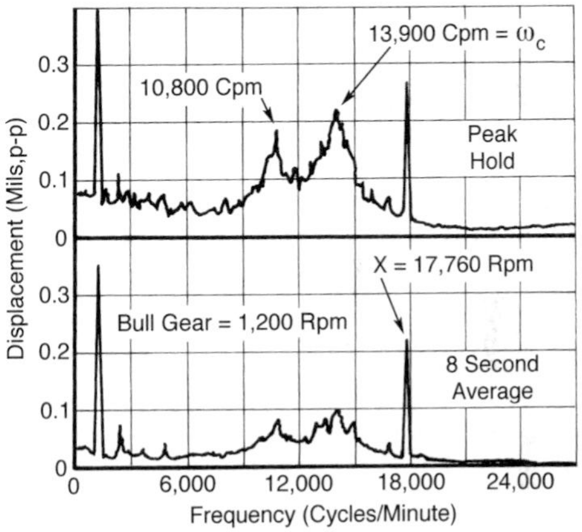

Fig. 14–20 Steady State Shaft Vibration After Pinion Modification And Trim Balance

During this project, the OEM was responsible for correcting the abnormal vibratory behavior. The basic diagnostic objectives were directed at reducing the subsynchronous instability and providing a machine with acceptable vibration amplitudes. Mechanical inspection was used to verify that all compressor elements were in agreement with the design drawings and associated tolerances. The developed test plan consisted of shaft vibration data acquisition during steady state conditions at variable loads, plus transient startup and coastdown information. Data processing consisted of standard formats for the measured vibration response data. This hard copy data was interpreted as a lower than desired pivotal resonance for the fourth stage pinion, plus excitation of this critical speed under full load conditions. Based on this information, the OEM performed an audit of the design rotor response calculations and suggested the fabrication of a stiffer overhung pinion assembly. This conclusion and associated recommendations were accepted by the end user

This modified pinion assembly was tested in a manner similar to the original fourth stage rotor design. The superiority of the modified design was demonstrated under all operating conditions. Hence, the final corrective action plan was simply to update drawings and part numbers to reflect the improved high speed pinion design. In this case, a responsible, and technically competent OEM provided a clean transition from a rough running initial design into a smooth operating modified design.

CONCLUSIONS ON DIAGNOSTIC METHODOLOGY

The seven steps associated with the identification and diagnosis of machinery malfunctions presented at the beginning of this paper are common to many types of problems. The actual sequence may not be the same for each problem, but eventually all the bases must be covered. The presented methodology will produce actionable results. As previously stated, there is little magic in the business of machinery diagnosis; it is primarily a function of an orderly plan, careful execution, and a lot of hard work by everyone along the way.

The final point is that complex machinery problems do not allow easily discovered solutions. In order to solve many of these problems a team effort is required. It takes a lot of work and input from a lot of folks. This includes operations and maintenance personnel, OEM designers and field representatives, instrumentation, controls, and measurement specialists, plus rotor analytical specialists. The associated disciplines of metallurgical, structural, and fluid experts may also be required. Finally, don't forget to talk to the guys that are turning valves and twisting nuts. They can provide a practical insight that is often lost as machinery problems migrate upwards through an organization.

BIBLIOGRAPHY

1. Eisenmann, Robert C., "Machinery Diagnostic Methodology." *Vibration Institute - Mini Course Notes - Machinery Vibration Monitoring and Analysis Meeting*, New Orleans, Louisiana (May 1985).
2. Gruber, J.M., and E.F. Hansen, "Electrostatic Shaft Voltage on Steam Turbine Rotors," *Transactions of the American Society of Mechanical Engineers*, Paper Number 58-SA-5, 1958.

CHAPTER **15**

Closing Thoughts and Comments

*I*n the early 1960s, the tools available to the machinery diagnostician were limited to a handful of vibration transducers, some analog instruments with filtration capabilities, and a minimal number of manual computations that could be successfully executed. The number of mechanical failures detectable by vibration analysis could be listed on one piece of paper, and the fundamental data acquisition device was the clipboard. A lot of credibility was placed in touching and feeling the machinery, listening to the various pitches and noises emitted by the machines, making accurate dimensional measurements of the stationary and rotating parts, and observing the oil flow through the sight glasses. On the other hand, the machinery was less complicated, operating speeds were slower, rotors were often rigid, machinery casings and supports were often massive, operating pressures and temperatures were lower, and routine maintenance was performed on the equipment whether it needed it or not. In most instances, a complete analysis of a machinery train could be easily completed within an eight hour time period. That included the physical inspection, vibration measurements, process correlations, review of the last overhaul report, plus plenty of time for lunch and three or four coffee breaks.

As demonstrated throughout this text, the technological advances achieved since that time have been somewhat incredible. Transducer technology has evolved from simple casing vibration probes with a finite life to highly sophisticated transducers systems with a virtually infinite life span. The analytical instruments that used to be dependent on vacuum tubes to perform a few basic tasks have given way to dynamic signal analyzers with internal computers. These devices perform an extraordinary range of functions and data manipulations. From a computational standpoint, the evolution and rapid development of the laptop computer has provided the field capability for complex vector calculations, transfer matrix calculations, finite element analysis, and computational fluid dynamics. The emergence of additional technologies, such as quantifiable structural measurements, thermography, oil analysis, laser alignment, and precision machining and balancing, have exponentially increased the available information on process machinery. Furthermore, digital control systems in the operating plants can provide volumes of data on the operational characteristics of the machinery and associated processes.

Certainly the complexity and sophistication of the machinery have grown, and a thorough analysis of a machinery train may not be completed in a single day. In reality, it may require several days or several weeks to provide a full assessment of all aspects of a machinery problem or failure. In some situations, an extended study is appropriate, and in many other cases, this approach is totally unacceptable. Whereas machinery problems have historically been approached from *best effort* basis, the current, and in all probability the future trend will be towards *economic justification*.

ECONOMIC REALITY

One of the key elements in the problem solving scenario is the ability to judge the true *net worth* of a problem. This should be matched by enough effort to solve the problem in a respectable amount of time, and for a reasonable amount of money. Clearly, the situation of a diagnostician devoting three hours presumably analyzing an increased vibration level on a critical turbomachinery train — and then casually increasing the vibration alarm setpoints, is totally inappropriate. Conversely, the diagnostician that spends a week analyzing an unusual frequency component on a fractional horsepower motor is wasting resources with no potential for payback.

In the harsh corporate realities of fiscal daylight, there must be an equilibrium maintained between the level and complexity of the technology, the time required and the cost incurred for performing the analysis, plus the cost for the necessary repairs or modifications. As suggested in Fig. 15-1, these three fundamental aspects form three different sides of an equilateral triangle.

Fig. 15–1 Problem Solving Equilibrium

The machinery diagnostician must be fully aware that the most elegant engineering solution provided in an inappropriately long time period is just about as useless as the casual *off the cuff* answer. Furthermore, as suggested in Fig. 15-1, the diagnostician must have the *knowledge* to properly access the overall situation, and the *experience* to establish a logical solution path. In addition to competency in a variety of technical disciplines, the machinery diagnostician must embrace the financial aspects of the work, plus have the wherewithal to fully appreciate the implications of the machinery problem upon the operating facility.

CORPORATE CONSIDERATIONS

During the construction and initial commissioning of a new plant, a variety of corporate entities are compelled to work together. In virtually all cases, these corporations function with two fundamental objectives. The publicly stated primary goal or objective would be to build the plant. Privately, the top objective is clearly the maximization of their individual corporate profits. Everyone generally understands this relationship, and manages to coexist in order to get the job done. If everyone goes home with a suitable paycheck, and maybe a little bonus or overtime pay every now and then, the project goes along fairly smoothly.

However, this delicate balance has a strong tendency to become unraveled and highly competitive when major technical problems occur in the new plant. When this occurs, schedules are disrupted, plans must be changed, and some profit margins may be threatened. When serious problems occur on critical machinery trains, the problem is compounded due to the number of people involved. For instance, if a major turbine-compressor train experiences a significant vibration problem during initial commissioning, the following list of individuals become immediately involved:

○ Field representative of the OEM that built the turbine.
○ Field representative of the OEM that built the compressor.
○ Various technical and operating representatives of the Architect and Engineering (A&E) company that designed the plant.
○ General contractor performing the installation work.
○ Mechanical contractor performing the actual millwright work.
○ Instrumentation contractor that installed the vibration transducers, field wiring, control room wiring, and monitoring instrumentation.
○ Computer and systems control contractor responsible for the physical interface with the Distributed Control System (DCS).
○ Representative of the company that provided the vibration instrumentation.
○ Operating personnel from the End User organization.
○ Maintenance personnel from the End User organization.
○ Vibration and/or machinery specialists from the End User organization.
○ Lab personnel responsible for water treatment analysis and steam quality.
○ Supplier of lubricant used in the main oil console.

With this initial array of interested individuals, it does not take very long to fill up the main conference room. Furthermore, if the problem cannot be immediately solved, the next hierarchy of engineers, technicians, managers, and purchasing agents are summoned to participate in this frenzy commonly referred to as *problem solving*. If brute force and the availability of large quantities of manpower were all that were required to solve machinery problems — then virtually all problems would be beat into submission at this stage. Unfortunately, some malfunctions do elude the masses, and additional personnel are invited to participate. This third wave of troops includes a variety of consultants, head office personnel from the major technical organizations, plus the initial appearance of lawyers with cameras and deposition forms. At this point, daily life becomes complicated, and some of the original participants begin to recognize that their major goal of financial profit is in serious jeopardy.

Although people try to appear cooperative, corporate *finger pointing* eventually evolves. In many instances, the absence of blame and liability is more important than a solution to the machine problem. Although a proper technical solution is highly desirable, the issue of *who pays for what* always creeps into the discussions. In fact, many good engineering solutions have succumbed to the overzealous input of purchasing managers and contract administrators. In these difficult situations, the diagnostician must really strive to address the technical issues and stay away from the politics. Although this very difficult to achieve, it is the most direct path towards solving the machinery problem.

The diagnostician must exercise diplomacy when dealing with other individuals and disciplines working on the same problem. On multifaceted malfunctions, it is often necessary to bring in various technical experts that have a specific field of expertise. These individuals are often interested in other test results to substantiate their work, or they may want to collaborate on common conclusions or test results. These types of activities are usually very beneficial, and they should be encouraged. In addition to the high-tech measurements, extensive calculations, and beautiful color displays, the diagnostician should not forget the basics. There are still things to be learned by touching and feeling the machinery, listening to the multiple pitches and noises emitted by the machines, plus making accurate dimensional measurements of the stationary and rotating parts. Furthermore, many baffling problems have been solved by checking the oil flow through the sight glasses, and other fundamental human observations.

It should always be recognized that people have good ideas, irrespective of the company that issues their paycheck. It is always beneficial to address these serious machinery problems with an open mind, and maintain a positive attitude throughout the duration of the problem. Sometimes it is necessary to *start with a clean sheet of paper*, and try to reconstruct the true list of causes and effects. In other cases, you need to surge the compressor once or twice to run all the non-contributors off of the compressor deck, and provide access to the machinery for those individuals that are really doing something significant. If you can scare off the people who are just there to *walk fast and look worried*, it makes life much more bearable for the actual technical investigators

If the problem is not solved by the resident team of experts, then the participants are destined to years of litigation. Occasionally, this situation does occur, but it is usually complicated by many technical issues. For instance, a prototype machine installed at several different operating facilities may have a generic design flaw. In some of the new installations, the people identify and correct the factory design problem. They then proceed to put the machine in service, and they enjoy some short-term economic advantages over their competitors who cannot correct the defect. For the companies that cannot solve the problem, they eventually head for the courts to obtain some type of financial settlement. In this case, the machinery manufacture can rightfully testify that some clients can make the machine run, and others cannot. Obviously, this type of situation drives the lawyers crazy, and leaves the engineers babbling to themselves.

However, with a little bit of luck, and a systematic analysis of the machinery problem, the difficulty is normally rectified. There are some enormously talented and intelligent people in this business, and they typically can identify and implement corrections for most machinery malfunctions. In the best possible scenario, the root cause of the problem is identified and corrected. The experts and consultants can then pack up and go home, and the local personnel can return to the business of building the plant. Although the new construction has been delayed, it now may be finished. Following plant completion, and the required commissioning activities, the facility is eventually turned over to the end user company. This simplifies dealing with any future machinery problems due to the fact that the number of involved individuals have been greatly reduced.

As a side note, the senior author of this text has long held the opinion that the original startup crew of operators and maintenance technicians are the most highly trained people in the intricacies of running and maintaining the new plant. They have to endure many startups and shutdowns during a short span of time, and these people either learn their jobs, or they don't survive. As the years pass, the startup crew are promoted, or transferred, or any number of other methods of attrition. Unfortunately, the replacement personnel can never be as fully qualified. Whereas the startup crew may experience two emergency crash downs, and three cold startups during one month, the replacement crew may never see a plant shutdown. This is particularly true for units that run five to ten years between major overhauls. Hence, as the operating facility matures into a continuously running process unit, the original startup difficulties often slip into vague and undocumented memories. The other danger is that some of the discipline and attention to detail that was instrumental in the original plant startup is either lost, ignored, or eliminated over the years.

In the evolution of many plants, the startup personnel are often rewarded with promotions into an ever increasing management structure. Some of this expansion is well overdue, and other facets of personnel expansions fall into the category of *bureaucratic spread*. Sometimes it's hard to tell the difference between the two categories. In general though, the ranks of the administrators tend to grow, and the number of productive workers tend to decrease. For discussion purposes, a typical organization chart for a mature chemical plant is presented in Fig. 15-2.

The facility is managed by a staff of department managers who report directly to the plant manager, or in some cases indirectly through an assistant plant manager. The normal departments of operations, maintenance, engineering, and administration are represented in Fig. 15-2. Some operating facilities also have research and development functions, plus specialized departments such as exploration and production, logistics and shipping, or environmental services. In this organization chart, the various department heads have a series of superintendents reporting to them. The superintendents have a group of supervisors that work directly for them. For functions that are staffed 24 hours a day, such as operations and certain maintenance jobs, there will be multiple supervi-

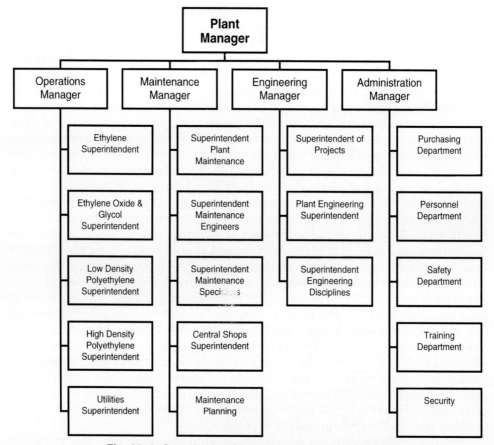

Fig. 15–2 Organization Chart In A Mature Process Plant

sors to cover all of the active and inactive shifts. At the bottom of this structure, the working troops report to a shift supervisor. Clearly this type of plant organization involves a lot of people, and a lot of resources.

In the early 1990s the trend in corporate management structures began to migrate towards *down-sizing* or *right-sizing*. People who lost their jobs due to corporate layoffs and early retirements tended to call the trend *down-sizing*. Those who retained their positions, or perhaps moved up a notch or two, generally referred to the changes as *right-sizing*. Whichever side of the politics an individual happens to be on, the fact remains that management structures are getting smaller. For instance, the organization chart shown in Fig. 15-2 can easily change into the streamlined or evolutionary version presented in Fig. 15-3.

Notably absent from the organization chart in Fig. 15-3 is the classical maintenance organization. This function has been subdivided and absorbed by the operations department and the engineering group. The number of functional superintendents have been reduced across the board, and some jobs have been

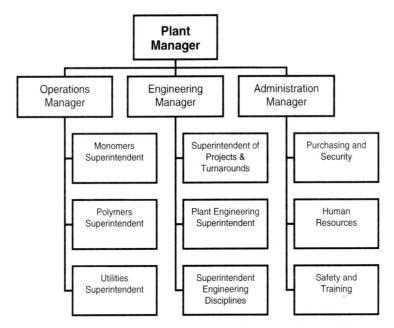

Fig. 15–3 Evolutionary Organization Chart In A Process Plant

combined under the administration manager. Although this evolutionary organization chart represents the same basic functions, the number of participants has been significantly reduced. In some operating plants even further reductions in personnel head count have occurred. There are operating facilities that only employ one or two technical people with applicable engineering degrees. And that starts to get to the real crux of the management problem. Specifically, it is unreasonable to expect one single individual to be responsible for all of the mechanical technology in a plant, plus optimization of the process, and the 101 tasks that must be accomplished on a daily basis.

In actuality, it takes a well-trained team of dedicated people to successfully run most industrial plants. A broad range of skills are required, and there is a definite need for knowledge and experience in a variety of technical disciplines. Hopefully, sanity and intelligence will prevail, and competent technical individuals will be retained in future organizations.

From the standpoint of the machinery diagnostician, contact and visibility within the formal plant organization will sometimes control the success or failure of the machinery analysis project. For instance, if a machine is operating with a significant subsynchronous vibration component, it may be desirable to change speed, load, or oil supply temperature to assist in identification of the specific mechanical malfunction. Any process change that reduces production rates is heavily frowned upon by the operations personnel. In some instances operations will refuse to change anything on the machine. This type of reply severely limits the ability of the diagnostician to identify the problem. Instead of being able to specifically conclude that: *the outboard compressor bearing is damaged, and is experiencing a 43% whirl due to excessive clearance* — the diagnostician is limited to say that: *the machine shakes at 43% of running speed, and there are a number of things that can cause that frequency.* This is the fundamental difference between solving a problem, and just providing reams of data that most people cannot properly understand nor interpret.

In this simple example, the machinery diagnostician is placed in a compromising position. If the only available option is to work with the local operators, the problem solving capability may be severely restricted. For this reason, it is necessary for the diagnostician to understand the plant organization, and solicit help from the available resources. In other words, don't be *stonewalled* by some jerk that is more interested in sitting in an air conditioned control room than getting out and working on the real problem.

PRESENTATION OF RESULTS

Regardless of the type of organization, or the number of independent companies involved in a machinery problem, there eventually comes a time for the diagnostician to render an opinion. Sometimes the conclusions of the analysis may be stated in an engineering report. In other cases involving difficult mechanical problems, a formal presentation is often requested in addition to the written report. The engineering report should begin with an *introduction* that identifies the specific machinery, the problem being addressed, and any general details or appropriate historical facts. The introduction should be immediately followed by the *summary and conclusions* that provide a management overview of the analysis project. Most folks, especially managers, have more paper cross their desks than they can stand. If you bury the project conclusions in the middle of a report, most people will never read the conclusions. However, a brief introduction, followed by a one or two page summary located at the front of the report, will stand the best chance of being read by the right people.

The engineering report should include a *discussion of results*. This section is a detailed review of the entire machinery analysis project, with appropriate discussion of the key points, specific references to the data, and explanation of the logic behind the conclusions. Due to the length and complexity of this report section it is often bypassed by the busy diagnostician. However, it should be realized that not preparing this section is a disservice to yourself. If the analysis needs to be repeated in the future, or if this report needs to be referenced at some later date, the discussion of results section provides the only coherent explanation of the work performed, and the logic path to the report conclusions.

The remainder of the engineering report should include topics such as a detailed *review of instrumentation* and *data presentation formats.* The machinery should be fully described, and any applicable sketches or drawings included. If *analytical calculations* are involved, the specific programs and software versions should be identified along with the summarized program outputs. Finally, the significant or pertinent *process and vibration data* should be enclosed. Sometimes this information may be summarized into various tables or other meaningful graphical formats. In all cases, the acquired and processed data must support the final project conclusions. People want to deal with engineering facts obtained from credible and repeatable sources.

Idle speculation has no place in an engineering report. If you cannot substantiate a statement, then either do not include the statement, or clearly identify it as speculation or conjecture. There are occasions when the data is inconclusive, and the diagnostician must reach an opinion based on reasoning and extrapolation. This type of conclusion is acceptable if it is clearly identified as speculation instead of a fully supported engineering fact.

Although the final report is typically quite detailed with numerous levels of rich technical information, the formal presentation of results normally will be less rigorous. The diagnostician must predetermine the technical makeup of the audience, and then prepare the formal presentation to match the audience. If the presentation is to a group of engineers, the contents might consist of fifty slides that require two hours to discuss. However, if the presentation is intended for upper management, the contents might include six slides to be used during a ten minute formal presentation. Often, the simplest format is the best, and direct statements are far superior to esoteric comments. Within any technical business, there are various jargons and acronyms that evolve over the years. While these technical languages may be necessary for the engineering discussions, they are totally inappropriate for the boardroom.

Many years ago, a considerate plant manager clearly stated that he needed to know two basic pieces of information. First, can the machine be safely run at full rated capacity plus an additional 5% or 10%. If not, then the second piece of information that he needs to know is how long will the current machine last, and what needs to be done to fix the machine. The ideal presentation for this plant manager is depicted in the final Fig. 15-4. Obviously, this is a facetious or tongue-in-cheek simplification of the real world. However, it does reinforce the concept that not all people are particularly enthralled with technology, and many individuals just want to know the basic *go* or *no-go* conclusion.

Machinery Analysis Summary

❏ Unit may be run up to 110% of normal process rates.

❏ Unit must be shutdown within the next _____ days for repair or replacement of the _____.

Instructions: Check Only One Box, and Fill In the Blanks.

Fig. 15–4 Idealized Final Data Presentation Format

SILVER BULLETS

This book was written for the machinery diagnostician who works diligently at his or her chosen profession. It is presumed that these individuals will learn about the mechanics and physical behavior of the machinery. They will learn the operation of the instrumentation, plus the use of measurement and analytical tools. If they keep working at this business they will gain both knowledge and experience. They will soon learn that there are no tricks in this business, and the *answers are in the details*. More specifically, *do accurate work and get excellent results, or do sloppy work and be mediocre or fail*. This formula for success has been repeated many times, and it is anticipated that it will be duplicated by dedicated individuals in the years to come.

The machinery diagnostician will also appreciate the fact that there are multiple ways to accomplish necessary tasks. Different balancing and alignment techniques coexist, a variety of calculations may be employed, a myriad of physical tests may be conducted, and many dynamic transducers and instruments are currently available. These working tools may be integrated or interleaved in numerous combinations to solve machinery problems. In many instances, the challenge to the diagnostician is to solve the problem with the available tools. It must also be recognized that there is no single unique solution for many problems. In fact, there may be several acceptable technical answers. In other cases, a variety of malfunctions will commingle to confuse the situation, and test the logic capabilities of the diagnostician. Actually, this is the fun part of the business. There is a great personal satisfaction in solving complicated mechanical problems. There is even more satisfaction to be gained from solving problems with the available tools, and providing management with more than one cost-effective solution path.

As proficiency increases, the machinery diagnostician will gain credibility with management, and his or her opinions will become more respected with each passing success. In some ways this is analogous to acquiring a *silver bullet* after each successful endeavor. However, with success and professional credibility, there comes a strong measure of responsibility. In all cases, the experienced machinery diagnostician carries an obligation to promote safe practices, and to foster growth and development of better techniques and methods to *make the machinery run better*. Furthermore, there are times when various situations may end up endangering the people working around the machines. In these instances, it is sincerely hoped that the machinery diagnostician will have the courage to relinquish some *silver bullets* to make management aware of the errors, and redirect things back into a safe and sane direction.

Machinery Diagnostic Glossary

A

absolute motion: Vibratory motion with respect to an inertial reference frame such as free space.

acceleration: The time rate of change of velocity. In the time domain, acceleration leads velocity by 90°, and displacement by 180°. Typical engineering units are $G's_{o-p}$, or $Inches/Second^2$.

accelerometer: A vibration transducer designed to measure acceleration. Generally, a piezoelectric crystal is used to generate a charge proportional to acceleration. The charge sensitive signal is converted to a voltage sensitive signal with an internal or external Charge Amplifier.

ADRE® for Windows: An acronym for Automated Diagnostics for Rotating Equipment that consists of hardware and software that is typically used for acquisition and processing of vibration data. ADRE® is a registered trademark of Bently Nevada Corporation of Minden, Nevada.

aim: In alignment terminology, to regulate or control the direction of a sighting device.

alidade: All of the upper part of an optical instrument that turns in azimuth with the sighting device (this is usually the telescope).

alignment: The relative position between machine elements. This is applied to the position or location of stationary elements within the machinery casing. It is also applied to the position of rotating elements (e.g., couplings) between machinery cases. In an ideal case, the shaft centerlines between two machines should be collinear when the machinery is fully loaded and heat soaked.

American Petroleum Institute (API): An organization for the development and distribution of technical standards, procedures, and certification programs for the petroleum industries. The main offices are located in Washington, D.C.

American Society of Mechanical Engineers (ASME): A technical organization devoted to promoting the arts and sciences in the field of mechanical engineering. As part of this charter, an extensive series of technical standards, procedures, and test codes have been developed for industry. The main offices are located in New York, NY.

amplification, signal: A uniform increase in amplitude of an electronic signal without variation of the wave form frequency or timing content. Normally performed with a voltage amplifier.

amplification factor: A measure of the susceptibility of a rotor to vibrate as it passes through a critical speed region. A high value indicates low damping, and a small value is a well-damped system. Also equal to π divided by the log decrement in analytical computations.

amplitude: The magnitude measurement of periodic dynamic motion. Displacement is measured in *Peak to Peak* values. Velocity and acceleration are presented as *Zero to Peak* amplitudes.

aperiodic motion: The motion of a body within a critically or highly damped system where all periodic motion is suppressed. During this type of non-vibratory motion, the body (following an initial disturbance) tends to creep back to its equilibrium position without oscillation.

Apple Computer, Inc.®: A manufacturer of the Macintosh personal computer, and associated hardware and operating systems, with main offices located in Cupertino, California.

asymmetrical support: Support systems that do not provide uniform restraint (stiffness) in all lateral directions.

asynchronous: For rotating machinery applications, frequency components which are not integer multiples, or integer fractions of shaft rotative speed — and remain at a fixed or constant frequency irrespective of speed changes.

attenuation, signal: A uniform reduction in the amplitude of a signal without variation of the wave form frequency content or timing. Normally performed with a voltage attenuator.

attitude angle: The included angle between a bearing centerline and a line connecting the geometric center of the bearing with the center of the shaft. This angle usually follows the direction of shaft rotation, and varies as a function of external plus internal preloads, and the particular bearing configuration.

averaging: Digital averaging of successive FFT samples to reduce random noise, and enhance fundamental spectral components and associated harmonics.

axial: Parallel to the rotor centerline. This applies to vibration and position measurements of the rotating shaft and the machine casing.

axial position: The average position, or the change in position, of a rotor in the axial direction with respect to an established reference. Normally measured with respect to the thrust bearing.

azimuth: The direction in a horizontal plane, parallel to the surface of the earth.

azimuth axis: The vertical axis. The axis of the bearing and spindle of an optical instrument which confines rotation to a horizontal plane.

azimuth motion: The clamp and tangent screw of an optical instrument which controls rotation in the horizontal plane.

B

balance resonance speed: A rotative speed corresponding to a natural resonant frequency of a rotor. The midpoint of the amplitude peak and phase shift that occurs as the rotor passes through a resonance. Balance resonances are usually referred to as rotor critical speeds.

balancing: A systematic procedure for adjusting the radial mass distribution of a rotor so that the mass centerline approaches the geometric centerline; thus, reducing the vibration and the lateral forces applied to the bearings and surrounding structure.

band-pass filter: An electronic filter that has a single transmission band extending from a finite lower to an upper cutoff frequency. Frequencies within the pass band are retained, and frequencies outside of the pass band are eliminated or attenuated. The cutoff frequencies are those points on either side of the center frequency where the amplitude is attenuated by 3 dB (or 0.707 of the value at the center frequency).

band-reject filter: An electronic filter that has a single rejection band extending from a finite lower to an upper cutoff frequency. Frequencies within the rejection band are eliminated or attenuated, and frequencies outside the rejection band are retained. This type of filter is also called a notch filter.

bandwidth: The span between frequencies at which a band-pass filter attenuates the signal by 3 dB to 0.707. Also, the frequency range over which a given electronic device is set to operate.

baseline: Vibration, position, and performance data acquired on a machine in good condition, and used as a reference for future trending and analysis of the machinery.

bearing clearance ratio (BCR): The relationship between the bearing diametrical clearance measured in Mils, and the shaft diameter measured in Inches. A value of 1.5 Mils/Inch is typical on many industrial machines. On units with nonsymmetrical (e.g., elliptical) bearings, a unique value will be associated with the vertical and the horizontal clearances separately.

bearing unit load (BUL): The static load placed on a journal bearing due to the weight of the shaft divided by the plane area of the bearing. Common engineering units are $Pounds/Inch^2$.

beats: The variation in amplitude due to the time domain superposition or summation of two periodic wave forms occurring at different frequencies. The differential between the two fundamental frequencies is known as the bear frequency.

bench marks: In machinery alignment work, the permanent fixtures used to identify reference points in the X, Y, and Z directions. Linear distance measurements are made with respect to these reference points to determine the three-dimensional location of a point in space.

Bently Nevada Corporation®: A manufacturer of vibration transducers, monitors, data acquisition and analysis systems, plus associated software and training activities. A worldwide organization with corporate headquarters located in Minden, Nevada.

bias voltage: The DC voltage at the output of a dynamic transducer on which the AC signal is superimposed. For an ICP® device this DC voltage has no mechanical significance. For a piezoresistive pressure probe, this voltage is proportional to the average static pressure. For a Proximitor® output from a displacement system transducer, this voltage is proportional to the distance between the probe tip and the observed surface. Variations in this voltage allow the computation of position changes between the proximity probe tip and the observed surface.

blade passing: For bladed machinery, a frequency equal to the number of blades times shaft rotative speed. This may be stationary or rotating blades, or geometric combinations thereof.

Bode plot: Cartesian plot of a vector (typically 1X) where speed is plotted on the abscissa, versus synchronous phase angle and vibration amplitude on the ordinate. This data is often combined with a polar plot of the same information. This plot may be constructed from measured vibration data, or calculated response information. Named for the frequency response plots of H.W. Bode.

bow, shaft: A shaft condition where the geometric shaft centerline is physically distorted by gravitational sag and/or thermal warpage.

broadband noise: The total noise at the output of an electronic circuit, or a transducer system.

buck-in: To place an optical instrument so that the line of sight satisfies two requirements (such as aiming at two targets) simultaneously. This is usually accomplished by trial and error.

C

Campbell diagram: A mathematically constructed plot used to predict the interference between system resonances and excitations. The abscissa displays speed, and the excitation frequencies (e.g., unbalance, misalignment, oil whirl, blade passing, etc.) are shown as multiples of this fundamental. The ordinate displays the lateral and torsional natural resonant frequencies.

cascade plot: A diagram used to observe frequency changes versus rotor speed. This plot consists of a series of spectra acquired at consecutive speeds. The abscissa displays frequency; and amplitude, incremented at various rotor speeds, is shown along the ordinate axis.

casing expansion: Measurement of axial growth of a machine casing relative to the foundation or the support structure. Common engineering units are *Mils* or *Inches*.

center frequency: For a band-pass filter, the frequency equal to the center of the transmission band. Common engineering units are *Cycles / Second* or *hertz*.

circular level: The round level attached to the alidade.

clipping: The distortion or truncation of a dynamic electronic signal due to the signal amplitude exceeding the limits of the amplifier or supply voltage.

coherence: The dimensionless ratio of coherent output power between two channels of an FFT. A high coherence (approaching 1.0) provides good confidence in the direct relationship of the data.

Compaq Computer Corporation: A manufacturer of personal computers and associated hardware and software. This equipment is distributed under a brand name of COMPAQ, with main offices located in Houston, Texas.

compliance: A frequency response function (FRF), also known as a transfer function measurement, where the output displacement response is divided by the input force. This measurement is the reciprocal of dynamic stiffness, with typical engineering units of *Inches / Pound*.

Computed Order Tracking® (COT®): A signal processing technique for identifying and tracking orders of a fundamental component. Computed Order Tracking® is a registered trademark of Hewlett Packard of Everett, Washington.

coupled modes: Vibratory modes which influence each other due to energy transfer between the respective modes. This could be lateral to lateral, or lateral to torsional interaction.

coupling, AC: A signal conditioning technique for eliminating low frequency and/or bias voltages from a transducer signal. This is often used on proximity probe signals to remove the DC gap voltage. It is achieved by inserting a coupling capacitor in series between the signal conditioner and the applied instrument. A coupling capacitor is installed only on the signal conductor, and not on the signal ground.

coupling, DC: A signal that has not been subjected to AC Coupling. A signal that contains AC signals riding upon a DC bias signal.

critical damping: The smallest amount of damping required to achieve aperiodic motion, and return the system to equilibrium in the shortest time without oscillation. The damping value which provides the most rapid response with minimal overshoot of the equilibrium position.

critical speed: A rotor balance resonance speed, or a speed which corresponds to a system resonance frequency. Also, an operating speed which is associated with high vibration amplitudes.

critical speed map: A machinery design diagram used to evaluate changes in resonant frequencies (plotted on the ordinate) versus variable support stiffness (on the abscissa).

CRITSPD: A computer program for determination of undamped response of flexible rotor-bearing systems using the complex matrix transfer method. Program coded in MS-DOS®, and distributed by Rodyn Vibration, Inc., located in Charlottesville, Virginia.

cross-coupled: A mechanical condition where a force applied in one plane will affect or influence a perpendicular plane, or another lateral plane.

cross talk: Electronic interference from one data channel superimposed upon another channel. This interference can occur between transducers, monitor channels, channels on a tape recorder or individual data channels on a digital data acquisition device.

Curvic®: A type of close tolerance coupling manufactured by The Gleason Works in Rochester, NY.

C_{xx}: Horizontal damping with common engineering units of *Pounds-Seconds/Inch*.

C_{xy} or C_{yx}: Cross-coupling damping between vertical and horizontal directions with common engineering units of *Pounds-Seconds/Inch*.

C_{yy}: Vertical damping with common engineering units of *Pounds-Seconds/Inch*.

D

damping: The energy converter in a vibrating mechanical system that restrains the amplitude of motion with each successive oscillation. As applied to shaft motion, damping is provided by the oil in bearings, seals, etc. This is the property that restrains motion through a resonance.

decibel: A ratio expressed as 20 times the log of the voltage ratio, or 10 times the log of the power ratio: $dB=20 \log (V/V_{ref})=10 \log (P/P_{ref})$.

degrees of freedom: Description of a mechanical system complexity. The number of independent variables describing the state of a vibrating mechanical system.

differentiation: An electronic circuit or calculation procedure that performs mathematical time differentiation. Converting displacement to velocity requires single differentiation, and displacement to acceleration would be considered as double differentiation.

differential expansion: Axial rotor position with respect to the casing, at the end of the machine opposite the thrust bearing. Common engineering units are *Mils* or *Inches*.

Digital Vector Filter (DVF): An electronic instrument that uses digital signal processing and band-pass tracking filters to extract vector data from complex dynamic signals. Devices such as the Bently Nevada DVF2 and DVF3 perform this function.

displacement: The change in distance or position of an object. Eddy current probes directly measure displacement. Single integration of a velocity signal, or double integration of an accelerometer are required to obtain displacement. Common units for displacement are *Mils*, or *Mils$_{p\text{-}p}$*.

dual probe: A transducer set consisting of a proximity probe plus a casing vibration transducer installed at the same location. The time summation of the shaft relative and the casing absolute signals allows the measurement of absolute shaft motion.

dynamic mass: A frequency response function (FRF), also known as a transfer function measurement, where input force is divided by the output acceleration response. This measurement is the reciprocal of inertance, with typical units of *Pounds / G's*.

dynamic motion: Vibratory motion due to forces that are active only when the rotor is turning at speeds above slow roll.

dynamic range: The ratio of the largest to the smallest signals that can be measured at the same time with typical engineering units of *dB*.

dynamic stiffness: A frequency response function (FRF), also known as a transfer function measurement where the input force is divided by output displacement. This measurement is the reciprocal of compliance, with typical units of *Pounds / Inch*.

Dynamic Signal Analyzer (DSA): An electronic instrument that uses digital signal processing and the Fast Fourier Transform to convert complex dynamic signals into frequency components, with associated parameters such as phase, coherence, and time records. Instruments such as the HP-3560A, HP-35665A, and the HP-35670A are DSAs.

DYROBES: A computer program for the statics and dynamics of rotor bearing systems to a variety of forcing functions using finite element analysis (FEA). Program coded in MS-DOS®, and distributed by Rodyn Vibration, Inc., located in Charlottesville, Virginia.

E

eccentricity: Variation of a shaft diameter when referenced to the true geometric shaft centerline. The measurement of shaft bow or runout at slow rotational speeds. When measured with a dial indicator, this is often referred to as total indicated runout (TIR).

eccentricity ratio: A dimensionless value obtained by dividing the change in radial shaft centerline position by the diametrical bearing clearance.

Eigenvalue: A root of the characteristic equation of a given matrix. As applied to rotor dynamics, a complex root that identifies natural frequencies, and the associated damping. The imaginary part of the root defines the calculated natural frequency.

Eigenvector: The calculated mode shape at each natural frequency (Eigenvalue). Multiplication of transfer matrices determine the mode shape by computation of displacement at each station.

electrical runout: A source of error on the output signal of a proximity probe system. Usually a function of the varying conductivity of the observed surface or localized shaft magnetism.

electro magnetic interference (EMI): A condition in which an electromagnetic field produces an unwanted signal or noise.

elevation: The direction of a line of sight in a vertical plane perpendicular to the earth.

elevation axis: The horizontal axis. The axis of the bearing and the journal of the optical telescope axle which confines rotation to a vertical plane.

elevation motion: The clamp and tangent screw of an optical instrument which controls rotation in elevation.

Endevco Corporation: A manufacturer of accelerometers, acoustic microphones, pressure and force transducers, plus associated hardware and calibration services. Corporate headquarters are located in San Juan Capistrano, California.

Excel, Microsoft®: A multi-platform spreadsheet program with numerical computation and data analysis capability by Microsoft Corporation located in Redmond, Washington.

F

focus: In alignment terminology, to move the optical parts so that a sharp image is observed.

Fourier Transform: A mathematical operation which converts a time-varying signal into a finite number of discrete frequency components with defined amplitudes.

Fast Fourier Transform (FFT): A mathematical operation which decomposes a time-varying signal into a finite number of discrete frequency components. The FFT approximates a true Fourier transform, and is performed by a digital computer-based instrument such as a DSA.

filter: For electronic signal processing, a circuit designed to pass or reject a specific frequency range. For machinery applications, the filters used for cleaning lubricants or other process fluids.

flat-top window: A time domain weighting function applied to the input signal of an FFT Analyzer. This window removes signals that are not periodic at both ends of the time record. It provides a flat filter shape to maximize amplitude accuracy (typically within ±0.005 dB).

forced vibration: System oscillation due to the action of a forcing function. Typically, forced vibration occurs at the frequency of the exciting force (e.g., unbalance occurs at running speed).

free vibration: Motion of a mechanical system following an initial perturbation. Depending on the system and the kind of perturbation, the system responds by free vibration at one or more of its natural frequencies.

frequency: The repetition rate of a periodic event within a specific unit of time. Typical units for frequency are *Cycles per Minute (CPM) or Cycles per Second (hertz)*.

frequency response: The variation of amplitude and phase characteristics of a mechanical or electronic system versus frequency.

frequency response function (FRF): A dynamic signal processing technique that provides a ratio between the FFT of the output divided by the FFT of the input signal. The results are displayed as an amplitude ratio, and a differential phase measurement. The validity of this measurement is verified by the coherence between signals. FRF is also called a transfer function.

G

G: The acceleration of gravity on the surface of the earth. It is equal to 32.174 *Feet per Second^2*, or 386.1 *Inches per Second^2*.

gear mesh: A normal gear box frequency equal to the number of gear teeth times each shaft frequency. For example, in a two element gear: Gear Mesh = Number Pinion Teeth x Pinion Speed = Number Bull Gear Teeth x Bull Gear Speed.

ground loop: System noise due to a circulating current between two or more electrical connections and a signal ground. This typically appears at frequencies such as 60, 120, or 180 hertz.

H

Hammer-3D: A computer program for determination of structural mode shapes based on impact hammer tests. Program runs exclusively on an HP-35670A in Hewlett Packard Instrument Basic. This was developed by Seattle Sound and Vibration, inc., located in Seattle, Washington.

Hann window: A time domain weighting function applied to the input signal of an FFT Analyzer. This window removes signals that are not periodic at both ends of the time record. It provides a good compromise between amplitude (+0, -1.5 dB), and frequency accuracy. Also known as a Hanning window on many instruments.

harmonic: A frequency which is an integer multiple of a specific fundamental frequency.

heavy spot: The angular location of the unbalance vector at a specific lateral location on a shaft.

hertz (Hz): A measurement of frequency, with engineering units of *Cycles per Second*.

Hewlett Packard®: A manufacturer of a wide array of electronic instrumentation, data acquisition and analysis systems, calibration equipment, plus associated software, service, and training activities. A worldwide organization with Test and Measurement headquarters in Palo Alto, CA.

high-pass filter: A filter with a single transmission band extending from a defined finite lower cutoff frequency (amplitude @ -3 dB) to the upper frequency limit of the device.

high spot: The angular location of the shaft directly under the vibration probe at the point in time when the shaft makes its closest approach to the probe.

horizontal: In alignment terminology, the direction perpendicular to the direction of gravity. As applied to the identification of vibration transducers, the right-hand or X-Axis transducer.

I

impact test: A mechanical test where an input force is provided by the impact from a force hammer or battering ram. The mechanical element vibration response is measured, and a frequency response function between force and vibration is computed with a dual channel DSA.

impedance: The mechanical properties that govern system response to periodic forces.

inertia, area: With respect to a given axis, the product of area and the distance from the axis squared. This is typically expressed in units of $Inches^4$.

inertia, mass: With respect to a given axis, the product of mass and the distance from the axis squared. For a rotating mechanical system, Inertia is often referred to as the WR^2 of the rotor, and is expressed in units of $Pound\text{-}Inch\text{-}Second^2$. Multiplication by the acceleration of gravity yields the common inertia units of $Pound\text{-}Inches^2$.

inertance: A frequency response function (FRF), also known as a transfer function measurement where output acceleration response is divided by the input force. This measurement is the reciprocal of dynamic mass, with typical units of $G's/Pound$.

in-phase component: The magnitude of the 1X vector that is in line with the transducer. This is expressed as: $In\text{-}Phase = A \cos{(\phi)}$; where A is the amplitude, and ϕ is the associated phase angle.

inertially referenced: Motion referred to free space, or a vibration transducer which measures such motion.

instability: A mechanical system that has a negative log decrement. Any perturbation to a unstable mechanical system will result in increasing system vibration as a function of time.

Integrated Circuit Piezoelectric® (ICP®): A dynamic transducer for acceleration, force, load, pressure pulsation, or shock measurements that incorporates signal conditioning electronics into the body of the field mounted transducer. This type of transducer produces a voltage sensitive output signal. ICP® is a registered trademark of PCB® Piezotronics, Inc. of Buffalo, NY.

integration: An electronic circuit or calculation procedure that performs mathematical time integration. Converting velocity to displacement is considered a single integration, and acceleration to displacement would be double integration.

isotropic support: Support systems that provide uniform stiffness in all radial directions.

J

J: Typical nomenclature for area or mass polar moment of inertia with common engineering units of $Inches^4$ or $Pound\text{-}Inch\text{-}Second^2$ respectively.

K

Keyphasor® (KeyØ®): A probe used for sensing a once-per-rev event. The resultant pulse signal that is used for measuring phase angle, rotational speed, and synchronous tracking. Keyphasor® is a registered trademark of Bently Nevada Corporation of Minden, Nevada.

K_{ax}: Axial stiffness with typical engineering units of $Pounds/Inch$.

Kingsbury: A manufacturer of oil film type journal and thrust bearings that are often self aligning and load equalizing. Kingsbury, Inc. is located in Philadelphia, PA.

Krohn-Hite Corporation: A manufacturer of electronic filters, signal sources, and other electronic instruments with corporate headquarters located in Avon, Massachusetts.

K_t: Torsional stiffness with typical engineering units of $Inch\text{-}Pounds/Radian$.

Kulite Semiconductor Products, Inc.: A manufacturer of accelerometers, strain gages, pressure and force transducers, plus associated hardware and calibration services. Corporate headquarters are located in Leonia, New Jersey.

K_{xx}: Horizontal lateral stiffness with typical engineering units of *Pounds / Inch*.

K_{xy} or K_{yx}: Cross-coupling lateral stiffness between vertical and horizontal directions with typical engineering units of *Pounds / Inch*.

K_{yy}: Vertical lateral stiffness with typical engineering units of *Pounds / Inch*.

L

lateral location: Various points identified along the axial length or centerline of a rotor.

linear system: A mechanical system in which the response of each element is proportional to the excitation.

Lissajous figure: The series of plane curves traced by an object executing two mutually perpendicular harmonic motions, forming a distinct pattern. For vibration measurements on rotating machinery, this is normally referred to as an orbit.

Loctite®: A series of commercially available products for locking threaded connections. Loctite® is a registered trademark of Loctite Corporation of Cleveland, Ohio.

log decrement: A measure of damping based upon the rate of decay of free oscillatory motion. A dimensionless quantity defined as the natural logarithm of the ratio of any two successive amplitude peaks in the decay cycle: *Log Dec=ln(X_2/X_1)*. Based upon an Eigenvalue, the calculated value: *Log Dec. = -2 π *(Real/Imaginary)*.

low-pass filter: A filter with a single transmission band extending from the lower frequency limit of the device to some finite upper cutoff frequency (amplitude @ -3 dB).

M

Machinery Diagnostics, Inc.: A machinery consulting and personnel training company with corporate and intergalactic headquarters located in Minden, Nevada.

magnetic center: The normal axial operating position for an electric machine where stator and rotor forces are generally balanced, and axial forces on the rotating element are minimal.

Mathematica®: A computer program for performing a wide range of calculations and associated graphical results on a variety of personal computer platforms. This program is by Wolfram Research, Inc., with head offices located in Champaign, Illinois.

Measurements Group, Inc.: A manufacturer of strain gages and accessories, signal conditioning, data acquisition systems, photoelastic materials, and training, with corporate headquarters located in Raleigh, North Carolina.

mechanical impedance: A frequency response function (FRF), also known as a transfer function measurement where input force is divided by the output velocity response. This is the reciprocal of mobility, with typical units of *Pounds / IPS*.

mechanical runout: A source of proximity probe signal error. This includes eccentric shafts, scratches, rust or other conductive metal buildup, plus variations in metallic properties.

Microsoft® Windows: A computer operating system that is a registered trademark of Microsoft Corporation located in Redmond, Washington.

mil: A measurement of length or distance equal to 0.001 *Inches*.

mobility: A frequency response function (FRF), also known as a transfer function measurement where the velocity response is divided by the input force. This is the reciprocal of mechanical impedance, with typical units of *IPS / Pound*.

mode of vibration: A pattern of a vibrating system in which the motion of every particle occurs at the same frequency. In a multiple degree of freedom system, two or more modes may exist concurrently.

mode shape: The deflected rotor shape at a specific speed to an applied forcing function. A two-dimensional or three-dimensional presentation of the shaft lateral deflection. Also applies to torsional shaft motion and the motion of all structures, including stationary support elements.

modulation, amplitude (**AM**): A signal where the amplitude of a carrier signal is varied by the amplitude of the modulating signal.

modulation, frequency (**FM**): A constant amplitude carrier signal, with a frequency varied by the modulating signal frequency.

modulus of elasticity (**E**): The ratio between a specified increment of tensile or compressive stress divided by a corresponding increment of tensile or compressive strain. Also known as elastic modulus, coefficient of elasticity, or Young's modulus, with typical units of $Pounds/Inch^2$.

modulus of rigidity (G_{shear}): The ratio between a specified increment of shearing stress divided by a corresponding increment of shearing strain. Also known as the shear modulus, torsion modulus, or modulus of elasticity in shear with typical engineering units of $Pounds/Inch^2$.

MS-DOS®: A computer operating system that is a registered trademark of Microsoft® Corporation located in Redmond, Washington.

N

NARF: An acronym for Natural Axial Resonant Frequency that is typically applied to axially compliant machine couplings such as disk or membrane couplings.

natural frequency: The free vibration frequency of a system. The frequency at which an undamped system with a single degree of freedom will oscillate upon momentary displacement from its rest position. The natural frequencies of a multiple degree of freedom system are the frequencies of the normal modes of vibration.

nodal point: A point of minimum shaft deflection at a specific mode shape. May change location along the shaft axis due to variations in balance, restraint, or forcing functions. Motion on either side of a node is out of phase by 180°.

noise: Any portion of a signal which does not represent the variable intended to be measured.

notch filter: A filter which eliminates or attenuates frequencies within the filter bandwidth; and retains all frequencies outside the rejection band. Also called a band-reject filter.

nulling: Vector subtraction at slow roll speed for rotational speed amplitude and phase electrical and/or mechanical runout correction.

O

objective lens: In alignment terminology, the optical lens at the front end of a telescope, and therefore the lens nearest the sighted object.

octave: The interval between two frequencies with a 2:1 ratio. Typically applied for noise analysis.

oil whirl: A vibratory state caused by insufficient bearing load. Various mechanisms, such as excessive bearing clearance, or a force counteracting bearing load can trigger this instability. The motion is generally circular, with a forward precession, and it occurs at a frequency equal to the average oil velocity within the bearing (typically 35 to 49% of shaft speed).

oil whip: A vibratory state in which a subsynchronous frequency, such as oil whirl, coincides with a rotor critical speed. The motion is often severe, it occurs with a forward precession, and the frequency remains generally constant, regardless of changes in operating shaft speed.

optical transducer: A transducer system that provides an optical output signal, and senses a reflected optical signal. This type of transducer is typically used as a temporary Keyphasor®.

orbit: The dynamic centerline motion of a rotating shaft. This is typically observed by X-Y proximity probes connected to an oscilloscope, or via a computer simulation of the analog signals.

orders: Multiple harmonics of a given fundamental frequency.

order tracking plot: A diagram used to observe individual frequency component amplitudes versus rotor speed or time. This plot is generated with Computed Order Tracking® on a DSA.

overlap processing: An FFT Signal handling technique that combines old data with new data samples. The selected frequency span and associated time record lengths are the main parameters for determining the percentage of overlap, and the resultant reduction in sample time.

P

PCB® Piezotronics, Inc.: A manufacturer of accelerometers, pressure and force transducers, plus associated hardware. Corporate headquarters are located in Buffalo, New York.

period: The time required for one complete oscillation, or for a single cycle. The reciprocal of frequency with typical units of *Seconds* or *Minutes*.

periodic vibration: Oscillatory motion whose amplitude pattern repeats with time.

Permatex®: A series of commercially available products for sealing machined joints. Permatex® is a registered trademark of Loctite Corporation of Cleveland, Ohio.

phase: A timing measurement between two signals, or between a vibration signal and a Keyphasor® pulse. This can be a relative or an absolute value depending on the specific measurement.

phase angle: The angular measurement from the Keyphasor® pulse to the next positive 1X peak signal. Since the pulse occurs first in time, this angle is always identified as a phase lag from the peak of the vibration signal.

piezoelectric: A material such as quartz that converts mechanical to electrical energy. For a piezoelectric crystal, application of stress (force) through a spring mass system produces an electrical charge proportional to the vibration. The charge sensitive signal is converted to a voltage sensitive signal for observation and analysis.

piezoresistive: A solid state silicone resistor material that converts mechanical to electrical energy. These devices are physically attached to cantilevered beams or diaphragms, and electrically connected to a Wheatstone bridge. The application of stress produces an electrical signal proportional to the vibration or pressure.

pivotal resonance: A balance resonance during which shaft motion pivots through the geometric centerline of the rotor, causing a zero axis crossing nodal point within the rotor span. Additional nodes may be produced as a function of bearing stiffness. Often called the conical mode.

Plastigage: A commercially available product for measuring assembly clearances. Plastigage is a product of Perfect Circle® which is a registered trademark of Dana Corporation of Toledo, Ohio.

plate, alignment: In alignment terminology, the base of a jig transit to which the standards are attached. It forms the connection between the standards of the azimuth spindle (journal) and carries both the plate level and the circular level.

plate level: In alignment terminology, a comparatively sensitive tubular level mounted on the plate, used to place the azimuth axis in the direction of gravity.

polar plot: Polar coordinate plot of the locus of the 1X vector at a specific lateral shaft location with variation of speed, or other parameters. This data is generally combined with a Bode plot. This plot may be constructed from measured vibration data, or calculated from analytical models.

position alignment: In alignment terminology, the position in a horizontal plane of the center of the half ball or the spherical adapter about which the instrument turns when it is being leveled. Once a jig transit is properly leveled, it is the point at which the vertical and horizontal axes intersect. The generic description for the initial step in machinery alignment where the proper three-dimensional physical position between foundations and cases are established.

preload, bearing: The dimensionless quantity that is expressed as a number from 0 to 1. A preload of 0 indicates no bearing load upon the shaft; whereas a value of 1 indicates maximum preload (i.e., shaft to bearing line contact).

preload, external: Any mechanism that can externally load a bearing. This includes soft preloads such as process fluids or gravitational forces, as well as hard pre-loads from misalignment, gear contact, piping loads, rubs, etc.

pressure pulsation: Dynamic variation in the static pressure of fluids. Typically measured with piezoelectric transducers that generate a charge proportional to pulsation. The charge is converted to a voltage sensitive signal with a charge amplifier.

Proximitor®: An oscillator-demodulator for conditioning signals from an eddy current proximity probe. This device sends a high frequency signal to the probe, and demodulates the output to provide an AC vibration signal, and a DC signal that is proportional to the average distance between the probe tip and the observed surface. Proximitor® is a registered trademark of Bently Nevada Corporation of Minden, Nevada.

pyroelectric noise: A distortion of the output of a sensor employing a piezoelectric crystal. This distortion is caused by a variation in the thermal environment of the crystal.

Q

Q, filter: A band of frequencies passed or rejected. A narrow band of frequencies has a high Q, whereas a broad band displays a low Q.

quadrature component: The value of the 1X vector that lags the in-phase portion by 90°. Expressed as: $Quad = A \cos(\phi-90°) = A\sin(\phi)$ where A is the amplitude, and ϕ is the phase angle.

R

radial: A direction on a machine which is perpendicular to the shaft centerline. Also referred to as the lateral direction.

radial position: The average position of the shaft dynamic motion within a journal bearing. This is determined by evaluating changes in DC output signals of X-Y proximity probes.

radial vibration: Shaft or casing vibration which is perpendicular to the axial shaft centerline. Also known as lateral vibration.

radio frequency interference (RFI): A condition in which unwanted signals or noise appear at radio frequencies.

real time analyzer: An instrument which performs a vibration frequency spectrum (e.g., an FFT). Typically, this is a device that employs digital signal processing of the time domain signals.

reference line: In alignment terminology, the line of sight from which measurements are made.

relative motion: Vibration measured relative to a chosen reference; for example, proximity probes measure shaft motion relative to the probe mounting location (e.g. bearing housing).

resonance: A vibration amplitude and phase change due to a system frequency sensitivity. The condition of a forcing frequency coinciding with a natural frequency of the mechanical system.

Reynolds number: A non-dimensional number that is the ratio between inertia and viscous forces. This dimensionless number includes fluid density, viscosity, speed, and physical geometry.

ROTSTB: A computer program for determination of damped rotor stability by the complex matrix transfer method. Program coded in Hewlett Packard Basic, and distributed by Rodyn Vibration, Inc., located in Charlottesville, Virginia.

rotor kit: A mechanical device consisting of a small rotor driven by a variable speed fractional horsepower motor. The rotor is generally user configurable with different masses, bearings, and attachments. This type of device is used to simulate larger machines, and it is widely applied as a training and demonstration tool.

runout compensation or **correction**: Correction of a proximity probe signal for the error resulting from shaft electrical or mechanical runout.

S

SAFE diagram: Acronym for Singh's Advanced Frequency Evaluation. This analytical tool combines the two-dimensional Campbell plot with a third dimension of nodal diameters or mode shapes. The three-dimensional intersection of natural resonant frequencies, excitation frequencies, and nodal diameters are then used to identify potential resonant conditions.

scale factor: The magnitude of the output signal change to a known change in the measured variable. Generally applied towards defining the measurement sensitivity of transducers.

seismic transducer: A transducer that measures vibration relative to free space. For example, an accelerometer mounted on a machine bearing housing will measure the seismic vibration of the bearing housing (i.e., with respect to free space).

sensor: A measurement device that transforms one type of physical behavior into a calibrated electrical signal. Also known as a pickup, or a transducer.

signature: A vibration signal documented at a specific condition via spectrum, orbit, time base and shaft centerline data. Also applied to various types of transient vibration data.

simple harmonic motion (SHM): Vibratory motion in which the amplitude varies in a sinusoidal manner with respect to time.

slow roll: A low shaft rotative speed at which dynamic effects such as unbalance are negligible.

soft foot or **leg**: The mechanical condition of a machine support foot (or leg) that is improperly milled, and it resides in a plane that is different from the remaining support feet. This is a primary consideration during initial installation of machinery, and alignment between cases.

Sommerfeld number: A non-dimensional number that is used as a characteristic number for journal bearing performance. Typical values range from 0.01 to 10.0. This dimensionless number relates oil viscosity, speed, load, and clearance, with the bearing geometry.

spectrum plot: A plot where vibration frequency is presented on the abscissa, and amplitude on the ordinate. This includes individual signal spectrum plots, plus related multiple plot formats.

stability: A mechanical system that has a positive log decrement. Any perturbation to a stable mechanical system will result in a decay of the system vibration as a function of time.

standards, alignment: Uprights which support the telescopic axle bearings of a jig transit.

Starrett®: A line of precision measuring tools manufactured by The L.S. Starrett Co. of Athol, MA.

station: In alignment terminology, the distance given in inches and decimal parts of an inch measured parallel to a chosen centerline from a single chosen point. For rotor analytical modeling, a discrete section of a rotor that is identified by unique physical properties and dimensions.

stiffness: The spring-like quality of mechanical and hydraulic elements to elastically deform under load. Applies to shafts, bearings, cases, and structures. Lateral stiffness units are normally expressed as *Pounds/Inch*. Torsional stiffness is presented as *Inch-Pounds/Radian*.

stonewall: A high flow, low head, phenomena that results in a choked flow through a centrifugal compressor. The high flow limit for a compressor at a specific operating speed.

subharmonic: A vibration component that is a fixed fraction of a fundamental frequency such as rotative speed.

subsynchronous: A vibration component that occurs at a frequency less than the fundamental frequency, such as machine rotative speed.

supersynchronous: A vibration component which occurs at a frequency greater than the machine rotational speed.

surge: A low flow, high head, phenomena that results in a reversal of flow through a centrifugal or axial flow compressor. The low flow limit for a compressor at a specific operating speed.

synchronous: Vibration components that change frequency in direct proportion to changes in speed. Also vibration components that occur exactly at shaft rotational speed.

T

tangent screw: In alignment terminology, a hand operated screw which changes the direction of the line of sight in either the azimuth or in elevation.

TEAC®: A manufacturer of instrumentation grade analog and digital recorders, plus data storage peripherals. North American headquarters are located in Montebello, California.

Tektronix®: A manufacturer of electronic instrumentation and measurement products. A worldwide organization with corporate headquarters located in Wilsonville, Oregon.

telescope axle: In alignment terminology, the horizontal axle which supports the telescope.

telescope direct: In alignment terminology, the normal position of a jig transit telescope, as opposed to telescope reversed.

telescope reversed: In alignment terminology, the position of a jig transit telescope when it is turned over (transited) so that it is upside down to its normal position.

telescope sight: In alignment terminology, an optical system that consists of an objective lens and a focusing device that form an image on a cross-line reticle which is viewed through an eyepiece that magnifies the image and the cross lines simultaneously.

tensioner: In wire alignment terminology, a spring loaded device for maintaining wire tension.

time averaging: A noise reduction technique for FFTs that averages successive time records to reduce asynchronous components.

time record: The block of time data that is converted to the frequency domain by an FFT. Typically 1,024 time samples.

torsional vibration: Angular oscillation of a mechanical system, or the amplitude modulation of torque. Units for Torsional Displacement are $Degrees_{,p-p}$, with $Degrees/Second_{,o-p}$ for Velocity.

TorXimitor®: A transducer system for measuring torque on a rotating machinery coupling. TorXimitor® is a registered trademark of Bently Nevada Corporation of Minden, Nevada.

transducer: A device for converting physical behavior into a calibrated electronic signal. Vibration transducers convert mechanical motion into electronic signals for enhanced viewing and observation. Also known as a pickup, or a sensor.

transfer function: A signal processing technique that provides a ratio of the cross power to the auto power spectrum of two dynamic signals. The results are displayed as an amplitude ratio, and a differential phase measurement. The validity of this measurement is verified by the coherence between signals. Also the mathematical relationship between the Laplace transform of the output signal divided by the Laplace transform of the input signal. Also known as a frequency response function (FRF).

transient data: Data obtained during changes in machine conditions such as speed, load, or time.

transient capture: A rapid data sampling technique where the dynamic data is digitized and stored in memory. At the conclusion of the sample event, the stored data is recalled, and post-processed into a variety of traditional transient data formats.

Transient Data Manager® (TDM): A computer-based system consisting of hardware and software for measuring and trending steady state and variable speed machinery behavior. Transient Data Manager® is a registered trademark of Bently Nevada Corporation of Minden, Nevada.

transient motion: Temporarily sustained vibration that occurs during changes in conditions such as speed, load, or time.

translational resonance: A balance resonance during which the shaft mode shape assumes a simple arc. At high support stiffness, a nodal point may exist at each end of the rotor.

transmissibility: The non-dimensional ratio of the response amplitude to a given excitation amplitude. This may also be expressed as a vector quantity with a non-dimensional magnitude.

U

UNBAL: A computer program for determination of rotor synchronous response to a variety of forcing functions using the complex matrix transfer method. Program coded in Hewlett Packard Basic, and distributed by Rodyn Vibration, Inc., located in Charlottesville, Virginia.

unbalance: Unequal radial weight distribution on a rotor. A condition where the mass centerline (principal inertial axis) does not coincide with the geometric centerline of the rotor.

uniform window: The absence of a time domain weighting function on the input signal of an FFT. This type of signal admission is used primarily for rapid transients that are self-windowing. This non-window has an amplitude accuracy of +0 to -4.0 dB.

unfiltered: Dynamic signals such as shaft orbital and time domain vibration data that has not been subjected to frequency filtration.

V

vane passing: In vaned machinery, a vibration frequency that is equal to the number of vanes times shaft rotational speed.

vector: A quantity with both magnitude and direction, For example, a rotational speed (1X) vector is normally expressed in engineering units of $Mils_{p-p}$ @ *Degrees*.

vector filter: A digital instrument that automatically adjusts a band-pass filter center frequency to coincide with the frequency of an external Keyphasor® (speed) pulse. Same as DVF.

velocity: The time rate of change of displacement. In the time domain, velocity leads displacement by 90°, and lags acceleration by 90°. Typical units for velocity are IPS_{o-p}.

velocity transducer: A mechanically activated vibration transducer used to measure relative velocity. Also an accelerometer that is internally integrated to yield a velocity output.

vertical: In alignment terminology, in the direction of gravity. As applied to the identification of vibration transducers, the left-hand or Y-Axis transducer.

W

waterfall plot: A diagram used to observe frequency changes versus time. This plot consists of a series of spectra acquired at consecutive times. The abscissa displays frequency; and amplitude, incremented at various times, is shown along the ordinate axis.

wave form: A display of the instantaneous dynamic signal with respect to time as observed on an oscilloscope.

white noise: A noise signal or source that displays a spectral density that is independent of frequency. A signal source that excites all frequencies within the measurement bandwidth.

X

X-probe: A vibration transducer mounted in a horizontal orientation. On *API 670* systems, the transducer mounted at 45° to the right of a true vertical centerline.

1X: Notation for the component in a dynamic signal that occurs at shaft rotational speed. This is also called the fundamental, or synchronous vibration.

X/2, X/3, X/4, etc.: Components in a dynamic signal having a frequency equal to a fixed fraction of rotative speed. Also called subharmonic and synchronous components.

2X, 3X, 4X, etc.: Components in a dynamic signal having a frequency equal to an exact multiple of rotative speed. Also called harmonics, superharmonics, and synchronous components.

X_{ls}, X_{hs}: Notation for components in a signal having a frequency equal to a specific rotational speed in the machinery train. For example: X_{ls} = low speed, X_{hs} = high speed rotational frequency.

Y

Y-probe: A vibration transducer mounted in a vertical orientation. On *API 670* systems, the transducer mounted at 45° to the left of a true vertical centerline.

Z

Zyglo: A brand name for fluorescent penetrant inspection equipment and materials. A product line of Magnaflux of Glenview, Illinois.

Physical Properties

Table B–1 Average Physical Properties For Common Industrial Metals[a]

Metal	Modulus of Elasticity (Pounds/Inches^2)	Shear Modulus of Rigidity (Pounds/Inches^2)	Density (Pounds/ Inch^3)	Coefficient of Expansion (Mils/Inch °F)
Aluminum	9,000,000	3,440,000	0.0975	0.0132
Aluminum Alloys	10,100,000	3,800,000	0.0995	0.0124
Brass - 70% Cu	15,900,000	6,000,000	0.308	0.0111
Copper	15,800,000	5,800,000	0.323	0.0092
Gold	11,600,000	4,100,000	0.698	0.0079
Cast Iron	28,500,000	—	0.285	0.0065
Inconel	31,000,000	11,000,000	0.307	0.0064
Invar - 36% Ni	21,400,000	—	0.289	0.0011
Monel	25,000,000	9,500,000	0.319	0.0078
Nickel	30,000,000	10,600,000	0.320	0.0074
Silver	11,000,000	3,770,000	0.379	0.0109
Steel - Cast	28,500,000	11,300,000	0.284	0.0063
Steel - Cold Rolled	29,500,000	11,500,000	0.283	0.0066
Steel - 4140 / 4340	29,100,000	11,100,000	0.283	0.0074
Stainless Steel 316	28,100,000	11,200,000	0.290	0.0092
Stainless 17-4 PH	28,400,000	11,300,000	0.280	0.0059
Titanium - 99%	15,900,000	6,100,000	0.163	0.00514
Tungsten	50,000,000	21,400,000	0.697	0.0026
Tungsten Carbide	84,500,000	34,800,000	0.522	0.0029

[a]For Critical Calculations, The Exact Physical Properties For The Specific Alloy Should Be Used.

Table B–2 Partial Physical Constants of Commonly Encountered Industrial Fluids[a]

Compound	Formula	Mol Weight	Ideal Gas Density[b] (Lb./Ft³)	Freezing Point[c] (°F)	Boiling Point (°F)	Specific Heat[d] (BTU/Lb.°F)
Hydrogen	H_2	2.016	0.00531	-434.8	-423.1	3.4066
Helium	He	4.003	0.01055	—	-452.1	1.2404
Methane	CH_4	16.043	0.04228	-296.4	-258.7	0.52676
Ammonia	NH_3	17.031	0.04488	-107.9	-28.0	0.49678
Water	H_2O	18.015	0.04747	32.0	212.0	0.44469
Acetylene	C_2H_2	26.038	0.06862	-113.4	-119.2	0.39754
Carbon Monoxide	CO	28.010	0.07381	-337.0	-312.6	0.24847
Nitrogen	N_2	28.013	0.07382	-346.0	-320.4	0.24833
Ethylene	C_2H_4	28.054	0.07393	-272.5	-154.7	0.35789
Air	N_2+O_2	28.962	0.07632	—	-317.8	0.23980
Ethane	C_2H_6	30.070	0.07924	-297.0	-127.5	0.40789
Oxygen	O_2	31.999	0.08432	-361.8	-297.3	0.21897
Methanol	CH_3OH	32.042	0.08444	-143.8	148.4	0.32429
Hydrogen Sulfide	H_2S	34.082	0.08981	-121.9	-76.5	0.23838
Propylene	C_3H_6	42.081	0.11089	-301.4	-53.8	0.35683
Carbon Dioxide	CO_2	44.010	0.11597	-69.8	-109.2	0.19909
Propane	C_3H_8	44.097	0.11620	-305.7	-43.7	0.38847
Ethanol	C_2H_5OH	46.069	0.12140	-173.4	172.9	0.33074
Butylene	C_4H_8	56.108	0.14785	-301.6	20.8	0.35535
n-Butane	C_4H_{10}	58.123	0.15316	-217.0	31.1	0.39500
Sulfur Dioxide	SO_2	64.065	0.16882	-103.8	14.1	0.14802
Chlorine	Cl_2	70.905	0.18685	-149.7	-29.1	0.11375
n-Pentane	C_5H_{12}	72.150	0.19013	-201.5	96.9	0.38831
Benzene	C_6H_6	78.114	0.20584	42.0	176.1	0.24295
Toluene	C_7H_8	92.141	0.24281	-139.0	231.1	0.26005
para-Xylene	C_8H_{10}	106.167	0.27977	55.9	281.0	0.27470

[a]Data from *Engineering Data Book* - 10th ed, Gas Processors Suppliers Association, Tulsa, OK, 1994.
[b]Ideal Gas Density at Standard Temperature of 60°F and Pressure of 14.696 Psia.
[c]Freezing and Boiling Point Temperature at Standard Pressure of 14.696 Psia.
[d]Specific Heat C_p of Ideal Gas at Standard Temperature of 60°F and Pressure of 14.696 Psia.

Conversion Factors

To Convert	Into	Multiply by
atmospheres	centimeters of Hg (0°C)	76.
atmospheres	feet of H_2O (4°C)	33.90
atmospheres	inches of Hg (0°C)	29.92
atmospheres	inches of H_2O (4°C)	406.83
atmospheres	kilograms/centimeter^2	1.0332
atmospheres	kilograms/meter^2	10,332.
atmospheres	Pascal	101,320.
atmospheres	pounds/foot^2	2,116.80
atmospheres	pounds/inch^2	14.696
barrels (liquid)	gallons	31.5
barrels (oil)	gallons (oil)	42.
BTU	ergs	1.055×10^{10}
BTU	foot-pounds	778.2
BTU	calories	252.
BTU	horsepower-hours	3.930×10^{-4}
BTU	kilowatt-hours	2.931×10^{-4}
BTU	joules	1,055.
BTU/hour	foot-pounds/second	0.2161
BTU/hour	calorie/second	0.0700
BTU/hour	horsepower	3.929×10^{-4}
BTU/hour	kilowatts	2.931×10^{-4}
BTU/hour	watts	0.2931
BTU/minute	horsepower	0.02358
BTU/minute	kilowatts	0.01757
BTU/minute	watts	17.57
calories	BTU	0.0039683
calories	ergs	4.1868×10^{7}
calories	foot-pounds	3.087
calories	kilowatt-hours	1.163×10^{-6}
calories/second	BTU/hour	14.286
calories/second	foot pounds/minute	185.2
calories/second	horsepower	5.613×10^{-3}
calories/second	kilowatts	4.186×10^{-3}
centimeters	feet	0.03281
centimeters	inches	0.3937
centimeters	kilometers	10^{-5}
centimeters	meters	0.01

To Convert	Into	Multiply by
centimeters	millimeters	10.
centimeters	mils	393.7
centimeters^2	feet^2	0.001076
centimeters^2	inches^2	0.1550
centimeters^2	kilometers^2	10^{-10}
centimeters^2	meters^2	10^{-4}
centimeters^2	millimeters^2	100.
centimeters^3	feet^3	3.531×10^{-5}
centimeters^3	inches^3	0.0610
centimeters^3	meters^3	10^{-6}
centimeters^3	gallons (U.S. liquid)	2.642×10^{-4}
centimeters^3	milliliters	1.
centimeters^3	liters	10^{-3}
centimeters^3	quarts (U.S. liquid)	1.057×10^{-3}
centimeters of Hg	atmospheres	0.01316
centimeters of Hg	feet of H_2O (4°C)	0.4461
centimeters of Hg	kilograms/meter^2	136.0
centimeters of Hg	pounds/feet^2	27.85
centimeters of Hg	pounds/inch^2	0.1934
centimeter/second	feet/minute	1.9685
centimeter/second	feet/second	0.03281
centimeter/second	meters/minutes	0.6
centipoise	poise	0.01
centipoise	microreyn	0.145
centipoise	pounds-second/inch^2	1.45×10^{-7}
centistokes	inch^2/second	1.550×10^{-3}
degrees	revolutions	0.002778
degrees	radians	0.017453
degrees/second	revolutions/minute	0.1667
degrees/second	revolutions/second	0.002778
dynes	grams	1.020×10^{-3}
dynes	newtons	10^{-5}
dynes	kilograms	1.020×10^{-6}
dynes	pounds	2.248×10^{-6}
dyne-second/cm^2	poise	1.
erg	BTU	9.481×10^{-11}
erg	foot-pounds	7.367×10^{-8}

To Convert	Into	Multiply by
erg	calorie	2.389×10^{-8}
erg	horsepower-hour	3.725×10^{-14}
erg	joule	10^{-7}
erg	kilowatt-hour	2.778×10^{-14}
erg	watt-hours	2.778×10^{-11}
erg/second	horsepower	1.341×10^{-10}
ergs/second	kilowatts	10^{-10}
ergs/second	watts	10^{-7}
farads	microfarads	10^6
feet	centimeters	30.48
feet	inches	12.
feet	kilometers	3.048×10^{-4}
feet	meters	0.3048
feet	millimeters	304.8
feet	mils	12,000.
feet^2	centimeters^2	929.
feet^2	inches^2	144.
feet^2	kilometers^2	9.29×10^{-8}
feet^2	meters^2	0.09290
feet^2	millimeters^2	9.290×10^4
feet^3	centimeters^3	2.832×10^4
feet^3	inches^3	1,728.
feet^3	meters^3	0.02832
feet^3	gallons (U.S. liquid)	7.48052
feet^3	liters	28.32
feet^3	quarts (U.S. liquid)	29.92
feet of H$_2$O (4°C)	atmospheres	0.02950
feet of H$_2$O (4°C)	centimeters of Hg (0°C)	2.2418
feet of H$_2$O (4°C)	inches of Hg (0°C)	0.8826
feet of H$_2$O (4°C)	kilograms/centimeter^2	0.03048
feet of H$_2$O (4°C)	kilograms/meter^2	304.8
feet of H$_2$O (4°C)	pounds/foot^2	62.43
feet of H$_2$O (4°C)	pounds/inch^2	0.4336
feet/minute	centimeters/second	0.508
feet/minute	feet/second	0.01667
feet/minute	kilometers/hour	0.01829
feet/minute	meters/minute	0.3048
feet/minute	meters/second	0.00508
feet/second	centimeters/second	30.48
feet/second	kilometers/hour	1.097
feet/second	meters/minute	18.29
feet/second	meters/second	0.3048
feet/second	feet/minute	60.
feet^3/hour	gallons/minute	0.1247
feet^3/minute	gallons/minute	7.48052
feet^3/second	gallons/minute	448.8
foot-pounds	BTU	1.2851×10^{-3}
foot-pounds	ergs	1.356×10^7
foot-pounds	calories	0.3239
foot-pounds	horsepower-hours	5.050×10^{-7}
foot-pounds	joules	1.3558
foot-pounds	kilowatt-hours	3.766×10^{-7}
foot-pounds	watt-hours	3.766×10^{-4}

To Convert	Into	Multiply by
foot-pounds/minute	horsepower	3.030×10^{-5}
foot-pounds/minute	kilowatts	2.260×10^{-5}
foot-pounds/minute	calories/second	5.40×10^{-3}
foot-pounds/second	BTU/hour	4.6263
foot-pounds/second	horsepower	0.001818
foot-pounds/second	kilowatts	0.001356
gallons (U.S. liquid)	barrels (U.S. liquid)	0.03175
gallons (U.S. liquid)	centimeters^3	3,785.4
gallons (U.S. liquid)	feet^3	0.1337
gallons (U.S. liquid)	inches^3	231.
gallons (U.S. liquid)	liters	3.785
gallons (U.S. liquid)	meters^3	0.003785
gallons (U.S. liquid)	quarts	4.
gallons (oil)	barrels (oil)	0.02381
gallons/minute	liters/second	0.06308
gallons/minute	feet^3/hour	8.0208
gallons/minute	feet^3/minute	0.13368
gallons/minute	feet^3/second	0.002228
grams	dynes	980.7
grams	newtons	9.807×10^{-3}
grams	kilograms	10^{-3}
grams	milligrams	10^3
grams	ounces (avdp)	0.03527
grams	pounds	0.002205
grams/cm^3	pounds/foot^3	62.42
grams/cm^3	pounds/inch^3	0.03613
horsepower	BTU/hour	2,545.
horsepower	BTU/minute	42.417
horsepower	calories/second	178.2
horsepower	ergs/second	7.457×10^9
horsepower	foot-pounds/minute	33,000.
horsepower	foot-pounds/second	550.
horsepower	kilowatts	0.7457
horsepower	watts	745.7
horsepower-hours	BTU	2,544.
horsepower-hours	ergs	2.685×10^{13}
horsepower-hours	foot-pounds	1.980×10^6
horsepower-hours	joules	2.685×10^6
horsepower-hours	kilocalories	641.2
horsepower-hours	kilogram-meters	2.737×10^5
horsepower-hours	kilowatt-hours	0.7457
horsepower-hours	watt-hours	745.7
inches	centimeters	2.54
inches	feet	0.08333
inches	kilometers	2.54×10^{-5}
inches	meters	0.0254
inches	millimeters	25.4
inches	mils	10^3
inches^2	centimeters^2	6.452
inches^2	feet^2	0.006944
inches^2	kilometers^2	6.452×10^{-10}

To Convert	Into	Multiply by	To Convert	Into	Multiply by
inches^2	meters^2	6.452x10^{-4}	kilometers	meters	1,000.
inches^2	millimeters^2	645.16	kilometers	millimeters	10^6
inches^2/second	centistokes	645.16	kilometers^2	centimeters^2	10^{10}
inches^3	centimeters^3	16.387	kilometers^2	feet^2	10.763x10^6
inches^3	feet^3	5.787x10^{-4}	kilometers^2	inches^2	1.550x10^9
inches^3	meters^3	1.639x10^{-5}	kilometers^2	meters^2	10^6
inches^3	gallons	4.329x10^{-3}	kilometer/hour	feet/minute	54.68
inches^3	liters	0.01639	kilometer/hour	feet/second	0.9113
inches^3	quarts (U.S. liquid)	0.01732	kilometer/hour	meters/minute	16.667
inches of Hg (0°C)	atmospheres	0.03342	kilometer/hour	meters/second	0.27778
inches of Hg (0°C)	feet of H$_2$O (4°C)	1.133	kilometer/minute	meters/second	16.667
inches of Hg (0°C)	inches of H$_2$O (4°C)	13.596	kilowatts	BTU/hour	3,412.
inches of Hg (0°C)	kilograms/centimeter^2	0.03453	kilowatts	BTU/minute	56.87
inches of Hg (0°C)	kilograms/meter^2	345.3	kilowatts	calories/second	238.9
inches of Hg (0°C)	pounds/foot^2	70.73	kilowatts	erg/second	10^{10}
inches of Hg (0°C)	pounds/inch^2	0.4912	kilowatts	foot-pounds/minute	4.425x10^4
inches of H$_2$O (4°C)	atmospheres	0.002458	kilowatts	foot-pounds/second	737.5
inches of H$_2$O (4°C)	inch of Hg (0°C)	0.07355	kilowatts	horsepower	1.341
inches of H$_2$O (4°C)	kilograms/centimeter^2	0.00254	kilowatts	watts	10^3
inches of H$_2$O (4°C)	pounds/foot^2	5.204	kilowatt-hours	BTU	3,412.
inches of H$_2$O (4°C)	pounds/inch^2	0.03613	kilowatt-hours	ergs	3.6x10^{13}
			kilowatt-hours	foot-pounds	2.655x10^6
joules	BTU	9.478x10^{-4}	kilowatt-hours	calories	8.599x10^5
joules	ergs	10^7	kilowatt-hours	horsepower-hours	1.341
joules	foot-pounds	0.7376	kilowatt-hours	joule	3.6x10^6
joules	horsepower-hours	3.725x10^{-7}			
joules	kilowatt-hour	2.778x10^{-7}	liters	centimeters^3	10^3
			liters	feet^3	0.03532
kilograms	dynes	9.807x10^5	liters	inches^3	61.02
kilograms	grams	10^3	liters	meters^3	10^{-3}
kilograms	newtons	9.807	liters	gallons (U.S. liquid)	0.2642
kilograms	pounds	2.2046	liters	milliliters	10^3
kilograms/meter^3	pounds/foot^3	0.06242	liters/minute	feet^3/second	5.886x10^{-4}
kilograms/meter^3	pounds/inch^3	3.613x10^{-5}	liters/second	gallons/minute	15.852
kilogram-meters	horsepower-hours	3.653x10^{-6}			
kilogram-meter/sec^2	newton	1.	megohms	ohms	10^6
kilograms/cm^2	atmospheres	0.9678	meters	centimeters	10^2
kilograms/cm^2	feet of H$_2$O (4°C)	32.81	meters	feet	3.281
kilograms/cm^2	inches of Hg (0°C)	28.96	meters	inches	39.37
kilograms/cm^2	inch of H$_2$O (4°C)	393.71	meters	kilometers	10^{-3}
kilograms/cm^2	pounds/foot^2	2,048.2	meters	microns	10^6
kilograms/cm^2	pounds/inch^2	14.22	meters	millimeters	1,000.
kilograms/meter^2	atmospheres	9.678x10^{-5}	meters^2	centimeters^2	10^4
kilograms/meter^2	centimeters of Hg (0°C)	0.007356	meters^2	feet^2	10.76
kilograms/meter^2	feet of H$_2$O (4°C)	0.003281	meters^2	inches^2	1,550.
kilograms/meter^2	inches of Hg (0°C)	0.002896	meters^2	kilometers^2	10^{-6}
kilograms/meter^2	pounds/foot^2	0.2048	meters^2	millimeters^2	10^6
kilograms/meter^2	pounds/inch^2	0.001422	meters^3	centimeters^3	10^6
kilograms/meter^2	kilograms/mm^2	10^{-6}	meters^3	feet^3	35.31
kilograms/mm^2	kilograms/meter^2	10^6	meters^3	inches^3	61,023.
kilocalories	horsepower/hours	0.001560	meters^3	gallons (U.S. liquid)	264.2
kilometers	centimeters	10^5	meters^3	liters	10^3
kilometers	feet	3,280.8	meters^3	quarts (U.S. liquid)	1,056.7
kilometers	inches	39,370.	meters/minute	centimeter/second	1.667

To Convert	Into	Multiply by
meters/minute	feet/minute	3.281
meters/minute	feet/second	0.05468
meters/minute	kilometers/hour	0.06
meters/second	feet/minute	196.8
meters/second	feet/second	3.281
meters/second	kilometers/hour	3.6
meters/second	kilometers/minute	0.06
microfarad	farads	10^{-6}
microhms	ohms	10^{-6}
microns	meters	10^{-6}
microreyn	centipoise	6.897
microreyn	reyn	10^{-6}
milligrams	grams	10^{-3}
milliliters	centimeters^3	1.
milliliters	liter	10^{-3}
millimeters	centimeters	0.1
millimeters	feet	0.003281
millimeters	inches	0.03937
millimeters	kilometers	10^{-6}
millimeters	meters	10^{-3}
millimeters	mils	39.37
millimeter^2	centimeters^2	0.01
millimeter^2	feet^2	1.076×10^{-5}
millimeter^2	inches^2	0.001550
millimeter^2	meters^2	10^{-6}
mils	centimeters	0.00254
mils	feet	8.333×10^{-5}
mils	inches	10^{-3}
mils	millimeters	0.0254
newtons	dynes	10^5
newtons	kilogram-meter/second^2	1.
newtons	grams	102.
newtons	kilograms	0.102
newtons	pounds	0.2248
ohms	megohms	10^{-6}
ohms	microhms	10^6
ounces	grams	28.35
ounces	pounds	0.0625
ounces	tons (short)	3.125×10^{-5}
Pascal	atmospheres	9.87×10^{-6}
poise	dyne-second/centimeter^2	1.
poise	centipoise	10^2
pounds	dynes	4.448×10^5
pounds	grams	453.59
pounds	kilograms	0.4536
pounds	newtons	4.448
pounds	ounces	16.
pounds	tons (short)	0.0005
pounds$_f$-sec./inch^2	centipoise	6.895×10^6
pounds$_f$-sec./inch^2	reyn	1.
pounds/foot^3	grams/centimeter^3	0.01602

To Convert	Into	Multiply by
pounds/foot^3	kilograms/meter^3	16.02
pounds/foot^3	pounds/inch^3	5.787×10^{-4}
pounds/inch^3	grams/centimeter^3	27.68
pounds/inch^3	kilograms/meter^3	2.768×10^4
pounds/inch^3	pounds/foot^3	1,728.
pounds/foot^2	atmospheres	4.724×10^{-4}
pounds/foot^2	centimeters of Hg (0°C)	0.0359
pounds/foot^2	feet of H_2O (4°C)	0.01602
pounds/foot^2	inches of Hg (0°C)	0.01414
pounds/foot^2	inches of H_2O (4°C)	0.1922
pounds/foot^2	kilograms/centimeter^2	4.882×10^{-4}
pounds/foot^2	kilograms/meter^2	4.882
pounds/foot^2	pounds/inch^2	0.006944
pounds/inch^2	atmospheres	0.06803
pounds/inch^2	centimeters of Hg (0°)	5.1715
pounds/inch^2	feet of H_2O (4°C)	2.307
pounds/inch^2	inches of Hg (0°C)	2.036
pounds/inch^2	inches of H_2O (4°C)	27.678
pounds/inch^2	kilograms/centimeter^2	0.07031
pounds/inch^2	kilograms/meter^2	703.1
pounds/inch^2	pounds/foot^2	144.
quarts (U.S. liquid)	centimeters^3	946.4
quarts (U.S. liquid)	feet^3	0.03342
quarts (U.S. liquid)	inches^3	57.75
quarts (U.S. liquid)	meters^3	9.464×10^{-4}
quarts (U.S. liquid)	gallons	0.25
radians	degrees	57.30
radians	revolutions	0.15916
radians/second	revolutions/minute	9.5493
radians/second	revolutions/second	0.15916
revolutions	degrees	360.
revolutions	radians	6.2831
revolutions/minute	degrees/second	6.
revolutions/minute	radians/second	0.10472
revolutions/minute	revolutions/second	0.01667
revolutions/second	revolutions/minute	60.
revolutions/second	radians/second	6.2831
reyn	pound$_f$-second/inch^2	1.
reyn	microreyn	10^6
tons (short)	ounces	32,000.
tons (short)	pounds	2,000.
watts	BTU/hour	3.413
watts	BTU/minute	0.05692
watts	ergs/second	10^7
watts	horsepower	1.341×10^{-3}
watts	kilowatts	10^{-3}
watt-hours	BTU	3.413
watt-hours	ergs	3.6×10^{10}
watt-hours	foot-pounds	2,655.3
watt-hours	horsepower-hours	1.341×10^{-3}

Index